# Nanotechnology in the Life Sciences

**Series Editor**

Ram Prasad
Department of Botany
Mahatma Gandhi Central University
Motihari, Bihar, India

Nano and biotechnology are two of the 21st century's most promising technologies. Nanotechnology is demarcated as the design, development, and application of materials and devices whose least functional make up is on a nanometer scale (1 to 100 nm). Meanwhile, biotechnology deals with metabolic and other physiological developments of biological subjects including microorganisms. These microbial processes have opened up new opportunities to explore novel applications, for example, the biosynthesis of metal nanomaterials, with the implication that these two technologies (i.e., thus nanobiotechnology) can play a vital role in developing and executing many valuable tools in the study of life. Nanotechnology is very diverse, ranging from extensions of conventional device physics to completely new approaches based upon molecular self-assembly, from developing new materials with dimensions on the nanoscale, to investigating whether we can directly control matters on/in the atomic scale level. This idea entails its application to diverse fields of science such as plant biology, organic chemistry, agriculture, the food industry, and more.

Nanobiotechnology offers a wide range of uses in medicine, agriculture, and the environment. Many diseases that do not have cures today may be cured by nanotechnology in the future. Use of nanotechnology in medical therapeutics needs adequate evaluation of its risk and safety factors. Scientists who are against the use of nanotechnology also agree that advancement in nanotechnology should continue because this field promises great benefits, but testing should be carried out to ensure its safety in people. It is possible that nanomedicine in the future will play a crucial role in the treatment of human and plant diseases, and also in the enhancement of normal human physiology and plant systems, respectively. If everything proceeds as expected, nanobiotechnology will, one day, become an inevitable part of our everyday life and will help save many lives.

More information about this series at http://www.springer.com/series/15921

Inamuddin • Abdullah M. Asiri
Editors

# Nanosensor Technologies for Environmental Monitoring

Springer

*Editors*
Inamuddin
Chemistry Department
King Abdulaziz University
Jeddah, Saudi Arabia

Department of Applied Chemistry
Aligarh Muslim University
Aligarh, India

Abdullah M. Asiri
Chemistry Department
King Abdulaziz University
Jeddah, Saudi Arabia

ISSN 2523-8027 ISSN 2523-8035 (electronic)
Nanotechnology in the Life Sciences
ISBN 978-3-030-45115-8 ISBN 978-3-030-45116-5 (eBook)
https://doi.org/10.1007/978-3-030-45116-5



This Springer imprint is published by the registered company Springer Nature Switzerland AG
The registered company address is: Gewerbestrasse 11, 6330 Cham, Switzerland

# Preface

Due to the rapid development of industrialization and urbanization, toxic chemicals are released, which leads to a greater risk to human health and the environment. Today environmental pollution issues exist all over the world at different levels and have attracted increasing attention. Environmental security is one of the major concerns for our well-being. Nanomaterials in nanotechnology are a game-changer which can resolve both issues and improve the quality of human life, especially in recent years. Based on various recognition elements, sensor technologies have been widely used as effective analytical tools to achieve the accurate detection of various pollutants. Therefore, it is necessary to understand how sensor technologies impact environmental monitoring applications.

*Nanosensor Technologies for Environmental Monitoring* describes a wide range of topics and methods, beginning with an overview of sensors for environmental applications. Chapters discuss the biosensors, gas sensors, and electrochemical sensors responsible for monitoring various environmental pollutants, such as pesticides, toxins, pathogens, heavy metals, and phenols. Green solvent-modified sensors, screen-printed sensors, lab-on-chip devices, and modern portable sensors are discussed in detail. This book should be useful for faculty, researchers, students from academics and laboratories which are linked to nanotechnologies, analytical chemistry, toxicology, and environmental sciences. Based on thematic topics, the book edition contains the following 17 chapters:

Chapter 1 summarizes the latest development of electrochemical sensors and biosensors for environmental contaminant detection in the past five years. It mainly discusses the synthesis of novel electrode materials and new proposed electrochemical (bio)sensing strategies for the detection of various pollutants.

Chapter 2 discusses the classification of different biosensors elaborately. It also focuses on electrochemical biosensors, namely, amperometric, cyclic voltammogram, potentiometric, conductometric and electrochemical impedance spectroscopy. It also futher discusses the application of biosensors in the detection of contamination in air, soil, and water.

Chapter 3 discusses the demand for consumer products and crises in energy sectors. The serious consequence of toxic gases and their adverse impact on our endangered health and environment over the long term. The major focus of this chapter reviews gas sensing and biosensor materials and their sensitivity to the operating conditions reported in the literature.

Chapter 4 discusses the various types of biosensors that can be implemented for monitoring various environmental parameters like heavy metals, pathogens, potentially toxic elements, toxins, endocrine disrupting chemicals, and pesticides. It focuses on various advantages of using a biosensor-based monitoring system.

Chapter 5 discusses the significance of disposable screen-printed electrodes for the electrochemical quantification of hazardous environmental contaminants. The fabrication of miniature sensing systems and the basic principles and methodologies reported in the literature for the on-site detection of major environmental contaminants is also discussed in detail.

Chapter 6 discusses the potential role of nanosensors as sustainable tools that can assist in the real-time monitoring of generated or released environmental contaminants within a process. Illustration of the nanosensors compositions with different recognition elements is presented. Besides, the green synthesis methodology of nanoparticles is discussed.

Chapter 7 offers an overview of the new trends in the employment of ultrasound and microwave technologies, as well as the use of biosynthesis routes and hybrid green synthetic pathways for the synthesis of (nano)materials. The applications of these (nano)materials for analytical (bio)sensing are also highlighted. Finally, some key points about the industrial application of (bio)sensing devices are reported.

Chapter 8 focuses on green and biological-assisted synthesis of plasmonic metal nanoparticles using various microorganisms as nano-factories. It also discusses the successful application of biosynthesized metal nanoparticles as an eco-friendly plasmonic sensing platform for the detection of several toxic heavy metal ions and organic pollutants in the environmental samples.

Chapter 9 discusses the ionic liquid (IL)-modified sensors and biosensors for the detection of environmental contaminants. ILs can act as both binder and conductor. Different materials have been modified with various ILs. The IL-modified sensors are potentially used for the detection of phenolic compounds, heavy metals, and pesticides.

Chapter 10 provides an overview of technical challenges and recent developments of biosensors for monitoring phenolic compounds. The chapter highlights research activities carried out on the preparation of nano-biosensors using nanomaterials and conducting polymers. Analytical parameters affecting the performance of phenolic compound-based biosensors are discussed.

Chapter 11 describes the potential application and superiority of hybrid nanostructures of noble metals and metal oxides for the detection of various analyte molecules, employing surface-enhanced Raman scattering technique. The role of charge transfer, coupled with the electromagnetic mechanism for improving the enhancement factor, is discussed. Additionally, degradation approaches of probe molecules towards renewable substrates are discussed.

Chapter 12 presents molecular imprinting-based sensing systems for the detection of microorganisms. Nanosensor applications for the detection of microorganisms are discussed by highlighting the principles, advantages, and limitations of sensing strategies. The main focus is given to offer directions to the future feasibility of nanosensors for the rapid detection of microorganisms.

Chapter 13 discusses various examples of nanomaterials that are used to fabricate sensors with the ability to detect toxic gases and biological molecules. Besides, the drawbacks of nanomaterials as sensors, the efficiency and limitations of green synthesized nanomaterials for sustainable toxic gas detection and biosensing applications are also discussed.

Chapter 14 discusses various types of flexible pressure, temperature, and pH sensors, which are mainly used for point care diagnosis.

Chapter 15 deals with lab-on-a-chip (LOC) technologies for water quality monitoring and highlights the advantages offered by LOC sensors over conventional techniques. Besides discussing fabrication materials for the sensors, monitoring of chemical and biological contaminants in water are discussed. Future directions and challenges of LOC technology are indicated.

Chapter 16 discusses the advance electrochemical sensors designed particularly for the sensitive detection of pharmaceutical drugs. The provided discussion emphasizes the surface-modification methods, where interfacial electrode engineering is comprehensively described as a straightforward approach to produce responsive and economically feasible sensors for the detection of antibiotics and therapeutic drugs.

Chapter 17 discusses in detail different types of green chemical sensors for environmental monitoring, including carbon quantum dots, greenly produced metal nanoparticles, and greenly functionalized metal nanoparticles. The basic principles of different environmental biosensors are also presented with an overview of recent review articles covering this type of sensor.

Jeddah, Saudi Arabia Inamuddin
Aligarh, India
Jeddah, Saudi Arabia Abdullah M. Asiri

# Contents

# Recent Advances in Electrochemical Sensor and Biosensors for Environmental Contaminants

**Li-Ping Mei, Pei Song, Yuan-Cheng Zhu, Yi-Fan Ruan, Xiao-Mei Shi, Wei-Wei Zhao, Jing-Juan Xu, and Hong-Yuan Chen**

## Contents



## 1 Introduction

Due to the development of society and technologies, the quality of human life has been obviously improved, especially in recent years. However, the following environmental issues are becoming increasingly serious. Numerous types of contaminants are released by modern factories, suburban farms, and even by urban communities, which are threatening human health and destroying the balance of

L.-P. Mei · P. Song · Y.-C. Zhu · Y.-F. Ruan · X.-M. Shi · W.-W. Zhao (✉)
J.-J. Xu · H.-Y. Chen (✉)
State Key Laboratory of Analytical Chemistry for Life Science, School of Chemistry and Chemical Engineering, Nanjing University, Nanjing, China
e-mail: zww@nju.edu.cn; xujj@nju.edu.cn; hychen@nju.edu.cn

 
Inamuddin, A. M. Asiri (eds.), *Nanosensor Technologies for Environmental Monitoring*, Nanotechnology in the Life Sciences,
https://doi.org/10.1007/978-3-030-45116-5_1

ecosystems. In September 1996, the European Union adopted the Integrated Pollution Prevention and Control Directive, which could combine prevention and control of pollution to ensure a high level of environmental protection (O'Malley 1999). Up to now, almost all of the countries have promulgated several related laws to control the emission of environmental pollution.

Common environmental contaminants usually include heavy metal ions, organic contaminants, antibiotics, pathogens, gas pollutants, and so on. Heavy metals refer to those metal elements with a density larger than 5 g $cm^{-3}$ and the atomic weights between 63.5 and 200.6 g $mol^{-1}$ (Fu and Wang 2011; Srivastava and Majumder 2008), which are not biodegradable and easily accumulate in living organisms through food chains. Usually, heavy metal ions, such as copper, nickel, zinc, cadmium, mercury, chromium, and lead are released along with industrial wastewater and car exhausts to the outer environment. Excessive heavy metal ions can cause series of health problems, such as skin irritations, gastrointestinal distress, pulmonary fibrosis, and damage to the central nervous system, or even death (Paulino et al. 2006; Njau et al. 2000; Zhu et al. 2009; Bojdi et al. 2014; Metters et al. 2012). In addition, organic pollutions are also considered as other important environmental contaminants. Aromatic compounds are the most harmful to the human health and ecological environment among organic pollutions, such as phenolic compounds and polyaromatic hydrocarbons. Phenolic compounds are formed by hydroxyl substitution of hydrogen atoms on benzene rings in aromatic hydrocarbons. According to the number of hydroxyl groups, phenolic compounds can be divided into monophenols and polyphenols. Phenolic compounds in the environment come from a wide range of sources, including wastewater from chemical and pharmaceutical industries, degradation of organic pesticides, automobile exhaust, and so on. Phenolic compounds are commonly transmitted through water and air and can remain in the soil for a long time. Due to their carcinogenic, teratogenic, mutagenic, and other potential toxicity, phenolic compounds will cause serious harm to the ecological environment, animals, plants, and human health (Teh and Mohamed 2011; Su et al. 2011; Aksu 2005; Samanta et al. 2002; Falahatpisheh et al. 2001). Similarly, antibiotics, as one of the common pollutions, are the most useful drug for bacterial infection. However, antibiotics are not only used for disease treatment but also extensively used for growth-promoting of animals in animal husbandry (Martinez 2009; Smith et al. 2002). The release of antibiotics in the natural environment will lead to the selection of resistant bacteria, which may cause the appearance of super bacteria (Davies 1994). In addition to the above contaminants, gas pollutants mainly include carbon monoxide, nitric oxides, oxides of sulfur, and volatile organic compounds, and are derived from gaseous industrial emissions, such as the burning of coal, petroleum, and other fossil fuels, and automobile exhaust (Li et al. 2012). These gas pollutants may cause acid rain and damage the respiratory systems of humans and animals.

Highly sensitive detection of environmental pollutants can help to guide the pollution control and also protect humans from the damage of pollutants. In recent years, environmentalists and analytical chemists have devoted plentiful of efforts to seek a sensitive detection method to measure and monitor the concentration of environmental pollutants. Up to now, much-advanced detection techniques have been developed, such as atomic spectroscopy (Hatch and Ott 1968), molecular spectroscopy (Xu et al. 2007), spectrophotometry (Baker et al. 2003), chromatography (Reemtsma 2003), mass spectrometry (Daughton 2001), electrochemistry method (Metters et al. 2012; Su et al. 2011), and so on. Although the accuracy and sensitivity of the previously reported methods are acceptable, the detection procedure of samples is usually expensive, complex, and time-consuming. These weaknesses are not suitable for the development toward civilianization and portability. Fortunately, electrochemistry methods, or electrochemical sensors, hold great promise to overcome the above disadvantage, due to their simplicity, rapid response, high sensitivity, small size, and low cost. Electrochemical sensors have been widely used to determine the amount of trace heavy metals ions in natural waters (Batley 1983), carcinogen (Barek et al. 2001), gas pollutants (Wan et al. 2018), and organic pollutants (Yang et al. 2018). Electrochemical biosensors have gradually emerged through an effective integration of specific biorecognition elements and electrochemical sensors. Benefit from the high selectivity of bio-recognition element, biosensors are worked without complex sample processing and can even be applied to multicomponent detection. Hence, the electrochemical sensors and biosensors have shown great potential for laboratory-based and on-site analysis of environmental contaminants.

This chapter has stated the important roles of electrochemical sensors and biosensors for environmental contaminants. We have summarized the common types, electrode materials, and composition of electrochemical sensors and biosensors. We have focused on the advances of electrochemical sensor technology in an update of electrode materials and fabricated strategies. In recent 5 years, applications of electrochemical sensors and biosensors in environmental problems and future trends have also been reviewed.

# 2 Fabrication

## *2.1 The Principles of Electrochemical Sensor and Biosensors*

Electrochemical sensors and biosensors are devices that can sense the physical, chemical, or biological changes of analytes and convert them into measurable amperometry, voltammetry, impedance, or other electrochemical signals (Fig. 1). The electrochemical sensing system usually contains three electrodes: working electrode, counter electrode, and reference electrode. The working electrode is responsible for generating redox reaction of analytes and sensing redox signal. And the counter electrode is used to establish a closed circuit through electrolytic

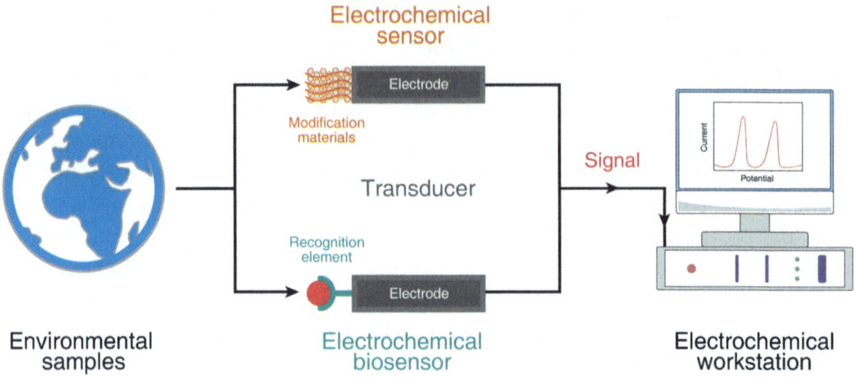


**Fig. 1** The principle of electrochemical sensors and biosensors for environmental pollutions

solution so that the voltage/current can be applied to the working electrode. The reference electrode serves as a benchmark to maintain a known and stable operating potential.

The fundamental of electrochemical detection is redox reaction that occurs on the surface of work electrodes. Therefore, environmental contaminants with electrochemical activities can be directly sensed by electrochemical sensors. Voltammetry is the most widespread technology for electrochemical sensors, which can measure the current response to achieve a precise quantitative analysis of target analytes under an applied potential. Voltammetry exhibits a wide linear range that is from sub-ppb to ppm. A high concentration of analytes with ppm or ppb can be analyzed by cyclic voltammetry, differential pulse, or square-wave voltammetry techniques, while the analytes with low concentrations with ppb or sub-ppb can be measured by the stripping voltammetry techniques. The sensing process is mainly described as follows: First, a constant or variational voltage is applied to the working electrode that is immersed in the electrolyte containing various analytes. Then, redox reactions will occur on the electrode surface, resulting in the change of the current signal. Finally, the number of analytes can be obtained through the corresponding variations of electrical signals.

In order to improve selectivity and sensitivity for environmental pollutants, electrochemical biosensors were gradually developed, which involved high specificity biorecognition elements during their fabrication process. The biosensor is composed of a molecular recognition element and a conversion section. For environmental monitoring, the recognition element must specifically recognize one or several particular contaminants from a complex real sample without any interference. Therefore, enzymes, antibodies, and nucleic acids with high specificity are often used as the recognition element to modify the working electrode. Once the target environmental pollutants are identified and captured by the recognition element, electric signals of modified electrodes will be immediately changed.

## *2.2 Materials for Electrochemical Sensor and Biosensors*

As one of the most important parts of electrochemical sensors and biosensors, novel electrode materials are widely studied to improve the electrochemical catalytic activity of as-fabricated sensors and biosensors. Most electrochemical reactions are normally carried out on the surface of working electrodes, so the electrode materials should be chemically inert and with good electrical conductivity. Up to now, gold electrode (Shen et al. 2008; Lupu et al. 2009), glassy carbon electrode (Zhang et al. 2019a, b), and indium tin oxide (Fu et al. 2018; Vaishnav et al. 2015) are commonly used as substrate electrode to design high sensitive electrochemical sensors. Shen and co-workers have developed an enzyme-based electrochemical biosensor based on a modified gold electrode for sensitive detection of trace lead ions (Shen et al. 2008). Li's group has used glass carbon electrode as the substrate electrode to coat a carbon sphere and fabricate electrochemical sensor for the determination of dihydroxybenzene isomers (Yang et al. 2019a, b). Vaishnav and co-workers developed an indium tin oxide film electrochemical sensor, which exhibits improved performance for sensitive and selective detection of benzene (Vaishnav et al. 2015).

Meanwhile, many emerging materials are explored to modify the surface of the electrode, such as conducting polymers (Hatchett and Josowicz 2008; Ates 2013), metal-based nanomaterials (Wu et al. 2019; Li et al. 2019a, b), carbon nanotubes (Şenocak et al. 2019; Alam et al. 2019), graphene (Yi et al. 2019; Shao et al. 2010), and metal–organic framework nanomaterials (Cao et al. 2019; Lu et al. 2019a, b). This leads to the electrodes with good stability, huge specific area, improved redox performance, and recyclability. In the following paragraphs we introduce the application of these electrode materials in the construction of electrochemical sensors and electrochemical biosensors.

### 2.2.1 Conducting Polymers

Because of the delocalization of π-bonded electrons over polymeric backbone, the conducting polymers exhibit unique electronic properties, such as low ionization potentials and high electron affinities (Ates 2013). Owing to excellent electronic conductivity properties, operability, and low cost, conducting polymers are deemed to be one of the most attractive materials to modify the electrode surface. Liu and co-workers have explored the over-oxidized poly(3,4-ethylenedioxythiophene) films to replace Nafion films as the fixed layer for the modification of electrodes (Liu et al. 2011). Polypyrrole was utilized to enhance electrocatalytic currents in this system, attributing to its good conducting performance. Manisankar and coworkers have prepared poly(3,4-ethylenedioxythiophene)-modified electrodes and fabricated electrochemical sensors for the detection of pesticides (Manisankar et al. 2005). The sensors show excellent performance toward the determination of pesticides. Recently, copolymer-based electrode was developed by in situ generation 3-poly(propylene thiophenoimine)-co-poly(3-hexylthiophene), which was applied to design a highly sensitive pyrene electrochemical sensor (Makelane et al. 2019). In addition, conducting polymers have been also applied to immobilize enzymes onto the electrode surface for the preparation of electrochemical biosensors, due to the unique cross-linking properties (Lin and Yan 2012).

### 2.2.2 Metal-Based Nanomaterials

With the rapid development of technologies, metal nanomaterials have been widely used in various fields of catalysis, biomedicine, energy, and environmental analysis. Currently, metal nanomaterials with different compositions and morphologies are successfully synthesized, in which some of them act as the electrode-modified materials used for the fabrication of electrochemical sensors and biosensors. Rahman and coworkers have synthesized the gold–silver alloy nanoparticles and fabricated an electrochemical sensor for the detecting of pyrene with the low detection limit of 0.1 μM (Latif ur et al. 2015). Poliana and coworkers have constructed an electrochemical sensor with high sensitivity for the quantitative analysis of phenolic compounds, which is on the strength of zinc oxide nanocomposites-modified glass carbon electrode (Freire et al. 2016). The synthesized zinc oxide nanoflowers can be used to effectively enhance the current response and reduce the separated peaks during the process of electrochemical oxidation of phenolic compounds. Magnetic ferroferric oxide nanoparticles were synthesized and utilized to fabricate an electrochemical sensor for simultaneous detection of multiple heavy metal ions (Wu et al. 2019). The integration between as-prepared magnetic ferroferric oxide nanoparticles and multiwalled carbon nanotubes has effectively improved the performance of electrochemical sensor. Meanwhile, the metal-based nanomaterials are served as electrode materials and also used to fabricate electrochemical biosensors. Gu's group has used multisegment gold/platinum nanowire and nanoparticle hybrid arrays as the electrode materials and developed antibiotics electrochemical biosensors with excellent analytical properties, in which L-cysteine and penicillinase were considered as biological recognition molecules for tetracycline and penicillin, respectively (Li et al. 2019a, b).

### 2.2.3 Carbon Nanotubes

The carbon nanotubes consist of $sp^2$ carbon units, which is different from the traditional carbon fibers. There are two categories in the family of carbon nanotubes: single-wall carbon nanotubes and multiwall carbon nanotubes. Carbon nanotubes possess excellent mechanical and electronic properties and can be used in the sensing field. Alam and coworkers have developed an electrochemical sensor for detecting of lead ions in drinking water using the modified multiwall carbon nanotubes (Alam et al. 2019). Benefit from high electron transfer efficiency of multiwall carbon nanotubes, the sensors exhibit high sensitivity and selectivity for the analysis of lead ion. Ahmet and coworkers have synthesized single-wall carbon nanotubes-based hybrid material and designed an electrochemical sensor with excellent analysis performance for the determination of serine (Şenocak et al. 2019). Sun's group has developed an ultrasensitive electrochemical aptasensor for chlorpyrifos using ordered mesoporous carbon/ferrocene hybrid multiwall carbon nanotubes as electrode materials, which could be provided to enhance the sensitivity of the developed biosensor (Jiao et al. 2016).

### 2.2.4 Graphene

Graphene and its derivatives have attracted tremendous attention from researchers in recent years, due to their special physical and chemical properties of good electric conductivity, strong mechanical strength, and large surface area (Shao et al. 2010). These excellent features make them as good candidate electrode materials for the construction of electrochemical sensors and biosensors. Although the basic building unit of graphene is similar to zero-dimension fullerenes and one-dimension carbon nanotubes, the two-dimensional structure of graphene and its derivatives result in an enlarged specific surface area. Surface modification of working electrodes through graphene can significantly increase the effective electrode area and enhance response signals of electrochemical sensors and biosensors. Furthermore, graphene oxide composite materials have been prepared and used as the modification electrode materials for the fabrication of sensitive electrochemical sensor, which have been applied to simultaneously determine a variety of heavy metal ions (Yi et al. 2019). Recently, porous three-dimensional graphene framework has also been investigated to enhance the mechanical property. Shi and coworkers have synthesized a novel three-dimensional graphene framework/bismuth nanoparticles film with rapid electron transfers ability, remarkably large active area, excellent structure stability, and high mass transfer efficiency, which have been utilized to design a heavy metal ions electrochemical sensor with enhanced analytical performance (Shi et al. 2017). Similarly, Sethuraman and coworkers have developed an electrochemical catechol biosensor based on reduced graphene oxide-metal oxide composite and enzyme, in which the introduced graphene nanocomposites were provided to improve the sensitivity of catechol biosensors (Sethuraman et al. 2016).

### 2.2.5 Metal–Organic Frameworks

The metal–organic framework is an organic-inorganic hybrid material with intramolecular pores formed by self-assembly of organic ligands and metal ions or clusters through coordination bonds. In recent years, metal–organic frame materials have captured considerable attention, because of their characteristics of porosity, large specific surface area, structural and functional diversity, and unsaturated metal sites. Various metal–organic framework materials have been prepared, and have important applications in the fields of hydrogen storage, gas adsorption and separation, sensors, drug release, catalytic reaction, and so on. With the increasing number of metal-organic framework materials and the gradual rise of composite metal–organic framework materials, metal–organic framework materials will have inestimable application prospects. However, there are only a few studies on the application of metal-organic framework materials as electrode materials in the fabrication of electrochemical sensors and biosensors. Based on as-prepared graphene aerogel–metal organic frameworks, Wang's group has developed an effective electrochemical sensor for simultaneous detection of heavy metal ions (Lu et al. 2019a, b). Owing to the synergy effect of graphene and metal-organic frameworks, the sensors show high sensitivity and selectivity. Cao and coworkers have utilized hierarchically porous copper-based metal–organic framework materials to construct an electrochemical glyphosate sensor, which exhibits high sensitivity with the ultralow detection limit of $1.4 \times 10^{-13}$ mol $L^{-1}$ (Cao et al. 2019).

## 3 Application of Environmental Contaminants

Currently, many efforts have been devoted to the research and developments of technologies toward a decrease or detection of the impacts of environmental pollutions (Maduraiveeran and Jin 2017). The potential of electrochemical sensors and biosensors was explored for detecting commonly environmental contaminants including heavy metal ions, phenolic compounds, polyaromatic hydrocarbons, pesticides, antibiotics, pathogens, gas pollutants, and so on (Fig. 2). Due to their simplicity, accuracy, and portability, electrochemical sensors and biosensors have been widely demonstrated to be useful for the recognition and quantitative analysis of specific compounds (Ramnani et al. 2016; Rotariu et al. 2016; Saidur et al. 2017). The detailed applications of electrochemical sensors and biosensors in the detection of environmental pollutions are depicted as follows.

### 3.1 *Heavy Metal Ions*

When the specific gravity of metals is larger than 5 and metals with a density greater than 4.5 g $cm^{-3}$, the metals can be generally defined as heavy metals, containing gold, silver, copper, iron, mercury, lead, cadmium, and so on (Bansod et al. 2017; Cui et al. 2015; Zhang et al. 2019a, b). Heavy metals can cause chronic poisoning when the accumulation reaches a certain extent in the living organism. However, heavy metals usually involve lead, cadmium, mercury, chromium, metalloid

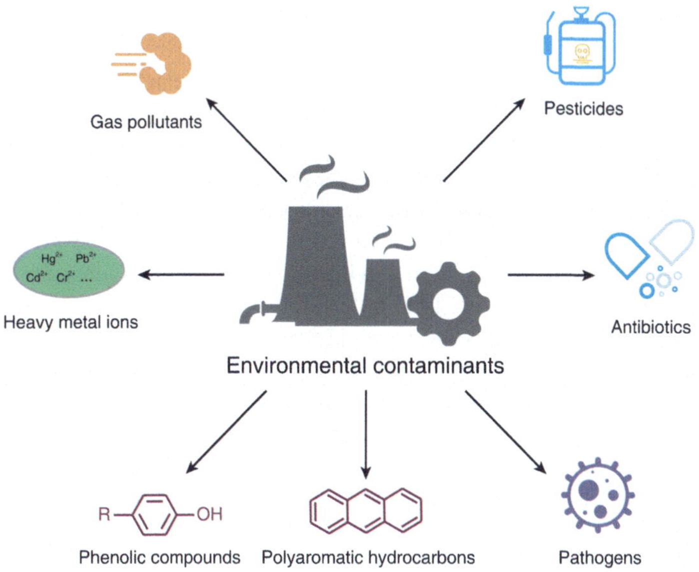


**Fig. 2** The classification of commonly environmental pollutions

arsenic, and other crucial biological poisonous heavy elements in the matter of environmental contaminants. Heavy metals are very hard to biodegrade and can be easily accumulated in vivo through the biomagnification effect of the food chain. Thus, the aggregated heavy metals can cause some irreversible damages to human beings, because they can lead to chronic poisoning and inactivation of enzymes and proteins in the human body. Due to the increasing industrial activities of human beings, heavy metal ions have become a significantly serious environmental problem that cannot be ignored. Therefore, it is urgent to develop an efficient, rapid, sensitive, accurate, and highly selective approach for the determination of heavy metal ions.

Compared with the traditional methods (Liu et al. 2017; Barbosa et al. 2016; Liang et al. 2000; Sun et al. 2017), electrochemical technique can be regarded as one of the most potential methods for the detection of heavy metal ions, due to its simplicity, rapid, high selectivity, and sensitivity. The process of electrochemical determination of heavy metal ions primarily involves two parts: recognition process and signal translation process. The detection mechanism of heavy metal ion electrochemical sensors and biosensors are both mainly based on the conversion of the specific recognition reaction between recognition element and target analytes to a measurable electrochemical signal (Saidur et al. 2017). And the generated signal response is related to the concentration of studied analyte, so heavy metal ions can be quantitatively analyzed by electrochemical sensors and biosensors.

### 3.1.1 Lead Ions

Lead ions, as one of the most toxic metallic pollutants, can accumulate in living organisms through the food chain and cause a series of adverse effects on human health (Wang et al. 2018; Dolati et al. 2017; Zhou et al. 2016). Due to the potential threats of lead ions to public health, many countries and organizations have strict hygiene regulations on the content of lead ions in drinking water. The maximum limit for lead ions contaminant in drinking water has been set to be 72 nM by the U.S. Environmental Protection Agency, while the International Agency for Research on Cancer has defined the safety limit of lead ion as 48.26 nM in food and drinking water (Guo et al. 2015). Thus the development of a novel analytical approach is significant for effective and routine detection of lead ions in the environment.

Currently, various lead ions electrochemical sensors and biosensors have been developed for highly selective and sensitive determination of lead ions. Zhang and coworkers have prepared $Mn_{1-x}Zn_xFe_2O_{4s}$-modified electrodes and used for directly electrochemical detection of lead ions, and the analytical performance of this sensor was enhanced by the introduction of zinc ion and Nafion (Zhang et al. 2019a, b). Additionally, several electrochemical biosensors have also been designed for the effective determination of lead ions. For example, Wei and coworkers have proposed a highly sensitive lead ion electrochemical aptasensor, which is based on gold nanoparticles-modified graphene nanocomposite as a signal probe to achieve ultrasensitive determination of lead ions (Fig. 3) (Wang et al. 2019a, b).

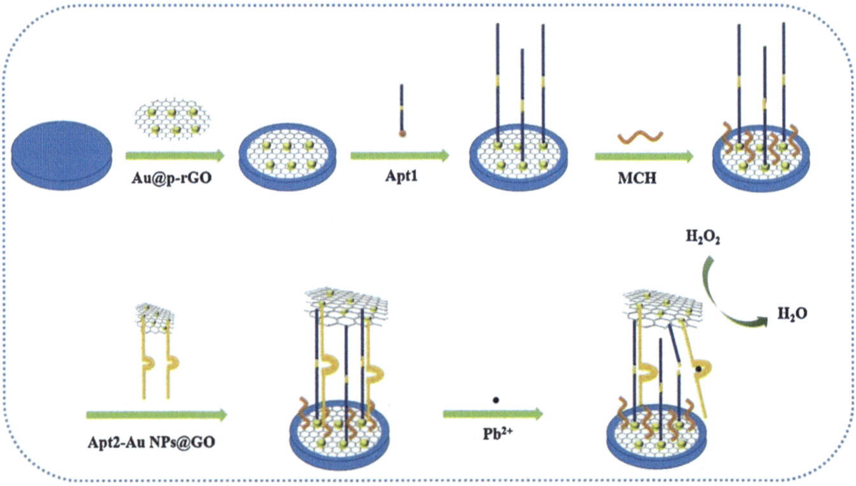


**Fig. 3** The scheme of fabrication of lead ion electrochemical aptasensor. Note the Au@p-rGO was prepared and supplied to immobilize the aptamer, which showed excellent electrocatalytic activity toward $H_2O_2$. Based on this, a highly sensitive electrochemical aptasensor was constructed and exhibited good performance for the detection of lead ions. *Au@p-rGO* Gold-modified porous reduced graphene oxide, *Apt1* sulfydryl-labeled substrate strand, *Apt2* sulfhydryl-labeled catalytic strand, *MCH* 6-mercapto-1-hexanol. Reprinted from Wang et al. (2019a, b), with permission from Elsevier

### 3.1.2 Cadmium Ions

Cadmium ion is also one of the heavy metal pollutions that easily accumulates in the environment through the food chain (Si et al. 2018). It is highly toxic and endangers the health of humans and the environment. Thus, it is of great importance for researchers to develop a rapid, convenient, low-cost, and high-efficiency technique to monitor and detect trace amounts of cadmium ions in the environment.

Due to their simplicity and cost-efficiency, electrochemical sensors have been frequently used in cadmium ion detection in recent 5 years (Si et al. 2018; Zhao et al. 2016). For example, a novel cadmium ions electrochemical sensor on account of reduced graphene oxide–gold nanoparticles–tetraphenylporphyrin nanoconjugates has been proposed by Si and coworkers (Si et al. 2018). The cadmium ions can be effectively detected, according to the coordination effect between cadmium ions and porphyrin. The sensitivity of the as-developed electrochemical sensor could be enhanced by the introduced Au nanoparticles.

### 3.1.3 Mercury Ion

As one of the most toxic heavy metal contaminants, accumulated mercury ion can cause serious adverse effects on human and environmental health. The mercury ions determination has attracted extensive attention to environmental monitoring. In recent 5 years, numerous literatures based on electrochemical detection mercury ions have been reported (Akbari Hasanjani and Zarei 2019; Xu et al. 2018; Isa et al. 2017; Wang et al. 2016). For example, Li and coworkers have constructed a highly sensitive electrochemical biosensor for the quantitative analysis of mercury ions, in which the electrical signals have been amplified by the synergistic effect between DNA-based hybridization chain and silver@gold core–shell nanoparticles with a positive charge (Li et al. 2016). This mercury ions electrochemical biosensor shows excellent performance with a low detection limit of 3.6 nM. Subsequently, He's group has utilized the specific thymine-$Hg^{2+}$-thymine base pair to fabricate a highly sensitive mercury ions electrochemical biosensor (Fig. 4) (Yu et al. 2019). Toluidine blue integrated with hybridization chain reaction could effectively realize mercury ions signal amplification. The developed biosensor exhibited a remarkable response for mercury ions, which includes a low detection limit of 0.2 pM and a wide linear range from 1 pM to 100 nM.

### 3.1.4 Arsenic Ions

Arsenic ion is also considered as a category of toxic environmental pollution, which can bring about various health and environmental issue, such as the skin, kidneys, urinary bladder, and lungs (Kato et al. 2016; Jaishankar et al. 2014). Therefore, it is necessary to develop a convenient, rapid, and efficient approach for quantitative analysis of arsenic ions in the environment. In contrast with traditional spectroscopy

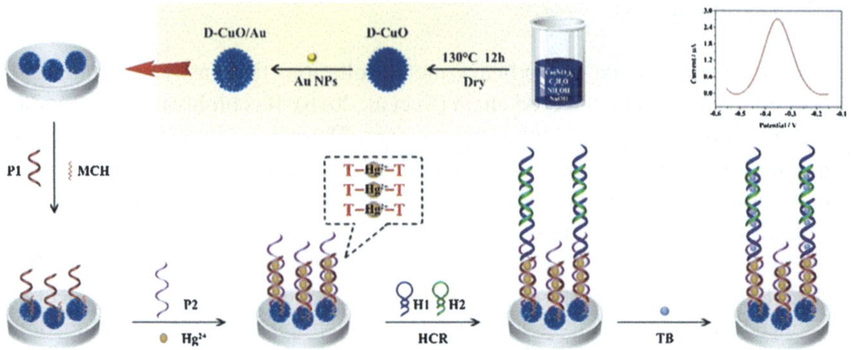


**Fig. 4** The fabrication of the mercury ions electrochemical biosensor. Note CuO dandelion-like microspheres modified with Au nanoparticles were prepared and used as electrode materials to enhance the analytical properties of biosensors. And toluidine blue, as redox indicator, was integrated with the hybridization chain reaction for signal amplification. *D-CuO-Au* gold nanoparticle modified dandelion-like copper oxide, *P1* thiolated probe; *P1: other oligonucleotide H1 and H2* two hairpin DNA, *HCR* hybridization chain reaction, *TB* toluidine blue. Reprinted from Yu et al. (2019), with permission from Elsevier

methods (Kempahanumakkagari et al. 2017), electrochemical methods are more suitable for the monitor of arsenic ion content in terms of their easy operation, low-cost, and high sensitivity. For example, Gumpu and coworkers have fabricated a preeminent sensor for electrochemical determination of arsenite and arsenate, on the basis of ruthenium bipyridine–graphene oxide nanocomposite modified electrode (Fig. 5) (Gumpu et al. 2018). This as-developed sensor exhibits excellent properties toward the determination of arsenic(III) and arsenic(V) with the detection limits of 21 nM and 34 nM, respectively.

### 3.1.5 Simultaneous Detection of Heavy Metal Ions

As well known, common heavy metal ions in the environment mainly include lead, cadmium, mercury, arsenic, and other metal ions. Although numerous literatures on electrochemical detection heavy ions have been reported (Cui et al. 2015; Si et al. 2018; Akbari Hasanjani and Zarei et al. 2019; Kato et al. 2016), it is still important in terms of research significance of the simultaneous monitor for various heavy metal ions due to the complexity of actual samples. As illustrated in Fig. 6, Ferreira's group has investigated the impacts of the incorporation of gold nanoparticles into layer-by-layer films of emeraldine salt polyaniline and sodium montmorillonite clay mineral and constructed an electrochemical sensor for simultaneous several heavy metal ions detection (de Barros et al. 2017). Electrochemical sensors exhibited high performance toward the detection of trace-level cadmium, lead, and copper ions with the induced gold nanoparticles.

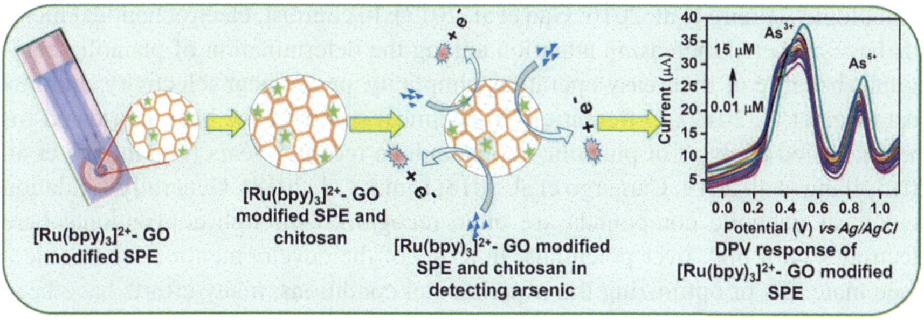


**Fig. 5** Schematic of electrochemical detection of arsenic ions. Note a preeminent sensor for electrochemical determination of arsenite and arsenate, on the basis of ruthenium bipyridine–graphene oxide nanocomposite ($[Ru(bpy)_3]^{2+}$-GO)-modified electrode. And $[Ru(bpy)_3]^{2+}$-GO was used as interface and promoted electron transfer rates. Reprinted from Gumpu et al. (2018), with permission from Elsevier

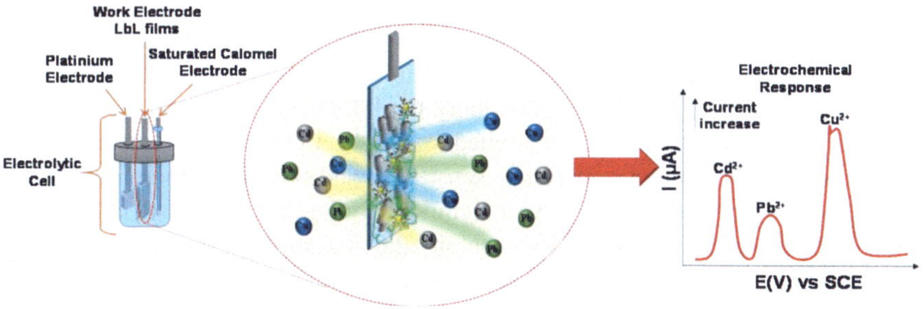


**Fig. 6** Schematic of simultaneous electrochemical detection of several heavy ions. Note the impacts of the incorporation of gold nanoparticles into layer-by-layer films of emeraldine salt polyaniline and sodium montmorillonite clay mineral have been investigated. On the basis of this, an electrochemical sensor was constructed for simultaneous detection of cadmium, lead, and copper ions by square wave anodic stripping voltammetry. Reprinted from Barros et al. (2017), with permission from Elsevier

## *3.2 Phenolic Compounds*

Phenolic compounds are prototype poisons substances and are as one of the 129 priority controlled pollutants identified by the US Environmental Protection Agency, which has great potential harm to human health and ecological environment (Huang et al. 2016; Xie et al. 2006). Phenolic compounds, as one of the main pollutants, are mainly derived from dye, pesticide, petrochemical, and other enterprises discharged wastewater, which are widely distributed in the natural environment (Gan et al. 2017). It is of great significance for precise analysis of trace phenolic compounds in the real sample.

Common analytical methods for phenolic compounds mainly involve spectrophotometry (Han et al. 2014), high-performance liquid chromatography (Hofmann et al. 2015), gas chromatography (Kovács et al. 2011), and electrochemical technologies (Freire et al. 2016; Gan et al. 2017). In contrast, electrochemical methods have attracted increasing attention among the determination of phenolic compounds because of their easy operation, simplicity, preeminent selectivity, and low cost (Wee et al. 2019). Different electrochemical sensors have been fabricated for the quantified analysis of phenolic compounds in recent 5 years (Govindhan et al. 2015; Jiang et al. 2019; Camargo et al. 2018; Lima et al. 2018). Generally, oxidation signals of phenolic compounds are often recognized through conventional bare electrodes with high over potentials. In terms of the development of various electrode materials or optimizing the experimental conditions, many efforts have been devoted to magnifying measurable signals for the electrochemical detection of phenolic compounds (Lima et al. 2018; Maikap et al. 2016; Huang et al. 2015). For example, a titanium dioxide-based electrochemical sensor was designed for the detection of bisphenol A, in which high-energy {0 0 1}-exposed titanium dioxide single crystals were served as the electrode material to amplify the signal (Fig. 7) (Pei et al. 2018). This as-fabricated sensitive sensor obtained excellent performance toward the bisphenol A with a low detection limit of 3.0 nM.

Phenolic catalytic enzymes are involved during the development of electrochemical biosensors, which can be used to effectively catalyze oxidation phenolic compounds (Apetrei et al. 2013; Roychoudhury et al. 2016). Tyrosinase, laccase, and polyphenol oxidase are commonly applied to construct phenolic biosensors, which have advantages of rapid response, high sensitivity, and selectivity, long-term stability (Andreescu and Sadik 2004). As a significant polyphenol oxidase, laccase can oxidize the phenolic hydroxyl to generate the measurable signal (Rodríguez-Delgado et al. 2015). For instance, Jiang and coworkers have constructed a new electrochemical biosensor for simultaneous determination of catechol and hydroquinone (Fig. 8) (Jiang et al. 2019). Meanwhile, tyrosinase has the capacity for the oxidization of mono and diphenol compounds among the most phenolic biosensors, which is due to its two copper atoms within the active sites (Sethuraman et al. 2016).

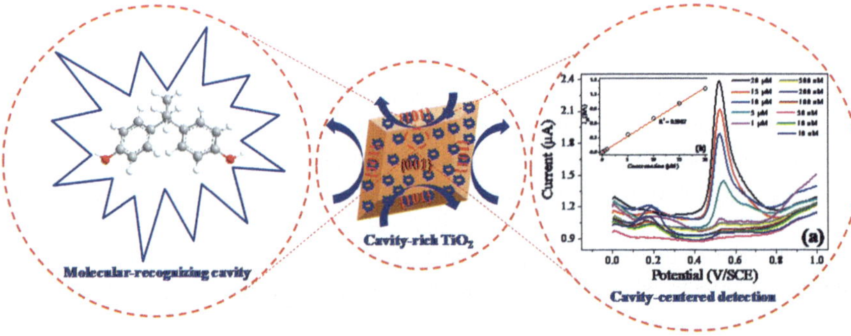


**Fig. 7** Schematic of electrochemical detection of bisphenol A. Note high-energy {0 0 1}-exposed titanium dioxide single crystals were prepared and served as the electrode material to amplify the signal. The fabricated electrochemical sensor was applied for the detection of bisphenol A with wide linear range. Reprinted from Pei et al. (2018), with permission from American Chemical Society

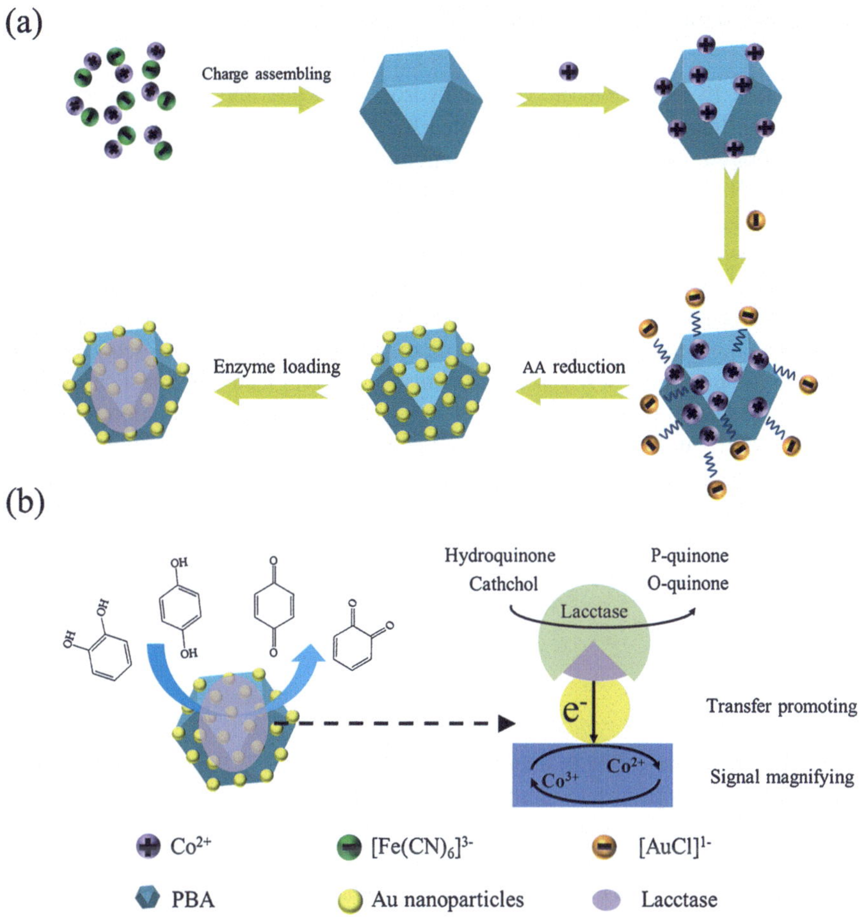


**Fig. 8** Lactase-based electrochemical biosensor for the simultaneous detection of catechol and hydroquinone. Note each prepared Prussian blue analog crystal with the high surface area can be a benefit for the combination of enzyme center and Prussian blue analog crystal, which can accelerate the electron transfer. Based on the synergy between electrode materials, the as-prepared phenolic biosensor was designed and obtained simultaneous detection for the trace hydroquinone and catechol with high sensitivity under various applied potentials. *PBA* Prussian blue analog. Reprinted from Jiang et al. (2019), with permission from Elsevier

A highly selective biosensor was built for the electrochemical determination of catechol on the basis of enzyme-modified electrode (Sethuraman et al. 2016).

Simultaneous determination of phenolic isomers is of great significance due to the coexistence of them in actual samples. However, similar physicochemical properties of phenolic isomer would result in the overlapping peak potentials, which is a key obstacle for their quantitative analysis (Huang et al. 2016). Some researchers have carried out electrochemical pretreatment on the surface of screen-printed electrode through different methods and obtained a fine peak separation of hydroquinone and catechol (Wang et al. 2010).

## 3.3 Polyaromatic Hydrocarbons

Polyaromatic hydrocarbons are volatile hydrocarbons and important environmental and food pollutants commonly generated by incomplete combustion of coal, oil, wood, tobacco, organic polymer compounds, and other organic compounds (Makelane et al. 2019). Up to now, more than 200 kinds of polyaromatic hydrocarbons, such as benzo α pyrene, benzo α anthracene, and so on, are easily carcinogenic, and widely distributed in the environment (Tovide et al. 2014). A large number of polyaromatic hydrocarbons in the environment are worrying because they are known to be highly toxic and have adverse health effects at low concentrations. Thus the determination of polyaromatic hydrocarbons in food, seawater, and lakes has aroused intensely research interest. To date, electrochemical sensor has been considered as a vital analysis tool for environmental determination, which is mainly based on the electrooxidation behavior of polyaromatic hydrocarbons (Makelane et al. 2019; Tovide et al. 2014). Iwuoha's group has exploited an anthracene electrochemical sensor based on graphenated polyaniline nanocomposite (Tovide et al. 2014). The amount of anthracene could be directly quantified through the electrooxidation behavior of anthracene on the surface of the modified electrode.

## 3.4 Pesticides

Generally, pesticides refer to chemicals extensively used in agriculture to control diseases and insect pests and regulate plant growth. Pesticides are the most affluent among the whole environmental contaminants, which are widely presented in atmosphere, water, soil, food, and plants (Uniyal and Sharma 2018; Kaur and Prabhakar 2017; Kumar et al. 2015). It is well known that pesticides are highly toxic and have an adverse impact on human's health and ecological balance. Therefore, it is vitally important to exploit a highly sensitive, facile, and low-cost approach for the accurate determination of pesticides in the environmental samples.

Traditionally, the detection methods of pesticides are general mass spectrometry (Chamkasem and Harmon 2016) and chromatography (Berijani et al. 2006), but these methods have some disadvantages of time-consuming, require a professional operation, and expensive. As a novel analytical method, electrochemical analytical technologies have been also extensively used due to their simplicity, sensitivity, and selectivity (Pérez-Fernández et al. 2019). In recent 5 years, various electrochemical sensors have been fabricated and applied for the quantitative analysis of pesticides through the direct redox reaction (Şenocak et al. 2019; Kaur and Prabhakar 2017; Xu et al. 2019). As shown in Fig. 9, a novel pesticide electrochemical sensor has been proposed based on the copper metal–organic frameworks modified electrode (Cao et al. 2019). Oxidation–reduction reaction could occur on the modified electrode surface, which would lead to an obvious signal response. When glyphosate combined with the modified electrode surface, glyphosate–copper ion complex

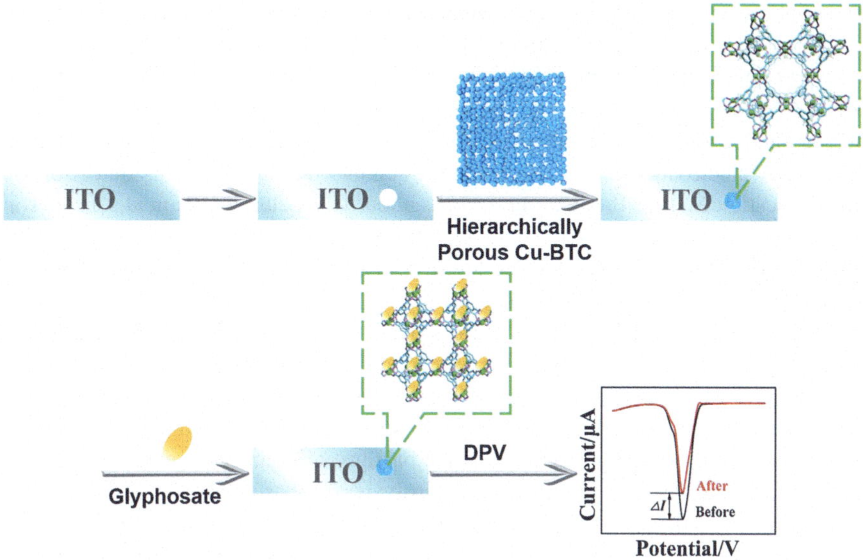


**Fig. 9** The principle of the glyphosate electrochemical sensor. Note a novel pesticide electrochemical sensor has been proposed based on the copper metal–organic frameworks modified electrode. When glyphosate combined with the modified electrode surface, glyphosate–copper ion complex would be produced because of adsorption of glyphosate on the metal center of copper metal–organic frameworks. And the generated complex could impede the electron transfer, causing a decrease of primary response current. The reduced current value is related to the reacted glyphosate concentration, so current changes before and after the reaction of a modified electrode with glyphosate could be utilized to detect the amount of glyphosate in real samples. *DPV* differential pulse voltammetry. Reprinted from Cao et al. (2019), with permission from Elsevier

would be produced because of adsorption of glyphosate on the metal center of copper metal–organic frameworks. Afterward, the generated complex could impede the electron transfer, causing a decrease of primary response current for the modified electrode. The reduced current value is related to the reacted glyphosate concentration, so current changes before and after the reaction of modified electrode with glyphosate were utilized to detect the amount of glyphosate in real samples. This as-developed electrochemical sensor exhibits excellent performance toward the determination of glyphosate.

In addition to the direct electrochemical sensor detection of pesticides, various electrochemical biosensors, including DNA-based biosensor (Eissa and Zourob 2017), enzyme sensors (Lu et al. 2018), immunosensors (El-Moghazy et al. 2018; Li et al. 2019a, b), and aptasensors (Roushani et al. 2018), have been also constructed and used for the highly sensitive detection of pesticides. According to the oxidation of guanine moieties, the electrochemical biosensors have been also developed for organophosphorus pesticide detection. As a recognition molecule, DNA is anchored on the modified electrodes surface in the fabrication process of DNA-based electrochemical biosensors (Eissa and Zourob 2017). The amount of organophosphorus pesticides can be quantified by monitoring the variation of redox characterizations of DNA that mainly refers to the oxidation of the guanine base. However, DNA-based electrochemical sensors are mainly based on the interactions between analytes and nitrogen bases, which can alter the electrochemical response of guanine base of DNA, so they have no specificity during the detection process. And DNA electrochemical sensors cannot achieve the simultaneous detection of several pesticides, thus there are few works about DNA electrochemical sensors for pesticide detection in recent 5 years. As prompted by the above limitations of DNA sensors, novel biomolecules and recognition strategies for the development of biosensors have been explored by researchers.

Enzymatic-based electrochemical biosensors for the detection of pesticides can be mainly based on the catalytic activity of organophosphorus hydrolase and the inhibition of several enzymes activity existed in pesticides, such as butyrylcholinesterase, acetylcholinesterase, tyrosinase, and so on (Cahuantzi-Muñoz et al. 2019). For the organophosphorus electrochemical biosensor, organophosphorus hydrolase enzyme is usually used to catalyze the hydrolysis of P-based bonds of organophosphorus, which will generate toxic products along with two protons. Because the released protons can result in a change of current or potential, organophosphorus hydrolase-based electrochemical biosensors have been fabricated and used for the detection of organophosphorus. Mulchandani and coworkers have proposed a highly sensitive electrochemical biosensor for paraoxon and methyl parathion based on organophosphorus hydrolase (Mulchandani et al. 2001). Based on the inhibition of acetylcholinesterase, Gao's group has fabricated and employed as a highly efficient electrochemical biosensor for the measurement of organophosphorus pesticides (Lu et al. 2019a, b).

As for immunosensors, measurable signals can be changed along with the organophosphorus concentrations because of the high specificity of organophosphorus antibodies to analytes (El-Moghazy et al. 2018). As shown in Fig. 10, an amino-modified metal-organic framework-based electrochemical immunosensor has been proposed for the simultaneous detection of pesticides triazophos and thiacloprid with high sensitivity (Yang et al. 2019a, b). With the assistance of specific monoclonal antibodies, Pérez-Fernández and coworkers have exploited a competitive immunosensor for the electrochemical determination of imidacloprid on screen-printed carbon electrodes (Pérez-Fernández et al. 2019).

Although enzymatic sensors and immunosensors have been widely used, they still possess several disadvantages of expensive. As a newly emerged method, aptasensors have appeared in the field of environmental monitoring to conquer the above limitations, which is mainly based on the folding of induced oligodeoxynucleotides aptamers. The combination of detected targets and nucleic acid aptamer on the electrode surface can alter its folding and flexibility, resulting in the electronic gain and loss of the redox-labeled substrates (Roushani et al. 2018). Consequently, an induced electrical signal will be generated and recorded, which is proportional to the concentration of pesticides. For instance, a high specific impedimetric biosensor was developed on the basis of aptamer modified platinum nanoparticles microwires for the detection of atrazine and acetamiprid (Fig. 11) (Madianos et al. 2018).

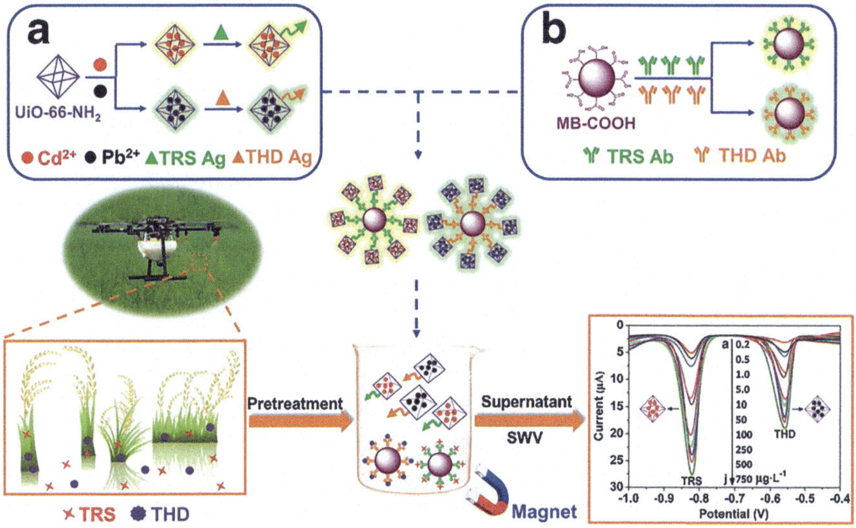


**Fig. 10** The fabrication of the glyphosate electrochemical immunosensor. Note amino-modified metal–organic framework nanoparticles were prepared to capture metal ions. Based on the specific binding of antigens to antibodies, a highly sensitive electrochemical immunosensor was constructed and for the simultaneous detection of pesticide triazophos and thiacloprid with a low limit of detection. *UiO-66-NH*$_2$ amino-modified metal–organic framework, *TRS* triazophos, *THD* thiacloprid, *Ag* antigen, *Ab* antibody, *MB-COOH* magnetic bead, *SWV* square wave voltammetry. Reprinted from Yang et al. (2019a, b), with permission from Springer

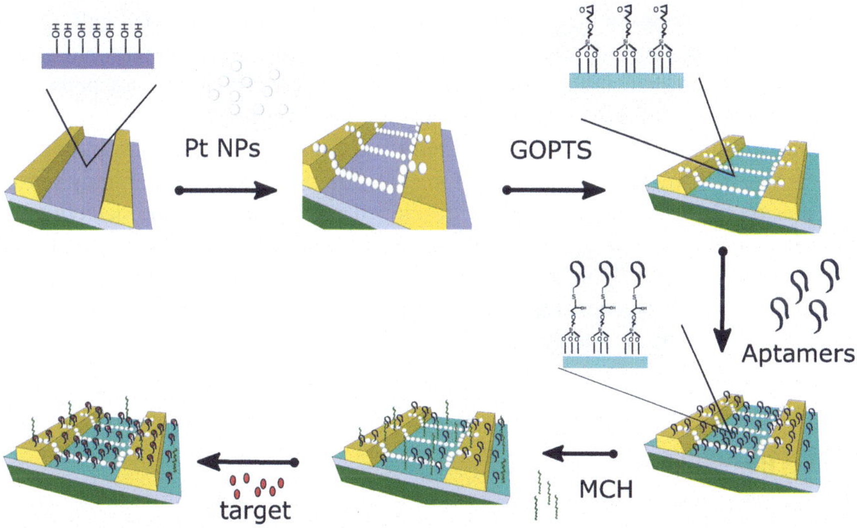


**Fig. 11** The fabrication and application of aptasensor. Note a high specific impedimetric biosensor was developed on the basis of aptamer modified platinum nanoparticles microwires. The aptasensor exhibited excellent performance toward the detection of atrazine and acetamiprid. *Pt NPs* platinum nanoparticles, *GOPTS* (3-glycidyloxypropyl)triethoxysilane, *MCH* 6-mercapto-1-hexanol. Reprinted from Madianos et al. (2018), with permission from Elsevier

## 3.5 Antibiotics

As a drug of inhibition and sterilizing bacteria, antibiotics have been widely used in medical and health, livestock and poultry breeding, agricultural production and other industries, and promoting the economic development of the society (Liu et al. 2018; Chen et al. 2019; Zhou et al. 2019). Antibiotics are mainly used as feed additives in livestock and poultry breeding, which are used to prevent and treat animal diseases and accelerate animal growth. However, the unreasonable use of antibiotics in livestock and poultry industry has been very common, highly sensitive detection of antibiotics is vitally necessary to regulate this phenomenon.

The detection of antibiotics is generally trace analysis due to their relatively low concentration in wastewater, which often needs highly sensitive instruments for accurate detection. According to the previously reported work (Batrawi et al. 2017), the main detection techniques of antibiotics are chromatography and its combination, enzyme immunoassay, capillary electrophoresis, and so on. The above-mentioned methods not only have the advantages of high sensitivity, accuracy, enhanced analytical efficiency, low detection limit, and strong specificity but possess the shortcomings of expensive instruments, tedious operation, need for professional operators, complex sample handling, and high cost. Therefore, these methods are difficult to become a conventional detection technique. At present, electrochemical sensors and biosensors have attracted increasing attention in antibiotics detection, based on their merit of low cost, easy miniaturization, biocompatibility, high sensitivity, and rapid response.

Electrochemical antibiotics sensors and biosensors are mainly based on the direct redox reactions of antibiotics or chemical reactions between targets and immobilized recognition elements, such as, antigen, enzyme, aptamer, cell, small molecular, macromolecule chemicals, antibody, and so on, which will cause the variations for the detectable properties of the work electrode, such as potential, response current, and resistance. In the recent 5 years, various electrochemical sensors and biosensors have been constructed for the detection of various antibiotics (Sun et al. 2019; Wang et al. 2019a, b; Liu et al. 2019). For example, a highly sensitive electrochemical sensor has been developed and applied for the detection of ciprofloxacin in view of the coordination interaction between ciprofloxacin and copper ions (Fig. 12a) (Fang et al. 2019) Lu's group has synthesized various metal–organic framework as label-free bioplatforms for the construction of sensitive electrochemical oxytetracycline aptasensors (Fig. 12b) (Zhou et al. 2019).

## 3.6 Pathogens

Pathogens refer to microorganisms, including bacteria, viruses, rickettsia, and fungi, parasites, or other vectors that can cause diseases in humans, animals, and plants (Silva et al. 2018). Pathogen infections are deemed to be a significantly serious

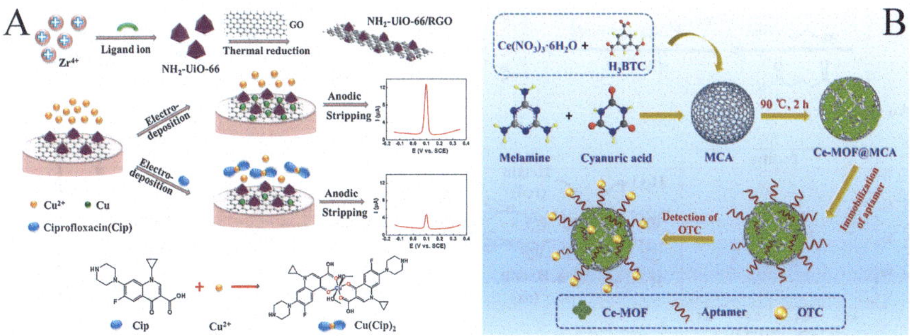


**Fig. 12** Electrochemical (**a**) sensor and (**b**) biosensor for different antibiotics. Note a highly sensitive electrochemical sensor has been developed and applied for the detection of ciprofloxacin based on the coordination interaction between ciprofloxacin and copper ions (**a**). Various metal–organic frameworks were synthesized and used as label-free bioplatforms for the construction of sensitive electrochemical oxytetracycline aptasensors (**b**). $Zr^{4+}$ Zirconium ions, *$NH_2$-UiO-66-RGO* amino-modified metal–organic framework supported on reduced graphene oxide, *MCA* melamine and cyanuric acid, *Ce-MOF* Ce-based metal–organic framework. A reprinted from Fang et al. (2019), with permission from American Chemical Society; B reprinted from Zhou et al. (2019), with permission from Elsevier

issue for global health, which can cause thousands of deaths and tremendous morbidity around the world (Campuzano et al. 2017). Therefore, it is urgent to develop an efficient method for the accurate detection of pathogens. Compared with the traditional approaches, diverse electrochemical biosensors have been also constructed and used to detect pathogens, due to their advantages of simplicity, inexpensive, sensitivity, and easy miniaturization (Hou et al. 2018; Jijie et al. 2018). The principle of pathogen electrochemical biosensors is mainly based on the specific recognition between various identification elements and targets, which can lead to the change of detectable signal. For instance, Guo and coworkers have fabricated a facile, label-free, cheap electrochemical *Escherichia coli* biosensor with satisfactory performance, while the electrochemical signals were amplified through rolling circle amplification and peroxidase-mimicking DNAzyme (Fig. 13) (Guo et al. 2016).

## 3.7 Gas Pollutants

Gas pollutants are one of the major pollutants, mainly including formaldehyde, nitrogen oxides, carbon monoxide, hydrogen sulfide, sulfur dioxide, ammonia, and so on (Wei et al. 2018; Trivedi et al. 2018). The released gas pollutants into the air not only cause environmental pollution but also pose a threat to human health. To date, various gas sensors have been fabricated and used for the detection of toxic gases. The electrochemical sensor is one of the most common sensors for detecting toxic and harmful gases, due to their serious of merits, such as easy miniaturization,

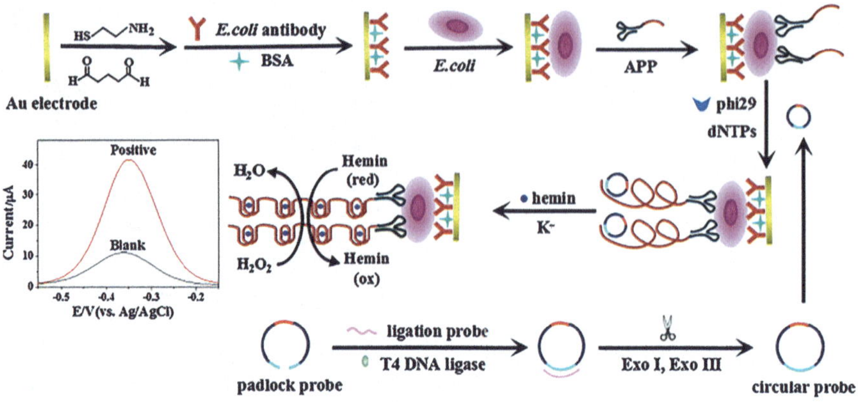


**Fig. 13** Fabrication and application of electrochemical *E. coli* biosensor. Note a facile, label-free, cheap electrochemical *E. coli* biosensor was fabricated on the gold electrode. The developed sensors indicated satisfactory performance, and the electrochemical signals were amplified through rolling circle amplification and peroxidase-mimicking DNAzyme. *BSA* bovine serum albumin, *APP* aptamer-primer probe, *Exo I* exonuclease I, *Exo III* exonuclease III, *red* reduction state, *ox* oxidation state. Reprinted from Guo et al. (2016), with permission from Elsevier

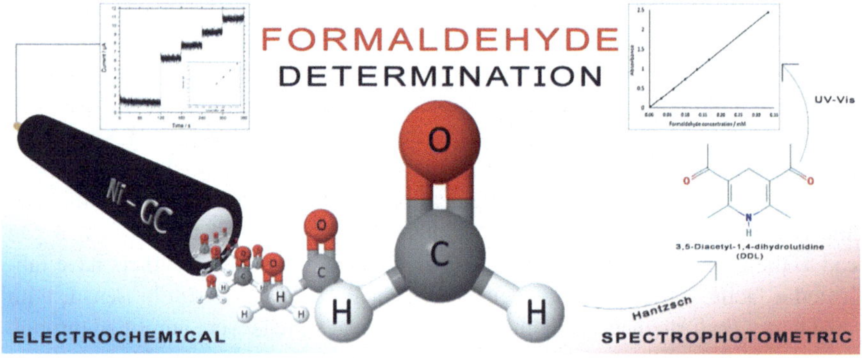


**Fig. 14** Detection of formaldehyde by electrochemical oxidation. Note nickel metal were prepared and severed as electrode materials. The as-designed electrochemical sensors showed excellent performance for the directly electrochemical oxidation of formaldehyde. Reprinted from Trivedi et al. (2018), with permission from Elsevier

inexpensive, good linearity and repeatability, long-term stability, and so on. The detecting principle of electrochemical gas sensors is based on electrochemical activity of the measured gas, which can be utilized to electrochemically oxidize or reduce the toxic gases, to distinguish the gas composition and detect the gas concentration. As depicted in Fig. 14, nickel–metal-modified electrodes were prepared and considered as work electrode for the determination of formaldehyde through directly electrochemical oxidation (Trivedi et al. 2018).

## 4 Conclusions and Perspectives

Electrochemical sensors and biosensors have been considered an effective analytical tool and have been applied for the detection of various environmental pollutants, owing to the remarkable performance of rapid response, high sensitivity, and selectivity. In this chapter, we have summarized their latest developments of electrochemical sensors and biosensors for environmental contaminants detection in the past 5 years. On the one hand, a number of novel electrode materials have been synthesized, which have significantly improved the sensitivity and selectivity of the electrochemical sensors. On the other hand, new biosensing strategies have been proposed and have been applied to fabricate electrochemical biosensors for the detection of various pollutants.

Although electrochemical sensors and biosensors have achieved satisfactory achievements in the detection of environmental pollutants, there are still some important challenges to be resolved. In order to better meet the needs of electrochemical sensors and biosensors in practical applications, some efforts should be made from the following aspects: (1) Novel electrode materials. The synthesis of new nanomaterials is of great significance for the fabrication of electrical sensors and biosensors, which can effectively enhance analytical performance. (2) Miniaturization. The development of portable sensors can improve work efficiency and reduce the consumption of reagents and manpower. (3) Combined with other spectral analysis techniques, such as Raman spectroscopy, the wide detection range and enhanced accuracy can be achieved. In brief, it is our research directions and goals to develop electrochemical sensors and biosensors with excellent performance and apply them to environmental and life analysis in the future.

**Acknowledgments** Financial support from the National Natural Science Foundation of China (Grant 21675080), the Natural Science Foundation of Jiangsu Province (Grant BK20170073) is appreciated.

## References

Akbari Hasanjani HR, Zarei K (2019) An electrochemical sensor for attomolar determination of mercury(II) using DNA/poly-L-methionine-gold nanoparticles/pencil graphite electrode. Biosens Bioelectron 128:1–8. https://doi.org/10.1016/j.bios.2018.12.039

Aksu Z (2005) Application of biosorption for the removal of organic pollutants: a review. Process Biochem 40:997–1026. https://doi.org/10.1016/j.procbio.2004.04.008

Alam AU, Howlader MMR, Hu N-X, Deen MJ (2019) Electrochemical sensing of lead in drinking water using β-cyclodextrin-modified MWCNTs. Sensors Actuators B Chem 296:126632. https://doi.org/10.1016/j.snb.2019.126632

Andreescu S, Sadik OA (2004) Correlation of analyte structures with biosensor responses using the detection of phenolic estrogens as a model. Anal Chem 76:552–560. https://doi.org/10.1021/ac034480z

Apetrei IM, Rodriguez-Mendez ML, Apetrei C, de Saja JA (2013) Enzyme sensor based on carbon nanotubes/cobalt(II) phthalocyanine and tyrosinase used in pharmaceutical analysis. Sensors Actuators B Chem 177:138–144. https://doi.org/10.1016/j.snb.2012.10.131

Ates M (2013) A review study of (bio)sensor systems based on conducting polymers. Mater Sci Eng C 33:1853–1859. https://doi.org/10.1016/j.msec.2013.01.035

Baker A, Inverarity R, Charlton M, Richmond S (2003) Detecting river pollution using fluorescence spectrophotometry: case studies from the Ouseburn, NE England. Environ Pollut 124:57–70. https://doi.org/10.1016/S0269-7491(02)00408-6

Bansod B, Kumar T, Thakur R, Rana S, Singh I (2017) A review on various electrochemical techniques for heavy metal ions detection with different sensing platforms. Biosens Bioelectron 94:443–455. https://doi.org/10.1016/j.bios.2017.03.031

Barbosa VMP, Barbosa AF, Bettini J, Luccas PO, Figueiredo EC (2016) Direct extraction of lead(II) from untreated human blood serum using restricted access carbon nanotubes and its determination by atomic absorption spectrometry. Talanta 147:478–484. https://doi.org/10.1016/j.talanta.2015.10.023

Barek J, Cvačka J, Muck A, Quaiserová V, Zima J (2001) Electrochemical methods for monitoring of environmental carcinogens. Fresen J Anal Chem 369:556–562. https://doi.org/10.1007/s002160100707

de Barros A, Constantino CJL, da Cruz NC, Bortoleto JRR, Ferreira M (2017) High performance of electrochemical sensors based on LbL films of gold nanoparticles, polyaniline and sodium montmorillonite clay mineral for simultaneous detection of metal ions. Electrochim Acta 235:700–708. https://doi.org/10.1016/j.electacta.2017.03.135

Batley GE (1983) Electroanalytical techniques for the determination of heavy metals in seawater. Mar Chem 12:107–117. https://doi.org/10.1016/0304-4203(83)90074-9

Batrawi N, Wahdan S, Al-Rimawi F (2017) A validated stability-indicating HPLC method for simultaneous determination of amoxicillin and enrofloxacin combination in an injectable suspension. Sci Pharm 85:6. https://doi.org/10.3390/scipharm85010006

Berijani S, Assadi Y, Anbia M, Milani Hosseini M-R, Aghaee E (2006) Dispersive liquid–liquid microextraction combined with gas chromatography-flame photometric detection: very simple, rapid and sensitive method for the determination of organophosphorus pesticides in water. J Chromatogr A 1123:1–9. https://doi.org/10.1016/j.chroma.2006.05.010

Bojdi MK et al (2014) Synthesis, characterization and application of novel lead imprinted polymer nanoparticles as a high selective electrochemical sensor for ultra-trace determination of lead ions in complex matrixes. Electrochim Acta 136:59–65. https://doi.org/10.1016/j.electacta.2014.05.095

Cahuantzi-Muñoz SL et al (2019) Electrochemical biosensor for sensitive quantification of glyphosate in maize kernels. Electroanalysis 31:927–935. https://doi.org/10.1002/elan.201800759

Camargo JR et al (2018) Electrochemical biosensor made with tyrosinase immobilized in a matrix of nanodiamonds and potato starch for detecting phenolic compounds. Anal Chim Acta 1034:137–143. https://doi.org/10.1016/j.aca.2018.06.001

Campuzano S, Yáñez-Sedeño P, Pingarrón MJ (2017) Molecular biosensors for electrochemical detection of infectious pathogens in liquid biopsies: current trends and challenges. Sensors 17:2533. https://doi.org/10.3390/s17112533

Cao Y et al (2019) An electrochemical sensor on the hierarchically porous cu-BTC MOF platform for glyphosate determination. Sensors Actuators B Chem 283:487–494. https://doi.org/10.1016/j.snb.2018.12.064

Chamkasem N, Harmon T (2016) Direct determination of glyphosate, glufosinate, and AMPA in soybean and corn by liquid chromatography/tandem mass spectrometry. Anal Bioanal Chem 408:4995–5004. https://doi.org/10.1007/s00216-016-9597-6

Chen C et al (2019) Amoxicillin on polyglutamic acid composite three-dimensional graphene modified electrode: reaction mechanism of amoxicillin insights by computational simulations. Anal Chim Acta 1073:22–29. https://doi.org/10.1016/j.aca.2019.04.052

Cui L, Wu J, Ju H (2015) Electrochemical sensing of heavy metal ions with inorganic, organic and bio-materials. Biosens Bioelectron 63:276–286. https://doi.org/10.1016/j.bios.2014.07.052

Daughton CG (2001) Emerging pollutants, and communicating the science of environmental chemistry and mass spectrometry: pharmaceuticals in the environment. J Am Soc Mass Spectrom 12:1067–1076. https://doi.org/10.1016/S1044-0305(01)00287-2

Davies J (1994) Inactivation of antibiotics and the dissemination of resistance genes. Science 264:375. https://doi.org/10.1126/science.8153624

Dolati S, Ramezani M, Abnous K, Taghdisi SM (2017) Recent nucleic acid based biosensors for $Pb^{2+}$ detection. Sensors Actuators B Chem 246:864–878. https://doi.org/10.1016/j.snb.2017.02.118

Eissa S, Zourob M (2017) Selection and characterization of DNA aptamers for electrochemical biosensing of carbendazim. Anal Chem 89:3138–3145. https://doi.org/10.1021/acs.analchem.6b04914

El-Moghazy AY et al (2018) Ultrasensitive label-free electrochemical immunosensor based on PVA-co-PE nanofibrous membrane for the detection of chloramphenicol residues in milk. Biosens Bioelectron 117:838–844. https://doi.org/10.1016/j.bios.2018.07.025

Falahatpisheh MH, Donnelly KC, Ramos KS (2001) Antagonistic interactions among nephrotoxic polycyclic aromatic hydrocarbons. J Toxicol Environ Health A 62:543–560. https://doi.org/10.1080/152873901300007833

Fang X et al (2019) Nanocomposites of Zr(IV)-based metal–organic frameworks and reduced graphene oxide for electrochemically sensing ciprofloxacin in water. ACS Appl Nano Mater 2:2367–2376. https://doi.org/10.1021/acsanm.9b00243

Freire PG et al (2016) Morphology of ZnO nanoparticles bound to carbon nanotubes affects electrocatalytic oxidation of phenolic compounds. Sensors Actuators B Chem 223:557–565. https://doi.org/10.1016/j.snb.2015.09.086

Fu F, Wang Q (2011) Removal of heavy metal ions from wastewaters: a review. J Environ Manag 92:407–418. https://doi.org/10.1016/j.jenvman.2010.11.011

Fu C et al (2018) A homogeneous electrochemical sensor for $Hg^{2+}$ determination in environmental water based on the T–$Hg^{2+}$–T structure and exonuclease III-assisted recycling amplification. Analyst 143:2122–2127. https://doi.org/10.1039/C8AN00462E

Gan T et al (2017) Flexible graphene oxide-wrapped $SnO_2$ hollow spheres with high electrochemical sensing performance in simultaneous determination of 4-aminophenol and 4-chlorophenol. Electrochim Acta 250:1–9. https://doi.org/10.1016/j.electacta.2017.08.043

Govindhan M, Lafleur T, Adhikari B-R, Chen A (2015) Electrochemical sensor based on carbon nanotubes for the simultaneous detection of phenolic pollutants. Electroanalysis 27:902–909. https://doi.org/10.1002/elan.201400608

Gumpu MB, Veerapandian M, Krishnan UM, Rayappan JBB (2018) Electrochemical sensing platform for the determination of arsenite and arsenate using electroactive nanocomposite electrode. Chem Eng J 351:319–327. https://doi.org/10.1016/j.cej.2018.06.097

Guo Y, Li J, Zhang X, Tang Y (2015) A sensitive biosensor with a DNAzyme for lead(II) detection based on fluorescence turn-on. Analyst 140:4642–4647. https://doi.org/10.1039/C5AN00677E

Guo Y et al (2016) Label-free and highly sensitive electrochemical detection of E. coli based on rolling circle amplifications coupled peroxidase-mimicking DNAzyme amplification. Biosens Bioelectron 75:315–319. https://doi.org/10.1016/j.bios.2015.08.031

Han XX et al (2014) Magnetic titanium dioxide nanocomposites for surface-enhanced resonance Raman spectroscopic determination and degradation of toxic anilines and phenols. Angew Chem Int Ed 53:2481–2484. https://doi.org/10.1002/anie.201310123

Hatch WR, Ott WL (1968) Determination of submicrogram quantities of mercury by atomic absorption spectrophotometry. Anal Chem 40:2085–2087. https://doi.org/10.1021/ac50158a025

Hatchett DW, Josowicz M (2008) Composites of intrinsically conducting polymers as sensing nanomaterials. Chem Rev 108:746–769. https://doi.org/10.1021/cr068112h

Hofmann T, Nebehaj E, Albert L (2015) The high-performance liquid chromatography/multistage electrospray mass spectrometric investigation and extraction optimization of beech (*Fagus sylvatica* L.) bark polyphenols. J Chromatogr A 1393:96–105. https://doi.org/10.1016/j.chroma.2015.03.030

Hou Y-H et al (2018) A colorimetric and electrochemical immunosensor for point-of-care detection of enterovirus 71. Biosens Bioelectron 99:186–192. https://doi.org/10.1016/j.bios.2017.07.035

Huang YH et al (2015) One-pot hydrothermal synthesis carbon nanocages-reduced graphene oxide composites for simultaneous electrochemical detection of catechol and hydroquinone. Sensors Actuators B Chem 212:165–173. https://doi.org/10.1016/j.snb.2015.02.013

Huang J, Zhang X, Zhou L, You T (2016) Simultaneous electrochemical determination of dihydroxybenzene isomers using electrospun nitrogen-doped carbon nanofiber film electrode. Sensors Actuators B Chem 224:568–576. https://doi.org/10.1016/j.snb.2015.10.102

Isa IM et al (2017) Chloroplatinum(II) complex-modified MWCNTs paste electrode for electrochemical determination of mercury in skin lightening cosmetics. Electrochim Acta 253:463–471. https://doi.org/10.1016/j.electacta.2017.09.092

Jaishankar M, Tseten T, Anbalagan N, Mathew BB, Beeregowda KN (2014) Toxicity, mechanism and health effects of some heavy metals. Interdiscip Toxicol 7:60–72. https://doi.org/10.2478/intox-2014-0009

Jiang D et al (2019) Simultaneous biosensing of catechol and hydroquinone via a truncated cube-shaped au/PBA nanocomposite. Biosens Bioelectron 124–125:260–267. https://doi.org/10.1016/j.bios.2018.09.094

Jiao Y et al (2016) An ultrasensitive aptasensor for chlorpyrifos based on ordered mesoporous carbon/ferrocene hybrid multiwalled carbon nanotubes. RSC Adv 6:58541–58548. https://doi.org/10.1039/C6RA07735H

Jijie R et al (2018) Reduced graphene oxide/polyethylenimine based immunosensor for the selective and sensitive electrochemical detection of uropathogenic *Escherichia coli*. Sensors Actuators B Chem 260:255–263. https://doi.org/10.1016/j.snb.2017.12.169

Kato D, Kamata T, Kato D, Yanagisawa H, Niwa O (2016) Au nanoparticle-embedded carbon films for electrochemical $As^{3+}$ detection with high sensitivity and stability. Anal Chem 88:2944–2951. https://doi.org/10.1021/acs.analchem.6b00136

Kaur N, Prabhakar N (2017) Current scenario in organophosphates detection using electrochemical biosensors. TrAC Trends Anal Chem 92:62–85. https://doi.org/10.1016/j.trac.2017.04.012

Kempahanumakkagari S, Deep A, Kim K-H, Kumar Kailasa S, Yoon H-O (2017) Nanomaterial-based electrochemical sensors for arsenic-a review. Biosens Bioelectron 95:106–116. https://doi.org/10.1016/j.bios.2017.04.013

Kovács Á, Mörtl M, Kende A (2011) Development and optimization of a method for the analysis of phenols and chlorophenols from aqueous samples by gas chromatography–mass spectrometry, after solid-phase extraction and trimethylsilylation. Microchem J 99:125–131. https://doi.org/10.1016/j.microc.2011.04.007

Kumar P, Kim K-H, Deep A (2015) Recent advancements in sensing techniques based on functional materials for organophosphate pesticides. Biosens Bioelectron 70:469–481. https://doi.org/10.1016/j.bios.2015.03.066

Latif ur R et al (2015) Synthesis, characterization, and application of Au–Ag alloy nanoparticles for the sensing of an environmental toxin, pyrene. J Appl Electrochem 45:463–472. https://doi.org/10.1007/s10800-015-0807-2

Li M, Li Y-T, Li D-W, Long Y-T (2012) Recent developments and applications of screen-printed electrodes in environmental assays—a review. Anal Chim Acta 734:31–44. https://doi.org/10.1016/j.aca.2012.05.018

Li Z et al (2016) Ultrasensitive electrochemical sensor for $Hg^{2+}$ by using hybridization chain reaction coupled with ag@au core–shell nanoparticles. Biosens Bioelectron 80:339–343. https://doi.org/10.1016/j.bios.2016.01.074

Li X, Yuan Y, Pan X, Zhang L, Gong J (2019a) Boosted photoelectrochemical immunosensing of metronidazole in tablet using coral-like $g\text{-}C_3N_4$ nanoarchitectures. Biosens Bioelectron 123:7–13. https://doi.org/10.1016/j.bios.2018.09.084

Li Z, Liu C, Sarpong V, Gu Z (2019b) Multisegment nanowire/nanoparticle hybrid arrays as electrochemical biosensors for simultaneous detection of antibiotics. Biosens Bioelectron 126:632–639. https://doi.org/10.1016/j.bios.2018.10.025

Liang Q, Jing H, Gregoire DC (2000) Determination of trace elements in granites by inductively coupled plasma mass spectrometry. Talanta 51:507–513. https://doi.org/10.1016/S0039-9140(99)00318-5

Lima AP et al (2018) Influence of $Al_2O_3$ nanoparticles structure immobilized upon glassy-carbon electrode on the electrocatalytic oxidation of phenolic compounds. Sensors Actuators B Chem 262:646–654. https://doi.org/10.1016/j.snb.2018.02.028

Lin P, Yan F (2012) Organic thin-film transistors for chemical and biological sensing. Adv Mater 24:34–51. https://doi.org/10.1002/adma.201103334

Liu M et al (2011) A stable sandwich-type amperometric biosensor based on poly(3,4-ethylenedioxythiophene)—single walled carbon nanotubes/ascorbate oxidase/nafion films for detection of L-ascorbic acid. Sensors Actuators B Chem 159:277–285. https://doi.org/10.1016/j.snb.2011.07.005

Liu X et al (2017) Liquid spray dielectric barrier discharge induced plasma–chemical vapor generation for the determination of lead by ICPMS. Anal Chem 89:6827–6833. https://doi.org/10.1021/acs.analchem.7b01255

Liu X et al (2018) Recent advances in sensors for tetracycline antibiotics and their applications. TrAC Trends Anal Chem 109:260–274. https://doi.org/10.1016/j.trac.2018.10.011

Liu P et al (2019) An ultrasensitive electrochemical immunosensor for procalcitonin detection based on the gold nanoparticles-enhanced tyramide signal amplification strategy. Biosens Bioelectron 126:543–550. https://doi.org/10.1016/j.bios.2018.10.048

Lu X, Tao L, Song D, Li Y, Gao F (2018) Bimetallic Pd@au nanorods based ultrasensitive acetylcholinesterase biosensor for determination of organophosphate pesticides. Sensors Actuators B Chem 255:2575–2581. https://doi.org/10.1016/j.snb.2017.09.063

Lu M et al (2019a) Graphene aerogel–metal–organic framework-based electrochemical method for simultaneous detection of multiple heavy-metal ions. Anal Chem 91:888–895. https://doi.org/10.1021/acs.analchem.8b03764

Lu X, Tao L, Li Y, Huang H, Gao F (2019b) A highly sensitive electrochemical platform based on the bimetallic Pd@au nanowires network for organophosphorus pesticides detection. Sensors Actuators B Chem 284:103–109. https://doi.org/10.1016/j.snb.2018.12.125

Lupu S, Lete C, Marin M, Totir N, Balaure PC (2009) Electrochemical sensors based on platinum electrodes modified with hybrid inorganic–organic coatings for determination of 4-nitrophenol and dopamine. Electrochim Acta 54:1932–1938. https://doi.org/10.1016/j.electacta.2008.07.051

Madianos L, Tsekenis G, Skotadis E, Patsiouras L, Tsoukalas D (2018) A highly sensitive impedimetric aptasensor for the selective detection of acetamiprid and atrazine based on microwires formed by platinum nanoparticles. Biosens Bioelectron 101:268–274. https://doi.org/10.1016/j.bios.2017.10.034

Maduraiveeran G, Jin W (2017) Nanomaterials based electrochemical sensor and biosensor platforms for environmental applications. Trends Environ Anal 13:10–23. https://doi.org/10.1016/j.teac.2017.02.001

Maikap A, Mukherjee K, Mondal B, Mandal N (2016) Zinc oxide thin film based nonenzymatic electrochemical sensor for the detection of trace level catechol. RSC Adv 6:64611–64616. https://doi.org/10.1039/C6RA09598D

Makelane H, Waryo T, Feleni U, Iwuoha E (2019) Dendritic copolymer electrode for second harmonic alternating current voltammetric signalling of pyrene in oil-polluted wastewater. Talanta 196:204–210. https://doi.org/10.1016/j.talanta.2018.12.038

Manisankar P, Viswanathan S, Pusphalatha AM, Rani C (2005) Electrochemical studies and square wave stripping voltammetry of five common pesticides on poly 3,4-ethylenedioxythiophene modified wall-jet electrode. Anal Chim Acta 528:157–163. https://doi.org/10.1016/j.aca.2004.08.027

Martinez JL (2009) Environmental pollution by antibiotics and by antibiotic resistance determinants. Environ Pollut 157:2893–2902. https://doi.org/10.1016/j.envpol.2009.05.051

Metters JP, Kadara RO, Banks CE (2012) Electroanalytical sensing of chromium (III) and (VI) utilising gold screen printed macro electrodes. Analyst 137:896–902. https://doi.org/10.1039/C2AN16054D

Mulchandani P, Chen W, Mulchandani A (2001) Flow injection amperometric enzyme biosensor for direct determination of organophosphate nerve agents. Environ Sci Technol 35:2562–2565. https://doi.org/10.1021/es001773q

Njau KN, Woude MV, Visser GJ, Janssen LJJ (2000) Electrochemical removal of nickel ions from industrial wastewater. Chem Eng J 79:187–195. https://doi.org/10.1016/S1385-8947(00)00210-2

O'Malley V (1999) The integrated pollution prevention and control (IPPC) directive and its implications for the environment and industrial activities in Europe. Sensors Actuators B Chem 59:78–82. https://doi.org/10.1016/S0925-4005(99)00199-9

Paulino AT et al (2006) Novel adsorbent based on silkworm chrysalides for removal of heavy metals from wastewaters. J Colloid Interface Sci 301:479–487. https://doi.org/10.1016/j.jcis.2006.05.032

Pei D-N, Zhang A-Y, Pan X-Q, Si Y, Yu H-Q (2018) Electrochemical sensing of bisphenol A on facet-tailored $TiO_2$ single crystals engineered by inorganic-framework molecular imprinting sites. Anal Chem 90:3165–3173. https://doi.org/10.1021/acs.analchem.7b04466

Pérez-Fernández B, Mercader JV, Checa-Orrego BI, de la Escosura-Muñiz A, Costa-García A (2019) A monoclonal antibody-based immunosensor for the electrochemical detection of imidacloprid pesticide. Analyst 144:2936–2941. https://doi.org/10.1039/C9AN00176J

Ramnani P, Saucedo NM, Mulchandani A (2016) Carbon nanomaterial-based electrochemical biosensors for label-free sensing of environmental pollutants. Chemosphere 143:85–98. https://doi.org/10.1016/j.chemosphere.2015.04.063

Reemtsma T (2003) Liquid chromatography–mass spectrometry and strategies for trace-level analysis of polar organic pollutants. J Chromatogr A 1000:477–501. https://doi.org/10.1016/S0021-9673(03)00507-7

Rodríguez-Delgado MM et al (2015) Laccase-based biosensors for detection of phenolic compounds. TrAC Trends Anal Chem 74:21–45. https://doi.org/10.1016/j.trac.2015.05.008

Rotariu L, Lagarde F, Jaffrezic-Renault N, Bala C (2016) Electrochemical biosensors for fast detection of food contaminants—trends and perspective. TrAC Trends Anal Chem 7:80–87. https://doi.org/10.1016/j.trac.2015.12.017

Roushani M, Nezhadali A, Jalilian Z (2018) An electrochemical chlorpyrifos aptasensor based on the use of a glassy carbon electrode modified with an electropolymerized aptamer-imprinted polymer and gold nanorods. Microchim Acta 185:551. https://doi.org/10.1007/s00604-018-3083-0

Roychoudhury A, Basu S, Jha SK (2016) Dopamine biosensor based on surface functionalized nanostructured nickel oxide platform. Biosens Bioelectron 84:72–81. https://doi.org/10.1016/j.bios.2015.11.061

Saidur MR, Aziz ARA, Basirun WJ (2017) Recent advances in DNA-based electrochemical biosensors for heavy metal ion detection: a review. Biosens Bioelectron 90:125–139. https://doi.org/10.1016/j.bios.2016.11.039

Samanta SK, Singh OV, Jain RK (2002) Polycyclic aromatic hydrocarbons: environmental pollution and bioremediation. Trends Biotechnol 20:243–248. https://doi.org/10.1016/S0167-7799(02)01943-1

Şenocak A et al (2019) Highly selective and ultra-sensitive electrochemical sensor behavior of 3D SWCNT-BODIPY hybrid material for eserine detection. Biosens Bioelectron 128:144–150. https://doi.org/10.1016/j.bios.2018.12.052

Sethuraman V, Muthuraja P, Anandha Raj J, Manisankar P (2016) A highly sensitive electrochemical biosensor for catechol using conducting polymer reduced graphene oxide–metal oxide enzyme modified electrode. Biosens Bioelectron 84:112–119. https://doi.org/10.1016/j.bios.2015.12.074

Shao Y et al (2010) Graphene based electrochemical sensors and biosensors: a review. Electroanalysis 22:1027–1036. https://doi.org/10.1002/elan.200900571

Shen L et al (2008) Electrochemical DNAzyme sensor for lead based on amplification of DNA–Au bio-bar codes. Anal Chem 80:6323–6328. https://doi.org/10.1021/ac800601y

Shi L, Li Y, Rong X, Wang Y, Ding S (2017) Facile fabrication of a novel 3D graphene framework/bi nanoparticle film for ultrasensitive electrochemical assays of heavy metal ions. Anal Chim Acta 968:21–29. https://doi.org/10.1016/j.aca.2017.03.013

Si Y et al (2018) rGO/AuNPs/tetraphenylporphyrin nanoconjugate-based electrochemical sensor for highly sensitive detection of cadmium ions. Anal Methods 10:3631–3636. https://doi.org/10.1039/C8AY01020J

Silva NFD, Magalhães JMCS, Freire C, Delerue-Matos C (2018) Electrochemical biosensors for Salmonella: state of the art and challenges in food safety assessment. Biosens Bioelectron 99:667–682. https://doi.org/10.1016/j.bios.2017.08.019

Smith DL, Harris AD, Johnson JA, Silbergeld EK, Morris JG (2002) Animal antibiotic use has an early but important impact on the emergence of antibiotic resistance in human commensal bacteria. Proc Natl Acad Sci 99:6434. https://doi.org/10.1073/pnas.082188899

Srivastava NK, Majumder CB (2008) Novel biofiltration methods for the treatment of heavy metals from industrial wastewater. J Hazard Mater 151:1–8. https://doi.org/10.1016/j.jhazmat.2007.09.101

Su W-Y, Wang S-M, Cheng S-H (2011) Electrochemically pretreated screen-printed carbon electrodes for the simultaneous determination of aminophenol isomers. J Electroanal Chem 651:166–172. https://doi.org/10.1016/j.jelechem.2010.11.028

Sun Q et al (2017) A new electrochemical system based on a flow-field shaped solid electrode and 3D-printed thin-layer flow cell: detection of $Pb^{2+}$ ions by continuous flow accumulation square-wave anodic stripping voltammetry. Anal Chem 89:5024–5029. https://doi.org/10.1021/acs.analchem.7b00383

Sun Y et al (2019) Novel three-dimensional electrochemical sensor with dual signal amplification based on $MoS_2$ nanosheets and high-conductive $NH_2$-MWCNT@COF for sulfamerazine determination. Sensors Actuators B Chem 281:107–114. https://doi.org/10.1016/j.snb.2018.10.055

Teh CM, Mohamed AR (2011) Roles of titanium dioxide and ion-doped titanium dioxide on photocatalytic degradation of organic pollutants (phenolic compounds and dyes) in aqueous solutions: a review. J Alloys Compd 509:1648–1660. https://doi.org/10.1016/j.jallcom.2010.10.181

Tovide O et al (2014) Electro-oxidation of anthracene on polyanilino-graphene composite electrode. Sensors Actuators B Chem 205:184–192. https://doi.org/10.1016/j.snb.2014.07.116

Trivedi D, Crosse J, Tanti J, Cass AJ, Toghill KE (2018) The electrochemical determination of formaldehyde in aqueous media using nickel modified electrodes. Sensors Actuators B Chem 270:298–303. https://doi.org/10.1016/j.snb.2018.05.035

Uniyal S, Sharma RK (2018) Technological advancement in electrochemical biosensor based detection of organophosphate pesticide chlorpyrifos in the environment: a review of status and prospects. Biosens Bioelectron 116:37–50. https://doi.org/10.1016/j.bios.2018.05.039

Vaishnav VS, Patel SG, Panchal JN (2015) Development of ITO thin film sensor for detection of benzene. Sensors Actuators B Chem 206:381–388. https://doi.org/10.1016/j.snb.2014.07.037

Wan H, Yin H, Lin L, Zeng X, Mason AJ (2018) Miniaturized planar room temperature ionic liquid electrochemical gas sensor for rapid multiple gas pollutants monitoring. Sensors Actuators B Chem 255:638–646. https://doi.org/10.1016/j.snb.2017.08.109

Wang S-M, Su W-Y, Cheng S-H (2010) A simultaneous and sensitive determination of hydroquinone and catechol at anodically pretreated screen-printed carbon electrodes. Int J Electrochem Sci 5:1649–1664

Wang N, Lin M, Dai H, Ma H (2016) Functionalized gold nanoparticles/reduced graphene oxide nanocomposites for ultrasensitive electrochemical sensing of mercury ions based on thymine–mercury–thymine structure. Biosens Bioelectron 79:320–326. https://doi.org/10.1016/j.bios.2015.12.056

Wang L et al (2018) Sensitive and label-free electrochemical lead ion biosensor based on a DNAzyme triggered G-quadruplex/hemin conformation. Biosens Bioelectron 115:91–96. https://doi.org/10.1016/j.bios.2018.04.054

Wang M et al (2019a) Covalent organic framework-based electrochemical aptasensors for the ultrasensitive detection of antibiotics. Biosens Bioelectron 132:8–16. https://doi.org/10.1016/j.bios.2019.02.040

Wang Y et al (2019b) Electrochemical aptasensor based on gold modified graphene nanocomposite with different morphologies for ultrasensitive detection of $Pb^{2+}$. Sensors Actuators B Chem 288:325–331. https://doi.org/10.1016/j.snb.2019.03.010

Wee Y et al (2019) Tyrosinase-immobilized CNT based biosensor for highly-sensitive detection of phenolic compounds. Biosens Bioelectron 132:279–285. https://doi.org/10.1016/j.bios.2019.03.008

Wei P et al (2018) Impact analysis of temperature and humidity conditions on electrochemical sensor response in ambient air quality monitoring. Sensors 18:59. https://doi.org/10.3390/s18020059

Wu W et al (2019) Sensitive, selective and simultaneous electrochemical detection of multiple heavy metals in environment and food using a lowcost $Fe_3O_4$ nanoparticles/fluorinated multi-walled carbon nanotubes sensor. Ecotoxicol Environ Saf 175:243–250. https://doi.org/10.1016/j.ecoenv.2019.03.037

Xie T, Liu Q, Shi Y, Liu Q (2006) Simultaneous determination of positional isomers of benzenediols by capillary zone electrophoresis with square wave amperometric detection. J Chromatogr A 1109:317–321. https://doi.org/10.1016/j.chroma.2006.01.135

Xu HL et al (2007) Simultaneous detection and identification of multigas pollutants using filament-induced nonlinear spectroscopy. Appl Phys Lett 90:101106. https://doi.org/10.1063/1.2711537

Xu A et al (2018) Ultrasensitive electrochemical sensing of $Hg^{2+}$ based on thymine-$Hg^{2+}$-thymine interaction and signal amplification of alkaline phosphatase catalyzed silver deposition. Biosens Bioelectron 104:95–101. https://doi.org/10.1016/j.bios.2018.01.005

Xu G et al (2019) Dual-signal aptamer sensor based on polydopamine-gold nanoparticles and exonuclease I for ultrasensitive malathion detection. Sensors Actuators B Chem 287:428–436. https://doi.org/10.1016/j.snb.2019.01.113

Yang S et al (2018) Advances in the use of carbonaceous materials for the electrochemical determination of persistent organic pollutants. A review. Microchim Acta 185:112. https://doi.org/10.1007/s00604-017-2638-9

Yang H et al (2019a) In situ construction of hollow carbon spheres with N, Co, and Fe co-doping as electrochemical sensors for simultaneous determination of dihydroxybenzene isomers. Nanoscale 11:8950–8958. https://doi.org/10.1039/C9NR01146C

Yang Y et al (2019b) An amino-modified metal-organic framework (type UiO-66-$NH_2$) loaded with cadmium(II) and lead(II) ions for simultaneous electrochemical immunosensing of triazophos and thiacloprid. Microchim Acta 186:101. https://doi.org/10.1007/s00604-018-3201-z

Yi W, He Z, Fei J, He X (2019) Sensitive electrochemical sensor based on poly(l-glutamic acid)/graphene oxide composite material for simultaneous detection of heavy metal ions. RSC Adv 9:17325–17334. https://doi.org/10.1039/C9RA01891C

Yu Y et al (2019) Dandelion-like CuO microspheres decorated with au nanoparticle modified biosensor for $Hg^{2+}$ detection using a T-$Hg^{2+}$-T triggered hybridization chain reaction amplification strategy. Biosens Bioelectron 131:207–213. https://doi.org/10.1016/j.bios.2019.01.063

Zhang J, Sun X, Wu J (2019a) Heavy metal ion detection platforms based on a glutathione probe: a mini review. Appl Sci 9:489. https://doi.org/10.3390/app9030489

Zhang Y et al (2019b) High-performance electrochemical sensor based on $Mn_{1-x}Zn_xFe_2O_4$ nanoparticle/nafion-modified glassy carbon electrode for $Pb^{2+}$ detection. J Electrochem Soc 166:B341–B348. https://doi.org/10.1149/2.0151904jes

Zhao G, Wang H, Liu G, Wang Z (2016) Box–Behnken response surface design for the optimization of electrochemical detection of cadmium by square wave anodic stripping voltammetry on bismuth film/glassy carbon electrode. Sensors Actuators B Chem 235:67–73. https://doi.org/10.1016/j.snb.2016.05.051

Zhou Q et al (2016) Highly sensitive electrochemical sensing platform for lead ion based on synergetic catalysis of DNAzyme and Au–Pd porous bimetallic nanostructures. Biosens Bioelectron 78:236–243. https://doi.org/10.1016/j.bios.2015.11.055

Zhou N et al (2019) Construction of Ce-MOF@COF hybrid nanostructure: label-free aptasensor for the ultrasensitive detection of oxytetracycline residues in aqueous solution environments. Biosens Bioelectron 127:92–100. https://doi.org/10.1016/j.bios.2018.12.024

Zhu Z et al (2009) Highly sensitive electrochemical sensor for mercury (II) ions by using a mercury-specific oligonucleotide probe and gold nanoparticle-based amplification. Anal Chem 81:7660–7666. https://doi.org/10.1021/ac9010809

# Research Insights on the Development of Biosensors

**Mohan Kumar Anand Raj, Rajasekar Rathanasamy, Gobinath Velu Kaliyannan, and Mohan Raj Thangamuthu**

## Contents



## Abbreviations

| | |
|---|---|
| CE | Counter electrode |
| EIS | Electrolyte-insulator-semiconductor |
| $E_{pa}$ | Potential of the anodic peak current |
| $E_{pc}$ | Potential of the cathodic peak current |
| IHP | Inner Helmholtz plane |
| $I_{pa}$ | Anodic peak current |
| $I_{pc}$ | Cathodic peak current |

M. K. Anand Raj (✉) · R. Rathanasamy · G. V. Kaliyannan
Department of Mechanical Engineering, Kongu Engineering College, Erode, Tamilnadu, India

M. R. Thangamuthu
Department of Mechanical Engineering, Amrita School of Engineering, Coimbatore, Amrita Vishwa Vidyapeetham, Coimbatore, India



Inamuddin, A. M. Asiri (eds.), *Nanosensor Technologies for Environmental Monitoring*, Nanotechnology in the Life Sciences,
https://doi.org/10.1007/978-3-030-45116-5_2

| | |
|---|---|
| LAPS | Light-addressable potentiometric sensor |
| OHP | Outer Helmholtz plane |
| $R_B$ | Bulk resistance |
| $R_C$ | Contact resistance |
| RE | Reference electrode |
| $R_I$ | Interface resistance |
| $R_S$ | Surface resistance |
| WE | Working electrode |

# 1 Introduction

Biosensors are applicable in such fields as the medical realm, food industries, agriculture, the treatment of industrial wastes, and armed defence (Kara 2012; Thévenot et al. 2001). Biosensors are devices that convert biological or biochemical signals to electrical responses (Mohanty and Kougianos 2006; Turner 2013). It is an advanced technology compared to traditional sensors. The multifunctional sensing system has been developed by combining many sensors. Various transducer types such as acoustical, optical, electronic, and mechanical have been developed by utilizing biosensors. The biologically active materials have more influence on the functions of biosensors. Factors to be considered when choosing biological materials include environmental conditions, storage, and operations. Annually, 60% of research work has been carried out with biosensors in the medical field (Malhotra 2017). Clark developed an enzyme electrode at the beginning of biosensor research (Pandey and Malhotra 2019). Thereafter, researchers from such fields as physics, materials, and medical worked together to develop multipurpose, higher-sensitivity biosensor devices (Hinze 1994; Malhotra et al. 2005; Vadgama and Crump 1992). Biosensors are utilized not only for commercial purposes but also in the field of defence during biowars (Song et al. 2006). Biosensors are defined based on the field of application and the purpose. In general, a biosensor is a device that has a biological sensing element for converting complex biochemical signals to electrical signals with sophisticated, understandable measurable formats.

The following factors or conditions are required when fabricating biosensors (Grieshaber et al. 2008; Turner 2013):

- The required sensing device should be stable under environmental conditions.
- The device should be accurate, precise, and have a higher degree of sensitivity.
- The sensor must be small in size and biocompatible so it can be used in medical settings.
- It should be easy to fabricate, portable, low in cost, and usable by less skilled laborers (Perumal and Hashim 2014).

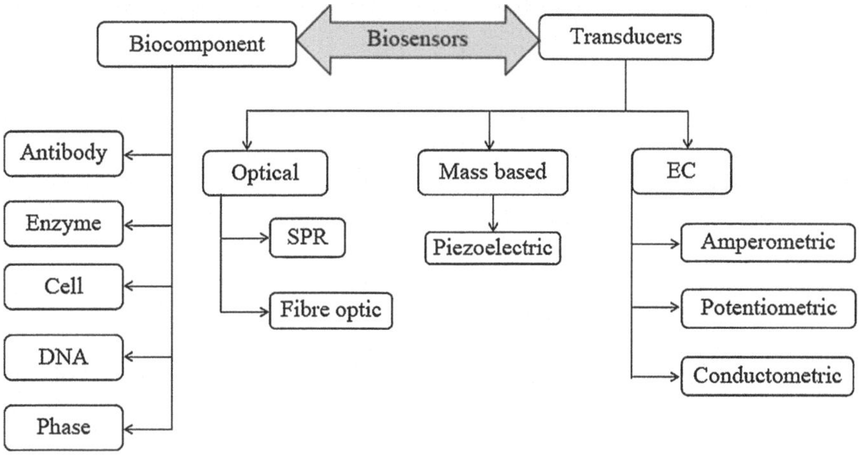


**Fig. 1** Classification of biosensors

The most important part of the biosensor is the transducer, which functions to convert complex measurable biochemical data to understandable electrical signals Biosensors are classified into two types: biocomponents and transducers (Fig. 1).

Biosensors are classified as optical, thermal, electrochemical, and piezoelectric according to the transducer element.

## 2 Electrochemical Biosensors

Electrochemical sensors are widely used in various fields because they allow biomolecules with high sensitivity (Thévenot et al. 2001). Biosensor response is based on the proximity of the three electrodes required for functioning: working electrode (WE), reference electrode (RE), and counter electrode (CE). The overall performance of biosensors depends on the type of electrode used. The factors that affect the detecting properties of electrodes are electrode material and dimensions (Grieshaber et al. 2008).

Figure 2 shows the electrodes present in electrochemical biosensors. Figure 3 is a diagrammatic representation of an electrochemical biosensor. The working electrode (WE) is a transducer element of the electrochemical biosensor, which is present in the biochemical reaction taking place. The counter electrode (CE) functions as a connection between the electrolytic solution and working electrode. To obtain the necessary stable potential, the reference electrode (RE) is kept at a distance from the reaction area. The materials silver and silver chloride have been used for the reference electrode. Electrons move from the analyte (the electroactive element) to the working electrode or from the working electrode to the analyte while the oxidation-reduction (redox) reaction is taking place. The direction of electron flow depends on the characteristics or properties of the analyte and the electric potential

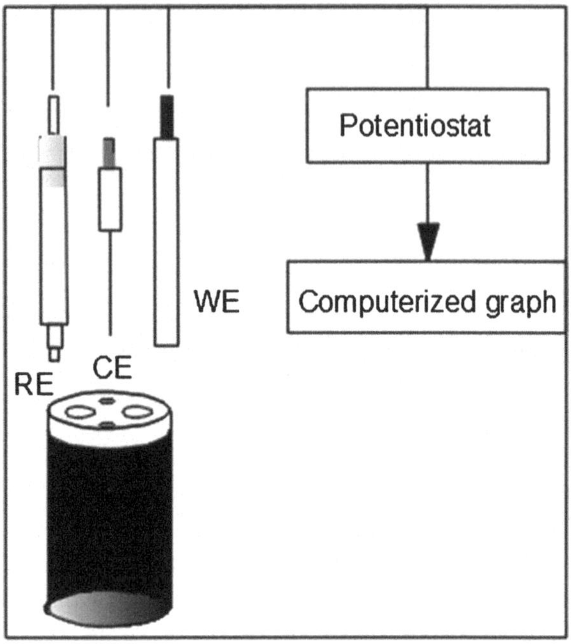


**Fig. 2** Electrodes present in an electrochemical biosensor

of the working electrode. The working electrode is stimulated as a positive potential when the oxidation reaction takes place. The potential difference that occurs depends on the concentration of the electroactive element dispersed on the surface of the working electrode (Grieshaber et al. 2008). The reduction reaction takes place when the working electrode is stimulated as a negative potential. The counter electrode is used to measure the potential difference when the working electrode and counter electrode are acting as anode and cathode, respectively. Such materials as gold, platinum, and carbon are used to fabricate auxiliary sensors.

Electrochemical biosensors have been used to measure current flow during oxidation and reduction reactions. Three electrodes are connected to the potentiostat: the working electrode, reference electrode, and conductive or auxiliary electrode. When a reaction takes place, a potential is applied on the working electrode, and the resulting current is displayed in the form of a graph with respect to time. A redox couple forms in solution from the equilibrium concentration of the oxidation and reduction reactions. The Nernst equation [Eq. (1)] shows the link between potential and redox couple:

$$E = E_0 + \frac{RT}{nF} \ln \frac{C_{\text{oxi}}}{C_{\text{red}}} \tag{1}$$

where

$E$ = potential

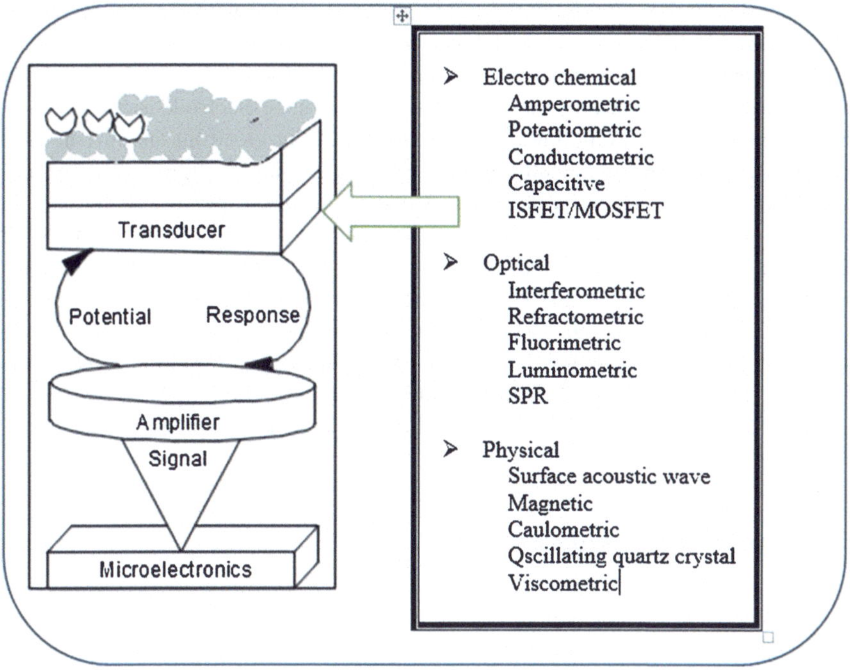


**Fig. 3** Diagrammatic representation of an electrochemical biosensor (*ISFET* ion-selective field-effective transistor, *MOSFET* metal oxide semiconductor field-effect transistor)

$E_0$ = standard half-cell potential
$F$ = Faraday constant.
$C_{oxi}$ = oxidation concentration
$C_{red}$ = reduction concentration
$T$ = absolute temperature.

The advantages of the electrochemical sensor are fast response, ease of use, and high sensitivity: it is economical and a comparatively simple device (Ronkainen et al. 2010; Sin et al. 2014). Electrochemical biosensors are classified into four types: amperometric, potentiometric, impedance, and conductometric (Aizawa 1991).

## 2.1 Amperometric Sensors

Amperometric sensors are electrochemical biosensors that are used to measure current variations resulting from redox of electroactive materials when the same intensity of potential is applied. The amount of the variation of current corresponds to the concentration of the electroactive materials in solution. In amperometric sensors three types of electrodes are used: working electrode, reference electrode, and counter electrode. Carbon, platinum, and gold are used for the working electrodes and silver or silver chloride is employed as a reference electrode. The potential of the working electrode is controlled by a fixed electrode, the reference electrode. The counter electrode or auxiliary electrode with the reference electrode is employed for the measurement of current flow. During redox reactions, electrons are moving from the analyte to the working electrode. The flow of electrons is controlled by the analyte characteristics. If the working electrode stimulates a positive potential, the results show an oxidation reaction has taken place (Dzyadevych et al. 2008). In other words, a reduction reaction takes place when the working electrode stimulates negative potential. Various analytes can be combined with the amperometric sensor for medical applications (Hasan et al. 2014). One of the major disadvantages of the amperometric sensor is showing false current readings; this limitation can be rectified by coating the electrode with conducting polymers and altering the type of analyte (Reinhardt et al. 2002). Figure 4 shows an amperometric biosensor.

## 2.2 *Cyclic Voltammogram*

Figure 5 depicts a cyclic voltammogram of a bare gold electrode in phosphate-buffered saline. In cyclic voltammogram graphs, current (μA) and potential (V) are drawn on the *x*- and *y*-axes, respectively. The sweep is gradual and denotes the potential of the cathodic peak current and anodic peak current as positive and negative on the *y*-axis, respectively; the potential of the working electrode decreases at a definite rate when the resulting electron flow is noted with respect to time. In general, the resulting current is inverted at a certain potential. The graph trajectory rate is constant, and the origin and potential are known; then, the time can be effortlessly

**Fig. 4** Amperometric biosensor

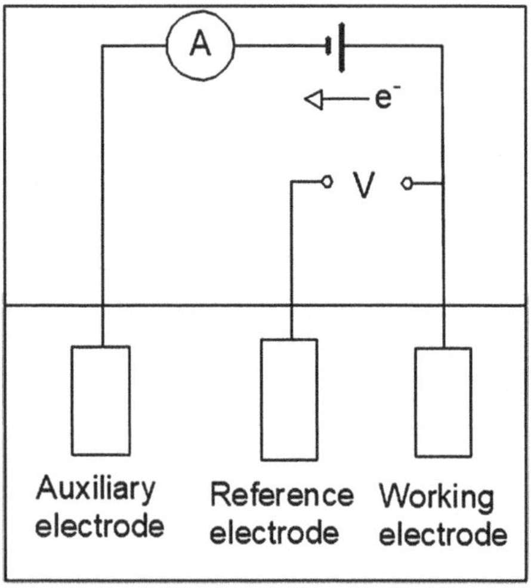

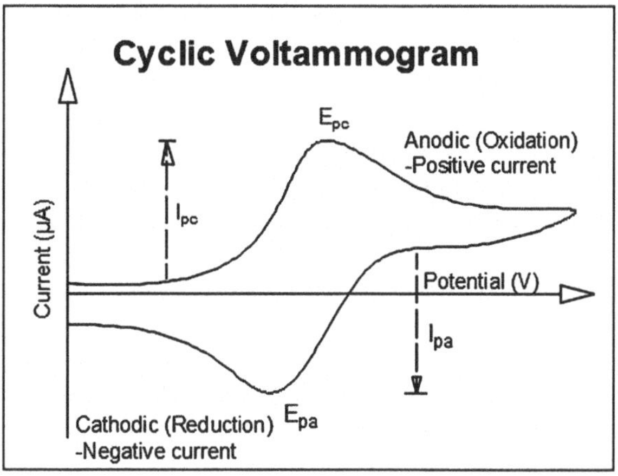


**Fig. 5** Cyclic voltammogram of a bare gold electrode in phosphate buffer saline ($E_{pa}$, potential of the anodic peak current; $E_{pc}$, potential of the cathodic peak current; $I_{pa}$, anodic peak current; $I_{pc}$, cathodic peak current)

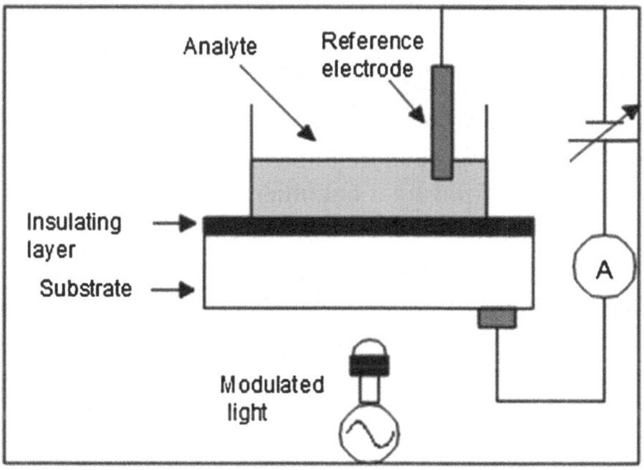


**Fig. 6** Light-addressable potentiometric sensor (LAPS)

converted into potential and the current versus potential graph is drawn easily. The current is noted at the working electrode through a potential test against a constant reference electrode potential.

The cyclic voltammogram is a multipurpose device used to analyse reversibility of the electrochemical reactions and its dispersion-controlled properties. Beyond the sensing application, it can be using to identify the processes carried out at the electrode (Mabbott 1983).

## 2.3 *Light-Addressable Potentiometric Sensor (LAPS)*

Figure 6 shows a schematic diagram of the light-addressable potentiometric sensor (LAPS) (Hafeman et al. 1988; Owicki et al. 1994). It is a kind of chemical sensor with a semiconductor device that is used to envision the three-dimensional distribution of ion concentrations in the solution. The sensor consists of an electrolyte-insulator-semiconductor (EIS) as shown in figure. A direct current bias voltage has passed over the EIS and the exhaust layer has formed between the insulator and semiconductor. The thickness of the exhaust layer depends on the concentration of ion and applied potential differences. The ion concentration-quantifying principle is the same as that of EIS capacitance sensors (Poghossian et al. 2001a, b). In LAPS, a photocurrent is induced by illuminating the semiconductor surface with modulated light. Unlike the EIS capacitance sensor, the LAPS sensor gives the mean value of the entire sensing surface, the quantifying area of the sensor defined by the illuminating surface. In a LAPS, the measured photocurrent value has been obtained by a laser beam focused on the sensing surface. The LAPS is also applicable as integrated multisensors, in which more than one measuring probe is used to the sensing surface by a light beam (Ermolenko et al. 2003; Schöning et al. 2005; Shimizu et al. 1994; Yoshinobu et al. 2005). The three-dimensional resolution is a significant factor in both electrochemical and multisensory applications. Three-dimensional resolution defines the smallest size of particles or elements that can be envisioned by the chemical sensor, and it restricts the concentration of the measuring point on the multisensor layer (George et al. 2000; Nakao et al. 1994; Sartore et al. 1992). The beam size and adjacent dispersion of photocarriers on the semiconductor surface have more influence on the three-dimensional resolution. The beam size is minimized by a one micrometer (1-μm) scale or is equivalent to the light wavelength with suitable focusing optics (Yoshinobu et al. 2004). The effect of light absorption coefficient and dispersion length is greater in lateral diffusion. More effort has been made by researchers to reduce the thickness of the silicon substrate for developing three-dimensional resolution of a chemical imaging and potentiometric sensor. Instead of minimizing the thickness of the sensor element, thickening of the semiconductor has been achieved by increasing the deposition rate and time from the vapor phase.

## 2.4 Conductometric Sensors

Conductometric sensors are generally bipolar instruments. A sample arranged with a selective layer in an adjacent surface is shown in Fig. 7.

In a conductometric sensor, the DC current is applied during measurement. The conducting samples sort the resistance from chemiresistors; the field of application may be a gas or a nonconducting liquid. Figure 7 shows chemiresistors and the corresponding circuit. The impedance is measured between the boundaries of conducting samples and selective layers, with AC current owing to the periodically altered exciting signals. The electrochemical sensor consists of complex arrangements of capacitance and resistance. The selective layer contains the primary interface between the conducting samples and the sensors. $R_S$, $R_B$, $R_I$, and $R_C$ represent surface, bulk, interface, and contact resistance, respectively. The points 1 and 2 or 3 and 4 show equivalent resistance by combining all the resistances (Bard et al. 1980).

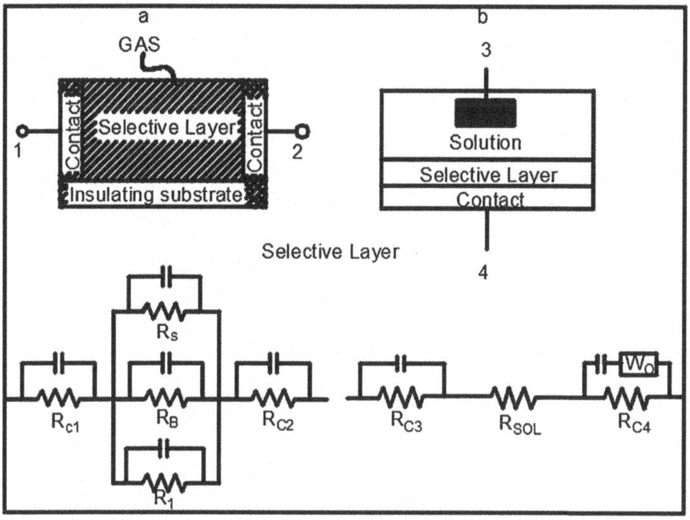


**Fig. 7** Common chemiresistors and their corresponding circuits: (**a**) adjacent structure in which any of the five resistances can be controlled by chemical interaction; (**b**) impedimetric chemiresistor in which capacitance $C_B$ is chemically modulated

In recent times, NA hybridisation and immunosensors have been detected by using micro- and nanofield effect sensors. One major advantage is that a reference electrode is not required; the cost is low, and it can be used for miniaturisation or for direct electrical pulses. Although these are specific advantages, sensitivity is less compared to other electrochemical sensors.

## 2.5 *Electrochemical Impedance Spectroscopy*

Electrochemical impedance is defined as resistance offered against the flow of electrical current in a circuit. The electrolyte solution, interfaces, and coatings are employed to measure the flow of ions in electrochemical impedance (Ianeselli et al. 2014). The impedance approach is very effective because it can move the test electron at higher frequency and mass movement at lower frequency (Bogomolova et al. 2009). During electrochemical reactions, measurement of impedance is done at an open circuit depending on which current flow occurs in an electrified boundary.

Electrons move from electrodes transversely to the electrified boundary (Fig. 8).

This method has a non-Faradaic component, which is written in the form of Eq. (2):

$$[\mathrm{O}] + n\mathrm{e}^{-} \rightarrow [\mathrm{R}] \quad (2)$$

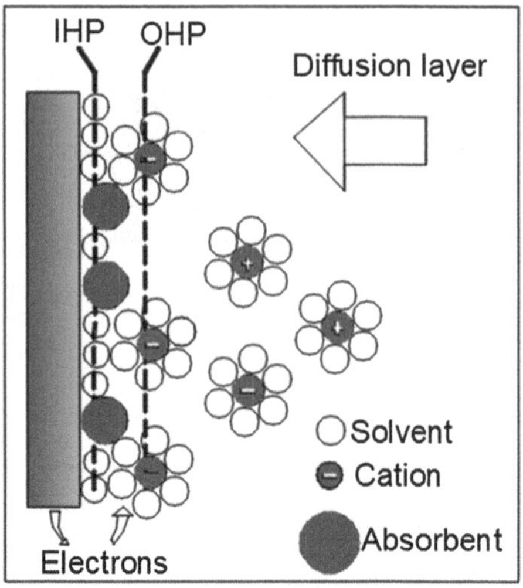


**Fig. 8** The electrified boundary at which the electrode is negatively charged; counter cations are arranged in line along the electrified layer. *IHP* inner Helmholtz plane, *OHP* outer Helmholtz plane

where $n$ denotes the number of electrons moved, O denotes oxidant, and R denotes reduction.

The activation barriers called polarisation resistance and uncompensated resistance are overwhelmed by electron movement across the boundary (Chang and Park 2010; Wolfbeis 2004). The rate of electron movement is defined by the mass movement of the reactant. It also depends on the depletion of the oxidant and the making of reductant close to the electrode surface. The mass movement of the reactants and the products give alternative impedance, which can be used to sense immunological binding measures such as antibody-binding occurrences on the surface of the electrode (Borisov and Wolfbeis 2008; McDonagh et al. 2008). The resulting in-phase and out-phase current reactions can be used for calculating capacitance and resistance in the circuit. Cell toxicology studies, cell movements, and morphology changes are monitored by this technology (Baronas and Kulys 2007).

# 3 Biosensor Application: Environmental Monitoring Biosensors

The use of biosensors in monitoring the concentrations of different contaminants present in the environment is an emerging field (Rinken 2013). Water, soil, and air are taken as major measurement areas for environmental monitoring. The major analytes identified are pesticides, heavy metals, herbicides, and phenolic compounds (Somerset 2011).

Biosensors for environmental monitoring consist of the analyte used for sensing biomaterials or chemical elements or combinations of these two (Serra 2011). The advantages of biosensors for environmental monitoring compared to other traditional methods are (1) the data can be transferred easily, and (2) minimal test specimens are sufficient for pollution measurements. The biosensors used for environmental monitoring are generally optical (Akkaya et al. 2016; Baro et al. 2016; Fojta et al. 2016; Giovanardi et al. 2017) or electrochemical biosensors (Dey and Goswami 2011; Fojta et al. 2016; Martins et al. 2013; Xiao et al. 2016).

Table 1 shows the different biosensors for environmental monitoring.

The table displays the different contaminants, the responding elements in the biosensors, the physical transducers, and different applications. Optical biosensors have more advantages compared to other traditional analytical methods (Borisov and Wolfbeis 2008; McDonagh et al. 2008; Wolfbeis 2004). Higher sensitivity, very small size, and low cost are the major advantages of optical biosensors (Rogers 2006). Optical biosensors are classified into two types: label-free and label-based biosensors. In the label-free method, a response signal that has been created by the analyte is exposed to the transducer. In the label-based method, luminescence is employed to generate the optical signal and label. Figures 9 and 10, respectively, depict optical and electrochemical biosensors for environmental monitoring.

## *3.1 Heavy Metals*

Heavy metals are dangerous environmental contaminants. Even though much less contaminating than other pollution sources, heavy metals are also harmful to human beings and the environment. Heavy metals are not eco-friendly. Copper, zinc, mercury, lead, and cadmium are some common heavy metal contaminants. Bacteria have been used as a sensing element for heavy metal detection in the environment. Bioluminescent protein is used as a cell biosensor.

**Table 1** Biosensor for environmental applications (from Gieva et al. 2014)

| Contamination | Biological responding element | Physical transducer | Applications |
|---|---|---|---|
| Pesticides | Antibody, enzyme, and microbe | Optical sensor, electrode | Water, soil, air |
| Herbicides | Antibody, enzyme, and microbe | Optical sensor, electrode | Water, soil, air |
| Dioxins | Microbe and slim mode | Optical sensor, electrode | Water, soil, air |
| Heavy metals | Enzyme, microbe | Optical sensor, electrode | Water, soil |
| Nitrogen compounds | Enzyme | Electrode | Water, soil, wastewater |
| Phenolic compounds | Enzyme, microbe | Optical sensor, electrode | Water, soil |

Fig. 9 Optical biosensor for environmental monitoring

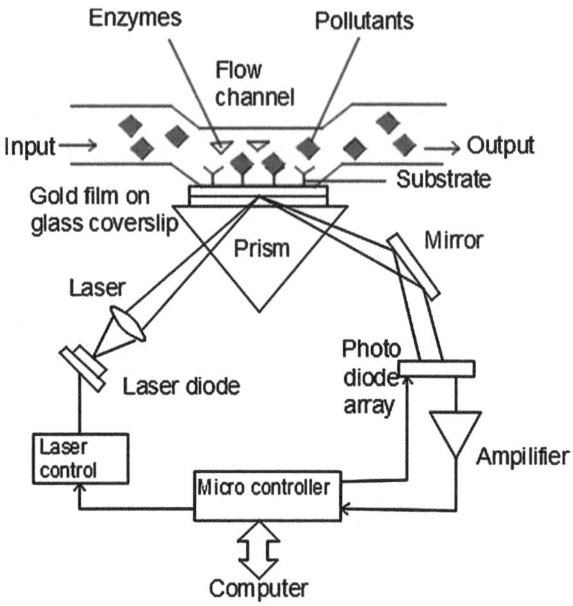


Fig. 10 Electrochemical biosensor for environmental monitoring

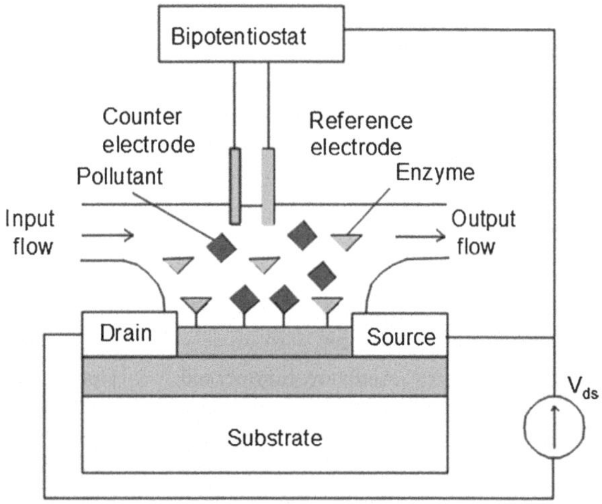

The enzyme technique has also been used to find heavy metal ions; in this regard, more enzymes have been employed for the detection of ions. Table 2 contains biosensors for the detection of heavy metals.

## 3.2 *Nitrites*

Nitrites are generally utilized to treat plants for protection from insects. However, this method is not suitable for human beings or environments; nitrites are very harmful to the human haemoglobin system. Cytochrome *c* protein is used as the biosensitive element in amperometric biosensors for detecting nitrites. This biosensor is highly sensitive and constant. Table 3 shows the determination of nitrites by two different techniques and sensing elements.

## 3.3 *Herbicides*

Herbicides are used to destroy weed plants and thus may damage crops. The toxicity of the herbicides has a wide range. Some herbicides are deleterious to bird populations and some prevent photosynthesis. Phenylureas and triazines are the biosensing elements utilized to detect herbicides in the environment. Two methods, amperometric and bioluminescence, are used to detect these herbicides. Table 4 shows the biosensors used for detection of herbicides.

**Table 2** Biosensors for heavy metals detection (from Gieva et al. 2014)

| Substances | Sensing biocatalyzer | Technique |
|---|---|---|
| Mercury, cadmium | Urease enzyme and microbe | Electrochemical |
| Cadmium | DNA | Optical, electrochemical |
| Cadmium, copper, lead | Sol-gel-immobilized urease | Electrochemical |
| Zinc, copper, cadmium, nickel | Enzyme | Optical |
| Mercury, lead | DNA | Optical |
| Copper | Fluorescent protein | Optical |

**Table 3** Biosensors for nitrite detection (from Gieva et al. 2014)

| Substances | Recognition biocatalyzer | Technique |
|---|---|---|
| Nitrite | Cytochrome *c* | Amperometric |
| | Viologen mediator | Electrochemical |

**Table 4** Biosensors used to detect herbicides (from Gieva et al. 2014)

| Analyte | Type of interaction | Recognition biocatalyzer | Technique |
|---|---|---|---|
| Dichlorofenoxiacetic | Immunoanalysis | Acetylcholinesterase | Amperometric |
| Diuron, Paraquat | Biocatalytic | Cyanobacterial | Bioluminescence |

**Table 5** Biosensors for dioxin detection (from Gieva et al. 2014)

| Analyte | Type of interaction | Recognition biocatalyzer | Technique |
|---|---|---|---|
| Dioxin | Immunoanalysis | Cell | Biomimetic |
| Dioxin-like polychlorinated biphenyls | Immunoanalysis | Cell | Biomimetic |

### 3.4 *Dioxins*

Dioxins are organic by-products from different industries. Dioxin contamination is present in water and soil. The waste discharged from industries travels long distances by wind, rivers, and seawater. Dioxins are reduced by various steps, including a recycling process. Table 5 shows biosensors for detection of dioxins.

## 4 Conclusion

Different types of electrochemical biosensors, namely, amperometric, cyclic voltammogram, potentiometric, conductometric, and electrochemical impedance spectroscopy, have been presented here. Environmental monitoring is presently an emerging trend; hence, the detection of different contaminants in air, soil, and water has been discussed in detail. The detection of different contaminants, such as heavy metals, herbicides, nitrites, and dioxins, is reported. The two methods of biosensors, optical and electrochemical, are discussed with diagrammatic representation.

## References

Aizawa M (1991) Principles and applications of electrochemical and optical biosensors. Anal Chim Acta 250:249–256. https://doi.org/10.1016/0003-2670(91)85073-2

Akkaya OC, Kilic O, Digonnet MJ, Kino G, Solgaard O (2016) Apparatus and methods utilizing optical sensors operating in the reflection mode. In: Google Patents

Bard AJ, Faulkner LR, Leddy J, Zoski CG (1980) Electrochemical methods: fundamentals and applications, vol 2. Wiley, New York

Baro JA, Nevares I, del Álamo Sanza M, Mayr T, Ehgartner J (2016) Biofilm monitoring of dissolved oxygen in wine aging barrel wood with optical chemical sensors. Paper presented at the 2016 IEEE international instrumentation and measurement technology conference proceedings. https://doi.org/10.1109/I2MTC.2016.7520505

Baronas R, Kulys J (2007) Modelling a peroxidase-based optical biosensor. Sensors (Basel) 7(11):2723–2740. https://doi.org/10.3390/s7112723

Bogomolova A, Komarova E, Reber K, Gerasimov T, Yavuz O, Bhatt S, Aldissi M (2009) Challenges of electrochemical impedance spectroscopy in protein biosensing. Anal Chem 81(10):3944–3949. https://doi.org/10.1021/ac9002358

Borisov SM, Wolfbeis OS (2008) Optical biosensors. Chem Rev 108(2):423–461

Chang B-Y, Park S-M (2010) Electrochemical impedance spectroscopy. Annu Rev Anal Chem (Palo Alto Calif) 3:207–229

Dey D, Goswami T (2011) Optical biosensors: a revolution towards quantum nanoscale electronics device fabrication. J Biomed Biotechnol 2011:348218. https://doi.org/10.1155/2011/348218

Dzyadevych S, Arkhypova V, Soldatkin A, El'Skaya A, Martelet C, Jaffrezic-Renault N (2008) Amperometric enzyme biosensors: past, present and future. IRBM 29(2–3):171–180. https://doi.org/10.1016/j.rbmret.2007.11.007

Ermolenko Y, Yoshinobu T, Mourzina Y, Furuichi K, Levichev S, Schöning M et al (2003) The double $K^+/Ca^{2+}$ sensor based on laser scanned silicon transducer (LSST) for multi-component analysis. Talanta 59(4):785–795

Fojta M, Daňhel A, Havran L, Vyskočil V (2016) Recent progress in electrochemical sensors and assays for DNA damage and repair. TrAC Trends Anal Chem 79:160–167. https://doi.org/10.1016/j.trac.2015.11.018

George M, Parak W, Gerhardt I, Moritz W, Kaesen F, Geiger H et al (2000) Investigation of the spatial resolution of the light-addressable potentiometric sensor. Sensors Actuators A Phys 86(3):187–196. https://doi.org/10.1016/S0924-4247(00)00455-6

Gieva E, Nikolov G, Nikolova B (2014) Biosensors for Environmental Monitoring. Challenges in Higher Education & Research, 12:123–127

Giovanardi F, Cucinotta A, Vincetti L (2017) Inhibited coupling guiding hollow fibers for label-free DNA detection. Opt Express 25(21):26215–26220. https://doi.org/10.1364/OE.25.026215

Grieshaber D, MacKenzie R, Vörös J, Reimhult E (2008) Electrochemical biosensors-sensor principles and architectures. Sensors (Basel) 8(3):1400–1458. https://doi.org/10.3390/s80314000

Hafeman DG, Parce JW, McConnell HM (1988) Light-addressable potentiometric sensor for biochemical systems. Science 240(4856):1182–1185. https://doi.org/10.1126/science.3375810

Hasan A, Nurunnabi M, Morshed M, Paul A, Polini A, Kuila T et al (2014, 2014) Recent advances in application of biosensors in tissue engineering. BioMed Res Int. https://doi.org/10.1155/2014/307519

Hinze S (1994) Bibliographical cartography of an emerging interdisciplinary discipline: the case of bioelectronics. Scientometrics 29(3):353–376. https://doi.org/10.1007/BF02033445

Ianeselli L, Grenci G, Callegari C, Tormen M, Casalis L (2014) Development of stable and reproducible biosensors based on electrochemical impedance spectroscopy: three-electrode versus two-electrode setup. Biosens Bioelectron 55:1–6. https://doi.org/10.1016/j.bios.2013.11.067

Kara S (2012) A roadmap of biomedical engineers and milestones: BoD–Books on Demand

Mabbott GA (1983) An introduction to cyclic voltammetry. J Chem Educ 60(9):697

Malhotra BD (2017) Biosensors: fundamentals and applications: Smithers Rapra

Malhotra BD, Singhal R, Chaubey A, Sharma SK, Kumar A (2005) Recent trends in biosensors. Curr Appl Phys 5(2):92–97

Martins TD, Ribeiro ACC, de Camargo HS, da Costa Filho PA, Cavalcante HPM, Dias DL (2013) New insights on optical biosensors: techniques, construction and application, State of the Art in Biosensors—General Aspects, pp 112–139. https://doi.org/10.5772/52330

McDonagh C, Burke CS, MacCraith BD (2008) Optical chemical sensors. Chem Rev 108(2):400–422. https://doi.org/10.1021/cr068102g

Mohanty SP, Kougianos E (2006) Biosensors: a tutorial review. IEEE Potentials 25(2):35–40. https://doi.org/10.1109/MP.2006.1649009

Nakao M, Yoshinobu T, Iwasaki H (1994) Improvement of spatial resolution of a laser-scanning pH-imaging sensor. Jpn J Appl Phys 33(3A):L394

Owicki JC, Bousse LJ, Hafeman DG, Kirk GL, Olson JD, Wada HG et al (1994) The light-addressable potentiometric sensor: principles and biological applications. Annu Rev Biophys Biomol Struct 23(1):87–114
Pandey CM, Malhotra BD (2019) Biosensors: fundamentals and applications. Walter de Gruyter GmbH & Co KG, Berlin/Boston, Germany.
Perumal V, Hashim U (2014) Advances in biosensors: principle, architecture and applications. J Appl Biomed 12(1):1–15. https://doi.org/10.1016/j.jab.2013.02.001
Poghossian A, Thust M, Schroth P, Steffen A, Lüth H, Schöning MJ (2001a) Penicillin detection by means of silicon-based field-effect structures. Sensors Mater 13(4):207–223
Poghossian A, Yoshinobu T, Simonis A, Ecken H, Lüth H, Schöning MJ (2001b) Penicillin detection by means of field-effect based sensors: EnFET, capacitive EIS sensor or LAPS? Sensors Actuators B Chem 78(1–3):237–242. https://doi.org/10.1016/S0925-4005(01)00819-X
Reinhardt G, Mayer R, Rösch M (2002) Sensing small molecules with amperometric sensors. Solid State Ionics 150(1–2):79–92. https://doi.org/10.1016/S0167-2738(02)00265-5
Rinken T (2013) State of the art in biosensors: general aspects: BoD–Books on Demand
Rogers KR (2006) Recent advances in biosensor techniques for environmental monitoring. Anal Chem Acta 568(1–2):222–231. https://doi.org/10.1016/j.aca.2005.12.067
Ronkainen NJ, Halsall HB, Heineman WR (2010) Electrochemical biosensors. Chem Soc Rev 39(5):1747–1763
Sartore M, Adami M, Nicolini C, Bousse L, Mostarshed S, Hafeman D (1992) Minority carrier diffusion length effects on light-addressable potentiometric sensor (LAPS) devices. Sensors Actuators A Phys 32(1–3):431–436. https://doi.org/10.1016/0924-4247(92)80025-X
Schöning MJ, Wagner T, Wang C, Otto R, Yoshinobu T (2005) Development of a handheld 16 channel pen-type LAPS for electrochemical sensing. Sensors Actuators B Chem 108(1–2):808–814. https://doi.org/10.1016/j.snb.2005.01.055
Serra PA (2011) Biosensors for health, environment and biosecurity: BoD–Books on Demand
Shimizu M, Kanai Y, Uchida H, Katsube T (1994) Integrated biosensor employing a surface photovoltage technique. Sensors Actuators B Chem 20(2–3):187–192. https://doi.org/10.1016/0925-4005(94)01176-1
Sin ML, Mach KE, Wong PK, Liao JC (2014) Advances and challenges in biosensor-based diagnosis of infectious diseases. Expert Rev Mol Diagn 14(2):225–244. https://doi.org/10.1586/14737159.2014.888313
Somerset V (2011) Environmental biosensors: BoD–Books on Demand
Song S, Xu H, Fan C (2006) Potential diagnostic applications of biosensors: current and future directions. Int J Nanomed 1(4):433
Thévenot DR, Toth K, Durst RA, Wilson GS (2001) Electrochemical biosensors: recommended definitions and classification. Biosens Bioelectron 34(5):635–659. https://doi.org/10.1351/pac199971122333
Turner AP (2013) Biosensors: sense and sensibility. Chem Soc Rev 42(8):3184–3196. https://doi.org/10.1039/C3CS35528D
Vadgama P, Crump PW (1992) Biosensors: recent trends. A review. Analyst 117(11):1657–1670. https://doi.org/10.1039/AN9921701657
Wolfbeis OS (2004) Fiber-optic chemical sensors and biosensors. Anal Chem 76(12):3269–3284. https://doi.org/10.1021/ac040049d
Xiao F, Wang L, Duan H (2016) Nanomaterial based electrochemical sensors for in vitro detection of small molecule metabolites. Biotechnol Adv 34(3):234–249. https://doi.org/10.1016/j.biotechadv.2016.01.006
Yoshinobu T, Schöning M, Finger F, Moritz W, Iwasaki H (2004) Fabrication of thin-film LAPS with amorphous silicon. Sensor (Basel) 4(10):163–169. https://doi.org/10.3390/s41000163
Yoshinobu T, Iwasaki H, Ui Y, Furuichi K, Ermolenko Y, Mourzina Y et al (2005) The light-addressable potentiometric sensor for multi-ion sensing and imaging. Methods 37(1):94–102. https://doi.org/10.1016/j.ymeth.2005.05.020

# Toxic Gas Sensors and Biosensors

**Umesh Fegade**

## Contents



## Abbreviations

| | |
|---|---|
| °C | Degree Celsius (temperature) |
| μÅ | Micro-angstrom |
| $C_2H_5OH$ | Ethanol |
| $C_3H_8$ | Propane |
| cm | Centimetre |
| CNF | Carbon nanofiber |
| CNTs | Carbon nanotubes |
| CO | Carbon monoxide |
| $Co_3O_4$ | Cobalt oxide |
| CuO | Copper oxide |
| CZC | CuO/ZnO composite |
| FET | Field effect transistor |
| $H_2$ | Hydrogen |
| $H_2S$ | Hydrogen sulfide |
| K | Kelvin (temperature) |
| mm | Millimetre |

U. Fegade (✉)
Bhusawal Arts Science and P. O. Nahata Commerce College,
Bhusawal, Jalgaon 425201, Maharashtra, India



Inamuddin, A. M. Asiri (eds.), *Nanosensor Technologies for Environmental Monitoring*, Nanotechnology in the Life Sciences,
https://doi.org/10.1007/978-3-030-45116-5_3

| | |
|---|---|
| MOF | Metal–organic framework |
| $MoTe_2$ | Molybdenum ditelluride |
| MWCNT | Multi-walled carbon nanotube |
| $Nb_2O_5$ | Niobium pentoxide |
| N-DWCNTs | Nitrogen-doped double-walled carbon nanotubes |
| NiO | Nickel oxide |
| $NO_2$ | Nitrogen dioxide |
| NWs | Nanowires |
| Pd | Palladium |
| Pd/ZnO | Palladium on zinc oxide |
| PdO | Palladium oxide |
| PPase | Pyrophosphatase |
| ppm | Parts per million |
| Pt | Platinum |
| s | Second |
| SWCN | Single-walled carbon nanotube |
| SWNT | Single-walled nanotube |
| TONTs | Tin dioxide nanocrystalline tubes |
| UV | Ultraviolet |
| $WO_3$ | Tungsten oxide |
| ZnO | Zinc oxide |

# 1 Introduction

Nowadays, the increase in human populations and their unsatisfied nature create an extreme demand for consumer products and cause crises in energy sectors. To increase the production of energy and consumer goods, industries release a number of toxic gases as by-products such as nitrogen dioxide, carbon monoxide, and hydrogen sulfide that significantly endanger our health and environment over the long term (Duy et al. 2015). The concern for environmental hygiene requires strict regulations on the emission of toxic and hazardous gases from automobiles/motor vehicles and from industries (Navale et al. 2017).

Gas sensors are the most important technology in our daily life. Typical applications are detection of toxic analyte gases for domestic safety and monitoring of environmental pollution, air quality, stylish houses and vehicles, etc. (Zhang et al. 2016). Because of their small price, high sensitivity, in situ detection ability, and portability, nanomaterials have an extraordinary capability for sensing 150 gases (Mirzaei et al. 2018; Vetter et al. 2015; Zhang et al. 2018; Liu et al. 2018).

Nanomaterials are the new era of science with many applications in the fields of chemistry, biology, and technology (Jethave et al. 2017; Jethave and Fegade 2018; Kondalkar et al. 2018; Fegade et al. 2018). Scientists have developed various methods for nanomaterial synthesis, including hydrothermal synthesis (Yang et al. 2001;

Carp al. 2004), chemical vapor deposition (CVD) (Wei et al. 2009; Xue et al. 2012), thermal decomposition (Sheng et al. 2011; Bhunia et al. 2012), solvothermal (Deng et al. 2011), templating (Bavykin et al. 2006; Jinsoo et al. 2005; Iwasaki et al. 2004), and the conventional sol–gel method (Pierre 1998; Kolen'ko et al. 2005; Fernandez-Garcia et al. 2004). The nanomaterials have shown various extraordinary properties greater than their own bulk. In the field of sensing, toxic gases in industry (Li et al. 2013; Poloju et al. 2018) and various metabolites and biological compounds in the living body (Wu et al. 2015) have been targeted (Niu et al. 2014).

## 2 Toxic Gas Sensors

The gas-sensing mechanism of a metal oxide is a surface phenomenon that involves a gas–solid interaction at the surface of the material (Sankar Ganesh et al. 2018; Sahay and Nath 2008; Lagowski et al. 1977; Hou et al. 2016; Morrison 1987).

In the adsorption mechanism of carbon monoxide, CO acts as a reducing agent; the oxidation of CO increases the electron concentration on the surface (Shankar and Rayappan 2015). In a dry atmosphere, CO reacts with chemisorbed oxygen to form carbon dioxide ($CO_2$). The detection of CO by a metal oxide semiconductor in an oxygen atmosphere is better than in a nitrogen or wet atmosphere

$$\mathrm{CO} + \mathrm{OH}^{-}\left(\text{adsorbed}\right) \rightleftharpoons \mathrm{HCO}_2^{-} \rightleftharpoons \mathrm{H}^{+}\left(\text{adsorbed}\right) + \mathrm{CO}_2 + 2\mathrm{e}^{-} \tag{1}$$

$$\mathrm{CO} + \mathrm{O}_2^{-}\left(\text{adsorbed}\right) \rightarrow \mathrm{CO}_2 + \mathrm{e}^{-} \tag{2}$$

The experimental setup is shown in Fig. 1. The sensor array is located in a chamber, and gases are injected into the chamber at different concentrations. The electrical responses of the sensor array are automatically measured using a data logger and are clearly seen in computer-generated data (Zhang et al. 2017).

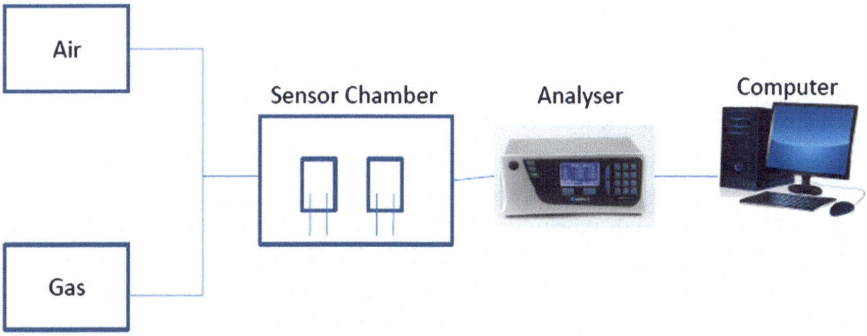


**Fig. 1** Experimental setup of sensor array

## 2.1 Carbon Monoxide (CO) Gas Sensor

Du et al. (2008) synthesized ultrathin $SnO_x$ ALD film with a thickness of ~10 Å and response time to $O_2$ within ~100 s and for CO within ~10 s at 300 °C. The gas sensor electrical conductivity was measured using in situ transmission Fourier transform infrared (FTIR) spectroscopy. Kuang et al. (2008) fabricated a single $SnO_2$ NW and NiO nanoparticle deposition on a $SnO_2$ NW surface. The sensitivity to CO was as much as 9.8 and to $CH_4$ only 3.3 on using $SnO_2$ NW without a NiO coating. After NiO surface functionalization, the sensitivity of $SnO_2$ NW to CO was magnified to 15.9, whereas the sensitivity of $SnO_2$ NW to $CH_4$ was not affected. Thus, NiO surface functionalization enhances the sensitivity of CO and $CH_4$ (Kuang et al. 2008). Huang et al. (2010) fabricated $SnO_2$ nanorod sensors that showed a small response to CO and $H_2$gases. Variations in the surface of Pt by dip coating up to 2 nm showed the best sensing performance to CO (Huang et al. 2010).

Her et al. (2012) synthesized an $In_2O_3$ nanomaterial on $SnO_2$ by a thermal vapour transfer process. $In_2O_3$ nanostructures displayed a greater sensitivity of 1.4 for CO than the film counterpart, which may be attributed to the existence of large quantities of active centres on $In_2O_3$ nanostructures (Her et al. 2012). Wang et al. (2012a) synthesized a PdO–NiO ring-like structure of diameter and thickness about 3.5 mm and 15 nm, respectively. It shows a low operating temperature (180 °C), a high response (46.3), and a rapid recovery (2–3 s) towards 50 ppm CO (Wang et al. 2012a). Lai and Chen (2012) developed a Pt or Pd nanomaterial coating on $In_2O_3$ using parallel nanoparticle chains. The uniform deposition enhanced CO gas sensitivity and response at 25 °C (Lai and Chen 2012).

Li et al. (2012) produced porous carbon nanofibers (CNF) from graphitic nanorolls by a simple electrospinning-assisted solid-phase graphitization method. Both CNF-800 and CNF-1000 as prepared showed remarkable gas-sensing properties for $H_2$, CO, $CH_4$, and ethanol at room temperature, with a detection limit of 500 ppm for CO (Li et al. 2012). Chang et al. (2012) synthesized Pd/ZnO nanorods by Pd nanoparticles that display an important CO gas-sensing performance. When the concentration of CO changed from 100 to 600 ppm, the gas sensitivity was enhanced by 4–6 fold, and the response and recovery times were reduced by 4–12- and 1–2 fold, respectively (Chang et al. 2012). Zhang et al. (2013a) fabricated $Co_3O_4$/PEI–CNTs composite materials by a noncovalent strategy and hydrothermal temperature at 190 °C for 4 h. The sensitivity and response time of composite thin film sensors are 24% and 8 s, respectively, for 700 ppm CO and limit of detection at 5 ppm (Zhang et al. 2013a).

Zhang et al. (2013b) synthesized p-type uniform $Cu_2O$–CuO microframes by the etching oxidation of $Cu_2O$ microcubes. The sensor exhibited an outstanding sensing performance with a high sensitivity and low detection limit for CO with response time of only 21 s (Zhang et al. 2013b). An Al-doped flower-like ZnO nanomaterial was synthesized by Bai et al. (2013). Its sensing response towards CO was appreciably raised from 79 to 464 when the temperature was decreased from 180° to 155 °C with doping of aluminum up to 0.3 wt% (Bai et al. 2013). Shi et al. (2014)

investigated CO sensing characteristics and mechanism for $La_{1-x}Ca_xFeO_3$ nanocrystalline powders. The optimum Ca doping concentration was $x = 0.2$ for obtaining a large response to 200 ppm CO among $La_{1-x}Ca_xFeO_3$ sensors (Shi et al. (2014). Bai et al. (2014) developed ZnO nanofibers using electrospinning and annealing method that showed enhancement in response from 95 to 300–400 ppm CO by 5.0 wt% Cd doping at 235 °C. The response and recovery time were about 10 s and 27 s, respectively (Bai et al. 2014).

Saberi et al. (2015) developed a novel dual selective Pt/$SnO_2$ sensor for CO and propane in the exhaust gases of a gasoline engine. In a mixture of CO, NO, and $C_3H_8$ in 1.0% oxygen, a Pt/$LaFeO_3$ catalytic filter–Pt/$SnO_2$ sensor combination is selective to CO at the temperature range of 150–250 °C, and the response time to CO is about 30 s (Saberi et al. 2015). Steinhauer et al. (2016) reported a CuO nanowire for in situ measurement of the electrical resistance for well-controlled integration into conductometric gas-sensing devices. Excellent gas sensor performance for the detection of CO concentrations down to 1 ppm was observed; on the other side a considerable decrease of CO response was found after exposure to humidity (Steinhauer et al. 2016). Shahid et al. (2018) developed a wireless E-nose using an array of commercially obtainable $SnO_2$ gas sensors. An ANN was designed using pattern recognition techniques for $CH_4$ and CO, with 98.7% accuracy. The LSR estimator achieved 94.4% minimum accuracy for CO (Shahid et al. 2018).

## 2.2 *Hydrogen Sulfide ($H_2S$) Gas Sensor*

Mubeen et al. (2011) developed metal nanoparticles decorated on SWNTs that were investigated towards $H_2S$ using FET transfer for gas sensing. At low concentrations (≤100 ppb), FET transfer characteristics showed that the gold nanoparticles at the surface of SWNTs acted as nano-Schottky barriers to predominately modulate trans-conductance (Mubeen et al. 2011). Zhang et al. (2011) synthesized highly aligned $SnO_2$ nanorods on graphene by a hydrothermal method that exhibits a sensitivity of 2.1 to $H_2S$ with a concentration as low as 1 ppm; the response and recovery time for 1 ppm $H_2S$ is 5 s and 10 s, respectively (Zhang et al. 2011). Mickelson et al. (2012) demonstrated a small, low-cost, low-power, highly sensitive, and selective tungsten oxide ($WO_3$) nanomaterial-based gas sensor for $H_2S$. The sensing mechanism is related to electron donation from $H_2S$ to $WO_3$, which causes the Fermi energy to increase and decreases the sensor resistance (Mickelson et al. 2012). Ma et al. (2012) developed cubic $ZnSnO_3$ and octahedral $Zn_2SnO_4$ microcrystals. The As-synthesized zinc stannate faceted microcrystals are converted to hollow structures through an acid etching process. The hollow zinc stannate was exploited as gas sensors and exhibited improved sensing performances to $H_2S$, $C_2H_5OH$, and HCHO (Ma et al. 2012).

$SnO_2$-based gas sensing films were synthesized by Kida et al. (2013) under hydrothermal conditions. Sensor reactions to $H_2$ were amplified with decrease in particle size. In contrast, the response to $H_2S$ increased with increasing particle size and showed excellent response to a very small concentration of $H_2S$ (5 ppm) (Kida et al. 2013). Tabassum et al. (2013) fabricated a probe by depositing thin films of copper and zinc oxide over the unclad portion of the fiber by the thermal evaporation technique. The sensor works on the principle of change in refractive index of the zinc oxide when molecules of zinc oxide come into contact with $H_2S$ gas. The sensitivity of the sensor decreases with increase in the concentration of $H_2S$ gas with the response and recovery time about 1 min (Tabassum et al. 2013).

Zhang et al. (2014) developed a Cu-doped and undoped $SnO_2$ porous gas sensor. The sensitivity of the Cu-doped $SnO_2$sensor was higher than that of the undoped $SnO_2$, the average response and recovery time to 100 ppm $H_2S$ being ~10.1 and ~42.4 s, respectively, at 180 °C (Zhang et al. 2014). Woo et al. (2014) prepared an Mo-doped ZnO nanowire by coating the $MoS_2$ layer during successive ionic layer adsorption at 600 °C for 2 h. The sensor confirmed a large response of 14.11–5 ppm $H_2S$ at 300 °C (Woo et al. 2014). Jiang et al. (2014) fabricated $Fe_2O_3$/graphene nanosheets that showed outstanding response to $H_2S$ gas with high selectivity at low cost and high efficiency. The sensor displayed an important CL emission at 450 absorption units in response to 15 ppm $H_2S$ at 190 °C. Furthermore, the sensor has a small response time, 500 ms, and a fast recovery time, less than 30 s, with a 10 ppm detection limit of $H_2S$ at 130 °C (Jiang et al. 2014). Yin et al. (2014) synthesized a novel hierarchical $In_2O_3$@$WO_3$ nanocomposite and explored its $H_2S$-sensing properties. The sensor displayed excellent $H_2S$-sensing performance at 150 °C; its response to 10 ppm $H_2S$ is as high as 143, 4 times greater than $WO_3$ nanoplates and 13 times greater than $In_2O_3$ nanocrystals. The enhance response of sensors for $H_2S$ resulted from the synergistic effect of $In_2O_3$ and $WO_3$ nanoplates (Yin et al. 2014).

The ZnS-decorated layer was fabricated by Qi et al. (2014) by passivating oriented ZnO nanorods in an $H_2S$ atmosphere. The ZnO nanorods decorated with a 2-nm-thick ZnS layer possessed a repeatable and better response to 10 ppm $H_2S$ for 200 s at 25 °C for the repeated detection of 1 ppm $H_2S$ (Qi et al. 2014). Sun et al. (2014) synthesized undoped and Cd-doped $SnO_2$ nanomaterial; the 3.0 wt% Cd-doped $SnO_2$-based sensor showed remarkable selectivity towards $H_2S$ at 275 °C, giving a response of about 31–10 ppm (Sun et al. 2014). Mendoza et al. (2014) developed $SnO_2$ particles distributed in the carbon nanotube (CNT) matrix that showed superior sensitivity to alcohol vapors and $H_2S$ than bare $SnO_2$ or CNTs (Mendoza et al. (2014). Li et al. (2016) prepared porous CuO nanosheets on alumina tubes using a facile hydrothermal method that were investigated for $H_2S$ gas-sensing properties. The sensor showed a good response sensitivity of 1.25 with the response/recovery time of 234 s and 76 s, respectively, at 10 ppb $H_2S$ (Li et al. 2016).

Yassine et al. (2016) developed a fumarate-based fcu-MOF sensor that showed a significant detection sensitivity for $H_2S$ at 100 ppb, with detection limit around 5 ppb (Yassine et al. 2016). Song et al. (2016) demonstrated $H_2S$ gas sensors based on $SnO_2$ quantum wires that were reduced graphene oxide (rGO) nanosheets. The sensor showed sensitivity to 50 ppm $H_2S$ was 33 in 2 s, and it was fully reversible upon $H_2S$ release at 22 °C (Song et al. 2016). Huang et al. (2017) prepared Ce-doped $BaTiO_3$ that showed an interesting gas-sensing performance for $H_2S$, including the ability to detect low concentrations of $H_2S$ (400 ppb), fast response and recovery speed, 45 s and $t_{rec}$ = 124 s, respectively, and 150 °C working temperatures (Huang et al. (2017).

Arafat et al. (2017) fabricated $TiO_2$-$Al_2O_3$ nanostructures that show n-type sensing behavior towards $H_2S$, $CH_3OH$, and $C_2H_5OH$ in an $N_2$ background with response values of 38.7, 349.6, and 1108.9, respectively (Arafat et al. 2017). Chu et al. (2018) reported tin oxide-modified reduced graphene oxide ($SnO_2$–rGO) that was used for the detection of $H_2S$ and $SOF_2$. The results showed that the $SnO_2$–rGO sensor shows better responses (34.31%) at 125 °C in the presence of 100 ppm $H_2S$ (Chu et al. 2018). Ding et al. (2018) developed a ZnO to ZnS shell over a single crystalline base that showed improvement of sensitivity to $H_2S$ gas. At 10 ppm $H_2S$ gas, the response of the sensor is 0.67, significantly higher than that of pure ZnO NWs (0.28) (Ding et al. 2018). Hussain et al. (2018) designed effective Ge-NS nanosensors based on the large surface area of the 2D monolayers. The sensing behavior of two gases, $H_2S$ and $SO_2$, on pristine, defected, and Ge-NS layers, has been studied by first-principles calculations of discrete Fourier transform (DFT) (Hussain et al. 2018).

## 2.3 *Nitrogen Dioxide ($NO_2$) Gas Sensor*

Sasaki et al. (2009) reported SWNT networks used for sensing of $NO_2$gas. The sensor annealed at 400 °C showed the highest detection sensitivity, indicating high sensitivity down to 25 ppb $NO_2$ (Sasaki et al. 2009). Offermans et al. (2010) presented gas-sensing InAs nanowire arrays. Noise measurements were performed to determine the measurement resolution for gas detection. These devices are sensitive to $NO_2$ concentrations well below 100 ppb at room temperature (Offermans et al. 2010). Bai et al. (2011) synthesized quantum-sized ZnO nanoparticles by the sol–gel process. The ZnO sensor exhibited the highest response, 264 to 40 ppm $NO_2$, and high selectivity, 8.8 and 13.7, compared to CO and $CH_4$, respectively, at 290 °C (Bai et al. 2011).

Jiang et al. (2012) developed tin dioxide nanocrystalline tubes (TONTs) that predominantly consist of 5–10 nm $SnO_2$ nanoparticles with outstanding sensing characteristics to $NO_x$ at <10 ppb. The possible gas-sensing mechanism of the TONTs is explained by DFT analysis in which the effects of $NO_x$ adsorption on nanoparticles and electronic transport properties are discussed (Jiang et al. 2012). Bai et al. (2012) fabricated $WO_3$ nanorods by hydrothermal reaction at 100 °C. The calcination conditions for $WO_3$ at 600 °C and 4 h and the As-prepared sensor exhibit high sensitivity to ppm-level $NO_2$ at 200 °C, and sensitivity increases nearly linearly with the increase of $NO_2$ concentration in the detection range from 5 to 40 ppm (Bai et al. 2012). Wang et al. (2012b) developed a-$Ni(OH)_2$ with a 3D hierarchical structure. The application of gas sensors at 25 °C showed that the obtained a-$Ni(OH)_2$ nanocrystals display quick response, excellent repeatability, and a lifetime longer than 35 days (Wang et al. 2012b). Pan et al. (2013) fabricated crystalline nanocombs via a chemical vapour deposition method. Nanobelt sensors demonstrated a significant resistance change from 2.5 ppm. For 5 ppm $NO_2$, the nanobelt sensitivity is estimated at only 2.45, much less than the nanocomb device (Pan et al. 2013).

Sonker et al. (2013) synthesized nanocrystalline composites (NCC) of zinc oxide and tin oxide (ZSO) using a chemical route for efficient sensing of $NO_2$ gas at a lower operating temperature. The sensor structure showed a better sensing response ($S \sim 6.64 \times 102$) at 70 °C for 20 ppm $NO_2$ gas with an average response time of about 2 min (Sonker et al. 2013). Jiang et al. (2013) fabricated CaO–$SnO_2$ rod-like structures by a facile electrospinning approach, followed by appropriate thermal treatment in ambient conditions. The 2 at% CaO–$SnO_2$ L-NRs display outstanding sensing properties with large response and good selectivity at 25 °C for $NO_x$. The 2 at% CaO–SnO2 L-NRs showed excellent sensitivity and detection to $NO_x$ even at less than 10 ppb (Jiang et al. 2013). Yang et al. (2013) synthesized $CeO_2$/graphene-like nanosheet composites (CeGNCs) via a facile solvothermal reaction. The CeGNCs with 46.7 wt% of $CeO_2$ showed high sensitivity of 10.39% for $NO_x$ gas with a detection limit of 5.0 ppm, and a short response time of 7.33 s, which is more than 4.6 times larger than pure $CeO_2$ (Yang et al. 2013). Xu et al. (2013) developed flower-like $In_2O_3$ through a easy hydrothermal process of precursor $In(OH)_3$ at 600 °C for 2 h. The sensor response is about 2.4–5 ppb $NO_2$ at 125 °C (Xu et al. 2013).

Gao et al. (2014) fabricated $Al_2O_3$–$In_2O_3$ composites by one-step electrospinning and thermal treatment. The sensor showed good response properties to $NO_x$ with LOD 291 ppb at 25 °C (Gao et al. 2014). Highly mesoporous hierarchical nickel and cobalt double hydroxide composites (NCDHs) have been synthesized via a simple reflux method by Ge et al. (2014). Furthermore, the sensor illustrated outstanding gas-sensing properties, having LOD at 0.97 ppm and a short response time of 0.6 s to 97 ppm $NO_x$, because of the single crystal structure (Ge et al. 2014).

Dai et al. (2015) synthesized an $NO_2$ chemiresistor, a monolayer a-$Fe_2O_3$. The porous $Fe_2O_3$sensor is capable of detecting low $NO_2$ at 10 ppb and shows good stability (Dai et al. 2015). Zhou et al. (2015a) designed ZnO nanowire sensors for $NO_2$ gases. The sensitivity and resolution of the sensor are clearly superior, together with a significant development in output current by 238.8% for $NO_2$ detection (Zhou et al. 2015a). $(La_{0.8}Sr_{0.2})_2FeMnO_6$ (LSFM) oxide powder with a double-perovskite structure has been prepared using the Pechini method by Zhou et al. (2015b), who fabricated a sensitive electrode (SE) for $NO_2$ potentiometric sensors. The selectivity to $NO_2$ for the sensor with a 1200 °C-sintered LSFM sensing electrode is good at 550 °C (Zhou et al. 2015b).

Kumar et al. (2017) demonstrated a $MoS_2$ ultrafast detection gas sensor at 25 °C. Sensor performance was investigated to $NO_2$ at 25 °C, under thermal and photo energy, showing an elevated response time, ~249 s. $MoS_2$ exhibited enhancement in response with a fast response time of ~29 s and outstanding recovery to $NO_2$ (100 ppm) at 25 °C (Kumar et al. 2017). Wu et al. (2017) reported a $NaHSO_3$ functionalized RGOH exhibited significant results 118.6 and 58.9 times superior to $NO_2$ and $NH_3$, respectively (Wu et al. 2017). $WO_3$ microbricks have been synthesized by Harale et al. (2018) using the hydrothermal route at 180 °C. The $WO_3$ sensor displayed 11.5 response towards $NO_2$ gas, with response time of 16 s and recovery time of 260 s at 100 ppm (Harale et al. 2018). Mahendraprabhu and Elumalai (2017) synthesized a niobium pentoxide ($Nb_2O_5$) powder by the citrate-based sol–gel method using niobium pentachloride as a precursor and citric acid as the gelling agent. The YSZ-based sensor attached with $Nb_2O_5$-SE exhibited selective and sensitive response to $NO_2$ at 800 °C. The sensor showed the highest sensitivity and selectivity to $NO_2$ in 5 vol% $O_2$ (Mahendraprabhu and Elumalai 2017).

Shim et al. (2018) prepared $MoS_2$ on 500-nm-thick $SiO_2$ NRs that showed 90 times greater gas-sensing response to 50 ppm $NO_2$ at 25 °C than the $MoS_2$ film prepared on flat $SiO_2$, and the theoretical LOD was ~2.3 ppb (Shim et al. 2018). Wu et al. (2018) reported a p-type molybdenum ditelluride ($MoTe_2$) gas sensor for $NO_2$ detection. The sensitivity of the sensor to $NO_2$ is dramatically enhanced under UV illumination as compared to that in the dark condition, with detection limit of 123 ppt (Wu et al. 2018). Ko et al. (2018) synthesized a $WSe_2$ gas sensor showing an appreciably high response (4140%) to $NO_2$. The $WSe_2$ recovery improved by reacting $NH_3$ and adsorbed $NO_2$ on the surface of $WSe_2$: the $NO_2$ are suddenly desorbed, with decrease in recovery time (Ko et al. 2018). Mutkule et al. (2018) synthesized a ZnO film sensor having high sensitivities from 176 to 610% towards $NO_2$ gas with 10 to 200 ppm at 200 °C. A large response, about 84.42%, on the 15th day showed chemical stability (Mutkule et al. 2018).

Muangrat et al. (2018) fabricated $NO_2$ gas sensors based on nitrogen-doped double-walled carbon nanotubes (N-DWCNTs). The N-DWCNTs at 900 °C exhibited a 2.7-fold improvement in the response to $NO_2$ gas compared to DWCNTs at 900 °C, with LOD of the best sensor being 0.14 ppm $NO_2$ (Muangrat et al. 2018). Wu et al. (2019) chemically derived ionic conductive polyacrylamide/carrageenan double-network (DN) hydrogels exploited to fabricate ultra-stretchable and transparent $NO_2$ and $NH_3$ sensors with high sensitivity (78.5 $ppm^{-1}$) and a low theoretical limit (1.2 ppb) in $NO_2$ detection (Wu et al. 2019). Bae et al. (2019) designed a graphene-ZnO heterostructure gas sensor. The charge interaction of the heterostructures was explored by monitoring changes in the transfer curves at RT and elevated temperature (250 °C) after introducing 20 ppm $NO_2$ (Bae et al. 2019). Casals et al. (2019) synthesized ZnO nanoparticles to achieve a $NO_2$-parts-per-billion conductimetric gas sensor operating at 25 °C. Responses of 94% to 25 ppb were achieved, corresponding to a lower detection limit of 1 ppb of $NO_2$ (Casals et al. 2019).

## 3 Biosensor

Zhang et al. (2015) developed a novel biosensor platform for detection of microRNAs based on graphene quantum dots (GQDs) and pyrene-functionalized molecular beacon probes (py-MBs) that showed sensing of miRNAs in the range 0.1–200 nM (Zhang et al. 2015). Bao et al. (2015) reported a novel esterase−chitosan/gold nanoparticle−graphene nanosheet (PLaE-CS/AuNPs-GNs) biosensor. The PLaE-CS/AuNPs-GNs composite-based biosensor measured as low as 50 ppt (0.19 nM) methyl parathion and 0.5 ppb (1.51 nM) malathion (Bao et al. 2015). Li et al. (2015) fabricated a biosensor that can detect various human metabolites. Our sensor displayed outstanding sensing response with good range (uric acid, 0.07–1 mM, cholesterol, 0.3–9 mM, triglycerides, 0.2–5 mM), high sensitivity, low LOD, and rapid response time (~3 s) (Li et al. 2015). Soikkeli et al. (2016) confirmed the idea of a modular and programmable label-free biosensor. The sensitivity of biorecognition in both model systems is high, with 3% response recorded for analyte concentrations less than 100 fM in large ionic strength buffers (Soikkeli et al. 2016).

Chong and Ching (2016) successfully engineered a transcription factor, DmpR (dimethyl phenol regulatory protein), to alter the expression level of RFP regulated by its cognate promoter, so that parathion could be detected at concentrations as low as 10 μM by simple colour visualization (Chong and Ching 2016). Park et al. (2017) demonstrated consistent and quantitative detection of a prostate cancer biomarker using the $MoS_2$-FET biosensor by consistent chemisorption of prostate cancer antigen (anti-PSA). Biosensor is able to detect PSA concentrations with a limit of 100 fg/mL (Park et al. 2017). Iitani et al. (2017) developed a fiberoptic biosensor for measurement of AcH by a combination of the NADH fluorescence detection system. It showed a dynamic range (1–500 μM) of the ADH-mediated biosensor greater than that of the ALDH (5–200 μM) (Iitani et al. 2017). Wen et al. (2017) reported a DNA-encoded biosensor used to measure a bacterial biomarker of Pseudomonas aeruginosa infection from human sputum samples. The concentration at which expression was equal to fluorescence was 1.55 nM (SD = 0.00124), and the LLOD for 3OC12-HSL was calculated as 1.56 nM (Wen et al. 2017).

Jett and Bonham (2017) designed an electrochemical DNA biosensor that when exposed to increasing concentrations of uranyl ions, ranging from 100 nM to 500 μM in an acetate buffer, showed a concentration-specific change in observed current and a KD of 15 ± 5 μM, with a limit of detection of 1.2 μM (Jett and Bonham 2017). Doran et al. (2017) designed a sensor using 200 U/mL superoxide dismutase (SOD), 0.5% glutaraldehyde, and 2%polyethylenimine (Sty-(SOD-0.5%GA-2%PEI)5). It had a response time of 1 s, sensitivity to $O^{2-}$, and in vitro LOD of 0.063 mM, thus signifying prospective utilization for monitoring $O^{2-}$ in vivo (Doran et al. 2017). Soylemez et al. (2017) designed two new water-soluble monomers that were successfully synthesized by solid-phase peptide synthesis via coupling reagents. The optimum biosensor exhibited a linear range of 0.01–0.75 mmol $L^{-1}$ ($R^2 = 0.998$), sensitivity 91.37 μA $mM^{-1}$ $cm^{-2}$, and KM app 0.208 mmol $L^{-1}$ (Soylemez et al. 2017).

Sriwichai et al. (2017) fabricated biomolecules on electropolymerized carboxylated conducting polymers, poly(3-aminobenzoic acid) and poly(3-pyrrole carboxylic acid), that were selective for simultaneous detection of two biomolecules using electrochemical-surface plasmon resonance spectroscopy (Sriwichai et al. 2017). Janissen et al. (2017) reported biomarkers using biocompatible ethanolamine and poly(ethylene glycol) derivate. The developed devices provide ultrahigh label-free detection sensitivities of about 1 fM for specific DNA sequences (Janissen et al. 2017). Tachibana et al. (2017) developed and implemented the wavelength-tunable excited state FSRS to explain the structural dynamics for the fluorescence modulation of a single-site P377R mutant of the emergent genetically encoded calcium indicators for optical imaging (GEM-GECO1) calcium biosensor. The transient Raman modes following 400 nm photoexcitation report the characteristic ESPT time constants of 36 ps in the $Ca^{2+}$-free biosensor versus 16–90 ps in the $Ca^{2+}$-bound state. Furthermore, from the integrated Raman peak intensity oscillations, a chromophore in-plane ring rocking mode at ~180 $cm^{-1}$ is found to dominate in the $Ca^{2+}$-free biosensor but is joined by two other out-of-plane low-frequency modes in gating the multidimensional ESPT reaction coordinate upon $Ca^{2+}$ binding (Tachibana et al. 2017). Wang et al. (2018a) reported a cocaine sensor based on a single nanochannel coupled with DNA aptamers. A linear relationship between target cocaine concentration and output ionic current is obtained in a wide concentration range of cocaine from 1 nM to 10 μM. The cocaine sensor showed a LOAD at 1 nM (Wang et al. 2018a).

Wang et al. (2018b) developed a specific single-stranded DNA embedded four PS-modified RNA (HgDPR) and Hg-DPR covalently linked with SWCN fruitfully obtained detection of Hg(II) as low as 10 pM, and the linear range of 50 pM to 100 nM and 100 nM to 10 μM (Wang et al. 2018b). Leonie Baumann et al. (2018) developed rapid detection of short- and medium-chain fatty acids (SMCFA). The sensor detected hexanoic, heptanoic, and octanoic acid over a linear range up to 2, 1.5, and 0.75 mM, respectively (Baumann et al. 2018). Han et al. (2018) designed enzyme-linked immunosorbent assay (ELISA)-on-a-chip biosensors for rapid analysis of Bacillus anthracis spores. The biosensor can detect a minimum threshold of $5 \times 10^3$ and $5 \times 10^2$ spores/mL for two different B. anthracis strains, respectively (Han et al. 2018). Soldatkina et al. (2018) developed a conductometric biosensor based on coimmobilized urease and arginase for arginine detection in medicine. The LOD was 2.5 μM, the linear range 2.5–500 μM, sensitivity to arginine 13.4 ± 2.4 μS/mM, and response time 20 s (Soldatkina et al. 2018).

Sun et al. (2019) developed a biosensor consisting of a prism, silver, barium titanate ($BaTiO_3$), and graphene layers with the theoretical analysis explanation that the maximum sensitivity can reach at 257°/RIU on an Ag thickness of 45 nm with 10 nm $BaTiO_3$ and one graphene layer (Sun et al. 2019). Narayanan and Slaughter (2019) reported sensitive and selective electrochemical tungsten and gold microwire electrodes decorated with gold nanoparticles (AuNPs) as a platform for the detection of glucose and $H_2O_2$. The sensor displayed a linear range of 0.5 mM to 8 mM glucose. Linearity for $H_2O_2$ was up to 70 μM (Narayanan and Slaughter 2019). Mallikarjunarao Ganesana et al. (2019) developed a poly-*o*-phenylenediamine layer electropolymerized onto a 50-μm Pt wire for sensing L-glutamate in vivo with a linear range from 5 to 150 μM, with a sensitivity of 0.097 ± 0.001 nA/μM and steady-state response to L-glutamate within 2 s, with a LOD at 0.044 μM (Ganesana et al. 2019). Tucci et al. (2019) developed a microbial biosensor based on the cyanobacteria Anabaena variabilis for in situ herbicide detection during inhibition of the generated photocurrent. The biosensor displayed sensitivity of 24.6 $\mu A\ \mu M^{-1}\ cm^{-2}$ towards atrazine at 0.56 μM (Tucci et al. 2019).

Tzouvadaki et al. (2019) developed biosensors for label-free detection of free prostate-specific antigen (fPSA) for prostate cancer (PC). This biosensor provides outstanding analytical sensitivity with a LOD low at 0.144 pg/mL on PSA

(Tzouvadaki et al. 2019). Malik et al. (2019) fabricated a $PO_x$NPs/Au electrode biosensor that displayed best response within 7.5 s, at a potential of 0.28 V, pH 5.5, and 35 °C with concentration in the range 0.01 μM–5000 μM, with LOD at 0.67 μM. The recovery of pyruvate was 99.0% and 99.5% and with CV was 0.045% and 0.040%, respectively (Malik et al. 2019). Solis-Tinoco et al. (2019) fabricated a novel plasmo-nanomechanical biosensor for studying cell contractile forces. In relationship to the plasmonic behaviour of our sensors, the 310 M sensor showed a bulk sensitivity of ≈204 nm/RIU whereas the OG142 biosensor showed a sensitivity of ≈164 nm/RIU (Solis-Tinoco et al. 2019).

## References

Arafat MM, Haseeb ASMA, Akbar SA, Quadir MZ (2017) In-situ fabricated gas sensors based on one dimensional core-shell $TiO_2$-$Al_2O_3$ nanostructures. Sensors Actuators B 238:972–984. https://doi.org/10.1016/j.snb.2016.07.135

Bae G, Jeon IS, Jang M, Song W, Myung S, Lim J, Lee SS, Jung HK, Park C-Y, An K-S (2019) Complementary dual channel gas sensor devices based on a role-allocated ZnO-graphene hybrid heterostructure. ACS Appl Mater Interfaces 11(18):16830–16837. https://doi.org/10.1021/acsami.9b01596

Bai S, Hu J, Li D, Luo R, Chen A, Liu CC (2011) Quantum-sized ZnO nanoparticles: synthesis, characterization and sensing properties for $NO_2$. J Mater Chem 21:12288–12294. https://doi.org/10.1039/c1jm11302j

Bai S, Zhang K, Luo R, Li D, Chen A, Liu CC (2012) Low-temperature hydrothermal synthesis of $WO_3$ nanorods and their sensing properties for $NO_2$. J Mater Chem 22:12643–12650. https://doi.org/10.1039/c2jm30997a

Bai S, Guo T, Zhao Y, Luo R, Li D, Chen A, Liu CC (2013) Mechanism enhancing gas sensing and first-principle calculations of Al-doped ZnO nanostructures. J Mater Chem A 1:11335–11342. https://doi.org/10.1039/c3ta11516j

Bai S, Chen S, Zhao Y, Guo T, Luo R, Li D, Chen A (2014) Gas sensing properties of Cd-doped ZnO nanofibers synthesized by the electrospinning method. J Mater Chem A 2:16697–16706. https://doi.org/10.1039/c4ta03665d

Bao J, Hou C, Chen M, Li J, Huo D, Yang M, Luo X, Yu L (2015) Plant esterase−chitosan/gold nanoparticles−graphene nanosheet composite-based biosensor for the ultrasensitive detection of organophosphate pesticides. J Agric Food Chem 63(47):10319–10326. https://doi.org/10.1021/acs.jafc.5b03971

Baumann L, Rajkumar AS, Morrissey JP, Boles E, Oreb M (2018) A yeast-based biosensor for screening of short- and medium-chain fatty acid production. ACS Synth Biol 71(1):2640–2646. https://doi.org/10.1021/acssynbio.8b00309

Bavykin DV, Friedrich JM, Walsh FC (2006) Protonated titanates and $TiO_2$ nanostructured materials: synthesis, properties, and applications. Adv Mater 18:2807–2824. https://doi.org/10.1002/adma.200502696

Bhunia P, Hwang E, Yoon Y, Lee E, Seo S, Lee H (2012) Synthesis of highly n-type graphene by using an ionic liquid. Chem Eur J 18:12207–12212. https://doi.org/10.1002/chem.201201593

Carp O, Huisman CL, Reller A (2004) Photoinduced reactivity of titanium dioxide. Prog Solid State Chem 32:33–177. https://doi.org/10.1016/j.progsolidstchem.2004.08.001

Casals O, Markiewicz N, Fabrega C, Gracia I, Cané C, Wasisto HS, Waag A, Prades JD (2019) A parts per billion (ppb) sensor for $NO_2$ with microwatt (μW) power requirements based on micro light plates. ACS Sensors 44:822–826. https://doi.org/10.1021/acssensors.9b00150

Chang C-M, Hon M-H, Leu I-C (2012) Improvement in CO sensing characteristics by decorating ZnO nanorod arrays with Pd nanoparticles and the related mechanisms. RSC Adv 2:2469–2475. https://doi.org/10.1039/c2ra01016j

Chong HH, Ching CB (2016) Development of colorimetric-based whole-cell biosensor for organophosphorus compounds by engineering transcription regulator DmpR. ACS Synth Biol 511:1290–1298. https://doi.org/10.1021/acssynbio.6b00061

Chu J, Wang X, Wang D, Aijun Yang A, Pinlei LV, Wu Y, Rong M, Gao L (2018) Highly selective detection of sulfur hexafluoride decomposition components $H_2S$ and $SOF_2$ employing sensors based on tin oxide modified reduced graphene oxide. Carbon 135:95–103. https://doi.org/10.1016/j.carbon.2018.04.037

Dai Z, Lee C-S, Tian Y, Kimb I-D, Lee J-H (2015) Highly reversible switching from P- to N-type $NO_2$ sensing in a monolayer $Fe_2O_3$ inverse opal film and the associated P–N transition phase diagram. J Mater Chem A 3:3372–3381. https://doi.org/10.1039/C4TA05438E

Deng DH, Pan XL, Yu L, Cui Y, Jiang YP, Qi J, Li WX, Fu Q, Ma XC, Xue QK, Sun GQ, Bao XH (2011) Toward N-doped graphene via solvothermal synthesis. Chem Mater 23:1188–1193. https://doi.org/10.1021/cm102666r

Ding P, Xu D, Huang H, Xu P, Zheng D (2018) ZnS modified ZnO nano-heterojunction for efficient trace $H_2S$ gas detection. Mater Sci Eng 452:022031. https://doi.org/10.1088/1757-899X/452/2/022031

Doran MM, Finnerty NJ, Lowry JP (2017) In–vitro development and characterisation of a superoxide dismutase-based biosensor. ChemistrySelect 2(14):4157–4164. https://doi.org/10.1002/slct.201700793

Du X, Du Y, George SM (2008) CO gas sensing by ultrathin tin oxide films grown by atomic layer deposition using transmission FTIR spectroscopy. J Phys Chem A 112:9211–9219. https://doi.org/10.1021/jp800518v

Duy LT, Kim D-J, Trung TQ, Dang VQ, Kim B-Y, Moon HK, Lee N-E (2015) High performance three-dimensional chemical sensor platform using reduced graphene oxide formed on high aspect-ratio micro-pillars. Adv Funct Mater 25:883–890. https://doi.org/10.1002/adfm.201401992

Fegade U, Jethave G, Kang-Yang S, Huang W-R, Wu R-J (2018) An multifunction $Zn_{0.3}Mn_{0.4}O_4$ nanospheres for carbon dioxide reduction to methane via photocatalysis and reused after five cycles for phosphate adsorption. J Environ Chem Eng 6:1918–1925. https://doi.org/10.1016/j.jece.2018.02.040

Fernandez-Garcia M, Martinez-Arias A, Hanson JC, Rodriguez JA (2004) Nanostructured oxides in chemistry: characterization and properties. Chem Rev 104(9):4063–4104. https://doi.org/10.1021/cr030032f

Ganesana M, Trikantzopoulos E, Maniar Y, Lee ST, Venton BJ (2019) Development of a novel micro biosensor for in vivo monitoring of glutamate release in the brain. Biosens Bioelectron 130:103–109. https://doi.org/10.1016/j.bios.2019.01.049

Gao J, Wang L, Kan K, Xu S, Jing L, Liu S, Shen P, Li L, Shi K (2014) One-step synthesis of mesoporous $Al_2O_3$–$In_2O_3$ nanofibres with remarkable gas-sensing performance to NOx at room temperature. J Mater Chem A 2:949–956. https://doi.org/10.1039/c3ta13943c

Ge Y, Kan K, Yang Y, Zhou L, Jing L, Shen P, Li L, Shi K (2014) Highly mesoporous hierarchical nickel and cobalt double hydroxide composite: fabrication, characterization and ultrafast NOx gas sensors at room temperature. J Mater Chem A 2:4961–4969. https://doi.org/10.1039/c3ta14607c

Han S-M, Kim Y-W, Kim Y-K, Chun J-H, Hee-Bok O, Paek S-H (2018) Performance characterization of two-dimensional paper chromatography-based biosensors for biodefense, exemplified by detection of *Bacillus anthracis* spores. Biochip J 12(1):59–68. https://doi.org/10.1007/s13206-017-2108-9

Harale NS, Dalavi DS, Mali SS, Tarwal NL, Vanalakar SA, Rao VK, Hong CK, Kim JH, Patil PS (2018) Single-step hydrothermally grown nanosheetassembled tungsten oxide thin films for sensitive and selective $NO_2$ gas detection. J Mater Sci 8. https://doi.org/10.1007/s10853-017-1905-9

Her Y-C, Chiang C-K, Jean S-T, Huang S-L (2012) Self-catalytic growth of hierarchical $In_2O_3$ nanostructures on $SnO_2$ nanowires and their CO sensing properties. CrystEngComm 14:1296–1300. https://doi.org/10.1039/c1ce06086d

Hou J, Huang H, Han Z, Pan H (2016) The role of oxygen adsorption and gas sensing mechanism for cerium vanadate ($CeVO_4$) nanorods. RSC Adv 6:14552–14558

Huang H, Ong CY, Guo J, White BT, Tse MS, Tan OK (2010) Pt surface modification of $SnO_2$ nanorod arrays for CO and $H_2$ sensors. Nanoscale 2:1203–1207. https://doi.org/10.1039/c0nr00159g

Huang H-M, Li H-Y, Wang X-X, Guo X (2017) Detecting low concentration of $H_2S$ gas by $BaTiO_3$ nanoparticle-based sensors. Sensors Actuators B 238:16–23. https://doi.org/10.1016/j.snb.2016.06.172

Hussain T, Kaewmaraya T, Chakraborty S, Vovusha H, Amornkitbamrung V, Ahuja R (2018) Defected and functionalized germanene-based nanosensors under sulfur comprising gas exposure. ACS Sensors 34:867–874. https://doi.org/10.1021/acssensors.8b00167

Iitani K, Chien P-J, Suzuki T, Toma K, Arakawa T, Iwasaki Y, Mitsubayashi K (2017) Improved sensitivity of acetaldehyde biosensor by detecting ADH reverse reaction-mediated NADH fluoro-quenching for wine evaluation. ACS Sensors 27:940–946. https://doi.org/10.1021/acssensors.7b00184

Iwasaki M, Davis SA, Mann S (2004) Spongelike macroporous $TiO_2$ monoliths prepared from starch gel template. J Sol-Gel Sci Technol 32:99–105. https://doi.org/10.1007/s10971-004-5772-x

Janissen R, Sahoo PK, Santos CA, da Silva AM, von Zuben AAG, Souto DEP, Costa ADT, Celedon P, Zanchin NIT, Almeida DB, Oliveira DS, Kubota LT, Cesar CL, de Souza AP, Cotta MA (2017) InP nanowire biosensor with tailored biofunctionalization: ultrasensitive and highly selective disease biomarker detection. Nano Lett 17(10):5938–5949. https://doi.org/10.1021/acs.nanolett.7b01803

Jethave G, Fegade U (2018) Design and synthesis of $Zn_{0.3}Fe_{0.45}O_3$ nanoparticle for efficient removal of Congo red dye and its kinetic and isotherm investigation. Int J Ind Chem 9:85–97. https://doi.org/10.1007/s40090-018-0140-9

Jethave G, Fegade U, Attarde S, Ingle S (2017) Facile synthesis of lead doped zinc-aluminum oxide nanoparticles (LD-ZAO-NPs) for efficient adsorption of anionic dye: kinetic, isotherm and thermodynamic behaviors. J Ind Eng Chem 53:294–306. https://doi.org/10.1016/j.jiec.2017.04.038

Jett SE, Bonham AJ (2017) Reusable E-DNA biosensor for the detection of water-borne uranium. ChemElectroChem 4(4):843–845. https://doi.org/10.1002/celc.201600617

Jiang C, Guo Z, Wu Y, Li L, Shi K (2012) Facile synthesis of $SnO_2$ nanocrystalline tubes by electrospinning and their fast response and high sensitivity to NOx at room temperature. CrystEngComm 14:2739–2747. https://doi.org/10.1039/c2ce06405g

Jiang C, Xu S, Guo Z, Li L, Yang Y, Shi K (2013) Facile synthesis of CaO–$SnO_2$ nanocrystalline composite rods by electrospinning method with enhanced gas sensitive performance at room temperature. CrystEngComm 15:2482–2489. https://doi.org/10.1039/c2ce26736e

Jiang Z, Li J, Aslan H, Li Q, Li Y, Chen M, Huang Y, Froning JP, Otyepka M, Zbŏril R, Besenbacher F, Dong M (2014) A high efficiency $H_2S$ gas sensor material: paper like $Fe_2O_3$/graphene nanosheets and structural alignment dependency of device efficiency. J Mater Chem A 2:6714–6717. https://doi.org/10.1039/c3ta15180h

Jinsoo K, Jae Won L, Tai Gyu L, Suk Woo N, Jonghee H (2005) Nanostructured titania membranes with improved thermal stability. J Mater Sci 40:1797–1799

Kida T, Fujiyama S, Suematsu K, Yuasa M, Shimanoe K (2013) Pore and particle size control of gas sensing films using $SnO_2$ nanoparticles synthesized by seed-mediated growth: design of highly sensitive gas sensors. J Phys Chem C 117:17574–17582. dx.doi.org. https://doi.org/10.1021/jp4045226

Ko KY, Park K, Lee S, Kim Y, Woo WJ, Kim D, Song J-G, Park J, Kim H (2018) Recovery improvement for large-area tungsten Diselenide gas sensor. ACS Appl Mater Interfaces 1028:23910–23917. https://doi.org/10.1021/acsami.8b07034

Kolen'ko YV, Kovnir KA, Gavrilov AI, Garshev AV, Meskin PE, Churagulov BR, Bouchard M, Colbeau-Justin C, Lebedev OI, Van Tendeloo G, Yoshimura M (2005) Structural, textural, and electronic properties of a nanosized mesoporous $Zn_xTi_{1-x}O_{2-x}$ solid solution prepared by a supercritical drying route. J Phys Chem B 109(43):20303–20309. https://doi.org/10.1021/jp0535341

Kondalkar M, Fegade U, Attarde S, Ingle S (2018) Experimental investigation on phosphate adsorption, mechanism and desorption properties of Mn-Zn-Ti oxide trimetal alloy nanocomposite. J Dispers Sci Technol 11:1635–1643. https://doi.org/10.1080/01932691.2018.1459678

Kuang Q, Lao C-S, Li Z, Liu Y-Z, Xie Z-X, Zheng L-S, Wang ZL (2008) Enhancing the photon- and gas-sensing properties of a single $SnO_2$ nanowire based nanodevice by nanoparticle surface functionalization. J Phys Chem C 112:11539–11544. https://doi.org/10.1021/jp802880c

Kumar R, Goel N, Kumar M (2017) UV-activated $MoS_2$ based fast and reversible $NO_2$ sensor at room temperature. ACS Sensors 2(11):1744–1752. https://doi.org/10.1021/acssensors.7b00731

Lagowski J, Sproles ES, Gatos HC (1977) Quantitative study of the charge transfer in chemisorption; oxygen chemisorption on ZnO. J Appl Phys 48:3566–3575

Lai H-Y, Chen C-H (2012) Highly sensitive room-temperature CO gas sensors: Pt and Pd nanoparticle-decorated $In_2O_3$ flower-like nanobundles. J Mater Chem 22:13204–13208. https://doi.org/10.1039/c2jm31180a

Li W, Zhang L-S, Qiong Wang YY, Chen Z, Cao C-Y, Song W-G (2012) Low-cost synthesis of graphitic carbon nanofibers as excellent room temperature sensors for explosive gases. J Mater Chem 22:15342–15347. https://doi.org/10.1039/c2jm32031b

Li P, Fan H, Yu C (2013) $In_2O_3/SnO_2$ heterojunction microstructures: facile room temperature solid-state synthesis and enhanced $Cl_2$ sensing performance. Sensors Actuators B 185:110–116. https://doi.org/10.1016/j.snb.2013.05.010

Li L, Wang Y, Pan L, Shi Y, Cheng W, Shi Y, Yu G (2015) A nanostructured conductive hydrogels-based biosensor platform for human metabolite detection. Nano Lett 152:1146–1151. https://doi.org/10.1021/nl504217p

Li Z, Wang N, Lin Z, Wang J, Liu W, Sun K, Yong Qing F, Wang Z (2016) Room-temperature high performance $H_2S$ sensor based on porous CuO nanosheets prepared by hydrothermal method. ACS Appl Mater Interfaces 83:220962–220968. https://doi.org/10.1021/acsami.6b02893

Liu N, Li T-T, Yu H, Xia L (2018) Fabrication of a disordered mesoporous ZnO matrix modified by CuO film as high-performance $NO_x$ sensor. ChemistrySelect 3:5377–5385. https://doi.org/10.1002/slct.201800429

Lei Y, Zhao W, Zhang Y, Jiang Q, He J-H, Baeumner AJ, Wolfbeis OS, Wang ZL, Salama KN, Alshareef HN (2019) A MXene-Based Wearable Biosensor System for High-Performance In Vitro PerspirationAnalysis, Small 15(19):1901190. https://doi.org/10.1002/smll.2019011

Lee CY, Liao CH, Tso JT, Hsieh Y-Z (2019) A pyrophosphatase biosensor with photocurrent analysis. Sensors and Actuators, B: Chemical, 284, 159–163. https://doi.org/10.1016/j.snb.2018.12.123

Ma G, Zou R, Lin J, Zhang Z, Xue Y, Yu L, Song G, Li W, Hu J (2012) Phase-controlled synthesis and gas-sensing properties of zinc stannate ($ZnSnO_3$ and $Zn_2SnO_4$) faceted solid and hollow microcrystals. CrystEngComm 14:2172–2179. https://doi.org/10.1039/c2ce06272k

Mahendraprabhu K, Elumalai P (2017) Stabilized zirconia-based selective $NO_2$ sensor using sol-gel derived $Nb_2O_5$ sensing-electrode. Sensors Actuators B 238:105–110. https://doi.org/10.1016/j.snb.2016.07.010

Malik M, Chaudhary R, Pundir CS (2019) An improved enzyme nanoparticles based amperometric pyruvate biosensor for detection of pyruvate in serum. Enzyme Microb Technol 123:30–38. https://doi.org/10.1016/j.enzmictec.2019.01.006

Mendoza F, Hernández DM, Makarov V, Febus E, Weiner BR, Morell G (2014) Room temperature gas sensor based on tin dioxide-carbon nanotubes composite films. Sensors Actuators B 190:227–233. https://doi.org/10.1016/j.snb.2013.08.050

Mickelson W, Sussman A, Zettl A (2012) Low-power, fast, selective nanoparticle-based hydrogen sulfide gas sensor. Appl Phys Lett 100(173110):173110. https://doi.org/10.1063/1.3703761

Mirzaei A, Kim SS, Kim HW (2018) Resistance-based $H_2S$ gas sensors using metal oxide nanostructures: a review of recent advances. J Hazard Mater 357:314–331. https://doi.org/10.1016/j.jhazmat.2018.06.015

Morrison SR (1987) Mechanism of semiconductor gas sensor operation. Sensors Actuators 11:283–287

Muangrat W, Wongwiriyapan W, Yordsri V, Chobsilp T, Inpaeng S, Issro C, Domanov O, Ayala P, Pichler T, Shi L (2018) Unravel the active site in nitrogen-doped double-walled carbon nanotubes for nitrogen dioxide gas sensor. Phys Status Solidi A 215:1800004–1800010. https://doi.org/10.1002/pssa.201800004

Mubeen S, Lim J-H, Srirangarajan A, Mulchandani A, Deshusses MA, Myung NV (2011) Gas sensing mechanism of gold nanoparticles decorated single-walled carbon nanotubes. Electroanalysis 23(11):2687–2692. https://doi.org/10.1002/elan.201100299

Mutkule S, Navale S, Patil V, Tehare K, Liu X, Gunturu KC, Mane R (2018) Chemical bath deposition of ZnO films at low pH for high chemoresistivity towards $NO_2$ gas. Mater Res Express 5:075021. https://doi.org/10.1088/2053-1591/aac918

Narayanan JS, Slaughter G (2019) Towards a dual in-line electrochemical biosensor for the determination of glucose and hydrogen peroxide. Bioelectrochemistry 128:56–65. https://doi.org/10.1016/j.bioelechem.2019.03.005

Navale ST, Jadhav VV, Tehare KK, Sagar RUR, Biswas CS, Galluzzi M, Liang W, Patil VB, Mane RS, Stadler FJ (2017) Solid-state synthesis strategy of ZnO nanoparticles for the rapid detection of hazardous $Cl_2$. Sensors Actuators B 283:1102–1110. https://doi.org/10.1016/j.snb.2016.07.136

Niu F, Tao L-M, Deng Y-C, Wang Q-H, Song W-G (2014) Phosphorus doped graphene nanosheets for room temperature $NH_3$ sensing. New J Chem 38:2269–2272. https://doi.org/10.1039/c4nj00162a

Offermans P, Crego-Calama M, Brongersma SH (2010) Gas detection with vertical InAs nanowire arrays. Nano Lett 10:2412–2415. https://doi.org/10.1021/nl1005405

Pan X, Liu X, Bermak A, Fan Z (2013) Self-gating effect induced large performance improvement of ZnO nanocomb gas sensors. ACS Nano 7:9318–9324. https://doi.org/10.1021/nn4040074

Park H, Han GC, Lee SW, Lee H, Jeong SH, Naqi M, AlMutairi AA, Kim YJ, Lee J, Kim W-J, Kim S, Yoon Y, Yoo G (2017) Label-free and recalibrated multilayer MoS2 biosensor for point-of-care diagnostics. ACS Appl Mater Interfaces 9(50):43490–43497. https://doi.org/10.1021/acsami.7b14479

Pierre AC (1998) Introduction to sol-gel processing. Kluwer Academic Publishers, Boston, p 394

Poloju M, Jayababu N, Reddy MVR (2018) Improved gas sensing performance of Al doped ZnO/CuO nanocomposite based ammonia gas sensor. Mater Sci Eng B 227:61–67. https://doi.org/10.1016/j.mseb.2017.10.012

Qi G, Zhang L, Yuan Z (2014) Improved $H_2S$ gas sensing properties of ZnO nanorods decorated by a several nm ZnS thin layer. Phys Chem Chem Phys 16:13434–13439. https://doi.org/10.1039/c4cp00906a

Saberi MH, Mortazavi Y, Khodadadi AA (2015) Dual selective Pt/$SnO_2$ sensor to CO and propane in exhaust gases of gasoline engines using Pt/LaFeO3 filter. Sensors Actuators B 206:617–623. https://doi.org/10.1016/j.snb.2014.10.007

Sahay PP, Nath RK (2008) Al-doped ZnO thin films as methanol sensors. Sensors Actuators B Chem 134:654–659

Sankar Ganesh R, Durgadevi E, Navaneethan M, Patil VL, Ponnusamy S, Muthamizhchelvan C, Kawasaki S, Patil PS, Hayakawa Y (2018) Tuning the selectivity of NH3 gas sensing response using Cu-doped ZnO nanostructures. Sensors Actuators A 269:331–341. https://doi.org/10.1016/j.sna.2017.11.042

Sasaki I, Minami N, Karthigeyan A, Iakoubovskii K (2009) Optimization and evaluation of networked single-wall carbon nanotubes as a $NO_2$ gas sensing material. Analyst 134:325–330. https://doi.org/10.1039/b813073f

Shahid A, Choi J-H, Rana A u HS, Kim H-S (2018) Least squares neural network-based wireless E-nose system using an $SnO_2$ sensor array. Sensors 18:1446–1461. https://doi.org/10.3390/s18051446

Shankar P, Rayappan JBB (2015) Gas sensing mechanism of metal oxides: the role of ambient atmosphere, type of semiconductor and gases—a review. ScienceJet 4:126

Sheng ZH, Shao L, Chen JJ, Bao WJ, Wang FB, Xia XH (2011) Catalyst-free synthesis of nitrogen-doped graphene via thermal annealing graphite oxide with melamine and its excellent electrocatalysis. ACS Nano 5:4350–4358. https://doi.org/10.1021/nn103584t

Shi C, Qin H, Zhao M, Wang X, Li L, Hu J (2014) Investigation on electrical transport, CO sensing characteristics and mechanism for nanocrystalline $La_{1-x}Ca_xFeO_3$ sensors. Sensors Actuators B 190:25–31. https://doi.org/10.1016/j.snb.2013.08.029

Shim Y-S, Kwon KC, Suh JM, Choi KS, Song YG, Sohn W, Choi S, Hong KT, Jeon J-M, Hong S-P, Kim S, Kim SY, Kang C-Y, Jang HW (2018) Synthesis of numerous edge sites in $MoS_2$ via $SiO_2$ nanorods platform for highly sensitive gas sensor. ACS Appl Mater Interfaces 10(37):31594–31602. https://doi.org/10.1021/acsami.8b08114

Soikkeli M, Kurppa K, Kainlauri M, Arpiainen S, Paananen A, Gunnarsson D, Joensuu JJ, Laaksonen P, Prunnila M, Linder MB, Ahopelto J (2016) Graphene biosensor programming with genetically engineered fusion protein monolayers. ACS Appl Mater Interfaces 8:8257–8264. https://doi.org/10.1021/acsami.6b00123

Soldatkina OV, Soldatkin OO, Velychko TP, Prilipko VO, Kuibida MA, Dzyadevych SV (2018) Conductometric biosensor for arginine determination in pharmaceutics. Bioelectrochemistry 124:40–46. https://doi.org/10.1016/j.bioelechem.2018.07.002

Solis-Tinoco V, Marquez S, Quesada-Lopez T, Villarroya F, Homs-Corbera A, Lechuga LM (2019) Building of a flexible microfluidic plasmo-nanomechanical biosensor for live cell analysis. Sensors Actuators B Chem 291:48–57. https://doi.org/10.1016/j.snb.2019.04.038

Song Z, Wei Z, Wang B, Luo Z, Xu S, Zhang W, Yu H, Li M, Huang Z, Zang J, Yi F, Liu H (2016) Sensitive room-temperature $H_2S$ gas sensors employing $SnO_2$ quantum wire/reduced graphene oxide nanocomposites. Chem Mater 284:1205–1212. https://doi.org/10.1021/acs.chemmater.5b04850

Sonker RK, Sharma A, Shahabuddin M, Tomar M, Gupta V (2013) Low temperature sensing of $NO_2$ gas using $SnO_2$-ZnO nanocomposite sensor. Adv Mater Lett 4(3):196–201. https://doi.org/10.5185/amlett.2012.7390

Soylemez S, Yılmaz T, Buber E, Udum YA, Özçubukçu S, Toppare L (2017) Polymerization and biosensor application of water soluble peptide-SNS type monomer conjugates. J Mater Chem B 5:7384–7392. https://doi.org/10.1039/C7TB01674C

Sriwichai S, Janmanee R, Phanichphant S, Shinbo K, Kato K, Kaneko F, Yamamoto T, Baba A (2017) Development of an electrochemical-surface plasmon dual biosensor based on carboxylated conducting polymer thin films. J Appl Polym Sci 134:45641–45651. https://doi.org/10.1002/APP.45641

Steinhauer S, Chapelle A, Menini P, Sowwan M (2016) Local CuO nanowire growth on microhotplates: in situ electrical measurements and gas sensing application. ACS Sensors 15:503–507. https://doi.org/10.1021/acssensors.6b00042

Sun P, Zhou X, Wang C, Wang B, Xu X, Lu G (2014) One-step synthesis and gas sensing properties of hierarchical Cd-doped $SnO_2$ nanostructures. Sensors Actuators B 190:32–39. https://doi.org/10.1016/j.snb.2013.08.045

Sun P, Wang M, Liu L, Jiao L, Wei D, Xia F, Liu M, Kong W, Dong L, Yun M (2019) Sensitivity enhancement of surface plasmon resonance biosensor based on graphene and barium titanate layers. Appl Surf Sci 475:342–347. https://doi.org/10.1016/j.apsusc.2018.12.283

Tabassum R, Mishra SK, Gupta BD (2013) Surface plasmon resonance-based fiber optic hydrogen sulphide gas sensor utilizing Cu–ZnO thin films. Phys Chem Chem Phys 15:11868–11874. https://doi.org/10.1039/c3cp51525g

Tachibana SR, Tang L, Wang Y, Zhu L, Liu W, Fang C (2017) Tuning calcium biosensors with a single-site mutation: structural dynamics insights from femtosecond Raman spectroscopy. Phys Chem Chem Phys 19:7138–7146. https://doi.org/10.1039/C6CP08821J

Tucci M, Grattieri M, Schievano A, Cristiani P, Minteer SD (2019) Microbial amperometric biosensor for online herbicide detection: photocurrent inhibition of *Anabaena variabilis*. Electrochim Acta 302:102–108. https://doi.org/10.1016/j.electacta.2019.02.007

Tzouvadaki I, Zapatero-Rodríguez J, Naus S, de Micheli G, O'Kennedy R, Carrara S (2019) Memristive biosensors based on full-size antibodies and antibody fragments. Sensors Actuators B Chem 286:346–352. https://doi.org/10.1016/j.snb.2019.02.001

Vetter S, Haffer S, Wagner T, Tiemann M (2015) Nanostructured $Co_3O_4$ as a CO gas sensor: temperature-dependent behaviour. Sensors Actuators B 206:133–138. https://doi.org/10.1016/j.snb.2014.09.025

Wang L, Zheng L, Wang R, Fei T, Zhang T (2012a) Ring-like PdO–NiO with lamellar structure for gas sensor application. J Mater Chem 22:12453–12456. https://doi.org/10.1039/c2jm16509k

Wang H, Gao J, Li Z, Ge Y, Kanab K, Shi K (2012b) One-step synthesis of hierarchical a-$Ni(OH)_2$ flowerlike architectures and their gas sensing properties for NOx at room temperature. CrystEngComm 14:6843–6852. https://doi.org/10.1039/c2ce25553g

Wang J, Hou J, Zhang H, Tian Y, Jiang L (2018a) Single nanochannel-aptamer-based biosensor for ultrasensitive and selective cocaine detection. ACS Appl Mater Interfaces 10(2):2033–2039. https://doi.org/10.1021/acsami.7b16539

Wang H, Liu Y, Liu G (2018b) Electrochemical biosensor using DNA embedded phosphorothioate modified RNA for mercury ions determination. ACS Sensors 33:624–631. https://doi.org/10.1021/acssensors.7b00892

Wei DC, Liu YQ, Wang Y, Zhang HL, Huang LP, Yu G (2009) Synthesis of N-doped graphene by chemical vapor deposition and its electrical properties. Nano Lett 9:1752–1758. https://doi.org/10.1021/nl803279t

Wen KY, Cameron L, Chappell J, Jensen K, Bell DJ, Kelwick R, Kopniczky M, Davies JC, Filloux A, Freemon PS (2017) A cell-free biosensor for detecting quorum sensing molecules in *P. aeruginosa*-infected respiratory samples. ACS Synth Biol 6:2293–2301. https://doi.org/10.1021/acssynbio.7b00219

Woo H-S, Kwak C-H, Kim I-D, Lee J-H (2014) Selective, sensitive, and reversible detection of $H_2S$ using Mo-doped ZnO nanowire network sensors. J Mater Chem A 2:6412–6418. https://doi.org/10.1039/c4ta00387j

Wu L, Lu X, Wang X, Song Y, Chen J (2015) An electrochemical deoxyribonucleic acid biosensor for rapid genotoxicity screening of chemicals. Anal Methods 7:3347–3352. https://doi.org/10.1039/c5ay00020c

Wu J, Tao K, Guo Y, Li Z, Wang X, Luo Z, Feng S, Chunlei D, Chen D, Miao J, Norford LK (2017) A 3D chemically modified graphene hydrogel for fast, highly sensitive, and selective gas sensor. Adv Sci 4:1600319–1600328. https://doi.org/10.1002/advs.201600319

Wu E, Xie Y, Yuan B, Zhang H, Hu X, Liu J, Zhang D (2018) Ultra-sensitive and fully reversible $NO_2$ gas sensing based on p-type MoTe2 under ultra-violet illumination. ACS Sensors 39:1719–1726. https://doi.org/10.1021/acssensors.8b00461

Wu J, Wu Z, Han S, Yang B-R, Gui X, Tao K, Liu C, Miao J, Norford LK, Deformable E (2019) Transparent, and high-performance gas sensor based on ionic conductive hydrogel. ACS Appl Mater Interfaces 112:2364–2373. https://doi.org/10.1021/acsami.8b17437

Xu X, Wang D, Liu J, Sun P, Guan Y, Zhang H, Sun Y, Liu F, Liang X, Gao Y, Lu G (2013) Template-free synthesis of novel $In_2O_3$ nanostructures and their application to gas sensors. Sensors Actuators B 185:32–38. https://doi.org/10.1016/j.snb.2013.04.078

Xue YZ, Wu B, Jiang L, Guo YL, Huang LP, Chen JY, Tan JH, Geng DC, Luo BR, Hu WP, Yu G, Liu YQ (2012) Low temperature growth of highly nitrogen-doped single crystal graphene arrays by chemical vapor deposition. J Am Chem Soc 134:11060–11063. https://doi.org/10.1021/ja302483t

Yang J, Mei S, Ferreira JMF (2001) Hydrothermal synthesis of $TiO_2$ nanopowders from tetraalkylammonium hydroxide peptized sols. Mater Sci Eng C 15:183–185. https://doi.org/10.1016/S0928-4931(01)00274-0

Yang Y, Tian C, Sun L, Lü R, Zhou W, Shi K, Kan K, Wang J, Fu H (2013) Growth of small sized $CeO_2$ particles in the interlayers of expanded graphite for high-performance room temperature NOx gas sensors. J Mater Chem A 1:12742–12749. https://doi.org/10.1039/c3ta12399e

Yassine O, Shekhah O, Assen AH, Belmabkhout Y, Salama KN, Eddaoudi M (2016) $H_2S$ sensors: fumarate-based fcu-MOF thin film grown on a capacitive interdigitated electrode. Angew Chem Int Ed 55:15879–15883. https://doi.org/10.1002/anie.201608780

Yin L, Chen D, Hu M, Shi H, Yang D, Fan B, Shao G, Zhang R, Shao G (2014) Microwave-assisted growth of $In_2O_3$ nanoparticles on $WO_3$ nanoplates to improve $H_2S$-sensing performance. J Mater Chem A 2:18867–18874. https://doi.org/10.1039/c4ta03426k

Zhang Z, Zou R, Song G, Yu L, Chen Z, Hu J (2011) Highly aligned $SnO_2$ nanorods on graphene sheets for gas sensors. J Mater Chem 21:17360–17365. https://doi.org/10.1039/c1jm12987b

Zhang G, Dang L, Li L, Wang R, Honggang F, Shi K (2013a) Design and construction of $Co_3O_4$/PEI–CNTs composite exhibiting fast responding CO sensor at room temperature. CrystEngComm 15:4730–4738. https://doi.org/10.1039/c3ce40206a

Zhang L, Cui Z, Wu Q, Guo D, Xu Y, Lin G (2013b) $Cu_2O$–CuO composite microframes with well-designed micro/nano structures fabricated via controllable etching of $Cu_2O$ microcubes for CO gas sensors. CrystEngComm 15:7462–7467. https://doi.org/10.1039/c3ce40595h

Zhang S, Zhang P, Wang Y, Ma Y, Zhong J, Sun X (2014) Facile fabrication of a well-ordered porous Cu-doped $SnO_2$ thin film for $H_2S$ sensing. ACS Appl Mater Interfaces 617:14975–14980. https://doi.org/10.1021/am502671s

Zhang H, Wang Y, Zhao D, Zeng D, Xia J, Aldalbahi A, Wang C, San L, Fan C, Zuo X, Xianqiang (2015) Universal fluorescence biosensor platform based on graphene quantum dots and pyrene-functionalized molecular beacons for detection of MicroRNAs. ACS Appl Mater Interfaces 7(30):16152–16156. https://doi.org/10.1021/acsami.5b04773

Zhang J, Liu X, Neri G, Pinna N (2016) Nanostructured materials for room-temperature gas sensors. Adv Mater 28:795–831. https://doi.org/10.1002/adma.201503825

Zhang D, Liu J, Jiang C, Liu A, Xia B (2017) Quantitative detection of formaldehyde and ammonia gas via metal oxide-modified graphene-based sensor array combining with neural network model. Sensors Actuators B 240:55–65. https://doi.org/10.1016/j.snb.2016.08.085

Zhang J, Zhu Z, Chen C, Chen Z, Cai M, Baihua Q, Wang T, Zhang M (2018) ZnO-carbon nanofibers for stable, high response, and selective $H_2S$ sensors. Nanotechnology 29:275501. https://doi.org/10.1088/1361-6528/aabd72

Zhou R, Hu G, Yu R, Pan C, Wang ZL (2015a) Piezotronic effect enhanced detection of flammable/toxic gases by ZnO micro/nanowire sensors. Nano Energy 12:588–596. https://doi.org/10.1016/j.nanoen.2015.01.036

Zhou L, Yuan Q, Li X, Xu J, Xia F, Xiao J (2015b) The effects of sintering temperature of $(La_{0.8}Sr_{0.2})_2FeMnO_6^{-1}$ on the $NO_2$ sensing property for YSZ-based potentiometric sensor. Sensors Actuators B 206:311–318. https://doi.org/10.1016/j.snb.2014.09.018

# Biosensors Used for Monitoring of Environmental Contaminants

**Naveen Patel, Pankaj Pathak, Dhananjai Rai, and Vinod Kumar Chaudhary**

## Contents



## Abbreviations

| | |
|---|---|
| AChE | Acetylcholine esterase |
| BOD | Biological oxygen demand |
| MBTH | 3-Methyl-2-benzothiazolinone hydrazine |
| SERS | Surface-enhanced Raman spectroscopy |
| SWCNT | Single-walled carbon nanotube |

N. Patel (✉)
Department of Civil Engineering, NIT, Agartala, Agartala, Tripura, India

P. Pathak
Department of Electrical Engineering, Indian Institute of Technology, Delhi, Delhi, New Delhi, India
e-mail: pankaj.pathak@ee.iitd.ac.in

D. Rai
Department of Civil Engineering, Bundelkhand Institute of Engineering and Technology, Jhansi, Jhansi, Uttar Pradesh, India

V. K. Chaudhary
Department of Environmental Sciences, Dr Ram Manohar Lohia Avadh University, Ayodhya, Uttar Pradesh, India
e-mail: vinodkrchaudhary@rmlau.ac.in



Inamuddin, A. M. Asiri (eds.), *Nanosensor Technologies for Environmental Monitoring*, Nanotechnology in the Life Sciences,
https://doi.org/10.1007/978-3-030-45116-5_4

# 1 Introduction

Environmental security is one of the major concerns for our well-being. However, environmental security still remains a major challenge as an increase in living standards along with an increase in demands have caused an increase in pollution of water with various types of chemicals, nutrients, and oil spills; of air with particulate matter, $CO_2$, and other greenhouse gases; and of soil with herbicides, pesticides, and other hazardous wastes and non-biodegradable materials (Nigam and Shukla 2015). Pesticides, pharmaceuticals, cosmetics, and personal care products are the chemical-based products being used worldwide, and are adding pollutants constantly to the environment (Claude et al. 2007). Along with these chemical pollutants, manmade activities have also caused pollution of water bodies with viruses, bacteria, and other biological micro-pollutants. Biological micro-pollutants are the main reason for the water-borne diseases and are the major cause of death worldwide (Silva et al. 2010). Toxicants present in environmental samples of soil, air, and water are needed to be analysed comprehensively along with their detection and regular monitoring for the overall security of living systems.

Two different methodologies can be used for the monitoring of the pollutants present in environment. The conventional methodology gives exact analysis of physical or chemical properties of the pollutant but proves to be costly as it requires specified instruments along with well-established laboratories for analysis of pollutant. This system still lacks in providing details about the bioavailability of pollutants and effects related to it on biological systems. The above-mentioned drawback can be eliminated by developing a new approach for environmental monitoring over the conventional one. Bacteria, fungi, and enzymes along with others are the biological systems that can be applied for such analyses (Nigam and Shukla 2015). Unicellular microorganisms, especially bacterial system, can be used for such purpose because of their large population size, rapid growth system, and minimum requirement for the growth and maintenance. Furthermore, cells of unicellular microorganisms contain proteins, and that too can be utilised for detection of specific analyte (Zourob 2010). As a result, there is a need for developing a method that is simple, rapid, and sensible for analysis of pollutants and making biosensors to be seen as a suitable choice (Podola et al. 2004).

On the basis of transduction principle, biosensors can be classified as electrochemical (includes impedance and amperometric biosensors), piezoelectric (includes quartz crystal microbalance biosensors), optical (includes surface plasmon resonance and optical fibre biosensors), or biosensors can be categorised on the basis of recognition element. Depending on recognition elements like aptamers, nucleic acids, antibodies, and enzymes, biosensors are classified as aptasensors, immunosensors, genosensors, and enzymatic biosensors, respectively. Figure 1 shows the block diagram of biosensor system.

Previously, immunosensors and enzymatic biosensors were commonly used for environmental monitoring but recently because of thermal stability, easiness to modify, to distinguish target with different functional groups, possibility to design their structure and rehybridisation, aptasensors are receiving more attention (Justino

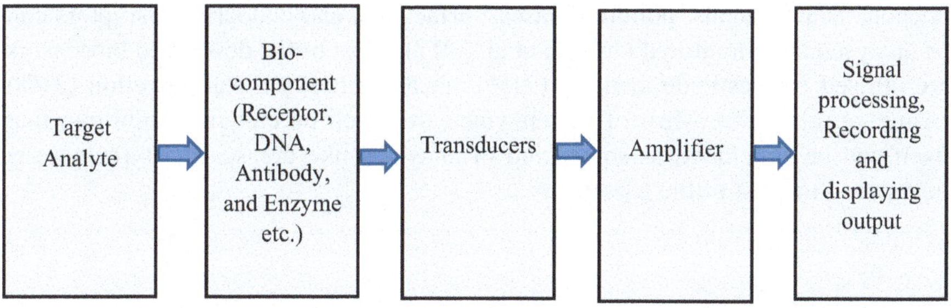


**Fig. 1** Block diagram of biosensing system

**Table 1** Different environmental application of biosensor

| Compound class | Biosensing element | Applications |
|---|---|---|
| Herbicides | Antibody, enzyme, and microbe | Soil, air, and water |
| Pesticides | Antibody, enzyme, and microbe | Soil, air, and water |
| Dioxins | Microbe | Soil, air, and water |
| Heavy metals | Enzyme and microbe | Water and soil |
| Phenolic compounds | Enzyme and microbe | Water and soil |
| Nitrogen compounds | Enzyme | Water and wastewater, soil |

et al. 2015). Role of nanotechnology for the development of biosensors has been increased as nanomaterials and novel nanocomposites based biosensors are more advantageous for the improvement of analytical properties like limit of detection and sensitivity (Maduraiveeran and Jin 2017). For example, gold nanostructure, because of its good electron mediation capability, high surface area, and stabilisation property of enzyme through electrostatic interaction, acts as a promising and versatile platform for enzyme immobilisation (Lang et al. 2016; Zhao et al. 2013). Commercialisation of the biosensors is one of the biggest problems because of its fabrication-related issues, in situ operation, and their reproducibility. Moreover, these sensors also find limitation in real samples as most of the sensors have been tested along with distilled water-contained pollutants or buffered solutions. However, few of the biosensors, with prime utilisation on biochemical oxygen demand biosensing, are available commercially (Bahadır and Sezgintürk 2015). This chapter will give an insight regarding the implementation of biosensors for the detection of pathogens, potentially toxic elements, toxins, endocrine disrupting chemicals, and pesticides for environmental monitoring as shown in Table 1.

## 2 Environmental Applications

Enormous number of biosensors including genosensors, immunosensors, and enzymatic biosensors, for the concern related to environmental monitoring, have been grown by the help of which a variety of pollutants that include pesticides, phenols, heavy metals, polluting gases, surfactants, and endocrine disruptors can be analysed and monitored (Justino et al. 2017). Most of the developed biosensors are utilised for pesticide analysis (71%), ahead of heavy metal detection (21%) (Amine et al. 2006). Most of the enzymes are based on enzyme inhibition; that itself can be concluded as some kind of enzymes like urease, and tyrosinase is being used in most of the biosensors.

## *2.1 Heavy Metals*

Heavy metals such as zinc, copper, mercury, arsenic, chromium, and cadmium are found to have difficulty in degradation and hence the tendency to accumulate. These above-mentioned heavy metals along with others enter the aquatic system via anthropogenic activities, industrial sources, and agriculture activities, creating imbalance in whole ecosystem. Many heavy metals pose adverse and toxic effects on environment that have the ability to get deposited in the living organisms and thus known to be one of the important classes of environmental pollutants (Rodriguez-Mozaz et al. 2004). Cadmium, chromium, lead, and mercury along with other heavy metals can be detected by immobilised urease and glucose oxidase biosensors. In case of mercury, higher affinity has been founded for the cysteine residue of urease; therefore, it leads to detection of mercury even to lower level that is up to 10 nm (Tsai and Doong 2005). Yang along with other researchers have developed a new renewable potentiometric biosensor and that was further investigated to have the ability to detect mercury ions even at low detection limit (Yang et al. 2006). A structure-switching DNA optical biosensor has been developed and are being utilised for detection of highly toxic and ubiquitous pollutants of heavy metals like mercury ions ($Hg^{2+}$). The main advantage of this biosensors were fast screening of $Hg^{2+}$ ions, i.e. within 10 min in natural water, portability, and cost-effectiveness (Long et al. 2013). The ability of some metal ions to form stable metal-mediated DNA duplexes after getting combined with some bases is the main principle for detection by such biosensor. Metal ions including zinc, copper, nickel, lead, and chromium along with others up to various concentration of 20 μM were tested using this type of biosensor, but no significant result has been observed. A new surface-enhanced biosensor known as surface-enhanced Raman spectroscopy (SERS) biosensor, having more simplicity and sensitivity, was designed for detection of $Hg^{2+}$, using single-walled carbon nanotubes (SWCNT) and magnetic substrate conjugated with single-stranded DNA. Magnetic substrate leads to magnetic aggregation of biosensor and leads to improvement in their detection limit and sensitivity (Yang et al. 2017; Madianos et al. 2018).

Enzymatic biosensors can be used to monitor other toxic metals like cadmium, lead, copper, and nickel, as these metals have the tendency to hinder the activity of various enzymes (Tsai et al. 2003; Guascito et al. 2008; Ghica et al. 2013; Ilangovan et al. 2006). Furthermore, the main drawbacks of these sensors are that they are not target specific and are generally utilised for the determination of the total inhibition effect induced in the sample by the heavy metals. In order to control the above-mentioned drawbacks, some recombinant luminescent bacterial sensors were developed (Ivask et al. 2004). Cytochrome c3 and metallothionein SmtA are the metal-binding proteins that are found in bacteria; they are purified and have been utilised in developing sensors for detection of heavy metals (Bontidean et al. 2000; Michel et al. 2006).

For analysis of $Pb^{2+}$ and $Cd^{2+}$, a multi-analyte biosensor using mesoporous carbon nitride/self-doped polyaniline nanofibers was developed, and the limit of detection for $Pb^{2+}$ and $Cd^{2+}$ was observed to be 0.2 nM and 0.7 nM, respectively (Zhang et al. 2016). Further, wireless biosensor based on magnetoelastic principle had shown similar detection limit for $Pb^{2+}$ and $Cd^{2+}$, i.e. 0.33 nM and 0.24 nM, respectively. The main advantage of such sensor involves its real-time monitoring in remote areas (Guo et al. 2018).

Electrochemical biosensor, paper-based channel with reduced chitosan and graphene, is developed for detection of $Zn^{2+}$ (Li et al. 2017). The biosensor has low detection limit of 0.03 nM and has the ability to analyse $Zn^{2+}$ even in complex environmental samples that contains other seven cations including: $Cu^{2+}$, $Fe^{3+}$, $Cd^{2+}$, $Hg^{2+}$, $Mn^{2+}$, $Mg^{2+}$, and $Ag^{2+}$ (Li et al. 2017).

Moreover, recently immunoassay, because of its advantage of high selectivity, sensitivity, and species specificity, has been used for heavy metal analysis. A KinExA-automated immunoassay has been developed by Blake and his co-workers for analysis of uranium within the range of 1–5 mm, and it can be used for analysis of other metals like cobalt, cadmium, and lead (Blake et al. 2001). Table 2 shows some biosensors along with their biorecognition element and transducer for heavy metal detection.

## 2.2 *Pathogens*

Bacteria, virus, and protozoa are the pathogenic form of microorganisms that are present in environmental matrices and founded to have the possibility to cause harmful diseases in humans. Hence, it is our prime importance to eliminate them and for the same, the role of biosensors becomes more important. The conventional methods of identification of pathogen involves cultivation of bacteria that itself is a time-consuming and difficult job; therefore, the role of biosensor for analysis of these disease-causing microorganism has been increasing immensely (Eriksson et al. 2009). A variety of immunosensors that are based on fluorescence, surface plasmon resonance, and impedance have been developed and are currently being used for detection of biological samples (Heyduk and Heyduk 2010; Baccar et al. 2010; Guo et al. 2012; Dos Santos et al. 2013). Optical biosensors based on surface plasmon resonance have been developed, and these sensors are known to be more rapid and specific in the case of analysis of Legionella pneumophila found in the water sample (Martín et al. 2015; Enrico et al. 2013). A whole cell imprinting biosensor based on optical and piezoelectric principles has been utilised for analysis

**Table 2** Biosensors along with their biorecognition element and transducer for heavy metal detection

| Heavy metal | Biorecognition element | Transducer | Detection limit | Reference |
|---|---|---|---|---|
| Cu(II) | Urease<br>Urease<br>Glucose oxidase<br>Glucose oxidase<br>Metallothionein SmA (*E. coli*) | Optical<br>Conductometric<br>Amperometric<br>Amperometric<br>Potentiometric | 10 μM<br>0.1–10 mM<br>5 μM<br>0.2 μM<br>$10^{-15}$ M | Tsai et al. (2003)<br>Ilangovan et al. 2006<br>Guascito et al. (2008)<br>Ghica et al. (2013)<br>Bontidean et al. (2000) |
| Hg(II) | Urease<br>Urease<br>Glucose oxidase<br>*E. coli* S30 extract<br>Metallothionein SmA (*E. coli*) | Potentiometric<br>Amperometric<br>Amperometric<br>Luminescence<br>Potentiometric | 0.05 μM<br>$7.4 * 10^{-6}$ M<br>2.5 μM<br>$5 * 10^{-9}$ M<br>$10^{-15}$ M | Yang et al. (2006)<br>Guascito et al. (2008)<br>Pellinen et al. (2004)<br>Bontidean et al. (2000) |
| Cd(II) | Urease<br>Urease<br>Glucose oxidase<br>Glucose oxidase<br>Metallothionein SmA (*E. coli*) | Optical<br>Conductometric<br>Amperometric<br>Amperometric<br>Potentiometric | 10 μM<br>0.1–10 mM<br>5 μM<br>2.4 μM<br>$10^{-15}$ M | Tsai et al. (2003)<br>Ilangovan et al. 2006<br>Guascito et al. (2008)<br>Ghica et al. (2013)<br>Bontidean et al. (2000) |
| Pb(II) | Urease | Conductometric | 0.1–10 mM | Ilangovan et al. 2006 |
| Ag(I) | Glucose oxidase | Amperometric | 0.05 μM | Guascito et al. (2008) |
| Co(II) | Glucose oxidase | Amperometric | 2.4 μM | Ghica et al. (2013) |
| Ni(II) | Glucose oxidase | Amperometric | 3.3 μM | Ghica et al. (2013) |
| Zn(II) | Metallothionein SmA (*E. coli*) | Potentiometric | $10^{-15}$ M | Bontidean et al. (2000) |
| Cr(VI) | Cytochrome $C_3$ (*D. norvegicum*) | Amperometric | 0.2 mg/L | Michel et al. (2006) |

*E. coli* provides real-time detection capability along with a decrease in total detection time (Yilmaz et al. 2015). $3.72 * 10^5$ and $1.54 * 10^6$ CFU/mL are the detection limit that have been observed for optical and piezoelectric biosensors, respectively. Use of capacitive biosensor for estimation of *Bacillus subtilis, Staphylococcus aureus,* and *Salmonella paratyphi* had shown 70% lower response as its application in the case of *E. coli*. Recently, an electrochemical immunosensor had shown an excellent limit of detection 8 CFU/mL, when applied for *E. coli* detection. The sensing surface of this biosensor was designed by using polydopamine imprinted polymer along with quantum dots (nitrogen-doped graphene oxide) (Chen et al. 2017).

Microfluidic culture-based biosensor can be used to detect mycobacteria present in the environmental samples which act as an agent of diseases like tuberculosis and leprosy (Jing et al. 2007).

## 2.3 *Biosensors for Pesticides Monitoring*

Pesticides because of their presence in significant concentration in the environment, because of their high pesticidal activity, have been listed as one of the most important classes of environmental pollutant around the world. Pesticides are being used as weed controller, insect, and fungi controller and therefore helps us in increasing the production of crop. However, extensive utilisation of these pesticides had created contamination in soil, water, and food with pesticide residues and its metabolites (Mostafa 2010). Insecticides, herbicides, and fungicides are the three different groups in which pesticides have been classified, based on their persistence, volatility, and polarity (Hernandez et al. 2005). Biosensors based on acetylcholine esterase (AChE) enzyme are mainly used for detection of pesticides. These biosensors are mostly used for detection of organophosphorus and carbamate pesticides. The property of inhibiting the activity of AChE of organophosphorus compounds by phosphorylating its serine group makes it detectable by various biosensors easily. AChE biosensors can be utilised to analyse carbaryl, carbofuran, aldicarb, paraoxon, malaoxon, chlorpyrifos ethyl oxon, diisopropyl fluorophosphates, and trichlorfon pesticides (Dzyadevych et al. 2005). AChE biosensors can be modified by coupling it with choline oxidase enzyme which makes it more effective in regard to linear range, sensitivity, and limit of detection. Carbamate pesticide with much lower detection limit of 0.01 ppb can be observed by using bienzymatic system of AChE–choline oxidase (Snejdarkova et al. 2003).

Organophosphorus hydrolase (OPH)-based biosensors can be used for direct monitoring of pesticides and rapid too. The OPH biosensor has been used to detect the paraoxon, and the limit of detection was founded to be as low as 0.1 μM (Wang et al. 2003). Organophosphorus contaminant can also be investigated using potentiometric biosensors (detection limit of 2 μM). A dual amperometric/potentiometric has been developed and being utilised to enhance the sensitivity and reliability of biosensors (Mulchandani et al. 1999; Schöning et al. 2003). A large number of immunosensors has been developed in recent years for detection and analysis of pesticides. These may be categorised as a labelled or label-free biosensors, where former utilises enzyme in order to quantify the amount of pesticide present in the environment whereas latter detects the pesticide binding on the surface of transducer. With the help of label-free electrochemical immunosensor, atrazine pesticide can be detected easily (Liu et al. 2014a; Tran et al. 2012). Liu along with his co-workers have developed a novel multi-analyte immunosensor that can be employed to detect endosulfan and paraoxon pesticides with more sensitivity (Liu et al. 2014b). Table 3 gives information about various biosensors used for analysis of pesticides.

**Table 3** Biosensors along with their biorecognition element and transducer for pesticide detection

| Pesticide | Recognition element | Transducer | Detection limit | References |
|---|---|---|---|---|
| Carbaryl | AChE | Amperometric<br>Impedimetric | $1 * 10^{-8}$ M<br>3.87 nM | Bucur et al. (2006)<br>Gong et al. (2014) |
| Atrazine | Aptamers<br>Antibodies | Immunosensor<br>Immunosensor<br>Impedimetric | 0.016 ng/mL<br>0.2 ng/L<br>0.01 ng/mL | Liu et al. (2014a)<br>Tran et al. (2012)<br>Belkhamssa et al. (2016) |
| Malaoxon | AChE | Amperometric | 0.6–1.2 μM | Jeanty et al. (2002) |
| Carbofuran | AChE | Amperometric<br>Voltammetric | $8 * 10^{-10}$ M<br>3.6 nM | Bucur et al. (2006)<br>Jeyapragasam and Saraswathi (2014) |
| Methyl parathion | Tyrosinase<br>*Sphingomonas* sp. cells<br>AChE | Amperometric<br>Optical<br>Impedimetric | 6–100 ppb<br>0.01 ppm<br>0.42 pg/mL | de Albuquerque and Ferreira (2007)<br>Mishra et al. (2017)<br>Peng et al. (2017) |
| Aldicarb | AChE | Amperometric | 24 ppb | Arduini et al. (2006) |
| Paraoxon | Butyrylcholinesterase<br>AChE<br>AChE and choline oxidase | Voltammetric<br>Amperometric<br>Optical | 5 μg /L<br>$1.91 * 10^{-8}$ M<br>4.7 ppb | Arduini et al. (2015)<br>Andreescu (2002)<br>Guo et al. (2017) |
| Parathion | Organophosphorus hydrolase | Amperometric | 10 nM | Tang et al. (2014) |
| Diisopropyl fluorophosphates | AChE | Conductometric | $5 * 10^{-11}$ M | Dzyadevych et al. (2005) |
| Acetamiprid | Aptamers<br>Aptamers | Optical<br>Impedimetric | 5 nM<br>1 pM | Shi et al. (2013)<br>Madianos et al. (2018) |

## 3 Other Environmental Pollutants

Along with the analysis and detection of pesticides, heavy metals, and pathogens, various other kind of environmental pollutants can also be analysed using biosensors. The other kind of pollutants are:

## 3.1 Toxins

Eutrophication of aquatic system is the main source of production of toxins. Algal bloom of cyanobacteria leads to the production of brevetoxins and microcystins toxins from algal bloom of cyanobacteria. As these toxins are highly poisonous, a more reliable, efficient, and cost-effective system is needed to be developed for detection of toxins at an early stage (Justino et al. 2017). Marine neurotoxin, a brevetoxin-2, can be detected by using an electrochemical aptasensor with gold electrode functionalised with cysteamine (Eissa et al. 2015). A much lower detection and improved selectivity was obtained in the case of brevetoxin-2 as compared to okadaic acid and microcystins (Eissa et al. 2015). Cardiomyocyte-based biosensor, a portable biosensor, has been developed for detection of saxitoxin and brevetoxin-2, with 0.35 ng/mL and 1.55 ng/mL as detection limits, respectively (Wang et al. 2015). Microelectrode arrays are used to grow cardiomyocytes, which makes this sensor as a tool for rapid and real-time monitoring of some of the environmental pollutants. Okadaic acid, a toxin, found in the real algal, seawater, and shellfish samples, can be detected by the help of sensors (McNamee et al. 2013; Pan et al. 2017). Multiplex surface plasmon resonance biosensor can be used for the detection of okadaic toxin obtained from algal samples (McNamee et al. 2013).

Graphene-based disposable electrochemical immunosensors can also be utilised for the investigation of okadaic acid that has been developed in real seawater samples (Antunes et al. 2018). Large surface area, high current density because of graphene leads to improvement over limit of detection, working range, and reproducibility (Antunes et al. 2018). Disposable carbon nanotube field effect transistor immunosensors is used to analyse domoic acid, a toxin, present in seawater (Marques et al. 2017).

## 3.2 Phenolic Compounds

Phenols and phenolics are being utilised in various kinds of industry that includes paint, mining, plastic, pharmaceutical, and plastic industries. These compounds are found to be highly toxic even at low concentration and hence have been classified as one of the priority pollutants. Phenol and its derivatives are founded to be persistent and bioaccumulative in nature, animals, and plants (Davì and Gnudi 1999). Most of the biosensors developed for determination of phenolic compound are of tyrosinase (an enzyme). A phenolic biosensor developed by using immobilised tyrosinase on magnetic nanoparticles helps it in detecting the phenolic compounds at much lower concentration. For detection of phenolic compounds, an optical biosensor can be fabricated by immobilising 3-methyl-2-benzothiazolinone hydrazone (MBTH) on silicate film, whereas chitosan film is used for immobilisation in the case of laccase enzyme (Patel et al. 2019a). Developing a bienzymatic biosensor by using titania gel matrix provides high sensitivity to cholorophenol, methylcatechol, and dimethoxyphenol (Kochana et al. 2008). Biosensors, based on polyphenol oxidase, can be utilised for analysis of *p*-cresol, *m*-cresol, phenol, and catechol (Xue and Shen 2002). Phenol and its derivatives can also be determined or analysed using amperometric biosensor that have horseradish peroxidase-modified electrodes (Korkut et al. 2008). Lipases are enzymes that also can be utilised for developing biosensors and hence can be utilised in pharmaceutical industries (Patel et al. 2019b).

### *3.3 Biological Oxygen Demand*

Biological oxygen demand (BOD) is used to describe the pollution in the water because of the organic pollutants. It is defined as the amount of oxygen required to degrade the organic matter that are present in wastewater by using aerobic microorganism. The conventional method for BOD estimation is very time consuming, and it further requires high skill along with experience (Wang et al. 2010). Therefore, application of biosensors for BOD analysis provides advantage of being rapid and more reliable as compared to conventional method. Biosensors are developed, in general, by using biomembrane, oxygen electrode, and that can be utilised for determining the respiration activity of the cells in order to BOD. Karube along with his co-workers were the first to develop the microbial biosensor for estimation of BOD, later on in order to enhance the various properties, a number of biosensors have been developed (Karube et al. 1977; Dhall et al. 2008). The development of biosensors using immobilized microbial cell (IMC) beads will help in eliminating various disadvantages of microbial membrane-type biosensor like stability, detection limit, and working range (Wang et al. 2010). Furthermore, BOD detection can be done by the help of a sensor developed by utilising thermally killed microbial culture at 300 °C because of its better stability, reproducibility, and sensitivity (Tan and Lim 2005).

## 4 Conclusion

In recent years, the development of biosensors along with their implementation in the field of environment management has been increased enormously. These sensors are known to be highly potential, but still they are lacking their application on commercial scale. The limitation of these sensors includes lower sensitivity, lack of application in real environmental sample, low lifetime stability, and high response time. With the help of better immobilisation techniques, a more comprehensive interface mechanism can be developed between advanced receptors and pollutants and that will help in resolving the above-mentioned problems. Further, this technique will be highly helpful in developing a biosensor of high reliability and precision.

**Acknowledgements** The authors wish to acknowledge the cooperation and assistance received from NIT Agartala. The corresponding author, Naveen Patel, is thankful to Assistant Professor Vinod Kumar Chaudhary, RMLAU, Ayodhya, India, and other anonymous reviewers for their valuable suggestions to improve this chapter.

## References

Amine A, Mohammadi H, Bourais I, Palleschi G (2006) Enzyme inhibition-based biosensors for food safety and environmental monitoring. Biosens Bioelectron 21:1405–1423. https://doi.org/10.1016/j.bios.2005.07.012

Andreescu S, Noguer T, Magearu V, Marty JL (2002) Screen-printed electrode based on AChE for the detection of pesticides in presence of organic solvents. Talanta 57:169–176. https://doi.org/10.1016/S0039-9140(02)00017-6

Antunes J, Justino C, da Costa JP, Cardoso S, Duarte AC, Rocha-Santos T (2018) Graphene immunosensors for okadaic acid detection in seawater. Microchem J 138:465–471. https://doi.org/10.1016/j.microc.2018.01.041

Arduini F, Ricci F, Tuta CS, Moscone D, Amine A, Palleschi G (2006) Detection of carbamic and organophosphorous pesticides in water samples using a cholinesterase biosensor based on Prussian Blue-modified screen-printed electrode. Anal Chim Acta 580:155–162. https://doi.org/10.1016/j.aca.2006.07.052

Arduini F, Forchielli M, Amine A, Neagu D, Cacciotti I, Nanni F, Moscone D, Palleschi G (2015) Screen-printed biosensor modified with carbon black nanoparticles for the determination of paraoxon based on the inhibition of butyrylcholinesterase. Microchim Acta 182:643–651. https://doi.org/10.1007/s00604-014-1370-y

Baccar H, Mejri MB, Hafaiedh I, Ktari T, Aouni M, Abdelghani A (2010) Surface plasmon resonance immunosensor for bacteria detection. Talanta 82:810–814. https://doi.org/10.1016/j.talanta.2010.05.060

Bahadır EB, Sezgintürk MK (2015) Applications of commercial biosensors in clinical, food, environmental, and biothreat/biowarfare analyses. Anal Biochem 478:107–120. https://doi.org/10.1016/j.ab.2015.03.011

Belkhamssa N, Justino CI, Santos PS, Cardoso S, Lopes I, Duarte AC, Rocha-Santos T, Ksibi M (2016) Label-free disposable immunosensor for detection of atrazine. Talanta 146:430–434. https://doi.org/10.1016/j.talanta.2015.09.015

Blake DA, Jones RM, Blake RC II, Pavlov AR, Darwish IA, Yu H (2001) Antibody-based sensors for heavy metal ions. Biosens Bioelectron 16:799–809. https://doi.org/10.1016/S0956-5663(01)00223-8

Bontidean I, Lloyd JR, Hobman JL, Wilson JR, Csöregi E, Mattiasson B, Brown NL (2000) Bacterial metal-resistance proteins and their use in biosensors for the detection of bioavailable heavy metals. J Inorg Biochem 79:225–229. https://doi.org/10.1016/S0162-0134(99)00234-2

Bucur B, Fournier D, Danet A, Marty JL (2006) Biosensors based on highly sensitive acetylcholinesterases for enhanced carbamate insecticides detection. Anal Chim Acta 562:115–121. https://doi.org/10.1016/j.aca.2005.12.060

Chen S, Chen X, Zhang L, Gao J, Ma Q (2017) Electrochemiluminescence detection of *Escherichia coli* O157: H7 based on a novel polydopamine surface imprinted polymer biosensor. ACS Appl Mater Interfaces 9:5430–5436. https://doi.org/10.1021/acsami.6b12455

Claude D, Houssemeddine G, Andriy B, Jean-Marc C (2007) Whole cell algal biosensors for urban waters monitoring. Ther Nova 7(3):1507–1514

de Albuquerque YDT, Ferreira LF (2007) Amperometric biosensing of carbamate and organophosphate pesticides utilizing screen-printed tyrosinase-modified electrodes. Anal Chim Acta 596:210–221. https://doi.org/10.1016/j.aca.2007.06.013

Davì ML, Gnudi F (1999) Phenolic compounds in surface water. Water Res 33:3213–3219. https://doi.org/10.1016/S0043-1354(99)00027-5

Dhall P, Kumar A, Joshi A, Saxsena TK, Manoharan A, Makhijani SD, Kumar R (2008) Quick and reliable estimation of BOD load of beverage industrial wastewater by developing BOD biosensor. Sensors Actuators B Chem 133:478–483. https://doi.org/10.1016/j.snb.2008.03.010

Dos Santos MB, Agusil JP, Prieto-Simón B, Sporer C, Teixeira V, Samitier J (2013) Highly sensitive detection of pathogen *Escherichia coli* O157: H7 by electrochemical impedance spectroscopy. Biosens Bioelectron 45:174–180. https://doi.org/10.1016/j.bios.2013.01.009

Dzyadevych SV, Soldatkin AP, Arkhypova VN, Anna V, Chovelon JM, Georgiou CA, Martelet C, Jaffrezic-Renault N (2005) Early-warning electrochemical biosensor system for environmental monitoring based on enzyme inhibition. Sensors Actuators B Chem 105:81–87. https://doi.org/10.1016/j.snb.2004.02.039

Eissa S, Siaj M, Zourob M (2015) Aptamer-based competitive electrochemical biosensor for brevetoxin-2. Biosens Bioelectron 69:148–154. https://doi.org/10.1016/j.bios.2015.01.055

Enrico DL, Manera MG, Montagna G, Cimaglia F, Chiesa M, Poltronieri P, Santino A, Rella R (2013) SPR based immunosensor for detection of *Legionella pneumophila* in water samples. Opt Commun 294:420–426. https://doi.org/10.1016/j.optcom.2012.12.064

Eriksson R, Jobs M, Ekstrand C, Ullberg M, Herrmann B, Landegren U, Nilsson M, Blomberg J (2009) Multiplex and quantifiable detection of nucleic acid from pathogenic fungi using padlock probes, generic real time PCR and specific suspension array readout. J Microbiol Methods 78:195–202. https://doi.org/10.1016/j.mimet.2009.05.016

Ghica ME, Carvalho RC, Amine A, Brett CM (2013) Glucose oxidase enzyme inhibition sensors for heavy metals at carbon film electrodes modified with cobalt or copper hexacyanoferrate. Sensors Actuators B Chem 178:270–278. https://doi.org/10.1016/j.snb.2012.12.113

Gong Z, Guo Y, Sun X, Cao Y, Wang X (2014) Acetylcholinesterase biosensor for carbaryl detection based on interdigitated array microelectrodes. Bioprocess Biosyst Eng 37:1929–1934. https://doi.org/10.1007/s00449-014-1195-4

Guascito MR, Malitesta C, Mazzotta E, Turco A (2008) Inhibitive determination of metal ions by an amperometric glucose oxidase biosensor. Sensors Actuators B Chem 131:394–402. https://doi.org/10.1016/j.snb.2007.11.049

Guo X, Lin CS, Chen SH, Ye R, Wu VC (2012) A piezoelectric immunosensor for specific capture and enrichment of viable pathogens by quartz crystal microbalance sensor, followed by detection with antibody-functionalized gold nanoparticles. Biosens Bioelectron 38:177–183. https://doi.org/10.1016/j.bios.2012.05.024

Guo L, Li Z, Chen H, Wu Y, Chen L, Song Z, Lin T (2017) Colorimetric biosensor for the assay of paraoxon in environmental water samples based on the iodine-starch color reaction. Anal Chim Acta 967:56–93. https://doi.org/10.1016/j.aca.2017.02.028

Guo X, Sang S, Jian A, Gao S, Duan Q, Ji J, Zhang Q, Zhang W (2018) A bovine serum albumin-coated magnetoelastic biosensor for the wireless detection of heavy metal ions. Sensors Actuators B Chem 256:318–324. https://doi.org/10.1016/j.snb.2017.10.040

Hernandez F, Sancho JV, Pozo OJ (2005) Critical review of the application of liquid chromatography/mass spectrometry to the determination of pesticide residues in biological samples. Anal Bioanal Chem 382:934–946. https://doi.org/10.1007/s00216-005-3185-5

Heyduk E, Heyduk T (2010) Fluorescent homogeneous immunosensors for detecting pathogenic bacteria. Anal Biochem 396:298–303. https://doi.org/10.1016/j.ab.2009.09.039

Ilangovan R, Daniel D, Krastanov A, Zachariah C, Elizabeth R (2006) Enzyme based biosensor for heavy metal ions determination. Biotechnol Biotechnol Equip 20:184–189. https://doi.org/10.1080/13102818.2006.10817330

Ivask A, François M, Kahru A, Dubourguier HC, Virta M, Douay F (2004) Recombinant luminescent bacterial sensors for the measurement of bioavailability of cadmium and lead in soils polluted by metal smelters. Chemosphere 55(2):147–156. https://doi.org/10.1016/j.chemosphere.2003.10.064

Jeanty G, Wojciechowska A, Marty JL, Trojanowicz M (2002) Flow-injection amperometric determination of pesticides on the basis of their inhibition of immobilized acetylcholinesterases of different origin. Anal Bioanal Chem 373:691–695. https://doi.org/10.1007/s00216-002-1336-5

Jeyapragasam T, Saraswathi R (2014) Electrochemical biosensing of carbofuran based on acetylcholinesterase immobilized onto iron oxide–chitosan nanocomposite. Sensors Actuators B Chem 191:681–687. https://doi.org/10.1016/j.snb.2013.10.054

Jing G, Polaczyk A, Oerther DB, Papautsky I (2007) Development of a microfluidic biosensor for detection of environmental mycobacteria. Sensors Actuators B Chem 123:614–621. https://doi.org/10.1016/j.snb.2006.07.029

Justino CI, Freitas AC, Pereira R, Duarte AC, Santos TAR (2015) Recent developments in recognition elements for chemical sensors and biosensors. Trends Anal Chem 68:2–17. https://doi.org/10.1016/j.trac.2015.03.006
Justino C, Duarte A, Rocha-Santos T (2017) Recent progress in biosensors for environmental monitoring: a review. Sensors 17(12):2918. https://doi.org/10.3390/s17122918
Karube I, Matsunaga T, Mitsuda S, Suzuki S (1977) Microbial biosensors. Biotechnol Bioeng 19:1535–1547
Kochana J, Nowak P, Jarosz-Wilkołazka A, Bieroń M (2008) Tyrosinase/laccase bienzyme biosensor for amperometric determination of phenolic compounds. Microchem J 89:171–174. https://doi.org/10.1016/j.microc.2008.02.004
Korkut S, Keskinler B, Erhan E (2008) An amperometric biosensor based on multiwalled carbon nanotube-poly (pyrrole)-horseradish peroxidase nanobiocomposite film for determination of phenol derivatives. Talanta 76:1147–1152. https://doi.org/10.1016/j.talanta.2008.05.016
Lang Q, Han L, Hou C, Wang F, Liu A (2016) A sensitive acetylcholinesterase biosensor based on gold nanorods modified electrode for detection of organophosphate pesticide. Talanta 156:34–41. https://doi.org/10.1016/j.talanta.2016.05.002
Li L, Zhang Y, Zhang L, Ge S, Yan M, Yu J (2017) Steric paper based ratio-type electrochemical biosensor with hollow-channel for sensitive detection of $Zn^{2+}$. Sci Bull 62(16):1114–1121. https://doi.org/10.1016/j.scib.2017.07.004
Liu X, Li WJ, Li L, Yang Y, Mao LG, Peng Z (2014a) A label-free electrochemical immunosensor based on gold nanoparticles for direct detection of atrazine. Sensors Actuators B Chem 191:408–414. https://doi.org/10.1016/j.snb.2013.10.033
Liu G, Guo W, Song D (2014b) A multianalyte electrochemical immunosensor based on patterned carbon nanotubes modified substrates for detection of pesticides. Biosens Bioelectron 52:360–366. https://doi.org/10.1016/j.bios.2013.09.009
Long F, Zhu A, Shi H, Wang H, Liu J (2013) Rapid on-site/in-situ detection of heavy metal ions in environmental water using a structure-switching DNA optical biosensor. Sci Rep 3:2308. https://doi.org/10.1038/srep02308
Madianos L, Tsekenis G, Skotadis E, Patsiouras L, Tsoukalas D (2018) A highly sensitive impedimetric aptasensor for the selective detection of acetamiprid and atrazine based on microwires formed by platinum nanoparticles. Biosens Bioelectron 101:268–274. https://doi.org/10.1016/j.bios.2017.10.034
Maduraiveeran G, Jin W (2017) Nanomaterials based electrochemical sensor and biosensor platforms for environmental applications. Trends Environ Anal Chem 13:10–23. https://doi.org/10.1016/j.teac.2017.02.001
Marques I, Pinto da Costa J, Justino C, Santos P, Duarte K, Freitas A, Cardoso S, Duarte A, Rocha-Santos T (2017) Carbon nanotube field effect transistor biosensor for the detection of toxins in seawater. J Environ Anal Chem 97:597–605. https://doi.org/10.1080/03067319.2017.1334056
Martín M, Salazar P, Jiménez C, Lecuona M, Ramos MJ, Ode J, Alcoba J, Roche R, Villalonga R, Campuzano S, Pingarrón JM (2015) Rapid Legionella pneumophila determination based on a disposable core–shell $Fe_3O_4$@ poly (dopamine) magnetic nanoparticles immunoplatform. Anal Chim Acta 887:51–58. https://doi.org/10.1016/j.aca.2015.05.048
McNamee SE, Elliott CT, Delahaut P, Campbell K (2013) Multiplex biotoxin surface plasmon resonance method for marine biotoxins in algal and seawater samples. Environ Sci Pollut Res 20:6794–6807. https://doi.org/10.1007/s11356-012-1329-7
Michel C, Ouerd A, Battaglia-Brunet F, Guigues N, Grasa JP, Bruschi M, Ignatiadis I (2006) Cr(VI) quantification using an amperometric enzyme-based sensor: interference and physical and chemical factors controlling the biosensor response in ground waters. Biosens Bioelectron 22:285–290. https://doi.org/10.1016/j.bios.2006.01.007
Mishra A, Kumar J, Melo JS (2017) An optical microplate biosensor for the detection of methyl parathion pesticide using a biohybrid of *Sphingomonas* sp. cells-silica nanoparticles. Biosens Bioelectron 87:332–338. https://doi.org/10.1016/j.bios.2016.08.048
Mostafa GA (2010) Electrochemical biosensors for the detection of pesticides. Open Electrochem J 2:22–42

Mulchandani P, Mulchandani A, Kaneva I, Chen W (1999) Biosensor for direct determination of organophosphate nerve agents. 1. Potentiometric enzyme electrode. Biosens Bioelectron 14:77–85. https://doi.org/10.1016/S0956-5663(98)00096-7

Nigam VK, Shukla P (2015) Enzyme based biosensors for detection of environmental pollutants-a review. J Microbiol Biotechnol 25(11):1773–1781. https://doi.org/10.4014/jmb.1504.04010

Pan Y, Zhou J, Su K, Hu N, Wang P (2017) A novel quantum dot fluorescence immunosensor based on magnetic beads and portable flow cytometry for detection of okadaic. Proc Technol 27:214–216

Patel N, Shahane S, Majumdar R, Mishra U (2019a) Mode of action, properties, production, and application of Laccase: a review. Recent Pat Biotechnol 13:19–32. https://doi.org/10.2174/1872208312666180821161015

Patel N, Rai D, Shahane S, Mishra U (2019b) Lipases: sources, production, purification, and applications. Recent Pat Biotechnol 13:45–56. https://doi.org/10.2174/1872208312666181029093333

Pellinen T, Huovinen T, Karp M (2004) A cell-free biosensor for the detection of transcriptional inducers using firefly luciferase as a reporter. Anal Biochem 330:52–57. https://doi.org/10.1016/j.ab.2004.03.064

Peng L, Dong S, Wei W, Yuan X, Huang T (2017) Synthesis of reticulated hollow spheres structure $NiCo_2S_4$ and its application in organophosphate pesticides biosensor. Biosens Bioelectron 92:563–569. https://doi.org/10.1016/j.bios.2016.10.059

Podola B, Nowack EC, Melkonian M (2004) The use of multiple-strain algal sensor chips for the detection and identification of volatile organic compounds. Biosens Bioelectron 19(10):1253–1260. https://doi.org/10.1016/j.bios.2003.11.015

Rodriguez-Mozaz S, Marco MP, De Alda ML, Barceló D (2004) Biosensors for environmental applications: Future development trends. Pure Appl Chem 76:723–752. https://doi.org/10.1351/pac200476040723

Schöning MJ, Krause R, Block K, Musahmeh M, Mulchandani A, Wang J (2003) A dual amperometric/potentiometric FIA-based biosensor for the distinctive detection of organophosphorus pesticides. Sensors Actuators B Chem 95:291–296. https://doi.org/10.1016/S0925-4005(03)00426-X

Shi H, Zhao G, Liu M, Fan L, Cao T (2013) Aptamer-based colorimetric sensing of acetamiprid in soil samples: sensitivity, selectivity and mechanism. J Hazard Mater 260:754–761. https://doi.org/10.1016/j.jhazmat.2013.06.031

Silva LMC, Salgado AM, Coelho MAZ (2010) Agaricus bisporus as a source of tyrosinase for phenol detection for future biosensor development. Environ Technol 31:611–616. https://doi.org/10.1080/09593331003592238

Snejdarkova M, Svobodova L, Nikolelis DP, Wang J, Hianik T (2003) Acetylcholine biosensor based on dendrimer layers for pesticides detection. Electroanalysis 15:1185–1191. https://doi.org/10.1002/elan.200390145

Tan TC, Lim EWC (2005) Thermally killed cells of complex microbial culture for biosensor measurement of BOD of wastewater. Sensors Actuators B Chem 107:546–551. https://doi.org/10.1016/j.snb.2004.11.013

Tang X, Zhang T, Liang B, Han D, Zeng L, Zheng C, Li T, Wei M, Liu A (2014) Sensitive electrochemical microbial biosensor for p-nitrophenylorganophosphates based on electrode modified with cell surface-displayed organophosphorus hydrolase and ordered mesopore carbons. Biosens Bioelectron 60:137–142. https://doi.org/10.1016/j.bios.2014.04.001

Tran HV, Yougnia R, Reisberg S, Piro B, Serradji N, Nguyen TD, Tran LD, Dong CZ, Pham MC (2012) A label-free electrochemical immunosensor for direct, signal-on and sensitive pesticide detection. Biosens Bioelectron 31:62–68. https://doi.org/10.1016/j.bios.2011.09.035

Tsai HC, Doong RA (2005) Simultaneous determination of pH, urea, acetylcholine and heavy metals using array-based enzymatic optical biosensor. Biosens Bioelectron 20:1796–1804. https://doi.org/10.1016/j.bios.2004.07.008

Tsai HC, Doong RA, Chiang HC, Chen KT (2003) Sol–gel derived urease-based optical biosensor for the rapid determination of heavy metals. Anal Chim Acta 481:75–84. https://doi.org/10.1016/S0003-2670(03)00066-7

Wang J, Krause R, Block K, Musameh M, Mulchandani A, Schöning MJ (2003) Flow injection amperometric detection of OP nerve agents based on an organophosphorus–hydrolase biosensor detector. Biosens Bioelectron 18:255–260. https://doi.org/10.1016/S0956-5663(02)00178-1

Wang J, Zhang Y, Wang Y, Xu R, Sun Z, Jie Z (2010) An innovative reactor-type biosensor for BOD rapid measurement. Biosens Bioelectron 25:1705–1709. https://doi.org/10.1016/j.bios.2009.12.018

Wang Q, Fang J, Cao D, Li H, Su K, Hu N, Wang P (2015) An improved functional assay for rapid detection of marine toxins, saxitoxin and brevetoxin using a portable cardiomyocyte-based potential biosensor. Biosens Bioelectron 72:10–17. https://doi.org/10.1016/j.bios.2015.04.028

Xue H, Shen Z (2002) A highly stable biosensor for phenols prepared by immobilizing polyphenol oxidase into polyaniline–polyacrylonitrile composite matrix. Talanta 57:289–295. https://doi.org/10.1016/S0039-9140(02)00028-0

Yang Y, Wang Z, Yang M, Guo M, Wu Z, Shen G, Yu R (2006) Inhibitive determination of mercury ion using a renewable urea biosensor based on self-assembled gold nanoparticles. Sensors Actuators B Chem 114:1–8. https://doi.org/10.1016/j.snb.2005.04.005

Yang X, He Y, Wang X, Yuan R (2017) A SERS biosensor with magnetic substrate $CoFe_2O_4$@Ag for sensitive detection of $Hg^{2+}$. Appl Surf Sci 416:581–586. https://doi.org/10.1016/j.apsusc.2017.04.106

Yilmaz E, Majidi D, Ozgur E, Denizli A (2015) Whole cell imprinting based *Escherichia coli* sensors: a study for SPR and QCM. Sensors Actuators B Chem 209:714–721. https://doi.org/10.1016/j.snb.2014.12.032

Zhang C, Zhou Y, Tang L, Zeng G, Zhang J, Peng B, Xie X, Lai C, Long B, Zhu J (2016) Determination of $Cd^{2+}$ and $Pb^{2+}$ based on mesoporous carbon nitride/self-doped polyaniline nanofibers and square wave anodic stripping voltammetry. Nanomaterials 6(1):7. https://doi.org/10.3390/nano6010007

Zhao Y, Zhang W, Lin Y, Du D (2013) The vital function of $Fe_3O_4$@Au nanocomposites for hydrolase biosensor design and its application in detection of methyl parathion. Nanoscale 5:1121–1126. https://doi.org/10.1039/C2NR33107A

Zourob M (2010) Recognition receptors in biosensors. Springer, New York, pp 415–448. https://doi.org/10.1007/978-1-4419-0919-0

# Screen-Printed Electrochemical Sensors for Environmental Contaminants

**A. M. Vinu Mohan**

**Contents**



## Abbreviations

| | |
|---|---|
| CV | Cyclic voltammetry |
| DPASV | Differential pulse anodic stripping voltammetry |
| PSA | Potentiometric stripping analysis |
| SWASV | Square wave anodic stripping voltammetry |

A. M. Vinu Mohan (✉)
Electrodics and Electrocatalysis Division, CSIR-Central Electrochemical Research Institute (CECRI), Karaikudi, Tamil Nadu, India
e-mail: vinumohan@cecri.res.in



Inamuddin, A. M. Asiri (eds.), *Nanosensor Technologies for Environmental Monitoring*, Nanotechnology in the Life Sciences,
https://doi.org/10.1007/978-3-030-45116-5_5

# 1 Introduction

Pollution is the effect of contamination that renders part of the environment unhealthy for intended or desired uses. Due to the rapid development of industrialization and urbanization, toxic chemicals are released into the environment, which leads to a greater risk to human health and the environment (Ebenstein 2012). The rapid growth of industrialization has been accompanied by substantial environmental hazardous effects including contamination in microbial and aquatic ecosystem (Mallin and Cahoon 2003; Leonard 2006). Other sources of pollution include emissions from steel and iron mills, lead, zinc, and copper smelters, oil refineries, municipal incinerators, and cement plants (Ettler et al. 2004). The large-scale pollution crises are deforestation, polluted water resources, and contaminated soils. Environmental monitoring of hazardous contaminants usually involves several steps such as sampling, sample treatment, and transportation to specialized laboratories (Worsfold et al. 2005). The in situ environmental monitoring requires precise analytical instruments featuring sensitivity, selectivity, rapidity, and ease of operation to detect changes in concentrations (Hanrahan et al. 2004). These devices must be simple to operate, portable, inexpensive, and able to supply reliable real-time information. Electrochemical sensors represent a key class of chemical sensors in which a functionalized electrode is used as the transduction element and are more suitable for meeting the cost, size, and power requirements of real-time environmental monitoring (Taillefert et al. 2000; Bakker and Telting-Diaz 2002). Also, electrochemical sensing platforms possess high sensitivity and selectivity, a wide detection range, minimal space and power requirements, and economical instrumentation.

Screen printing technology is a versatile tool for electrochemical analysis of environmental hazards and for point of care testing in health care management (Hayat and Marty 2014). The use of screen-printed electrodes allows remote detection of hazardous pollutants or environmental contaminants or toxic warfare agents outside the centralized laboratory (Wang et al. 1998a, b). The screen printing technique allows the mass production of sensing electrodes with precision and consistent analytical performance (Wang et al. 1998c). The scalable process is not dependent on highly skilled personnel or complicated methods and is devoid of tedious pretreatment steps (Metters et al. 2011). The surface of the electrodes can be easily functionalized to realize various purposes and to enhance signal sensitivity and selectivity. Several modifiers have been used for environmental analyses, including enzymes, inorganic nanomaterials, noble metals, and DNA sequences. Hence, such disposable electrodes can be effectively applied for in situ assessment of water quality and environmental monitoring of toxic contaminants like heavy metal ions, hazardous organic compounds, gases, pathogens, and trace radioactive species (Fig. 1).

Also, such thick film fabrication technology extends to a suitable platform for constructing flexible and stretchable electrochemical devices for wearable

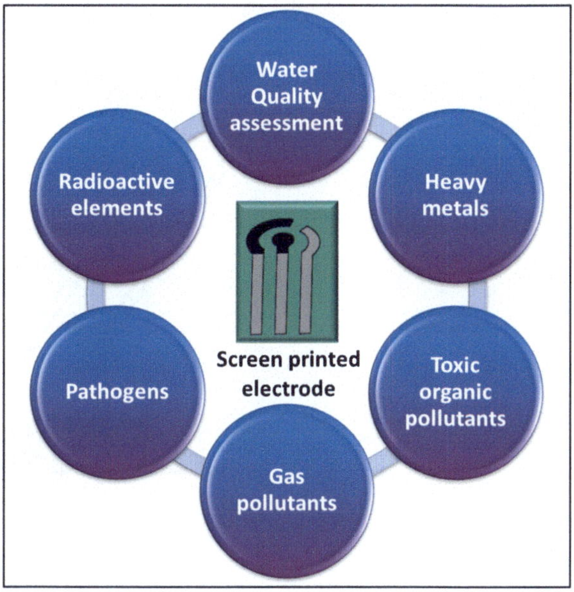


**Fig. 1** Applications of screen-printed electrodes for environmental analyses

applications. Figure 2 reveals the stretchable carbon electrodes printed on flexible polyester sheet, which is ideal for skin-based sweat monitoring application. Such wearable electronics showed potential application in defense-related sensing technologies for chemical warfare agents (Seto et al. 2005; Mishra et al. 2018; Barfidokht et al. 2019). The screen printing technique also facilitates to build temporary tattoo like sensing modalities for skin-based continuous monitoring applications.

## 2 Fabrication of Screen-Printed Electrodes

The thick film electrodes enable the direct sensing of analytes in its 'environment' itself without changing normal surrounding conditions. The sensor development involves mainly two stages: (1) construction of screen-printed electrodes and (2) functionalization of the electrode surface. Initially, a design of the electrode has to be created with specific geometry and unique specifications suitable for application and the corresponding stencil needs to be prepared on chemically etched mesh screen or laser-cut stainless-steel stencils. The various steps involved in the development of screen-printed electrodes are demonstrated in Fig. 3. The substrate (A) and the stencil (B) for the electrodes need to be attached on the respective spots in the

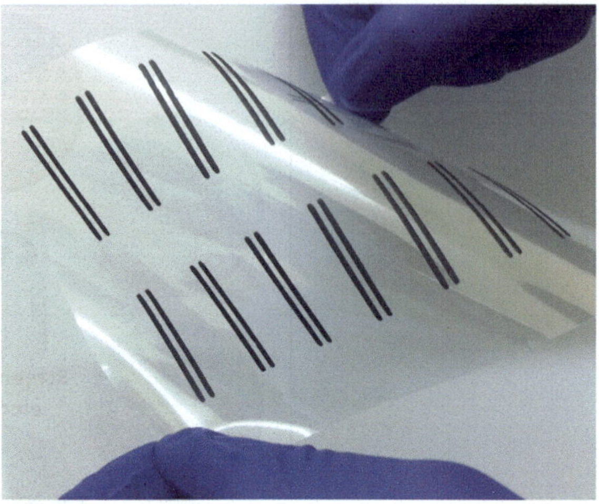

**Fig. 2** Screen-printed carbon electrodes on flexible polyester substrate suitable for wearable sensing application

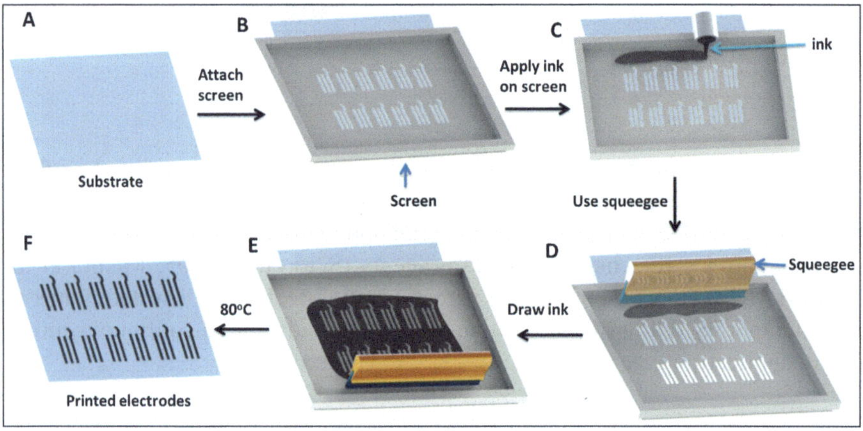


**Fig. 3** Schematic representation of the fabrication process of screen-printed electrodes

screen printer. The printing ink has to be injected on the stencil close to the printing pattern (C). The movement of ink over the stencil using a squeegee facilitates printing of the stencil design on the underlying substrate (D–F).

A three stage printing is the most common that consists of (a) conductive silver (Ag/AgCl) underlayer, (b) a conductive carbon-based active layer, and an (c) insulator layer (Fig. 4). Subsequent to each printing step, the patterned substrate is annealed in a convection oven at suitable temperature in order to evaporate the volatile organic solvent so as to get solid printed traces in a three-electrode configuration. The printing ink consists of conductive fillers, polymeric binder and other

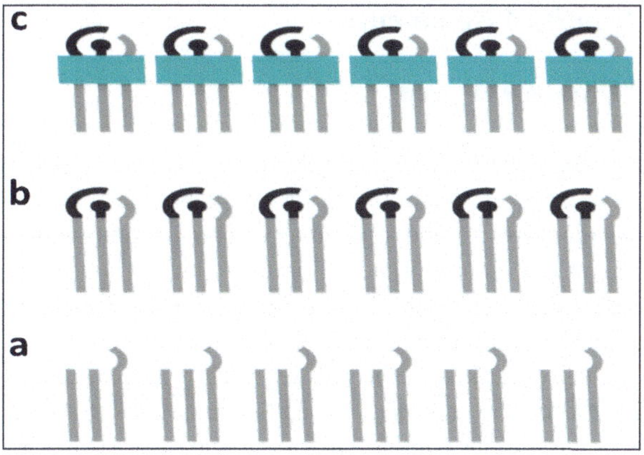


**Fig. 4** Three-layer printing of (**a**) silver (Ag/AgCl), (**b**) carbon, and (**c**) insulator to constitute screen-printed electrodes (three-electrode configuration)

constituents required for homogenous dispersion, suitable viscosity and adhesion. Conductive graphitic carbon and silver-based inks are commonly using to realize working electrode and reference electrodes, respectively.

The stable immobilization of bioreceptor on the transducer surface is crucial for biorecognition of the specific target molecule. The bioreceptor translates information from the target into a chemical or physical output response with a characteristic sensitivity for quantitatively monitoring the analytes of interest. Various methods are being employed for immobilizing a thin layer of recognition element at the transducer surface such as bulk modification along with the electrode material, entrapment within a polymeric matrix, behind a membrane or within self-assembled monolayers or sandwich type bilayer lipid membranes and covalent bonding on functionalized surfaces (Sassolas et al. 2012; Garcia-Galan et al. 2011; Mateo et al. 2007; Costa et al. 2011).

## 3 Application of Screen-Printed Electrodes for Environmental Analysis

Screen-printed electrodes–based electrochemical sensors are uniquely positioned to enable the miniaturization of a laboratory for on-site detection of analytes. The combination of these electrodes with portable hand held analyzer can realize sensitive platform for heavy metal trace analysis (Ahmed et al. 2016). Such portable system facilitate exceptional platform to perform environmental assessment for water quality, heavy metals, toxic organic pollutants, pathogens, hazardous gas contaminants, pathogens, and other radioactive components.

## 3.1 *Water Quality Assessment*

### 3.1.1 pH and Dissolved Oxygen

There is an increasing need for portable, robust, inexpensive, and sensitive pH sensors that have extensive applications in environments such as agriculture, food industry, clinical diagnostics, and oil refinery. The conventional glass electrode necessitates frequent and inappropriate calibration which is not adequate for in-field use. The electrochemical detection of pH using screen-printed electrodes can simplify the electronic sensing components. Potentiometric determination of pH based on ion selective electrodes have been well reported for environmental monitoring (Vonau and Guth 2006). The presence of specific ionophores in the membrane facilitates the selective analysis in complex samples. Ruthenium dioxide–modified disposable electrodes were used for amperometric sensing of pH with printed silver electrode as reference electrode (Koncki and Mascini 1997). The changes in conductivity of the metal oxide with respect to change in pH were monitored even in presence of interfering alkaline cations. Screen-printed electrodes with pH sensitive phenanthraquinone moiety along with pH insensitive dimethylferrocene as an internal reference showed Nernstian potential shift with pH (Kampouris et al. 2009).

Monitoring the level of dissolved oxygen is also crucial for water quality assessment. The chemical oxygen demand can be defined as the indicative parameter in environmental chemistry which indicates the quantity of oxygen that could be consumed by reactions in a measured medium. The chemical oxygen demand is normally expressed in terms of mass of oxygen consumed over volume of solution. The most common application is in measuring the amount of oxidizable pollutants found in surface water like lakes and rivers or wastewater. A miniaturized amperometric sensor with immobilized electrolyte gel and oxygen permeable membrane has been utilized for dissolved oxygen monitoring (Glasspool and Atkinson 1998). $RuO_2$-based potentiometric sensors with $TiO_2$ or polyisoftalamide diphenylsulfone showed a linear electrode response as a function of the logarithm of the dissolved oxygen concentration in the 0.8–8 ppm range (Martínez-Máñez et al. 2004).

### 3.1.2 Nitrite and Phosphate

Nitrite is present in variety of food samples like a preservative or a natural component which results in the formation of carcinogenic nitrosoamines in the stomach. The drinking water can be contaminated by the formation of nitrite via chloramination and other microbial activities. A reliable, inexpensive unmodified screen-printed graphite microelectrode array is reported for the sensing of nitrite in aqueous solutions (Khairy et al. 2010). The approach is shown to be feasible for the sensing of nitrite in river water samples at optimum levels indicated by the World Health Organization. The electrochemically activated graphite-modified screen-printed carbon electrode has been developed for nitrite detection that showed a lowest detection limit of 38 nM and exhibits appreciable selectivity in the presence of common interfering ions (Palanisamy et al. 2014). Anion exchanger–modified carbon electrode was developed for pre-concentration followed by the voltammetric detection of nitrite (Neuhold et al. 1995). A short accumulation period of 30 s provides a detection limit of 30 μg/L nitrite.

Phosphate is an inorganic form of phosphorus and a necessary nutrient for the entirety of aquatic organism. Though plant growth is stimulated due to increase in phosphate content, its augmented level adversely affects plants and animals by lowering the level of dissolved oxygen (Zhou et al. 2003). Hence, the development of a portable sensor for remote monitoring of phosphate level is critical for judicious phosphate removal to maintain a healthy aquatic environment. Pyruvate oxidase–immobilized screen-printed electrode was developed for real time analysis of phosphate content in a sequential batch reactor (Kwan et al. 2005). The entrapment of enzyme on the electrode surface was accomplished by a Nafion layer and protected with a hydrogel of poly(carbamoyl)sulfonate. The sensor showed an appreciable response time and short recovery time of 2 s and 2 min, respectively. Reagentless paper-based electrochemical phosphate sensor has been fabricated by wax patterning and functionalizing the paper substrate (Cinti et al. 2016). The sensor possessed a detection limit of 4 μM and offers in situ determination of phosphate levels from river water samples.

## *3.2 Heavy Metal Detection*

The required quantities of heavy metal ions like zinc, manganese, copper, cobalt and iron are essential for the maintenance of metabolic activities. The exposure of major heavy metal ions such as cadmium, lead, mercury, and arsenic are the main hazard to the human health and environment. The accumulation of these metals on the human body causes serious impact on human health including impaired neurologic development, cardiovascular, gastrointestinal tract, pulmonary diseases, kidney damage, decreased bone growth, behavioral issues, and hyperirritability. (Caito and Aschner 2017; Jarup 2003; Sabath and Robles-Osorio 2012). According to a report from World Resources Institute, more than 100 million people are living in areas of heavy metal ion contaminated water and 0.4 million lives are in danger every year. Hence, there is a high need to develop a highly sensitive and cost-effective sensor for continuous monitoring of heavy metal ions in water. In this scenario, the development of portable electrochemical sensor for simultaneous monitoring of heavy metal ions is extremely important for preventing environmental pollution (Hughes et al. 2016).

The typical analyses of heavy metal concentrations have been based on typical spectroscopic techniques, such as atomic absorption spectrometry (Siaka et al. 1998), inductively coupled plasma optical mass spectrometry (Yuan et al. 2004, Sereshti et al. 2012), X-ray fluorescence spectrometry (Melquiades and Appoloni 2004), and neutron activation analysis (Landsberger and Wu 1995) Majority of these techniques depend on bulky, expensive instruments demanding complex experimental protocols. Due to this, researchers are extensively focused on the development of portable, compact, easy-to-use analytical instruments for trace level quantification of heavy metal ions from various environmental samples. Electrochemical methods are superior to optical techniques due to the benefits associated with portability, sensitivity, selectivity, and the ability of in-field analysis. Various electroanalytical techniques like voltammetry and potentiometry are extensively employed for heavy metals analysis (Barton et al. 2016).

Stripping voltammetry is the most sensitive technique for analysis of toxic heavy metals from solution at trace levels. The technique offers picomolar level detection of metal ions and allows the selective detection from a mixture of trace metal ions. Stripping voltammetry is further distinguished into anodic or cathodic stripping voltammetry depending on oxidative and reductive scan, respectively. The studies revealed that printed electrodes modified with mercury, gold, silver, and bismuth can enhance the selectivity and sensitivity of heavy metal ion determination. There are three major steps involved in the stripping analysis of toxic heavy metals such as (1) pre-concentration, (2) equilibration, and (3) anodic stripping.

1. Pre-concentration

   The metal ions of interest ($M^{n+}$) needs to be electrochemically deposited on the Hg or Bi film–modified electrode surface as illustrated in Eq. (1). The stripping performance of mercury-based electrodes is based on amalgam formation, but bismuth film electrodes depend on the development of multicomponent alloys. The electrodeposition of trace heavy metals can be carried out by applying suitable negative potential for the reduction of metal ions. The solution needs to be stirred throughout the pre-concentration process, which enriches the deposition of metal ions on the transducer and enhances the sensitivity of the detection.

$$\mathrm{M}^{n+} + ne^{-} \rightarrow \mathrm{M}(\mathrm{Hg}) \tag{1}$$

2. Equilibration

   The equilibration step is to obtain a uniform metal-amalgam/metal-alloy composition at the electrode surface. The solution is allowed to be quiescent, during which the applied potential can be remains unchanged.

3. Anodic stripping

   After the equilibration step the deposited metal can be oxidized from the electrode during the stripping scan as metal ions to the solution (Eq. 2). The current can be measured during the stripping step that must be proportional to time of electrolysis, stirring rate and concentration of metal ions in solution. The oxidation of metals can be registered as a current response at the potential at which the species starts to be oxidized.

$$\mathrm{M}(\mathrm{Hg}) \rightarrow \mathrm{M}^{n+} + ne^{-} \tag{2}$$

### 3.2.1 Mercury-Modified Screen-Printed Electrodes

Mercury possesses several advantages as an electrode material (Kolthoff 1952) like purity at ambient temperatures, regeneration of reproducible surfaces, and sensitivity of analysis. Mercury shows a wide potential window from 0.1 V to low as −2.6 V (vs. SCE) at aqueous medium and −3.0 V (vs. SCE) at non-aqueous medium (Mikkelsen and Schroder 2003). Hanging mercury drop electrode was successfully used for measuring cadmium (Cd) at trace levels by anodic stripping voltammetry (DeMars and Shain 1957). But the low surface-to-volume ratios at the mercury drop electrode adversely affects the plating efficiency and results broadening of the stripping peak. Hg thin film on the carbon electrode surface has been found to be superior without the risk of dislodging the film with improved sensitivity and selectivity compared to the mercury drop electrode.

Researchers explored the application of Hg film–modified screen-printed carbon disposable electrodes for selective and sensitive detection of $Pb^{2+}$ and $Cd^{2+}$ (Wang and Tian 1992). Here, glucose biosensors available in the market (ExacTech) were utilized for the potentiometric stripping analysis (PSA) and differential pulse anodic stripping voltammetry (DPASV) analysis of metals at trace levels. Subsequently, a great number of reports have been published, utilizing Hg films–coated screen-printed electrodes. A thin film of Hg can be either ex-situ deposited from a separate solution, or co-deposited with the specific analyte using soluble Hg salt with the sample solution. Table 1 summarizes a number of these sensors based on DPASV,

**Table 1** Mercury-modified screen-printed electrodes for metal ion determination

| Heavy metal | Modifier | Technique | Limit of detection | Real sample | References |
|---|---|---|---|---|---|
| $Cd^{2+}$ | Ex situ Hg thin film | SWASV, −1.0 V | 0.2 ng/mL | River water | Zaouak et al. (2010) |
| $Pb^{2+}$<br>$Cd^{2+}$<br>$Cu^{2+}$ | Thin film of Hg | SWASV, −1.1 V | 7.0 ng/mL, 0.31 ng/mL, 0.53 ng/mL | Seawater | Aragay et al. (2010) |
| $Cd^{2+}$<br>$Pb^{2+}$<br>$Zn^{2+}$ | Thin film of hg on carbon electrode | CV, DPASV | 25 ng/mL, 08 ng/mL, 5 ng/mL | Wastewater, lake water | Ensafi et al. (2010) |
| $Pb^{2+}$ | Pre-anodized hg | SWASV | 0.23 ng/mL | Blood | Yang et al. (2005) |
| $Pb^{2+}$<br>$Cd^{2+}$ | Thin film of Hg | SWASV, −1.1 V | – | Herbal drugs | Palchetti et al. (2003) |
| $Pb^{2+}$<br>$Cd^{2+}$ | Thin film of Hg | SWASV, −1.1 V | 1.8 ng/mL, 2.9 ng/mL | Seawater | Guell et al. (2008) |
| $Sb^{3+}$ | Thin film of Hg | DPASV, −0.9 V | $1.27 \times 10^{-8}$ M | Seawater | Dominguez-Renedo et al. (2009b) |
| $Cd^{2+}$ | Hg film–modified microelectrode array | SWASV | 1.3 ng/mL | River water | Cugnet et al. (2009) |

square wave anodic stripping voltammetry (SWASV), or cyclic voltammetry (CV), illustrating the low limits of detection and versatility for practical application.

### 3.2.2 Bismuth-Modified Screen-Printed Electrodes

Bismuth film–modified screen-printed electrodes have been successfully utilized for heavy metal detection as an efficient alternative to the more toxic Hg film. The unique performance of bismuth film electrodes is featured by the construction of binary or multicomponent alloys. Such alloys facilitate the nucleation process during the pre-concentration of metal ions. The screen-printed carbon is a robust platform for maintaining the stability of Bi film. Bi films can be prepared either by ex situ or in situ approach. In ex situ method the film can be pre-plated on the electrode surface, while in situ method involves the addition of trace level of bismuth(III) to the sample solution, and simultaneously electrodepositing the target heavy metals along with bismuth. The Bi plating bath is prepared in acetate buffer solution with pH 4.5. Equations (3), (4), and (5) depict the Bi film formation from the plating solution and the alloy formation with heavy metals, respectively.

$$\mathrm{Bi}^{3+} + 3\mathrm{e}^{-} \rightarrow \mathrm{Bi}^{0} \tag{3}$$

$$\mathrm{M}^{n+} + n\mathrm{e}^{-} \rightarrow \mathrm{M}(\mathrm{Bi}) \tag{4}$$

$$\mathrm{M}(\mathrm{Bi}) \rightarrow \mathrm{M}(\mathrm{Bi})\mathrm{M}^{n+} + n\mathrm{e}^{-} + \mathrm{Bi}^{0} \tag{5}$$

Several elements including zinc, chromium, and lead have been determined, using Bi film–coated or bulk-modified screen-printed electrodes (Table 2).

### 3.2.3 Gold Film-Coated Screen-Printed Electrodes

Gold nanoparticle film-deposited screen-printed electrodes will be utilized as transducers for the detection of heavy metals. Apart from the detection of cadmium and lead, other metals such as arsenic and mercury can be detected using gold film-coated disposable sensors (McNerney et al. 1972; Punrat et al. 2014; Some et al. 2016; Punrat et al. 2013; Song and Swain 2007). Furthermore, mercury and arsenic metals can be subjected to a fundamental electrochemical process called under potential deposition (Kolb et al. 1974). The method involves the formation of an adlayer by the strong interaction between the metal and the gold electrode followed by the reduction of metal ions. The process is judiciously crucial for Hg due to the distinct properties like amalgam formation, equilibrium and steady-state potentials and discharge ionization (Salie and Bartels 1994; Li et al. 1998).

**Table 2** Bismuth-modified screen-printed electrodes for trace metal monitoring

| Heavy metal | Modifier | Technique | Limit of detection | Real sample | References |
|---|---|---|---|---|---|
| $Pb^{2+}$ | Bi film | SWASV, −1.0 V | 0.3 ng/mL | Drinking water | Wang et al. (2001) |
| $Pb^{2+}$ | Bi, 0.5% Nafion | SWASV, −1.0 V | 4 ng/mL | Leachates from cooking vessels | Mandil and Amine (2009) |
| $Zn^{2+}$ $Pb^{2+}$ $Cd^{2+}$ | Ex situ deposited bismuth film | DPASV, −1.5 V | 3.5 ng/mL, 0.5 ng/mL, 3.9 ng/mL | Barcelona tap water | Serrano et al. (2010) |
| $Cd^{2+}$ $Pb^{2+}$ | Bismuth oxide–modified ink | Chronopotentiometric | 8.0 ng/mL, 16 ng/mL | Soil and water samples | Kadara and Tothill (2008) |
| $Zn^{2+}$ $Cd^{2+}$ $Pb^{2+}$ | Hydrogel-modified Bi ink | SWASV, −1.2 V | 1 ng | Metal vapors | Lu et al. (2010) |
| $Zn^{2+}$ $Cd^{2+}$ $Pb^{2+}$ | In situ plated Bi film | ASV (sequential injection analysis) | 0.2 ng/mL, 0.8 ng/mL, 11 ng/mL | Herbs | Injang et al. (2010) |
| $Pb^{2+}$ | In situ plated Bi lab on a chip | SWASV, −1.2 V | 1.0 ng/mL | – | Nie et al. (2010) |
| $Cd^{2+}$ $Pb^{2+}$ | Bismuth oxide–modified ink | SWASV, −1.2 V | 2.3 ng/mL, 1.5 ng/mL | River water | Hwang et al. (2008) |

### 3.2.4 Ion Selective Membrane–Based Screen-Printed Electrodes

The potentiometric sensors are based on the tracking of the recognition process in the form of a potential response, which is logarithmically proportional to the concentration or activity of the analyte of interest generated or consumed (Bratov et al. 2010). Ion selective electrodes are usually used for selective detection of analytes. Such electrodes are constructed by modifying with a perm selective membrane with suitable charge carriers and ionophores. Calixarene or crown ether–based ion selective membranes can be constructed on screen-printed electrodes using polyvinyl chloride. Such membrane immobilized electrodes have been utilized for potentiometric determination of heavy metal ions (Gupta et al. 1997, 2013). The use of target specific ionophores for each heavy metal ions will impart the appreciable selectivity towards the real sample analysis (Buhlmann et al. 1998).

### 3.2.5 Biosensors-Based Metal Monitoring

Screen-printed electrodes modified by biological components like enzymes, nucleotides, and bacteria. They have recently emerged as a tool for metal ion determination. Enzymes like urease and Acetylcholine esterase has been immobilized by cross-linking with bovine serum albumin and glutaraldehyde and used for the determination of $Hg^{2+}$ and $As^{3+}$ (Dominguez-Renedo et al. 2009a; Sanllorente-Mendez et al. 2010). The steady-state current was recorded at suitable potential in the presence of urea or acetylthiocholine iodide and further additions of metal ions resulted in a decrease in current response, proportional to the concentration.

DNA-modified gold screen-printed electrode was reported for $Hg^{2+}$ determination by tracking the induced hybridization with respect to metal ion concentration (Niu et al. 2011). Ferrocene derivative tagged oligonucleotide was used as an electroactive indicator which can able to hybridize with other oligonucleotide by forming thymine–$Hg^{2+}$–thymine complexes. Peptide nanofibrils self-assembled at screen-printed Au electrodes have been utilized for the determination of $Cu^{2+}$ ions (Viguier et al. 2011). Copper ions were pre-concentrated onto the gold electrode surface functionalized with cysteine containing octapeptide at open circuit potential and SWVs were recorded for measuring the analytical signal.

## 3.3 *Detection of Toxic Organic Pollutants*

Several toxic synthetic organic compounds are contaminating the environment through either industrial waste or pesticide contamination. The increased concentration of such hazardous organic pollutants in the aquatic environment contaminate food and the accumulation of those compounds in human body have been correlated to severe health threats including cancer (McGuinness and Dowling 2009).

### 3.3.1 Pesticides

The exposure of low level of organophosphate pesticides can cause cholinergic dysfunction in both humans and animals (Sanchez-Santed et al. 2016). Sensitive quantification of concentration of pesticides in the environment and water is crucial to track the hazardous effect of such compounds (Li et al. 2011). Several enzymes are being used for monitoring such pesticides by monitoring the inhibition of enzyme activity. Screen-printed electrodes modified with acetylcholinesterase (Andreescu et al. 2002; Lin et al. 2004), butyrylcholinesterase (Arduini et al. 2015a, b), organophosphate hydrolase (Mulchandani et al. 1999), and tyrosinase (de Albuquerque and Ferreira 2007) are promising inhibition based biosensor systems for in-field detection of pesticides. The stable immobilization of the enzyme layer is crucial to measure the decrease in enzyme activity and the resulting decrease in current density. Acetylcholinesterase immobilized via $Al_2O_3$ sol–gel matrix on the sonogel–carbon was utilized for developing a disposable organophosphate biosensor (Zejli et al. 2008). Other immobilization strategies such as cross-linking and physical entrapment have been used for the production of reproducible biosensors. Magnetic materials–based enzyme immobilization has been widely utilized for the production of electrochemical transducers for pesticides (Won et al. 2009; Gan et al. 2010).

Magnetic materials provide stable and reproducible microenvironment without affecting the bioactivity of the electrode. The use of electron transfer mediator is suitable for reducing the overpotential required for oxidizing the electroactive by-products during the enzymatic reaction (Schuhmann et al. 1991). Prussian blue, poly(3,4-ethylenedioxythiophene), cobalt phthalocyanine, and carbon nanotubes have been employed for reducing the overpotential and thus eliminate the interference response (Istamboulie et al. 2010; Arduini et al. 2006; Cai and Du 2008; Crew et al. 2011).

### 3.3.2 Herbicides

Herbicides are chemical substances used to control the growth of unwanted plants. The increase in level of herbicides contaminates soil, drinking water and affects crop safety. Disposable immunoassay-based detection of herbicide is a common approach which allows the rapid quantification and free from cleaning and reusing the active materials. A bienzyme immunoassay method has been developed for the electrochemical detection of chlorsulfuron (Dzantiev et al. 2004). A membrane having entrapped anti-chlorsulfuron antibodies was immobilized onto the screen-printedelectrode.Thesamplecontainingfreechlorsulfuronandachlorsulfuron–glucose oxidase conjugate competed for the accessible recognition sites. The electric signal corresponds to the generated hydrogen peroxide during the addition of glucose was detected. Molecularly imprinted polymer with specific recognition sites has been used for the selective detection of 2,4-dichlorophenoxyacetic acid (Kroger et al. 1999). Poly(ethylene glycol dimethacrylate-co-4-vinylpyridine was modified on electrodes for competitive detection of target molecules. Isolated photosystem(II) complexes–based biosensor was constructed for the detection of triazine and phenylurea type herbicides (Koblížek et al. 2002). Photosystem(II) was isolated from thermophilic cyanobacterium *Synechococcus elongatus* and immobilized on screen-printed electrode deposited on a polymeric substrate. The herbicides selectively block the photosystem electron transport kinetics in a concentration-dependent manner and the changes of the activity were registered by the amperometric technique.

### 3.3.3 Phenolic Compounds

Phenolic derivatives are one of the predominant environmental contaminants. The presence of even trace levels of such compounds are toxic carcinogens and can exert negative effects on different biological processes (Dąbrowski et al. 2005). The rapid, in situ monitoring of toxic phenolic compounds and the derivatives is crucial because of the easy penetration of such compounds through membranes or skins of humans, animals, and plants (Su et al. 2011). Screen-printed electrodes–based electrochemical sensors have established as an inexpensive, easy-to-use, sensitive, and selective analytical tool for the detection of phenolic compounds. The phenolic compounds can be directly detected by unmodified electrodes using electrochemical techniques (Mersal 2009; Brugnera et al. 2010). In some cases, the oxidation product of the phenolic compound attaches and passivates the electrode surface and obstructs the electron transfer process (Kawde et al. 2013). This can be avoided by the detection of such analytes in the presence of certain surfactants which can prevent the surface fouling by micelle formation (Fernandez et al. 2006). Such methodology was employed for the detection of bisphenol A (endocrine disruptor) on screen-printed electrode using cetyltrimethylammonium bromide as a cationic surfactant (Brugnera et al. 2010; Rubin 2011).

The simultaneous monitoring of structurally similar phenolic compounds from environmental samples is challenging due to the overlapping of anodic peaks. The pretreatment of the electrode enhances the electrochemically active area by removing the organic residues from the electrode surface. Such studies were reported for the simultaneous quantification of hydroquinone and catechol with appreciable peak separation in river water samples (Wang et al. 2010; Su et al. 2011). Screen-printed electrodes functionalized with various nanomaterials like multiwalled carbon nanotubes and gold nanoparticles were also used for monitoring phenolic compounds to have sensitive and selective detection due to the unique chemical, catalytic and electronic properties (Li et al. 2010; Wang 2005). Enzyme-based transducers for quantification of phenolic electroactive compounds have been reported which can perform measurement at a relatively low applied potential, thus significantly reducing the effect of interference. Enzymes (polyphenol oxidase, tyrosinase, laccase), can catalytically oxidize benzenediols and phenols to quinones by reaction with dissolved oxygen (Renedo et al. 2007; Montereali et al. 2010). The incorporation of magnetic nanoparticles with bioreceptors provides unique abilities and enhances the sensitivity of detection. A disposable bisphenol A sensor was developed with magnetic nickel nanoparticles with good sensitivity and short response time of 30 s (Alkasir et al. 2010). The amperometric sensor showed reproducible responses for more than 100 assays.

## *3.4 Gas Pollutants*

The presence of excessive quantities of harmful gases caused air pollution at the earth's atmosphere which is dangerous to living organisms, animals, and food crops. Electrochemical gas sensors offer simple and low-cost production of reusable gas sensors for environmental monitoring application.

### 3.4.1 Volatile Organic Compounds

Volatile organic compounds are organic chemicals that possess high vapor pressure at room temperature. The low boiling point of these compounds causes large numbers of molecules to evaporate or sublimate from the liquid or solid state of the compound and enter the surrounding air. Some volatile organic compounds are dangerous to human, animals and harm to the environment (Li et al. 2012). $SnO_2$–$TiO_2$ composite oxide doped with Ag ion was used for quantitatively measuring various volatile organic compounds such as methanol, ethanol, acetone, and formaldehyde (Wen and Tian-Mo 2010). Sol–gel method was used to fabricate the sensing materials, and the powder samples were mixed with diethanolamine. The paste was screen-printed on an alumina substrate attached with a pair of Au electrodes. The approach presents a possibility for selective detection of volatile organic compounds by controlling the operating temperature.

### 3.4.2 Carbon Monoxide

The incomplete combustion of fuels such as natural gas, coal, or wood generates carbon monoxide, which is a toxic, odorless, colorless, and nonirritating gas. Carbon monoxide complexes with hemoglobin to produce carboxyhemoglobin, that adversely affects the oxygen delivering mechanism to body tissues. Tin oxide and platinum-modified tin oxide–based sensor array was developed for simultaneously monitoring carbon monoxide and methane in humid air (Huyberechts et al. 1997). Both gases are commonly considered as hazardous gases in a domestic environment and with the developed artificial neural network, the microsystem allows the quantitative monitoring even in the presence of varying humidity levels. Platinum-deposited screen-printed edge band ultra-microelectrodes have been employed for amperometric detection of carbon monoxide (Chou et al. 2009). Nafion membrane was used as a solid polymer electrolyte and carbon monoxide detection was carried out at an applied potential of 0.45 V against pseudo Ag reference electrode. Disposable porous $SnO_2$ thin film on alumina substrate was used as a carbon monoxide gas sensor (Khadayate and Patil 2010). The semiconductor film transfers electrons from the conduction band to the pre-adsorbed oxygen atoms, which leads to the construction of ionic groups like $O^{2-}$ or $O^{-}$. The exposure of the $SnO_2$ layer to carbon monoxide oxidizes the gas molecule, followed by the release of oxygen and the discharged electrons are transferred to the conduction band.

### 3.4.3 Nitrogen Oxide

The high temperature combustions expel nitrogen oxides as a by-product. In the atmosphere, nitric oxide converts into nitric acid which leads to the development of acid rain. Nitrogen dioxide is a reddish-brown colored poisonous gas and one of the prominent air pollutants. The nitrogen dioxide poisoning causes pulmonary edema, which leads to impaired gas exchange and may cause respiratory failure. Thick films of $LaFeO_3$ and $SmFeO_3$ were fabricated by screen-printing technology on alumina substrates with comb-type Au electrodes (Martinelli et al. 1999). These perovskite-type oxide powders have been prepared by thermal decomposition of hexacyanocomplexes, $Ln[Fe(CN)_6]\ nH_2O$, at 700 °C. The conductivity of the film is not significantly affected by the presence of humid air. These p-type semiconducting oxides are ideal for the analysis of nitrogen dioxide. Disposable tin-doped indium oxide layer on glass surface was developed to detect nitrogen oxide gases (Mbarek et al. 2007). The transparent and conductive ITO films are granular and porous in nature and found to be robust and proficient platform for nitrogen oxide gas monitoring.

## *3.5 Pathogens*

Pathogens are infectious microorganism as a bacterium, virus, protozoan, or fungus. Disposable screen-printed electrodes offer versatile substrate for the sensitive quantification of pathogens to diminish the threats of contamination and avoid harmful diseases and environmental hazards (Hayat and Marty 2014). An electrochemical immunosensor was developed for the simultaneous detection of food pathogenic bacteria, namely, *Escherichia coli* O157:H7, campylobacter and salmonella (Viswanathan et al. 2012). Specific antibodies which have releasable metal ions were immobilized on multiwall carbon nanotube polyallylamine-modified screen-printed electrode and SWASV was employed to measure the stripping current response. Aptamer-based viable impedimetric sensor for bacteria was developed by immobilizing selective DNA sequences onto a gold nanoparticle–modified screen-printed carbon electrode (Labib et al. 2012). This method is envisaged to open a new platform for the aptamer-based label-free electrochemical sensing of different viable microorganisms. A sensitive, phage-based amperometric biosensor was developed for the detection of extremely low levels of *Bacillus cereus* and *Mycobacterium smegmatis* as simulants for *Bacillus anthracis* and *Mycobacterium tuberculosis*, respectively (Yemini et al. 2007). P-AP-α-GLU and p-AP-β-GLU were used as the substrate and the product of the enzymatic reaction, *p*-aminophenol, was analyzed at screen-printed carbon electrode at constant potential.

## *3.6 Radioactive Elements*

Radioactive contamination is the presence or accumulation of radioactive substances on surfaces or within solids, liquids, or gases, where the presence of those materials is unintended or undesirable. The radioactive decay of contaminants produces hazardous ionizing radiations such as α, β, and γ rays. Recently, screen-printed electrodes have been explored for the analysis of trace radioactive elements in the ground water. A disposable electrode grafted with 4-carboxyphenyl moiety has been utilized for extremely low-level detection of uranium(VI) using square wave voltammetric technique (Betelu et al. 2009). This approach provides interference-free detection even in the presence of high concentrations of Cd(II), Pb(II), and Zn(II). Unmodified screen-printed electrodes were also explored for trace-level uranium monitoring in environmental samples with a detection limit of 4.5 nM (Kostaki et al. 2011).

## 3.7 Conclusion

As elaborated in this chapter, there have been enormous developments in the application of screen-printed electrodes for effective analysis of environmental contaminants. The combination of screen printing technology along with microelectronics realizes miniaturized analytical tools for pollution monitoring. Electrochemical sensors are ideal for remote monitoring of environmental contaminants. Screen-printed electrodes facilitates the production of disposable, portable, and economical analytical devices with improved sensitivity and selectivity. Such disposable electrodes have been applied to determine water quality and trace heavy metals, toxic organic pollutants, hazardous gases, pathogens, and radioactive compounds. Besides, the use of microchip components can improve the miniaturization procedure to improve response time, reduce sample volumes and reagent consumption, and enhance portability for on-site detection of pollutants.

## References

Ahmed MU, Hossain MM, Safavieh M, Wong YL, Rahman IA, Zourob M, Tamiya E (2016) Toward the development of smart and low cost point-of-care biosensors based on screen printed electrodes. Crit Rev Biotechnol 36(3):495–505. https://doi.org/10.3109/07388551.2014.992387

de Albuquerque YD, Ferreira LF (2007) Amperometric biosensing of carbamate and organophosphate pesticides utilizing screen-printed tyrosinase-modified electrodes. Anal Chim Acta 596(2):210–221. https://doi.org/10.1016/j.aca.2007.06.013

Alkasir RS, Ganesana M, Won YH, Stanciu L, Andreescu S (2010) Enzyme functionalized nanoparticles for electrochemical biosensors: a comparative study with applications for the detection of bisphenol A. Biosens Bioelectron 26(1):43–49. https://doi.org/10.1016/j.bios.2010.05.001

Andreescu S, Barthelmebs L, Marty JL (2002) Immobilization of acetylcholinesterase on screen-printed electrodes: comparative study between three immobilization methods and applications to the detection of organophosphorus insecticides. Anal Chim Acta 464(2):171–180. https://doi.org/10.1016/S0003-2670(02)00518-4

Aragay G, Puig-Font A, Cadevall M, Merkoci A (2010) Surface characterizations of mercury-based electrodes with the resulting micro and nano amalgam wires and spheres formations may reveal both gained sensitivity and faced nonstability in heavy metal detection. J Phys Chem C 114(19):9049–9055. https://doi.org/10.1021/jp102123w

Arduini F, Ricci F, Tuta CS, Moscone D, Amine A, Palleschi G (2006) Detection of carbamic and organophosphorous pesticides in water samples using a cholinesterase biosensor based on Prussian Blue-modified screen-printed electrode. Anal Chim Acta 580(2):155–162. https://doi.org/10.1016/j.aca.2006.07.052

Arduini F, Forchielli M, Amine A, Neagu D, Cacciotti I, Nanni F, Moscone D, Palleschi G (2015a) Screen-printed biosensor modified with carbon black nanoparticles for the determination of paraoxon based on the inhibition of butyrylcholinesterase. Microchim Acta 182(3–4):643–651. https://doi.org/10.1007/s00604-014-1370-y

Arduini F, Neagu D, Scognamiglio V, Patarino S, Moscone D, Palleschi G (2015b) Automatable flow system for paraoxon detection with an embedded screen-printed electrode tailored with butyrylcholinesterase and prussian blue nanoparticles. Chemosensors 3(2):129–145. https://doi.org/10.3390/chemosensors3020129

Bakker E, Telting-Diaz M (2002) Electrochemical sensors. Anal Chem 74(12):2781–2800. https://doi.org/10.1021/ac0202278

Barfidokht A, Mishra RK, Seenivasan R, Liu SY, Hubble LJ, Wang J, Hall DA (2019) Wearable electrochemical glove-based sensor for rapid and on-site detection of fentanyl. Sensors Actuators B Chem:296. https://doi.org/10.1016/J.Snb.2019.04.053

Barton J, García MBG, Santos DH, Fanjul-Bolado P, Ribotti A, McCaul M, Diamond D, Magni P (2016) Screen-printed electrodes for environmental monitoring of heavy metal ions: a review. Microchim Acta 183(2):503–517. https://doi.org/10.1007/s00604-015-1651-0

Betelu S, Vautrin-Ul C, Ly J, Chausse A (2009) Screen-printed electrografted electrode for trace uranium analysis. Talanta 80(1):372–376. https://doi.org/10.1016/j.talanta.2009.06.076

Bratov A, Abramova N, Ipatov A (2010) Recent trends in potentiometric sensor arrays—a review. Anal Chim Acta 678(2):149–159. https://doi.org/10.1016/j.aca.2010.08.035

Brugnera MF, Trindade MAG, Zanoni MVB (2010) Detection of bisphenol a on a screen-printed carbon electrode in Ctab micellar medium. Anal Lett 43(18):2823–2836. https://doi.org/10.1080/00032711003731332

Buhlmann P, Pretsch E, Bakker E (1998) Carrier-based ion-selective electrodes and bulk optodes. 2. Ionophores for potentiometric and optical sensors. Chem Rev 98(4):1593–1687. https://doi.org/10.1021/Cr970113+

Cai J, Du D (2008) A disposable sensor based on immobilization of acetylcholinesterase to multi-wall carbon nanotube modified screen-printed electrode for determination of carbaryl. J Appl Electrochem 38(9):1217–1222. https://doi.org/10.1007/s10800-008-9540-4

Caito S, Aschner M (2017) Developmental neurotoxicity of lead. In: Neurotoxicity of metals. Springer, pp 3–12. https://doi.org/10.1007/978-3-319-60189-2_1

Chou CH, Chang JL, Zen JM (2009) Homogeneous platinum-deposited screen-printed edge band ultramicroelectrodes for amperometric sensing of carbon monoxide. Electroanalysis 21(2):206–209. https://doi.org/10.1002/elan.200804376

Cinti S, Talarico D, Palleschi G, Moscone D, Arduini F (2016) Novel reagentless paper-based screen-printed electrochemical sensor to detect phosphate. Anal Chim Acta 919:78–84. https://doi.org/10.1016/j.aca.2016.03.011

Costa F, Carvalho IF, Montelaro RC, Gomes P, Martins MCL (2011) Covalent immobilization of antimicrobial peptides (AMPs) onto biomaterial surfaces. Acta Biomater 7(4):1431–1440. https://doi.org/10.1016/j.actbio.2010.11.005

Crew A, Lonsdale D, Byrd N, Pittson R, Hart JP (2011) A screen-printed, amperometric biosensor array incorporated into a novel automated system for the simultaneous determination of organophosphate pesticides. Biosens Bioelectron 26(6):2847–2851. https://doi.org/10.1016/j.bios.2010.11.018

Cugnet C, Zaouak O, Rene A, Pecheyran C, Potin-Gautier M, Authier L (2009) A novel microelectrode array combining screen-printing and femtosecond laser ablation technologies: development, characterization and application to cadmium detection. Sensors Actuators B Chem 143(1):158–163. https://doi.org/10.1016/j.snb.2009.07.059

Dąbrowski A, Podkościelny P, Hubicki Z, Barczak M (2005) Adsorption of phenolic compounds by activated carbon—a critical review. Chemosphere 58(8):1049–1070. https://doi.org/10.1016/j.chemosphere.2004.09.067

DeMars R, Shain I (1957) Anodic stripping voltammetry using the hanging mercury drop electrode. Anal Chem 29(12):1825–1827. https://doi.org/10.1021/ac60132a047

Dominguez-Renedo O, Alonso-Lomillo MA, Ferreira-Goncalves L, Arcos-Martinez MJ (2009a) Development of urease based amperometric biosensors for the inhibitive determination of Hg(II). Talanta 79(5):1306–1310. https://doi.org/10.1016/j.talanta.2009.05.043

Dominguez-Renedo O, Gonzalez MJG, Arcos-Martinez MJ (2009b) Determination of antimony (III) in real samples by anodic stripping voltammetry using a mercury film screen-printed electrode. Sensors 9(1):219–231. https://doi.org/10.3390/s90100219

Dzantiev BB, Yazynina EV, Zherdev AV, Plekhanova YV, Reshetilov AN, Chang SC, McNeil CJ (2004) Determination of the herbicide chlorsulfuron by amperometric sensor based on separation-free bienzyme immunoassay. Sensors Actuators B Chem 98(2–3):254–261. https://doi.org/10.1016/j.snb.2003.10.021

Ebenstein A (2012) The consequences of industrialization: evidence from water pollution and digestive cancers in China. Rev Econ Stat 94(1):186–201. https://doi.org/10.1162/REST_a_00150

Ensafi AA, Nazari Z, Fritsch I (2010) Highly sensitive differential pulse voltammetric determination of cd, Zn and Pb ions in water samples using stable carbon-based mercury thin-film electrode. Electroanalysis 22(21):2551–2557. https://doi.org/10.1002/elan.201000246

Ettler V, Mihaljevič M, Komárek M (2004) ICP-MS measurements of lead isotopic ratios in soils heavily contaminated by lead smelting: tracing the sources of pollution. Anal Bioanal Chem 378(2):311–317. https://doi.org/10.1007/s00216-003-2229-y

Fernandez L, Borras C, Carrero H (2006) Electrochemical behavior of phenol in alkaline media at hydrotalcite-like clay/anionic surfactants/glassy carbon modified electrode. Electrochim Acta 52(3):872–884. https://doi.org/10.1016/j.electacta.2006.06.021

Gan N, Yang X, Xie D, Wu Y, Wen W (2010) A disposable organophosphorus pesticides enzyme biosensor based on magnetic composite nano-particles modified screen printed carbon electrode. Sensors (Basel) 10(1):625–638. https://doi.org/10.3390/s100100625

Garcia-Galan C, Berenguer-Murcia A, Fernandez-Lafuente R, Rodrigues RC (2011) Potential of different enzyme immobilization strategies to improve enzyme performance. Adv Synth Catal 353(16):2885–2904. https://doi.org/10.1002/adsc.201100534

Glasspool W, Atkinson J (1998) A screen-printed amperometric dissolved oxygen sensor utilising an immobilised electrolyte gel and membrane. Sensors Actuators B Chem 48(1–3):308–317. https://doi.org/10.1016/S0925-4005(98)00063-X

Guell R, Aragay G, Fontas C, Antico E, Merkoci A (2008) Sensitive and stable monitoring of lead and cadmium in seawater using screen-printed electrode and electrochemical stripping analysis. Anal Chim Acta 627(2):219–224. https://doi.org/10.1016/j.aca.2008.08.017

Gupta VK, Jain S, Khurana U (1997) A PVC-based pentathia-15-crown-5 membrane potentiometric sensor for mercury(II). Electroanalysis 9(6):478–480. https://doi.org/10.1002/elan.1140090609

Gupta VK, Sethi B, Sharma RA, Agarwal S, Bharti A (2013) Mercury selective potentiometric sensor based on low rim functionalized thiacalix [4]-arene as a cationic receptor. J Mol Liq 177:114–118. https://doi.org/10.1016/j.molliq.2012.10.008

Hanrahan G, Patil DG, Wang J (2004) Electrochemical sensors for environmental monitoring: design, development and applications. J Environ Monit 6(8):657–664. https://doi.org/10.1039/b403975k

Hayat A, Marty JL (2014) Disposable screen printed electrochemical sensors: tools for environmental monitoring. Sensors (Basel) 14(6):10432–10453. https://doi.org/10.3390/s140610432

Hughes G, Westmacott K, Honeychurch KC, Crew A, Pemberton RM, Hart JP (2016) Recent advances in the fabrication and application of screen-printed electrochemical (bio)sensors based on carbon materials for biomedical, agri-food and environmental analyses. Biosensors (Basel) 6(4):50. https://doi.org/10.3390/bios6040050

Huyberechts G, Szecowka P, Roggen J, Licznerski BW (1997) Simultaneous quantification of carbon monoxide and methane in humid air using a sensor array and an artificial neural network. Sensors Actuators B Chem 45(2):123–130. https://doi.org/10.1016/S0925-4005(97)00283-9

Hwang GH, Han WK, Park JS, Kang SG (2008) An electrochemical sensor based on the reduction of screen-printed bismuth oxide for the determination of trace lead and cadmium. Sensors Actuators B Chem 135(1):309–316. https://doi.org/10.1016/j.snb.2008.08.039

Injang U, Noyrod P, Siangproh W, Dungchai W, Motomizu S, Chailapakul O (2010) Determination of trace heavy metals in herbs by sequential injection analysis-anodic stripping voltammetry using screen-printed carbon nanotubes electrodes. Anal Chim Acta 668(1):54–60. https://doi.org/10.1016/j.aca.2010.01.018

Istamboulie G, Sikora T, Jubete E, Ochoteco E, Marty JL, Noguer T (2010) Screen-printed poly(3,4-ethylenedioxythiophene) (PEDOT): a new electrochemical mediator for acetylcholinesterase-based biosensors. Talanta 82(3):957–961. https://doi.org/10.1016/j.talanta.2010.05.070

Jarup L (2003) Hazards of heavy metal contamination. Br Med Bull 68(1):167–182. https://doi.org/10.1093/bmb/ldg032

Kadara RO, Tothill IE (2008) Development of disposable bulk-modified screen-printed electrode based on bismuth oxide for stripping chronopotentiometric analysis of lead(II) and cadmium(II) in soil and water samples. Anal Chim Acta 623(1):76–81. https://doi.org/10.1016/j.aca.2008.06.010

Kampouris DK, Kadara RO, Jenkinson N, Banks CE (2009) Screen printed electrochemical platforms for pH sensing. Anal Methods 1(1):25–28. https://doi.org/10.1039/b9ay00025a

Kawde AN, Morsy MA, Odewunmi N, Mahfouz W (2013) From electrode surface fouling to sensitive electroanalytical determination of phenols. Electroanalysis 25(6):1547–1555. https://doi.org/10.1002/elan.201300101

Khadayate RS, Patil PP (2010) CO gas sensing properties of screen printed $SnO_2$ thick films. J Optoelectron Adv Mater 12(6):1338–1342

Khairy M, Kadara RO, Banks CE (2010) Electroanalytical sensing of nitrite at shallow recessed screen printed microelectrode arrays. Anal Methods 2(7):851–854. https://doi.org/10.1039/c0ay00142b

Koblížek M, Malý J, Masojídek J, Komenda J, Kučera T, Giardi MT, Mattoo AK, Pilloton R (2002) A biosensor for the detection of triazine and phenylurea herbicides designed using Photosystem II coupled to a screen-printed electrode. Biotechnol Bioeng 78(1):110–116. https://doi.org/10.1002/bit.10190

Kolb D, Przasnyski M, Gerischer H (1974) Underpotential deposition of metals and work function differences. J Electroanal Chem Interfacial Electrochem 54(1):25–38. https://doi.org/10.1016/S0022-0728(74)80377-3

Kolthoff IM (1952) Polarography, vol 2. Interscience Publishers, London

Koncki R, Mascini M (1997) Screen-printed ruthenium dioxide electrodes for pH measurements. Anal Chim Acta 351(1–3):143–149. https://doi.org/10.1016/S0003-2670(97)00367-X

Kostaki VT, Florou AB, Prodromidis MI (2011) Electrochemically induced chemical sensor properties in graphite screen-printed electrodes: the case of a chemical sensor for uranium. Electrochim Acta 56(24):8857–8860. https://doi.org/10.1016/j.electacta.2011.07.092

Kroger S, Turner APF, Mosbach K, Haupt K (1999) Imprinted polymer based sensor system for herbicides using differential-pulse voltammetry on screen printed electrodes. Anal Chem 71(17):3698–3702. https://doi.org/10.1021/Ac9811827

Kwan RCH, Leung HF, Hon PYT, Barford JP, Renneberg R (2005) A screen-printed biosensor using pyruvate oxidase for rapid determination of phosphate in synthetic wastewater. Appl Microbiol Biotechnol 66(4):377–383. https://doi.org/10.1007/s00253-004-1701-8

Labib M, Zamay AS, Kolovskaya OS, Reshetneva IT, Zamay GS, Kibbee RJ, Sattar SA, Zamay TN, Berezovski MV (2012) Aptamer-based viability impedimetric sensor for bacteria. Anal Chem 84(21):8966–8969. https://doi.org/10.1021/ac302902s

Landsberger S, Wu D (1995) The impact of heavy metals from environmental tobacco smoke on indoor air quality as determined by Compton suppression neutron activation analysis. Sci Total Environ 173(1–6):323–337. https://doi.org/10.1016/0048-9697(95)04755-7

Leonard HJ (2006) Pollution and the struggle for the world product: multinational corporations, environment, and international comparative advantage. Cambridge University Press, Cambridge

Li J, Herrero E, Abruna HD (1998) The effects of anions on the underpotential deposition of Hg on Au(111)—an electrochemical and in situ surface X-ray diffraction study. Colloids Surf Physicochem Eng Asp 134(1–2):113–131. https://doi.org/10.1016/S0927-7757(97)00341-5

Li DW, Li YT, Song W, Long YT (2010) Simultaneous determination of dihydroxybenzene isomers using disposable screen-printed electrode modified by multiwalled carbon nanotubes and gold nanoparticles. Anal Methods 2(7):837–843. https://doi.org/10.1039/c0ay00076k

Li H, Li J, Yang Z, Xu Q, Hu X (2011) A novel photoelectrochemical sensor for the organophosphorus pesticide dichlofenthion based on nanometer-sized titania coupled with a screen-printed electrode. Anal Chem 83(13):5290–5295. https://doi.org/10.1021/ac200706k

Li M, Li Y-T, Li D-W, Long Y-T (2012) Recent developments and applications of screen-printed electrodes in environmental assays—a review. Anal Chim Acta 734:31–44. https://doi.org/10.1016/j.aca.2012.05.018

Lin Y, Lu F, Wang J (2004) Disposable carbon nanotube modified screen-printed biosensor for amperometric detection of organophosphorus pesticides and nerve agents. Electroanalysis 16(1–2):145–149. https://doi.org/10.1002/elan.200302933

Lu DL, La Belle J, Le Ninivin C, Mabic S, Dimitrakopoulos T (2010) In situ electrochemical detection of trace metal vapors at bismuth doped carbon screen printed electrodes. J Electroanal Chem 642(2):157–159. https://doi.org/10.1016/j.jelechem.2010.02.022

Mallin MA, Cahoon LB (2003) Industrialized animal production—a major source of nutrient and microbial pollution to aquatic ecosystems. Popul Environ 24(5):369–385. https://doi.org/10.1023/A:1023690824045

Mandil A, Amine A (2009) Screen-printed electrodes modified by bismuth film for the determination of released lead in moroccan ceramics. Anal Lett 42(9):1245–1257. https://doi.org/10.1080/00032710902901772

Martinelli G, Carotta MC, Ferroni M, Sadaoka Y, Traversa E (1999) Screen-printed perovskite-type thick films as gas sensors for environmental monitoring. Sensors Actuators B Chem 55(2–3):99–110. https://doi.org/10.1016/S0925-4005(99)00054-4

Martínez-Máñez R, Soto J, Lizondo-Sabater J, García-Breijo E, Gil L, Ibáñez J, Alcaina I, Alvarez S (2004) New potentiomentric dissolved oxygen sensors in thick film technology. Sensors Actuators B Chem 101(3):295–301. https://doi.org/10.1016/j.snb.2004.03.008

Mateo C, Palomo JM, Fernandez-Lorente G, Guisan JM, Fernandez-Lafuente R (2007) Improvement of enzyme activity, stability and selectivity via immobilization techniques. Enzym Microb Technol 40(6):1451–1463. https://doi.org/10.1016/j.enzmictec.2007.01.018

Mbarek H, Saadoun M, Bessais B (2007) Porous screen printed indium tin oxide (ITO) for NOx gas sensing. Phys Status Solidi C 4(6):1903. https://doi.org/10.1002/pssc.200674315

McGuinness M, Dowling D (2009) Plant-associated bacterial degradation of toxic organic compounds in soil. Int J Environ Res Public Health 6(8):2226–2247. https://doi.org/10.3390/ijerph6082226

McNerney JJ, Buseck PR, Hanson RC (1972) Mercury detection by means of thin gold films. Science 178(4061):611–612. https://doi.org/10.1126/science.178.4061.611

Melquiades FL, Appoloni CR (2004) Application of XRF and field portable XRF for environmental analysis. J Radioanal Nucl Chem 262(2):533–541. https://doi.org/10.1023/B:Jrnc.0000046792.52385.B2

Mersal GAM (2009) Electrochemical sensor for Voltammetric determination of catechol based on screen printed graphite electrode. Int J Electrochem Sci 4(8):1167–1177

Metters JP, Kadara RO, Banks CE (2011) New directions in screen printed electroanalytical sensors: an overview of recent developments. Analyst 136(6):1067–1076. https://doi.org/10.1039/c0an00894j

Mikkelsen O, Schroder KH (2003) Amalgam electrodes for electroanalysis. Electroanalysis 15(8):679–687. https://doi.org/10.1002/elan.200390085

Mishra RK, Barfidokht A, Karajic A, Sempionatto JR, Wang J, Wang J (2018) Wearable potentiometric tattoo biosensor for on-body detection of G-type nerve agents simulants. Sensors Actuators B Chem 273:966–972. https://doi.org/10.1016/j.snb.2018.07.001

Montereali M, Della Seta L, Vastarella W, Pilloton R (2010) A disposable laccase–tyrosinase based biosensor for amperometric detection of phenolic compounds in must and wine. J Mol Catal B Enzym 64(3–4):189–194. https://doi.org/10.1016/j.molcatb.2009.07.014

Mulchandani A, Mulchandani P, Chen W, Wang J, Chen L (1999) Amperometric thick-film strip electrodes for monitoring organophosphate nerve agents based on immobilized organophosphorus hydrolase. Anal Chem 71(11):2246–2249. https://doi.org/10.1021/ac9813179

Neuhold CG, Wang J, Cai XH, Kalcher K (1995) Screen-printed electrodes for nitrite based on anion-exchanger-doped carbon inks. Analyst 120(9):2377–2380. https://doi.org/10.1039/An9952002377

Nie Z, Nijhuis CA, Gong J, Chen X, Kumachev A, Martinez AW, Narovlyansky M, Whitesides GM (2010) Electrochemical sensing in paper-based microfluidic devices. Lab Chip 10(4):477–483. https://doi.org/10.1039/b917150a

Niu XH, Ding YL, Chen C, Zhao HL, Lan MB (2011) A novel electrochemical biosensor for $Hg^{2+}$ determination based on $Hg^{2+}$-induced DNA hybridization. Sensors Actuators B Chem 158(1):383–387. https://doi.org/10.1016/j.snb.2011.06.040

Palanisamy S, Karuppiah C, Chen SM, Periakaruppan P (2014) Highly sensitive and selective amperometric nitrite sensor based on electrochemically activated graphite modified screen printed carbon electrode. J Electroanal Chem 727:34–38. https://doi.org/10.1016/j.jelechem.2014.05.025

Palchetti I, Mascini M, Minunni M, Bilia AR, Vincieri FF (2003) Disposable electrochemical sensor for rapid determination of heavy metals in herbal drugs. J Pharm Biomed Anal 32(2):251–256. https://doi.org/10.1016/S0731-7085(03)00132-8

Punrat E, Chuanuwatanakul S, Kaneta T, Motomizu S, Chailapakul O (2013) Method development for the determination of arsenic by sequential injection/anodic stripping voltammetry using long-lasting gold-modified screen-printed carbon electrode. Talanta 116:1018–1025. https://doi.org/10.1016/j.talanta.2013.08.030

Punrat E, Chuanuwatanakul S, Kaneta T, Motomizu S, Chailapakul O (2014) Method development for the determination of mercury(II) by sequential injection/anodic stripping voltammetry using an in situ gold-film screen-printed carbon electrode. J Electroanal Chem 727:78–83. https://doi.org/10.1016/j.jelechem.2014.05.026

Renedo OD, Alonso-Lomillo MA, Martinez MJA (2007) Recent developments in the field of screen-printed electrodes and their related applications. Talanta 73(2):202–219. https://doi.org/10.1016/j.talanta.2007.03.050

Rubin BS (2011) Bisphenol a: an endocrine disruptor with widespread exposure and multiple effects. J Steroid Biochem Mol Biol 127(1–2):27–34. https://doi.org/10.1016/j.jsbmb.2011.05.002

Sabath E, Robles-Osorio ML (2012) Renal health and the environment: heavy metal nephrotoxicity. Nefrologia 32(3):279–286. https://doi.org/10.3265/Nefrologia.pre2012.Jan.10928

Salie G, Bartels K (1994) Partial charge-transfer and adsorption at metal-electrodes—the underpotential deposition of Hg(I), Tl(I), Bi(III) and Cu(II) on polycrystalline gold electrodes. Electrochim Acta 39(8–9):1057–1065. https://doi.org/10.1016/0013-4686(94)E0020-Z

Sanchez-Santed F, Colomina MT, Herrero Hernandez E (2016) Organophosphate pesticide exposure and neurodegeneration. Cortex 74:417–426. https://doi.org/10.1016/j.cortex.2015.10.003

Sanllorente-Mendez S, Dominguez-Renedo O, Arcos-Martinez MJ (2010) Immobilization of acetylcholinesterase on screen-printed electrodes. Application to the determination of arsenic(III). Sensors 10(3):2119–2128. https://doi.org/10.3390/s100302119

Sassolas A, Blum LJ, Leca-Bouvier BD (2012) Immobilization strategies to develop enzymatic biosensors. Biotechnol Adv 30(3):489–511. https://doi.org/10.1016/j.biotechadv.2011.09.003

Schuhmann W, Ohara TJ, Schmidt HL, Heller A (1991) Electron-transfer between glucose-oxidase and electrodes via redox mediators bound with flexible chains to the enzyme surface. J Am Chem Soc 113(4):1394–1397. https://doi.org/10.1021/Ja00004a048

Sereshti H, Heravi YE, Samadi S (2012) Optimized ultrasound-assisted emulsification microextraction for simultaneous trace multielement determination of heavy metals in real water samples by ICP-OES. Talanta 97:235–241. https://doi.org/10.1016/j.talanta.2012.04.024

Serrano N, Diaz-Cruz JM, Arino C, Esteban M (2010) Stripping analysis of heavy metals in tap water using the bismuth film electrode. Anal Bioanal Chem 396(3):1365–1369. https://doi.org/10.1007/s00216-009-3294-7

Seto Y, Kanamori-Kataoka M, Tsuge K, Ohsawa I, Matsushita K, Sekiguchi H, Itoi T, Iura K, Sano Y, Yamashiro S (2005) Sensing technology for chemical-warfare agents and its evaluation using authentic agents. Sensors Actuators B Chem 108(1–2):193–197. https://doi.org/10.1016/j.snb.2004.12.084

Siaka M, Owens CM, Birch GF (1998) Evaluation of some digestion methods for the determination of heavy metals in sediment samples by flame-AAS. Anal Lett 31(4):703–718. https://doi.org/10.1080/00032719808001873

Some IT, Sakira AK, Mertens D, Ronkart SN, Kauffmann JM (2016) Determination of groundwater mercury (II) content using a disposable gold modified screen printed carbon electrode. Talanta 152:335–340. https://doi.org/10.1016/j.talanta.2016.02.033

Song Y, Swain GM (2007) Development of a method for total inorganic arsenic analysis using anodic stripping voltammetry and a Au-coated, diamond thin-film electrode. Anal Chem 79(6):2412–2420. https://doi.org/10.1021/ac061543f

Su WY, Wang SM, Cheng SH (2011) Electrochemically pretreated screen-printed carbon electrodes for the simultaneous determination of aminophenol isomers. J Electroanal Chem 651(2):166–172. https://doi.org/10.1016/j.jelechem.2010.11.028

Taillefert M, Luther GW, Nuzzio DB (2000) The application of electrochemical tools for in situ measurements in aquatic systems. Electroanalysis 12(6):401–412. https://doi.org/10.1002/(Sici)1521-4109(20000401)12:6<401::Aid-Elan401>3.0.Co;2-U

Viguier B, Zór K, Kasotakis E, Mitraki A, Clausen CH, Svendsen WE, Castillo-León J (2011) Development of an electrochemical metal-ion biosensor using self-assembled peptide nanofibrils. ACS Appl Mater Interfaces 3(5):1594–1600. https://doi.org/10.1021/am200149h

Viswanathan S, Rani C, Ho JAA (2012) Electrochemical immunosensor for multiplexed detection of food-borne pathogens using nanocrystal bioconjugates and MWCNT screen-printed electrode. Talanta 94:315–319. https://doi.org/10.1016/j.talanta.2012.03.049

Vonau W, Guth U (2006) pH monitoring: a review. J Solid State Electrochem 10(9):746–752. https://doi.org/10.1007/s10008-006-0120-4

Wang J (2005) Nanomaterial-based electrochemical biosensors. Analyst 130(4):421–426. https://doi.org/10.1039/B414248A

Wang J, Tian BM (1992) Screen-printed stripping voltammetric potentiometric electrodes for decentralized testing of trace lead. Anal Chem 64(15):1706–1709. https://doi.org/10.1021/Ac00039a015

Wang J, Bhada RK, Lu JM, MacDonald D (1998a) Remote electrochemical sensor for monitoring TNT in natural waters. Anal Chim Acta 361(1–2):85–91. https://doi.org/10.1016/S0003-2670(97)00702-2

Wang J, Tian BM, Lu JM, Wang JY, Luo DB, MacDonald D (1998b) Remote electrochemical sensor for monitoring trace mercury. Electroanalysis 10(6):399–402. https://doi.org/10.1002/(Sici)1521-4109(199805)10:6<399::Aid-Elan399>3.0.Co;2-K

Wang J, Tian BM, Nascimento VB, Angnes L (1998c) Performance of screen-printed carbon electrodes fabricated from different carbon inks. Electrochim Acta 43(23):3459–3465. https://doi.org/10.1016/S0013-4686(98)00092-9

Wang J, Lu J, Hocevar SB, Ogorevc B (2001) Bismuth-coated screen-printed electrodes for stripping voltammetric measurements of trace lead. Electroanalysis 13(1):13–16. https://doi.org/10.10002/1521-4109

Wang SM, Su WY, Cheng SH (2010) A simultaneous and sensitive determination of hydroquinone and catechol at anodically pretreated screen-printed carbon electrodes. Int J Electrochem Sci 5(11):1649–1664

Wen Z, Tian-Mo L (2010) Gas-sensing properties of $SnO_2$–$TiO_2$-based sensor for volatile organic compound gas and its sensing mechanism. Phys B Condens Matter 405(5):1345–1348. https://doi.org/10.1016/j.physb.2009.11.086

Won Y-H, Jang HS, Kim SM, Stach E, Ganesana M, Andreescu S, Stanciu LA (2009) Biomagnetic glasses: preparation, characterization, and biosensor applications. Langmuir 26(6):4320–4326. https://doi.org/10.1021/la903422q

Worsfold PJ, Gimbert LJ, Mankasingh U, Omaka ON, Hanrahan G, Gardolinski PC, Haygarth PM, Turner BL, Keith-Roach MJ, McKelvie ID (2005) Sampling, sample treatment and quality assurance issues for the determination of phosphorus species in natural waters and soils. Talanta 66(2):273–293. https://doi.org/10.1016/j.talanta.2004.09.006

Yang CC, Kumar AS, Zen JM (2005) Precise blood lead analysis using a combined internal standard and standard addition approach with disposable screen-printed electrodes. Anal Biochem 338(2):278–283. https://doi.org/10.1016/j.ab.2004.12.015

Yemini M, Levi Y, Yagil E, Rishpon J (2007) Specific electrochemical phage sensing for *Bacillus cereus* and *Mycobacterium smegmatis*. Bioelectrochemistry 70(1):180–184. https://doi.org/10.1016/j.bioelechem.2006.03.014

Yuan CG, Shi JB, He B, Liu JF, Liang LN, Jiang GB (2004) Speciation of heavy metals in marine sediments from the East China Sea by ICP-MS with sequential extraction. Environ Int 30(6):769–783. https://doi.org/10.1016/j.envint.2004.01.001

Zaouak O, Authier L, Cugnet C, Castetbon A, Potin-Gautier M (2010) Electroanalytical device for cadmium speciation in waters. Part 1: development and characterization of a reliable screen-printed sensor. Electroanalysis 22(11):1151–1158. https://doi.org/10.1002/elan.200900474

Zejli H, de Cisneros JLH-H, Naranjo-Rodriguez I, Liu B, Temsamani KR, Marty J-L (2008) Alumina sol–gel/sonogel-carbon electrode based on acetylcholinesterase for detection of organophosphorus pesticides. Talanta 77(1):217–221. https://doi.org/10.1016/j.talanta.2008.06.010

Zhou Y, Zhang F, Yang H, Zhang S, Ma X (2003) Comparison of effectiveness of different ashing auxiliaries for determination of phosphorus in natural waters, aquatic organisms and sediments by ignition method. Water Res 37(16):3875–3882. https://doi.org/10.1016/S0043-1354(03)00267-7

# Sensors and Biosensors for Environment Contaminants

**Heba M. Mohamed**

## Contents



## 1 Introduction: Sensors/Biosensors as Green Analytical Tools

Many years' worth of effort has been dedicated by researchers to investigate and develop technologies toward both detection and reduction of the environmental impact of hazardous compounds. Electrochemical sensors prove to have several potentialities to detect widespread environmental pollutants like pesticides, heavy metals, polycyclic aromatic hydrocarbons, and toxins, and other emerging contaminants including gasoline additives, pharmaceuticals, hormones, personal

H. M. Mohamed (✉)
Faculty of Pharmacy, Cairo University, Cairo, Egypt

Higher Colleges of Technology, Abu Dhabi, UAE



Inamuddin, A. M. Asiri (eds.), *Nanosensor Technologies for Environmental Monitoring*, Nanotechnology in the Life Sciences,
https://doi.org/10.1007/978-3-030-45116-5_6

care products, endocrine-disrupting agents, organometallic compounds, disinfection by-products, plasticizers, perfluorinated compounds, and surfactants that are considered a massive threat for living things and ecosystems. Nowadays, monitoring the environment for various pollutants has become a fundamental factor to attain sustainability goals (Azmuddin et al. 2017). The principles of sustainable and green chemistry have a great focus on new eco-friendly synthetic paths and promoting development of analytical processes for real-time and in situ monitoring of perilous substances. Pioneering green technologies and tools are substantial to diminish or abolish the usage or production of perilous materials and minimize energy consumption. In this context, sensors/biosensors are expansively recognized to be highly useful for identification, monitoring, and analysis of various substances owing to their ease of use, simplicity of construction, portability, sustainability, and considerably cost-effective development (Kimmel et al. 2011).

The application of nano-sensors/biosensors in environmental monitoring and analysis has become increasingly imperative, and many researchers show a particular interest on the different types of fabricated structure encompassing those made by self-assembly, which can be adapted for monitoring of chemical processes (Riu et al. 2006).

In the last decades, novel and green synthesis procedures for nanomaterials and miniaturization approaches used for fabrication of sensors and biosensors to improve their eco-friendliness level toward more sustainability have emerged as promising tools for environmental monitoring and prompt warning. Green synthesis is essential to evade hazardous by-product production by using sustainable, reliable, and environment-friendly synthesis processes (Singh et al. 2018). Eco-friendly synthesis of nanoparticles is implemented to lodge several biological components (e.g., bacteria, fungi, algae, and different plant parts and extracts). The use of plant extracts has been proven to be a quite easy and simple route for large-scale nanoparticles production when compared to either bacteria or fungi-based synthesis. Green nanoparticles synthesized using biological components are collectively recognized as biogenic nanoparticles.

## 2 Miniaturization and Microfabrication

The current trend is heading toward using disposable and portable sensors that are pliable to miniaturization. A sensor is a tiny device which transforms chemical or biochemical data into a signal; it commonly comprises a recognition element, transducer, and signal processor, as three main parts, as shown in Fig. 1. The transducer functioned by transforming the signal obtained by the sensor into an electrical signal. Due to current progress in instrumentation and electronics very small electrical signals can now be measured to enable on/in site pollution monitoring. Different

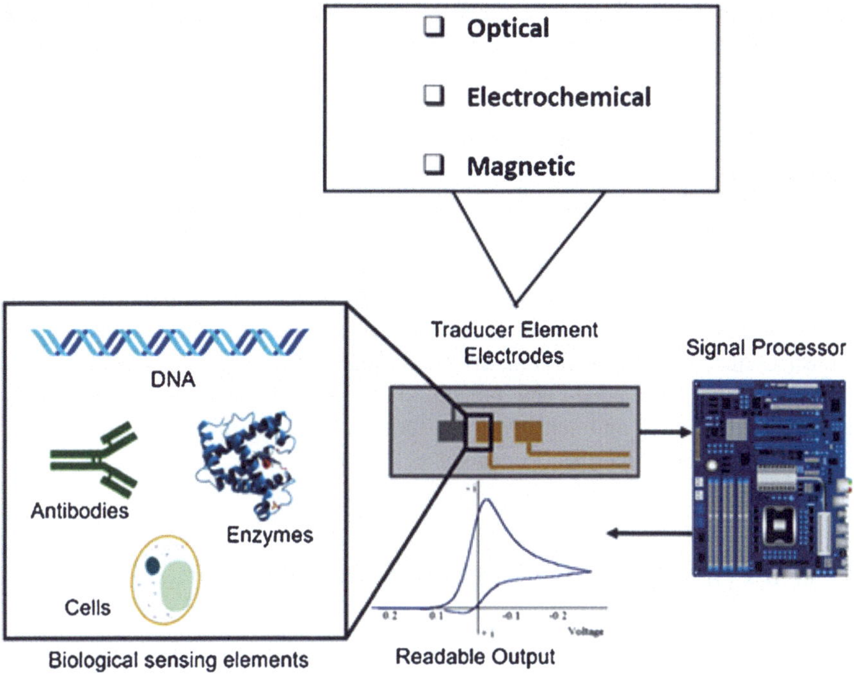


**Fig. 1** Scheme of an electrochemical biosensor composition

**Table 1** Types of sensors based on the transducer used for detection of chemical species

| Type of sensors | Example(s) |
|---|---|
| Electrochemical | Biosensors, potentiometric sensors, and voltammetric sensors |
| Calorimetric | Thermistor |
| Optical | Spectrophotometric, colorimetric, fiber optic |
| Surface plasmon resonance | Biomolecule concentration |

types of sensors based on transducers are mentioned in Table 1. Miniaturization along with automation, have been the emphasis of growing efforts recently. Such miniaturization, implicit by the new terminology "lab on a chip," leads to less reagent consumption, less in-flow injection analysis approaches, and the ability to use the analysis system outside the laboratory. The aim of "lab on a chip" is that the whole process is conducted by means of a microfluidic system on the same device throughout, integrating all the steps and phases of pretreatment followed by separation and finally detection. Their fabrication is usually accompanied by some special challenges in terms of accuracy, precision, reproducibility, calibration, and so on that cannot be figured out in similar ways for conventional ones. (Rios et al. 2006). Many criteria have to be fulfilled within miniaturized systems to ensure their successful implantation including being probably disposable and being robust, and they require a minimum direct operator intervention, especially for environmental applications.

Microfabrication of lab-on-chip assays use clearly less reagents compared to classical bench methodologies and improve reaction kinetics and reduce overall reaction cost (Shitanda et al. 2011 and Das et al. 2011). Recently, a massive range of electrochemical devices has been developed for detection and monitoring of various organic and inorganic toxins, for instance, heavy metals (Choi and Kim 2009, Cooper et al. 2007 and Beni et al. 2005). Similarly, the use of disposable screen-printed electrodes has added new breadth to electrochemical analysis, particularly in support of speedy and sensitive monitoring of several materials with various characteristics and properties (Renedo et al. 2007).

## 3 Eco-Friendly Sensor/Biosensor Development

The reliability and efficiency of an electrochemical sensor highly depend on the constitution of the detection platform. The synergistic effect between the technology of electrochemical sensors and nanomaterials has offered many merits in novel transducing context, alongside signal enhancement. Carbon paste electrodes (CPE) have been modified by linking CPE with other distinctive substances to produce chemically modified carbon electrodes showing very high selectivity. They have valuable advantages like being easy to manufacture, low cost, wider operational window, stability, and flexibility in composition to fit for different purposes (Svancara et al. 2009). Modifiers like bismuth (Bi) nanoparticles (Rico et al. 2009), hydroxyapatite (HA) (El Mhammedi et al. 2007), and Bi-HA (Khan and Abdullah 2014) have been used, with higher sensitivity, to enhance cadmium and lead deposition through HA ion-exchange. There has also been an increasing trend of using plant tissues to prepare chemically modified carbon electrodes (CMEs) with many merits including simplicity of construction, being environmentally friendly, and being less harmful. Plant tissues CME was initially fabricated and used for L-glutamate determination. Table 2 shows some plant and animal materials used as modifiers for heavy metal detection (Kwon et al. 2000).

Sensors with plant extract modifiers contain several chemical components that function as active constituents for analyte binding. Amino acids can serve as a ligand for a variety of metal ions due to the large number of donor atoms they contain. Likewise, lignin and lignocellulosic materials can be used as binding sites for metals due to the oxygen-containing functional groups, for example, alcohol,

**Table 2** Examples of some modifiers of plant and animal origin

| Modifier | Detection limit (ppm) |
|---|---|
| *Plant origin* | |
| Kapok fiber | 1000 |
| Apple peelings | – |
| Grass weed | 10 |
| *Solanum tuberosum* (potato) | – |
| Stems from cabbage | – |
| Banana | 100 |
| *Animal origin* | |
| Feather | 121 |

phenol, and carboxylic acid structures in lignin that are in control for highly stable lignin–metal complexes via different bonding interactions. Likewise, cellulose, by its carbonyl and hydroxyl functional groups, can work as binding sites for different metals (Nazir et al. 2013).

## 4 Sensors/Biosensors Composition

Sensors technologies based on nanomaterials have been developed over the last two decades targeting the sensitive and highly specific detection and monitoring of environmental pollutants with many superior advantages of easy use, low cost, field-deployable technology. In general, sensors are composed of three main components: nanomaterials, recognition element (to increase selectivity), and a signal transduction means for analyte detection, as summarized in Fig. 1. Sensors are generally characterized on the basis of these three elements.

### *4.1 Nanomaterials*

Nanomaterials have reinforced improvements and advances in sensor design in the direction of more sustainability, for instance, miniaturization, portability, disposability, and rapid signal response. Simplified surface functionalization and the high surface area-to-volume ratio support better sensors sensitivity and improve selectivity and reduce the detection limits to extreme low values. Graphene and carbon nanotubes are frequently used in nanosensors due to their large surface area, perfect electrical ad thermal conductivity, and improved mechanical strength (Yang et al. 2010). Metal and metal oxides nanoparticles have extensive uses in sensor fabrications for various applications as they can be produced in many different shapes with different extinction coefficients (Link and El-Sayed 1999) and easily modified by surface functionalization. Colloidal solutions of both silver and gold nanoparticles are widely used and due to their distinctive characteristics and color changes that make them useful as visual colorimetric sensors. Added to that, their nanoparticles excitation can cause the uniform oscillation of conduction electrons that leads to localized surface plasmon resonance–based spectroscopies, for example, surface enhanced Raman spectroscopy and surface plasmon resonance (Romo-Herrera et al. 2011). Gold nanoparticles are stable, biocompatible, and have been widely used in sensing applications (Saha et al. 2012). Surface coatings can be used for modification purposes and simplify the addition of recognition elements. Thioglycolic acid and 3-mercaptopropionic acid, thiol capping agents, are frequently used to afford chemical functionality and colloidal stability. A wide range of nanostructured metal oxides, for example, iron oxides, zinc oxides, titanium oxides, zirconium oxides, and others, have been tried for sensing uses. Quantum dots (QDs) are semiconductor nanocrystals and normally have broad absorption bands, yet narrow fluorescence emission bands; therefore they can superbly serve as optical transducers.

### 4.1.1 Green Synthesis of Metal/Metals Oxide Nanoparticles

An innovative era of green synthesis methodologies is attaining prodigious focus in the modern research and material improvement. Nanoparticle green synthesis is composed of single bio-reduction step methodology that necessitates fairly low consumption of energy and is also cost-effective, and allows for large-scale production of nanoparticles (Wadhwani et al. 2016), Fig. 2 illustrates the key merits of green synthesis. Ultimately, regulated and controlled green synthesis of materials/nanomaterials will directly assist elevating their environmental friendliness. Prevention/minimization of waste, using safer solvent/reagents and reduction of pollution are all considered some of the main principles for green synthesis. Solvents are an essential element in the greening of synthesis methods. Ideally, water is always the solvent of choice for green synthesis processes, for example, synthesis of Ag and Au nanoparticles using gallic acid in an aqueous medium at room temperature (Yoosaf et al. 2007). Ionic liquids are also acknowledged for being eco-friendly and can be used for synthesis of various metal nanoparticles (Vollmer et al. 2010). Ionic liquids are able to work both as a reducing and a protecting agent; this facilitates and simplifies the nanoparticle synthesis process.

Various reaction parameters like temperature, solvent, pH and pressure can affect and control green synthesis methodologies. One of the highly considered factor in metal/metal oxide nanoparticles synthesis is biodiversity in plants because of the presence of different useful phytochemicals within different plants, for instance, aldehydes, ketones, phenols, carboxylic acids, amides, ascorbic acids, flavones, and terpenoids in different percentages based on cultivation location, season and species,

**Fig. 2** Significant virtues of green synthesis

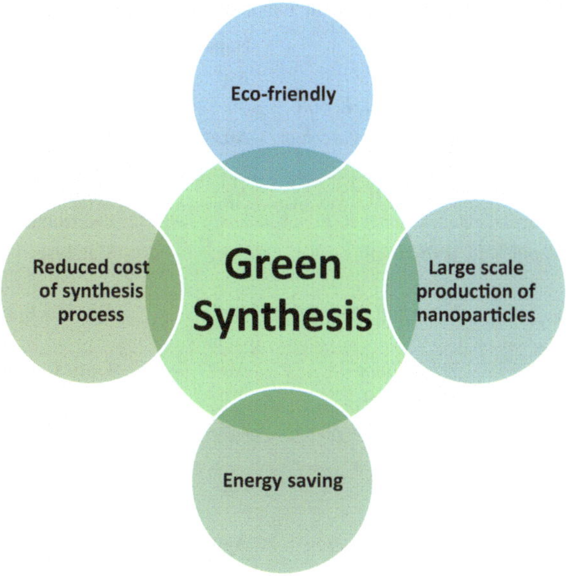


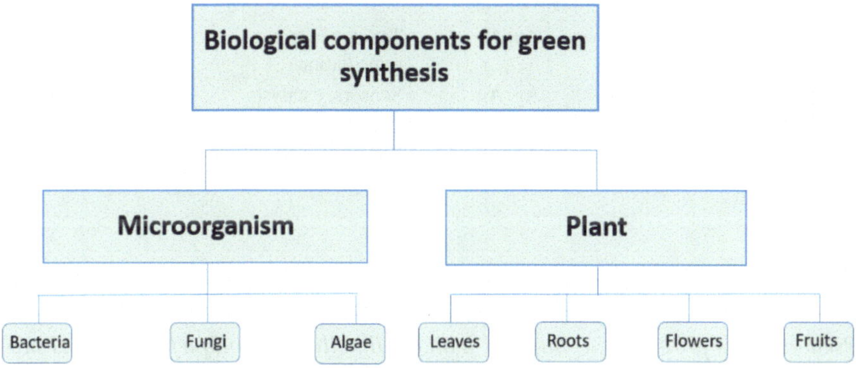


**Fig. 3** Different biological components used for green synthesis

and other factors. These components reduce metal salts into metal nanoparticles. Figure 3 summarizes the currently used biological components for green synthesis.

### 4.1.2 Bacteria

Bacterial species, mainly prokaryotic and actinomycetes, have been broadly used for biotechnological applications (Gericke and Pinches 2006). With a relatively easy manipulation, bacteria owns the capability of reducing metal ions and considered as significant parameter in the preparation of various nanoparticles (Iravani 2014; Thakkar et al. 2010). Some strains of bacteria have expansively been

**Table 3** Examples of metallic nanoparticles prepared from biological components

| Organism | Species | Nanoparticles | Applications | Ref. |
|---|---|---|---|---|
| Bacteria | *Klebsiella pneumonia*, *E. coli* | Ag | Electrical batteries, optical receptors | Ahmad et al. (2010) |
| | *Bacillus megaterium* D01 | Au | Bio sensing, catalysis | Wen et al. (2009) |
| | *E. coli* DH 5α | Ag | Hemoglobin electrochemistry | Du et al. (2007) |
| Fungi | *Verticillium* | Ag | Catalysis | Mukherjee et al. (2001) |
| | *Fusarium semitectum* | Ag | Bio-labeling | Basavaraja et al. (2008) |
| | *Verticillium luteoalbum* | Au | Optics, sensor, coatings | Gericke and Pinches (2006) |
| | *Aspergillus terreus* | ZnO | Bio sensing, catalysis | Raliya and Tarafdar (2014) |
| Yeast | MKY3 | Ag | Coatings, electrical batteries | Kowshik et al. (2002) |
| | *Saccharomyces cerevisiae* broth | Ag, Au | Catalysis | Mourato et al. (2011) |
| Plant | *Aloe barbadensis*, Miller | Au, Ag | Optical coatings | Chandran et al. (2006) |
| | *Azadirachta indica* | Ag, Au | Toxic metals remediation | Shankar et al. (2004) |
| | *Camellia sinensis* | Ag, Au | Sensors, catalysts | Raliya and Tarafdar (2014) |
| | *Medicago sativa* | Au | Labeling in structural biology | Gardea-Torresdey et al. (2002) |

investigated for the green production of silver nanoparticles, for instance, *Escherichia coli*, *Enterobacter cloacae*, *Lactobacillus casei*, *Bacillus cereus*, *Pseudomonas proteolytica*, *Bacillus indicus*, *Bacillus amyloliquefaciens*, and others. Similarly, gold nanoparticles are bio-reduced using several bacterial species, for instance, *Desulfovibrio desulfuricans*, *E. coli* DH5a, *Bacillus megaterium*, and others. Distinct size/shape and morphology can be obtained as summarized in Table 3.

### 4.1.3 Fungi

Biosynthesis of metal/metal oxide nanoparticles based on fungi as a biological element is a very competent development for green nanoparticles generation. Owing to the presence of intracellular enzyme in addition to enzymes, proteins, and other reducing elements on the fungi cell surfaces, make them a better biological agents for the nanoparticles preparation (Chen et al. 2009; Narayanan and Sakthivel 2011). Efficient fungi can produce greater amounts of nanoparticles rather than bacteria

(Mohanpuria et al. 2008). Reductase enzymatic reduction is the most probable mechanism for the metallic nanoparticle formation. Various fungal species can be used for the synthesis of nanoparticles of several metals and metals oxides like gold, silver, zinc oxide, and titanium dioxide, and synthesis of other components, Table 3.

### 4.1.4 Yeast

There are around 1500 yeast species that have been recognized. Many research groups have reported successful synthesis trials of nanoparticles/nanomaterials by the use of yeast. *Saccharomyces cerevisiae* broth and silver-tolerant yeast strain have been reported to successfully produce silver and gold nanoparticles (Yurkov et al. 2011). Various species are incorporated for numerous metallic nanoparticles preparation (Table 3).

### 4.1.5 Plants

Plants can store specific amounts of heavy metals. Consequently, biosynthesis practices using plant extracts have acquired amplified attention as a simple, feasible, efficient, and cost-effective methods along with being an excellent alternate to traditional methods of nanoparticles preparation, as illustrated in Fig. 4. Plants contain organic biomolecules, for instance, proteins, carbohydrates, and coenzymes, that are capable of reducing metal salts to their nanoparticles in a single step process. Many plants including neem, aloe vera, oat, tulsi, coriander, mustard, lemon grass, and lemon have been extensively used for silver and gold nanoparticle synthesis, as described in Table 3.

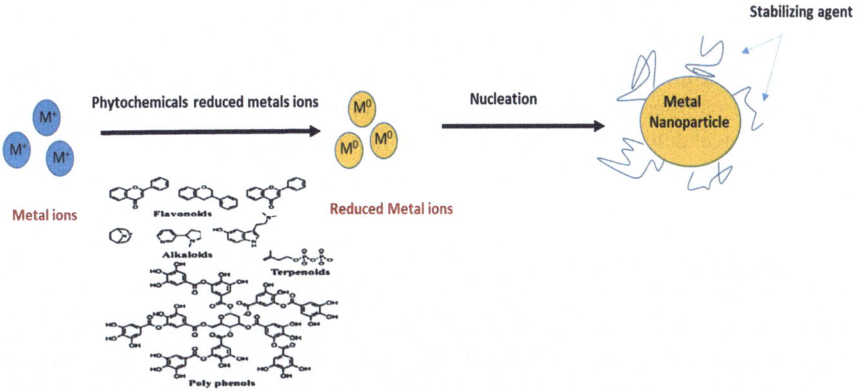


**Fig. 4** Mechanism of nanoparticle formation by plant leaf extracts

## *4.2 Recognition Elements*

Selectivity is crucial property in the design of an efficacious sensor. A wide range of recognition elements have been executed in the design of nanosensor that includes antibodies (Jiajie et al. 2014), aptamers (Ma et al. 2013) functional proteins (Bies et al. 2004), enzymes (Evtugyn et al. 1998), and whole cells (Olaniran et al. 2011).

### 4.2.1 Antibodies

Antibodies bind specifically to a distinct antigen, and they are commonly used for capturing and labeling microorganisms or substances that evoke an immune response (Ellington et al. 2010). Three types of antibodies used to recognize the analytes: polyclonal, monoclonal, and antibody fragments. Though antibodies are commonly utilized in biosensor fabrication, they have some drawbacks such as high costs, pH sensitivity, short shelf-lives, temperature, and batch-to batch variations (Bordeaux et al. 2010). Even with these weaknesses, antibodies are regularly the most selective recognition means for immunogenic analytes (Huang et al. 2005). Many immunosensors and screen-printed electrodes are designed and used for the detection of environmental pollutants as multi-analytes in different settings and are offered both as a laboratory device and a portable system.

### 4.2.2 Aptamers

Aptamers can be defined as short oligonucleotide of RNA or ssDNA that can bind to certain molecules. Aptamers have long shelf-lives, have low variability between batches, are thermally stable, and have low cost production compared to antibodies (Low et al. 2009; Hamula et al. 2011). Aptamers of nucleic acid are known for their high specificity (Hoinka et al. 2015). The oligonucleotide sequences are isolated, identified and amplified through polymerase chain reaction and affinity testing, and after de novo synthesis they can be integrated into biosensors (Tombelli et al. 2002). Bacteria and pathogenic microorganisms were detected using aptamers-based sensors with optical detection in different environment matrices, air and drinking water, and food samples too.

### 4.2.3 Enzymes

Enzyme-based biosensors are having increasing significance and abundance in environmental and food applications. Improvement of enzyme biosensors using different strategies were done, for example, electrochemical interfaces to enzymes (Chen and Gorski 2001), application of quinoproteins (Matsushita et al. 2002) and metalloproteins (Sinibaldi et al. 2001), and investigations on immobilization (Gill and Ballesteros 2000). Acetylcholine esterase was used to detect organophosphates and carbamates with either single-use devices (Schulze et al. 2002) or in traditional graphite electrodes. For instant detecting phenols in environmental matrices was done using continuous electrochemical sensor (Freire et al. 2002) going up to biosensor arrays (Young et al. 2001). Toxic gas can also be detected using enzyme-based sensors. For example, $SO_2$ was detected by screen-printed electrodes (Hart et al. 2002). Nitrite reductase-based optical biosensor was used for monitoring purposes. In addition heavy metals and nitrite can be detected in potable water using several biosensors (Lee and Lee 2002).

### 4.2.4 Whole Cells

Whole cell-mediated biosensors are founded on the use of biosensing cells, for instance, microorganisms, protozoa, algae, and plant cells. Being cheaper gives an advantage to whole cell-based biosensors compared to enzyme-based biosensors. Multistep reactions are usually possible since all the enzymes and cofactors needed are present in one cell. Many microbial-mediated optical biosensors have been designed and used for toxicity and pollutants detection (Olaniran et al. 2011). *A rapid and effective heavy metals monitoring in waste water was achievable by* a whole-cell bacterial biosensors (Olaniran et al. 2011). Because of the use of *Shigella sonnei* and *Escherichia coli*, the sensors were recording high sensitivity. An integrated fluorescence-based sensor was reported to monitor bacterial respiratory activity by measuring the reduction in oxygen partial pressure and pH value reduction (Arain et al. 2006).

## *4.3 Signal Transduction*

There are three chief signal transduction means used in nanomaterials-mediated sensors, namely, optical, electrochemical, and magnetic methods.

### 4.3.1 Optical

Optical transduction takes place because of the interaction between electromagnetic radiation (ultraviolet, visible, or infrared light) and the sensing element. The two widely used optical methods employed in the design of nanosensor are surface plasmon resonance and fluorescence. Quantum dots or polymer nanoparticle probes or dye-doped silicon is commonly used in fluorescent nanosensors due to their photostability and robustness (Vikesland and Wigginton 2010).

### 4.3.2 Electrochemical

Electrochemical methods of detection works by measuring the change in potential or current resulting from the reaction between the analyte and the electrode. Different methods can be used for change detection including amperometry, cyclic voltammetry, chronopotentiometry, chronoamperometry, and impedance spectroscopy (Grieshaber et al. 2008).

### 4.3.3 Magnetic

Magnetic transduction has been incorporated to detect signals upon analyzing biological samples due to the low background magnetic signal (Koets et al. 2009).

## 5 Environmental Applications for Sensors/Biosensors

There are vast of environment contaminants that can be accurately detected and monitored via sensors/biosensors using antibodies, enzymes, and aptamers as recognition elements. Different environmental applications are summarized in Fig. 5.

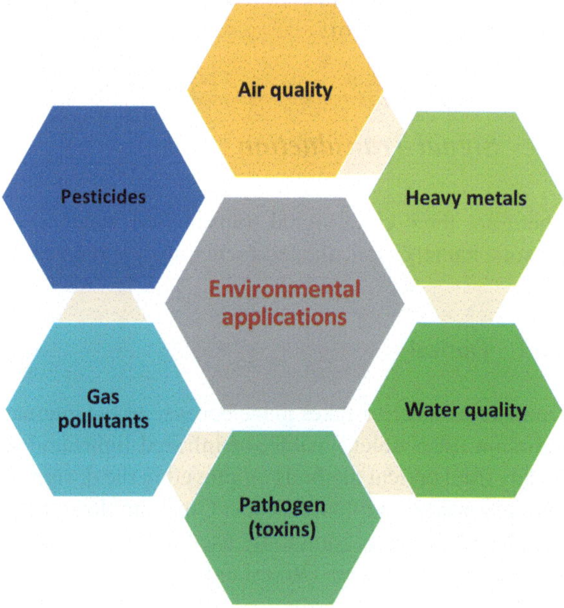


**Fig. 5** Environmental applications of sensors/biosensors

## *5.1 Pesticides*

Owing to their significant environmental existence, pesticides are among the utmost crucial environmental pollutants. Currently, over 800 active ingredients are present in pesticides (Liu et al. 2013). Organophosphates (OP), carbamates, neonicotinoids, and triazines are the prevailing ones. The organophosphates insecticides are broadly used in agriculture but they have the topmost environmental concern because of their unlimited harmfulness. Consequently, simple, sensitive, reliable, and sustainable approaches such as biosensors have been used to detect and monitor the pesticides with minimum sample pretreatment procedures. Table 4 describes a summary of some latest biosensors for pesticides detection and monitoring.

### 5.1.1 Organophosphates

Paraoxon has been detected by the aid of disposable amperometric acetylcholinesterase-based biosensors on gold screen-printed electrodes in water samples (Arduini et al. 2013). The disposable biosensors showed a good analytical performance in regard to linearity, sensitivity, and detection limits due to the proper enzyme immobilization by the self-assembled monolayer. Other biosensors for paroxon detection were fabricated (Arduini et al. 2015) using butyrylcholinesterase and carbon black nanoparticles that added some advantages such as low applied potential, cost-effectiveness, and ease of preparing a stable dispersion (Arduini et al. 2015). Gold nanorods were also used in colorimetric acetylcholinesterase biosensor (Guo et al. 2017) to detect paraoxon in real water samples. Besides, the colorimetric biosensor allowed paraoxon detection in irrigation water with good recoveries.

Methyl parathion, was determined by hydrolase- based biosensor using magnetic $Fe_3O_4$ nanocomposite and gold nanoparticles with high sensitivity and selectivity (Zhao et al. 2013). The advantages of using hydrolase were; no poisoning by organophosphates, produces a reusable biosensor and allows the continuous measurement (Zhao et al. 2013). Due to great catalytic effectiveness, high conductivity, and being exceptionally biocompatible, biosensors using gold nanoparticles showed wide linear range and high sensitivity. Another acetylcholinesterase amperometric biosensor was used for methyl parathion detection in contaminated lake, for immobilization purpose a graphite working electrode and macroalgae were employed (Nunes et al. 2014). Another acetylcholinesterase-based biosensor for methyl parathion detection was fabricated using a nanoporous carbon paste electrode with gold nanoparticles, chitosan, and Nafion (Deng et al. 2016). Determination of chlorpyrifos in river water samples was done at low cost by disposable tyrosinase based biosensor on screen printed electrodes (Mayorga-Martinez et al. 2014) and aptasensor using a new composite film (Wei et al. 2014). Dichlorvos detection in environmental samples was achieved using several biosensors that constructed via bi-enzyme system composed of acetylcholinesterase in addition to choline oxidase

**Table 4** Biosensors for pesticide monitoring

| Pesticide | Biosensor type | Sensing material | Recognition element | Ref. |
|---|---|---|---|---|
| Paraoxon | Electrochemical | Gold SPE[a] + cysteamine SAM[b] | Acetylcholinesterase enzyme | Arduini et al. (2013) |
| | Electrochemical | SPE[a] + carbon black NP[c] | Butyrylcholinesterase enzyme | Arduini et al. (2015) |
| | Optical | Iodine-starch | Acetylcholinesterase and choline oxidase enzymes | Gao et al. (2012) |
| Methyl parathion | Electrochemical | SPE[a] with $Fe_3O_4$ + gold NP[c] | Hydrolase enzyme | Zhao et al. (2013) |
| | Electrochemical | Graphite + macroalgae | Acetylcholinesterase enzyme | Nunes et al. (2014) |
| | Electrochemical | $NiCo_2S_4$ reticulated spheres in carbon electrode | Acetylcholinesterase enzyme | Deng et al. (2016) |
| Acetamiprid | Electrochemical | Gold NP[c] + MWCNT[d] + rGO[e] nanoribbons | Aptamers | Fei et al. (2015) |
| | Electrochemical | Silver NP[c] + nitrogen doped GO[f] | Aptamers | Jiang et al. (2015) |
| | Optical | Gold NP[c] | Aptamers | Shi et al. (2013) |
| Dichlorvos | Optical | QD[g] + acetylcholine | Enzyme (AChE[c] + ChO[e]) | Meng et al. (2013) |
| | Electrochemical | Platinum electrode + ZnO | Acetylcholinesterase enzyme | Sundarmurugasan et al. (2016) |
| | Electrochemical | Ionic liquids-gold NP[c] + porous carbon comp | Acetylcholinesterase enzyme | Wei et al. (2014) |
| Chlorpyrifos | Electrochemical | SPCE[h] and IrOx NP[c] | Enzyme (tyrosinase) | Mayorga-Martinez et al. (2014) |
| | Electrochemical | Boron-doped diamond electrode + gold NP[c] | Acetylcholinesterase enzyme | Wei et al. (2014)<br>Chai et al. (2013) |
| Pirimicarb | Electrochemical | Prussian blue + MWCNT[d] SPE[a] | Acetylcholinesterase enzyme | Oliveira et al. (2013) |
| | Electrochemical | Carbon paste electrode + MWCNT[d] | Laccase enzyme | |
| Atrazine | Electrochemical | Gold NP[c] | Monoclonal antibodies | Liu et al. (2014a, b) |
| | Electrochemical | SWCNT[i] | Monoclonal antibodies | Belkhamssa et al. (2016a, b) |
| | Electrochemical | Magnetic beads + G protein | Phage–antibody complex | González-Techera et al. (2015) |

| | | | | |
|---|---|---|---|---|
| Carbofuran | Electrochemical | IrOx-chitosan nanocomposite | Acetylcholinesterase enzyme | Jeyapragasam and Saraswathi (2014) |
| | Electrochemical | GCE[j] with GO[f] + MWCNT[d] | Acetylcholinesterase enzyme | Li et al. (2017) |
| | Electrochemical | GCE[j] + NiO NP[c] + COOH graphene-Nafion | Acetylcholinesterase enzyme | Yang et al. (2013) |
| Fenitrothion | Electrochemical | Carbon paste electrode | Whole cell (Pseudomonas putida JS444) | Lei et al. (2007) |

[a]*SPE* screen printed electrode
[b]*SAM* self-assembled monolayer
[c]NP nanoparticles
[d]*MWCNT* multiwalled carbon nanotubes
[e]*rGO* reduced graphene oxide
[f]*GO* graphene oxide
[g]*QD* quantum dots
[h]*SPCE* screen-printed carbon electrode
[i]*SWCNT* single-walled carbon nanotubes
[j]*GCE* glassy carbon electrode

(Meng et al. 2013), a platinum electrode modified with acetylcholinesterase-zinc oxide (Sundarmurugasan et al. 2016) and a composite of ionic liquids, gold nanoparticles, and porous carbon (Peng et al. 2017). The combination of ionic liquid, porous carbon, and gold nanoparticles improves the adsorption of enzyme, preserves the activity of enzyme, and enhances the sensitivity of the analysis, Table 4.

### 5.1.2 Other Types of Pesticides

Acetamiprid analysis in water samples has been achieved by impedimetric aptasensors (Fei et al. 2015). Gold nanoparticles, a composite composed of reduced graphene oxide nanoribbons along with multiwalled carbon nanotubes were utilized, which cause higher electron transfer with overall improvement of analytical performance (Fei et al. 2015). Atrazine was analyzed in crop samples using an electrochemical immunosensor based on gold nanoparticles (Liu et al. 2014a, b) and in seawater/river water samples using a disposable immunosensor with single-walled carbon nanotubes (Belkhamssa et al. 2016a, b). A novel recognition element formed by recombinant complex of antibody/ M13 phage and G protein functionalized magnetic beads was used in fabrication of atrazine electrochemical immunosensor (González-Techera et al. 2015).The biosensor displayed a boosted detection limit due to high sensitivity of phage/antibody complex (González-Techera et al. 2015).

Enzymatic biosensors were used to determine pirimicarb using enzymatic laccase and multiwalled carbon nanotubes on carbon paste electrode composite (Chai et al. 2013). Carbofuran, a carbamate insecticide, was determined by acetylcholinesterase biosensor immobilized onto iron oxide–chitosan nanocomposite with square wave voltammetry (Jeyapragasam and Saraswathi 2014). A superior limit of detection was achieved by immobilizing acetylcholinesterase using modified electrode with (nickel oxide +Nafion+ carboxylic graphene) to detect carbofuran in a mixture with methyl parathion and chlorpyrifos (Yang et al. 2013). It was suggested that the conjugation of nickel oxide nanoparticles and carboxylic graphene decrease oxidation peak potential and increase the electron transfer (Yang et al. 2013), Table 4.

## *5.2 Metals*

Nano-enabled sensors have been effectively utilized to detect and monitor a number of heavy metals in different environmental matrices. From these metals are mercury, lead, chromium, and cadmium.

### 5.2.1 Mercury

There is usually an extensive research concern for Mercury detection due to its known negative neurological effects to humans (Selid et al. 2009). Production of DNA-based probes is a main emphasis of mercury nanosensor development (Liu

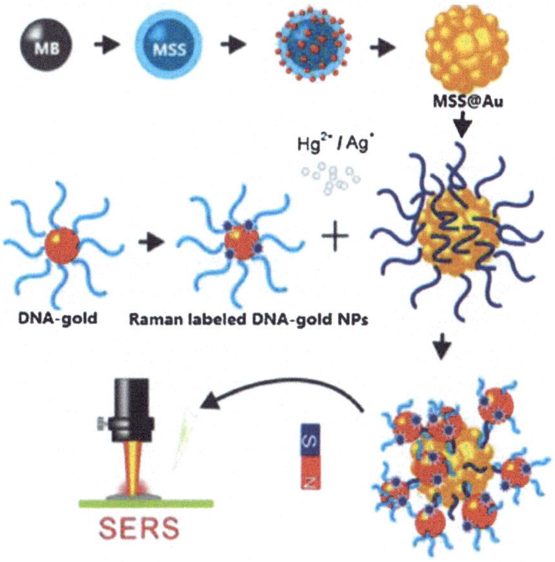


**Fig. 6** Schematic of surface enhanced Raman spectroscopy (SERS)-active system for Hg*II* ion detection based on T–Hg–T bridges using DNA-Au NPs and DNA-MSS@Au NPs (Reprinted with permission from Liu et al. 2014a, b Copyright 2014 American Chemical Society)

et al. 2014a, b). In DNA, the base mismatch of thymine–thymine are of significant finding in case of mercury presence due to metal base pairs formation. Many mercury sensors have been fabricated with several nanomaterials, for example using mercury sandwich type assay (Liu et al. 2014a, b) where spheres of magnetic silica are captured into gold shell along with gold nanoparticles attached to complementary sequences of DNA containing five positions of mismatched thymine with an insufficient binding energy not to allow complete hybridization, as shown in Fig. 6. In the presence of mercury, full hybridization took place, hence reducing the inter-probe spacing and generating a plasmonic hotspot. Thiol based assays for the detection of mercury have been severally using various nanoparticles, for instance, gold (Chansuvarn et al. 2015), silver (Alam et al. 2015) or quantum dots (Ke et al. 2012). Likewise, a competition-based reaction assay wherein surface coating is replaced with mercury has been designated (Huang and Chang 2006). It was reported that thiol coatings improved the assay specificity.

### 5.2.2 Lead

Lead can result in higher risk of different types of cancer and neurological problems (Goyer 1990) consequently, lead detection is considered as a chief concern. Various nanosensors have been constructed for lead analysis. 8–17 DNAzyme and catalytic nucleic acid were used for label-based lead detection (Tang et al. 2013), in addition to a class of oligonucleotides capable of formation of G-quadruplexes in lead presence (Li et al. 2010). An AlOOH–graphene oxide nanocomposite for detection of lead and other metals like cadmium by voltammetry was investigated (Gao et al. 2012), the great adsorption capacity of AlOOH to form a composite results in the electron transfer kinetics. The AlOOH does not show good selectivity for a single metal; thus, the AlOOH–graphene oxide nanocomposite is considered a good option for multiplex detection.

### 5.2.3 Chromium

High chromium absorption can result in several health problems, for example, airway hypersensitivity, nasal and lung cancer, and other types of tumors and fibroproliferative diseases, (Gibb et al. 2000). Several immunoassays have been suggested for chromium detection (Liu et al. 2012). A new anti-CrIII-EDTA antibody was developed for chromium detection via immunochromatographic assay. In order to ensure evoking an immune response, chromium ions were mixed with bifunctional chelating agent, isothiocyanobenzyl-EDTA then linked to a carrier protein to reach an appropriate size for ultimate, sensitive, and rapid detection (Liu et al. 2012).

### 5.2.4 Cadmium

A diversity of nanomaterials have been investigated for cadmium detection, this includes quantum dots (Gui et al. 2012, 2013), SWCNT (Sun et al. 2007), and antimony nanoparticles (Toghill et al. 2009). An off/on- fluorescence sensor for cadmium detection was designed (Gui et al. 2012,). First, Photoluminescent CdTe/CdS quantum dots were quenched using ammonium pyrrolidine dithiocarbamate, upon introducing cadmium ions, it displaced the quenching agent from the QD surface and reinstated the photoluminescence; therefore, turning on the sensor. These sensors showed high selectivity for cadmium, and to improve their accuracy, a ratiometric sensor was later developed (Luan et al. 2012).

## 5.3 *Toxins*

Many sensors were constructed for detection and monitoring of various toxins (Eissa et al. 2015). Brvetoxin was detected with a sensitive and selective aptasensor by gold electrodes modified via cysteamine SAM (Eissa et al. 2015). A further detection of brevetoxin and saxitoxin was reported using cardiomyocyte-based portable biosensor (Wang et al. 2015). Cardiomyocytes provided real-time screening of pathogens in a rapid manner detection. Microcystin-LR detection in water samples was performed using graphene electrochemical impedance spectroscopy immunosensor and showed good reproducibility and repeatability results (Zhang et al. 2017a, b). A better detection limit for microcystin-LR detection in the studies samples of lake water was achieved using immunosensor with gold electrodes modified with a molybdenum disulfide–gold nanorod composite (Zhang et al. 2017a, b). This improvement could be explained by the augmented effect of gold nanorods and

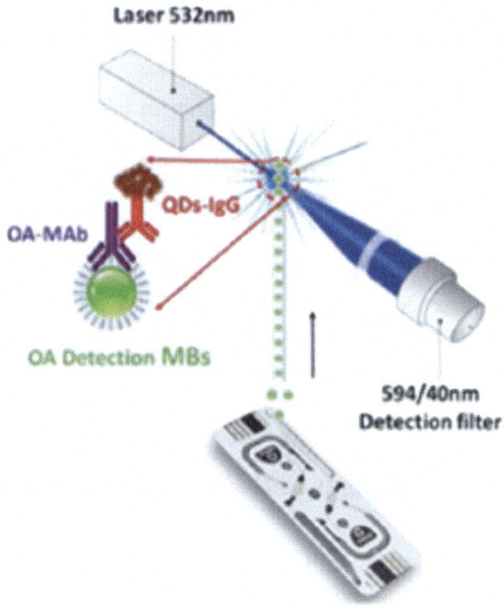


**Fig. 7** Working principle of fluorescence immunosensor for detection of okadaic acid reprinted from Pan et al. (2017)

molybdenum disulfide that endorsed larger surface area, increasing electrical conductivity and overall electrochemical performance.

Okadaic acid toxin in algal, shellfish and seawater can be detected using different biosensors (McNamee et al. 2013, Pan et al. 2017). For okadaic acid, saxitoxin, and domoic acid determination in seawater and algal samples, fabrication of multiplex surface plasmon resonance biosensor was reported (McNamee et al. 2013). A multiplex immunological method was developed to be used as an early warning tool for variable marine biotoxin monitoring in seawater samples. A highly sensitive fluorescence-based immunosensor using CdTe QDs and carboxylic acid modified magnetic beads was investigated for the analysis of okadaic acid in mussel (Pan et al. 2017). An illustration of the fluorescence immunosensor working principle is shown in Fig. 7.

To detect domoic acid toxin, carbon nanotube disposable immunosensors were fabricated and showed good reproducibility and low limit of detection when analyzing the seawater samples (Marques et al. 2017). For improving the detection limit an underwater surface plasmon resonance biosensor was designed (Colas et al. 2016); in addition, it also allows for in situ quantitative determination of domoic acid in seawater.

## 5.4 *Endocrine Disrupting Chemicals*

Detection of bisphenol A, an endocrine disrupting chemical, in water samples was achieved by using fluorescence aptasensors (functionalized by fluorescein amidite) and gold nanoparticles (Ragavan et al. 2013). A portable, fast, and cost-effective evanescent-wave optical fiber aptasensor was successfully used for sensitive detection of bisphenol A in water samples with no need of pretreatment procedures (Yildirim et al. 2014).

17β-estradiol was monitored in lake water using CdSe nanoparticles and $TiO_2$ nanotube arrays aptasensor with an outstanding selectivity and femtomolar level of detection. The particular recognition reaction takes place between the 17-estradiol and aptamer and leads to an increase in the steric hindrances to the electron donor diffusion, which leads to a decrease of the photocurrent, thereby resulting in in situ complex formation. The superior selectivity toward 17-estradiol could be credited to the excellent photoelectrical activity, the tubular microstructure of sensing interface, high affinity of the aptamer to 17β-estradiol, and large packing density of aptamer (Fan et al. 2014).

## 6 Challenges and Future Perspectives

It is obvious that the future trends and the forthcoming advances of sensors/biosensors will count on the achievements of evolving micro- and nano-level technologies including electronics, materials science, physics, and biochemistry. Since environmental pollution within various media is growing incredibly fast and becoming a severe global fear, designing and developing novel biosensor-based techniques capable of precisely identifying and analyzing different pollutants from a larger spectrum is of high significance. Nevertheless, sensors/biosensors for environmental monitoring have some limitations including response time, selectivity, sensitivity, stability, and lifetime. Researchers have to collaborate in work to eliminate these limitations for effective eco-friendly, reliable, competitive tools of analysis.

It is also clear that need for speedy detecting tools like biosensors will grow in the upcoming future. Despite the current and past significant research and efforts in electrochemical sensor development field, a challenge to improve devices to evade instrumental errors and improve reliability in complex matrices is still crucial. Sensors have to fulfil all demands for proper, robust, and sustainable monitoring by being integrated networks, offering screening and analysis of complex mixtures of multi-analytes and providing remote sensing through combining with wireless signal transmitters. In this context, comprehensive studies and research are required in the fields of biosensors, bioelectronics, and bionanotechnology that will definitely have a manifest impact on expanding innovative biosensing approaches, putting into consideration being "smart" and user friendly to fit more into the future.

## 7 Conclusion

Researchers' effort is being focused in the direction of designing and developing reliable, sustainable, and more sensitive methods of analysis that are able to detect, monitor, and eliminate noxious environmental contaminants. Being able to give more sensitive, selective, robust, and low-cost results in addition of being eco-friendly, sensors are rapidly becoming a vital deliberation in all environmental screening and monitoring programs. Shifts from classical harmful mercury-based electrodes with low sensitivity and stability and lifetime problems to disposable electrodes with inert green materials, improved stability are among the essential routes explored. Consequently, selection of appropriate sensing and modifying materials, miniaturization and green synthesis approaches are of supreme importance toward sustainability of environmental analysis. Together with methods optimization, reliable analyses should guard public health and improve the quality of the environment.

## References


Ahmad N, Sharma S, Alam MK et al (2010) Rapid synthesis of silver nanoparticles using dried medicinal plant of basil. Colloids Surf B Biointerfaces 81:81–86. https://doi.org/10.1016/j.colsurfb.2010.06.029

Alam A, Ravindran A, Chandran P, Sudheer Khan S (2015) Highly selective colorimetric detection and estimation of $Hg^{2+}$ at nano-molar concentration by silver nanoparticles in the presence of glutathione. Spectrochim Acta A Mol Biomol Spectrosc 137:503–508

Arain S, John GT, Kranse C, Gerlach J, Wolfbeis OS, Klimant I (2006) Characterization of microtiterplates with integrated optical sensors for oxygen and pH, and their applications to enzyme activity screening, respirometry, and toxicological assays. Sensors Actuators B 113:639–648

Arduini F, Guidone S, Amine A, Palleschi G, Moscone D (2013) Acetylcholinesterase biosensor based on self-assembled monolayer-modified gold-screen printed electrodes for organophosphorus insecticide detection. Sensors Actuators B Chem 179:201–208

Arduini F, Forchielli M, Amine A, Neagu D, Cacciotti I, Nanni F, Moscone D, Palleschi G (2015) Screen-printed biosensor modified with carbon black nanoparticles for the determination of paraoxon based on the inhibition of butyrylcholinesterase. Microchim Acta 182:643–651

Azmuddin AM, Khan AA, Ajab H (2017) Environmental monitoring by eco-friendly fabricated carbon-modified electrode sensor. Int J Biosens Bioelectron 2(5):141–144

Basavaraja S, Balaji SD, Lagashetty A et al (2008) Extracellular biosynthesis of silver nanoparticles using the fungus Fusarium semitectum. Mater Res Bull 43:1164–1170. https://doi.org/10.1016/j.materresbull.2007.06.020

Belkhamssa N, da Costa JP, Justino CIL et al (2016a) Development of an electrochemical biosensor for alkylphenol detection. Talanta 158:30–34

Belkhamssa N, Justino CIL, Santos PSM, Cardoso S et al (2016b) Label-free disposable immunosensor for detection of atrazine. Talanta 146:430–434

Beni V, Ogurtsov V, Bakunin N et al (2005) Development of a portable electroanalytical system for the stripping voltammetry of metals: determination of copper in acetic acid soil extracts. Anal Chim Acta 552:190–200

Bies C, Lehr CM, Woodley JF (2004) Lectin-mediated drug targeting: history and applications. Adv Drug Deliv Rev 56:425–435

Bordeaux J, Welsh AW, Agarwal S, Killiam E, Baquero MT, Hanna JA, Anagnostou VK, Rimm DL (2010) Antibody validation. BioTechniques 48:197–209

Chai Y, Niu X, Chen C, Zhao H, Lan M (2013) Carbamate insecticide sensing based on acetylcholinesterase/Prussian blue-multi-walled carbon nanotubes/screen-printed electrodes. Anal Lett 46:803–817

Chandran SP, Chaudhary M, Pasricha R et al (2006) Synthesis of gold nanotriangles and silver nanoparticles using aloe vera plant extract. Biotechnol Prog. https://doi.org/10.1021/bp0501423

Chansuvarn W, Tuntulani T, Imyim A (2015) Colorimetric detection of mercury(II) based on gold nanoparticles, fluorescent gold nanoclusters and other gold-based nanomaterials. TrAC Trends Anal Chem 65:83–96

Chen L, Gorski W (2001) Bioinorganic composites for enzyme electrodes. Anal Chem 73(13):2862–2868

Chen Y-L, Tuan H-Y, Tien C-W et al (2009) Augmented biosynthesis of cadmium sulfide nanoparticles by genetically engineered *Escherichia coli*. Biotechnol Prog 25:1260–1266. https://doi.org/10.1002/btpr.199

Choi HS, Kim HD (2009) Development of a portable heavy metal ion analyzer using disposable screen-printed electrodes. Bull Kor Chem Soc 30(8):1881–1883

Colas F, Crassous M-P, Laurent S et al (2016) A surface plasmon resonance system for the underwater detection of domoic acid. Limnol Oceanogr Methods 14:456–465

Cooper J, Bolbot J, Saini S et al (2007) Electrochemical method for the rapid on site screening of cadmium and lead in soil and water samples. Water Air Soil Pollut 79(1):183–195

Das RN, Lin HT, Lauffer JM et al (2011) Printable electronics: towards materials development and device fabrication. Circuit World 37:38–45

Deng Y, Liu K, Liu Y, Dong H, Li S (2016) An novel acetylcholinesterase biosensor based on nano-porous pseudo carbon paste electrode modified with gold nanoparticles for detection of methyl parathion. J Nanosci Nanotechnol 16:9460–9467

Du L, Jiang H, Liu X, Wang E (2007) Biosynthesis of gold nanoparticles assisted by *Escherichia coli* DH5α and its application on direct electrochemistry of hemoglobin. Electrochem Commun 9:1165–1170. https://doi.org/10.1016/j.elecom.2007.01.007

Eissa S, Siaj M, Zourob M (2015) Aptamer-based competitive electrochemical biosensor for brevetoxin-2. Biosens Bioelectron 69:148–154

El Mhammedi MA, Bakasse M, Chtaini A (2007) Electrochemical studies and square wave voltammetry of paraquat at natural phosphate modified carbon paste electrode. J Hazard Mater 145(1–2):1–7

Ellington AA, Kullo IJ, Bailey KR, Klee GG (2010) Antibody-based protein multiplex platforms: technical and operational challenges. Clin Chem 56:186–193

Evtugyn GA, Budnikov HC, Nikolskaya EB (1998) Sensitivity and selectivity of electrochemical enzyme sensors for inhibitor determination. Talanta 46:465–484

Fan L, Zhao G, Shi H et al (2014) A femtomolar level and highly selective 17-estradiol photoelectrochemical aptasensor applied in environmental water samples analysis. Environ Sci Technol 48:5754–5761

Fei A, Liu Q, Huan J, Qian J, Dong X et al (2015) Label-free impedimetric aptasensor for detection of femtomole level acetamiprid using gold nanoparticles decorated multiwalled carbon nanotube-reduced graphene oxide nanoribbon composites. Biosens Bioelectron 70:122–129

Freire R, Duran N, Kubota L (2002) Electrochemical biosensor-based devices for continuous phenols monitoring in environmental matrices. J Braz Chem Soc 13(4):456–462

Gao C, Yu XY, Xu RX, Liu JH, Huang XJ (2012) AlOOH-reduced graphene oxide nanocomposites: one-pot hydrothermal synthesis and their enhanced electrochemical activity for heavy metal ions. ACS Appl Mater Interfaces 4:4672–4682

Gardea-Torresdey JL, Parsons JG, Gomez E et al (2002) Formation and growth of Au nanoparticles inside live alfalfa plants. Nano Lett 2:397–401. https://doi.org/10.1021/nl015673

Gericke M, Pinches A (2006) Microbial production of gold nanoparticles. Gold Bull 9:22–28. https://doi.org/10.1007/BF03215529

Gibb HJ, Lees PS, Pinsky PF, Rooney BC (2000) Lung cancer among workers in chromium chemical production. Am J Ind Med 38:115–126

Gill I, Ballesteros A (2000) Bioencapsulation within synthetic polymers (Part 2): non-sol-gel protein-polymer biocomposites. Trends Biotechnol 18(11):469–479

González-Techera A, Zon MA, Molina PG et al (2015) Development of a highly sensitive noncompetitive electrochemical immunosensor for the detection of atrazine by phage anti-immunocomplex assay. Biosens Bioelectron 64:650–656

Goyer RA (1990) Lead toxicity: from overt to subclinical to subtle health effects. Environ Health Perspect 86:177–181

Grieshaber D, MacKenzie R, Vörös J, Reimhult E (2008) Electrochemical biosensors—sensor principles and architectures. Sensors 8:1400–1458

Gui R, An X, Su H, Shen W, Chen Z, Wang X (2012) A near-infrared-emitting CdTe/CdS core/shell quantum dots-based OFF–ON fluorescence sensor for highly selective and sensitive detection of $Cd^{2+}$. Talanta 94:257–262

Gui R, An X, Huang W (2013) An improved method for ratiometric fluorescence detection of pH and $Cd^{2+}$ using fluorescein isothiocyanatequantum dots conjugates. Anal Chim Acta 767:134–140

Guo L, Li Z, Chen H, Wu Y, Chen L, Song Z, Lin T (2017) Colorimetric biosensor for the assay of paraoxon in environmental water samples based on the iodine-starch color reaction. Anal Chim Acta 967:59–63

Hamula CLA, Zhang H, Li F, Wang Z, Le Chris X, Li X-F (2011) Selection and analytical applications of aptamers binding microbial pathogens. TrAC Trends Anal Chem 30:1587–1597

Hart J, Abass A, Cowell D (2002) Development of disposable amperometric sulfur dioxide biosensors based on screen printed electrodes. Biosens Bioelectron 17(5):389–394

Hoinka J, Berezhnoy A, Dao P, Sauna ZE, Gilboa E, Przytycka TM (2015) Large scale analysis of the mutational landscape in HT-SELEX improves aptamer discovery. Nucleic Acids Res 43:5699–5707

Huang CC, Chang H (2006) Selective gold-nanoparticle-based "turn-on" fluorescent sensors for detection of mercury (II) in aqueous solution. Anal Chem 78:8332–8338

Huang CC, Huang YF, Cao Z, Tan W, Chang HT (2005) Aptamer-modified gold nanoparticles for colorimetric determination of platelet-derived growth factors and their receptors. Anal Chem 77:5735–5741

Iravani S (2014, 2014) Bacteria in nanoparticle synthesis: current status and future prospects. Int Sch Res Not:1–18. https://doi.org/10.1155/2014/359316

Jeyapragasam T, Saraswathi R (2014) Electrochemical biosensing of carbofuran based on acetylcholinesterase immobilized onto iron oxide-chitosan nanocomposite. Sensors Actuators B Chem 191:681–687

Jiajie L, Hongwu L, Caifeng L, Qiangqiang F, Caihong H, Zhi L, Tianjiu J, Yong T (2014) Silver nanoparticle enhanced Raman scattering-based lateral flow immunoassays for ultra-sensitive detection of the heavy metal chromium. Nanotechnology 25:495501

Jiang D, Du X, Liu Q et al (2015) Silver nanoparticles anchored on nitrogen-doped graphene as a novel electrochemical biosensing platform with enhanced sensitivity for aptamer-based pesticide assay. Analyst 140:6404–6411

Ke J, Li X, Shi Y, Zhao Q, Jiang X (2012) A facile and highly sensitive probe for Hg(II) based on metal-induced aggregation of ZnSe/ZnS quantum dots. Nanoscale 4:4996–5001

Khan AAA, Abdullah MA (2014) Bismuth-modified hydroxyapatite carbon electrode for simultaneous *in-situ* cadmium and lead analysis. Int J Electrochem Sci 8:195–203

Kimmel DW, LeBlanc G, Meschievitz ME, Cliffel DE (2011) Electrochemical sensors and biosensors. Anal Chem 84:685–707

Koets M, van der Wijk T, van Eemeren JTWM, van Amerongen A, Prins MWJ (2009) Rapid DNA multi-analyte immunoassay on a magneto-resistance biosensor. Biosens Bioelectron 24:1893–1898

Kowshik M, Vogel W, Urban J et al (2002) Microbial synthesis of semiconductor PbS nanocrystallites. Adv Mater 14:815–818. https://doi.org/10.1002/1521-4095(20020605)14:11%3c815:AID-ADMA815%3e3.0.CO;2-K

Kwon HSP, Kil Yoont J, Seo ML (2000) Plant tissue-based amperometric sensor for determination of phenols in methylene chloride. J Korean Chem Soc 44(4):376–379

Lee S-M, Lee W-Y (2002) Determination of heavy metal ions using conductometric biosensor based on sol-gel-immobilized urease. Bull Kor Chem Soc 23(8):1169–1172

Lei Y, Mulchandani P et al (2007) Biosensor for direct determination of fenitrothion and EPN using recombinant Pseudomonas putida JS444 with surface-expressed organophosphorous hydrolase. 2. Modified carbon paste electrode. Appl Biochem Biotechnol 136:243–245

Li T, Wang E, Dong S (2010) Lead(II)-induced allosteric G-quadruplex DNAzyme as a colorimetric and chemiluminescence sensor for highly sensitive and selective $Pb^{2+}$ detection. Anal Chem 82:1515–1520

Li Z, Qu S, Cui L, Zhang S (2017) Detection of carbofuran pesticide in seawater by using an enzyme biosensor. J Coast Res 80:1–5

Link S, El-Sayed MA (1999) Spectral properties and relaxation dynamics of surface plasmon electronic oscillations in gold and silver nanodots and nanorods. J Phys Chem B 103:8410–8426

Liu X, Xiang JJ, Tang Y, Zhang XL, Fu QQ, Zou JH, Lin Y (2012) Colloidal gold nanoparticle probe-based immunochromatographic assay for the rapid detection of chromium ions in water and serum samples. Anal Chim Acta 745:99–105

Liu S, Zheng Z, Li X (2013) Advances in pesticide biosensors: current status, challenges, and future perspectives. Anal Bioanal Chem 405:63–90

Liu X, Li W-J, Yang Y, Mao L-G, Peng ZA (2014a) Label-free electrochemical immunosensor based on gold nanoparticles for direct detection of atrazine. Sensors Actuators B Chem 191:408–414

Liu M, Wang Z, Zong S, Chen H, Zhu D, Wu L, Hu G, Cui Y (2014b) SERS detection and removal of mercury(II)/silver(I) using oligonucleotide-functionalized core/shell magnetic silica Sphere@Au nanoparticles. ACS Appl Mater Interfaces 6:7371–7379

Low SY, Hill JE, Peccia J (2009) DNA aptamers bind specifically and selectively to (1→3)-β-d-glucans. Biochem Biophys Res Commun 378:701–705

Luan W, Yang H, Wan Z, Yuan B, Yu X, Tu S-T (2012) Mercaptopropionic acid capped CdSe/ZnS quantum dots as fluorescence probe for lead(II). J Nanopart Res 14:1–8

Ma J, Chen Y, Hou Z, Jiang W, Wang L (2013) Selective and sensitive mercuric(II) ion detection based on quantum dots and nicking endonuclease assisted signal amplification. Biosens Bioelectron 43:84–87

Marques I, da Costa JP, Justino C, Santos P, Duarte K et al (2017) Carbon nanotube field effect biosensor for the detection of toxins in seawater. J Environ Anal Chem 97:597–605

Matsushita K, Toyama H, Yamada M, Adachi O (2002) Quinoproteins: structure, function, and biotechnological applications. Appl Microbiol Biotechnol 58(1):13–22

Mayorga-Martinez C, Pino F, Kurbanoglua S, Rivas L, Ozkan SA, Merkoci A (2014) Iridium oxide nanoparticles induced dual catalytic/inhibition based detection of phenol and pesticide compounds. J Mater Chem B 2:2233–2239

McNamee SE, Elliott CT, Delahaut P, Campbell K (2013) Multiplex biotoxin surface plasmon resonance method for marine biotoxins in algal and seawater samples. Environ Sci Pollut Res 20:6794–6807

Meng X, Wei J, Ren X, Ren J, Tang F (2013) A simple and sensitive fluorescence biosensor for detection of organophosphorus pesticides using $H_2O_2$-sensitive quantum dots/bi-enzyme. Biosens Bioelectron 47:402–407

Mohanpuria P, Rana NK, Yadav SK (2008) Biosynthesis of nanoparticles: technological concepts and future applications. J Nanopart Res 10:507–517

Mourato A, Gadanho M, Lino AR, Tenreiro R (2011) Biosynthesis of crystalline silver and gold nanoparticles by extremophilic yeasts. Bioinorg Chem Appl 1:1. https://doi.org/10.1155/2011/546074

Mukherjee P, Ahmad A, Mandal D et al (2001) Fungus-mediated synthesis of silver nanoparticles and their immobilization in the mycelial matrix: a novel biological approach to nanoparticle synthesis. Nano Lett 1:515–519. https://doi.org/10.1021/nl0155274

Narayanan KB, Sakthivel N (2011) Synthesis and characterization of nanogold composite using Cylindrocladium floridanum and its heterogeneous catalysis in the degradation of 4-nitrophenol. J Hazard Mater 189:519–525. https://doi.org/10.1016/j.jhazmat.2011.02.069

Nazir MS, Wahjoedi BA, Yussof AW, Abdullah MA (2013) Eco-friendly extraction, characterization and modification of microcrystalline cellulose from oil palm empty fruit bunches. BioRes 8(2):2161–2172

Nunes GS, Lins JAP, Silva FGS, Araujo LC et al (2014) Design of a macroalgae amperometric biosensor; application to the rapid monitoring of organophosphate insecticides in an agroecosystem. Chemosphere 111:623–630

Olaniran AO, Hiralal L, Pillay B (2011) Whole-cell bacterial biosensors for rapid and effective monitoring of heavy metals and inorganic pollutants in wastewater. J Environ Monit 13:2914–2920

Oliveira TMBF, Barroso MF, Morais S et al (2013) Biosensor based on multi-walled carbon nanotubes paste electrode modified with laccase for pirimicarb pesticide quantification. Talanta 106:137–143

Pan Y, Zhou J, Su K, Hu N, Wang P (2017) A novel quantum dot fluorescence immunosensor based on magnetic beads and portable flow cytometry for detection of okadaic acid. Procedia Technol 27:214–216

Peng L, Dong S, Wei W, Yuan X, Huang T (2017) Synthesis of reticulated hollow spheres structure $NiCo_2S_4$ and its application in organophosphate pesticides biosensor. Biosens Bioelectron 92:563–569

Ragavan KV, Selvakumar LS, Thakur MS (2013) Functionalized aptamers as nano-bioprobes for ultrasensitive detection of bisphenol-A. Chem Commun 49:5960–5962

Raliya R, Tarafdar JC (2014) Biosynthesis and characterization of zinc, magnesium and titanium nanoparticles: an eco-friendly approach. Int Nano Lett 4:93. https://doi.org/10.1007/s40089-014-0093-8

Renedo OD, Alonso-Lomillo MA, Martínez MJ (2007) Recent developments in the field of screen-printed electrodes and their related applications. Talanta 73(2):202–219

Rico MA, Olivares-Marín M, Gil EP (2009) Modification of carbon screen-printed electrodes by adsorption of chemically synthesized Bi nanoparticles for the voltammetric stripping detection of Zn(II), Cd(II) and Pb(II). Talanta 80(2):631–635

Rios A, Escarpa A, González MC, Crevillén AG (2006) Challenges of analytical microsystems. Trends Anal Chem 25:467–479

Riu J, Maroto A, Ruis FX (2006) Nanosensors in environmental analysis. Talanta 69:288–301

Romo-Herrera JM, Alvarez-Puebla RA, Liz-Marzan LM (2011) Controlled assembly of plasmonic colloidal nanoparticle clusters. Nanoscale 3:1304–1315

Saha K, Agasti SS, Kim C, Li X, Rotello VM (2012) Gold nanoparticles in chemical and biological sensing. Chem Rev 112:2739–2779

Schulze H, Schmid R, Bachmann T (2002) Rapid detection of neurotoxic insecticides in food using disposable acetylcholinesterase biosensors and simple solvent extraction. Anal Bioanal Chem 372(2):268–272

Selid P, Xu H, Collins EM, Striped Face-Collins M, Zhao JX (2009) Sensing mercury for biomedical and environmental monitoring. Sensors 9:5446–5459

Shankar SS, Rai A, Ahmad A, Sastry M (2004) Rapid synthesis of Au, Ag, and bimetallic Au core Ag shell nanoparticles using Neem (Azadirachta indica) leaf broth. J Colloid Interface Sci 1:1. https://doi.org/10.1016/j.jcis.2004.03.003

Shi H, Zhao G, Liu M, Fan L, Cao T (2013) Aptamer-based colorimetric sensing of acetamiprid in soil samples: sensitivity, selectivity and mechanism. J Hazard Mater 260:754–761

Shitanda I, Irisako T, Itagaki M (2011) Three-electrode type micro-electrochemical cell fabricated by screen-printing. Sensors Actuators B Chem 160:1606–1609

Singh J, Dutta T, Kim K, Rawat M, Samddar P, Kumar P (2018) Green' synthesis of metals and their oxide nanoparticles: applications for environmental remediation. J Nanobiotechnol 16:84

Sinibaldi F, Bongiovanni C, Ferri T, Santucci R (2001) Trends Inorg Chem 7:77–87

Sun D, Xie X, Cai Y, Zhang H, Wu K (2007) Voltammetric determination of $Cd^{2+}$ based on the bifunctionality of single-walled carbon nanotubes-Nafion film. Anal Chim Acta 581:27–31

Sundarmurugasan R, Gumpu MB, Ramachandra BL et al (2016) Simultaneous detection of monocrotophos and dichlorvos in orange samples using acetylcholinesterase-zinc oxide modified platinum electrode with linear regression calibration. Sensors Actuators B Chem 230:306–313

Svancara I, Walcarius A, Kalcher K et al (2009) Carbon paste electrodes in the new millennium. Cent Eur J Chem 7(4):598–656

Tang S, Tong P, Li H, Tang J, Zhang L (2013) Ultrasensitive electrochemical detection of $Pb^{2+}$ based on rolling circle amplification and quantum dots tagging. Biosens Bioelectron 42:608–611

Thakkar KN, Mhatre SS, Parikh RY (2010) Biological synthesis of metallic nanoparticles. Nanomed Nanotechnol Biol Med 6:257–262

Toghill KE, Xiao L, Wildgoose GG, Compton RG (2009) Electroanalytical determination of cadmium(II) and lead(II) using an antimony nanoparticle modified boron-doped diamond electrode. Electroanalysis 21:1113–1118

Tombelli S, Mascini M, Turner A (2002) Improved procedures for immobilization of oligonucleotides on gold coated piezoelectric quartz crystals. Biosens Bioelectron 17(11–12):929–936

Vikesland PJ, Wigginton KR (2010) Nanomaterial enabled biosensors for pathogen monitoring—a review. Environ Sci Technol 44:3656–3669

Vollmer C, Redel E, Abu-Shandi K et al (2010) Microwave irradiation for the facile synthesis of transition-metal nanoparticles (NPs) in ionic liquids (ILs) from metal-carbonyl precursors and Ru-, Rh-, and Ir-NP/IL dispersions as biphasic liquid-liquid hydrogenation nanocatalysts for cyclohexene. Chem A Eur J 16:3849–3858. https://doi.org/10.1002/chem.200903214

Wadhwani SA, Shedbalkar UU, Singh R, Chopade BA (2016) Biogenic selenium nanoparticles: current status and future prospects. Appl Microbiol Biotechnol 100:2555–2566

Wang Q, Fang J, Cao D, Li H et al (2015) An improved functional assay for rapid detection of marine toxins, saxitoxin and brevetoxin using a portable cardiomyocyte-based potential biosensor. Biosens Bioelectron 72:10–17

Wei M, Zeng G, Lu Q (2014) Determination of organophosphate pesticides using an acetylcholinesterase-based biosensor based on a boron-doped diamond electrode modified with gold nanoparticles and carbon spheres. Microchim Acta 181:121–127

Wen L, Lin Z, Gu P et al (2009) Extracellular biosynthesis of monodispersed gold nanoparticles by a SAM capping route. J Nanoparticle Res 11:279–288. https://doi.org/10.1007/s11051-008-9378-z

Yang W, Ratinac KR, Ringer SP, Thordarson P, Gooding JJ, Braet F (2010) Carbon nanomaterials in biosensors: should you use nanotubes or graphene? Angew Chem Int Ed Engl 49:2114–2138

Yang L, Wang G, Liu Y, Wang M (2013) Development of a biosensor based on immobilization of acetylcholinesterase on NiO nanoparticles-carboxylic graphene-nafion modified electrode for detection of pesticides. Talanta 113:135–141

Yildirim N, Long F, He M et al (2014) A portable optic fiber aptasensor for sensitive, specific and rapid detection of bisphenol-A in water samples. Environ Sci Process Impacts 16:1379–1386

Yoosaf K, Ipe BI, Suresh CH, Thomas KG (2007) In situ synthesis of metal nanoparticles and selective naked-eye detection of lead ions from aqueous media. J Phys Chem C 111:12839–12847. https://doi.org/10.1021/jp073923q

Young S, Hart J, Dowman A, Cowell D (2001) The non-specific inhibition of enzymes by environmental pollutants: a study of a model system towards the development of electrochemical biosensor arrays. Biosens Bioelectron 16(9–12):887–894

Yurkov AM, Kemler M, Begerow D (2011) Species accumulation curves and incidence-based species richness estimators to appraise the diversity of cultivable yeasts from beech forest soils. PLoS One 1:1. https://doi.org/10.1371/journal.pone.0023671

Zhang Y, Chen M, Li H et al (2017a) A molybdenum disulfide/gold nanorod composite-based electrochemical immunosensor for sensitive and quantitative detection of microcystin-LR in environmental samples. Sensors Actuators B Chem 244:606–615

Zhang W, Han C, Jia B et al (2017b) A 3D graphene-based biosensor as an early microcystin-LR screening tool in sources of drinking water supply. Electrochim Acta 236:319–327

Zhao Y, Zhang W, Lin Y, Du D (2013) The vital function of $Fe_3O_4$@Au nanocomposites for hydrolase biosensor design and its application in detection of methyl parathion. Nanoscale 5:1121–1126

# Green Synthesis of NanoMaterials for BioSensing

**Juan José García-Guzmán, David López-Iglesias, Dolores Bellido-Milla, José María Palacios-Santander, and Laura Cubillana-Aguilera**

## Contents



J. J. García-Guzmán · D. López-Iglesias · D. Bellido-Milla
J. M. Palacios-Santander (✉) · L. Cubillana-Aguilera (✉)
Faculty of Sciences, Department of Analytical Chemistry, Campus de Excelencia Internacional del Mar (CEIMAR), Institute of Research on Electron Microscopy and Materials (IMEYMAT), University of Cadiz, Puerto Real, Cádiz, Spain
e-mail: juanjo.garciaguzman@uca.es; david.lopeziglesias@uca.es; dolores.milla@uca.es; josem.palacios@uca.es; laura.cubillana@uca.es



Inamuddin, A. M. Asiri (eds.), *Nanosensor Technologies for Environmental Monitoring*, Nanotechnology in the Life Sciences,
https://doi.org/10.1007/978-3-030-45116-5_7

# 1 Introduction

"Chemophobia" (Mckinnon 1981; Kauffman 1991; Chalupa and Nesměrák 2018) (or chemphobia or chemonoia) (Ropeik 2015), an irrational panic or preconception against chemical compounds or chemistry, is a relatively very new term quite widespread, with rather repercussions in food chemistry mainly in both the Western world and Asia (Gribble 2013) but founded in conceptions and apprehensions that have been coming for centuries. This mistrust, fed in part by news media's one-sided stories, natural and technological disasters of different magnitude (Cherry et al. 2018; Liu and Wang 2019) and current environmental problems of anthropogenic origin, such as extreme plastics pollution (Eriksen et al. 2014; Bergmann et al. 2017; Belontz et al. 2019), the greenhouse effect (Stephens et al. 2016; Büntgen et al. 2019), global warming (Trenberth et al. 2014), the erosion of ozone layer (Kuttippurath et al. 2018), and the proliferation of emerging pollutants and wastes (Gavrilescu et al. 2015; Calvo-Flores et al. 2018), among many others, has spread to industry, in general, and to chemical industry, in particular.

Arriving at this point, someone may wonders, would it be possible to imagine a life without chemicals or without things manufactured with them? Would it be possible to imagine our life without the enormous amount of goods generated by Chemical Industry? The answers to these questions are very simple: absolutely not and there would be no life at all, at least as we know it now, respectively. This is due to most manufactured goods involve at least one and in most cases many chemical processes. Chemistry and chemical industry have always had a central role in the provision of food and energy, materials, and medicines. Hence, they form part of our daily life (Fortineau 2004; Roy 2016). Nevertheless, chemical industry must follow a pathway that allows it to recuperate the lost confidence, to claim itself as an essential part of the engine that makes the world go round. The sustainability of these industries and hence of our planet depends on strategic choices made by governments. These entities intend to speed up the development and growth of industries to the transition towards a low-carbon economy by implementing green and strategic industrial policies (Cosbey 2013; Schmitz et al. 2015). That is why during the past few decades, scientific community and industry have been cooperating to make it possible, transmitting a new conception based on concerns that are more positive: that of a much greener, ecological, and sustainable chemical industry that generates environmentally benign products, which would be rather more appreciated by Governments, ONGs, and society in general.

For this purpose, the philosophy of "Green Chemistry" saw the light in the 1980s (Clark 2005). Anastas and Warner defined this term at the end of 1990s as "the design of chemical products and processes that reduce or eliminate the use and generation of hazardous substances" (Anastas and Warner 1998). This term was reformulated and completed by Manley et al. 10 years later as "the design, development, and implementation of chemical products and processes to reduce or eliminate the use and generation of toxic compounds to human health and the environment" (Manley et al. 2008). The Green Chemistry idea was accepted and worldwide extended as an essential development of chemistry and chemical industry. Despite the general acceptance of this philosophy for the sustainable development, its application, not only in developed countries but also in India and China, was fragmented at the beginning, representing only a small part of the chemistry done in the past decades. Currently this concept has influenced education, research, and industrial practice in such a manner that an increasing and significant part of the most environmentally aware society daily demands greener and sustainable products, processes, and methodologies. In fact, Green Chemistry is understood now as a group of actions and attitudes rather than being constrained to chemical analyses based on nontoxic solvents; in other words, it must be considered as multidimensional (Kogawa and Salgado 2015). It is thinking about the process as a whole and reducing steps, energy, reagents, and costs to the minimum (de Marco et al. 2019). This sustainability even reaches Academia, since it is responsible for teaching new university students to gain knowledge on how to safeguard the earth for future generations (Płotka-Wasylka et al. 2018). However, as these authors state, there exist several false "greenness" in chemical literature and teaching practices, what suggests that this proper knowledge must be carefully obtained through what we call a "sustainable and green education," "environmental education" as United States Environmental Protection Agency reports (U.S. Environmental Protection Agency 1992), or "education for sustainable development" as UNESCO calls (Leicht et al. 2018).

As it is well known, the increasing industrialization process was a keystone for world economic evolution. During the second half of the twentieth century, social movements promoted a revolution in Green Chemistry and provoked changes that affected all the industry and its processes, trying making them more sustainable. As a consequence of this, environmental impact and companies and population awareness increased. That is why Paul Anastas and John Warner in the 1990s (Anastas and Warner 1998; Anastas and Eghbali 2010) listed the 12 Green Chemistry principles (Table 1), mainly based on the reduction or rejection of toxic solvents and the non-generation of residues in analyses and chemical processes.

In fact, the core of Green Chemistry can be understood as a set of reduction processes (Fig. 1). The reductions shown in the figure lead to many economic, environmental, and social benefits (ENDS 2003; European Commission 2003). Costs can be saved by reducing by-products and energy use, as well as increasing the efficiency of the whole process due to decreasing materials consumption. The abovementioned reductions also drive to environmental profit in terms of both feedstock depletion and end-of-life disposal. Moreover, the growing employment of renewable resources will make the manufacturing industry more sustainable (ENDS 2004). The minimization of hazardous events and the handling of dangerous substances offers additional social benefits to plant operators and local communities (Clark 2005).

Figure 2 summarizes the main important aspects about Green Chemistry and the role that all actors play for reaching global sustainability.

One of the most important documents regarding industrial sustainability is the Green Industry Initiative, where sustainable industrial development was tried to be

**Table 1** The 12 principles of green chemistry (Anastas and Warner 1998; Anastas and Eghbali 2010)

| N° | Description |
|---|---|
| 1 | **Prevention.** It is better to prevent waste than to treat of clean up waste after it is formed |
| 2 | **Atom economy.** Synthetic methods should be designed to maximize the incorporation of all materials used in the process into the final product |
| 3 | **Less hazardous chemical synthesis.** Whenever practicable, synthetic methodologies should be designed to use and generate substances that pose little or no toxicity to human health and the environment |
| 4 | **Designing safer chemicals.** Chemical products should be designed to preserve efficacy of the function while reducing toxicity |
| 5 | **Safer solvents and auxiliaries.** The use of auxiliary substances (e.g., solvents, separation agents) should be made unnecessary whenever possible and, when used, innocuous |
| 6 | **Design for Energy Efficiency.** Energy requirements of chemical processes should be recognized for their environmental and economic impacts and should be minimized. If possible, synthetic methods should be conducted at ambient temperature and pressure |
| 7 | **Use of renewable Feedstocks.** A raw material of feedstock should be renewable rather than depleting whenever technically and economically practicable |
| 8 | **Reduce derivatives.** Unnecessary derivatization (use of blocking groups, protection/deprotection, temporary modification of physical/chemical processes) should be minimized or avoided if possible, because such steps require additional reagents and can generate waste |
| 9 | **Catalysis.** Catalytic reagents (as selective as possible) are superior to stoichiometric reagents |
| 10 | **Design for Degradation.** Chemical products should be designed so that at the end of their function they break down into innocuous degradation products and do not persist in the environment |
| 11 | **Real-time analysis for pollution prevention.** Analytical methodologies need to be further developed to allow for real-time, in process monitoring and control prior to the formation of hazardous substances |
| 12 | **Inherently safer chemistry for accident prevention.** Substances and the form of a substance used in chemical process should be chosen to minimize the potential for chemical accidents, including releases, explosions, and fires |

placed in the context of new global sustainable development challenges through Green Industry (UNIDO 2011). In fact, in this document the main points in which this initiative is based are highlighted.

In this declaration, firstly the need for Green Industry is discussed; later, Green Industry as a tool for implementing sustainable development is exposed; then, the benefits of Green Industry are described: economic, social, and environmental; in fourth place, the opportunities that Green Industry entails regarding mitigation of climate change and chemical pollution are presented. Finally, the existing hindrances towards the evolution of Green Industry in developing countries are outlined.

As previously affirmed, most goods and products that industry, especially Chemical Industry, manufactures and people use involve one or more chemical processes. Our life without them would not be the same for sure. However, Green

**Fig. 1** "Reducing" as the core of green chemistry

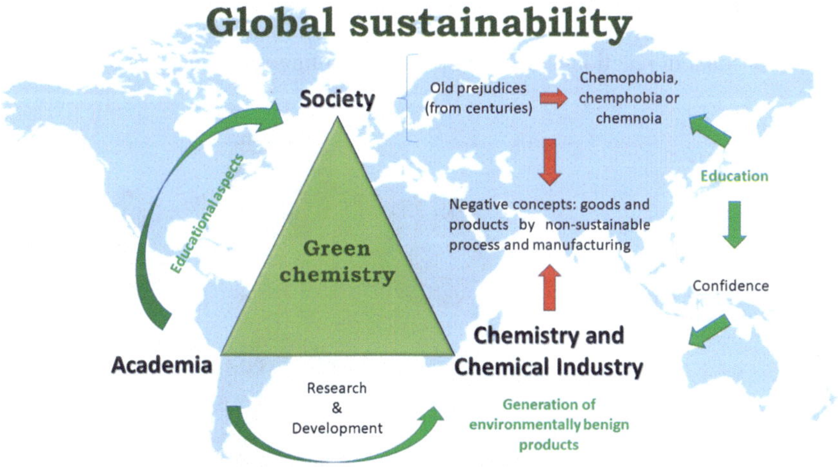


**Fig. 2** Concepts related to green chemistry: philosophical aspects

Chemistry helps industry to become more ecological and sustainable, generating in this way environmentally benign products. Green Chemistry is considered as a multidisciplinary field that encompasses many areas like catalysis, solvents, synthesis, raw materials, products, and efficient processes (Song and Han 2015). Green synthesis constitutes one of the pillars of Green Chemistry and that is why the search for efficient synthetic routes is of vital relevancy to achieve the sustainable industrial production, like in healthcare industry (Morgon 2015). Generation of chemists, mainly organic chemists, have been trained to formulate synthetic reactions in order to maximize yield and purity. Nevertheless, many chemical production processes lack efficiency in the employment of feedstocks and generate large amount of side products. A crucial point to reduce both is increasing atom economy, premise that would satisfy 4 out of the 12 Green Chemistry principles, particularly, principles of Prevention, Atom Economy, Less Hazardous Chemical Synthesis, and Use of Renewable Feedstocks (see Principles 1, 2, 3, and 7 in Table 1). Ideally, all the atoms in reactants should be turned into the desired products. The perfect synthesis, according to Green Chemistry, can be represented in Fig. 3.

Despite that in all industrial chemical processes 100% of atom economy is an utopia. Another form of minimizing the formation of secondary products is the integration of different reactions and processes, being the by-product in a certain reaction the feedstock of another (Song and Han 2015). Of course, yield, product isolation easiness, and purity needs, among other factors, should not be replaced by the concept of atom economy when implementing a chemical synthesis; in fact, it should be considered as an additional aspect (Lancaster 2002). It is noteworthy to mention that there are some reaction types that, due to their nature, would probably minimize waste because of being inherently atom efficient. A3 coupling (Alkyne, Aldehyde, and Amine) (Wei and Li 2002; Wei et al. 2004) and mainly Diels-Alder reaction are two typical and excellent examples of atom-economical reactions (Trost 1991, 1995) (Fig. 4).

Other either well-known or new efficient synthetic tools can be found in Anastas and Eghbali's work (Anastas and Eghbali 2010): cycloadditions, rearrangements, cascade or tandem reactions, multicomponent coupling reactions, metathesis, C-H

**Fig. 3** Scheme and features of the perfect synthesis

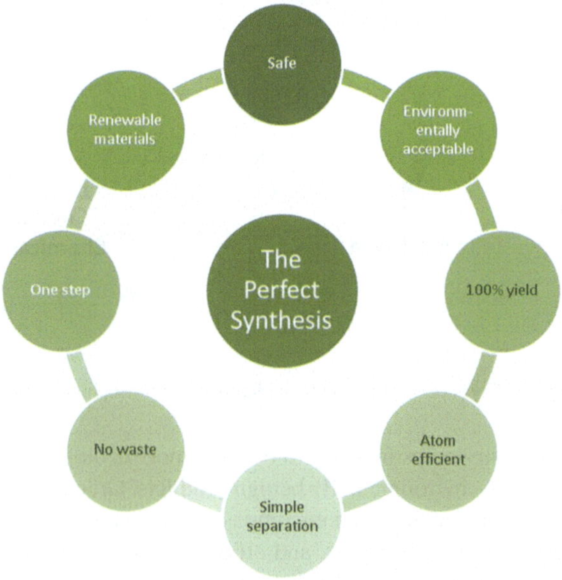

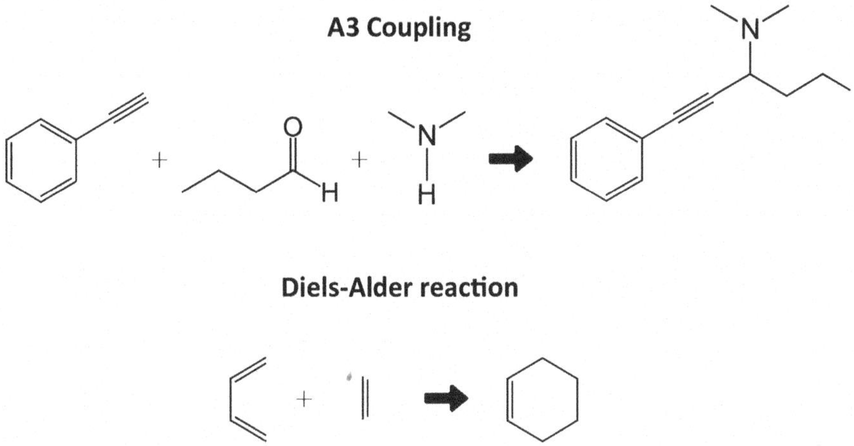


**Fig. 4** Examples of two atom-economical reactions

activation, and enzymatic reactions, for citing some examples. Hence, these reaction types are necessary to be taken into account when implementing a synthetic strategy. However, other elements must obviously be considered as responsibles of the most competitive, efficient, and eco-friendly route: cost and feedstocks availability; toxicity/hazardous nature of feedstocks; yield; product isolation and purification easiness; energy, solvent, and cost-effective equipment exigencies; process times; and waste materials nature (Lancaster 2002).

On the other hand, green synthesis cannot be only circumscribed to organic chemistry, as it is formerly suggested, but to other disciplines as well, such as inorganic chemistry, materials science, or even analytical chemistry. In the latter area, it is also possible to point out Green Analytical Chemistry (De la Guardia and Garrigues 2012; Koel and Kaljurand 2019). This recent discipline focuses on the elaboration of new, green, and sustainable analytical procedures for organic and inorganic compounds determination in different kinds of samples characterized by complex matrices composition (Płotka-Wasylka and Namieśnik 2019). In fact, according to experts, the attention should be focused on making sample-pretreatment and analytical methods much greener thanks to the development of new strategies and tools (Armenta et al. 2008). To these ones, other complementary practices can be added, such as minimization of wastes, recovery of reagents, on-line decontamination of wastes, and the use of reagent-free methodologies. Hence, it is mandatory to fix a group of clear and concise reccomendations constituting the principles of Green Analytical Chemistry, susceptible to be applied in laboratory practices. In this way, the 12 principles of Green Analytical Chemistry were proposed in 2013 (Gałuszka et al. 2013) (see Table 2). They supposed a reformulation of the existing principles of Green Chemistry and Green Engineering since they did not fully fulfill the needs of analytical chemistry. From the 12 principles of Green Chemistry, only four can be directly made suitable for analytical chemistry: (1) residues prevention

**Table 2** The principles of green analytical chemistry expressed as the mnemonic significance

| | |
|---|---|
| S | Select direct analytical technique. |
| I | Integrate analytical processes and operations. |
| G | Generate as little waste as possible and treat it properly |
| N | Never waste energy. |
| I | Implement automation and miniaturization of methods. |
| F | Favor reagents obtained from renewable source. |
| I | Increase safety for operator. |
| C | Carry on in situ measurements. |
| A | Avoid derivatization. |
| N | Note that the sample number and size should be minimal. |
| C | Choose multi-analyte or multi-parameter method. |
| E | Eliminate or replace toxic reagents. |

(principle 1); (2) environmentally friendly solvents and reagents (principle 5); (3) energetically efficient designs (principle 6); and (4) minimization of derivatization (principle 8). These four principles constitute the core of synthesis in some areas of analytical chemistry. Besides, two more key goals should be considered to achieve sustainability in analytical synthesis: (5) removing or reducing chemical substances whatever the purpose they are used for and (6) increased safety for the operator. As it can be seen, most of these topics demand reductions (Clark 2005) (see Fig. 1). However, the necessity of reaching an agreement between the performance parameters and Green Analytical Chemistry exigencies constitutes one of the disadvantages of green laboratory practices. Most of the strategies for analytical chemists projected in the 12 principles collected in Table 2 may worsen some quality analytical parameters: sensitivity, precision, accuracy, selectivity, detectability, or representativeness (Gałuszka et al. 2013). Regardless, the fast technological advance and knowledge about current problems will lead to an enhancement of green analytical methods.

According to Gałuszka et al. (2013), many green alternatives existing in different fields of analytical chemistry versus conventional methods imply the use of sensors and biosensors, i.e., bisphenol A or 17β-estradiol determination in urban wastewater (instead of gas chromatography/mass spectrometry); folic acid determination in medicines, blood glucose or atrazine in water (instead of spectrophotometry); and lead in water (instead of graphite furnace/atomic absorption spectrometry). (Bio) sensors show several advantages versus other commonly used analytical techniques like chromatography or mass spectrometry, including no or simple sample treatment, non-complex instrumentation, low-cost, high specificity, sensitivity, fast response, relatively compact size, multiparameter analysis, in situ determination, and ease of implementation to detect biomolecules (Bahadir and Sezgintürk 2015; Amine et al. 2016; García-Guzmán et al. 2019; Vogiazi et al. 2019). All these advantages are in agreement with the principles of Green Chemistry and Green Analytical

Chemistry (Arduini et al. 2019). Hence, (bio)sensors are a good and serious alternative to determine different kinds of analytes from biomedical, agrifood, or environmental interest versus other typical analytical techniques, which require expensive instrumentations, laboratory setup, skilled personnel, and usually the employment of organic solvents, producing hazardous waste. Due to the aforementioned advantages, (bio)sensing devices are being recently considered of great importance for being used in industry (Siontorou 2019), mainly in food processing industry (Murugaboopathi et al. 2013; Thakur and Ragavan 2013; Mehrotra 2016; Mustafa and Andreescu 2018; Neethirajan et al. 2018a), and also in pharmaceutical (Macdonald 2019), environmental (Patil et al. 2019), and biomedical industries (Cifric et al. 2020).

Nevertheless, most times these (bio)sensors are employed in combination with different kind of materials and nanomaterials in order to enhance analytical quality parameters, like sensitivity, limit of detection, and selectivity, among others, and, thus, their analytical performance (Attar et al. 2015; Bernardo-Boongaling et al. 2019; Shafiei-Irannejad et al. 2019), which make them quite competitive versus other analytical techniques. Among the materials and nanomaterials employed in (bio)sensing, metal (Cubillana-Aguilera et al. 2011; Franco-Romano et al. 2014; Zarzuela et al. 2018; Shukla and Iravani 2019) and metal oxide nanoparticles (González-Álvarez et al. 2016; Henam et al. 2019; Sundaresan et al. 2019), nanowires (Liu et al. 2012; Luo et al. 2019), nanocarbon-based materials (Roh et al. 2019), graphene (Chang et al. 2019; Dong et al. 2019; Hafeez et al. 2019) and magnetic nanostructured molecularly imprinted polymers (Lahcen et al. 2019), among others (Kumar 2007; Merkoçi 2009), as well as new electrode materials stand out (Cordero-Rando et al. 2002; Hidalgo-Hidalgo de Cisneros et al. 2003; Cubillana-Aguilera et al. 2006; López-Iglesias et al. 2016, 2018; Palacios-Santander et al. 2017). There are many different technologies for preparing them, but in this chapter special attention will be paid to two of the most commonly used environmentally friendly technologies, from the energy cost savings point of view: microwave and ultrasound (Strauss 2002; Timothy and Cintas 2002; Chatel 2016). Both techniques are based on the use of focused radiation that reduces reaction times, increases product yields, and also makes reactions more selective (Clark 2005); that is why ultrasound and microwave can be considered the base of the most powerful, ecological, and interesting technologies developed for green analytical synthesis (Lodeiro and Capelo-Martínez 2009). Their advantages versus other synthetic routes can be summarized as follows: (1) environmentally friendliness, (2) very low energy requirements, (3) drastically reduced time of synthesis: from days/hours to few minutes, and (4) simple and (5) low cost instrumentation compared to other technologies.

Moreover, synthesis routes of functional materials and nanomaterials through biomineralization and biotemplating, typically known as biosynthetic routes, by using biopolymers (Khomand and Afsharpour 2019), plant extracts (Agarwal et al. 2019) and other biomolecular structures (bacteria and fungi, among many others) (Gahlawat and Choudhury 2019) play also an important role in Green Analytical

Chemistry and attract tremendous amount of interest. The reason for this attention is the promise to achieve enhanced control over positioning and linking different functional nanostructures to give place to complex nanodevices (Padalkar et al. 2010). In the last decade, the exploitation of natural biopolymers fibers, like cellulose, and plant extracts (olive, geranium, *Aloe vera*, and a much wide etcetera) for synthesizing inorganic nanoparticles, nanoparticle chains, and nanowires has increased a lot. The advantages of using these biostructures in synthetic routes of (nano)materials are evident. For example, in the case of cellulose or other biopolymers-based materials, they are relatively inexpensive, renewable, abundant in many different forms, and have hydroxyl and other functional groups that are accessible for chemical modification (Azizi Samir et al. 2005). With respect to plant extracts (Makarov et al. 2014; Shah et al. 2015), the enormous diversity of reactive chemical compounds possessing different functional groups (alcohols, ketones, esters, etc.) may also offer, in some cases, other advantages, such as abundancy, long lifetime, low or zero-cost requirements for their culture, the easiness of extract preparation, antioxidant activity, antimicrobial properties, and bacterial growth inhibition of some compounds present in the extracts of most plants (Franco-Romano et al. 2014). That is why biosynthesis routes for obtaining (nano)materials will also be paid special attention in this work.

This chapter intends to provide a summary of the most relevant green synthesis or biosynthesis routes, mainly based on the clean ultrasound and microwave technologies, or even in hybrid techniques (Sect. 2), and how they can be employed to synthesize nanomaterials or materials for building (bio)sensing devices (Sect. 3). Currently some of these materials are directly or can be susceptible of being produced by industry, due to scaling-up processes and employed in the industry itself for detection and/or determination purposes and/or for quality control, among other applications in different kinds of industrial companies: food, environmental, biopharmaceutical, and biomedical industries (Sects. 4 and 5). Finally, a critical discussion about the abovementioned topics and their relationships with sustainable development of chemical industry is reported.

## 2 Green Synthesis Routes

Synthesis routes followed by Green Chemistry principles were given particularly relevance in the last decades. Many research efforts are focused on their development for many reasons, such as short reaction times, low cost, easy workup, and low energy requirements.

Three main synthetic routes can be distinguished: microwave-assisted, ultrasound-assisted, and biosynthesis. A schematic representation is shown in Fig. 5.

In addition to their importance in Green Chemistry, a brief overview for each one will be discussed.

## 2.1 Ultrasound-Assisted Method

Nowadays, sonochemistry constitutes a broad research field with a growing interest, especially for synthesis purposes. Thus, the employment of ultrasound has been developed as an emergent powerful tool for obtaining an extensive set of organic and inorganic compounds.

### 2.1.1 Basic Aspects of Ultrasound: Cavitation Process

Sonochemical reactions involves the ultrasonic cavitation, phenomenon based on the formation, growth, and implosion of air bubbles in the liquid phase (Cravotto and Cintas, 2006). Several theories concerning the cavitation phenomena have been proposed: electrical, plasma discharge, supercritical, and hot spots. According to the last one mentioned, high local pressures and temperatures are produced inside the air bubbles and at their interfaces after collapsing, reaching values around 5000 K (Leong et al. 2011). Under these extreme conditions, short-live species from solvent and/or substrate molecules pyrolysis are produced. Hydrogen peroxide and oxygen are generated as by-products by coupling of radical species ($H^{\bullet}$ and $OH^{\bullet}$). This process, known as sonolysis (Torres-Palma and Serna-Galvis 2018), is represented as follows.

*Decomposition of water molecules by ultrasound*

$$H_2O \rightarrow H^{\bullet} + OH^{\bullet}$$

*Formation of secondary radicals*

$$H^{\bullet} + O_2 \rightarrow HO_2^{\bullet}$$

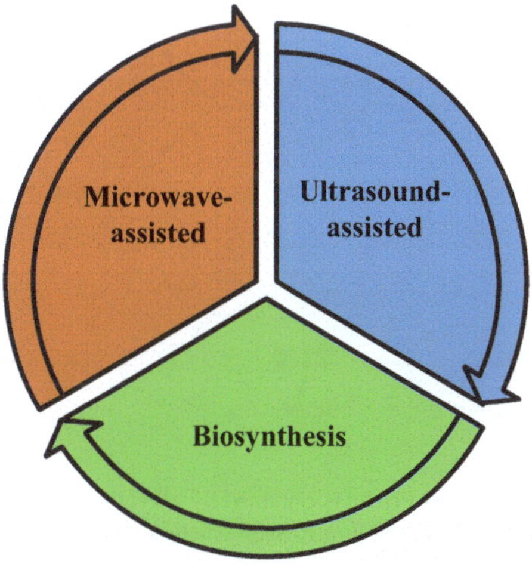


**Fig. 5** Schematic representation of the major green synthetic routes

*Generation of hydrogen peroxide by radical coupling*

$$HO_2^{\bullet} + HO_2^{\bullet} \rightarrow H_2O_2 + O_2$$

$$OH^{\bullet} + OH^{\bullet} \rightarrow H_2O_2$$

The radical species derived from ultrasonic waves can react with other reagents, following a single-electron transfer pathway (SET). A representative example is the alkylation of 4-nitrobenzylbromide, which led to a different product under sonication with respect to the one obtained under silent conditions (Fig. 6a). This result can be explained with the involvement of radicals species from reagents' cleavage under sonication (sonochemical switching), as shown in Fig. 6b (Dickens and Luche 1991).

The ultrasound approach has been extensively applied to assist the synthesis of inorganic and organic compounds. In Sects. 2.1.2 and 2.1.3, numerous examples are exposed.

### 2.1.2 Ultrasound-Assisted Synthesis of Organic Compounds

There are several reviews in the bibliography related to the synthesis of organic compounds assisted by ultrasound, providing a detailed discussion including the reaction pathways proposed for each case (Baig and Varma 2012; Banerjee 2017). Ando's reaction is an illustrative example of organic synthesis in which different product was obtained under ultrasound. The reaction between benzylbromide, potassium cyanide, and alumina in toluene gave the benzylcyanide under sonication, instead of the diphenylmethane derivative obtained under mechanical stirring conditions. The authors proposed the inactivation of the acidic sites on the alumina

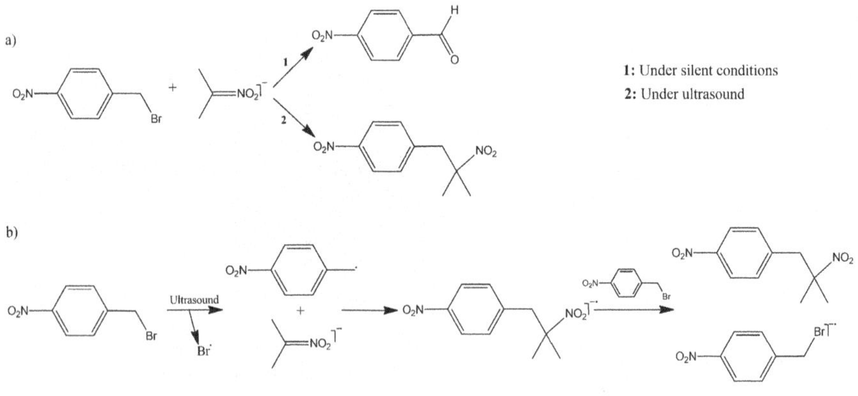


**Fig. 6** (**a**) Products obtained by alkylation of benzylbromide under sonication and silent conditions; (**b**) Kornblum-Russell reaction mechanism under ultrasound

surface by the strong adsorption of potassium cyanide under ultrasound conditions (Ando and Kimura 1990).

Not all the ultrasound-assisted organic reactions involve the sonochemical switching phenomena. Despite this fact, two important advantages are ascribed to the use of ultrasound: high yields and low reaction times, among others. All of them lead to several features in terms of Green Chemistry, summarized in Gregory Chatel's work (Chatel 2018): reduction of waste products, easy synthesis workup, and minimization of energy requirements.

Table 3 shows the reaction time and yield obtained for different organic products obtained synthesized under silent conditions and assisted by ultrasound.

As shown in the previous table, shorter reaction times, as well as higher yields were achieved under ultrasound conditions. This demonstrates the benefits of ultrasound in organic synthesis.

Other important aspect in some conventional organic syntheses is the use of toxic reagents, harmful for the environmental and human health. By using ultrasound, organic hazardous solvents can be replaced by greener ones or totally removed (solvent-free conditions). The selective oxidation of sulfides to sulfoxides can be carried out by using hydrogen peroxide as solvent (Mahamuni et al. 2006), instead of methanol. Polymerization can also be mediated by ultrasound, taking place by polymer or solvent-derived radical processes. Several examples of these types of reactions are reported by McKenzie and coworkers (McKenzie et al. 2019).

### 2.1.3 Ultrasound-Assisted Synthesis of Inorganic Compounds

Inorganic synthesis can also be ultrasound-assisted, leading to several advantages with respect to conventional routes. Suslick and coworkers summarized the influence of chemical and physical effects of ultrasound in the synthesis of nanostructured materials (Bang and Suslick 2010; Xu et al. 2013). Among all of them, metallic and metallic-oxide nanoparticles have received wide attention in the past few decades due to their optical properties, as well as their unique reactivity, very useful in many applications (Christian et al. 2008).

For noble metals, radical species generated by sonolysis of the water ($H^{\bullet}$ and $OH^{\bullet}$) can act as reducing agents, avoiding the use of any additional reducing compound in the synthesis. However, secondary species can be added to produce secondary radical species, promoting the rate of the process.

The overall process was detailed in Suslick and coworkers' work (Xu et al. 2013) and can be summarized as follows (recombination of radicals derived from the sonolysis of water was not included).

*Decomposition of water molecules by ultrasound*

$$H_2O \rightarrow H^{\bullet} + OH^{\bullet}$$

*Formation of secondary radicals ($R^{\bullet}$)*

**Table 3** Reaction times and yields of different organic compounds obtained under silent conditions and assisted by ultrasound

| | | Silent conditions | | Ultrasound-assisted | | |
|---|---|---|---|---|---|---|
| Synthesis process | Solvent | Time | Yield (%) | Time | Yield (%) | References |
| Synthesis of 2-organoselanyl-naphthalene | EtOH | 72 h | 78 | 50 min | 84 | Perin et al. (2018) |
| 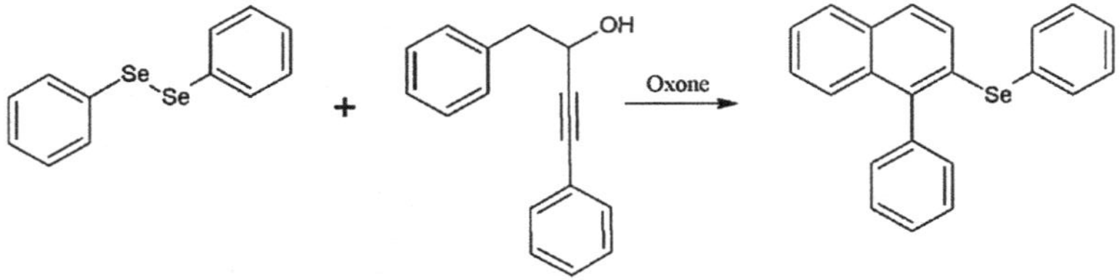 | | | | | | |
| Synthesis of quinolone derivative | Water | 160 min | 60 | 15 min | 90 | Bhardwaj et al. (2019) |
| 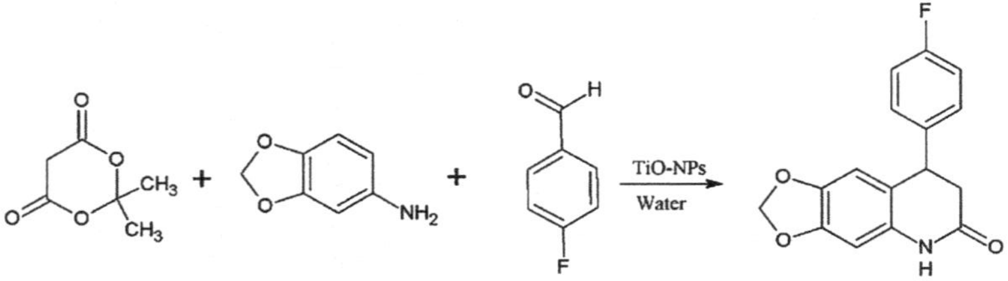 | | | | | | |
| Opening of ring epoxide | Free | 150 min | 72 | 60 min | 80 | Nasef et al. (2016) |
| 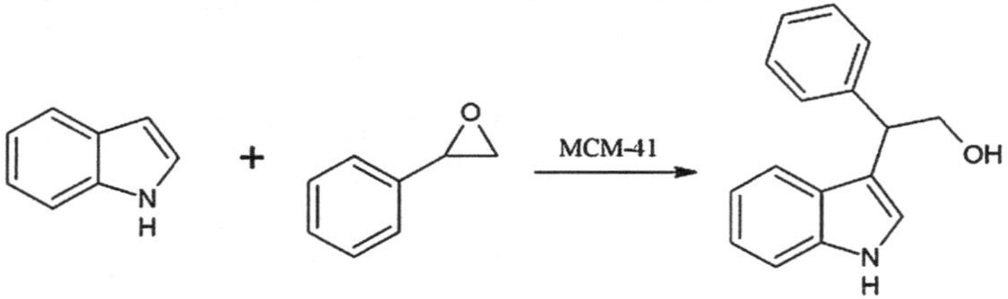 | | | | | | |

| | | | | | | |
|---|---|---|---|---|---|---|
| Aza-Michael reaction | Water | 16 h | 73 | 30 min | 98 | Yang et al. (2005) |
| 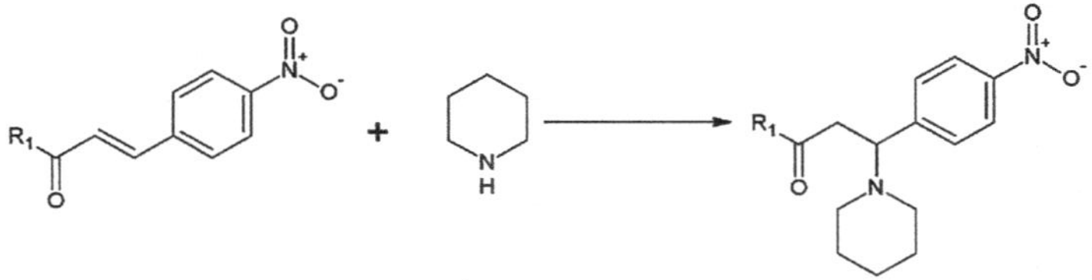 $R_1 = F_c$ | | | | | | |
| Multicomponent reaction of N-substituted pyrrole | Free | 6–9 h | 74 | 30 min | 90 | Shahvelayati et al. (2017) |
| 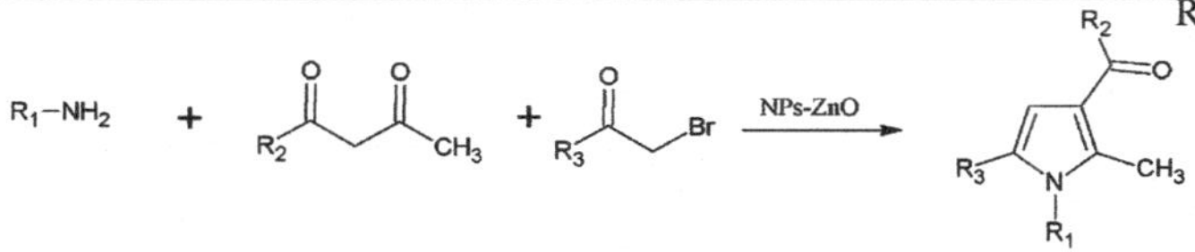 $R_1$=$CH_2CH_3$ $R_2$=$CH_3$ $R_3$=$C_6H_5$ | | | | | | |
| Cycloaddition reaction | EtOH | 13 h | 87 | 5 h | 93 | Hu et al. (2012b) |
| | MeOH | 15 h | 80 | 6 h | 85 | |
| 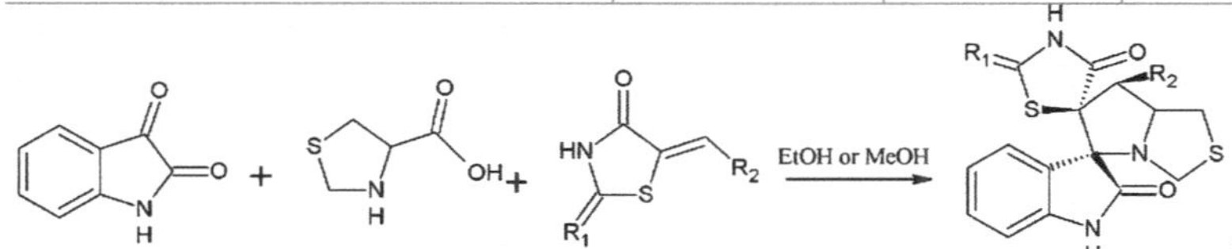 $R_1 = S$ $R_2 = C_6H_5$ | | | | | | |
| Synthesis of 2-sulfonylquinoline derivative | Water | 8 h | 67 | 25 min | 86 | Xie et al. (2017) |
| 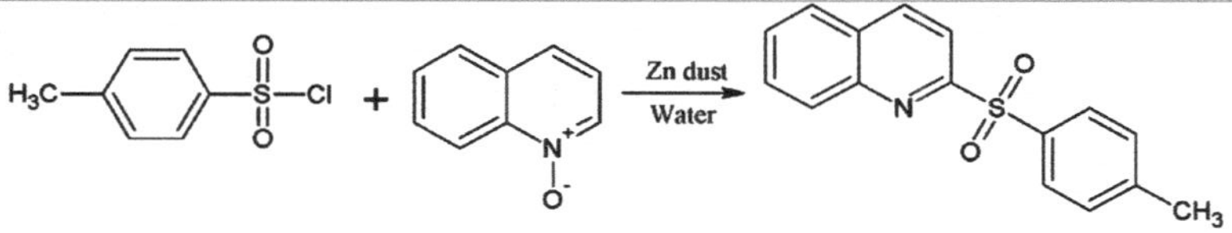 | | | | | | |

(continued)

**Table 3** (continued)

| Synthesis process | Solvent | Silent conditions | | Ultrasound-assisted | | References |
|---|---|---|---|---|---|---|
| | | Time | Yield (%) | Time | Yield (%) | |
| Synthesis of coumarin (Knoevenagel condensation) | EtOH | 7 h | 80 | 40 min | 88 | Da Silveira Pinto and De Souza (2017) |
| 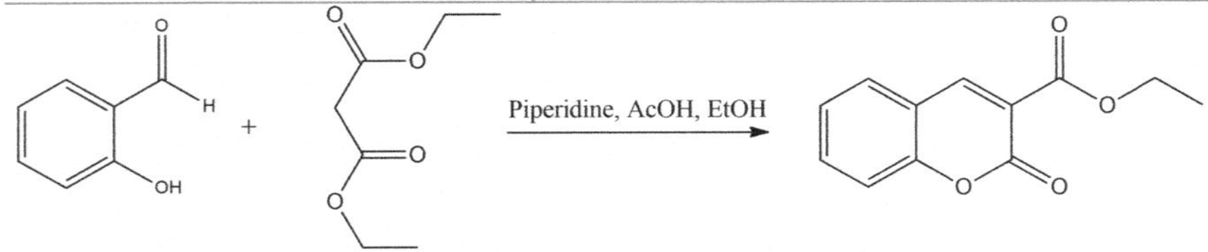 | | | | | | |

$$RH + OH^{\bullet} \rightarrow R^{\bullet} + H_2O$$

$$RH + H^{\bullet} \rightarrow R^{\bullet} + H_2$$

*Chemical reduction of noble metallic salt (M)*

$$M^{n+} + nH^{\bullet} \rightarrow M^0$$

$$M^{n+} + nR^{\bullet} \rightarrow M^0$$

*Formation of noble metallic salt nanoparticles ($M_m$)*

$$mM^0 \rightarrow M_m$$

By using this approach, gold (Okitsu et al. 2007), platinum (Mizukoshi et al. 1999), silver (Elsupikhe et al. 2015), and palladium (Nemamcha et al. 2006) nanoparticles were synthesized.

The sonochemical method provides mainly spherical nanoparticles, although other nanomaterial shapes can be obtained under ultrasound. Different gold nanoparticles morphologies were formed from the reduction of the gold salt under ultrasound by setting the sodium dodecyl sulfate concentration and the ultrasonic intensity. According to this research work (Park et al. 2006), sodium dodecyl sulfate concentration values lower than critical micelle concentration drove to the obtention of nonspherical nanoparticles. With respect to the influence of the ultrasonic power, low ultrasonic irradiation produces lesser radical species in comparison with those produced at higher values, and hence the generation of nanodisks, nanoprisms, and nanorods, among other structural shapes, is promoted due to the slow reduction rate of the gold salt. Other examples about the obtaining of nonspherical gold nanoparticles from the chemical reduction of the gold salt by ultrasound can be stated. Gold nanorods were obtained when using cetyltimethylammonium bromide as stabilizing agent and ascorbic acid. In this report, the pH of the solution plays a significant role in the aspect ratio of nanorods (length/width ratio): the average aspect ratio decreases as pH value increases (Okitsu et al. 2009). Gold nanobelts were formed in the presence of α-D glucose by using a green, nonhazardous, and rapid synthesis method. The effect of the concentration of α-D glucose was discussed in this research work. Furthermore, the use of ultrasound was proposed to accelerate the process, as well as to enhance the reorganization of the molecules of glucose on gold crystals (Zhang et al. 2006).

Nonspherical nanoparticles constituted by other metals were synthesized following the sonochemical route. Cubic silver nanoparticles were formed by self-arrangement of dodecylbenzenesulfonic acid sodium salt as surfactant. The nanoparticle morphology was also influenced by the concentration of poly(vinylpyrrolidinone), the ripening time, and the ultrasound (Moghimi-Rad et al. 2011). The formation of copper nanowhiskers was reported in Min Xu and coworkers' report (Xu et al. 2015). According to this work, ultrasound plays an essential role in the synthesis of nanowhiskers: low ultrasonic power favors the growth of the particles in one dimension, while higher values lead to irregular particles.

Gold nanoparticles can also be synthesized using sodium citrate as reductant under high-power ultrasound conditions (Cubillana-Aguilera et al. 2011). Their characteristic features, optical properties and size, can be easily characterized by analytical routine techniques, dynamic light scattering, and UV–vis spectrophotometry. The characterization of nanoparticles' size by dynamic light scattering, considered as a no complex and low-time consuming methodology, was also performed in the silicon oxide nanoparticles synthesis by ultrasound (González-Álvarez et al. 2016), as an alternative versus the classical Stöber method, and successfully correlated to the results obtained with transmission electron microscopy, the most used technique for evidencing particle shape and size. Thus, the utility of routine analysis in the monitoring of optical properties and sizes of metal or metal oxide nanoparticles is demonstrated.

Therefore, the development of ultrasound-assisted syntheses as an alternative to conventional methods is demonstrated. The production of radical species by ultrasonic irradiation enhances the reaction rate, as well as reduces the formation of by-products in some cases by sonochemical switching phenomena. Thus, shorter reaction times, higher yields, and high selectivity were ascribed to the employment of ultrasound, leading to several improvements in terms of Green Chemistry. Furthermore, the role of ultrasound in the formation of shape-controlled nanomaterials can also be stated, allowing tailor-made structures to reach unique and useful properties for sensing and catalysis, among other applications.

## 2.2 *Microwave-Assisted Synthesis*

In Sect. 2.1, ultrasound has been exposed as an interesting approach towards Green Chemistry, although it is not the only one. In this section, microwaves assistance will be discussed as an excellent alternative to conventional synthesis routes.

### 2.2.1 Microwave Heating: Fundamentals and Mechanisms

Microwaves is a kind of electromagnetic radiation comprised between radio waves and infrared, with a wavelength range from 1 mm to 1 m (0.3–300 GHz). The interaction between this radiation and the matter has different physical effects, such as mobilizing the electric charges in liquids or conducting ions in solids. The electromagnetic energy produced by microwave irradiation is transformed into heat, promoting the conversion of reagents to products. As an example, a schematic representation of a reaction assisted by microwaves radiation appears in Fig. 7. The heating mechanism of this phenomena will be explained deeply later (Sekhon, 2010).

Regarding the interaction between microwave and the compound, three different situations can be established:

1. Microwaves have no effects in the system, e.g., sulfur. The radiation travels through the species without altering it.
2. Microwave is reflected by the system, e.g., metals such as copper. The radiation cannot travel through it and it is reflected.
3. Microwaves are absorbed by the system, e.g., water. The absorption leads to an increase of the temperature.

Based on the previous classification, compounds which absorb the microwaves will be considered in this section due to their predominance in microwave chemistry. However, the other materials have interesting application as well; some of them will be explained later.

There are three main mechanisms about the role of the microwave radiation in the heating process: dipolar polarization, conduction mechanism, and interfacial polarization. All of them will be described in this section.

## Dipolar Polarization

A polar molecule, which is affected by a varying electric field (microwave radiation), try to reorientate itself. The molecular friction effect causes a release of heat to the system. Thus, it is necessary to possess a dipole to cause heat with this mechanism. In this sense, it is possible to take advantage of this phenomenon by using polar solvent, such as ethanol, methanol, and water, among others, or polar solutes such as ammonia or formic acid. The most notorious factor is the application of a nonconstant field with a suitable frequency, which facilitates the interaction of the particle. This frequency must not be too high because the molecule will not be fast enough to follow the changes in the field, leading to a premature stop of the

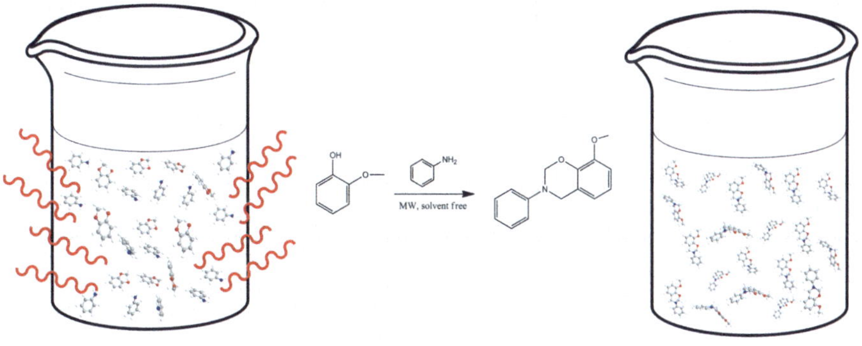


**Fig. 7** Example of reaction promoted by microwave

molecule. In another way, if the frequency is too low, the molecule will have enough time to realign itself with no effective interaction between molecules. By using the microwave radiation (frequency 0.3–30 GHz) it is possible to provoke an effective particle interaction (Lidström et al. 2001). An example of dipolar polarization can be observed in Fig. 8a.

### Conduction Mechanism

Electrons moved throughout a resistance provide heat to the system. This principle is the basis of the conduction mechanism. A conductor submitted into an electric field generates a flow of electron, or ions, which is against the internal resistance, leading to the heating of the conductor. Therefore, charge carriers (electrons, ions, etc.) in a sample can be moved using an electric field. The inner resistance will heat the sample due to the induced currents formed. It is important to clarify that high conductive materials will not be heated in this way due to the reflective properties of these ones (Wathey et al. 2002). A scheme of the conduction mechanism is presented in Fig. 8b.

### Interfacial Polarization

The combination of the previous mechanisms gives, as a result, a third heating mechanism. It is relevant when a conducting material is dispersed in a non-conductive medium such as metal powder in sulfur. Even if the sulfur does not absorb the radiation and the metal reflects it, the whole system is an appropriate microwave absorbing material. This heating mechanism is based on the dipolar polarization but it is slightly different. In this case, the metal powder limits the movement of the ions by forces that are similar to the cases of polar solvents. The

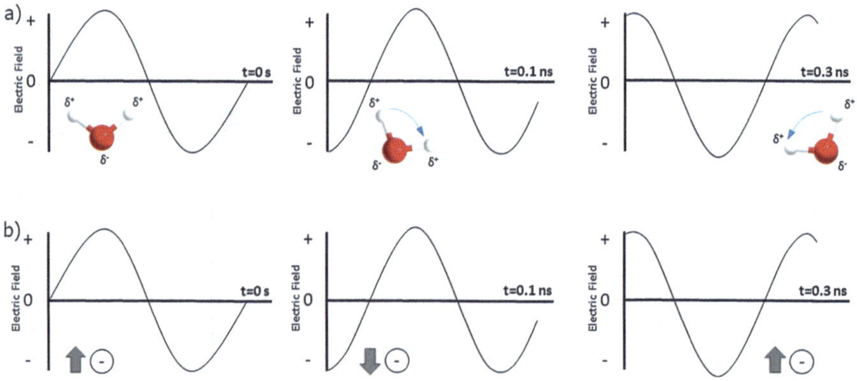


**Fig. 8** Heating mechanism of microwave radiation: (**a**) dipolar polarization and (**b**) conduction mechanism

restriction of these ions results in a random motion and, afterwards, the heating of the sample (Gabriel et al. 1998).

### 2.2.2 The Use of Microwaves as a Green Approach in Organic and Inorganic Syntheses

Classically, the heating methods used in organic and inorganic reactions were based on convection, which involves complex workup, such as oil baths, Bunsen burner, or furnace, among others. Other drawbacks in relation with green terms can be ascribed to conventional synthesis methods. They involved petrochemical ingredients, several catalyst and separation and purification processes. Furthermore, organic synthesis implied many health issues and risk for workers, long reaction times, high cost, and inefficiency to heat the system. The last one is based on the convection mechanism, leading to energy losses due to the heat dissipation phenomena. Microwave assistance is proposed as a suitable alternative to conventional routes. The microwave radiation affects only target molecules or solvent molecules, involving less energy. Thus, the heating procedure is more focused and effective.

Besides, the microwave provides other advantages like higher reaction rates, fast and easy optimization, and more reproducible syntheses. In addition to the more efficient heating mechanism, the microwave also reaches higher temperatures, decreasing even more the time needed for certain syntheses. For example, the fluorescein synthesis lasts 10 h with conventional procedures; otherwise, it can be performed in 35 min using a microwave approach (Charde et al. 2012). Moreover, in this kind of synthesis higher final yields of the desired product were reached, minimizing the formation of side products. Thus, the purification procedure is easier and takes shorter times.

In Table 4, the synthesis parameters obtained for synthesis procedures assisted by microwaves and by conventional heating are exposed.

In the previous table the reduction in the time reaction and the general increase in yields are demonstrated. It is also established that the microwave approach is highly versatile in organic synthesis. On the other hand, as it has been mentioned previously, there are several reactions where the organic solvents are minimized or completely removed (solvent free). A mineral support is used instead of the non-eco-friendly solvent and the microwave is adjusted to focus only on the sample (Mordini and Faigl 2005; Algul et al. 2008; Gaba and Dhingra 2011). These reactions will be much greener than their conventional analogous process due to saving in toxic solvents. Other aspect that can be stated is the possibility to carry out several reactions at the same time by using a multimode microwave device.

Microwave approach is also used in polymer syntheses. The energy saving in these procedures and the heating efficiency make this technique an economic and suitable option. In addition, the employment of microwave radiation in the curing process has greatly shortened the reaction time. It has been established that this process is strongly dependent on how the pulse is applied; on the contrary, the power applied is not so important. A higher volume of product can be produced by using

**Table 4** Reaction promoted by microwaves and their time and yield in comparison with the ones obtained by using conventional heating

| Reaction promoted | Microwave-assisted | | Conventional heating | | References |
|---|---|---|---|---|---|
| | Time | Yield (%) | Time | Yield (%) | |
| Synthesis of 6-allylbenzo[d][1,3] dioxol-5-ol | 5 min | 97 | 36 h | 72 | Majetich and Hicks (1995) |
| 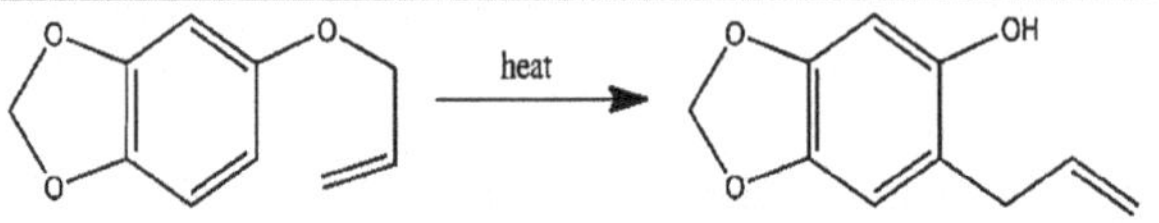  | | | | | |
| Synthesis of 2,3-dihydroxylpropyl decanoate | 1 min | 100 | 100 min | 77 | Mhanna et al. (2018) |
| 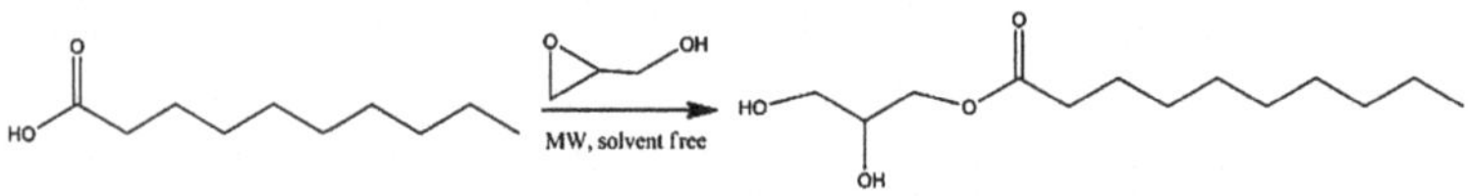  | | | | | |
| Synthesis of benzoxazines | 6 min | 55–82 | 90–180 min | 55–75 | Oliveira et al. (2017) |
| 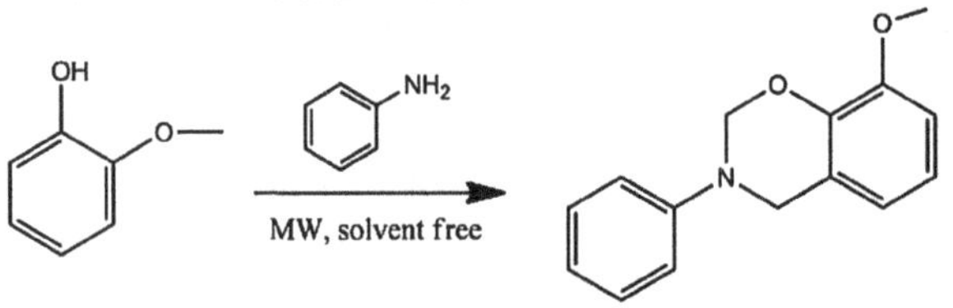  | | | | | |
| Synthesis of a protected *keto*-Lysidine | 1 h | 68 | 72 h | 71 | Sweeney et al. (2019) |
| 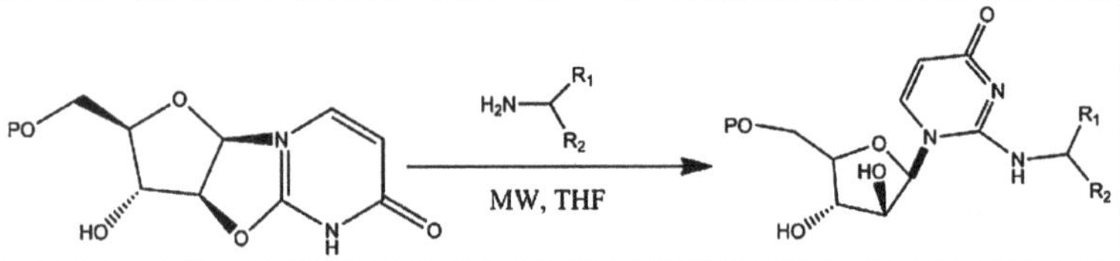  | | | | | |
| Diels-Alder reactions | 20 min | 58 | 6 h | 67 | Majetich and Hicks (1995) |
| 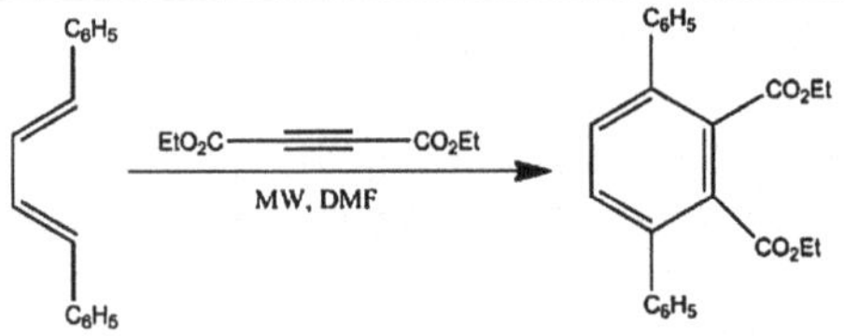  | | | | | |
| Finkelstein reactions | 4 min | 83 | 40 min | 85 | Majetich and Hicks (1995) |
| 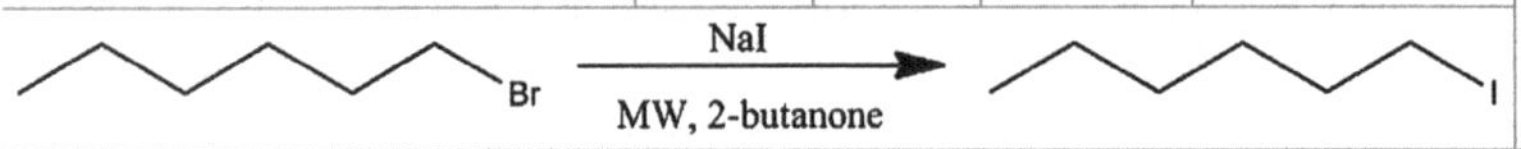  | | | | | |

(continued)

**Table 4** (continued)

| Reaction promoted | Microwave-assisted | | Conventional heating | | References |
|---|---|---|---|---|---|
| | Time | Yield (%) | Time | Yield (%) | |
| Oxidation reactions | 7 min | 52 | 8 h | 20 | Majetich and Hicks (1995) |
| 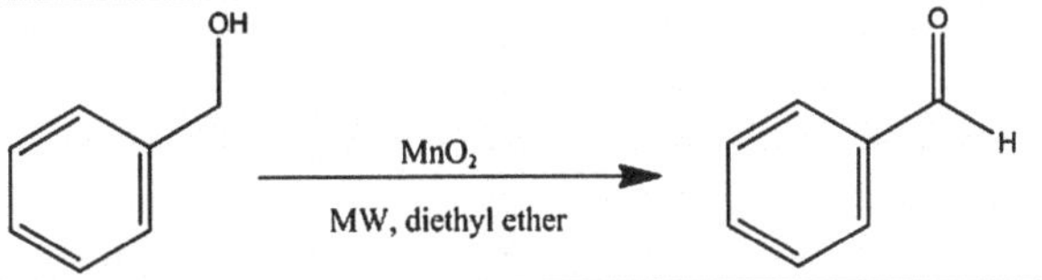 | | | | | |
| Esterifications | 2 min | 86 | 90 min | 84 | Majetich and Hicks (1995) |
| 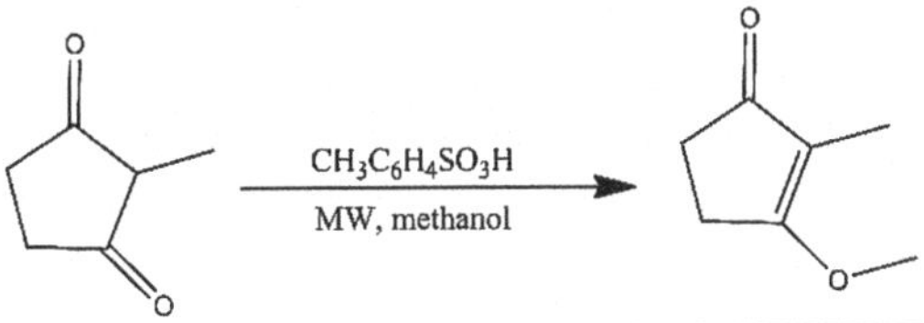 | | | | | |

large reactors and a controlled solvent-free synthesis with microwave heating. For instance, it has been reported that the peptides production leads high yields and purity. This process implies the assembly of peptide chains of about 30 amino acids carried out only in one night with an automated microwave system (Nayak et al. 2016).

Despite its wide use in organic syntheses, there is a growing increase in the use of microwave in inorganic compound syntheses (Darvishi et al. 2017; Zhao et al. 2017; Li et al. 2019; Rossini et al. 2019; Xu et al. 2019), especially in the nanomaterial field. In these cases, the microwave radiation provides advantages, such as narrow particle size distribution, energetic efficiency, particle size controllability, and fast crystallization rate, among others. The use of microwave was proposed for generating some inorganic compounds by hydrothermal and solvothermal method, leading to some improvements in terms of yields and reaction times, in comparison with the conventional method (Gaikwad and Han 2019). Zhou et al. (2014) proposed a microwave-assisted hydrothermal method to obtain CuO spheres, which are wrapped and linked by graphene nanosheets. Furthermore, it is possible to combine a redox system with a microwave-assisted hydrothermal method. Chen et al. (2013) obtained polymorphic $MnO_2$, being able to grow the material with different crystallographic phases (α, β and γ) modifying the synthesis parameters. Zhang et al. (2019) have obtained $CoFe_2O_4$ nanoparticles, by a microwave-assisted solvothermal method, which even possess better properties than the nanoparticles resulting from a conventional solvothermal method. Palma-Goyes et al. (2018) provide a facile and rapid microwave-assisted solvothermal method to synthesize $RuO_2$ nanoparticles employing citric acid in ethylene glycol as stabilizing agent, $H_2O_2$ as oxidizing agent, and the metallic precursor.

Moreover, it is also possible to use the microwave radiation to assist a sol-gel process. A great number of studies have been carried out in this direction. Ghule et al. (2011) informed the synthesis of zinc oxide nanorods taking advantage of the microwave radiation employing zinc, nitrate, ethylene glycol, and sodium hydroxide as precursors. Shrike et al. (2011) synthesized pure anatase $TiO_2$ nanoparticles with a particular porous structure by employing a sol-gel microwave-assisted method. Garadkar et al. (2013) developed an easy procedure to synthesize $ZnWO_4$ nanoparticles controlling successfully their size by using a microwave-assisted procedure. This approach can also be used to synthesize magnetic nanoparticles as well. Obaidullah et al. (2019) reported an easy and fast method to make $Fe_2O_3$ nanoparticles coated by a shell of $SiO_2$ retaining their magnetic properties even at high temperatures. Thus, the microwave radiation has greatly supported conventional materials syntheses methods enlarging the possibilities in this field. Consequently, the number of green process has been considerably increased over the last few years.

Therefore, it is noteworthy to mention some aspects regarding the use of microwave as green approach in synthesis. Unlike the monotonous conventional heating, which is very time consuming, microwave chemistry opens a wide range of new possibilities for the development of new methods of synthesis. The time saving, replacing the toxic solvents to greener ones, greater selectivity, enhancement of reaction yields, and easier setup are also highlighted. For all these reasons, microwaves assistance is widely considered as a very effective and promising tool in Green Chemistry (Ravichandran and Karthikeyan 2011).

## 2.3 *Biosynthesis*

The use of biological compounds is currently considered by the scientific community as an attractive way to synthesize complex molecules by one-pot method (Shamaila et al. 2016). In this manner, biosynthesis constitutes an environmental-friendly, economic, and low-time consuming synthesis route. In this chapter, the main groups used in biosynthesis are presented, providing some information about their role in syntheses.

### 2.3.1 Plant Extracts

Plants are a rich source of chemical compounds, commonly known as phytochemicals, such as polyphenols, terpenoids, alkaloids, and proteins, for instance. They can be easily extracted by using different solvents, such as water, methanol, ethanol, and dimethylformamide, among others (Altemimi et al. 2017). To follow the green rules, water-based and low-toxicity solvents are required for extraction purposes. Concerning the extraction method, microwave and ultrasound constitutes economic, low cost, low time-consuming, and viable tools in the extraction of phytochemicals (Tiwari 2015; Dhanani et al. 2017), and hence their use is encouraged in Green Chemistry.

The phytochemical compounds from plant extracts contain functionalized groups able to reduce, as well as to stabilize, the final product. Based on this dual role, plant extracts are excellent candidates to synthesize metallic nanoparticles, allowing controlling many properties, e.g., size and morphology. The influence of several parameters to obtain tailor-made nanoparticles, such as temperature, pH, and type of extract, will be commented in Sect. 3.4.1 of this chapter.

Plant extracts are very complex matrices; therefore, a detailed study about the bioreduction mechanism is required. Several research groups investigated the reducing role of flavonoids, proteins, and amino acids available in plant extracts (Makarov et al. 2014). In the case of flavonoids, the reactive hydrogen was released by keto-enol tautomerism, leading to the keto-form. As an example, A. K. Singh and coworkers summarizes a possible mechanism for the synthesis of metallic nanoparticles mediated by the keto-enol tautomeric process of eugenol, shown in Fig. 9 (Singh et al. 2010).

The participation of amino acids and proteins in the reduction process was also reported owing to the existence of functionalized amino groups. Based on the above, the overall process of metal nanoparticles formation implies reducing the metal precursor, followed by the nucleation and growth processes.

### 2.3.2 Biopolymers

Biopolymers can be obtained via microbial synthesis from biological resources as starting products (Ahmad et al. 2015). Among all of them, polysaccharide biopolymers have some excellent properties, enabling the development of advance functionalized materials for several applications (Wróblewska-Krepsztul et al. 2019). Figure 10 shows some representative polysaccharide biopolymeric structures.

The high reducing properties of metallic salts, together with their ability to coordinate metal ions, make them promising compounds for the biosynthesis of metallic nanoparticles. The reducing/stabilizing dual role enables the formation of metal nanoparticle with specific features, such as tailored sizes, biocompatibility, and nontoxicity, among others (Wang et al. 2017).

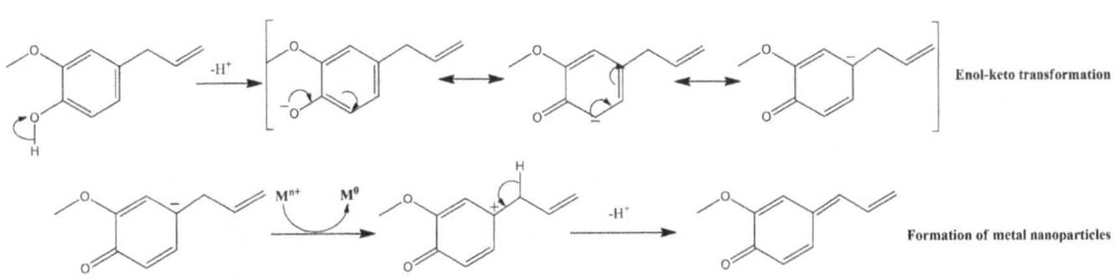


**Fig. 9** Schematic representation of the reducing role of eugenol in the bioreduction process

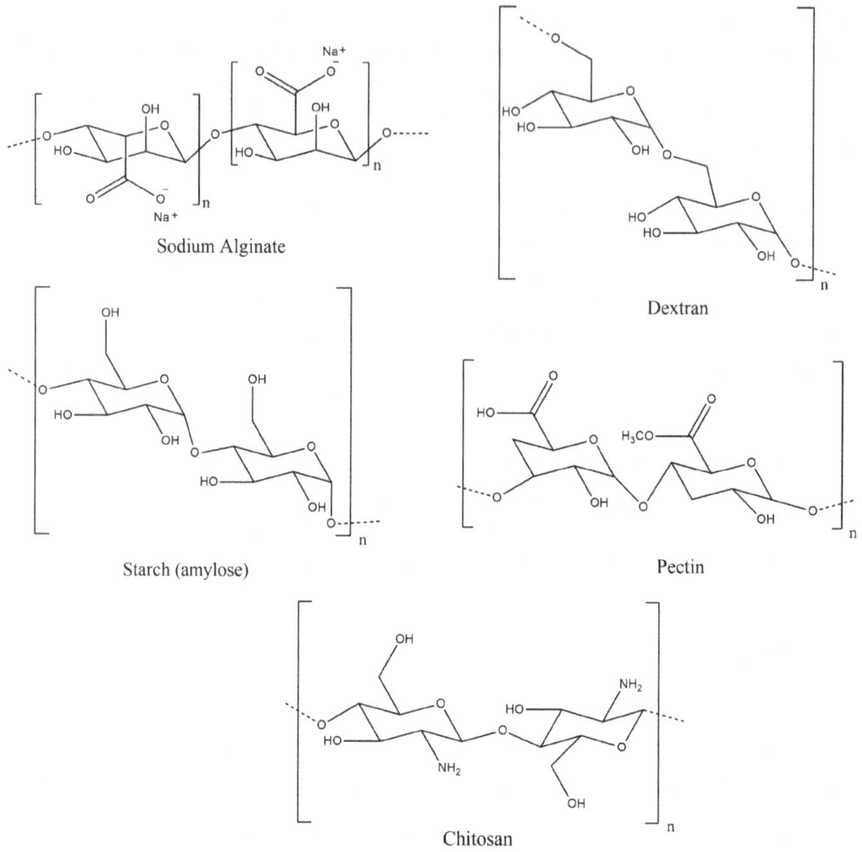


**Fig. 10** Representative polysaccharide biopolymers

As happened with plant extracts, the influence of some synthesis parameters on the size and morphology of the generated nanoparticles, as well as more details about the synthesis approaches, will be given in Sect. 3.4.2.

### 2.3.3 Microorganisms

Living microscopic organisms are involved in several synthesis routes, usually performed at mild conditions (Wang et al. 2016). The use of several microorganisms, such as algae, bacteria, and fungi, should be highlighted for biosynthetic purposes (Schmid et al. 2015; Dahoumane et al. 2017; Skellam 2019).

The metabolic engineering constitutes a powerful tool in biosynthetic routes using microorganisms, by enhancing endogenous metabolic pathways or by introducing exogenous pathways (Grunwald 2012). R. Kumar and S. Prasad summarized the fundamentals of metabolic engineering of bacteria, including three steps required in the metabolic process: understanding the metabolic pathway, use of a computational approach, and its application at experimental level using different engineering approaches (Kumar and Prasad 2011).

Thus, biosynthesis of several compounds mediated by engineered microorganisms was recently reported. Biodiesel was produced using oleaginous microorganisms from organic wastes by cost-effective approaches (Cho and Park 2018). The production of biofuels and biofuel feedstocks was reported using engineered microorganisms, such as yeasts, cyanobacteria, bacteria, and microalgae (Majidian et al. 2018). The synthesis of 2-phenylethanol, an important aromatic compound, was performed by metabolic engineering in yeast and bacteria, reducing the formation of toxic by-products, usually obtained by chemical synthesis route (Wang et al. 2019b). Lycopene was synthesized from cytosolic isoprenoid precursors using a viral vector (Majer et al. 2017). Biosynthesis of polyhydroxyalkanoate (PHA) seems to be improved by engineered microorganisms (Chen and Jiang 2018). Competing pathways of PHA were minimized, channeling the resources to the PHA biosynthesis pathways.

Microorganisms are also entailed in the synthesis of metal nanoparticles (Ovais et al. 2018). However, the biosynthesis of metal nanoparticles using plants and biopolymers will be presented in Sects. 3.4.1 and 3.4.2.

## 2.4 *Hybrid Green Synthetic Routes*

In Sects. 2.1, 2.2, and 2.3, several green approaches have been overviewed individually. However, a new generation of assisted green methods is recently arising based on the combination of some of the advance methods previously discussed. Therefore, new tandems have been proposed as green synthesis advances, such as microwave-ultrasound, microwave-biosynthesis, and ultrasound-biosynthesis assisted methods. These new methodologies take advantages of the synergistic effect involved using different techniques. In this section, they will be briefly presented, discussing some of the most notorious and promising results already available, and exposing their high potential applicability in the green synthesis.

### 2.4.1 Microwave-Ultrasound-Assisted Methods

The use of ultrasound and microwave radiation has been extensively commented in Sects. 2.1 and 2.2, respectively. Regarding ultrasound, the cavitation phenomena are the core of this approach, which implies physical-chemical process and the generation, growth, and collapse of bubbles. With respect to the microwave radiation, the high effective transference of heat (creation of "hot spots") leads to higher yields and faster processes than those obtained in conventional heating. However, the limitations of this approach with respect to the use of nonpolar compounds should be mentioned (Martina et al., 2016).

Despite microwaves and ultrasounds are based on two different phenomena, their combination leads to better results than the ones provided by the individual use of each technique. In this way, synergistic effect should be remarkable by using the hybrid approach. This synergistic effect is mainly explained by two different contributions: the highly effective heating of the system and the efficient temperature stimulation supplied by the cavitation process. Besides, the microwave heating prevents the loss of energy and the thermal pollution of the environment. In addition, ultrasounds allow the suitable mixing of substrates, even if they are found in different phases, avoiding the need of any additives, such as surfactants (Pawełczyk et al. 2018).

Many researchers have tried to exploit this binomial in order to develop even more efficient and fast synthesis of organic and inorganic compounds (Cravotto et al. 2015). According to the organic synthesis reactions, several kinds of processes have been improved by this approach. Firstly, the transesterification reactions, Martinez-Guerra et al. reported a protocol to convert waste vegetable oil into biodiesel with a yield of 98% in only 82 min (Martinez-Guerra and Gude 2014). The Heck reaction has also been improved with the simultaneous microwave/ultrasound irradiation. Saaco et al. have performed one of the most recent advances: the creation of carbon-carbon bonds by olefin metathesis under microwaves/ultrasound with 86% of yield (Sacco et al. 2015). The microwave/ultrasound irradiation approach has also provided very good alternatives in the C-Heteroatom Bond formation reactions, such as the ethers synthesis and aromatic azo compounds, among others. For instance, the hydrazinolysis of methyl salicylate has greatly improved investing 40 s instead of 9 h with a yield of 84% (Wu et al. 2008). On the other hand, the inorganic synthesis has been a matter of interest from the point of view of microwave/ultrasound irradiation approach, especially in the case of nanoparticles and nanomaterials (Cravotto and Boffa 2014). It is noteworthy to mention that in the case of metallic nanoparticles the application of microwave irradiation has no dangerous implication due to excessive metal dilution in the solution. Following the microwave/ultrasound irradiation methodology, copper nanoparticles has been successfully generated reducing $Cu(OAc)_2$ with hydrazine in ethylene glycol. Transmission electron microscopy and X-ray diffraction were used to characterize the nanoparticles, resulting in a highly pure and spherical nanomaterial. The yields and reaction time were 97% and 4.5 min, respectively, in comparison with those obtained by using the conventional method: 52% and 12 h, respectively (Feng et al. 2014). Cherkasov et al. obtained a solid supported Pd catalyst. In this case, the irradiation was applied sequentially. Palladium was firstly reduced by ultrasound radiation, and after then, it was deposited on the solid surface by microwave irradiation; the clusters obtained were around 100 nm. One of the most notorious milestone of this method is the absence of any surfactants, which make the process much greener than the previously made (Wu et al. 2015). The surface morphology of the nanoparticles can be selected altering the conditions of the synthesis. Mesoporous hydroxyapatite nanoparticles were obtained by Liang et al. through microwave/ultrasound irradiation. However, the most relevant results were the change of the porous structure given for different synthesis conditions. A flake-like non-mesoporous structure was generated at low temperature (10–50 °C) and microwave power. On the other hand, a clear mesoporous derivative was obtained by using higher temperatures (50–90 °C). Besides, if the microwave is increased until 200 W, a more mesoporous structure is found (Liang et al. 2013). Thus, due to all the reasons previously mentioned, the microwave/ultrasound irradiation approach can be established as an alternative route in order to obtain greener and more efficient procedures.

### 2.4.2 Microwave-Assisted Biosynthesis

The next approach to discuss is the employment of microwave radiation in a synthesis performed using biological extracts. Both methods have been widely discussed in Sects. 2.2 and 2.3, respectively. In this case, the synergistic effect is more evident than in the previous case. On the one hand, hazardous reagents were replaced to more ecofriendly reducing agents. On the other hand, the microwave radiation highly increases the efficiency and greatly reduces the time necessary for the synthesis process.

This approach has been employed to synthesize majorly metallic nanoparticles. Bhagavanth et al. described a gold nanoparticles synthesis employing an extract of *Annosa Squamosal* L. assisted with microwave radiation. In this work, the synthesis was carried out in 5 min, obtaining spherical-shaped forms and a distribution size of 11 ± 2 nm. In addition, the nanoparticles generated possess good stability due to the carbonyl and hydroxyl groups, which surround the nanoparticles (Reddy et al. 2018). Jahan et al. used an extract of *Rosa Santana* and the microwave radiation to generate silver nanoparticles. It should be noticed that the stability of the nanoparticles was outstanding, being able to maintain their characteristics during 9 months. The size distribution of the nanoparticles provided an average of 14.48 nm with a roughly spherical shape (Jahan et al. 2019). Other researchers have also reported their results in the silver nanoparticles synthesis employing this approach (Eshghi et al. 2018; Francis et al. 2018; Ukkund et al. 2019). Besides, metallic oxide nanoparticles can also be obtained by this hybrid technique. Chankaew et al. reported the synthesis of ZnO nanoparticles and their application in solar cell technology. They used a crude water extract of *Dimocarpus longan* as biological component. The nanoparticles were obtained in about 30 min using cycles of 1 min-off 1 min-on. The distribution size was about 10–100 nm and amorphous shape was observed. However, it is noteworthy to mention that the final nanoparticles possessed a pure hexagonal phase (Chankaew et al. 2019). $SnO_2$ quantum dots have been obtained using this approach as well. In this case, the extract of *Parkia speciosa* was employed with a microwave program consisting in 30 shots of 10 s each one. Highly pure crystalline and tetragonal rutile polycrystalline structure was observed. Besides, morphology of this nanoparticles was spherical with an average diameter of 1.9 nm (Begum and Ahmaruzzaman 2018). Biopolymers can be employed in the metallic nanoparticles synthesis assisted by microwave as well. Torabfam et al. described the silver nanoparticles generation by using chitosan and microwave radiation. Experimental design was carried out to study the best conditions for the chitosan solution. The final synthesis was performed in 100 s leading to an average size of 37 nm. In addition, a spherical shape was noted. The high zeta value obtained (+50 mV) indicated the high stability of the nanoparticles obtained (Torabfam and Jafarizadeh-Malmiri 2018). Naggar et al. performed similar process for the obtention of gold and bimetallic gold/silver nanoparticles. The biopolymer used in this work was curdlan, firstly reported for this purpose. The synthesis of both nanomaterials was done in 10 min with microwave assistance. AuNPs with 52 nm as average size were synthesized. Concerning the nanoparticles built as silver core and gold shell, lower size was appreciated, around 45 nm. Furthermore, despite the fact that no important differences were observed in Ag and Au X-ray diffraction patterns due to their close lattice constant, the structure face-centered cubic was confirmed (El-Naggar et al. 2016). Thus, as it can be noticed, this tandem seems to offer good stability, fast reaction times, and good distribution size of the nanomaterials obtained.

### 2.4.3 Ultrasound-Assisted Biosynthesis

The use of ultrasound as support in the biosynthesis will be briefly discussed in this section. Each individual component has been exposed in Sects. 2.1 and 2.3, respectively. As it has been commented in Sect. 2.4.1, the interest of these hybrid techniques lays in the synergistic effect gained through their simultaneous use. In this case, the cavitation phenomena creates nano-reactors and aids the chemical reaction, meanwhile the selected biological material will interact with the precursor, stabilizing afterwards the final product formed. This is translated into an easier formation of highly stable nanoparticles, much reduced reaction times, and higher yields. As in Sect. 2.4.2, this approach has been mainly exploited in the synthesis of nanoparticles. Manjamadha et al. have informed the synthesis of Ag nanoparticles using an extract of *Lantana camara* and the assistance of ultrasounds. The spherical AgNPs obtained had an average size of 33.8 nm. The process was done in 10 min obtaining pure crystalline phases. However, the nanoparticles obtained exhibit a wide size distribution (Manjamadha and Muthukumar 2016). The synthesis of gold nanoparticles was performed by Franco-Romano et al. They employed geranium extract (*Pelargonium zonale*) and a high-power ultrasound probe. The whole process was done in 3.5 min obtaining nanoparticles of 12 ± 3 nm. In this study, the synthesis conditions were optimized by means of an experimental design, demonstrating that higher volume of reducing agent and precursor led to better synthesis process. On the other hand, it was also exposed that higher volumes of the metallic precursor were related with nonspherical-shaped nanoparticles. The nanoparticles stability was also assessed, showing 8 weeks of lifetime (Franco-Romano et al. 2014). Gu et al. reported the ultrasound-assisted generation of CuO nanoparticles using an alga extract (*Cystoseira trinodis*). The process was performed in 90 min, obtaining a mean size of 9 nm, and a pure nanocrystalline phase was observed too. Nevertheless, several agglomeration were exposed in transmission electron microscopy analysis; these agglomeration increased the range of the size distribution (Gu et al. 2018). Bayrami et al. synthesized another metal oxide nanoparticle, $ZnO_2$, by using a leaf extract of *Vaccinium arctostaphylos*. The ultrasound application lasted 15 min, obtaining nonspherical nanoparticles with a size about 100 nm. A hexagonal wurtzite with a high grade of crystallinity was checked in X-ray diffraction assays. In addition, assays for the medical applications of these nanoparticles were done, exposing the improvement in their characteristics, antidiabetic and antibacterial, in comparison with the conventional chemical route of synthesis (Bayrami et al. 2019). Not only can nanoparticles be made employing this approach, but also biopolymers. Zhu et al. described the synthesis of chitosan employing a biological agent (*Ganoderma lucidum* spore powder) and the assistance of ultrasound. This synthesis was based on the ultrasound-assisted deacetylation approach, which reduce the temperatures required, diminish the time needed, and lesser severely the depolymerization risk. X-ray diffraction, thermogravimetric analysis, and Fourier transformed infrared spectroscopy characterization confirmed the successful generation of this polymer. Finally, the antibacterial properties were examined and the chitosan prepared via ultrasound-assisted deacetylation showed better antibacterial properties (Zhu et al. 2018a). Therefore, all the data previously exposed indicates that the ultrasound-biosynthesis binomial is an interesting approach which can offer several advantages over the traditional methods.

## 3 Trends in the Green Synthesis of (Nano)Materials

The development of (nano)materials with specific features has received much attention in the last decades. Many researchers have made great efforts on the modification of bare devices in order to improve their (bio)sensing properties following the Green Chemistry rules. In addition to the properties of the resulting material, other characteristics, such as low cost, easy setup, low toxicity, and high scalability, are desirable to obtain competitive materials with high applicability in several fields.

In this section, the obtention of diverse materials following green approaches will be summarized, while their application in (bio)sensing will be discussed in Sect. 4. Furthermore, the green synthesis of electrode materials used as bare electrochemical transducers is also reported.

### *3.1 Graphene-Based Materials*

Graphene is a two-dimensional material constituted by a single or few layers of carbon atoms comprised in hexagonal rings. Its promising optical, magnetic, electric, and thermal properties, such as high electrical and thermal conductivity, tunable band-gap, and high tensile strength, are suitable for sensing purposes (Rao et al. 2009). According to all these remarkable reasons, the modification of bare electrodes by graphene layer deposition was performed to constitute electrochemical devices with excellent analytical features (Kim et al. 2010).

Bottom-up and top-down approaches can be used to obtain pristine graphene. It is noteworthy to mention that only liquid-phase exfoliation, chemical reduction of graphene oxide (top-down approaches), and chemical vapor deposition (bottom-up approach) will be considered due to their high scalability, allowing obtaining graphene layers at industrial scale (Backes et al. 2017; Zhu et al. 2018b).

### 3.1.1 Graphene Layer by Liquid-Phase Exfoliation

Graphene flakes were produced via liquid-phase exfoliation of graphite by sonication or shear forces mixing. Figure 11 shows a schematic representation of this process.

Xu and coworkers summarized the main aspects regarding the exfoliation process, highlighting the use of solvents as dispersing agents of graphite and the subsequent graphene sheets (Xu et al. 2018). In their work, the selection of the solvent constitutes major relevance, since the dispersibility of the solid in the liquid strongly depends on their interfacial tension: high values lead to poor dispersibility, favoring the agglomeration of the graphene flakes. Organic solvents with surface tension values from 40 to 50 mJ $m^{-2}$ seem to be the best choices for obtaining stable graphene dispersions by liquid-phase exfoliation (Du et al. 2013). In Table 5 common organic solvents employed in exfoliation, together with their surfaces tension values reported in the literature, are summarized.

Various research papers were devoted to the replacement of organic solvents to water-based systems. The use of a wide variety of surfactants in the liquid-phase exfoliation of graphite was tested, since surface tension of water could be reduced, making feasible the exfoliation of graphite (Narayan and Kim 2015). A recent work regarding the role of surfactant in liquid-phase exfoliation of graphite could be highlighted, demonstrating its current research interest (Sukumaran et al. 2019).

The employment of ionic liquids was reported as promising solvents for liquid exfoliation of graphite in some research works owing to their extraordinary thermal stability, low vapor pressure, and low flammability (Bari et al. 2014; Godoy et al.

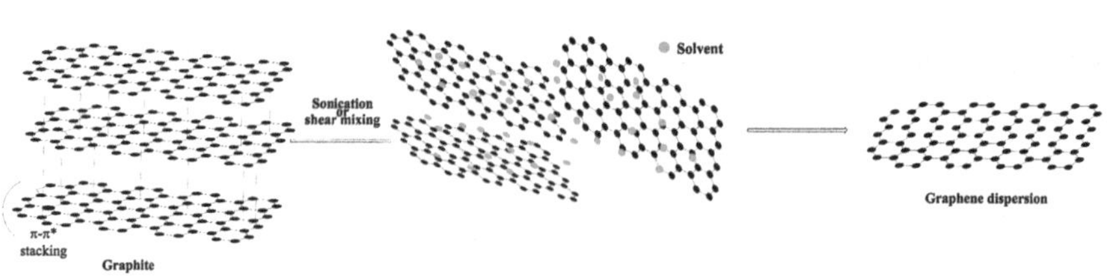


**Fig. 11** Liquid-phase exfoliation of graphite

**Table 5** Common organic solvents employed for liquid-phase exfoliation of graphite

| Organic solvent | Chemical structure | Surface tension, $\gamma^a$ (dyn·cm$^{-1}$) | Method | References |
|---|---|---|---|---|
| *N*-methyl-2-pyrrolidone (NMP) | 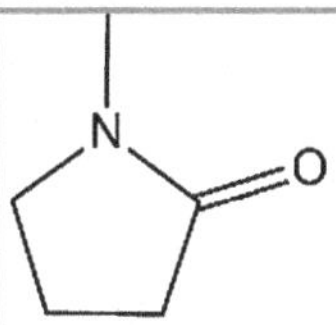 | 44.6 | Tip-sonication | Khan et al. (2012) |
| | | | Bath sonication | Bracamonte et al. (2014) |
| | | | High shear mixing | Tran et al. (2016) |
| Ortho-dichlorobenzene (O-DCB) | 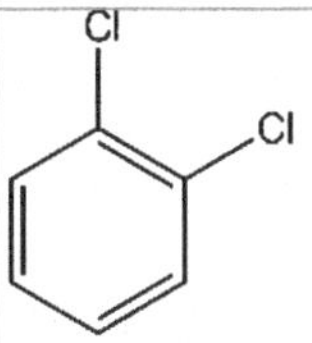 | 35.7 | Bath sonication | Sahoo et al. (2013) |
| N,N-Dimethylformamide (DMF) | 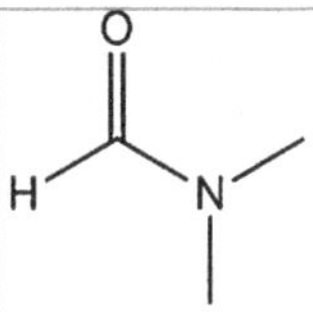 | 34.4 | Tip-sonication | Durge et al. (2014) |
| γ-Butyrolactone (GBL) | 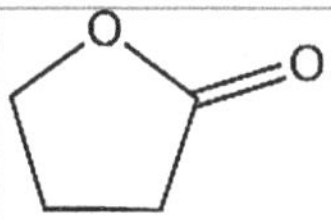 | 53.2 | Bath sonication | Hernandez et al. (2008) |

[a]Values reported at 25 °C from (Yaws 2009)

2019). However, their toxicity constitutes a high controversy nowadays (Bystrzanowska et al. 2019).

The use of natural extracts was also proposed for the exfoliation of graphene. Chitosan and alginate were studied by Uysal and coworkers as alternatives to organic solvents in the exfoliation of graphite (Uysal Unalan et al. 2015). In their work, the stability of the graphene dispersion was higher when chitosan-assisted, which can be explained in terms of the affinity between the biopolymer and the graphene sheets. Regarding the use of chitosan, the nonpolar segments have a solid affinity with the graphene surface. Moreover, the ionic repulsion between the amine groups and the sheets prevents their agglomeration, leading to stable graphene dispersion. In the case of alginate, the compatibility with the graphene sheets is thermodynamically unfavorable, leading to the restacking and precipitation of the graphene sheets. Black tea was employed to produce graphene in one-step exfoliation method by using a kitchen mixer (Ismail et al. 2017). Other research work reported the use of instant coffee to produce few-layer graphene by ultrasound, proposing the chlorogenic acid as the chemical active component for the graphene functionalization (Abdullah et al. 2019).

The manufacturing of graphene nanoplatelets by high temperature vapor exfoliation of graphite should be mentioned. In this work, no chemicals or surfactants were required for the exfoliation or dispersion of graphene, reaching good dispersibility even at higher concentrations (Ding et al. 2018).

### 3.1.2 Graphene Layer By Chemical Reduction of Graphene Oxide

The reduction of graphene oxide (GO) constitutes another scalable top-down approach to get graphene flakes for mass production (Lavin-Lopez et al. 2017). The first step consists of the oxidation and exfoliation of graphite to graphene oxide, which forms a stable colloid dispersion in water due to the presence of hydroxyl groups. The reduction of graphene oxide leads to reduced-graphene oxide (r-GO) by removing these groups, restoring partially the $\pi$-$\pi*$ conjugation (Chua and Pumera 2014) (Fig. 12).

The graphene oxide has poor electrical conductivity due to the presence of high content of hydroxyl groups on the edge and basal planes. After the reduction process, some residual oxygen groups still remain in the structure, leading to lower electrical conductivity in comparison with the pristine graphene layer (Pei and Cheng 2012).

C/O ratio should be a critical parameter to evaluate the electrical conductivity, since lower content of oxygen should lead to graphene sheets with higher electrical conductivity. However, this factor is also affected by other parameters, such as sheet orientation or percolation effects (Guex et al. 2017). In spite of this fact, C/O ratio could be used to evaluate the efficient removal of oxygen in resulting graphene sheets, by comparing this value with the one obtained for the initial GO.

The chemical GO reduction involves the addition of a reductant, which plays a key factor in the obtention of graphene layers with the desirable properties for sensing purposes. Several research works were focused on the generation of reduced-graphene oxide by employing diverse reductants, evaluating their C/O ratio. In this sense, hydrazine, hydrazine hydrate, and sodium borohydride were employed for obtaining high-quality graphene layer (Luo et al. 2011; Guex et al. 2017). Due to their toxicity for the living organisms and the environment, their replacing to greener ones is well studied in the last years. With this purpose, less hazardous chemicals and plant extracts, among others, were proposed as alternative reducing agents (De Silva et al. 2017). In Table 6 some examples reported in literature are summarized.

As observed in the previous table, the C/O ratio values of r-GO were higher than the one obtained for GO, indicating the removing of oxygen after reduction.

Chemical reduction of GO can also be assisted by microwave irradiation. Hassan and coworkers performed the chemical reduction of GO assisted by microwave

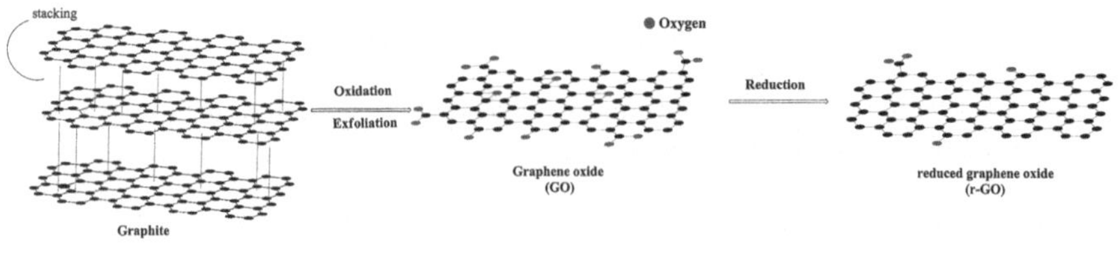


**Fig. 12** Chemical reduction of GO to generate reduced-graphene oxide

**Table 6** Compounds employed in the chemical reduction of graphene oxide

| Reducing agent | C/O ratio | References |
|---|---|---|
| *GO* | | |
| n.a. | 1.80–2.37 | Peng et al. (2016) |
| *rGO-chemicals* | | |
| Ascorbic acid | 4.70 | De Silva et al. (2018) |
| Sodium sulfite | 4.80 | Yin et al. (2019) |
| Caffeine | 6.50 | Vu et al. (2015) |
| *rGO-peel extract* | | |
| Lemon extract | 4.66 | Dandan et al. (2017) |
| *rGO-plant extract* | | |
| *Artemisin* | 11.7 | Hou et al. (2018) |
| *Phaseolusaureus L* | 6.60 | Jana et al. (2014) |
| *Ocinum sanctum* | 3.10 | Mahata et al. (2018) |
| *Colocasia esculenta* | 7.11 | Thakur and Karak (2012) |

*GO* graphene oxide, *rGO* reduced graphene oxide

employing hydrazine hydrate as reductant, obtaining r-GO in few minutes (Hassan et al. 2009), in comparison with the one required by conventional method, about 24 h (Stankovich et al. 2007). Hence, shorter reaction times seem to be endorsed to the application of microwave due to the effective transference of energy to the precursors (Hu et al. 2012a).

The sequential chemical reduction and microwave irradiation of GO was performed by Wen and coworkers (Wen et al. 2014). The electrical conductivity of the rGO after microwave treatment was higher than those obtained by using either single microwave-assisted or chemical reduction of GO. Besides, defects of the rGO obtained by chemical reduction seem to be repaired by microwave.

Microwave-assisted GO reduction was performed with ascorbic acid as reductant under $N_2$ atmosphere, leading to rGO in 3 min (Iskandar et al. 2017). The flow of nitrogen was found to be more effective, since gases formed as side products during reduction process were removed. Furthermore, the electrical conductivity of the resulting graphene flakes was increased after performing the annealing treatment. Another recent work shows novel approach to obtain graphene patterns onto graphene oxide film by using an rGO template assisted by microwave. This method provides shape-controlled graphene patterns with excellent electrical conductivity (Zhao and He 2019).

Therefore, microwave can be employed as a green powerful tool to obtain graphene flakes from the chemical reduction of GO at short reaction times. However, this technique is yet to be explored, since oxygen and some structural defects are still retained, disrupting the electronic conjugation and, hence, decreasing the electrical conductivity in comparison with pristine graphene (Xie et al. 2019).

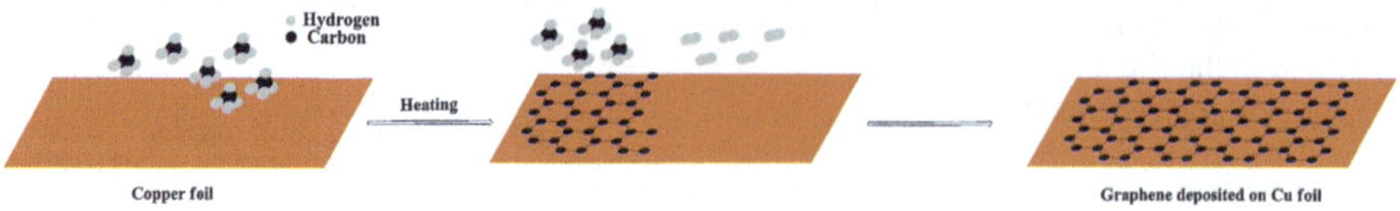


**Fig. 13** Chemical vapor deposition of graphene

### 3.1.3 Chemical Vapor Deposition

Chemical vapor deposition has attracted huge attention for the obtention of uniform and large area graphene layers with low defects. The first step involves the decomposition of the carbon precursor, carried in the vapor phase with an inert gas, e.g., argon. Afterwards, carbon was deposited and grown on a metal substrate surface, such as copper, nickel, palladium, or iridium, among others, which plays a catalytic role in the nucleation and growth process (Zhang et al. 2013a) (Fig. 13).

The first industrial chemical vapor deposition process of graphene deposited on copper foil was reported by Ruoff and coworkers using methane as carbon source. The graphene film can be moved to another substrate, like $SiO_2$/Si (Li et al. 2009).

The employment of high temperatures in the synthesis of graphene via chemical vapor deposition, around 1000 °C, is usually required (Zhang et al. 2013a). Other chemical vapor deposition derivatives, such as plasma enhanced (Li et al. 2016), inductively coupled plasma (Pekdemir et al. 2017), microwave plasma (Fang et al. 2016), and photo-induced, can be excellent alternatives to reduce the working temperature. Son and Ham summarized the recent progress in the employment of these techniques for synthesis of graphene, highlighting its application in electronic devices (Son and Ham 2017).

New alternative carbon sources were investigated as green precursors. The deposition of graphene on a copper substrate via microwave plasma-chemical vapor deposition using a camphor precursor was reported by Uchida and coworkers (Hideo Uchida et al. 2016). In their work, graphene sheets using camphor showed superior qualities in terms of lower sheet resistance than those obtained using methane. Other natural resources, such as palm oil (Salifairus et al. 2016) and tea tree extract (Jacob et al. 2015), were also employed as green carbon sources due to its high availability and low cost.

A recent work reported a new chemical vapor deposition method using a solid waste plastic as carbon solid precursor (You et al. 2017). A waste material was transformed into a functional material by following a green and sustainable approach at atmospheric pressure, since no chemicals and reagents were used. Therefore, this method seems to be a promising way to obtain high-quality monolayer graphene flakes at industrial scale.

## 3.2 *Carbon and Graphene-Quantum Dots*

Carbon-based dots are fluorescent particles with high applicability in analytical and biomedical fields. Cayuela and coworkers summarized the three main types, carbon nanodots, graphene-quantum dots, and carbon-quantum dots, defining their properties and providing a detailed discussion about their applications (Cayuela et al. 2016).

Carbon quantum dots are small carbon particles lower than 10 nm. Graphene quantum dots are constituted by graphene disks from 2 to 20 nm. Both materials have promising applications in sensing and bioimaging owing to their tunable fluorescent properties, small sizes, and low toxicity. Regarding the sensing applications, their deposition onto bare surfaces improves the electrochemical features of the resulting devices (Sun et al. 2013; Algarra et al. 2018).

Carbon-quantum dots can be doped or co-doped with heteroatoms to improve their fluorescence efficacy in terms of quantum yields. Hence, the doping with boron, nitrogen, sulfur, phosphorous, and fluorine was reported in literature (Kandasamy 2019). Regarding graphene-quantum dots, the doping with heteroatoms is also reported (Feng et al. 2018; Kaur et al. 2018).

Several ecofriendly methods have been recently developed for the obtaining of carbon-quantum dots and graphene-quantum dots and their doped derivatives by using different carbon precursors, including waste products and low-toxic chemical reagents (Das et al. 2018). In this sense, microwave and hydrothermal methods emerged in the production of carbon-quantum dots and graphene-quantum dots from green precursors. In Table 7 some recent examples found in the literature are summarized.

Graphene-quantum dots can also be obtained from bulk carbon sources. They were synthesized using different carbon precursors by hydrothermal route (De Xie et al. 2007). In this work, hydrogen peroxide was used as oxidizing reagent, instead of harsh chemicals used in conventional hydrothermal route, such as nitric acid and sulfuric acid. Another hydrothermal method using graphene oxide as carbon source and hydrogen peroxide was reported (Tian et al. 2016). The acid-free synthesis of graphene-quantum dots by sonochemical method with intermittent microwave heating was performed using graphene oxide as carbon source. The resulting graphene-quantum dots exhibited a high quantum and product yields (Nair et al. 2017). Other acid-free approach by microwave was performed to obtain boron-doped graphene-quantum dots from graphene oxide (Hai et al. 2015).

## 3.3 *Multi-Walled Carbon Nanotubes*

Carbon nanotubes are carbon allotropes consisting of rolled-up sheets of carbon atoms. Their outstanding optical, mechanical, thermal, and electrical properties make them suitable candidates in gas sensing (Mao et al. 2014) and electrochemical biosensing (Du et al. 2017), among other applications.

**Table 7** Obtention of C-QDs and G-QDs assisted by microwave and hydrothermal methods using different precursors

| Method | Precursors | Size (nm) | References |
|---|---|---|---|
| *Carbon-quantum dots* | | | |
| Microwave | Lysine | 5–10 | Park et al. (2017) |
| | Xylan | ≈7.9 | Yang et al. (2018) |
| | Phthalic acid and triethylenediamine | ≈3.5 | Yu et al. (2018) |
| | Chitosan and lysine | ≈5.5 | Janus et al. (2019) |
| Hydrothermal | Cabbage | 2–6 | Alam et al. (2015) |
| | Fresh lemon Ethanol | ≈3.1 | He et al. (2018) |
| | Citric acid | 2–4 | Shen et al. (2018) |
| | Glucose | 3–6 | |
| | Tofu yellow serofluid and Sodium hydroxide | 3.5–5.5 | Zhang et al. (2017) |
| Microwave-assisted hydrothermal | Arginine and glycerin | ≈4.4 | Huang et al. (2019) |
| *Graphene-quantum dots* | | | |
| Hydrothermal | Citric acid and thiourea | ≈2.69 | Qu et al. (2013a) |
| Microwave | Aspartic acid and ammonium bicarbonate | 1.8–2.4 | Zhang et al. (2016) |
| | Glucose and urea | <15 | Fresco-Cala et al. (2018) |
| Microwave-assisted hydrothermal | Citric acid and urea | ≈5 | Nguyen et al. (2019) |
| | Glucose and urea | ≈3 | Hou et al. (2016) |

Regarding the synthesis of carbon nanotubes, chemical vapor deposition, arc discharge, and laser ablation constitute the most popular methods for production of MWCNTs nanotubes (Rahman et al. 2019). Several drawbacks are ascribed to them, such as high temperatures required, low energy efficiency, and low yields. Furthermore, fossil carbon sources were employed, e.g., methane (Li et al. 2004) and ethylene (Weizhong et al. 2003). Several green attempts using biomass and other ecological products as raw precursors were carried out. In this section, chemical vapor deposition and microwave pyrolysis will be considered.

### 3.3.1 Carbon Nanotubes Production by Chemical Vapor Deposition from Ecological Sources

As previously commented, chemical vapor deposition raised as a well-established synthesis method of carbon nanomaterials at both laboratory and industrial scale (Pang et al. 2016).

The conventional carbon sources used in chemical vapor deposition were replaced to green precursors in the last decade. An environmental-friendly hydrocarbon, namely camphor, was reported in the obtention of carbon nanotubes via chemical vapor deposition (Pandey et al. 2013). This precursor was also used as carbon source using rice straws hydrothermally treated by a carbonization hydrothermal method with either ferrocene or ferrocene mixed with nickel nitrate (Fathy 2017). Neem oil was also employed in the synthesis of carbon nanotubes by spray pyrolysis-assisted chemical vapor deposition using ferrocene as catalyst (Kumar et al. 2011).

The use of catalysts, commonly transition metals, was required for the catalytic chemical vapor deposition of carbon nanotubes (Rashid et al. 2015). New alternative catalysts obtained from natural sources were proposed over metallic ones. The catalytic role of several plant leaves extracts, garden grass, rose, kaner, and walnut was proposed in the carbon nanotubes growth on silicon substrate (Tripathi et al. 2017). In this work, nontoxicity, high availability, low temperature, and low cost, among other advantages, are highlighted. Natural laterite powder was also recently employed as a catalyst source for the carbon nanotubes growth through chemical vapor deposition, proposing a growth mechanism involving the fragmentation of the pristine iron-containing material (Kumar et al. 2018). The minerals present in natural resources were reported to play a catalytic role in the carbon nanotubes grown process. The minerals obtained from bamboo charcoal, especially calcium silicate and magnesium metasilicate, were found to be responsible of the nucleation and growth of carbon nanotubes carried out by chemical vapor deposition (Zhu et al. 2012). Another work reports the participation of minerals present in the coconut shell in the growth of carbon nanotubes using radio frequency plasma-enhanced chemical vapor deposition (Araga and Sharma 2017).

Thus, new carbon sources and catalysts were investigated for carbon nanotubes production by means of a chemical vapor deposition process. The green methodologies proposed aimed to reduce the overall costs, energy requirements, reaction times, and the formation of toxic side products.

### 3.3.2 Carbon Nanotubes Production by Microwave Pyrolysis of Biomass

The carbon nanotubes production by microwave pyrolysis of the biomass has attracted huge attention nowadays. This method involves the decomposition of biomass by microwave heating, leading to higher yields of pyrolytic products in comparison with those obtained by conventional heating (Dhyani and Bhaskar 2018). The principles and mechanisms of pyrolysis by microwave were reported in the bibliography (Nomanbhay et al. 2017; Nizamuddin et al. 2018).

Gumwood was used as carbon source in the carbon nanotubes synthesis via microwave pyrolysis (Shi et al. 2014). The decomposition rate of the biomass by using microwave was found to be greater than the one obtained by means of conventional heating, leading to higher yields of pyrolytic products, as commented before.

The microwave pyrolysis of bagasse was performed by using iron and cobalt as metal susceptors (Debalina et al. 2017). Another recent work reported the use of palm kerner shell as raw material, obtaining carbon nanotubes in mild conditions and without adding any catalyst (Omoriyekomwan et al. 2019). The role of cellulose in the synthesis and growth of carbon nanotubes was also reported.

Therefore, microwave heating of biomass constitutes a promising synthesis method in terms of Green Chemistry. However, the mechanism of formation and growth of carbon nanotubes are not fully understood, and hence further investigations are required.

### 3.3.3 Functionalization of Carbon Nanotubes

Although this subsection is focused on the obtention of carbon nanotubes by green methodologies, their functionalization is also relevant, since their sensing properties seem to be enhanced (Setaro 2017). Among all the functionalization approaches, the decoration of carbon nanotubes with metal nanoparticles constitutes one of the most important methods (Kharisov et al. 2016).

Several decoration procedures were reported in the bibliography. A cost-effective and ecofriendly method using focused solar irradiation over metal precursor and carbon nanotubes dispersion was reported. Different metal, metal oxide, and metal alloy nanoparticles, such as Au, Pt, Ag, NiO, ZnO, and $Pt_3Co$, were successfully attached on carbon nanotubes surface by using this approach (Baro et al. 2013). Palladium nanoparticles were deposited onto the multi-walled carbon nanotubes surface by drop casting of palladium salt precursor, followed by ultraviolet irradiation (Yoo et al. 2019). The decoration of carbon nanotubes with bimetallic nanoparticles was mediated by aqueous plant extracts (Mendoza-Cachú et al. 2018). Silver nanoparticles were deposited on carboxylated and hydroxylated carbon nanotubes by using a modified Tollens process, which involves the chemical reduction of $[Ag(NH_3)_2]^+$ complex with sugars (Dinh et al. 2015). A one-pot microwave synthesis of copper oxide nanoparticle-carbon nanotubes was performed in a solvent-free system using copper acetate as precursor (Rudd et al. 2019). Figure 14 represents a scheme of each functionalization procedure described.

## *3.4 Metal and Metal Oxide Nanoparticles-Supported Materials*

The one-pot synthesis of metallic nanoparticles mediated by biological compounds is currently considered by the scientific community (Shamaila et al. 2016). A brief overview of the metallic nanoparticles synthesis using natural extracts and biopolymers will be stated in this section.

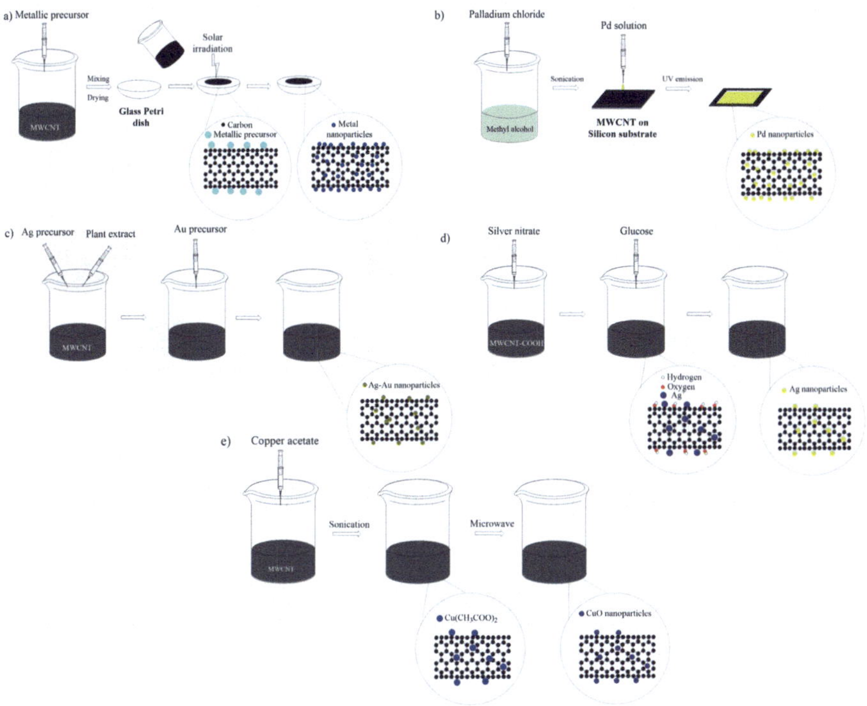


**Fig. 14** Different decoration methods of carbon nanotubes with metal nanoparticles: (**a**) Deposition of metal/metal oxide NPs by solar radiation, (**b**) deposition of PdNPs by drop casting on MWCNT/Si substrate, (**c**) deposition of Ag-AuNPs mediated by plant extract, (**d**) deposition of AgNPs by using a modified Tollens method, and (**e**) deposition of CuONPs mediated by the sonication and subsequent microwave treatment of copper acetate

### 3.4.1 Synthesis of Metal Nanoparticles Mediated by Plant Extracts

The employment of several organisms, such as fungi (Vágó et al. 2016; Gudikandula et al. 2017), bacteria (Ahmad et al. 2017), and plants (Makarov et al. 2014), for one-pot metal nanoparticles synthesis has been extensively studied. Among all of them, plant leaf extracts seems to be a good choice based on green terms: scalable process, lower time required, and eco-friendly waste products (Mittal et al. 2013; Yadi et al. 2018).

The last one is a crucial point to reduce the presence of pollutants, improving the human health. Furthermore, the synthesis procedure is very easy, allowing the generation of metal nanoparticles by direct mixing of the plant extract and the metal solution at room temperature (Fig. 15). Thus, high temperatures and additives are not required for the synthesis step.

The synthesis mechanism of metal nanoparticles using plant extracts is very complex owing to the high amount of phytochemicals able to reduce the metal salt, as well as to avoid the aggregation of the resulting nanoparticles. These compounds,

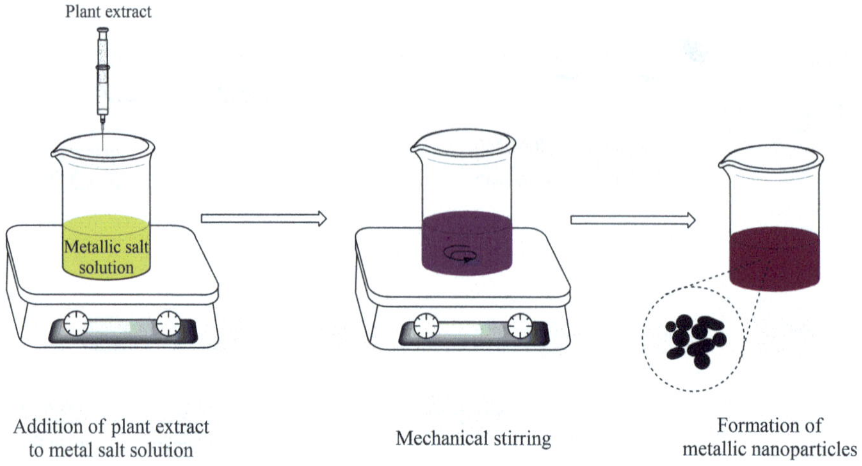


**Fig. 15** Synthesis of metal nanoparticles mediated by plant extracts

like flavonoids, sugars, and amino acids, among others, contain functionalized groups responsible for the bioreduction and the subsequent stabilization processes (Rai et al. 2013; Jain and Mehata 2017; Zarzuela et al. 2018).

According to the biochemical reduction mechanism proposed in several reports, the plant extract composition is a key factor in the shape and size of the metal nanoparticles obtained. Table 8 summarizes several examples reported in the bibliography in the last years.

As observed in Table 8, the morphology of AuNPs is strongly dependent on the composition of the extract. Particularly, Lee and coworkers carried out a detailed study about the reducing role of the phytochemicals contained in the extract for the gold nanoparticles synthesis (Lee et al. 2016). This research work revealed different morphologies depending on the fraction of the extract considered. The crude extract led a heterogeneous mixture of gold nanoparticles, while hexane fraction displayed spherical gold nanoparticles with a few proportion of anisotropic nanoparticles. Regarding the chloroform fraction, gold discs lower than 200 nm were mostly obtained. The aqueous and *n*-butanol extracts led to Au platelets and irregular nanoparticles, respectively. A similar study was performed for gold nanoparticles synthesized using *Zostera noltii* extract (an aquatic plant). The buthanolic extract led to mostly spherical nanoparticles with bimodal bell-shaped distribution, while the aqueous/dimethylsulfoxide flavone fraction of buthanolic extract led to nanoparticles with triangular, spherical, and polyhedral shapes (Zarzuela et al. 2018).

The concentration of the phytocompounds present in the extract can influence the shape and size of metal nanoparticles. Smitha and coworkers reported higher proportion of gold anisotropic particles at lower volumes, while spherical shape dominates at higher ones (Smitha et al. 2009). The change in the AuNPs shape by varying the concentration of the extract can be stated on the biosynthesis of gold using other sources. Gold nanotriangles were obtained at low concentration of *Aloe*

**Table 8** Size and shape of several metal and metal oxide nanoparticles obtained by using different plant extracts

| Metal nanoparticle | Plant specie | Origin of the extract | Size (nm) | Morphology | References |
|---|---|---|---|---|---|
| Gold nanoparticles (AuNPs) | *Pelargonium zonale* | Leaf | 8–20 | Spherical | Franco-Romano et al. (2014) |
| | *Ocimum sanctum* | Leaf | 10–300 | Spherical triangular hexagonal platelets | Lee et al. (2016) |
| | *Coleus aromaticus* | Leaf | <20 | Spherical triangular hexagonal | Boomi et al. (2019) |
| | *Solanum nigrum* | Leaf | 5–35 | Spherical | Muthuvel et al. (2014) |
| | *Sanseviera roxburghiana* | Leaf | 5–31 | Spherical triangular hexagonal rod | Kumar et al. (2019) |
| | *Croton caudatus Geisel* | Leaf | 20–50 | Spherical | Vijaya Kumar et al. (2019) |
| | *Pueraria lobata* | Root | 5–36 | Spherical | Zhou et al. (2019) |
| | *Chenopodium aristatum* L. | Stem | 7–19 | Spherical Triangular Pentagonal Hexagonal | Golinska et al. (2017) |
| | *Zostera noltii* | Leaf | <11; 20–35 (Buthanolic extract) | Spherical | Zarzuela et al. (2018) |
| | | | 25–65 (flavonoid fraction) | Spherical Triangular | |
| Silver nanoparticles (AgNPs) | *Ocimum basilicum* | Leaf | ≈23 | Spherical | Pirtarighat et al. (2019) |
| | *Satujera hortensis* | Leaf | 2.9–3.4 | Spherical | Rasaee et al. (2018) |
| | *Artemisia vulgaris* | Leaf | ≈25 nm | Irregular | Rasheed et al. (2017) |
| | *Capparis spinosa* | Leaf | 5–30 | Spherical | Benakashani et al. (2016) |
| | *Beberis vulgaris* | Leaf | 30–70 | Spherical | Behravan et al. (2019) |
| | *Cynara cardunculus* | Leaf | <45 | Semi-spherical | Ruíz-Baltazar et al. (2018) |

(continued)

**Table 8** (continued)

| Metal nanoparticle | Plant specie | Origin of the extract | Size (nm) | Morphology | References |
|---|---|---|---|---|---|
| | *Ocinum sanctum* | Roots | 10 | Spherical | Ahmad et al. (2010) |
| | *Ocinum sanctum* | Stem | 5 | Spherical | |
| | *Avicennia marina* | Leaf roots stem | 20–40 | Spherical | Abdi et al. (2018) |
| Zinc oxide nanoparticles (ZnONPs) | *Cassia fistula* | Leaf | 5–15 | Irregular | Suresh et al. (2015) |
| | *Aloe barbadensis miller* | Leaf | 25–55 | Spherical Hexagonal | Sangeetha et al. (2011) |
| Palladium nanoparticles (PdNPs) | *Camellia sinensis* | Leaf | 6–18 | Spherical | Azizi et al. (2017) |
| | *Catharanthus roseus* | Leaf | ≈38 | Spherical | Kalaiselvi et al. (2015) |
| Copper nanoparticles (CuNPs) | *Tilia extract* | Leaf | 4.7–17.4 | Spherical | Hassanien et al. (2018) |
| | *Plantago asiatica* | Leaf | 7–35 | Spherical | Nasrollahzadeh et al. (2017) |

*vera*, while spherical nanoparticles were obtained at higher values (Chandran et al. 2006). Anisotropic gold nanoparticles were also obtained using low concentration of *Magnolia kobus* (Song et al. 2009).

The nanoparticle's morphology is influenced by other elements, like pH, reduction time, and temperature, all of them summarized in Baranwal and coworkers review (Baranwal et al. 2016).

### 3.4.2 Synthesis of Metal Nanoparticles Mediated by Polysaccharide-Based Biopolymers

The green synthesis of metallic nanoparticles also includes the use of biopolymers, mainly characterized by their biodegradability, biocompatibility, and nontoxicity (Hernández et al. 2014). Among all of them, polysaccharides are mainly used for biosynthesis (Wang et al. 2017).

As in the case of the plant extracts use, the metallic nanoparticles synthesis using biopolymers is of great interest from the Green Chemistry point of view. The role of the biopolymer as reducing and/or capping agent avoids the use of any additive, reducing the employment of hazardous reagents. Furthermore, the formation of metal nanoparticles can be reached at short times, minimizing the energy requirements (Balachandran et al. 2015). The main synthesis approach consists of the direct mixing of the biopolymer and the metal solution, obtaining the metal nanoparticles after stirring in one-step.

The gold nanoparticles synthesis using pectin was investigated by Ahmed and coworkers (Ahmed et al. 2016). In their report, the fragmented units of pectin obtained after alkali treatment reduce $Au^{3+}$ to $Au^0$ and promote the nucleation process. In addition, the binding between the reducing sugar units and the gold nanoparticles was proposed, preventing their agglomeration. Consequently, small nanoparticles distribution was obtained.

The chemical bonding metal-biopolymer was also proposed in other reports related to the synthesis of metal nanoparticles using polysaccharide biopolymers. Guibal and coworkers summarized the mechanisms involved in metal ion binding on chitosan, proposing the formation of different metallic chitosan complexes (Guibal et al. 2014). The employment of this biopolymer in the synthesis of copper nanoparticles is reported, leading to improvements in their stability (Muthukrishnan 2015). Gold and silver nanoparticles with high stability were also mediated by chitosan (Wei and Qian 2008). The chelating role of soluble starch was also reported by the linkage between the aldehyde terminal of amylose units and silver in the silver nanoparticles synthesis (Yakout and Mostafa 2015).

Table 9 shows some examples of silver nanoparticles' size and morphology synthesized using biopolymers.

The synthesis of metallic nanoparticles can also be mediated by biopolymeric nanostructures using two approaches, according to several research works (Preiss et al. 2014): incorporation of metallic nanoparticles into a polymeric matrix (ex situ) and the synthesis of metallic nanoparticles in the biopolymeric matrix (in situ).

Boury and Plumejeau summarized the main aspects concerning the in situ synthesis of metal oxide nanocomposites (Boury and Plumejeau 2015). The combination between metal precursor and the biopolymer lead to biopolymer-metallic composite. Afterwards, the biotemplate was removed by thermal treatments, obtaining the metallic oxide nanocomposite. A general scheme is shown in Fig. 16.

The metal oxide nanoparticles can be formed and included into the polymeric scaffold using the sol-gel process, controlling their morphology. Catalytic effects in the sol-gel kinetics are ascribed to the bipolymer, as well as a strong chemical affinity with metals due to the presence of hydroxyl and amine groups (Plumejeau et al. 2015). Zlotski and Uglov reported the synthesis and immobilization of different oxide nanoparticles on cellulose fibers template by sol-gel process (Zlotski and Uglov 2017). A flake-like structure was observed after calcination with irregular

**Table 9** Size and morphology of silver nanoparticles using different polysaccharide biopolymers

| Biopolymer | Size(nm) | Morphology | Ref |
|---|---|---|---|
| Starch | 4–14 | Spherical | Božanić et al. (2007) |
| Sodium alginate | 3–12 | Spherical | Chunfa et al. (2016) |
| Chitosan | 10–60 | Spherical | Kalaivani et al. (2018) |
| Pectin | 5–10 | Spherical | Zahran et al. (2014) |
| Dextran | 5–10 | Spherical | Bankura et al. (2012) |

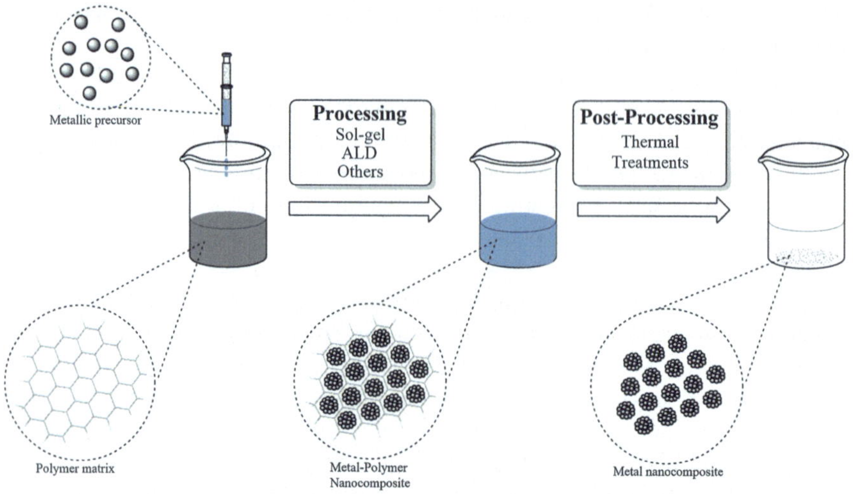


**Fig. 16** Synthesis of metallic nanocomposites assisted by biopolymer templating

pores, testifying that cellulose fibers provided seeding points of metal oxide nanoparticles nucleation and growth. Calcium alginate is also used as biotemplate for the synthesis of titanium oxide beads by sol-gel route (Kimling and Caruso 2012).

β-Cyclodextrin, chitosan, and starch were used for the synthesis of titanium oxide nanoparticles by Bao and coworkers using a simple mineralization process (Bao et al. 2013). In their work, the polysaccharide biotemplate plays a key role on the final nanostructure: small rods for β-cyclodextrin, chestnut-like for chitosan, and small nanoparticles for soluble starch. The influence of the polysaccharide on the crystalline structure is also discussed. The rutile phase was obtained mediated by chitosan, while anatase was obtained mediated by starch and β-cyclodextrin.

Therefore, the biosynthesis of metal and metal oxide nanoparticles was carried out by plant extracts and biopolymers, obtained from renewable and clean sources. Their reducing and capping agent role can be highlighted, reducing the waste products. Furthermore, the nanoparticles shape and size can be monitored by the synthesis conditions, such as phytocompounds composition and concentration of the extract and the nature of the polysaccharide, among others.

## *3.5 Polymers*

Polymers are compounds constituted by several monomeric units linked by covalent chemical bondings. They can be employed as immobilization matrices of several enzymatic compounds, increasing the selectivity of the system with respect to a target analyte (García-Guzmán et al. 2018; Bilal and Iqbal 2019). Their use as bulk transducers in some electronic devices should also be highlighted, improving their electronic transference and, hence, their electrochemical performance (Gautam et al. 2018).

Polymers can be ordered into two groups: intrinsic conducting polymers and nonconducting polymers. Several synthesis routes to produce some of them will be shown in this subsection.

### 3.5.1 Intrinsic Conducting Polymers: Polythiophene Derivatives, Polyaniline, and Polypyrrole

Conducting polymers have emerged in the last decade due to their high electrical properties, among others. Their high conductivity, attributed to the π-conjugation and electronic doping, enables their use in electrochemical sensing (Kenry and Liu 2018).

Moreover, their redox properties make them suitable for the fabrication of electrochromic and sensing devices (Naveen et al. 2017). Figure 17 shows the chemical structure of the most representative intrinsic conducting polymers, poly-(3,4-ethylenedioxythiophene) (PEDOT), polyaniline (PANI), and polypyrrole (Ppy).

The chemical polymerization is a high-scalable approach based on the chemical oxidation of the monomer. Many oxidizing agents and dopants were investigated to control the electrical properties of the resulting polymer and increase their oxidation degree (Nguyen and Yoon 2016). Several synthesis strategies were investigated as alternative polymerization routes.

The **ultrasound-assisted synthesis** led to high polymerization rates, as well as an improvement in the electrical features of the resulting polymer in comparison

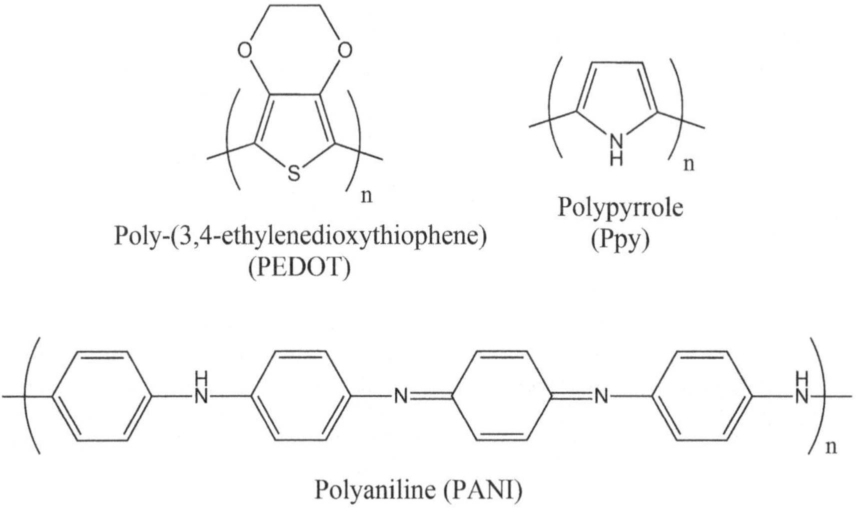


**Fig. 17** Chemical structure of PEDOT, PANI, and Ppy

with conventional route (Ali Mohsin et al. 2016). The influence of frequency waves was studied in the acid polymerization of aniline with ammonium persulfate (Husin et al. 2014). PANI nanofibers' diameter decreased when ultrasonic frequency takes higher values. Thus, the role of the ultrasound in the morphology of PANI was demonstrated. Uniform spherical PEDOT nanoparticles were produced under ultrasound, in contrast with irregular-shaped nanoparticles obtained by conventional stirring (Zhong et al. 2010). PEDOT microspheres with controlled morphology were obtained via ultrasonic spray polymerization by using different oxidants and solvents (Zhang and Suslick 2015). The **microwave-assisted synthesis** of conducting polymers was also proposed as an alternative to conventional synthesis, reducing significantly the reaction time and leading to improvements in product yields. Polyaniline was obtained under microwave in 5 min, with a similar yield value than the one obtained under conventional synthesis at 5 h, around 76%. In addition to the reaction time, the morphology was also affected by microwave heating. Nanofibrillar morphology was observed under microwave, while granular morphology was obtained under conventional conditions (Gizdavic-Nikolaidis et al. 2010). A more recent work shows the influence of the concentration of hydrochloric acid in the morphology of PANI, synthesized by oxidative microwave-assisted polymerization: low acid concentration values provided oligomeric chains with flat structure, while high acid concentration led to nanofibers (Qiu et al. 2017). It should be noted that longer oligomeric chains were observed in microwave-assisted polymerization with respect to those obtained under conventional synthesis at same reaction times, demonstrating the role of microwave heating in the improvement of reaction rates and yields.

Enzyme-catalyzed synthesis is another environmental friendly route to produce conducting polymers with desirable electrical and morphological features at mild conditions. The enzymatic polymerization of thiophene, aniline, and pyrrole with embedded glucose oxidase was performed (German et al. 2019). Several synthesis parameters, such as pH, monomer concentration, and ratio enzyme/substrate, were investigated by spectrophotometric assays. Laccase produced by *Aspergillus oryzae* was used as biocatalyst in the polymerization of aniline (De Salas et al. 2016). The polymerization of EDOT was achieved mediated by horseradish peroxidase with high efficiency, by using hydrogen peroxide as clean oxidant (Wang et al. 2014). Two *Aspergillus niger* strains were employed in the polymerization of pyrrole with hydrogen peroxide (Apetrei et al. 2018).

Therefore, biosynthetic routes provide several advantages with respect to the conventional polymerization process, like improvements in product yields, reduction of reaction times, tailored morphological features, and high electrical conductivities. However, some instrumental parameters required a specific control in order to obtain conducting polymers with the desired electrical properties and nanosized morphology.

### 3.5.2 Nonconducting Polymers: Polysaccharide Biopolymers

Polysaccharide biopolymers are obtained by natural resources. Several pieces of research were focused on their biosynthesis in order to understand the different mechanism pathways. The metabolic engineering offers some improvements in the green synthesis of tailored biopolymers, and thus different synthetic biologic strategies were reported in the literature (Anderson et al. 2018). The biosynthetic routes for three relevant polysaccharide biopolymers are listed in this subsection.

Chitosan and chitin are located in the shell of crustaceans, exoskeleton of mollusks, and fungi's cell walls. For industrial applications, crustacean shells constitute the main source for the production of chitosan and chitin. The chemical extraction procedure to obtain the pure product, shown in a recent review (Abo Elsoud and El Kady 2019), involves five steps: demineralization, discoloration, deproteinization, acid reflux, and deacetylation. The fungal biosynthesis of chitosan was also proposed in the same work for industrial applications over crustacean sources due to several reasons: simple extraction procedure, low cost waste management, and high availability, among others.

Alginate was initially isolated from farmed brown seaweeds for commercial production. However, these algal alginates suffer from heterogeneity in composition and material properties. Other alternative biosynthetic route is the microbial synthesis, summarized in the following review (Hay et al. 2013).

Two types of bacteria, pseudomonas and azotobacter, were used in the alginate biosynthesis (Hay et al. 2013). In this way, the production of alginate was reported using *Pseudomonas mandelii* (Vásquez-Ponce et al. 2017), *aeruginosa* (McCaslin et al. 2015), and *fluorescens* (Maleki et al. 2017). Regarding azotobacter, *Azotobacter vinelandii* was widely employed for alginate biosynthesis (Saeed et al. 2016). Starch is mainly located in the endosperm of cereal grains. Its biosynthesis via ADP-glucose pathway involves the use of sucrose as carbon source, imported from leaves, and the participation of three primary enzymes, ADP-glucose pyrophosphorylase, starch synthase, and starch branching (Thitisaksakul et al. 2012). The starch biosynthesis was investigated in the last decade by using different cereals, such as rice (Fujita 2014), wheat (Chen et al. 2016), maize (Jiang et al. 2013), and grass (Tetlow and Emes 2017). The starch biosynthesis in plants, such as *Cassava* (Tappiban et al. 2019) and *Arabidopsis thaliana* (Malinova et al. 2018), was also investigated.

## *3.6 Bare Electrochemical Devices and Their Modifications: Carbon Ceramic Materials*

Although the majority of reports related to chemical sensors were based on the deposition onto glassy carbon, metallic or ITO surfaces, the development of silicon and carbon ceramic materials by using sol-gel technology and assisted by high-power ultrasound was carried out by some research groups (Hidalgo-Hidalgo-de-Cisneros et al. 2001).

The sol-gel process has been extensively applied to obtain silicon oxide and silicon oxide-derived monolithic substrates at room temperature from silane alkoxides or metallic precursors (Kajihara 2013). The employment of organic solvents, like methanol or ethanol, is required due to the immiscibility of the silane/metallic precursors in aqueous solution. Their use can be suppressed by high-power ultrasound, leading to a new type of silica monolith with higher density matrix. This also implies the reduction of waste products. The structural and mechanical properties of these new materials, namely Sonogels, were extensively studied (Blanco et al. 1999).

The addition of carbon powder after the sonication process leads to a conducting material able to be used as electrochemical device. In the first formulations, Sonogel-Carbon, graphite powder was employed owing to its low cost and high availability, as well as its high electrical conductivity (Cordero-Rando et al. 2002; Cubillana-Aguilera et al. 2006). Figure 18 shows a schematic representation of the fabrication of the Sonogel-Carbon electrodes.

The synthesis process of Sonogel-Carbon is characterized by its high versatility, allowing the obtention of new bare electrode materials. The graphite powder added to the sonosol after sonication can be replaced totally by multi-walled carbon nanotubes and nanocarbon (Palacios-Santander et al. 2017) or partially by 1-furoylthiourea (Cubillana-Aguilera et al. 2010), β-cyclodextrin (Izaoumen et al. 2009), and

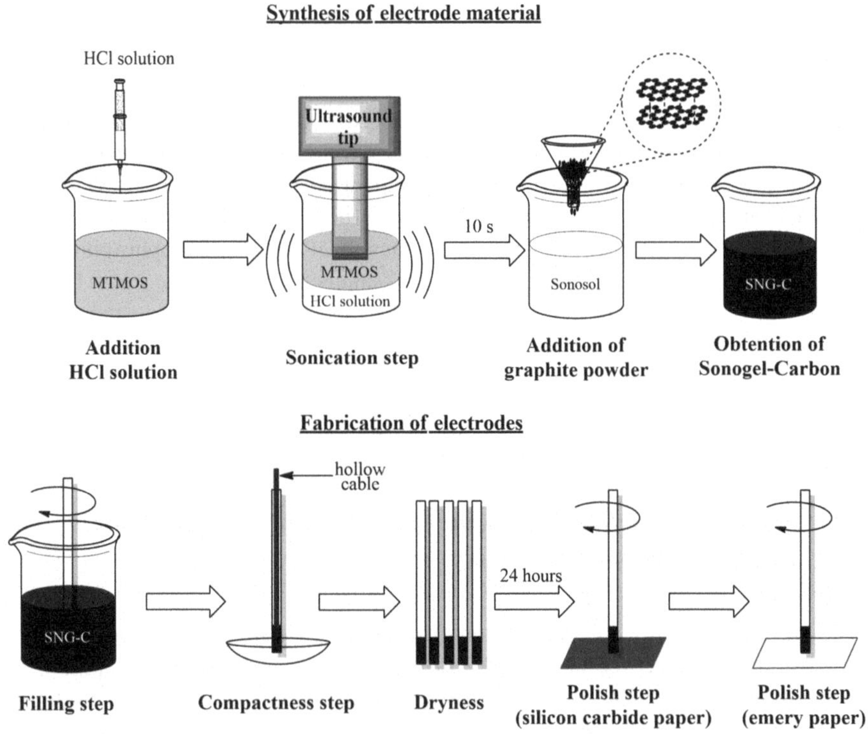


**Fig. 18** Fabrication of Sonogel-Carbon electrodes

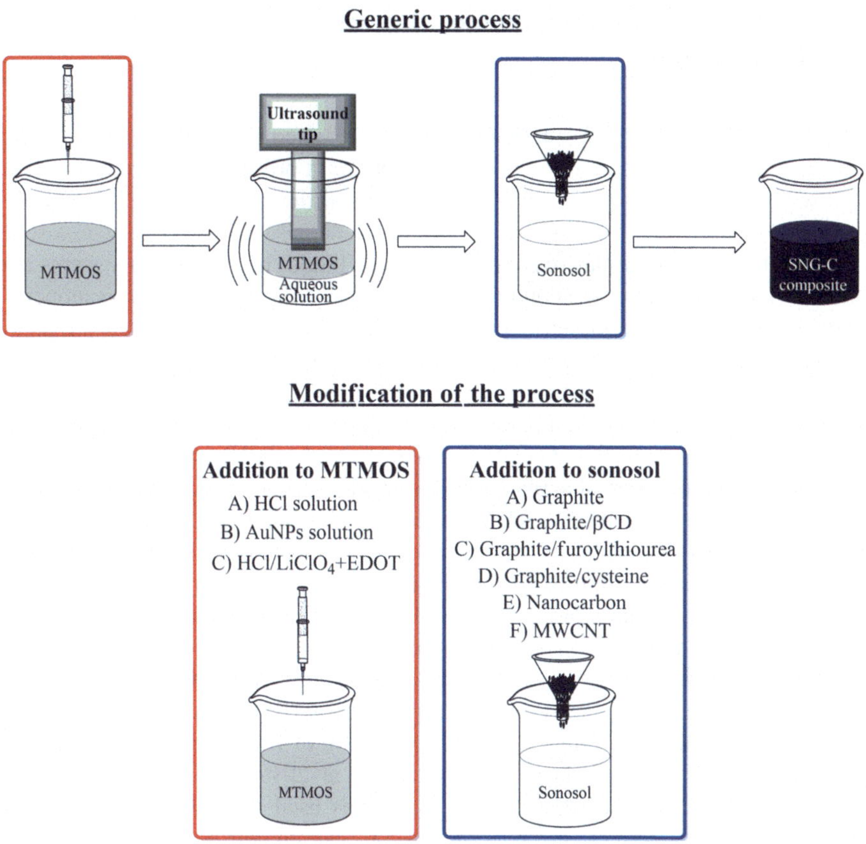


**Fig. 19** Different synthesis routes for the fabrication of Sonogel-Carbon derived composites

L-cysteine (El Bouhouti et al. 2009), among other modifiers (Bellido-Milla et al. 2013). The substitution of the acid solution used as catalyst by gold nanoparticles colloidal solution obtained from a green route is under investigation (Franco-Romano et al. 2013). Based on their acid pH, they can catalyze the formation of the silicon oxide network from the silane precursor. Finally, the inclusion of a conducting polymer in the silicon oxide network by using high-power ultrasound was reported in other work (López-Iglesias et al. 2018). Figure 19 summarizes all the synthesis routes mentioned in this subsection.

Therefore, the synthesis of several conducting ceramic materials assisted by high-power ultrasound in one-step was reported in the bibliography. The easy, low-cost and low-time consuming methods are highlighted, together with the use of high-power ultrasound, allowing decrease in the reaction time, as well as minimize the use of hazardous solvents. Moreover, the versatility and greenness of the synthesis scheme, as shown in Figs. 18 and 19, may increase, in the near future, the number of materials susceptible of being employed in the fabrication of ceramic-based electrode devices (López-Iglesias et al. 2016).

# 4 Application of Green (Nano)Materials in Analytical (Bio)Sensing

The applications of the different (nano)materials obtained by a green synthetic route for (bio)sensing are the main focus of this section. In order to clarify the structure that will be followed, several tables collecting the different materials by type have been included.

In Table 10, a representative collection of ionic liquids used in sensor systems are shown. Some of them correspond to enzymatic biosensors used either as amperometric sensors or in colorimetric tests. The half of them employs nanomaterials to enhance sensitivity, such as graphene and carbon nanotubes.

The amount of analytes that can be detected by these devices based on ionic liquids is quite diverse, including analytes of interest in food industry (like ascorbic and caffeic acid), medical and healthcare industry (i.e., cholesterol and neurotransmitters), and environmental applications (pesticides). They were determined in biological and agrifood matrices such as water, juices, human serum, and blood. The performance of ILs-based sensor systems can be evaluated by several quality analytical parameters, i.e., limit of detection and reproducibility. In general, the figures of merits are quite good in most cases for the linear ranges indicated.

The original interest and applications of ILs in (bio)sensing started decreasing when the scientific community realized the possible problematic related to the elimination of the ILs after their use from the environment (Bystrzanowska et al. 2019). More information concerning analytical application of ILs in sensing systems can be found in Yavir and coworkers' review (Yavir et al. 2019).

The use of other kinds of green nanomaterials used for building sensing and biosensing devices, like graphene, quantum dots, carbon dots, and metal and metal oxide nanoparticles, principally, may also be assessed. Table 11 reports optical devices and Table 12 electrochemical ones for multiple sensing applications.

With respect to the optical sensors described in Table 11, most of them are based on carbon or quantum dots, in some cases doped with lantanide metals to enhance the luminescence signal, and applied directly in buffer solutions. The variety of target analytes is much higher than in the previous case: nitrophenols, cations, anions, metals, amino acids, enzymes, guanine phosphorylated derivatives, and/or their nanocomposites. Some of them are considered as priority environmental pollutants and others have high applicability in biomedical or pharmaceutical fields. According to this classification, the type of samples measured are highly diverse, including cellular systems, such as MCF-7 and HeLa cells, soil samples, natural water samples, or serum samples. The limits of detection obtained for the analytes mentioned is quite low, most of them in the nanomolar range for more or less wide dynamic ranges.

On the other hand, Table 12 shows some examples of electrochemical sensing and (bio)sensing devices mainly based on carbon transducers modified with the aforementioned green nanomaterials. Their application is fundamentally focused on biomedical analysis, including some pharmaceuticals. That is why the real samples

**Table 10** Sensors and biosensor systems based on different ionic liquids and most relevant figures of merits

| Analyte (s) | Applied ILs | Composite components | Sample type | Reproducibility (RSD %) | LOD (μM) | Linear range (mM) | Sensitivity ($A \cdot M^{-1}\ cm^{-2}$) | References |
|---|---|---|---|---|---|---|---|---|
| $H_2O_2$ | [C4C1Im][BF4] | HRP, TEOS | Water | 3.1 | 1.1 | 0.02–0.26 | 7.2 | Liu et al. (2005) |
| Glucose | [C4C1Im][BF4] | HRP, GOD, PTMSPA, Nafion® | Serum | 3.6 | 10 | 0.05–7.0 | 0.0379 | Chen et al. (2012) |
| Carbaryl, monocrotphos | [(NH2C3)C1Im][BF4] | GN, AchE, | Tomatojuice | – | $5.3 \times 10^{-9}$; $4.6 \times 10^{-8}$ | $1.0 \times 10^{-11}$-$1.0 \times 10^{-5}$; $1.0 \times 10^{-10}$-$5.0 \times 10^{-5}$ | – | Zheng et al. (2015) |
| Eserine, neostigmine | [C2C1Im][TCB] | TTF, TCNQ | Tapwater | 6 | $2.6 \times 10^{-7}$; $0.3 \times 10^{-3}$ | $(0.1–1000) \times 10^{-6}$; $(1–500) \times 10^{-6}$ | – | Zamfir et al. (2013) |
| Caffeic acid | [C4C1Im][Br]; [C4C1Im][cl] | GN | Plasma | 3.2; 1.4 | 18.0; 5.0 | 0.025–2.0; 0.025–2.0 | 1.657; 3.389 | Valentini et al. (2015) |
| Cholesterol | [C8C1Im][TFA] | ChOx, Prussian blue | Water sample | – | 4.4 | 0.01–0.4 | – | Liu et al. (2013) |
| Dopamine | [C8C1Im][PF6] | MWCNTs | Human blood; serum | – | 0.1 | 0.001–0.1 | – | Zhao et al. (2005) |
| Glucose | [C8Pyrr][PF6] | MWCNTs | – | – | – | Up to 6 | 0.002 | Kachoosangi et al. (2009) |
| Hb, HRP | [C4C1Im][BF4] | MWCNTs | – | 6.8; 7.5 | – | – | – | Tao et al. (2006) |
| Cholesterol | [C4C1IM][BF4] | Chi, ChOx, MWCNTs, au | Serum sample | 1.9 (repeatability) | 0.5 | 0.5–5.0 | 0.0002 A/M | Gopalan et al. (2009) |

*IL* ionic liquid, *LOD* limit of detection, *Hb* hemoglobin, *HRP* horseradish peroxidase, *GOD* glucose oxidase, *[C4C1Im][BF4]* 1-butyl-3-methylimidazolium tetrafluoroborate, *[(NH2C3)C1Im][BF4]* 1-(3-aminopropyl)-3-methylimidazolium tetrafluoroborate, *[C2C1Im][TCB]* 1-ethyl-3-methylimidazolium tetracyanoborate, *[C4C1Im][Br]* 1-butyl-3-methylimidazoliumbromide, *[C4C1Im][Cl]* 1-butyl-3-methylimidazoliumchloride, *[C8C1Im][TFA]* 1-butyl-3-methylimidazolium trifluoroacetate, *[C8C1Im][PF6]* 1-methyl-3-octylimidazolium hexafluorophosphate, *[C8Pyrr][PF6]* N-octylpyrrolidiniumhexafluorophosphate, *TEOS* tetra-ethyl-ortho-silicate, *PTMSPA* poly[1-phenyl-2-(p-trimethylsilyl) phenylacetylene], *GN* graphene, *AChE* acetylcholinesterase, *TTF* tetrathiafulvalene, *TCNQ* tetracyanoquinodimethane, *ChOx* cholesterol oxidase, *MWCNTs* multi-walled carbon nanotubes, *Chi* chitosa

**Table 11** Optical sensors and biosensor devices based on green nanomaterials and most relevant figures of merits

| Analyte(s) | Green (nano)material | Sample type | LOD (nM) | Linear range (μM) | References |
|---|---|---|---|---|---|
| 2-Nitrophenol, 4-nitrophenol | Carbon dots | – | 1060; 500 | 0.001–1.0 | Ren et al. (2018) |
| Fe(III) | Graphitic carbon quantum dots | – | 2 | 0–1 | Zhang et al. (2013b) |
| Phosphate | Eu-adjusted carbon dots | Artificial wetland water | 51 | 0.4–15 | Zhao et al. (2011) |
| Thallium | Mn-doped ZnSeQDs&carbondots | Serum, water, soil | ≈ 4.9 | ≈ 24–490 | Lu et al. (2018) |
| Fe(III), pyrophosphate | Carbon dots | – | 60; 300 | 0.2–100; 1–100 | Chen and Tseng (2017) |
| Lysine | Carbon dots | Inside cellular systems | 94 | 0.5–260 | Song et al. (2017) |
| Thioredoxin reductase | Carbon dots | MCF-7 and HeLa cells | 20 | – | Sidhu et al. (2018) |
| Guanosine 3′-diphosphate-5′-diphosphate (ppGpp) | Tb(III)-modified carbon dots | – | 50 | 0.5–15 | Chen and Jiang (2018) |
| Ascorbic acid | Graphene QDs | HeLa cells | 270 | 0–800 | Feng et al. (2017) |
| Fe(III) | Carbon dots | Lake water | – | 0.01–46 | Qu et al. (2013b) |
| 2,4,6-Trinitrophenol | Carbon dots/ $Fe_3O_4$NPs-MIPs | Tap/lake water | 0.5 | 0–100 | Wang et al. (2019a) |

*LOD* limit of detection, *QDs* quantum dots, *$Fe_3O_4$NPs* $Fe_3O_4$ nanoparticles, *MIPs* molecularly imprinted polymers

where the determination has been accomplished correspond to human samples (i.e., serum, blood, and urine). Only in two cases food and environmental samples are measured. As observed in the previous devices reported, limits of detection are at nanomolar level; besides, reproducibility is excellent (lower than 5–7%) and, when applicable, sensitivity is also quite good. Other examples can be seen in Table 12, but exclusively referred to real environmental, biomedical, and food industrial uses. Excellent figures of merits are reported in all cases.

Hence, most of these (bio)sensing devices can be or are susceptible of being produced and/or used directly in industry. As discussed in the previous paragraphs and tables, the performance of these analytical systems are more than adequate for fulfilling the industry requirements whatever the purpose they are used for: detection and/or determination, quality, and waste control, in different kinds of industrial companies, such as food, environmental, biopharmaceutical, and biomedical industries (Siontorou 2019).

**Table 12** Electrochemical sensors and biosensor devices based on green nanomaterials and most relevant figures of merits

| Analyte(s) | Green (nano) material | Composite components | Sample type | Reproducibility (RSD %) | LOD (μM) | Linear range (mM) | Sensitivity ($A \cdot M^{-1}$ $cm^{-2}$) | References |
|---|---|---|---|---|---|---|---|---|
| p-nitrophenol | MnO NPs | BCA/AuE | – | – | 15.65 | 0.2–0.55 | 0.16 | Kumar et al. (2017) |
| Hydrogen peroxide | rGO/FeNPs | GCE | – | – | 0.056 | $1 \times 10^{-4}$–2.15 | 0.2085 | Amanulla et al. (2017) |
| Glucose | rGO-PtNPs-GOx | GCE | – | – | 1.21 | 2–10 | 0.0275 | Akkaya et al. (2018) |
| Glucose, *Escherichia coli* | fPtNCF-GN paper | | – | – | 0.08 ± 0.02; ≈4 $CFU \cdot mL^{-1}$ | 0–0.5; 4–$10^5$ $CFU \cdot mL^{-1}$ | 3.59 ± 0.68 $A \cdot M$; 16.1 $\Omega \cdot$log $CFU^{-1} \cdot mL^{-1}$ | Burrs et al. (2016) |
| Uric acid, tyrosine | MIP/rGO | GCE | Human serum and urine samples | 4.68 | 0.0032; 0.046 | $1 \times 10^{-5}$- 0.1;$1 \times 10^{-4}$-0.4 | – | Zheng et al. (2018) |
| Hydrogen peroxide | rGO-Nafion/AuNPs | GCE | – | 0.47 | 2 | 0.02–23 | 0.575 | Lv et al. (2016) |
| Glucose | Electro-chemically rGO/DexP | SPE | Human serum | 6.7 | 20 | 0.05–100 | – | Li et al. (2018) |
| Cathecol | PEDOT-rGO-$Fe_2O_3$-PPO | GCE | Green tea | 3.2 | 0.007 | $4 \times 10^{-5}$–$6.2 \times 10^{-3}$ | 0.0108 $A \cdot M^{-1}$ | Sethuraman et al. (2016) |
| Nitromethane | Hb-CS/rGO-CS | GCE | Fresh water samples | 3.7 | 1.5 | $5 \times 10^{-3}$–1.46 | – | Wen et al. (2016) |
| Chloramphenicol | TiN–rGO | GCE | – | 4.39 | 0.02 | $5 \times 10^{-5}$–0.1 | – | Kong et al. (2016) |
| Glucose | GelMA:NiPs-RGO | GCE | Human blood serum | 2.1 | 0.005 | $1.5 \times 10^{-4}$-10 | 0.056 $A \cdot M^{-1}$ | Darvishi et al. (2018) |

(continued)

**Table 12** (continued)

| Analyte(s) | Green (nano) material | Composite components | Sample type | Reproducibility (RSD %) | LOD (μM) | Linear range (mM) | Sensitivity ($A \cdot M^{-1} cm^{-2}$) | References |
|---|---|---|---|---|---|---|---|---|
| CEA, PSA | Anti-CEA/EDC/NHS/PdAuPt/COOH-rGO/au; anti-PSA/EDC/NHS/PdAuPt/COOH-rGO/au | Immunosensor platform | Human serum | 5.6 | 8; 2 $pg \cdot ml^{-1}$ | 12–8.5 × $10^4$; 3–6 × $10^4$ pg $mL^{-1}$ | 212.1; 258.7 $A \cdot L \cdot g^{-1}$ | Barman et al. (2018) |
| Glycerol | CuONPs/Pe-MWCNTs | GCE | Biodiesel | 10 | 5.8 × $10^{-6}$ $g \cdot L^{-1}$ | 9 × $10^{-6}$–1 × $10^{-3}$ $g \cdot L^{-1}$ | 5.5 ± 0.3 × $10^{-5}$ $A \cdot L \cdot g^{-1}$ | Arévalo et al. (2017) |
| Triglycerides | CNP-L/CuONPs/MWCNTs/Pe | GCE | Human serum | 5.9 | 3.2 × $10^{-3}$-3.6 × $10^{-3}$ $g \cdot L^{-1}$ | 0.001–0.53 $g \cdot L^{-1}$ | 1.64 × $10^{-6}$ $A \cdot L \cdot g^{-1}$ | Di Tocco et al. (2018) |
| Organophosphorous pesticides: Paraoxon and carbaryl | MWCNT–(PEI/DNA)$_2$/OPH/AChE | GCE | Apples (agricultural industry) | – | 0.5–10 | 0.5 × $10^{-3}$–0.04;0.01–0.08 | 0.021–0.00941 $A \cdot M^{-1}$ | Zhang et al. (2015) |
| Organophosphorous pesticide: Dimethoate | CNTs/$ZrO_2$/PB/NfGMP-AChE | SPCE | Cabbage (Agricultural industry) | 5.2 | 5.6 × $10^{-4}$ $ng \cdot mL^{-1}$ | 1.0 × $10^{-3}$-10 $ng \cdot mL^{-1}$ | – | Gan et al. (2010) |
| Choline (in food industry) | Glass/ZnONR/COD-HRP/PHA | Optical platform (chemilumini-scence) | Milk | 4.1 ± 0.5 | 0.5 | 0.006–2 | – | Pal et al. (2014) |
| Xanthine (in food industry) | XOD/ZnO-NP/CHIT/c-MWCNT/PANI | PtE | Fish meat | 5.30 | 0.1 | 1 × $10^{-4}$–0.1 | – | Devi et al. (2012) |

| | | | | | | | | |
|---|---|---|---|---|---|---|---|---|
| PSA | Ti/Pt sputtered (microfluidic channel) | Slide glass by photolithographic system | Human plasma | – | 100 fg·mL$^{-1}$ | 10–1 × 10$^{5}$ fg·mL$^{-1}$ | 5.3 MΩ·dec imalpoint$^{-1}$ | Shin et al. (2016) |

*CEA* carcinoembryonic antigen, *PSA* prostate specific antigen, *NPs* nanoparticles, *rGO* reduced graphene oxide, *GOx* glucose oxidase, *fPtNCF* fractal Pt nano cauliflower, *GN* Graphene, *MIP* molecularly imprinted polymer, *DexP,* phenoxyl-dextran, *PEDOT* poly(3,4-ethylenedioxythiophene), *PPO* polyphenol oxidase, *Hb* hemoglobin, *CS* chitosan, *TiN* titanium nitride, *GelMA* gelatin methacryloyl hybrid hydrogel, *EDC/NHS* 1-Ethyl-3-(3-dimethylaminopropyl) carbodiimide/N-hydroxysuccinimide, *Pe* Pectin, *MWCNT* multi-walled carbon nanotubes, *CNP-L* lipase-modified magnetic nanoparticles, *BCA* butyl carbitol acetate, *AuE* gold electrode, *GCE* glassy carbon electrode, *SPE* screen-printed electrode, *PSA* prostate specific antigen, *PEI* polyethyleneimine, *OPH* organo phosphate hydrolase, *AChE* acetylcholine esterase, *CNTs* carbon nanotubes, *PB* Prussian blue, *Nf GMP* Nafion gold magnetic nanoparticles ($Fe_3O_4$/Au), *ZnONR* ZnO nanorods, *COD* choline oxidase, *HRP* horseradish peroxidase, *PHA* phosphonohexadecanoicacid, *XOD* xanthine oxidase, *CHIT* chitosan, *PANI* polyaniline, *c-MWCNT* carboxylated-MWCNT, *GCE* glassy carbon electrode, *SPCE* screen-printed carbon electrode, *PtE* platinum electrode, *LOD* limit of detection

## 5 Industrial Application of (Bio)Sensing Devices

As mentioned in Sect. 4, analytical sensing constitutes high interest in the chemical industry, attracting great attention nowadays. Thus, the employment of robust analytical systems with suitable performance in the in situ analytical detection/determination or quality control is being implemented in industry. In this section, some applications of (bio)sensing devices able to be used at industrial scale are reported.

Several recent reports concerning the application of sensing devices in several fields of interest such as food, clinical, and medicines at industrial scale are highlighted. A recent review is focused on the application of biosensors for whole-cell bacterial detection. The interest of portable stand-alone biosensors in the rapid detection and diagnosis of critical illnesses (meningitis, food-borne pathogens, sexually transmitted diseases, anthrax detection) is reported. Furthermore, the review discusses recent progress in the use of the biosensors without the need for sample processing compared with current methods of bacterial detection. A particular focus on electrochemical biosensors is made (Ahmed et al. 2014). Another review encompasses recent developments in the technological innovations concerning the detection of food allergens. Hypersensitivity towards this kind of allergens is increasing worldwide. Therefore, improving the methodology for food allergens detection will permit the identification of individual immune triggers, the prediction of the response severity, and the monitoring of allergen tolerance over time. Neethirajan and coworkers reported several biosensors as innovative analytical devices for enzymes, antibodies, aptamers, and single-stranded DNA detection. Optical, electromechanical, and electrochemical biosensors employed in the detection of food allergens are also reviewed (Neethirajan et al. 2018b).

A review about nanogenerators for self-powered gas sensing discusses their ability as technological and economical driver for global industries. Nanogenerators are applied as self-powered environmental sensors and the paper summarized 24 references of nanogenerators-based self-powered gas sensors (Wen et al. 2017).

Another paper reported the virtual sensing technology, known as soft-sensors in the area of chemical engineering. This technology is a key to estimate successfully product quality when online analyzers are not available. The applications are broad and extendable to fields such as petrochemical, steel, and pharmaceutical industries (Kano and Fujiwara 2013).

Therefore, with some effort made by all the societal sectors: public in general, academia and industries, and always thinking on as much global as possible sustainability, some of these (bio)sensing devices might reach the widespread usefulness and success of the currently most extended biosensor, the glucometer, which in many cases is also complemented with the determination of other related analytes, such as cholesterol and lactic acid, among others (Calabria et al. 2017; Pilas et al. 2017; Cunha-Silva and Arcos-Martinez 2018; Kotanen et al. 2018). In this way, not only would biomedical point-of-care diagnoses be possible, but also environmental, pharmaceutical, and food ones.

## 6 Conclusions

As it has been highlighted throughout this chapter, Chemical Industry, with the aim of Academia and increasingly with the growing support of Society, is currently looking for and reaching, in many cases, sustainability thanks to Green Chemistry. This multidimensional philosophy can offer many different alternatives and tools to redirect the modern industry to a more ecological and efficient manufacturing and production. The concept of "reducing" (costs, energy, wastes, and materials, among others), sustainable and green education, and government policies are considered important milestones to bring this matter to a successful conclusion.

Particularly, green synthesis is part of the foundations of Green Chemistry, since the search for the perfect synthesis with all its implications (safe, atom economy, high yield, etc.) is of vital importance to achieve sustainable industrial production. Moreover, efficient synthetic routes can be extended to different areas, especially to Analytical Chemistry, where the use of sensors and biosensors can represent a green choice with respect to conventional methodologies, as widely reported in literature during the last decade. Their many advantages among which stand out rapidity, analytical performance, on line and in situ measurements and cost-effectiveness, make them of great relevancy for being used in many kinds of industries as for analyte detection/determination as for quality control. That is why it is essential to know the different possibilities regarding the green synthesis of materials and nanomaterials that can be used for building sensors and biosensors.

In this chapter, a summary of the most relevant green materials synthesis or biosynthesis routes is provided. Special attention has been paid to ultrasound and microwave technologies. During the past few decades, both has been widely recognized as ecological and clean, as well as cost and time effective, having been demonstrated that their use fulfills most of the 12 Principles of Green (Analytical) Chemistry and is quite sustainable for synthesis purposes of both organic and inorganic materials. Moreover, the use of biological compounds to synthesize complex molecules or nanostructures, preferably by a one-pot method, is also emphasized. The focus is put on the employment of plant extracts, biopolymers, and microorganisms, such as algae, bacteria, and fungi. The possibility of combining microwave or ultrasound technologies and any of them with some biosynthetic route is also discussed through the most recent and representative achievements reported in literature within the field of green synthesis of (nano)materials. It is also necessary to mention that the most important and current trends in green (nano)materials production affects to graphene, carbon and graphene quantum dots, multi-walled carbon nanotubes, metal and metal oxide nanoparticles and polymers. A very wide set of examples concerning green synthesized materials is reported in Sect. 3, most of them already applied or with high potential applicability in industry. Furthermore, the possibility of replacing actual and more expensive metal, glassy carbon or ITO-based materials to new ceramic electrode materials, mainly for electrochemical devices, is also explored. The versatility and greenness of their scheme of synthesis and their excellent analytical performance makes them a strong alternative to typical and conventional electrode materials. It has been demonstrated as well that the applicability of the mentioned green (nano)materials for analytical (bio)sensing is widely extended even at industrial level. An extensive set of optical or electrochemical devices employed for that purpose are described on depth according to their most relevant quality analytical parameters: limit of detection, sensitivity, linear range, and reproducibility. As reported, the kind of analytes that can be determined with (bio)sensing devices includes the fields of food, environment and medicine and healthcare.

Despite all, the path of industry towards sustainability has just begun and there is still too much work to do in order to completely erase the prejudices and the fear to chemistry, in general; in other words to banish chemophobia. However, one thing is absolutely clear: our life without chemistry or chemicals would be unthinkable, since life itself is chemistry.

**Acknowledgments** J.J. García-Guzmán thanks University of Cadiz for his post-doc contract (Plan Propio 2018/2019-UCA). D. López-Iglesias greatly acknowledge financial support of ESF funds (Sistema de Garantía Juvenil depending on Ministerio de Empleo y Seguridad Social and Junta de Andalucía, Spain) for his contract (E-11-20180043137). Authors thank Institute of Research on Electron Microscopy and Materials (IMEYMAT) for their economic support as well.

## References

Abdi V, Sourinejad I, Yousefzadi M, Ghasemi Z (2018) Mangrove-mediated synthesis of silver nanoparticles using native *Avicennia marina* plant extract from southern Iran. Chem Eng Commun 205:1069–1076. https://doi.org/10.1080/00986445.2018.1431624

Abdullah AH, Ismail Z, Zainal Abidin AS, Yusoh K (2019) Green sonochemical synthesis of few-layer graphene in instant coffee. Mater Chem Phys 222:11–19. https://doi.org/10.1016/j.matchemphys.2018.09.085

Abo Elsoud MM, El Kady EM (2019) Current trends in fungal biosynthesis of chitin and chitosan. Bull Natl Res Cent 43:59. https://doi.org/10.1186/s42269-019-0105-y

Agarwal H, Nakara A, Shanmugam VK (2019) Anti-inflammatory mechanism of various metal and metal oxide nanoparticles synthesized using plant extracts: a review. Biomed Pharmacother 109:2561–2572. https://doi.org/10.1016/j.biopha.2018.11.116

Ahmad N, Sharma S, Alam MK et al (2010) Rapid synthesis of silver nanoparticles using dried medicinal plant of basil. Colloids Surf B Biointerfaces 81:81–86. https://doi.org/10.1016/j.colsurfb.2010.06.029

Ahmad NH, Mustafa S, Man YBC (2015) Microbial polysaccharides and their modification approaches: a review. Int J Food Prop 18:332–347. https://doi.org/10.1080/10942912.2012.693561

Ahmad A, Wei Y, Syed F et al (2017) The effects of bacteria-nanoparticles interface on the antibacterial activity of green synthesized silver nanoparticles. Microb Pathog. https://doi.org/10.1016/j.micpath.2016.11.030

Ahmed A, Rushworth JV, Hirst NA, Millner PA (2014) Biosensors for whole-cell bacterial detection. Clin Microbiol Rev 27:631–646. https://doi.org/10.1128/CMR.00120-13

Ahmed HB, Zahran MK, Emam HE (2016) Heatless synthesis of well dispersible Au nanoparticles using pectin biopolymer. Int J Biol Macromol 91:208–219. https://doi.org/10.1016/j.ijbiomac.2016.05.060

Akkaya B, Çakiroğlu B, Özacar M (2018) Tannic acid-reduced graphene oxide deposited with Pt nanoparticles for switchable bioelectronics and biosensors based on direct electrochemistry. ACS Sustain Chem Eng 6:3805–3814. https://doi.org/10.1021/acssuschemeng.7b04164

Alam AM, Park BY, Ghouri ZK et al (2015) Synthesis of carbon quantum dots from cabbage with down- and up-conversion photoluminescence properties: excellent imaging agent for biomedical applications. Green Chem 17:3791–3797. https://doi.org/10.1039/c5gc00686d

Algarra M, González-Calabuig A, Radotić K et al (2018) Enhanced electrochemical response of carbon quantum dot modified electrodes. Talanta 178:679–685. https://doi.org/10.1016/j.talanta.2017.09.082

Algul O, Kaessler A, Apcin Y et al (2008) Comparative studies on conventional and microwave synthesis of some benzimidazole, benzothiazole and indole derivatives and testing on inhibition of hyaluronidase. Molecules 13:736–748. https://doi.org/10.3390/molecules13040736

Ali Mohsin ME, Elias M, Arsad A et al (2016) Ultrasonic irradiation: a novel approach for conductive polymer. J Eng Appl Sci 11:2557–2560. https://doi.org/10.3923/jeasci.2016.2557.2560

Altemimi A, Lakhssassi N, Baharlouei A et al (2017) Phytochemicals: extraction, isolation, and identification of bioactive compounds from plant extracts. Plan Theory 6:42. https://doi.org/10.3390/plants6040042

Amanulla B, Palanisamy S, Chen SM et al (2017) A non-enzymatic amperometric hydrogen peroxide sensor based on iron nanoparticles decorated reduced graphene oxide nanocomposite. J Colloid Interface Sci 487:370–377. https://doi.org/10.1016/j.jcis.2016.10.050

Amine A, Arduini F, Moscone D, Palleschi G (2016) Recent advances in biosensors based on enzyme inhibition. Biosens Bioelectron 76:180–194. https://doi.org/10.1016/j.bios.2015.07.010

Anastas P, Eghbali N (2010) Green chemistry: principles and practice. Chem Soc Rev 39:301–312. https://doi.org/10.1039/b918763b

Anastas PT, Warner JC (1998) Green chemistry: theory and practice. Oxford University Press, New York

Anderson LA, Islam MA, Prather KLJ (2018) Synthetic biology strategies for improving microbial synthesis of "green" biopolymers. J Biol Chem 293:5053–5061. https://doi.org/10.1074/jbc.TM117.000368

Ando T, Kimura T (1990) Reactivity and selectivity in organic sonochemical reactions involving inorganic solids. Ultrasonics 28:326–332. https://doi.org/10.1016/0041-624X(90)90040-U

Apetrei RM, Carac G, Bahrim G et al (2018) Modification of *Aspergillus niger* by conducting polymer, polypyrrole, and the evaluation of electrochemical properties of modified cells. Bioelectrochemistry 121:46–55. https://doi.org/10.1016/j.bioelechem.2018.01.001

Araga R, Sharma CS (2017) One step direct synthesis of multiwalled carbon nanotubes from coconut shell derived charcoal. Mater Lett 188:205–207. https://doi.org/10.1016/j.matlet.2016.11.014

Arduini F, Cinti S, Caratelli V et al (2019) Origami multiple paper-based electrochemical biosensors for pesticide detection. Biosens Bioelectron 126:346–354. https://doi.org/10.1016/j.bios.2018.10.014

Arévalo FJ, Osuna-Sánchez Y, Sandoval-Cortés J et al (2017) Development of an electrochemical sensor for the determination of glycerol based on glassy carbon electrodes modified with a copper oxide nanoparticles/multiwalled carbon nanotubes/pectin composite. Sensors Actuators B Chem 244:949–957. https://doi.org/10.1016/j.snb.2017.01.093

Armenta S, Garrigues S, de la Guardia M (2008) Green analytical chemistry. TrAC Trends Anal Chem 27:497–511. https://doi.org/10.1016/j.trac.2008.05.003

Attar A, Cubillana-Aguilera L, Naranjo-Rodriguez I et al (2015) Amperometric inhibition biosensors based on horseradish peroxidase and gold sononanoparticles immobilized onto different electrodes for cyanide measurements. Bioelectrochemistry 101:84–91. https://doi.org/10.1016/j.bioelechem.2014.08.003

Azizi Samir MAS, Alloin F, Dufresne A (2005) Review of recent research into cellulosic whiskers, their properties and their application in nanocomposite field. Biomacromolecules 6:612–626. https://doi.org/10.1021/bm0493685

Azizi S, Shahri MM, Rahman HS et al (2017) Green synthesis palladium nanoparticles mediated by white tea (*Camellia sinensis*) extract with antioxidant, antibacterial, and antiproliferative activities toward the human leukemia (MOLT-4) cell line. Int J Nanomedicine 12:8841–8853. https://doi.org/10.2147/IJN.S149371

Backes C, Higgins TM, Kelly A et al (2017) Guidelines for exfoliation, characterization and processing of layered materials produced by liquid exfoliation. Chem Mater 29:243–255. https://doi.org/10.1021/acs.chemmater.6b03335

Bahadir EB, Sezgintürk MK (2015) Applications of commercial biosensors in clinical, food, environmental, and biothreat/biowarfare analyses. Anal Biochem 478:107–120. https://doi.org/10.1016/j.ab.2015.03.011

Baig RBN, Varma RS (2012) Alternative energy input: mechanochemical, microwave and ultrasound-assisted organic synthesis. Chem Soc Rev 41:1559–1584. https://doi.org/10.1039/c1cs15204a

Balachandran YL, Panarin AY, Khodasevich IA et al (2015) Environmentally friendly preparation of gold and silver nanoparticles for Sers applications using biopolymer pectin. J Appl Spectrosc 81:962–968. https://doi.org/10.1007/s10812-015-0036-9

Banerjee B (2017) Recent developments on ultrasound assisted catalyst-free organic synthesis. Ultrason Sonochem 35:1–14. https://doi.org/10.1016/j.ultsonch.2016.09.023

Bang JH, Suslick KS (2010) Applications of ultrasound to the synthesis of nanostructured materials. Adv Mater 22:1039–1059. https://doi.org/10.1002/adma.200904093

Bankura KP, Maity D, Mollick MMR et al (2012) Synthesis, characterization and antimicrobial activity of dextran stabilized silver nanoparticles in aqueous medium. Carbohydr Polym 89:1159–1165. https://doi.org/10.1016/j.carbpol.2012.03.089

Bao SJ, Lei C, Xu MW et al (2013) Environmentally-friendly biomimicking synthesis of $TiO_2$ nanomaterials using saccharides to tailor morphology, crystal phase and photocatalytic activity. CrystEngComm 15:4694–4699. https://doi.org/10.1039/c3ce40310f

Baranwal A, Mahato K, Srivastava A et al (2016) Phytofabricated metallic nanoparticles and their clinical applications. RSC Adv 6:105996–106010. https://doi.org/10.1039/c6ra23411a

Bari R, Tamas G, Irin F et al (2014) Direct exfoliation of graphene in ionic liquids with aromatic groups. Colloids Surf A Physicochem Eng Asp 463:63–69. https://doi.org/10.1016/j.colsurfa.2014.09.024

Barman SC, Hossain MF, Yoon H, Park JY (2018) Trimetallic Pd@Au@Pt nanocomposites platform on -COOH terminated reduced graphene oxide for highly sensitive CEA and PSA biomarkers detection. Biosens Bioelectron 100:16–22. https://doi.org/10.1016/j.bios.2017.08.045

Baro M, Nayak P, Baby TT, Ramaprabhu S (2013) Green approach for the large-scale synthesis of metal/metal oxide nanoparticle decorated multiwalled carbon nanotubes. J Mater Chem A 1:482–486. https://doi.org/10.1039/c2ta00483f

Bayrami A, Alioghli S, Rahim S, Habibi-yangjeh A (2019) A facile ultrasonic-aided biosynthesis of ZnO nanoparticles using *Vaccinium arctostaphylos* L. leaf extract and its antidiabetic, antibacterial, and oxidative activity evaluation. Ultrason Sonochem 55:57–66. doi: https://doi.org/10.1016/j.ultsonch.2019.03.010

Begum S, Ahmaruzzaman M (2018) Green synthesis of $SnO_2$ quantum dots using *Parkia speciosa* Hassk pods extract for the evaluation of anti-oxidant and photocatalytic properties. J Photochem Photobiol B Biol 184:44–53. https://doi.org/10.1016/j.jphotobiol.2018.04.041

Behravan M, Hossein Panahi A, Naghizadeh A et al (2019) Facile green synthesis of silver nanoparticles using *Berberis vulgaris* leaf and root aqueous extract and its antibacterial activity. Int J Biol Macromol 124:148–154. https://doi.org/10.1016/j.ijbiomac.2018.11.101

Bellido-Milla D, Cubillana-Aguilera LM, El Kaoutit M et al (2013) Recent advances in graphite powder-based electrodes. Anal Bioanal Chem 405:3525–3539. https://doi.org/10.1007/s00216-013-6816-2

Belontz SL, Corcoran PL, Davis H et al (2019) Embracing an interdisciplinary approach to plastics pollution awareness and action. Ambio 48:855–866. https://doi.org/10.1007/s13280-018-1126-8

Benakashani F, Allafchian AR, Jalali SAH (2016) Biosynthesis of silver nanoparticles using *Capparis spinosa* L. leaf extract and their antibacterial activity. Karbala Int J Mod Sci 2:251–258. https://doi.org/10.1016/j.kijoms.2016.08.004

Bergmann M, Tekman M, Gutow L (2017) Marine liter. Sea change for plastin pollution. Nature 544:297. https://doi.org/10.1038/544297b

Bernardo-Boongaling VR, Serrano N, García-Guzmán J et al (2019) Screen-printed electrodes modified with green-synthesized gold nanoparticles for the electrochemical determination of aminothiols. J Electroanal Chem 847:113184. https://doi.org/10.1016/j.jelechem.2019.05.066

Bhardwaj D, Singh A, Singh R (2019) Eco-compatible sonochemical synthesis of 8-aryl-7,8-dihydro-[1,3]-dioxolo[4,5-g]quinolin-6(5H)-ones using green $TiO_2$. Heliyon 5:e01256. https://doi.org/10.1016/j.heliyon.2019.e01256

Bilal M, Iqbal HMN (2019) Naturally-derived biopolymers: potential platforms for enzyme immobilization. Int J Biol Macromol 130:462–482. https://doi.org/10.1016/j.ijbiomac.2019.02.152

Blanco E, Esquivias L, Litrán R et al (1999) Sonogels and derived materials. Appl Organomet Chem 13:399–418. https://doi.org/10.1002/(SICI)1099-0739(199905)13:5<399::AID-AOC825>3.0.CO;2-A

Boomi P, Ganesan RM, Poorani G et al (2019) Biological synergy of greener gold nanoparticles by using *Coleus aromaticus* leaf extract. Mater Sci Eng C 99:202–210. https://doi.org/10.1016/j.msec.2019.01.105

Boury B, Plumejeau S (2015) Metal oxides and polysaccharides: an efficient hybrid association for materials chemistry. Green Chem 17:72–88. https://doi.org/10.1039/c4gc00957f

Božanić DK, Djoković V, Blanuša J et al (2007) Preparation and properties of nano-sized Ag and $Ag_2S$ particles in biopolymer matrix. Eur Phys J E 22:51–59. https://doi.org/10.1140/epje/e2007-00008-y

Bracamonte MV, Lacconi GI, Urreta SE, Foa Torres LEF (2014) On the nature of defects in liquid-phase exfoliated graphene. J Phys Chem C 118:15455–15459. https://doi.org/10.1021/jp501930a

Büntgen U, Krusic PJ, Piermattei A et al (2019) Limited capacity of tree growth to mitigate the global greenhouse effect under predicted warming. Nat Commun 10:2171. https://doi.org/10.1038/s41467-019-10174-4

Burrs SL, Bhargava M, Sidhu R et al (2016) A paper based graphene-nanocauliflower hybrid composite for point of care biosensing. Biosens Bioelectron 85:479–487. https://doi.org/10.1016/j.bios.2016.05.037

Bystrzanowska M, Pena-Pereira F, Marcinkowski Ł, Tobiszewski M (2019) How green are ionic liquids?—A multicriteria decision analysis approach. Ecotoxicol Environ Saf 174:455–458. https://doi.org/10.1016/j.ecoenv.2019.03.014

Calabria D, Caliceti C, Zangheri M et al (2017) Smartphone–based enzymatic biosensor for oral fluid L-lactate detection in one minute using confined multilayer paper reflectometry. Biosens Bioelectron 94:124–130. https://doi.org/10.1016/j.bios.2017.02.053

Calvo-Flores F, Isac-Garcia J, Dobado JA (2018) Emerging pollutants: origin, structure and properties, 1st edn. Wiley-VCH, Weinheim

Cayuela A, Soriano ML, Carrillo-Carrión C, Valcárcel M (2016) Semiconductor and carbon-based fluorescent nanodots: the need for consistency. Chem Commun 52:1311–1326. https://doi.org/10.1039/c5cc07754k

Chalupa R, Nesměrák K (2018) Analytical chemistry as a tool for suppressing chemophobia: an introduction to the 5E-principle. Monatsh Chem 149:1527–1534. https://doi.org/10.1007/s00706-018-2224-9

Chandran SP, Chaudhary M, Pasricha R et al (2006) Synthesis of gold nanotriangles and silver nanoparticles using Aloe vera plant extract. Biotechnol Prog 22:577–583. https://doi.org/10.1021/bp0501423

Chang WT, Chao YH, Li CW et al (2019) Graphene oxide synthesis using microwave-assisted vs. modified Hummer's methods: efficient fillers for improved ionic conductivity and suppressed methanol permeability in alkaline methanol fuel cell electrolytes. J Power Sources 414:86–95. https://doi.org/10.1016/j.jpowsour.2018.12.020

Chankaew C, Tapala W, Grudpan K, Rujiwatra A (2019) Microwave synthesis of ZnO nanoparticles using longan seeds biowaste and their efficiencies in photocatalytic decolorization of organic dyes. Environ Sci Pollut Res. 26:17548–17554. https://doi.org/10.1007/s11356-019-05099-w

Charde MS, Shukla A, Bukhariya V, Chakole RD (2012) A review on: a significance of microwave assist technique in green chemistry. Int J Phytopharm 2:39–50. https://doi.org/10.7439/ijpp.v2i2.441

Chatel G (2016) Sonochemistry: new opportunities for green chemistry. World Scientific Publishing Europe Ltd, London

Chatel G (2018) How sonochemistry contributes to green chemistry? Ultrason Sonochem 40:117–122. https://doi.org/10.1016/j.ultsonch.2017.03.029

Chen GQ, Jiang XR (2018) Engineering microorganisms for improving polyhydroxyalkanoate biosynthesis. Curr Opin Biotechnol 53:20–25. https://doi.org/10.1016/j.copbio.2017.10.008

Chen TH, Tseng WL (2017) Self-assembly of monodisperse carbon dots into high-brightness nanoaggregates for cellular uptake imaging and iron(III) sensing. Anal Chem 89:11348–11356. https://doi.org/10.1021/acs.analchem.7b02193

Chen X, Zhu J, Tian R, Yao C (2012) Bienzymatic glucose biosensor based on three dimensional macroporous ionic liquid doped sol-gel organic-inorganic composite. Sensors Actuators B Chem 163:272–280. https://doi.org/10.1016/j.snb.2012.01.053

Chen K, Dong Noh Y, Li K et al (2013) Microwave-hydrothermal crystallization of polymorphic $MnO_2$ for electrochemical energy storage. J Phys Chem C 117:10770–10779. https://doi.org/10.1021/jp4018025

Chen GX, Zhou JW, Liu YL et al (2016) Biosynthesis and regulation of wheat amylose and amylopectin from proteomic and phosphoproteomic characterization of granule-binding proteins. Sci Rep 6:33111. https://doi.org/10.1038/srep33111

Chen BB, Liu ML, Zhan L et al (2018) Terbium(III) modified fluorescent carbon dots for highly selective and sensitive ratiometry of stringent. Anal Chem 90:4003–4009. https://doi.org/10.1021/acs.analchem.7b05149

Cherry KE, Sampson L, Galea S et al (2018) Spirituality, humor, and resilience after natural and technological disasters. J Nurs Scholarsh 50:492–501. https://doi.org/10.1111/jnu.12400

Cho HU, Park JM (2018) Biodiesel production by various oleaginous microorganisms from organic wastes. Bioresour Technol 256:502–508. https://doi.org/10.1016/j.biortech.2018.02.010

Christian P, Von Der Kammer F, Baalousha M, Hofmann T (2008) Nanoparticles: structure, properties, preparation and behaviour in environmental media. Ecotoxicology 17:326–343. https://doi.org/10.1007/s10646-008-0213-1

Chua CK, Pumera M (2014) Chemical reduction of graphene oxide: a synthetic chemistry viewpoint. Chem Soc Rev 43:291–312. https://doi.org/10.1039/c3cs60303b

Chunfa D, Xianglin Z, Hao C, Chuanliang C (2016) Sodium alginate mediated route for the synthesis of monodisperse silver nanoparticles using glucose as reducing agents. Rare Met Mater Eng 45:261–266. https://doi.org/10.1016/s1875-5372(16)30051-0

Cifric S, Nuhic J, Osmanovic D, Kisija E (2020) Review of electrochemical biosensors for hormone detection. In: Badnjevic A, Škrbić R, GurbetaPokvić L (eds) CMBEBIH 2019, IFMBE proceedings, vol 73. Springer, Cham, pp 173–177

Clark JH (2005) Green chemistry and environmentally friendly technologies. In: Afonso CAM, Crespo JG (eds) Green separation processes: fundamentals and applications. Wiley-VCH Verlag GmbH & Co. KGaA, Wienheim, pp 3–18

Cordero-Rando MM, Hidalgo-Hidalgo de Cisneros JL, Blanco E, Naranjo-Rodriguez I (2002) The sonogel-carbon electrode as a sol-gel graphite-based electrode. Anal Chem 74:2423–2427. https://doi.org/10.1021/ac010782u

Cosbey A (2013) Green industrial policy and the world trading system. SSRN Electron J 17:12. https://doi.org/10.2139/ssrn.2344558

Cravotto G, Boffa L (2014) Combined ultrasound-microwave irradiation for the preparation of nanomaterials. Standford Publishing, Redwood City

Cravotto G, Cintas P (2006) Power ultrasound in organic synthesis: moving cavitational chemistry from academia to innovative and large-scale applications. Chem Soc Rev 35:180–196. https://doi.org/10.1039/b503848k

Cravotto G, Rinaldi L, Carnaroglio D (2015) Efficient catalysis by combining microwaves with other enabling technologies. In: Horikoshi S, Serpone N (eds) Microwaves in catalysis. pp 155–170. Wiley-VCH, Weinheim

Cubillana-Aguilera LM, Palacios-Santander JM, Naranjo-Rodriguez I, Hidalgo-Hidalgo-De-Cisneros JL (2006) Study of the influence of the graphite powder particle size on the structure of the sonogel-carbon materials. J Sol-Gel Sci Technol 40:55–64. https://doi.org/10.1007/s10971-006-9151-7

Cubillana-Aguilera LM, Palacios-Santander JM, Estevez-Hernandez OL et al (2010) 1-Furoylthiourea-sonogel-carbon electrodes: structural and electrochemical characterization. Talanta 82:129–136. https://doi.org/10.1016/j.talanta.2010.04.005

Cubillana-Aguilera LM, Franco-Romano M, Gil MLA et al (2011) New, fast and green procedure for the synthesis of gold nanoparticles based on sonocatalysis. Ultrason Sonochem 18:789–794. https://doi.org/10.1016/j.ultsonch.2010.10.009

Cunha-Silva H, Arcos-Martinez MJ (2018) Dual range lactate oxidase-based screen printed amperometric biosensor for analysis of lactate in diversified samples. Talanta 188:779–787. https://doi.org/10.1016/j.talanta.2018.06.054

Da Silveira Pinto LS, De Souza MVN (2017) Sonochemistry as a general procedure for the synthesis of coumarins, including multigram synthesis. Synthesis 49:2677–2682. https://doi.org/10.1055/s-0036-1590201

Dahoumane SA, Mechouet M, Wijesekera K et al (2017) Algae-mediated biosynthesis of inorganic nanomaterials as a promising route in nanobiotechnology—a review. Green Chem 19:552. https://doi.org/10.1039/c6gc02346k

Dandan H, Quinfu L, Hongfei C, Kuo L (2017) Graphene synthesis via chemical reduction of graphene oxide using lemon extract. J Nanosci Nanotechnol 17:6518–6523

Darvishi M, Mohseni-Asgerani G, Seyed-Yazdi J (2017) Simple microwave irradiation procedure for the synthesis of CuO/graphene hybrid composite with significant photocatalytic enhancement. Surf Interfaces 7:69–73. https://doi.org/10.1016/j.surfin.2017.02.007

Darvishi S, Souissi M, Kharaziha M et al (2018) Gelatin methacryloyl hydrogel for glucose biosensing using Ni nanoparticles-reduced graphene oxide: an experimental and modeling study. Electrochim Acta 261:275–283. https://doi.org/10.1016/j.electacta.2017.12.126

Das R, Bandyopadhyay R, Pramanik P (2018) Carbon quantum dots from natural resource: a review. Mater Today Chem 8:96–109. https://doi.org/10.1016/j.mtchem.2018.03.003

De la Guardia M, Garrigues S (eds) (2012) Handbook of green analytical chemistry, 1st edn. Wiley, Chichester

De Salas F, Pardo I, Salavagione HJ et al (2016) Advanced synthesis of conductive polyaniline using laccase as biocatalyst. PLoS One 11:e0164958. https://doi.org/10.1371/journal.pone.0164958

De Silva KKH, Huang HH, Joshi RK, Yoshimura M (2017) Chemical reduction of graphene oxide using green reductants. Carbon NY 119:190–199. https://doi.org/10.1016/j.carbon.2017.04.025

De Silva KKH, Huang H, Yoshimura M (2018) Progress of reduction of graphene oxide by ascorbic acid. Appl Surf Sci 447:338–346. https://doi.org/10.1016/j.apsusc.2018.03.243

De Xie J, Lai GW, Huq MM (2007) Hydrothermal route to graphene quantum dots: effects of precursor and temperature. Diam Relat Mater 79:112–118. https://doi.org/10.1016/j.diamond.2017.08.014

Debalina B, Reddy RB, Vinu R (2017) Production of carbon nanostructures in biochar, bio-oil and gases from bagasse via microwave assisted pyrolysis using Fe and co as susceptors. J Anal Appl Pyrolysis 124:310–318. https://doi.org/10.1016/j.jaap.2017.01.018

Devi R, Yadav S, Pundir CS (2012) Amperometric determination of xanthine in fish meat by zinc oxide nanoparticle/chitosan/multiwalled carbon nanotube/polyaniline composite film bound xanthine oxidase. Analyst 137:754–759. https://doi.org/10.1039/c1an15838d

Dhanani T, Shah S, Gajbhiye NA, Kumar S (2017) Effect of extraction methods on yield, phytochemical constituents and antioxidant activity of *Withania somnifera*. Arab J Chem 10:S1193–S1199. https://doi.org/10.1016/j.arabjc.2013.02.015

Dhyani V, Bhaskar T (2018) A comprehensive review on the pyrolysis of lignocellulosic biomass. Renew Energy 129:695–716. https://doi.org/10.1016/j.renene.2017.04.035

Di Tocco A, Robledo SN, Osuna Y et al (2018) Development of an electrochemical biosensor for the determination of triglycerides in serum samples based on a lipase/magnetite-chitosan/copper oxide nanoparticles/multiwalled carbon nanotubes/pectin composite. Talanta 190:30–37. https://doi.org/10.1016/j.talanta.2018.07.028

Dickens MJ, Luche JL (1991) Further evidence for the effect of ultrasonic waves on electron transfer processes—the case of the Kornblum-Russell reaction. Tetrahedron Lett 32:4709–4712. https://doi.org/10.1016/S0040-4039(00)92288-3

Ding JH, Zhao HR, Bin YH (2018) A water-based green approach to large-scale production of aqueous compatible graphene nanoplatelets. Sci Rep 8:5567–5572. https://doi.org/10.1038/s41598-018-23859-5

Dinh NX, Van Quy N, Huy TQ, Le A-T (2015) Decoration of silver nanoparticles on multiwalled carbon nanotubes: antibacterial mechanism and ultrastructural analysis. J Nanomater. https://doi.org/10.1155/2015/814379

Dong LL, Ding YC, Huo WT et al (2019) A green and facile synthesis for rGO/Ag nanocomposites using one-step chemical co-reduction route at ambient temperature and combined first principles theoretical analyze. Ultrason Sonochem 53:152–163. https://doi.org/10.1016/j.ultsonch.2019.01.002

Du W, Jiang X, Zhu L (2013) From graphite to graphene: direct liquid-phase exfoliation of graphite to produce single- and few-layered pristine graphene. J Mater Chem A 1:10592–10606. https://doi.org/10.1039/c3ta12212c

Du F, Zhu L, Dai L (2017) Carbon nanotube-based electrochemical biosensors. Biosens Nanomater Nanodev 3:59. https://doi.org/10.1201/b16234

Durge R, Kshirsagar RV, Tambe P (2014) Effect of sonication energy on the yield of graphene nanosheets by liquid-phase exfoliation of graphite. Procedia Eng 97:1457–1465. https://doi.org/10.1016/j.proeng.2014.12.429

El Bouhouti H, Naranjo-Rodríguez I, Hidalgo-Hidalgo de Cisneros JL et al (2009) Electrochemical behaviour of epinephrine and uric acid at a sonogel-carbon l-cysteine modified electrode. Talanta 79:22–26. https://doi.org/10.1016/j.talanta.2009.02.057

El-Naggar ME, Shaheen TI, Fouda MMG, Hebeish AA (2016) Eco-friendly microwave-assisted green and rapid synthesis of well-stabilized gold and core-shell silver-gold nanoparticles. Carbohydr Polym 136:1128–1136. https://doi.org/10.1016/j.carbpol.2015.10.003

Elsupikhe RF, Shameli K, Ahmad MB et al (2015) Green sonochemical synthesis of silver nanoparticles at varying concentrations of κ-carrageenan. Nanoscale Res Lett 10:302–309. https://doi.org/10.1186/s11671-015-0916-1

ENDS (2003) REACH caught up in EU's competitiveness agenda

ENDS (2004) Retailers voice support for REACH chemicals reform

Eriksen M, Lebreton LCM, Carson HS et al (2014) Plastic pollution in the World's oceans: more than 5 trillion plastic pieces weighing over 250,000 tons afloat at sea. PLoS One 9:e111913. https://doi.org/10.1371/journal.pone.0111913

Eshghi M, Vaghari H, Najian Y, Najian MJ (2018) Microwave-assisted green synthesis of silver nanoparticles using *Juglans regia* leaf extract and evaluation of their physico-chemical and antibacterial properties. Antibiotics 7:68. https://doi.org/10.3390/antibiotics7030068

European Comission (2001). White paper: "Strategy for a future chemicals policy". COM 88. Available at https://eur-lex.europa.eu/legal-content/EN/TXT/PDF/?uri=CELEX:52001DC0088&from=EN

Fang L, Yuan W, Wang B, Xiong Y (2016) Growth of graphene on Cu foils by microwave plasma chemical vapor deposition: the effect of in-situ hydrogen plasma post-treatment. Appl Surf Sci 383:28–32. https://doi.org/10.1016/j.apsusc.2016.04.148

Fathy NA (2017) Carbon nanotubes synthesis using carbonization of pretreated rice straw through chemical vapor deposition of camphor. RSC Adv 7:28535–28541. https://doi.org/10.1039/c7ra04882c

Feng H, Li Y, Lin S et al (2014) Nano Cu-catalyzed efficient and selective reduction of nitroarenes under combined microwave and ultrasound irradiation. Sustain Chem Process 2:14. https://doi.org/10.1186/2043-7129-2-14

Feng LL, Wu YX, Zhang DL et al (2017) Near infrared graphene quantum dots-based two-photon nanoprobe for direct bioimaging of endogenous ascorbic acid in living cells. Anal Chem 89:4077–4084. https://doi.org/10.1021/acs.analchem.6b04943

Feng J, Pang B, Chen Y et al (2018) Investigation on tunable optical properties and structures of graphene quantum dots doped with sulfur-containing groups. ECS J Solid State Sci Technol 7:M180–M185. https://doi.org/10.1149/2.0171811jss

Fortineau A-D (2004) Chemistry perfumes in your daily life. J Chem Educ 81:45–50. https://doi.org/10.1021/ed081p45

Francis S, Joseph S, Koshy EP, B. M (2018) Microwave assisted green synthesis of silver nanoparticles using leaf extract of elephantopus scaber and its environmental and biological applications. Artif Cells Nanomed Biotechnol 46:795–804. doi: https://doi.org/10.1080/21691401.2017.1345921

Franco-Romano M, Cubillana-Aguilera L, Gil-Montero M, Palacios-Santander JM, Naranjo-Rodríguez I, Hidalgo-Hidalgo-de-Cisneros J (2013) Method for producing materials by means of sonogel technology catalysed by plant extracts and material produced thereby, p 13. Patent number WO2015018951. University of Cadiz (Spain)

Franco-Romano M, Gil MLA, Palacios-Santander JM et al (2014) Sonosynthesis of gold nanoparticles from a geranium leaf extract. Ultrason Sonochem 21:1570–1577. https://doi.org/10.1016/j.ultsonch.2014.01.017

Fresco-Cala B, Soriano ML, Sciortino A et al (2018) One-pot synthesis of graphene quantum dots and simultaneous nanostructured self-assembly via a novel microwave-assisted method: impact on triazine removal and efficiency monitoring. RSC Adv 8:29939–29946. https://doi.org/10.1039/c8ra04286a

Fujita N (2014) Starch biosynthesis in rice endosperm. AGri-Biosci Monogr 4:1–18. https://doi.org/10.5047/agbm.2014.00401.0001

Gaba M, Dhingra N (2011) Microwave chemistry: general features and applications. Indian J Pharm Educ Res 45:175–183

Gabriel C, Gabriel S, Grant EH et al (1998) Dielectric parameters relevant to microwave dielectric heating. Chem Soc Rev 27:213. https://doi.org/10.1039/a827213z

Gahlawat G, Choudhury AR (2019) A review on the biosynthesis of metal and metal salt nanoparticles by microbes. RSC Adv 9:12944–12967. https://doi.org/10.1039/c8ra10483b

Gaikwad S, Han S (2019) A microwave method for the rapid crystallization of UTSA-16 with improved performance for $CO_2$ capture. Chem Eng J 371:813–820. https://doi.org/10.1016/j.cej.2019.04.112

Gałuszka A, Migaszewski Z, Namieśnik J (2013) The 12 principles of green analytical chemistry and the SIGNIFICANCE mnemonic of green analytical practices. TrAC Trends Anal Chem 50:78–84. https://doi.org/10.1016/j.trac.2013.04.010

Gan N, Yang X, Xie D et al (2010) A disposable organophosphorus pesticides enzyme biosensor based on magnetic composite nano-particles modified screen printed carbon electrode. Sensors 10:625–638. https://doi.org/10.3390/s100100625

Garadkar KM, Ghule LA, Sapnar KB, Dhole SD (2013) A facile synthesis of $ZnWO_4$ nanoparticles by microwave assisted technique and its application in photocatalysis. Mater Res Bull 48:1105–1109. https://doi.org/10.1016/j.materresbull.2012.12.002

García-Guzmán JJ, López-Iglesias D, Cubillana-Aguilera L et al (2018) Assessment of the polyphenol indices and antioxidant capacity for beers and wines using a tyrosinase-based biosensor prepared by sinusoidal current method. Sensors (Basel) 19:66. https://doi.org/10.3390/s19010066

García-Guzmán J, López-Iglesias D, Marin M et al (2019) Electrochemical biosensors for antioxidants. In: Inamuddin D, Khan R, Mohammad A, Asiri A (eds) Advanced biosensors for health care applications, 1st edn. Elsevier, Amsterdam, pp 105–146

Gautam V, Singh KP, Yadav VL (2018) Polyaniline/MWCNTs/starch modified carbon paste electrode for non-enzymatic detection of cholesterol: application to real sample (cow milk). Anal Bioanal Chem 410:2173–2181. https://doi.org/10.1007/s00216-018-0880-6

Gavrilescu M, Demnerová K, Aamand J et al (2015) Emerging pollutants in the environment: present and future challenges in biomonitoring, ecological risks and bioremediation. New Biotechnol 32:147–156. https://doi.org/10.1016/j.nbt.2014.01.001

German N, Ramanaviciene A, Ramanavicius A (2019) Formation of polyaniline and polypyrrole nanocomposites with embedded glucose oxidase and gold nanoparticles. Polymers (Basel) 9:806. https://doi.org/10.3390/polym11020377

Ghule LA, Shirke BS, Sapnar KB et al (2011) Preparation of zinc oxide nanorods by microwave assisted technique using ethylene glycol as a stabilizing agent. J Mater Sci Mater Electron 22:1120–1123. https://doi.org/10.1007/s10854-010-0270-0

Gizdavic-Nikolaidis MR, Stanisavljev DR, Easteal AJ, Zujovic ZD (2010) A rapid and facile synthesis of nanofibrillar polyaniline using microwave radiation. Macromol Rapid Commun 31:657–661. https://doi.org/10.1002/marc.200900800

Godoy AP, Ecorchard P, Beneš H et al (2019) Ultrasound exfoliation of graphite in biphasic liquid systems containing ionic liquids: a study on the conditions for obtaining large few-layers graphene. Ultrason Sonochem 55:279–288. https://doi.org/10.1016/j.ultsonch.2019.01.016

Golinska P, Rathod D, Wypij M et al (2017) Mycoendophytes as efficient synthesizers of bionanoparticles: nanoantimicrobials, mechanism, and cytotoxicity. Crit Rev Biotechnol 37:765–778. https://doi.org/10.1080/07388551.2016.1235011

González-Álvarez RJ, Naranjo-Rodríguez I, Hernández-Artiga MP et al (2016) Experimental design applied to optimisation of silica nanoparticles size obtained by sonosynthesis. J Sol-Gel Sci Technol 80:378–388. https://doi.org/10.1007/s10971-016-4129-6

Gopalan AI, Lee KP, Ragupathy D (2009) Development of a stable cholesterol biosensor based on multi-walled carbon nanotubes-gold nanoparticles composite covered with a layer of chitosan-room-temperature ionic liquid network. Biosens Bioelectron 24:2211–2217. https://doi.org/10.1016/j.bios.2008.11.034

Gribble GW (2013) Food chemistry and chemophobia. Food Secur 5:177–187. https://doi.org/10.1007/s12571-013-0251-2

Grunwald P (2012) Metabolic pathway engineering. Biocatalysis 3:830–856. https://doi.org/10.1142/9781848162310_0013

Gu H, Chen X, Chen F et al (2018) Ultrasound-assisted biosynthesis of CuO-NPs using brown alga *Cystoseira trinodis*: characterization, photocatalytic AOP, DPPH scavenging and antibacterial investigations. Ultrason Sonochem 41:109–119. https://doi.org/10.1016/j.ultsonch.2017.09.006

Gudikandula K, Vadapally P, Singara Charya MA (2017) Biogenic synthesis of silver nanoparticles from white rot fungi: their characterization and antibacterial studies. OpenNano 2:64–78. https://doi.org/10.1016/j.onano.2017.07.002

Guex LG, Sacchi B, Peuvot KF et al (2017) Experimental review: chemical reduction of graphene oxide (GO) to reduced graphene oxide (rGO) by aqueous chemistry. Nanoscale 9:9562–9571. https://doi.org/10.1039/c7nr02943h

Guibal E, Vincent T, Navarro R (2014) Metal ion biosorption on chitosan for the synthesis of advanced materials. J Mater Sci 49:5505–5518. https://doi.org/10.1007/s10853-014-8301-5

Hafeez HY, Lakhera SK, Ashokkumar M, Neppolian B (2019) Ultrasound assisted synthesis of reduced graphene oxide (rGO) supported $InVO_4$-$TiO_2$ nanocomposite for efficient hydrogen production. Ultrason Sonochem 53:1–10. https://doi.org/10.1016/j.ultsonch.2018.12.009

Hai X, Mao QX, Wang WJ et al (2015) An acid-free microwave approach to prepare highly luminescent boron-doped graphene quantum dots for cell imaging. J Mater Chem B 3:9109–9114. https://doi.org/10.1039/c5tb01954k

Hassan HMA, Abdelsayed V, Khder AERS et al (2009) Microwave synthesis of graphene sheets supporting metal nanocrystals in aqueous and organic media. J Mater Chem 19:3832–3837. https://doi.org/10.1039/b906253j

Hassanien R, Husein DZ, Al-Hakkani MF (2018) Biosynthesis of copper nanoparticles using aqueous Tilia extract: antimicrobial and anticancer activities. Heliyon 4:e01077. https://doi.org/10.1016/j.heliyon.2018.e01077

Hay ID, Rehman ZU, Moradali MF et al (2013) Microbial alginate production, modification and its applications. Microb Biotechnol 6:637–650. https://doi.org/10.1111/1751-7915.12076

He M, Zhang J, Wang H et al (2018) Material and optical properties of fluorescent carbon quantum dots fabricated from lemon juice via hydrothermal reaction. Nanoscale Res Lett 13:175. https://doi.org/10.1186/s11671-018-2581-7

Henam SD, Ahmad F, Shah MA et al (2019) Microwave synthesis of nanoparticles and their antifungal activities. Spectrochim Acta A Mol Biomol Spectrosc 213:337–341. https://doi.org/10.1016/j.saa.2019.01.071

Hernandez Y, Nicolosi V, Lotya M et al (2008) High-yield production of graphene by liquid-phase exfoliation of graphite. Nat Nanotechnol 3:563–568. https://doi.org/10.1038/nnano.2008.215

Hernández N, Williams RC, Cochran EW (2014) The battle for the “green” polymer. Different approaches for biopolymer synthesis: bioadvantaged vs. bioreplacement. Org Biomol Chem 12:2834–2849. https://doi.org/10.1039/c3ob42339e

Hidalgo-Hidalgo de Cisneros J, Cordero-Rando M, Naranjo Rodríguez I, et al (2003) Materiales compuestos sonogel-carbono y sonogel-carbono modificados, un procedimiento para su preparación y su aplicación a la fabricación de electrodos y sensores amperométricos, p 7. Patent number ES2195715. University of Cadiz (Spain)

Hidalgo-Hidalgo-de-Cisneros JL, Cordero-Rando M., Naranjo Rodríguez I, et al. (2001) Materiales compuestos sonogel-carbono y sonogel-carbono modificados, un procedimiento para su preparación y su aplicación a la fabricación de electrodos y sensores amperométricos

Hou X, Li Y, Zhao C (2016) Microwave-assisted synthesis of nitrogen-doped multi-layer graphene quantum dots with oxygen-rich functional groups. Aust J Chem 69:357–360. https://doi.org/10.1071/CH15431

Hou D, Liu Q, Wang X et al (2018) Facile synthesis of graphene via reduction of graphene oxide by artemisinin in ethanol. J Mater 4:256–265. https://doi.org/10.1016/j.jmat.2018.01.002

Hu H, Zhao Z, Zhou Q et al (2012a) The role of microwave absorption on formation of graphene from graphite oxide. Carbon N Y 50:3267–3273. https://doi.org/10.1016/j.carbon.2011.12.005

Hu Y, Zou Y, Wu H, Shi D (2012b) A facile and efficient ultrasound-assisted synthesis of novel dispiroheterocycles through 1,3-dipolar cycloaddition reactions. Ultrason Sonochem 19:264–269. https://doi.org/10.1016/j.ultsonch.2011.07.006

Huang H, Liang C, Sha H et al (2019) Microwave assisted hydrothermal way towards highly crystalized N-doped carbon quantum dots and their oxygen reduction performance. Chem Res Chinese Univ 35:171–178. https://doi.org/10.1007/s40242-019-8343-y

Husin MR, Arsad A, Hassan A, Hassan O (2014) Influence of different ultrasonic wave on polymerization of polyaniline nanofiber. Appl Mech Mater 618:50–54. https://doi.org/10.4028/www.scientific.net/amm.618.50

Iskandar F, Hikmah U, Stavila E, Aimon AH (2017) Microwave-assisted reduction method under nitrogen atmosphere for synthesis and electrical conductivity improvement of reduced graphene oxide (rGO). RSC Adv 7:52391–52397. https://doi.org/10.1039/c7ra10013b

Ismail Z, Kassim NFA, Abdullah AH et al (2017) Black tea assisted exfoliation using a kitchen mixer allowing one-step production of graphene. Mater Res Express 4:1–11. https://doi.org/10.1088/2053-1591/aa7ae2

Izaoumen N, Cubillana-Aguilera LM, Naranjo-Rodríguez I et al (2009) β-Sonogel-carbon electrodes: a new alternative for the electrochemical determination of catecholamines. Talanta 78:370–376. https://doi.org/10.1016/j.talanta.2008.11.027

Jacob MV, Rawat RS, Ouyang B et al (2015) Catalyst-free plasma enhanced growth of graphene from sustainable sources. Nano Lett 15:5702–5708. https://doi.org/10.1021/acs.nanolett.5b01363

Jahan I, Erci F, Isildak I (2019) Microwave-assisted green synthesis of non-cytotoxic silver nanoparticles using the aqueous extract of Rosa santana (rose) petals and their antimicrobial activity. Anal Lett 52:1860–1873. https://doi.org/10.1080/00032719.2019.1572179

Jain S, Mehata MS (2017) Medicinal plant leaf extract and pure flavonoid mediated green synthesis of silver nanoparticles and their enhanced antibacterial property. Sci Rep 7:1–13. https://doi.org/10.1038/s41598-017-15724-8

Jana M, Saha S, Khanra P et al (2014) Bio-reduction of graphene oxide using drained water from soaked mung beans (*Phaseolus aureus* L.) and its application as energy storage electrode material. Mater Sci Eng B Solid-State Mater Adv Technol 186:33–40. https://doi.org/10.1016/j.mseb.2014.03.004

Janus Ł, Piątkowski M, Radwan-Pragłowska J et al (2019) Chitosan-based carbon quantum dots for biomedical applications: synthesis and characterization. Nano 9:274. https://doi.org/10.3390/nano9020274

Jiang L, Yu X, Qi X et al (2013) Multigene engineering of starch biosynthesis in maize endosperm increases the total starch content and the proportion of amylose. Transgenic Res 22:1133–1142. https://doi.org/10.1007/s11248-013-9717-4

Kachoosangi RT, Musameh MM, Abu-Yousef I et al (2009) Carbon nanotube-ionic liquid composite sensors and biosensors. Anal Chem 81:435–442. https://doi.org/10.1021/ac801853r

Kajihara K (2013) Recent advances in sol-gel synthesis of monolithic silica and silica-based glasses. J Asian Ceramic Soc 1:121–133. https://doi.org/10.1016/j.jascer.2013.04.002

Kalaiselvi A, Roopan SM, Madhumitha G et al (2015) Synthesis and characterization of palladium nanoparticles using *Catharanthus roseus* leaf extract and its application in the photo-catalytic degradation. Spectrochim Acta Part A Mol Biomol Spectrosc 135:116–119. https://doi.org/10.1016/j.saa.2014.07.010

Kalaivani R, Maruthupandy M, Muneeswaran T et al (2018) Synthesis of chitosan mediated silver nanoparticles (Ag NPs) for potential antimicrobial applications. Front Lab Med 2:30–35. https://doi.org/10.1016/j.flm.2018.04.002

Kandasamy G (2019) Recent advancements in doped/Co-doped carbon quantum dots for multi-potential applications. C 5:24

Kano M, Fujiwara K (2013) Virtual sensing technology in process industries: trends and challenges revealed by recent industrial applications. J Chem Eng Japan 46:1–17. https://doi.org/10.1252/jcej.12we167

Kauffman GB (1991) Chemophobia. Chem Br 27:512–516

Kaur M, Kaur M, Sharma VK (2018) Nitrogen-doped graphene and graphene quantum dots: a review onsynthesis and applications in energy, sensors and environment. Adv Colloid Interf Sci 259:44–64. https://doi.org/10.1016/j.cis.2018.07.001

Kenry K, Liu B (2018) Recent advances in biodegradable conducting polymers and their biomedical applications. Biomacromolecules 19:1783–1803. https://doi.org/10.1021/acs.biomac.8b00275

Khan U, O'Neill A, Porwal H et al (2012) Size selection of dispersed, exfoliated graphene flakes by controlled centrifugation. Carbon N Y 50:470–475. https://doi.org/10.1016/j.carbon.2011.09.001

Kharisov BI, Kharissova OV, Ortiz Méndez U, De La Fuente IG (2016) Decoration of carbon nanotubes with metal nanoparticles: recent trends. Synth React Inorg Met Nano-Metal Chem 46:55–76. https://doi.org/10.1080/15533174.2014.900635

Khomand E, Afsharpour M (2019) Green synthesis of nanostructured SiCs by using natural biopolymers (guar, tragacanth, Arabic, and xanthan gums) for oxidative desulfurization of model fuel. Int J Environ Sci Technol 16:2359–2372. https://doi.org/10.1007/s13762-018-1678-y

Kim Y-R, Bong S, Kang Y-J et al (2010) Electrochemical detection of dopamine in the presence of ascorbic acid using graphene modified electrodes. Biosens Bioelectron 25:2366–2369. https://doi.org/10.1016/j.talanta.2012.05.013

Kimling MC, Caruso RA (2012) Sol-gel synthesis of hierarchically porous $TiO_2$ beads using calcium alginate beads as sacrificial templates. J Mater Chem 22:4073–4082. https://doi.org/10.1039/c2jm15720a

Koel M, Kaljurand M (2019) Green analytical chemistry, 2nd edn. Royal Society of Chemistry, London

Kogawa AC, Salgado HRN (2015) Comparative study over methods developed for quantification of darunavir in tablets by environmental friendly infrared and capillary electrophoretic techniques. J Int Res Med Pharm Sci 2:99–105

Kong FY, Chen TT, Wang JY et al (2016) UV-assisted synthesis of tetrapods-like titanium nitride-reduced graphene oxide nanohybrids for electrochemical determination of chloramphenicol. Sensors Actuators B Chem 225:298–304. https://doi.org/10.1016/j.snb.2015.11.041

Kotanen CR, Karunwi O, Alam F et al (2018) Fabrication and in vitro performance of a dual responsive lactate and glucose biosensor. Electrochim Acta 267:71–79. https://doi.org/10.1016/j.electacta.2018.02.042

Kumar C (ed) (2007) Nanomaterials for biosensors, 1st edn. Wiley-VCH Verlag GmbH & Co. KGaA, Weinheim

Kumar RR, Prasad S (2011) Metabolic engineering of bacteria. Indian J Microbiol 51:403–409. https://doi.org/10.1007/s12088-011-0172-8

Kumar R, Tiwari RS, Srivastava ON (2011) Scalable synthesis of aligned carbon nanotubes bundles using green natural precursor: neem oil. Nanoscale Res Lett 6:92. https://doi.org/10.1186/1556-276X-6-92

Kumar V, Singh K, Panwar S, Mehta SK (2017) Green synthesis of manganese oxide nanoparticles for the electrochemical sensing of p-nitrophenol. Int Nano Lett 7:123–131. doi: https://doi.org/10.1007/s40089-017-0205-3

Kumar A, Kostikov Y, Orberger B et al (2018) Natural laterite as a catalyst source for the growth of carbon nanotubes and nanospheres. ACS Appl Nano Mater 1:6046–6054. https://doi.org/10.1021/acsanm.8b01117

Kumar I, Mondal M, Meyappan V, Sakthivel N (2019) Green one-pot synthesis of gold nanoparticles using *Sansevieria roxburghiana* leaf extract for the catalytic degradation of toxic organic pollutants. Mater Res Bull 117:18–27. https://doi.org/10.1016/j.materresbull.2019.04.029

Kuttippurath J, Kumar P, Nair PJ, Pandey PC (2018) Emergence of ozone recovery evidenced by reduction in the occurrence of Antarctic ozone loss saturation. Clim Atmos Sci 1:42. https://doi.org/10.1038/s41612-018-0052-6

Lahcen AA, García-Guzmán JJ, Palacios-Santander JM et al (2019) Fast route for the synthesis of decorated nanostructured magnetic molecularly imprinted polymers using an ultrasound probe. Ultrason Sonochem 53:226–236. https://doi.org/10.1016/j.ultsonch.2019.01.008

Lancaster M (2002) Principles of sustainable and green chemistry. In: Clark JH, Macquarrie D (eds) Handbook of green chemistry and technology. Blackwell Science Ltd, London, pp 10–27

Lavin-Lopez MP, Paton-Carrero A, Sanchez-Silva L et al (2017) Influence of the reduction strategy in the synthesis of reduced graphene oxide. Adv Powder Technol 28:3195–3203. https://doi.org/10.1016/j.apt.2017.09.032

Lee SY, Krishnamurthy S, Cho CW, Yun YS (2016) Biosynthesis of gold nanoparticles using *Ocimum sanctum* extracts by solvents with different polarity. ACS Sustain Chem Eng 4:2651–2659. https://doi.org/10.1021/acssuschemeng.6b00161

Leicht A, Heiss J, Wong JB (2018) Issues and trends in education for sustainable development. UNESCO Publishing, Paris

Leong T, Ashokkumar M, Sandra K (2011) The fundamentals of power ultrasound—a review. Acoust Aust. 39:54–63

Li YL, Kinloch IA, Shaffer MSP et al (2004) Synthesis of single-walled carbon nanotubes by a fluidized-bed method. Chem Phys Lett 384:98–102. https://doi.org/10.1016/j.cplett.2003.11.070

Li X, Cai W, An J et al (2009) Large-area synthesis of high-quality and uniform graphene films on copper foils. Science 324:1312–1314. https://doi.org/10.1126/science.1171245

Li M, Liu D, Wei D et al (2016) Controllable synthesis of graphene by plasma-enhanced chemical vapor deposition and its related applications. Adv Sci 3:1600003. https://doi.org/10.1002/advs.201600003

Li B, Yu A, Lai G (2018) Self-assembly of phenoxyl-dextran on electrochemically reduced graphene oxide for nonenzymatic biosensing of glucose. Carbon N Y 127:202–208. https://doi.org/10.1016/j.carbon.2017.10.096

Li S, Li G, Chen Q, Wang F (2019) Facile green synthesis of degraded-PVA coated $TiO_2$ nanoparticles with enhanced photocatalytic activity under visible light. J Phys Chem Solids 129:92–98. https://doi.org/10.1016/j.jpcs.2019.01.002

Liang T, Qian J, Yuan Y, Liu C (2013) Synthesis of mesoporous hydroxyapatite nanoparticles using a template-free sonochemistry-assisted microwave method. J Mater Sci 48:5334–5341. https://doi.org/10.1007/s10853-013-7328-3

Lidström P, Tierney J, Wathey B, Westman J (2001) Microwave-assisted green organic synthesis. Tetrahedron 57:9225–9283. https://doi.org/10.1039/9781782623632-00001

Liu P, Wang R (2019) Public attitudes toward technological hazards after a technological disaster. Disaster Prev Manag 28:216–227. https://doi.org/10.1108/dpm-08-2018-0244

Liu Y, Wang M, Li J et al (2005) Highly active horseradish peroxidase immobilized in 1-butyl-3-methylimidazolium tetrafluoroborate room-temperature ionic liquid based sol-gel host materials. Chem Commun:1778–1780. https://doi.org/10.1039/b417680d

Liu YQ, Zhang M, Wang FX, Pan GB (2012) Facile microwave-assisted synthesis of uniform single-crystal copper nanowires with excellent electrical conductivity. RSC Adv 2:11235–11237. https://doi.org/10.1039/c2ra21578k

Liu X, Nan Z, Qiu Y et al (2013) Hydrophobic ionic liquid immoblizing cholesterol oxidase on the electrodeposited Prussian blue on glassy carbon electrode for detection of cholesterol. Electrochim Acta 90:203–209. https://doi.org/10.1016/j.electacta.2012.11.119

Lodeiro C, Capelo-Martínez J-L (2009) Beyond analytical chemistry. In: Capelo-Martínez J-L (ed) Ultrasound in chemistry: analytical applications. Weinheim, Wiley-VCH Verlag GmbH, pp 129–149

López-Iglesias D, García-Romero M, Cubillana-Aguilera LM, et al (2016) Materiales compuestos sonogel-carbono-polímeros conductores y sus variantes: procedimiento de fabricación y su aplicación en la constitución de (bio) sensores electroquímicos, p 41. Patent number ES2670985. University of Cadiz (Spain)

López-Iglesias D, García-Guzmán JJ, Bellido-Milla D et al (2018) The Sonogel-carbon-PEDOT material: an innovative bulk material for sensor devices. J Electrochem Soc 165:B906–B915. https://doi.org/10.1149/2.1021816jes

Lu X, Zhang J, Xie YN et al (2018) Ratiometric phosphorescent probe for thallium in serum, water, and soil samples based on long-lived, spectrally resolved, Mn-doped ZnSe quantum dots and carbon dots. Anal Chem 90:2939–2945. https://doi.org/10.1021/acs.analchem.7b05365

Luo D, Zhang G, Liu J, Sun X (2011) Evaluation criteria for reduced graphene oxide. J Phys Chem C 115:11327–11335. https://doi.org/10.1021/jp110001y

Luo W, Gaumet JJ, Magri P et al (2019) Fast, green microwave-assisted synthesis of single crystalline Sb 2 Se 3 nanowires towards promising lithium storage. J Energy Chem 30:27–33. https://doi.org/10.1016/j.jechem.2018.03.013

Lv Y, Wang F, Zhu H et al (2016) Electrochemically reduced graphene oxide-Nafion/Au nanoparticle modified electrode for hydrogen peroxide sensing. Nanomater Nanotechnol 6:30. https://doi.org/10.5772/63519

Macdonald GJ (2019) Keeping up with bioprocess changes. Genet Eng Biotechnol News 39:37–38, 40. https://doi.org/10.1089/gen.39.03.10

Mahamuni NN, Gogate PR, Pandit AB (2006) Ultrasound-accelerated green and selective oxidation of sulfides to sulfoxides. Ind Eng Chem Res 45:8829–8836. https://doi.org/10.1021/ie061006l

Mahata S, Sahu A, Shukla P et al (2018) The novel and efficient reduction of graphene oxide using *Ocimum sanctum* L. leaf extract as an alternative renewable bio-resource. New J Chem 42:19945–19952. https://doi.org/10.1039/c8nj04086a

Majer E, Llorente B, Rodríguez-Concepción M, Daròs JA (2017) Rewiring carotenoid biosynthesis in plants using a viral vector. Sci Rep 7:41645. https://doi.org/10.1038/srep41645

Majetich G, Hicks R (1995) Application of microwave-accelerated organic synthesis. Radiat Phys Chem 45:567–579. https://doi.org/10.1016/0969-806X(94)00071-Q
Majidian P, Tabatabaei M, Zeinolabedini M et al (2018) Metabolic engineering of microorganisms for biofuel production. Renew Sust Energ Rev 82:3863–3885. https://doi.org/10.1016/j.rser.2017.10.085
Makarov VV, Love AJ, Sinitsyna OV et al (2014) "Green" nanotechnologies: synthesis of metal nanoparticles using plants. Acta Nat 6:35–44
Maleki S, Mærk M, Hrudikova R et al (2017) New insights into *Pseudomonas fluorescens* alginate biosynthesis relevant for the establishment of an efficient production process for microbial alginates. New Biotechnol 37:2–8. https://doi.org/10.1016/j.nbt.2016.08.005
Malinova I, Qasim HM, Brust H, Fettke J (2018) Parameters of starch granule genesis in chloroplasts of *Arabidopsis thaliana*. Front Plant Sci 9:761. https://doi.org/10.3389/fpls.2018.00761
Manjamadha VP, Muthukumar K (2016) Ultrasound assisted green synthesis of silver nanoparticles using weed plant. Bioprocess Biosyst Eng 39:401–411. https://doi.org/10.1007/s00449-015-1523-3
Manley JB, Anastas PT, Cue BW (2008) Frontiers in green chemistry: meeting the grand challenges for sustainability in R&D and manufacturing. J Clean Prod 16:743–750. https://doi.org/10.1016/j.jclepro.2007.02.025
Mao S, Lu G, Chen J (2014) Nanocarbon-based gas sensors: progress and challenges. J Mater Chem A 2:5573–5579. https://doi.org/10.1039/c3ta13823b
de Marco BA, Rechelo BS, Tótoli EG et al (2019) Evolution of green chemistry and its multidimensional impacts: a review. Saudi Pharm J 27:1–8. https://doi.org/10.1016/j.jsps.2018.07.011
Martina K, Tagliapietra S, Barge A, Cravotto G (2016) Combined microwaves/ultrasound, a hybrid technology. Top Curr Chem 374:1–27. https://doi.org/10.1007/s41061-016-0082-7
Martinez-Guerra E, Gude VG (2014) Synergistic effect of simultaneous microwave and ultrasound irradiations on transesterification of waste vegetable oil. Fuel 137:100–108. https://doi.org/10.1016/j.fuel.2014.07.087
McCaslin CA, Petrusca DN, Poirier C et al (2015) Impact of alginate-producing *Pseudomonas aeruginosa* on alveolar macrophage apoptotic cell clearance. J Cyst Fibros 14:70–77. https://doi.org/10.1016/j.jcf.2014.06.009
McKenzie TG, Karimi F, Ashokkumar M, Qiao GG (2019) Ultrasound and sonochemistry for radical polymerization: sound synthesis. Chem Eur J 25:5372–5388. https://doi.org/10.1002/chem.201803771
Mckinnon D (1981) Chemophobia. Chem Eng News 59:5
Mehrotra P (2016) Biosensors and their applications—a review. J Oral Biol Craniofacial Res 6:153–159. https://doi.org/10.1016/j.jobcr.2015.12.002
Mendoza-Cachú D, López-Miranda JL, Mercado-Zúñiga C, Rosas G (2018) Functionalization of MWCNTs with Ag-AuNPs by a green method and their catalytic properties. Diam Relat Mater 84:26–31. https://doi.org/10.1016/j.diamond.2018.03.004
Merkoçi A (ed) (2009) Biosensing using nanomaterials. Wiley, New York
Mhanna A, Chupin L, Brachais CH et al (2018) Efficient microwave-assisted synthesis of glycerol monodecanoate. Eur J Lipid Sci Technol 120:1–8. https://doi.org/10.1002/ejlt.201700133
Mittal AK, Chisti Y, Banerjee UC (2013) Synthesis of metallic nanoparticles using plant extracts. Biotechnol Adv 31:346–356. https://doi.org/10.1016/j.biotechadv.2013.01.003
Mizukoshi Y, Oshima R, Maeda Y, Nagata Y (1999) Preparation of platinum nanoparticles by sonochemical reduction of the Pt(II) ion. Langmuir 15:2733–2737. https://doi.org/10.1021/la9812121
Moghimi-Rad J, Isfahani TD, Hadi I et al (2011) Shape-controlled synthesis of silver particles by surfactant self-assembly under ultrasound radiation. Appl Nanosci 1:27–35. https://doi.org/10.1007/s13204-011-0004-5
Mordini A, Faigl F (2008) New methodologies and techniques for a sustainable organic chemistry. Intech Open, Siena
Morgon PA (ed) (2015) Sustainable development for the healthcare industry: reprogramming the healthcare value chain, 1st edn. Springer International Publishing, Cham

Murugaboopathi G, Parthasarathy V, Chellaram C et al (2013) Applications of biosensors in food industry. Biosci Biotechnol Res Asia 10:711–714. https://doi.org/10.13005/bbra/1185

Mustafa F, Andreescu S (2018) Chemical and biological sensors for food-quality monitoring and smart packaging. Foods 7:168. https://doi.org/10.3390/foods7100168

Muthukrishnan AM (2015) Green synthesis of copper-chitosan nanoparticles and study of its antibacterial activity. J Nanomed Nanotechnol 6:251–255. https://doi.org/10.4172/2157-7439.1000251

Muthuvel A, Adavallan K, Balamurugan K, Krishnakumar N (2014) Biosynthesis of gold nanoparticles using *Solanum nigrum* leaf extract and screening their free radical scavenging and antibacterial properties. Biomed Prev Nutr 4:325–332. https://doi.org/10.1016/j.bionut.2014.03.004

Nair RV, Thomas RT, Sankar V et al (2017) Rapid, acid-free synthesis of high-quality graphene quantum dots for aggregation induced sensing of metal ions and bioimaging. ACS Omega 2:8051–8061. https://doi.org/10.1021/acsomega.7b01262

Narayan R, Kim SO (2015) Surfactant mediated liquid phase exfoliation of graphene. Nano Converg 2:1–19. https://doi.org/10.1186/s40580-015-0050-x

Nasef MM, Zakeri M, Asadi J et al (2016) Environmentally benign and highly regioselective ring opening of epoxides accelerated by ultrasound irradiation. Green Chem Lett Rev 9:76–84. https://doi.org/10.1080/17518253.2015.1137975

Nasrollahzadeh M, Momeni SS, Sajadi SM (2017) Green synthesis of copper nanoparticles using *Plantago asiatica* leaf extract and their application for the cyanation of aldehydes using K4Fe(CN)6. J Colloid Interface Sci 506:471–477. https://doi.org/10.1016/j.jcis.2017.07.072

Naveen MH, Gurudatt NG, Shim Y-B (2017) Applications of conducting polymer composites to electrochemical sensors: a review. Appl Mater Today 9:419–433. https://doi.org/10.1016/j.apmt.2017.09.001

Nayak J, Devi C, Vidyapeeth L (2016) Microwave assisted synthesis: a green chemistry approach. Int Res J Pharm Appl Sci 3:278–285

Neethirajan S, Ragavan V, Weng X, Chand R (2018a) Biosensors for sustainable food engineering: challenges and perspectives. Biosensors 8:23. https://doi.org/10.3390/bios8010023

Neethirajan S, Weng X, Tah A et al (2018b) Nano-biosensor platforms for detecting food allergens—new trends. Sens Bio-Sensing Res 18:13–30. https://doi.org/10.1016/j.sbsr.2018.02.005

Nemamcha A, Rehspringer JL, Khatmi D (2006) Synthesis of palladium nanoparticles by sonochemical reduction of palladium(II) nitrate in aqueous solution. J Phys Chem B 110:383–387. https://doi.org/10.1021/jp0535801

Nguyen DN, Yoon H (2016) Recent advances in nanostructured conducting polymers: from synthesis to practical applications. Polymers (Basel) 8:118. https://doi.org/10.3390/polym8040118

Nguyen HY, Le XH, Dao NT et al (2019) Microwave-assisted synthesis of graphene quantum dots and nitrogen-doped graphene quantum dots: Raman characterization and their optical properties. Adv Nat Sci Nanosci Nanotechnol 10:025005. https://doi.org/10.1088/2043-6254/ab1b73

Nizamuddin S, Baloch HA, Siddiqui MTH et al (2018) An overview of microwave hydrothermal carbonization and microwave pyrolysis of biomass. Rev Environ Sci Biotechnol 17:813–837. https://doi.org/10.1007/s11157-018-9476-z

Nomanbhay S, Salman B, Hussain R, Ong MY (2017) Microwave pyrolysis of lignocellulosic biomass—a contribution to power Africa. Energy Sustain Soc 7:23. https://doi.org/10.1186/s13705-017-0126-z

Obaidullah M, Bahadur NM, Furusawa T et al (2019) Microwave assisted rapid synthesis of $Fe_2O_3$@$SiO_2$ core-shell nanocomposite for the persistence of magnetic property at high temperature. Colloids Surf A Physicochem Eng Asp 572:138–146. https://doi.org/10.1016/j.colsurfa.2019.03.062

Okitsu K, Mizukoshi Y, Yamamoto TA et al (2007) Sonochemical synthesis of gold nanoparticles on chitosan. Mater Lett 61:3429–3431. https://doi.org/10.1016/j.matlet.2006.11.090

Okitsu K, Sharyo K, Nishimura R (2009) One-pot synthesis of gold nanorods by ultrasonic irradiation: the effect of pH on the shape of the gold nanorods and nanoparticles. Langmuir 25:7786–7790. https://doi.org/10.1021/la9017739

Oliveira JR, Kotzebue LRV, Ribeiro FWM et al (2017) Microwave-assisted solvent-free synthesis of novel benzoxazines: a faster and environmentally friendly route to the development of bio-based thermosetting resins. J Polym Sci Part A Polym Chem 55:3534–3544. https://doi.org/10.1002/pola.28755

Omoriyekomwan JE, Tahmasebi A, Zhang J, Yu J (2019) Mechanistic study on direct synthesis of carbon nanotubes from cellulose by means of microwave pyrolysis. Energy Convers Manag 192:88–99. https://doi.org/10.1016/j.enconman.2019.04.042

Ovais M, Khalil AT, Ayaz M et al (2018) Biosynthesis of metal nanoparticles via microbial enzymes: a mechanistic approach. Int J Mol Sci 19:4100. https://doi.org/10.3390/ijms19124100

Padalkar S, Capadona JR, Rowan SJ et al (2010) Natural biopolymers: novel templates for the synthesis of nanostructures. Langmuir 26:8497–8502. https://doi.org/10.1021/la904439p

Pal S, Sharma MK, Danielsson B et al (2014) A miniaturized nanobiosensor for choline analysis. Biosens Bioelectron 54:558–564. https://doi.org/10.1016/j.bios.2013.11.057

Palacios-Santander JM, Cabeza-Saucedo J, García-Romero M, et al (2017) Materiales compuestos Sonogel-Nanotubos de Carbono y Sonogel-Nanocarbono: procedimiento de fabricación y su aplicación para la construcción de electrodos y (bio)sensores electroquímicos, p 38

Palma-Goyes RE, Vazquez-Arenas J, Romero-Ibarra IC, Ostos C (2018) Microwave-assisted solvothermal one-pot synthesis of $RuO_2$ nanoparticles: first insights of its activity towards oxygen and chlorine evolution reactions. ChemistrySelect 3:12937–12945. https://doi.org/10.1002/slct.201802695

Pandey K, Jain A, Meena AK et al (2013) Camphor based carbon nano tubes: a recent advancement in green chemistry. Bull Environ Pharmacol Life Sci 3:3–6

Pang J, Bachmatiuk A, Ibrahim I et al (2016) CVD growth of 1D and 2D sp2carbon nanomaterials. J Mater Sci 51:640–667. https://doi.org/10.1007/s10853-015-9440-z

Park JE, Atobe M, Fuchigami T (2006) Synthesis of multiple shapes of gold nanoparticles with controlled sizes in aqueous solution using ultrasound. Ultrason Sonochem 13:237–241. https://doi.org/10.1016/j.ultsonch.2005.04.003

Park SY, Thongsai N, Chae A et al (2017) Microwave-assisted synthesis of luminescent and biocompatible lysine-based carbon quantum dots. J Ind Eng Chem 47:329–335. https://doi.org/10.1016/j.jiec.2016.12.002

Patil SB, Annese VF, Cumming DRS (2019) Commercial aspects of biosensors for diagnostics and environmental monitoring. In: Deep A, Kumar S (eds) Advances in nanosensors for biological and environmental analysis, 1st edn. Elsevier, Amsterdam, pp 133–142

Pawełczyk A, Sowa-kasprzak K, Olender D (2018) Microwave (MW), ultrasound (US) and combined synergic MW-US strategies for rapid functionalization of pharmaceutical use phenols. Molecules 23:2360. https://doi.org/10.3390/molecules23092360

Pei S, Cheng HM (2012) The reduction of graphene oxide. Carbon N Y 50:3210–3228. https://doi.org/10.1016/j.carbon.2011.11.010

Pekdemir S, Onses MS, Hancer M (2017) Low temperature growth of graphene using inductively-coupled plasma chemical vapor deposition. Surf Coat Technol 309:814–819. https://doi.org/10.1016/j.surfcoat.2016.10.081

Peng W, Li H, Liu Y, Song S (2016) Comparison of Pb(II) adsorption onto graphene oxide prepared from natural graphites: diagramming the Pb(II) adsorption sites. Appl Surf Sci 364:620–627. https://doi.org/10.1016/j.apsusc.2015.12.208

Perin G, Araujo DR, Nobre PC et al (2018) Ultrasound-promoted synthesis of 2-organoselanyl-naphthalenes using Oxone® in aqueous medium as an oxidizing agent. PeerJ 6:e4706. https://doi.org/10.7717/peerj.4706

Pilas J, Yazici Y, Selmer T et al (2017) Optimization of an amperometric biosensor array for simultaneous measurement of ethanol, formate, D- and L-lactate. Electrochim Acta 251:256–262. https://doi.org/10.1016/j.electacta.2017.07.119

Pirtarighat S, Ghannadnia M, Baghshahi S (2019) Biosynthesis of silver nanoparticles using *Ocimum basilicum* cultured under controlled conditions for bactericidal application. Mater Sci Eng C 98:250–255. https://doi.org/10.1016/j.msec.2018.12.090

Płotka-Wasylka J, Namieśnik J (eds) (2019) Green analytical chemistry: past, present and perspectives, 1st edn. Springer Nature Singapore, Singapore

Płotka-Wasylka J, Kurowska-Susdorf A, Sajid M et al (2018) Green chemistry in higher education: state of the art, challenges, and future trends. ChemSusChem 11:2845–2858. https://doi.org/10.1002/cssc.201801109

Plumejeau S, Alauzun JG, Boury B (2015) Hybrid metal oxide@biopolymer materials precursors of metal oxides and metal oxide-carbon composites. J Ceram Soc Japan 123:695–708. https://doi.org/10.2109/jcersj2.123.695

Preiss LC, Landfester K, Muñoz-Espí R (2014) Biopolymer colloids for controlling and templating inorganic synthesis. Beilstein J Nanotechnol 5:2129–2138. https://doi.org/10.3762/bjnano.5.222

Qiu B, Li Z, Wang X et al (2017) Exploration on the microwave-assisted synthesis and formation mechanism of polyaniline nanostructures synthesized in different hydrochloric acid concentrations. J Polym Sci Part A Polym Chem 55:3357–3369. https://doi.org/10.1002/pola.28707

Qu D, Zheng M, Du P et al (2013a) Highly luminescent S, N co-doped graphene quantum dots with broad visible absorption bands for visible light photocatalysts. Nanoscale 5:12272–12277. https://doi.org/10.1039/c3nr04402e

Qu S, Chen H, Zheng X et al (2013b) Ratiometric fluorescent nanosensor based on water soluble carbon nanodots with multiple sensing capacities. Nanoscale 5:5514–5518. https://doi.org/10.1039/c3nr00619k

Rahman G, Najaf Z, Mehmood A et al (2019) An overview of the recent Progress in the synthesis and applications of carbon nanotubes. C 5:3. https://doi.org/10.3390/c5010003

Rai M, Ingle A, Gupta I et al (2013) Potential role of biological systems in formation of nanoparticles: mechanism of synthesis and biomedical applications. Curr Nanosci 9:576–587. https://doi.org/10.2174/1573413711309990092

Rao CNR, Sood AK, Subrahmanyam KS, Govindaraj A (2009) Graphene: the new two-dimensional nanomaterial. Angew Chem Int Ed 48:7752–7777. https://doi.org/10.1002/anie.200901678

Rasaee I, Ghannadnia M, Baghshahi S (2018) Biosynthesis of silver nanoparticles using leaf extract of *Satureja hortensis* treated with NaCl and its antibacterial properties. Microporous Mesoporous Mater 264:240–247. https://doi.org/10.1016/j.micromeso.2018.01.032

Rasheed T, Bilal M, Iqbal HMN, Li C (2017) Green biosynthesis of silver nanoparticles using leaves extract of *Artemisia vulgaris* and their potential biomedical applications. Colloids Surf B Biointerfaces 158:408–415. https://doi.org/10.1016/j.colsurfb.2017.07.020

Rashid HU, Yu K, Umar MN et al (2015) Catalyst role in chemical vapor deposition (CVD) process: a review. Rev Adv Mater Sci 40:235–248

Ravichandran S, Karthikeyan E (2011) Microwave synthesis - a potential tool for green chemistry. Int J ChemTech Res 3:466–470

Reddy B, Bandi R, Alle M et al (2018) Microwave assisted rapid green synthesis of gold nanoparticles using *Annona squamosa* L. peel extract for the efficient catalytic reduction of organic pollutants. J Mol Struct 1167:305–315. https://doi.org/10.1016/j.molstruc.2018.05.004

Ren G, Tang M, Chai F, Wu H (2018) One-pot synthesis of highly fluorescent carbon dots from spinach and multipurpose applications. Eur J Inorg Chem 2018:153–158. https://doi.org/10.1002/ejic.201701080

Roh HK, Lee GW, Chung KY, Kim KB (2019) Revisiting $NaTi_2(PO_4)_3$/nanocarbon composites prepared using nanocarbons with different dimensions for high-rate sodium-ion batteries: the surface properties of nanocarbons. J Alloys Compd 787:728–737. https://doi.org/10.1016/j.jallcom.2019.02.167

Ropeik D (2015) On the roots of, and solutions to, the persistent battle between "chemonoia" and rationalist denialism of the subjective nature of human cognition. Hum Exp Toxicol 34:1272–1278. https://doi.org/10.1177/0960327115603592

Rossini EL, Milani MI, Pezza HR (2019) Green synthesis of fluorescent carbon dots for determination of glucose in biofluids using a paper platform. Talanta 201:503–510. https://doi.org/10.1016/j.talanta.2019.04.045

Roy S (2016) Chemistry in our daily life: preliminary information. Int J Home Sci 2:361–366

Rudd JA, Gowenlock CE, Gomez V et al (2019) Solvent-free microwave-assisted synthesis of tenorite nanoparticle-decorated multi-walled carbon nanotubes. J Mater Sci Technol 35:1121–1127. https://doi.org/10.1016/j.jmst.2019.01.002

Ruíz-Baltazar Á d J, Reyes-López SY, Mondragón-Sánchez M d L et al (2018) Biosynthesis of Ag nanoparticles using *Cynara cardunculus* leaf extract: evaluation of their antibacterial and electrochemical activity. Results Phys 11:1142–1149. https://doi.org/10.1016/j.rinp.2018.11.032

Sacco M, Charnay C, De Angelis F et al (2015) Microwave-ultrasound simultaneous irradiation: a hybrid technology applied to ring closing metathesis. RSC Adv 5:16878–16885. https://doi.org/10.1039/c4ra14938f

Saeed S, Hashmi AS, Ikram-ul-Haq et al (2016) Bioconversion of agricultural by-products to alginate by Azotobacter vinelandii and physico-chemical optimization for hyper-production. J Anim Plant Sci 26:1516–1521

Sahoo S, Hatui G, Bhattacharya P et al (2013) One pot synthesis of graphene by exfoliation of graphite in ODCB. Graphene 2:42–48. https://doi.org/10.4236/graphene.2013.21006

Salifairus MJ, Abd Hamid SB, Soga T et al (2016) Structural and optical properties of graphene from green carbon source via thermal chemical vapor deposition. J Mater Res 31:1947–1956. https://doi.org/10.1557/jmr.2016.200

Sangeetha G, Rajeshwari S, Venckatesh R (2011) Green synthesis of zinc oxide nanoparticles by aloe barbadensis miller leaf extract: structure and optical properties. Mater Res Bull 46:2560–2566. https://doi.org/10.1016/j.materresbull.2011.07.046

Schmid J, Sieber V, Rehm B (2015) Bacterial exopolysaccharides: biosynthesis pathways and engineering strategies. Front Microbiol 6:496. https://doi.org/10.3389/fmicb.2015.00496

Schmitz H, Johnson O, Altenburg T (2015) Rent management—the heart of green industrial policy. New Polit Econ 20:812–831. https://doi.org/10.1080/13563467.2015.1079170

Sekhon BS (2010) Microwave-assisted pharmaceutical synthesis: an overview. Int J PharmTech Res 2:827–833

Setaro A (2017) Advanced carbon nanotubes functionalization. J Phys Condens Matter 29:423003

Sethuraman V, Muthuraja P, Anandha Raj J, Manisankar P (2016) A highly sensitive electrochemical biosensor for catechol using conducting polymer reduced graphene oxide–metal oxide enzyme modified electrode. Biosens Bioelectron 84:112–119. https://doi.org/10.1016/j.bios.2015.12.074

Shafiei-Irannejad V, Soleymani J, Azizi S et al (2019) Advanced nanomaterials towards biosensing of insulin: analytical approaches. TrAC Trends Anal Chem 116:1–12. https://doi.org/10.1016/j.trac.2019.04.020

Shah M, Fawcett D, Sharma S, et al (2015) Green synthesis of metallic nanoparticles via biological entities. Materials 8:7278–7308. https://doi.org/10.3390/ma8115377

Shahvelayati AS, Sabbaghan M, Banihashem S (2017) Sonochemically assisted synthesis of N-substituted pyrroles catalyzed by ZnO nanoparticles under solvent-free conditions. Monatsh Chem 148:1123–1129. https://doi.org/10.1007/s00706-016-1904-6

Shamaila S, Sajjad AKL, Ryma N u A et al (2016) Advancements in nanoparticle fabrication by hazard free eco-friendly green routes. Appl Mater Today 5:150–199. https://doi.org/10.1016/j.apmt.2016.09.009

Shen T, Wang Q, Guo Z et al (2018) Hydrothermal synthesis of carbon quantum dots using different precursors and their combination with $TiO_2$ for enhanced photocatalytic activity. Ceram Int 44:11828–11834. https://doi.org/10.1016/j.ceramint.2018.03.271

Shi K, Yan J, Lester E, Wu T (2014) Catalyst-free synthesis of multiwalled carbon nanotubes via microwave-induced processing of biomass. Ind Eng Chem Res 53:15012–15019. https://doi.org/10.1021/ie503076n

Shin KS, Ji JH, Hwang KS et al (2016) Sensitivity enhancement of bead-based electrochemical impedance spectroscopy (BEIS) biosensor by electric field-focusing in microwells. Biosens Bioelectron 85:16–24. https://doi.org/10.1016/j.bios.2016.04.086

Shirke BS, Korake PV, Hankare PP et al (2011) Synthesis and characterization of pure anatase $TiO_2$ nanoparticles. J Mater Sci Mater Electron 22:821–824. https://doi.org/10.1007/s10854-010-0218-4

Shukla AK, Iravani S (eds) (2019) Green synthesis, characterization and applications of nanoparticles. Elsevier, Amsterdam

Sidhu JS, Singh A, Garg N et al (2018) Carbon dots as analytical tools for sensing of thioredoxin reductase and screening of cancer cells. Analyst 143:1853–1861. https://doi.org/10.1039/c7an02040f

Singh AK, Talat M, Singh DP, Srivastava ON (2010) Biosynthesis of gold and silver nanoparticles by natural precursor clove and their functionalization with amine group. J Nanopart Res 12:1667–1675. https://doi.org/10.1007/s11051-009-9835-3

Siontorou CG (2019) University-industry relationships for the development and commercialization of biosensors. In: Thouand G (ed) Handbook of cell biosensors. Springer Nature, Cham, pp 1–16

Skellam E (2019) Strategies for engineering natural product biosynthesis in fungi. Trends Biotechnol 37:416–427. https://doi.org/10.1016/j.tibtech.2018.09.003

Smitha SL, Philip D, Gopchandran KG (2009) Green synthesis of gold nanoparticles using *Cinnamomum zeylanicum* leaf broth. Spectrochim Acta Part A Mol Biomol Spectrosc 74:735–739. https://doi.org/10.1016/j.saa.2009.08.007

Son M, Ham MH (2017) Low-temperature synthesis of graphene by chemical vapor deposition and its applications. FlatChem 5:40–49. https://doi.org/10.1016/j.flatc.2017.07.002

Song J, Han B (2015) Green chemistry: a tool for the sustainable development of the chemical industry. Natl Sci Rev 2:255–258. https://doi.org/10.1093/nsr/nwu076

Song JY, Jang HK, Kim BS (2009) Biological synthesis of gold nanoparticles using *Magnolia kobus* and *Diopyros kaki* leaf extracts. Process Biochem 44:1133–1138. https://doi.org/10.1016/j.procbio.2009.06.005

Song W, Duan W, Liu Y et al (2017) Ratiometric detection of intracellular lysine and pH with one-pot synthesized dual emissive carbon dots. Anal Chem 89:13626–13633. https://doi.org/10.1021/acs.analchem.7b04211

Stankovich S, Dikin DA, Piner RD et al (2007) Synthesis of graphene-based nanosheets via chemical reduction of exfoliated graphite oxide. Carbon N Y 45:1558–1565. https://doi.org/10.1016/j.carbon.2007.02.034

Stephens GL, Kahn BH, Richardson M (2016) The super greenhouse effect in a changing climate. J Clim 29:5469–5482. https://doi.org/10.1175/JCLI-D-15-0234.1

Strauss CR (2002) Applications of microwaves for environmentally benign organic chemistry. In: Clark JH, Macquarrie D (eds) Handbook of green chemistry and technology. Blackwell Science Ltd, London, pp 397–415

Sukumaran SS, Tripathi S, Resmi AN et al (2019) Influence of surfactants on the electronic properties of liquid-phase exfoliated graphene. Mater Sci Eng B Solid-State Mater Adv Technol 240:62–68. https://doi.org/10.1016/j.mseb.2019.01.003

Sun H, Wu L, Wei W, Qu X (2013) Recent advances in graphene quantum dots for sensing. Mater Today 16:433–442. https://doi.org/10.1016/j.mattod.2013.10.020

Sundaresan P, Karthik R, Chen SM et al (2019) Ultrasonication-assisted synthesis of sphere-like strontium cerate nanoparticles (SrCeO3 NPs) for the selective electrochemical detection of calcium channel antagonists nifedipine. Ultrason Sonochem 53:44–54. https://doi.org/10.1016/j.ultsonch.2018.12.013

Suresh D, Nethravathi PC, Udayabhanu, et al. (2015) Green synthesis of multifunctional zinc oxide (ZnO) nanoparticles using *Cassia fistula* plant extract and their photodegradative, antioxidant and antibacterial activities. Mater Sci Semicond Process 31:446–454. doi: https://doi.org/10.1016/j.mssp.2014.12.023

Sweeney JB, Bethel PA, Gill DM et al (2019) Synthesis of a protected keto-lysidine analogue via improved preparation of arabino-isocytosine nucleosides. Org Lett 21:2004–2007. https://doi.org/10.1021/acs.orglett.9b00086

Tao W, Pan D, Liu Q, Yao S, Nie Z, Han B (2006) Optical and Bioelectrochemical Characterization of Water-Miscible Ionic Liquids Based Composites of Multiwalled Carbon Nanotubes. Electroanalysis 18(17):1681–1688

Tappiban P, Smith DR, Triwitayakorn K, Bao J (2019) Recent understanding of starch biosynthesis in cassava for quality improvement: a review. Trends Food Sci Technol 83:167–180. https://doi.org/10.1016/j.tifs.2018.11.019

Tetlow IJ, Emes MJ (2017) Starch biosynthesis in the developing endosperms of grasses and cereals. Agronomy 7:81. https://doi.org/10.3390/agronomy7040081

Thakur S, Karak N (2012) Green reduction of graphene oxide by aqueous phytoextracts. Carbon N Y 50:5331–5339. https://doi.org/10.1016/j.carbon.2012.07.023

Thakur MS, Ragavan KV (2013) Biosensors in food processing. J Food Sci Technol 50:625–641. https://doi.org/10.1007/s13197-012-0783-z

Thitisaksakul M, Jiménez RC, Arias MC, Beckles DM (2012) Effects of environmental factors on cereal starch biosynthesis and composition. J Cereal Sci 56:67–80. https://doi.org/10.1016/j.jcs.2012.04.002

Tian R, Zhong S, Wu J et al (2016) Facile hydrothermal method to prepare graphene quantum dots from graphene oxide with different photoluminescences. RSC Adv 6:40422–40426. https://doi.org/10.1039/c6ra00780e

Timothy J, Cintas P (2002) Sonochemistry. In: Clark JH, Macquarrie D (eds) Handbook of green chemistry and technology. Blackwell Science Ltd, London, pp 372–396

Tiwari BK (2015) Ultrasound: a clean, green extraction technology. TrAC Trends Anal Chem 71:100–109. https://doi.org/10.1016/j.trac.2015.04.013

Torabfam M, Jafarizadeh-Malmiri H (2018) Microwave-enhanced silver nanoparticle synthesis using chitosan biopolymer: optimization of the process conditions and evaluation of their characteristics. Green Process Synth 7:530–537. https://doi.org/10.1515/gps-2017-0139

Torres-Palma RA, Serna-Galvis EA (2018) Sonolysis. In: Advanced oxidation processes for wastewater treatment: emerging green chemical technology. Elsevier Inc., Medellín, pp 177–213

Tran TS, Park SJ, Yoo SS et al (2016) High shear-induced exfoliation of graphite into high quality graphene by Taylor-Couette flow. RSC Adv 6:12003–12008. https://doi.org/10.1039/c5ra22273g

Trenberth KE, Dai A, Van der Schrier G et al (2014) Golbal warming and changes in drought. Nat Clim Chang 4:17–22

Tripathi N, Pavelyev V, Islam SS (2017) Synthesis of carbon nanotubes using green plant extract as catalyst: unconventional concept and its realization. Appl Nanosci 7:557–566. https://doi.org/10.1007/s13204-017-0598-3

Trost BM (1991) The atom economy—a search for synthetic efficiency. Science 254:1471–1477. https://doi.org/10.1126/science.1962206

Trost BM (1995) Atom economy—a challenge for organic synthesis: homogeneous catalysis leads the way. Angew Chem Int Ed Engl 34:259–281. https://doi.org/10.1002/anie.199502591

U.S. Environmental Protection Agency (1992) Environmental Education and The National Environmental Education Act of 1990. https://www.epa.gov.tw/public/Attachment/78716464493.pdf. Accessed 3 Jul 2019

Uchida H, Aryal HR, Adhikari S, Umeno M (2016) Low temperature plasma CVD grown graphene by microwave surface-wave plasma CVD using camphor precursor. J Phys Sci Appl 6:34–38. https://doi.org/10.17265/2159-5348/2016.02.005

Ukkund SJ, Darshanram KZ et al (2019) Microwave assisted green synthesis and characterization of silver nanoparticles from *Hibiscus* leaf extract and investigation of their antimicrobial activities. AIP Conf Proc:2080. https://doi.org/10.1063/1.5092885

UNIDO (2011) Green industry initiative for sustainable industrial development. Vienna. (Avaliable at: https://www.unido.org/our-focus/cross-cutting-services/green-industry/green-industry-initiative)

Uysal Unalan I, Wan C, Trabattoni S et al (2015) Polysaccharide-assisted rapid exfoliation of graphite platelets into high quality water-dispersible graphene sheets. RSC Adv 5:26482–26490. https://doi.org/10.1039/c4ra16947f

Vágó A, Szakacs G, Sáfrán G et al (2016) One-step green synthesis of gold nanoparticles by mesophilic filamentous fungi. Chem Phys Lett 645:1–4. https://doi.org/10.1016/j.cplett.2015.12.019
Valentini F, Roscioli D, Carbone M et al (2015) Graphene and ionic liquids new gel paste electrodes for caffeic acid quantification. Sensors Actuators B Chem 212:248–255. https://doi.org/10.1016/j.snb.2015.02.033
Vásquez-Ponce F, Higuera-Llantén S, Pavlov MS et al (2017) Alginate overproduction and biofilm formation by psychrotolerant *Pseudomonas mandelii* depend on temperature in Antarctic marine sediments. Electron J Biotechnol 28:27–34. https://doi.org/10.1016/j.ejbt.2017.05.001
Vijaya Kumar P, Mary Jelastin Kala S, Prakash KS (2019) Green synthesis of gold nanoparticles using Croton Caudatus Geisel leaf extract and their biological studies. Mater Lett 236:19–22. https://doi.org/10.1016/j.matlet.2018.10.025
Vogiazi V, De La Cruz A, Mishra S et al (2019) A comprehensive review: development of electrochemical biosensors for detection of cyanotoxins in freshwater. ACS Sensors 4:1151–1173. https://doi.org/10.1021/acssensors.9b00376
Vu THT, Tran TTT, Le HNT et al (2015) A new green approach for the reduction of graphene oxide nanosheets using caffeine. Bull Mater Sci 38:667–671
Wang J, Fang BS, Chou KY et al (2014) A two-stage enzymatic synthesis of conductive poly(3,4-ethylenedioxythiophene). Enzym Microb Technol 54:45–50. https://doi.org/10.1016/j.enzmictec.2013.10.002
Wang J, Guleria S, Koffas MAG, Yan Y (2016) Microbial production of value-added nutraceuticals. Curr Opin Biotechnol 37:97–104. https://doi.org/10.1016/j.copbio.2015.11.003
Wang C, Gao X, Chen Z et al (2017) Preparation, characterization and application of polysaccharide-based metallic nanoparticles: a review. Polymers (Basel) 9:689–722. https://doi.org/10.3390/polym9120689
Wang M, Fu Q, Zhang K et al (2019a) A magnetic and carbon dot based molecularly imprinted composite for fluorometric detection of 2,4,6-trinitrophenol. Microchim Acta 186:86. https://doi.org/10.1007/s00604-018-3200-0
Wang Y, Zhang H, Lu X et al (2019b) Advances in 2-phenylethanol production from engineered microorganisms. Biotechnol Adv 37:403–409. https://doi.org/10.1016/j.biotechadv.2019.02.005
Wathey B, Tierney J, Lidström P, Westman J (2002) The impact of microwave-assisted organic chemistry on drug discovery. Drug Discov Today 7:373–380. https://doi.org/10.1016/S1359-6446(02)02178-5
Wei C, Li CJ (2002) Grignard type reaction via C-H bond activation in water. Green Chem 4:39–41. https://doi.org/10.1039/b110102c
Wei D, Qian W (2008) Facile synthesis of Ag and Au nanoparticles utilizing chitosan as a mediator agent. Colloids Surf B Biointerfaces 62:136–142. https://doi.org/10.1016/j.colsurfb.2007.09.030
Wei C, Li Z, Li CJ (2004) The development of A3-coupling (aldehyde-alkyne-amine) and AA3-coupling (asymmetric aldehyde-alkyne-amine). Synlett 9:1472–1483. https://doi.org/10.1055/s-2004-829531
Weizhong Q, Fei W, Zhanwen W et al (2003) Production of carbon nanotubes in a packed bed and a fluidized bed. AICHE J 49:619–625. https://doi.org/10.1002/aic.690490308
Wen C, Zhao N, Zhang DW et al (2014) Efficient reduction and exfoliation of graphite oxide by sequential chemical reduction and microwave irradiation. Synth Met 194:71–76. https://doi.org/10.1016/j.synthmet.2014.04.023
Wen Y, Wen W, Zhang X, Wang S (2016) Highly sensitive amperometric biosensor based on electrochemically-reduced graphene oxide-chitosan/hemoglobin nanocomposite for nitromethane determination. Biosens Bioelectron 79:894–900. https://doi.org/10.1016/j.bios.2016.01.028
Wen Z, Shen Q, Sun X (2017) Nanogenerators for self-powered gas sensing. Nano-Micro Lett 9:45. https://doi.org/10.1007/s40820-017-0146-4

Wróblewska-Krepsztul J, Rydzkowski T, Michalska-Pożoga I, Thakur VK (2019) Biopolymers for biomedical and pharmaceutical applications: recent advances and overview of alginate electrospinning. Nano 9:404. https://doi.org/10.3390/nano9030404

Wu Z, Ondruschka B, Cravotto G et al (2008) Oxidation of primary aromatic amines under irradiation with ultrasound and/or microwaves. Synth Commun 38:2619–2624

Wu Z, Cherkasov N, Cravotto G et al (2015) Ultrasound- and microwave-assisted preparation of lead-free palladium catalysts: effects on the kinetics of diphenylacetylene semi-hydrogenation. ChemCatChem 7:952–959. https://doi.org/10.1002/cctc.201402999

Xie LY, Li YJ, Qu J et al (2017) A base-free, ultrasound accelerated one-pot synthesis of 2-sulfonylquinolines in water. Green Chem 19:5642–5646. https://doi.org/10.1039/c7gc02304a

Xie X, Zhou Y, Huang K (2019) Advances in microwave-assisted production of reduced graphene oxide. Front Chem 7:355. https://doi.org/10.3389/fchem.2019.00355

Xu H, Zeiger BW, Suslick KS (2013) Sonochemical synthesis of nanomaterials. Chem Soc Rev 42:2555–2567. https://doi.org/10.1039/c2cs35282f

Xu M, Peng W, Cai J et al (2015) Ultrasound-assisted synthesis and characterization of ultrathin copper nanowhiskers. Mater Lett 161:164–167. https://doi.org/10.1016/j.matlet.2015.08.092

Xu Y, Cao H, Xue Y et al (2018) Liquid-phase exfoliation of Graphene: An overview on exfoliation media, techniques, and challenges. Nano 8:942–974. https://doi.org/10.3390/nano8110942

Xu Y, Du C, Steinkruger JD et al (2019) Microwave-assisted synthesis of AuNPs/CdS composite nanorods for enhanced photocatalytic hydrogen evolution. J Mater Sci 54:6930–6942. https://doi.org/10.1007/s10853-018-03294-7

Yadi M, Mostafavi E, Saleh B et al (2018) Current developments in green synthesis of metallic nanoparticles using plant extracts: a review. Artif Cells Nanomed Biotechnol 46:S336–S343. https://doi.org/10.1080/21691401.2018.1492931

Yakout SM, Mostafa AA (2015) A novel green synthesis of silver nanoparticles using soluble starch and its antibacterial activity. Int J Clin Exp Med 8:3538–3544

Yang JM, Ji SJ, Gu DG et al (2005) Ultrasound-irradiated Michael addition of amines to ferrocenylenones under solvent-free and catalyst-free conditions at room temperature. J Organomet Chem 690:2989–2995. https://doi.org/10.1016/j.jorganchem.2005.03.030

Yang P, Zhu Z, Chen M et al (2018) Microwave-assisted synthesis of xylan-derived carbon quantum dots for tetracycline sensing. Opt Mater (Amst) 85:329–336. https://doi.org/10.1016/j.optmat.2018.06.034

Yavir K, Marcinkowski Ł, Marcinkowska R et al (2019) Analytical applications and physicochemical properties of ionic liquid-based hybrid materials: a review. Anal Chim Acta 1054:1–16. https://doi.org/10.1016/j.aca.2018.10.061

Yaws CL (2009) Thermophysical properties of chemicals and hydrocarbons, 1st edn. Elsevier Science, New York

Yin R, Shen P, Lu Z (2019) A green approach for the reduction of graphene oxide by the ultraviolet/sulfite process. J Colloid Interface Sci 550:110–116. https://doi.org/10.1016/j.jcis.2019.04.073

Yoo IH, Kalanur SS, Seo H (2019) Deposition of Pd nanoparticles on MWCNTs: green approach and application to hydrogen sensing. J Alloys Compd 788:936–943. https://doi.org/10.1016/j.jallcom.2019.02.298

You Y, Mayyas M, Xu S et al (2017) Growth of NiO nanorods, SiC nanowires and monolayer graphene: via a CVD method. Green Chem 19:5599–5607. https://doi.org/10.1039/c7gc02523h

Yu T, Wang H, Guo C et al (2018) A rapid microwave synthesis of green-emissive carbon dots with solid-state fluorescence and pH-sensitive properties. R Soc Open Sci 5:180245. https://doi.org/10.1098/rsos.180245

Zahran MK, Ahmed HB, El-Rafie MH (2014) Facile size-regulated synthesis of silver nanoparticles using pectin. Carbohydr Polym 111:971–978. https://doi.org/10.1016/j.carbpol.2014.05.028

Zamfir LG, Rotariu L, Bala C (2013) Acetylcholinesterase biosensor for carbamate drugs based on tetrathiafulvalene-tetracyanoquinodimethane/ionic liquid conductive gels. Biosens Bioelectron 46:61–67. https://doi.org/10.1016/j.bios.2013.02.018

Zarzuela R, Luna MJ, Gil MLA et al (2018) Analytical determination of the reducing and stabilization agents present in different *Zostera noltii* extracts used for the biosynthesis of gold nanoparticles. J Photochem Photobiol B Biol 179:32–38. https://doi.org/10.1016/j.jphotobiol.2017.12.025

Zhang Y, Suslick KS (2015) Synthesis of poly(3,4-ethylenedioxythiophene) microspheres by ultrasonic spray polymerization (USPo). Chem Mater 27:7559–7563. https://doi.org/10.1021/acs.chemmater.5b03423

Zhang J, Du J, Han B et al (2006) Sonochemical formation of single-crystalline gold nanobelts. Angew Chem Int Ed 45:1116–1119. https://doi.org/10.1002/anie.200503762

Zhang Y, Zhang L, Zhou C (2013a) Review of chemical vapor deposition of graphene and related applications. Acc Chem Res 40:2329–2339. https://doi.org/10.1021/ar300203n

Zhang YL, Wang L, Zhang HC et al (2013b) Graphitic carbon quantum dots as a fluorescent sensing platform for highly efficient detection of $Fe^{3+}$ ions. RSC Adv 3:3733–3738. https://doi.org/10.1039/c3ra23410j

Zhang Y, Arugula MA, Wales M et al (2015) A novel layer-by-layer assembled multi-enzyme/CNT biosensor for discriminative detection between organophosphorus and non-organophosphrus pesticides. Biosens Bioelectron 67:287–295. https://doi.org/10.1016/j.bios.2014.08.036

Zhang C, Cui Y, Song L et al (2016) Microwave assisted one-pot synthesis of graphene quantum dots as highly sensitive fluorescent probes for detection of iron ions and pH value. Talanta 150:54–60. https://doi.org/10.1016/j.talanta.2015.12.015

Zhang J, Wang H, Xiao Y et al (2017) A simple approach for synthesizing of fluorescent carbon quantum dots from tofu wastewater. Nanoscale Res Lett 12:611. https://doi.org/10.1186/s11671-017-2369-1

Zhang HJ, Liu LZ, Zhang XR et al (2019) Microwave-assisted solvothermal synthesis of shape-controlled $CoFe_2O_4$ nanoparticles for acetone sensor. J Alloys Compd 788:1103–1112. https://doi.org/10.1016/j.jallcom.2019.03.009

Zhao Y, He J (2019) Novel template-assisted microwave conversion of graphene oxide to graphene patterns: a reduction transfer mechanism. Carbon N Y 148:159–163. https://doi.org/10.1016/j.carbon.2019.03.081

Zhao Y, Gao Y, Zhan D et al (2005) Selective detection of dopamine in the presence of ascorbic acid and uric acid by a carbon nanotubes-ionic liquid gel modified electrode. Talanta 66:51–57. https://doi.org/10.1016/j.talanta.2004.09.019

Zhao HX, Liu LQ, De Liu Z et al (2011) Highly selective detection of phosphate in very complicated matrixes with an off-on fluorescent probe of europium-adjusted carbon dots. Chem Commun 47:2604–2606. https://doi.org/10.1039/c0cc04399k

Zhao Y, Huang Z, Chang W et al (2017) Microwave-assisted solvothermal synthesis of hierarchical $TiO_2$ microspheres for efficient electro-field-assisted-photocatalytic removal of tributyltin in tannery wastewater. Chemosphere 179:75–83. https://doi.org/10.1016/j.chemosphere.2017.03.084

Zheng Y, Liu Z, Jing Y et al (2015) An acetylcholinesterase biosensor based on ionic liquid functionalized graphene-gelatin-modified electrode for sensitive detection of pesticides. Sensors Actuators B Chem 210:389–397. https://doi.org/10.1016/j.snb.2015.01.003

Zheng W, Zhao M, Liu W et al (2018) Electrochemical sensor based on molecularly imprinted polymer/reduced graphene oxide composite for simultaneous determination of uric acid and tyrosine. J Electroanal Chem 813:75–82. https://doi.org/10.1016/j.jelechem.2018.02.022

Zhong X, Fei G, Xia H (2010) Synthesis and characterization of poly(3,4-ethylenedioxythiophene) nanoparticles obtained through ultrasonic irradiation. J Appl Polym Sci 118:2146–2152. https://doi.org/10.1002/app.32531

Zhou W, Zhang F, Liu S et al (2014) Microwave-assisted hydrothermal synthesis of graphene-wrapped CuO hybrids for lithium ion batteries. RSC Adv 4:51362–51365. https://doi.org/10.1039/c4ra09144b

Zhou Q, Zhou M, Li Q et al (2019) Facile biosynthesis and grown mechanism of gold nanoparticles in *Pueraria lobata* extract. Colloids Surf A Physicochem Eng Asp 567:69–75. https://doi.org/10.1016/j.colsurfa.2019.01.039

Zhu J, Jia J, Kwong FL et al (2012) Synthesis of multiwalled carbon nanotubes from bamboo charcoal and the roles of minerals on their growth. Biomass Bioenergy 36:12–19. https://doi.org/10.1016/j.biombioe.2011.08.023

Zhu L-F, Yao Z-C, Ahmad Z et al (2018a) Synthesis and evaluation of herbal chitosan from ganoderma lucidum spore powder for biomedical applications. Sci Rep 8:14608. https://doi.org/10.1038/s41598-018-33088-5

Zhu Y, Ji H, Cheng HM, Ruoff RS (2018b) Mass production and industrial applications of graphene materials. Natl Sci Rev 5:90–101. https://doi.org/10.1093/nsr/nwx055

Zlotski S, Uglov V (2017) Facile sol-gel synthesis of metaloxide nanoparticles in a cellulose paper template. J Nanomed Nanotechnol S8:002. https://doi.org/10.4172/2157-7439.S8-002

# Green Synthesis of Plasmonic Metal Nanoparticles and Their Application as Enviromental Sensors

**Ali Mehdinia and Simindokht Rostami**

## Contents



# Abbreviations

| | |
|---|---|
| AgNPs | Silver nanoparticles |
| AuNPs | Gold nanoparticles |
| FTIR | Fourier transform infrared spectroscopy |
| GSH | Glutathione |
| LSPR | Localized surface plasmon resonance |
| NADH | Nicotinamide adenine dinucleotide |
| SERS | Surface-enhanced Raman spectroscopy |

A. Mehdinia (✉)
Department of Marine Living Science, Ocean Sciences Research Center, Iranian National Institute for Oceanography and Atmospheric Science, Tehran, Iran
e-mail: mehdinia@inio.ac.ir

S. Rostami
Faculty of Chemistry, Department of Analytical Chemistry, K.N. Toosi University of Technology, Tehran, Iran

 
Inamuddin, A. M. Asiri (eds.), *Nanosensor Technologies for Environmental Monitoring*, Nanotechnology in the Life Sciences,
https://doi.org/10.1007/978-3-030-45116-5_8

# 1 Introduction

Plasmonic metal nanoparticles have drawn wide attention due to their unique optical properties, size, and shape-dependent properties and tunable optical response over a spectral range from visible to the near-infrared region (Rao et al. 2002; Tao et al. 2008). Among metal nanoparticles, silver and gold nanoparticles (AgNPs and AuNPs) exhibit unique optical features and strong significant absorption in the visible region. This behavior refers to a collective oscillation of the surface conduction electrons. The collective oscillation of the surface conduction electrons is called surface plasmons. The resonance of surface plasmons with certain frequencies of incident electromagnetic radiation of light is known as localized surface plasmon resonance (LSPR) (Willets and Van Duyne 2007; Mayer and Hafner 2011). These extraordinary properties have made plasmonic metal nanoparticles considerable interest and fantastic nanoscale platform in biological and chemical sensing approaches in the area of food safety (Homola 2004; Narsaiah et al. 2012), disease diagnosis (Barizuddin et al. 2016), and environmental analysis (Wei et al. 2015).

High-quality plasmonic metal nanoparticles can be fabricated by various physical and chemical synthetic methods (Nisbet and Weiss 2010). Besides, green and biosynthesis routes as alternatives to the conventional physical and chemical synthesis strategies offer clean, nontoxic, and environmentally friendly synthetic approaches for the large-scale, more feasible, cost-effective, and faster production of metal nanoparticles (Gahlawat and Choudhury 2019). By utilizing green synthesis techniques biocompatible nanoparticles are produced by natural agents (Babu et al. 2013). The green synthesis methods do not require expensive, toxic, and harmful chemical reagents. Therefore, green synthesis results in eliminating using toxic chemicals. Also, green synthesis can significantly reduce environmental contamination and harmful residues. These advantages make the produced nanoparticles safe for both therapeutic uses and environment (Abdeen et al. 2014). Moreover, biosynthesis of nanoparticles using the natural, pure, and green substances is mostly carried out at room temperature (Gao et al. 2014). Therefore, this synthetic strategy has other several advantages over the traditional chemical methods due to low energy consumption and moderate operation condition. Owing to these advantages, the green synthesis of nanoparticles has gained significant attention in recent years.

Green synthesis approaches using various microorganisms such as yeasts, molds, fungi, algae, and bacteria, and plants due to their natural bioactive molecules including proteins, enzymes, amino acids, polysaccharides, phenolic compounds, amines, alkaloids, vitamins, and pigments have been developed for eco-friendly synthesis of metal nanoparticles (Adil et al. 2015; Gahlawat and Choudhury 2019; Ahmad et al. 2019; Hulkoti and Taranath 2014). Most of these viable natural biomolecules can be used as reducing agents, stabilizers, or both. The bioactive molecules can bind to metal ions through their functional groups such as hydroxyl and carbonyl . Then these biomolecules can fabricate nanoparticles and also prevent the aggregation of nanoparticles (Duan et al. 2015; Mohammadlou et al. 2016). Figure 1 illustrates a graphical green synthetic route of metal nanoparticles from the extract of microor-

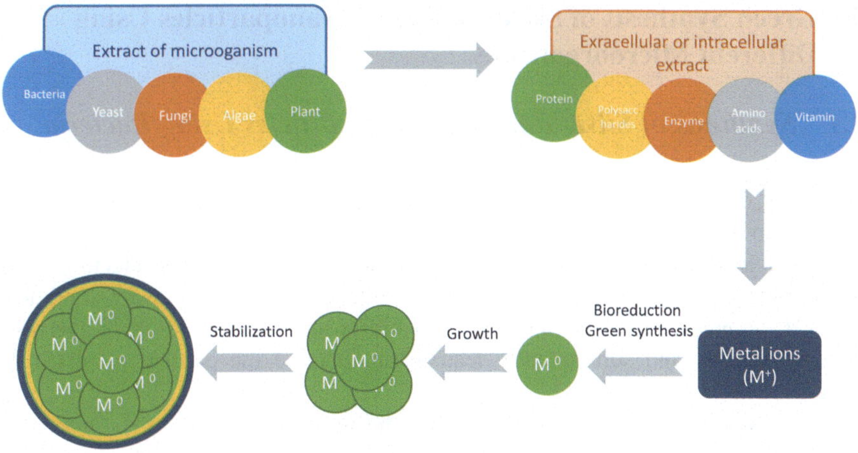


**Fig. 1** Graphical representation of green synthesis of metal nanoparticles using microorganisms and plant extracts and mechanism of metal nanoparticle formation

ganisms. Different green synthesis methods of metal nanoparticles are classified into the mentioned categories which are described in more detail in the following sections.

Depending on the area where nanoparticles are generated, the biosynthesis may occur either intracellular or extracellular. The exact mechanism of biosynthesis of metal nanoparticles using bioreagents is not very clear yet. Different microorganisms and different biomolecules have displayed different pathways for the formation of metal nanoparticles. However, in most of biosynthetic routes metal ions get trapped on the surface of microbial cell walls or diffuse inside the cells and then further get reduced to metal nanoparticles via an enzymatic or nonenzymatic method (Ali et al. 2019). In intracellular synthesis process, positively charged metal ions get attracted to the negatively charged cell wall based on electrostatic interaction between metal ions and the cell wall. Entrapped metal ions could be reduced to metal nanoparticles due to the presence of enzymes on the cell wall. Then the produced nanoparticles get diffused off through the cell wall (Hulkoti and Taranath 2014). In extracellular synthesis methods, nitrate reductase enzyme is mostly responsible for bioreduction of metal ions (Hulkoti and Taranath 2014). The nitrate reductase enzyme which is involved in the cellular nitrogen cycle is able to reduce nitrate to nitrite. In several studies on the mechanism of extracellular biofabrication of nanoparticle, it was revealed that nicotinamide adenine dinucleotide (NADH)-dependent nitrate reductase enzyme is capable to reduce metal ions to metal nanoparticles via an electron transfer process, followed by metal nanoparticle formation and stabilization (Shah et al. 2015). Indeed, NADH provides required electrons for reductase enzyme and is then oxidized to $NAD^+$. During the electron transfer from NADH to nitrate-dependent reductase, each metal ion receives an electron and gets reduced to a metal atom (Waghmare et al. 2014).

# 2 Green Synthesis of Silver and Gold Nanoparticles Using Different Microorganisms

## *2.1 Synthesis of Silver and Gold Nanoparticles Using Bacteria*

Generally, different parameters such as pH, temperature, metallic salt, and culturing media can significantly influence on size, shape, and properties of produced nanoparticles during bacterial-mediated synthesis methods. Also, the selection of an appropriate culturing method is very important. Because the culturing condition like pH, temperature, light, nutrients, and buffer strength can result to increase enzyme activity and subsequently lead to an increase the particle yield (Shah et al. 2015).

By considering synthesis condition, some bacterial strains such as *Actinobacter* sp. (Wadhwani et al. 2016), *Escherichia coli* (Srivastava et al. 2013), *Klebsiella pneumonia* (Prema et al. 2016), *Lactobacillus casei* (Kato et al. 2019), *Bacillus methylotrophicus* (Wang et al. 2016a), *Corynebacterium glutamicum* (Gowramma et al. 2015), and *Pseudomonas* sp. (Klaus et al. 1999) have been reported as potential biofactories for production of silver and gold nanoparticles either intracellularly or extracellularly. *Pseudomonas stutzeri AG259* has been applied for intracellular biosynthesis of triangles and hexagonal silver nanocrystals up to 200 nm (Klaus et al. 1999). *Pseudomonas mandelii* was used for fabrication of highly stable AgNPs with small diameter of 1.9–10 nm under low temperature conditions (Mageswari et al. 2015). Spherical AgNPs with the size of 5–50 nm were obtained from alkaliphilic actinobacterium *Nocardiopsis valliformis* (Rathod et al. 2016). Green synthesis of Ag and AuNPs and Au–Ag alloy crystal was assisted by *Lactobacillus* which is present in buttermilk (Nair and Pradeep 2002). Biosynthesis of 15 nm AgNPs has been achieved using native isolates of *Corynebacterium glutamicum* (Gowramma et al. 2015). *Acinetobacter* sp. SW30 (Wadhwani et al. 2016), *Bacillus methylotrophicus* DC3 (Wang et al. 2016a), and *Bacillus licheniformis* M09 supernatant (Momin et al. 2019) were used for biosynthesis of AgNPs with antimicrobial activity and a mean size of 19 nm, 10–30 nm, and 10–30 nm, respectively.

AuNPs were synthesized with the average size of 22.2 nm using cell extracts of *Labrys* sp. (Shen et al. 2018). pH-dependent extracellular biosynthesis of AuNPs with different sizes and shapes was reported using *Rhodopseudomonas capsulata*. Spherical AuNPs in the size range of 10–20 nm were formed at pH 7 while some nanoplates were obtained at pH 4 (He et al. 2007). Also in another study, cell free extract of *Rhodopseudomonas capsulata* was used for the synthesis of gold nanowires with the size of 10–20 nm. It was elucidated that proteins were responsible for gold nanowire formation (He et al. 2008). Also, another bacterium, *Stenotrophomonas maltophilia* synthesized well-dispersed AuNPs with an average diameter of 40 nm (Nangia et al. 2009). Circular AuNPs with the average size of 50 nm were obtained at room temperature using *Escherichia coli* K12. It was revealed that some peptides in the membrane of *Escherichia coli* K12 were involved in synthesis and stabilization of AuNPs (Srivastava et al. 2013). *Geobacillus* sp. strain ID17 was also used for biosynthesis of spherical AuNPs with the size of 5–50 nm. Also, quasi-hexagonal AuNPs with the size between 10 and 20 nm were obtained using *Geobacillus* sp. strain ID17 (Correa-Llantén et al. 2013). Microbial biosynthesis of AuNPs was evaluated using *Klebsiella pneumonia*. The extract of *Klebsiella pneumonia* yielded to the formation of spherical AuNPs with the size between 10 and 15 nm (Prema et al. 2016). In another study, glycolipids present in the cell membrane of *Lactobacillus casei* were involved in bioreducing gold ions into AuNPs (Kato et al. 2019).

## 2.2 Synthesis of Silver and Gold Nanoparticles Using Yeast

All yeast genera are capable of accumulation of different heavy metals. Cell membrane transportation of metals is controlled by enzymatic oxidation or reduction, chelating with extracellular polysaccharides and peptides, and sorption at the cell wall. The yeast species are known as "semiconductor crystals" or "quantum semiconductor crystals" because of their strong potential in the synthesis of semiconductor nanoparticles (Dameron et al. 1989). However, they can also produce other nanoparticles including metal nanoparticles. Spherical silver nanoparticles with the size of 6–20 nm were obtained using *Saccharomyces cerevisiae* as a bioreducing agent (Jha and Prasad 2008). AgNPs were synthesized on the yeast cells *Saccharomyces cerevisiae* BU-MBT-CY1 isolated from coconut cell sap, with an average diameter of 19 ± 9 nm. The prepared AgNPs were applied for As (V) removal (Selvakumar et al. 2011). Green and rapid synthesis of highly stable AgNPs with the size of 3–10 nm was reported using cell free extract of *Saccharomyces boulardii*. It was elucidated that proteins and peptides of this yeast were involved in the formation and stabilization of AgNPs. The obtained AgNPs exhibited anticancerous activity (Kaler et al. 2013). Also, the culture extract of *Rhodotorula* sp. was appeared as a novel biocatalyst for bioreduction of silver ions to AgNPs with an average size of 40 nm (Ashengroph 2014). The psychrotrophic marine yeast *Yarrowia lipolytica* containing brown pigment (melanin) synthesized AgNPs, and the obtained AgNPs were used as antibiofilm agents (Apte et al. 2013). Proteins present in marine yeast *Candida* sp. were used as a bioreducing agent for producing AgNPs with antimicrobial activity (Kumar et al. 2011). Yeast *Trichosporon montevideense* was used for biosynthesis of AuNPs (Shen et al. 2018). Spherical AuNPs with the size below 100 nm were generated using *Pichia jadini* (Gericke and Pinches 2006). Also, both intracellularly (Pimprikar et al. 2009) and extracellularly (Agnihotri et al. 2009) biosynthesis of AuNPs have been reported using marine yeast *Yarrowia lipolytica* (*Candida lipolytica*).

## 2.3 Synthesis of Metal Nanoparticles Using Fungi

Most of the fungal genera contain a high amount of proteins and enzymes. Therefore, fungi are biosynthetic productive agents for green synthesis of metal nanoparticles. In addition, they have high intracellular metal uptake capacity, high binding, and metal bioaccumulation capacity and specific enzymes such as reductase. The existence of these enzymes facilitates the biosynthesis of metal nanoparticles (Hulkoti and Taranath 2014). Using fungi in the synthesis process, highly monodisperse nanoparticles can be produced. Based on reports, different fungi have been evaluated for the production of metal nanoparticles. In this regard, stable 5–15 nm AgNPs were synthesized using *Fusarium oxysporum* (Ahmad et al. 2003). The fungus *Pestalotiopsis pauciseta* was used for biological synthesis of AgNPs with the size of 123–195 nm (Vardhana and Kathiravan 2015). Spherical AgNPs with the average diameter of 12–20 nm were produced biologically using the endophytic fungus *Fusarium* sp. (Singh et al. 2015). Fast and simple extracellular synthesis of AgNPs was carried out using fungal biomass of *Aspergillus fumigatus* and the obtained AgNPs were in the range of 5–25 nm (Bhainsa and D'souza 2006). In one study, biosynthesis of AgNPs using three endophytic fungi *Aspergillus tamarii* PFL2, *Aspergillus niger* PFR6, and *Penicillium ochrochloron* PFR8 were considered and the size of obtained nanoparticles was compared. According to observations, using the fungi *A. tamarii* PFL2 smaller particle size (3.5 ± 3 nm) was obtained compared to the generated AgNPs by the other two fungi (Devi and Joshi 2015). *Arthroderma fulvum* HT77 was employed for biosynthesis of antifungal AgNPs with the average diameter of 15.5 nm (Xue et al. 2016). Also, biosynthesis of AgNPs was investigated using four different fungal species such as *Rhizopus nigricans*, *Fusarium semitectum*, *Colletotrichum gloeosporioides*, and *Aspergillus nidulans* (Ravindra and Rajasab 2014). Biofabrication of AgNPs was studied using an endophytic fungus identified as *Botryosphaeria rhodina* which was secreted from the medicinal plant *Catharanthus roseus* (Linn.) (Akther et al. 2019).

Using *Fusarium acuminatum* MTCC-1983, AuNPs were obtained with the size of 17 nm via enzymes secreted by the fungus (Tidke et al. 2014). Moreover, different shapes of AuNPs were synthesized with various fungi. Triangle, pentagon, and hexagon-shaped AuNPs with the average size of 10–60 nm were produced using cells and biomass of *Aspergillus oryzae var. viridis* (Binupriya et al. 2010). Biomass of *Trichothecium* spp. *Link*. synthesized triangle and hexagonal AuNPs with the size in the range of 5–200 nm under stationary or shaking conditions which caused extracellular and intracellular formation of nanoparticles, respectively (Ahmad et al. 2005). Spherical and triangular AuNPs were synthesized using *Fusarium oxysporum* in the size range of 20–40 nm (Mukherjee et al. 2002). Gold nanocrystals with the mean size of 10 nm were biofabricated using *Rhizopus oryzae*. The gold nanocrystals were synthesized on the surface of *Rhizopus oryzae* cell (Das et al. 2009). *Fusarium oxysporum* was also used for generation of highly stable Au–Ag alloy with a diameter between 8 and 14 nm (Senapati et al. 2005). Ag-Au alloy was obtained using fungal xylanases extracted from *Aspergillus niger* L3 and *Trichoderma longibrachiatum* L2. Biosynthesized alloy nanoparticles displayed potential biomedical applications (Elegbede et al. 2019).

## 2.4 Synthesis of Silver and Gold Nanoparticles Using Algae

Metal nanoparticles can be fabricated using algae which are known as a source of biomaterials such as proteins, amino acids, carbohydrates, pigments, fatty acids, and nucleic acids. Several algae have been employed for green synthesis of silver and gold nanoparticles. Extracellular biosynthesis of AgNPs was evaluated using a brown seaweed *Sargassum wightii*. The obtained AgNPs displayed antibacterial activity (Govindaraju et al. 2009). Also, the edible blue green alga *Spirulina platensis* was studied for extracellular synthesis of biocompatible AgNPs (Govindaraju et al. 2009). In one study, different strains of microalgae such as *Botryococcus braunii, Coelastrum* sp., *Spirulina* sp., and *Limnothix* sp. produced AgNPs with the size of about 13–25 nm. Also, different shapes of AgNPs such as spherical, elongated, and irregular nanoparticles were synthesized. Based on the obtained data it was indicated that polysaccharides found in extracellular cell-free cultural liquid of algae and protein-based pigments of cyanobacteria were involved in nanoparticle formation (Patel et al. 2015). The biosynthesis of antibacterial AgNPs was assisted using the aqueous extract of the red marine macroalga *Amphiroa fragilissima*. Also, AgNPs with antibacterial activity were synthesized via marine alga *Caulerpa racemosa* (Kathiraven et al. 2015). In one report, an aqueous extract of *Chlorella vulgaris* proved to be suitable for the biosynthesis of AgNPs with the size range of 15–47 nm (Annamalai and Nallamuthu 2016). The synthesis of spherical AgNPs was carried out using *Ulva fasciata* extract as a reducing and capping agent (Rajesh et al. 2012). Recently, it was reported that the green alga *Botryococcus braunii* was used for the synthesis of spherical, cubic and truncated triangular AgNPs with the size of 40–90 nm. The synthesized AgNPs exhibited catalytic activity in the conversion of 2-nitroaniline to biologically important 2-arylbenzimidazoles. According to the FTIR spectrum, it was revealed that bioactive molecules such as proteins, amides, polysaccharides, and long chain fatty acids found in *Botryococcus braunii* extract were responsible for the synthesis and stabilizing of AgNPs (Arya et al. 2019). Another green alga, *Chlorella vulgaris*, resulted in the biosynthesis of AgNPs. Based on FTIR obtained spectrum, it was revealed that polysaccharides, amides, proteins, and chain fatty acids of *Chlorella vulgaris* are responsible for AgNPs formation. Green synthesis using *Chlorella vulgaris* yielded triangular AgNPs in the range of 40–90 nm (Mahajan et al. 2019). The whole-cell aqueous extract of *Neochloris oleoabundans* was employed for the synthesis of quasi-spherical AgNPs with a mean diameter of 16.63 nm under light (Bao et al. 2019). A sulfated polysaccharide present in the extract of the green alga *Ulva armoricana* appeared to be a reducing and stabilizing agent for synthesis of AgNPs with a thick polysaccharide shell. The synthesis was carried out under mild conditions (Massironi et al. 2019).

The marine alga *Sargassum wightii* has been applied for the biological formation of stable AuNPs with the size of 8–12 nm (Singaravelu et al. 2007). In one report, AuNPs were produced by blue green algae with antibacterial activity (Suganya et al. 2015). The biomass of a green microalga, *Chlorella vulgaris*, was used for the synthesis of gold nanoplates (Dahoumane et al. 2017). Highly stable small AuNPs with a mean diameter of 5.42 nm were yielded using brown marine macroalga *Sargassum muticum*. Moreover, the aqueous extract of *S. muticum* appeared to act as both reducing and stabilizing agent (Namvar et al. 2015). The brown alga *Cystoseira baccata* was employed for the biosynthesis of AuNPs with the size of 8.4 nm and cytotoxic activity against colon cancer cells (González-Ballesteros et al. 2017). In another study, green synthesis of AuNPs with the size of 15 and 47 nm was considered using the green alga *Chlorella vulgaris* (Annamalai and Nallamuthu 2016). Marine red seaweed *Gracilaria verrucosa* was applied for facile one-pot synthesis of biocompatible AuNPs with the size range of 20–80 nm. Different isotropic and anisotropic AuNPs such as spherical, oval, triangular, octahedral, pentagonal, and rhomboid nanoparticles were obtained. Proteins, phenolic and aromatic compounds of the studied seaweed were responsible for the synthesis of AuNPs (Chellapandian et al. 2019). The extract of the red marine alga *Gelidiella acerosa* fabricated spherical and hexagonal AuNPs with antibacterial, antioxidant, and antidiabetic activity. The particle size of the synthesized AuNPs was estimated in the range of 5.81–117.59 nm. Terpenoids, cardiac glycosides, alkaloids, and tannins present in *Gracilaria verrucosa* extract acted as bioreducing and capping agent (Senthilkumar et al. 2019). Aqueous extracts of two marine brown algae, *Turbinaria conoides* and *Sargassum tenerrimum*, were used as both reducing and capping agent for green synthesis of spherical AuNPs with the size of 27–35 nm. It appeared that the brown alga *T. conoides* is rich in polysaccharides, sulfated polysaccharides, and polyphenolic compounds and *S. tenerrimum* extract is containing various secondary metabolites such as proteins, amino acids, carbohydrates, phenolic acids, sterols, alkaloids, flavonoids, and tannins. These biomolecules can act as reducing as well as stabilizing agents. The biosynthesized nanoparticles using the applied brown algae exhibited catalytic activity (Ramakrishna et al. 2016).

Au and Ag nanoparticles were biosynthesized using the red seaweed *Chondrus crispus* and the green alga *Spirogyra insignia* as reducing agents. *C. crispus* is containing sulfated polysaccharides which are able to bind to gold nanoparticles surface and stabilize the produced AuNPs. Using *C. crispus* at different pH, AuNPs with different shapes were obtained. In acidic media (pH 2) mostly triangle and hexagonal AuNPs were produced and at pH 4 spherical nanoparticles with the average size of 30 nm were formed. Also, *Spirogyra insignia* is containing pectins which are polysaccharides rich in galacturonic acid. There are huge amount of hydroxyl, ketone, aldehyde and carboxylic acid groups in the structure of these biomolecules which are appeared as reducing and capping agent in synthesis of AgNPs (Castro et al. 2013). Also, it has been reported that a red alga, *Gracilaria* sp., was used for the extracellular biosynthesis of Au, Ag, and Au/Ag bimetallic nanoparticles with antibacterial activity (Ramakritinan et al. 2013). Some related studies on algae-mediated synthesis of Ag and AuNPs last recent years are summarized in Table 1.

**Table 1** Algae-assisted synthesis of silver and gold nanoparticles

| Algae | Biomolecules | Metal nanoparticle | Size and shape | Ref |
|---|---|---|---|---|
| *Green microalgae* | | | | |
| *Botryococcus braunii* | Proteins, amides, polysaccharides, and long chain fatty acids | Ag | 40–90 nm, spherical, cubic, truncated, triangular | Arya et al. (2019) |
| *Chlorella vulgaris* | Protein | Ag | 5–50 nm | Annamalai and Nallamuthu (2016) |
| *Caulerpa serrulata* | Caulerpenyne and/or its derivatives | Ag | 10 ± 2 nm, spherical | Aboelfetoh et al. (2017) |
| *Rhizoclonium fontinale* | Chlorophyll, protein, and carbohydrate | Au | pH 5: 5–20 nm spherical, 15–88 nm, nanotriangles 34 nm, nanohexagons, rod-shaped (~100 × 51.5 nm); pH 7: Spherical 13–22 nm, pH 9: 16 nm, nanospheres | Parial and Pal (2015) |
| *Brown microalgae* | | | | |
| *Sargassum vulgare* | Alginate moieties, secondary OH groups | Ag | 10 nm, spherical | Govindaraju et al. (2015) |
| *Padina pavonia* | Protein Polysaccharides | Ag | 49.58–86.37 nm, spherical, triangular, rectangle, polyhedral and hexagonal | Abdel-Raouf et al. (2018) |
| *Cystoseira baccata* | Polyphenols Polysaccharides | Au | 8.4 ± 2.2 nm, spherical | González-Ballesteros et al. (2017) |
| *Sargassum tenerrimum* | Amino acids Alkaloids Carbohydrates Flavonoids Saponins Sterols Tannins Proteins Phenolic acids | Au | 5–45 nm, anisotropic | Ramakrishna et al. (2016) |
| *Red microalgae* | | | | |
| *Acanthophora specifera* | Monosaccharide, polysaccharide, uronic acids and secondary metabolites | Ag | 33–81 nm, cubic | Ibraheem et al. (2016) |
| *Portieria hornemannii* | Protein Phenolic compounds | Ag | 9–80 nm, spherical | Ramamoorthy et al. (2019) |

(continued)

**Table 1** (continued)

| Algae | Biomolecules | Metal nanoparticle | Size and shape | Ref |
|---|---|---|---|---|
| *Gracilaria verrucosa* | Protein Phenolic and aromatic compounds | Au | 20–80 nm, as spherical, oval, triangular, octahedral, pentagonal and rhombus shapes | Chellapandian et al. (2019) |
| *Galaxaura elongata* | Palmitic acid Polyphenolic compounds | Au | 3.85–77.13 nm, rod, triangular, truncated, triangular, hexagonal | Abdel-Raouf et al. (2017) |

## 2.5 Synthesis of Metal Nanoparticles Using Plants

Plant-based synthesis method has attracted much more attention rather than the other green agents because it provides rapid single step synthesis process. Green synthesis of metal nanoparticles using reactive plant derivative reagents and phytochemicals has been greatly popularized nowadays. The extracts of different parts of plants like leaves, roots, stems, bark, pods, peel, seeds, flowers, and fruits have been used as reducing reagent for the synthesis of metal nanoparticles (Iravani 2011; Gour and Jain 2019). Phytochemicals present in the plant extracts such as polysaccharides, polyol and heterocyclic compounds, essential oils, sugar, flavonoids, proteins, enzymes, alkaloids, tannins, terpenoids, vitamins, and organic acids are mainly responsible for the bioreduction of metal ions into metal nanoparticles and also can act as a capping agent of the fabricated nanoparticles (Iravani 2011; Makarov et al. 2014). For instance, terpenoids are a group of organic polymers with strong antioxidant activity and are able to reduce metal ions. Flavonoids are polyphenolic compounds with various functional groups which are capable to actively chelate to metal ions and reduce them to form nanoparticles. Moreover, sugar, monosaccharides like glucose, disaccharides, and polysaccharides can act as reducing agents due to their active sites, free aldehyde groups, and their type and concentrations (Makarov et al. 2014). In comparison with using microorganisms for the synthesis of nanoparticles, plant-based synthesis is more rapid and has some advantages over microorganisms. Plants have a large range of active reagents. Plant extracts do not need to complex treatment process like isolation, culturing and maintenance. Plants are easily accessible and due to having great potential in detoxification and heavy metal accumulation are more appropriate candidates for the biosynthesis of metal nanoparticles. Also, more stable nanoparticles are produced using plant extracts. Moreover, size and morphology controlled nanoparticles can be obtained considerably more feasible when plant extracts were used for biosynthesis of metal nanoparticles (Mohammadlou et al. 2016; Iravani 2011; Shah et al. 2015; Jha et al. 2009). Several affecting parameters including type of plant and phytochemicals, concentration of metal ions and extracts, time, pH and temperature are the key factors on the metal nanoparticles generation and also can influence on the size and morphology of the produced nanoparticles (Shah et al. 2015).

It has been reported that different parts of plants have been employed for the biosynthesis of metal nanoparticles. For instance, in some studies the leaf extract of *Geranium* (Rivera-Rangel et al. 2018; Shankar et al. 2003), *Helianthus annuus, Basella alba,* and *Saccharum officinarum* (Leela and Vivekanandan 2008), *Glycine max* (soybean) (Vivekanandhan et al. 2009), *Syzygium cumini* (Kumar et al. 2010), *Coriandrum sativum* (Khan et al. 2018a), *Pinus densiflora* (Basiri et al. 2018), and *Casuarina equisetifolia L.* (Muthu and Rathika 2016) have appeared suitable for bioreduction and biosynthesis of AgNPs with different sizes. Green synthesis of AgNPs using banana peel extract has been demonstrated. The banana peel extract was used as both reducing and capping agent (Ibrahim 2015). Moreover, ethanolic extract of rose petals (*Rosa indica*) (Manikandan et al. 2015), aqueous seed extract of *Pistacia atlantica* (Sadeghi et al. 2015), aqueous stem bark extract of *Syzygium alternifolium* (Yugandhar et al. 2015), and fruit extracts like kiwi fruit juice (Gao et al. 2014), blackberry (Kumar et al. 2017a), *Cucumis melo* (Basiri et al. 2017), and lingonberry and cranberry juices (Puišo et al. 2014) were evaluated for synthesis of AgNPs.

The root extract of *Cucurbita pepo* L. was used for bio-catalyzed synthesis of AuNPs (Gonnelli et al. 2015). Also, *Cucurbita pepo* L. leaf extract was applied for fabrication of morphology controlled AuNPs (Gonnelli et al. 2018). Using mango peel extract, AuNPs in particle size of 6–18 nm were obtained at pH 9 and 2 (Yang et al. 2014). Leaf extract of *Sesbania grandiflora* was led to the formation of well-dispersed 7–34 nm AuNPs (Das and Velusamy 2014). Phytochemical compounds present in the seed, skin, and stalk of grape such as catechin, epicatechin, anthocyanidin, proanthocyanidin, and condensed tannins were able to produce stabilized AuNPs with the average size of 20 nm (Krishnaswamy et al. 2014). In another study, it was revealed that carbohydrates, tannins, flavonoids, and phenolic acids present in the fruit of *Phoenix dactylifera* (date palm) were responsible for the formation and stabilization of spherical AuNPs in the size range of 32–45 nm (Zayed and Eisa 2014).

Several plant extracts such as seed extract of *Madhuca longifolia* (Sharma et al. 2019), leaf extract of gold rod (*Solidago canadensis*) (Botha et al. 2019) and *Stigmaphyllon ovatum* (Elemike et al. 2019) have been used for green synthesis of both Ag and Au nanoparticles and also bimetallic Ag/AuNPs. Also, the extract of medicinal plants including the root extract of ginger (*Zingiber officinale*) (Velmurugan et al. 2014) and leaf extract of *Cinnamomum camphora* (Huang et al. 2007) have been evaluated for synthesis of both Ag and AuNPs. In one study, Ag and AuNPs were synthesized using the extract of blueberry, blackberry, turmeric, and pomegranate. In this study, more uniform metal nanoparticles were obtained using pomegranate extract (Nadagouda et al. 2014).

Different morphologies of silver and gold nanoparticles have been fabricated by plant extracts. The extract of *Aloe vera* have been utilized for biofabrication of triangular AuNPs and spherical AgNPs (Chandran et al. 2006) and also octahedron AgNPs (Logaranjan et al. 2016). Ag nanorod and cubes were synthesized biologically using sundried *Stevia rebaudiana* leaves (Varshney et al. 2010). Spherical, rod-like, prism, triangular, pentagonal, and hexagonal AgNPs were synthesized by

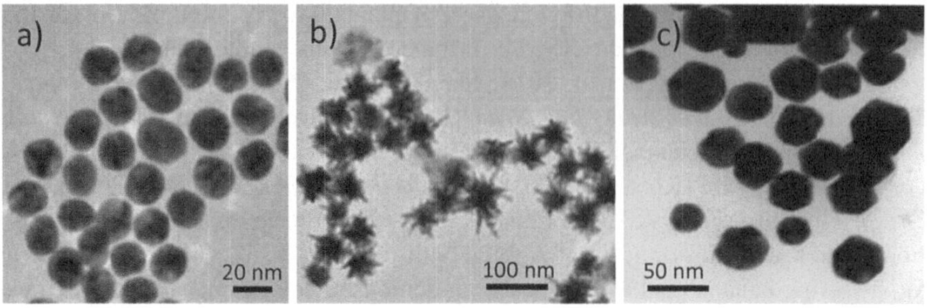


**Fig. 2** The TEM image of (**a**) spherical AuNPs and (**b**) star-shaped AuNPs synthesized using *Cistus incanus* extract (Klekotko et al. 2019) and (**c**) spherical/hexagonal AuNPs synthesized using *Acacia nilotica* twig bark extract (Emmanuel et al. 2014)

employing aqueous extract of *Salicornia brachiata* as a bioreducing agent (Seralathan et al. 2014). Spherical and triangular AgNPs in the size range of 6–60 nm were also formed using lingonberry and cranberry juices which contain anthocyanins, benzoic acid, and phenolic compounds (Puišo et al. 2014). Aqueous leaf extract of *Euphorbia prostrata* was resulted in Ag nanorods biofabrication (Zahir and Rahuman 2012). Spherical, triangular, and hexagonal AgNPs were obtained from *Caesalpinia coriaria* leaf extract while its boiling leaf extract yielded only triangular nanoparticles (Jeeva et al. 2014). Cylindrical shaped AgNPs with an average diameter of 250 nm were formed using aqueous bark extract of *Ficus racemosa* (Velayutham et al. 2013). Spherical and triangular AuNPs with the mean diameter of 12–38 nm were yielded using aqueous root extract of *Morinda citrifolia L.* (Suman et al. 2014). Green synthesis of hexagonal AuNPs was carried out utilizing essential oils extracted from fresh leaves of *Anacardium occidentale* (Sheny et al. 2012). It can be seen some TEM images of plant-mediated AuNPs in Fig. 2. Recently, plant-based biosynthesis of metal nanoparticles appeared to be an important branch of green synthesis route which could be an attractive and appropriate alternative to chemical methods and also microorganism-mediated synthetic methods. Hence, some recent studies for the biosynthesis of Ag and AuNPs using plants have been summarized in Table 2.

## 3 Assisted Green Synthesis of Metal Nanoparticles Using Sunlight, Microwave, and Ultrasound

In an effort to develop environmentally benign green synthesis approaches and reduce energy consumption, alternative assisting energy resources have emerged. The assisting resources are effective for reducing time and temperature of synthesis procedure and yield to create nanoparticles with a higher degree of crystallinity compared to the traditional heating methods (Kahrilas et al. 2013). Sunlight as the

**Table 2** Plant-based synthesis of silver and gold nanoparticles

| Plant | Metal nanoparticle | Size and shape | Ref. |
|---|---|---|---|
| *Camellia sinensis* (green tea and black tea leaf extracts) | Ag<br>Au | Au 10 nm<br>Ag 30 nm | Onitsuka et al. (2019) |
| *Amomum villosum* (Fructus Amomi (cardamom)) (aqueous extract of dried fruits) | Ag<br>Au | 5–10 nm (Au), spherical<br>5–15 nm (Ag), spherical | Soshnikova et al. (2018) |
| *Aglaia elaeagnoidea* (flower extract) | Ag<br>Au | 17 nm (Ag), spherical<br>25 nm (Au), spherical | Manjari et al. (2017) |
| *Camellia sinensis* (green tea extract) | Ag | 34.68 ± 4.95 nm, spherical | Rolim et al. (2019) |
| *Berberis vulgaris* (leaf and root extracts) | Ag | 30–70 nm, spherical | Behravan et al. (2019) |
| *Enicostemma axillare* (Lam.) (leaf extract) | Ag | 15–20 nm, spherical | Raj et al. (2018) |
| Geranium (*P. hortorum*) (leaf extract) | Ag | 25–150 nm | Rivera-Rangel et al. (2018) |
| *Cratoxylum formosum* (leaf extract)<br>*Phoebe lanceolata* (leaf extract)<br>*Scurrula parasitica* (aerial parts)<br>*Ceratostigma minus* (stem and root extracts)<br>*Mucuna birdwoodiana* (stem ectract)<br>*Myrsine africana* (root extract)<br>*Lindera strychnifolia* (root extract) | Ag | 8.8 ± 0.3 nm ~ 35.4 ± 5.9 nm, spherical | Ahn et al. (2019) |
| *Dodonaea viscosa* (leaf extract) | Ag | 20–50 nm for nano worms, 50–100 nm for flowers, 70–100 nm for spherical particles and micro sized dendrites (with a diameter about 0.7–2.5 μm and length about 3.3–30 μm) | Anandan et al. (2019) |
| *Rosa brunonii* Lindl. | Ag | Less than 100 nm | Bhagat et al. (2019) |
| *Capparis decidua* | Ag | 1–20 nm, spherical | Ahlawat and Sehrawat (2017) |

(continued)

**Table 2** (continued)

| Plant | Metal nanoparticle | Size and shape | Ref. |
|---|---|---|---|
| *Rheum palmatum* (root extract) | Ag | 44–113 nm, spherical and hexagonal | Arokiyaraj et al. (2017) |
| *Prunus persica* L. (outer peel extract) | Ag | 28.27 nm, spherical | Patra and Baek (2016) |
| *Cleome viscosa* (fruit extract) | Ag | 20–50 nm, spherical | Lakshmanan et al. (2018) |
| *Andean blackberry* (fruit extract) | Ag | 12–50 nm, spherical | Kumar et al. (2017a) |
| *Alpinia Katsumadai* (seed extract) | Ag | 12.6 nm, quasi-spherical | He et al. (2017) |
| *Nigella arvensis* (seed extract) | Ag | 2–15 nm, spherical | Chahardoli et al. (2017) |
| *Cucurbita pepo* L. (leaf extract) | Au | 10–15 nm, spherical | Gonnelli et al. (2018) |
| *Simarouba glauca* (leaf extract) | Au | Prism and spherical like particles | Thangamani and Bhuvaneshwari (2019) |
| *Cistus incanus* (dried, powdered leaves) | Au | 45–85 nm, popcorn-shape | Klekotko et al. (2019) |
| *Sansevieria roxburghiana* (leaf extract) | Au | 5–31.11 nm (ave 17.48 nm), spherical with a few triangle, hexagonal, rod and decahedral shaped particles | Kumar et al. (2019) |
| Waste macadamia nut shells | Au | 50 nm–2 μm, spherical, triangular and hexagonal morphology | Dang et al. (2019) |
| *Euphrasia officinalis* (leaf extract) | Au | 5–30 nm, spherical or hexagonal | Liu et al. (2019) |
| *Terminalia arjuna* (leaf extract) | Au | 15–30 nm, spherical | Dudhane et al. (2019) |
| *Dalbergia coromandeliana* (root extract) | Au | 10.5 nm, spherical | Umamaheswari et al. (2018) |
| Waste *Citrullis lanatus var* (watermelon) | Au | 100 nm–2.5 μm, spheres and hexagonal plates | Chums-ard et al. (2019) |
| *Gnidia glauca* | Au | 10–60 nm, spherical | Ghosh et al. (2016) |
| *Pterocarpus santalinus* L. (red Sanders) (bark extract) | Au | 13–26 nm, spherical | Keshavamurthy et al. (2018) |

largest available renewable energy source (Annadhasan et al. 2014; Annadhasan et al. 2015), microwave irradiation (Francis et al. 2018; Shore 2018), and ultrasound radiation (Elsupikhe et al. 2015) as nontoxic, clean, and safe sources have been studied for assisted green synthesis of different kind of nanostructures including plasmonic metal nanoparticles. These alternative sources can assist microorganisms, plant extracts, and active biomolecules to catalyze the formation of nanoparticles.

Microwave-assisted synthesis of metal nanoparticles provides rapid and uniform heating and thus results in homogeneous nucleation and growth condition and consequently narrow size distribution and controlled morphology (Francis et al. 2018; Shore 2018; Kahrilas et al. 2013). Ultrasound-assisted synthesis method is another preferred route for green synthesis of metal nanoparticles based on the application of powerful ultrasound radiation. The ultrasound radiation generates cavitation microbubbles. The bubbles grow in the solution and then collapse after reaching the maximum size. The collapse of bubbles results in extremely high pressure and high temperature. Consequently, this condition causes formation of highly reactive free radicals. It has been reported that using the ultrasound-mediated synthesis method can produce monodispersed nanoparticles with different shapes (Elsupikhe et al. 2015).

## 4 Application of Green-Synthesized Metal Nanoparticles in Sensing Approaches

Green-synthesized metal nanoparticles have displayed great potential for application in different fields including industry, biotechnology, medicine, food safety, and environmental studies. According to most of the reports, produced metal nanoparticles using green methods exhibit strong antibacterial activity (Koduru et al. 2018) and can be also applied as carriers for drug delivery (Saratale et al. 2018a), biosensor (Gayda et al. 2019), nanocatalyst (Palomo and Filice 2016), and environmental monitoring platform (Saratale et al. 2018b). Due to industrial development and intensive use of toxic chemicals, the level of environmental pollution is increasing. Therefore, there is an environmental concern to monitoring harmful chemicals and removal of contaminations from environmental resources. Moreover, the development of nanotechnology has led to excessive use of chemicals and can also result in a new class of hazardous materials and environmental concern (Masciangioli and Zhang 2003). Therefore, green synthesis for clean production of nanoparticles and monitoring of the potential environmental hazards such as heavy metal ions and organic contaminants have become a vital need to explore the sustainable method for new remedial technologies (Das et al. 2018). To this aim, various novel sensing platforms have been emerged for detection and sensing of toxic chemicals and pollutants by the nontoxic green-synthesized metal nanoparticles. It has been proved that utilizing green-synthesized plasmonic metal nanoparticles as colorimetric sensors can open up a new window for simple, rapid, and low-cost detection of toxic metal ions and organic compounds in environmental samples (Annadhasan et al. 2014; Ragam and Mathew 2019; Sebastian et al. 2019).

## *4.1 Colorimetric Detection of Environmental Pollutions*

Heavy metals such as mercury (Hg), cadmium (Cd), arsenic (As), chromium (Cr), cobalt (Co), nickel (Ni), magnesium (Mn), zinc (Zn), and lead (Pb) have been found highly toxic even in trace level concentrations. So far, several green-synthesized plasmonic metal nanoparticles have shown a promising development in sensing and detection of environmentally toxic heavy metal ions. As mercury ion ($Hg^{2+}$) and its related compounds are highly toxic, it is of great concern among various toxic heavy metals. Therefore, several sensors have been investigated for the detection of $Hg^{2+}$. In one work, AgNPs were produced using *Syzygium aromaticum* commonly known as clove. During AgNPs biosynthesis, it was revealed that the size of AgNPs increased by increasing the concentration of Ag ions and temperature. Also, a red-shift was observed in the LSPR absorption band of AgNPs. The biosynthesized AgNPs has been adopted for detection of $Hg^{2+}$ ions in water with a minimum detection level of 2.0 μM. Electrostatic interaction of $Hg^{2+}$ ions with AgNPs led to form Ag–Hg amalgam and consequently caused the color solution to vanish (Sangar et al. 2019). Monodispersed, quasi-spherical AgNPs with an average size of ~11 nm and antibacterial activity were produced using the aqueous extract of an agrowaste: *Terminalia catappa* leaves containing flavonoids, phenolic compounds, and antioxidants which acted as reducing and capping agents. The green-synthesized AgNPs were used as a sensor for the colorimetric detection of trace levels of $Hg^{2+}$ based on the color change of the solution of AgNPs from deep brown to colorless due to redox reaction between $Hg^{2+}$ ions and AgNPs. Moreover, a blue shift and decrease in the LSPR peak intensity were observed in the presence of $Hg^{2+}$ ions (Devadiga et al. 2017). Biologically green-synthesized AgNPs were obtained using a freshly prepared extract of the soap-root plant and aqueous extract of manna of *Hedysarum* plant. The biofabricated yellowish-brown AgNPs were used as a colorimetric sensor for detection $Hg^{2+}$ in water samples with the limit of detection of 2.2 μM. The color of AgNPs solution turned to pale yellow in the presence of $Hg^{2+}$. Also, the color of the solution decreased gradually with the increase of $Hg^{2+}$ concentration. Finally, it turned to colorless at concentration of 0.001 M of $Hg^{2+}$ ions (Farhadi et al. 2012). Figure 3 shows the changes of the LSPR absorption band and color change of AgNPs solution in the presence of $Hg^{2+}$ and some other heavy metal ions.

Green crystalline AgNPs were synthesized from *Allium sativum* (garlic) extract. The extract of *Allium sativum* is containing amino acids, carbohydrates, and vitamin (vitamin B6). These biomolecules acted as stabilizer. The produced AgNPs were employed as a colorimetric sensor for sensitive detection of $Hg^{2+}$ in the presence of $Fe^{3+}$ in water samples. It was observed that $Fe^{3+}$ was not able to oxidize and etch AgNPs in phosphate buffer. Therefore, the color change from yellow to colorless

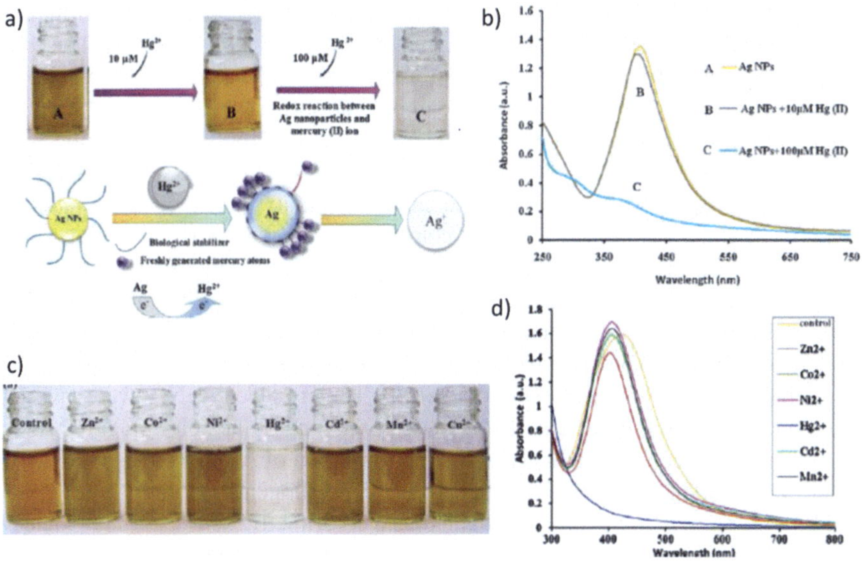


**Fig. 3** (**a**) Schematic representation of possible mechanism of $Hg^{2+}$ detection, (**b**) UV-Vis spectra of (*A*) biosynthesized AgNPs, (*B*) AGNPs in after addition of 10 μM of $Hg^{2+}$ and (*C*) 100 μM of $Hg^{2+}$; (**c**) image of biologically synthesized AgNPs in the presence of $Hg^{2+}$ and other heavy metal ions such as $Zn^{2+}$, $Co^{2+}$, $Ni^{2+}$, $Cd^{2+}$, $Mn^{2+}$, and $Cu^{2+}$ which confirms sensor selectivity; (**d**) UV-Vis spectra of the solution of AgNPs in the presence of the mentioned metal ions (Farhadi et al. 2012)

and the observed blue shift could be due to the interaction of $Hg^{2+}$ and AgNPs. This colorimetric sensor was also useful for colorimetric detection of $Pb^{2+}$ ions in the concentration range of 0.05–1 mM (Ghosh et al. 2018). Biosynthesized AgNPs using an aqueous extract of *Murraya koenigii* were capable of sensitive detection of $Hg^{2+}$ ions based on a redox reaction between $Hg^{2+}$ and AgNPs. Indeed, this sensing probe was evaluated separately for colorimetric detection of various metal ions such as $Fe^{3+}$, $Fe^{2+}$, $Pb^{2+}$, $Cd^{2+}$, $Co^{2+}$, $Zn^{2+}$, $Ni^{2+}$, $Al^{3+}$, $As^{3+}$, $As^{5+}$, $Cu^{2+}$, $Hg^{2+}$, and $Mn^{2+}$. Only in the presence of $Hg^{2+}$ ions, color change was occurred from dark brown to colorless, the LSPR peak shifted shorter wavelength and its intensity decreased. No color change was observed in the case of other studied metal ions and the intensity of the LSPR absorption band changed slightly (Kumar et al. 2017b). For colorimetric detection of $Hg^{2+}$ in various groundwater samples, a colorimetric sensor was designed based on green-synthesized AgNPs by gum kondagogu. In this study, a highly sensitive colorimetric sensor was proposed for detection of $Hg^{2+}$ in low concentration of 0.05 μM. This enhanced sensitivity could be due to small green-synthesized AgNPs with the average size of 5 ± 2.8 nm (Rastogi et al. 2014). Also, green-synthesized crystalline AgNPs using citrus fruit extracts (lemon, *Citrus limon* and sweet orange, *Citrus limetta*) has been demonstrated as a selective sensing probe for detection of $Hg^{2+}$ ions at wide pH range (3.2–8.5) (Ravi et al. 2013). Synthesized AgNPs using leaf extract of *Dahlia pinnata* showed the ability to selective sensing of hazardous $Hg^{2+}$ ions at wide pH range (3–8) (Roy et al. 2015).

A fully eco-friendly colorimetric assay was proposed for $Hg^{2+}$ detection in drinking water using AgNPs capped with carboxymethyl cellulose which were synthesized by carboxymethylation of cellulose waste isolated from Tunisian pal date petiole. This sensitive sensor exhibited great potential in the successful detection of hazardous $Hg^{2+}$ at a low concentration. The high observed sensitivity can make the proposed sensor as a powerful detection probe applied for water safety control (Sakly et al. 2017). AgNPs with anticancer and catalytic activity generated from aqueous mango leaf extract exhibited selective colorimetric sensing of $Hg^{2+}$ ions in water due to oxidation of AgNPs to $Ag^{+}$ ions and discoloration of solution (Samari et al. 2018). Green-synthesized AgNPs using *Euphorbia geniculate* extract were also evaluated for sensing of $Hg^{2+}$ ions according to the formation of Hg–Ag amalgam and etching of AgNPs (Santhosh et al. 2019). Similarly, green synthesis of AgNPs was achieved using *Panax ginseng* root extract. The produced AgNPs were investigated for colorimetric detection of $Hg^{2+}$ based on fading of the color of AgNP solution due to the dissolution of AgNPs and the formation of Ag–Hg amalgam (Tagad et al. 2017). The synthesized AuNPs by nonpathogenic fungal biomass of *Trichoderma harzianum* were investigated as a colorimetric sensor for $Hg^{2+}$ detection. The sensing strategy was based on aggregation of AuNPs in the presence of $Hg^{2+}$ ions. This interaction led to color change from pink-red to grayish blue. It was also observed that the plasmonic absorption band of AuNPs shifted from 532 to 540 nm through a red shift and a new LSPR peak was appeared at 720 nm. These changes in the color of solution and the LSPR absorption band indicated the $Hg^{2+}$ binding to AuNPs and their aggregation (Tripathi et al. 2014). In another study, a highly selective colorimetric sensor for the detection of $Hg^{2+}$ was reported. Well-dispersed quasi-spherical biofabricated AgNPs using plant extract of *Matricaria recutita* (Babunah) were used for detection of $Hg^{2+}$ ions (Uddin et al. 2017).

Besides the studied nanosensors for detection of $Hg^{2+}$, several plasmonic sensing platforms have emerged for detection of some other potential toxic heavy metals. Functionalized AuNPs were obtained using *Mangifera indica* leaf extract. The prepared AuNPs were used for colorimetric detection of $As^{3+}$ ions in an aqueous medium with limit of detection of 1.2 ppb. It was observed that $As^{3+}$ ions caused to aggregation of AuNPs and subsequent color change from red to blue. Also, a red shift occurred in the LSPR peak and a new absorption band was appeared ad 525 and 720 nm. These changes indicated aggregation of AuNPs in the presence of $As^{3+}$ ions (Boruah et al. 2019). In another study, AgNPs and AuNPs were prepared using L-tyrosine as reducing and stabilizing agent under ambient sunlight irradiation in aqueous medium and were used as a sensitive plasmonic nanoprobe for detection of metal ions. It was found that green-synthesized AgNPs were sensitive to $Hg^{2+}$ and $Mn^{2+}$ ions. The obtained AuNPs were used for highly sensitive naked-eye detection of $Hg^{2+}$ and $Pb^{2+}$ in drinking water and tap water samples with the detection limit in nM concentrations. Different strategies were reported for detection of $Hg^{2+}$, $Mn^{2+}$, and $Pb^{2+}$ ions. A blue shift was observed in the LSPR plasmonic absorption band of AgNPs in the presence of $Hg^{2+}$ ions and the yellow color of the solution turned to colorless. Moreover, the interaction of $Hg^{2+}$ ions with AuNPs led to a blue shift in the LSPR spectrum and the color of the solution remained pink. These observations

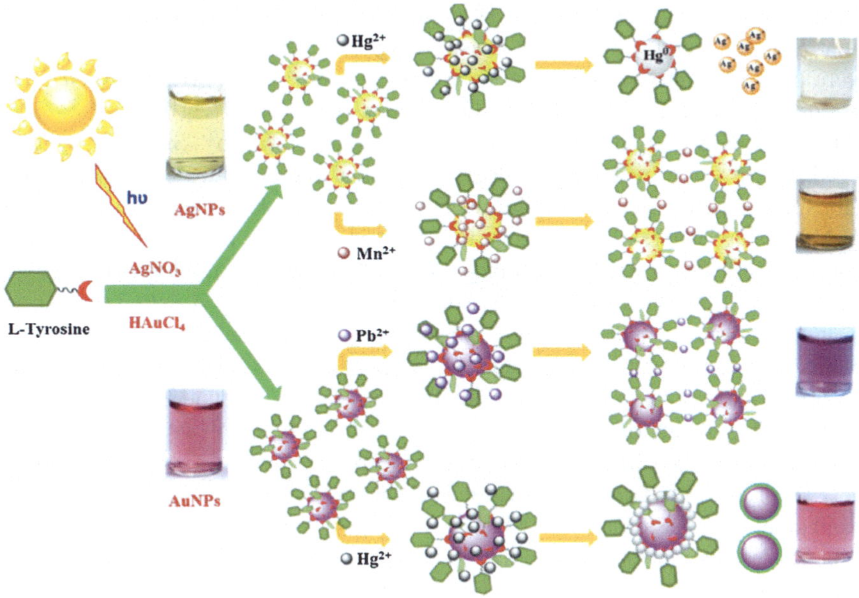


**Fig. 4** Colorimetric detection of $Hg^{2+}$ and $Mn^{2+}$ ions using green-synthesized AgNPs and $Hg^{2+}$ and $Pb^{2+}$ ions based on green-synthesized AuNPs (Annadhasan et al. 2014)

confirmed the formation of a core shell structure. Also, aggregation of nanoparticles occurred due to complex formation between $Mn^{2+}$ and $Pb^{2+}$ with L-tyrosine present on the surface of Ag and AuNPs, respectively. As it is known, these aggregations resulted in red shift in the LSPR spectra of nanoparticles (Fig. 4) (Annadhasan et al. 2014).

In a similar report, AuNPs were synthesized using N-cholyl-L-valine under natural sunlight irradiation. The green-synthesized AuNPs were used for colorimetric detection of $Co^{2+}$ and $Ni^{2+}$ in tap water and drinking water. A red shift from 525 nm towards 543 nm was observed in the LSPR absorption band of AuNPs after addition of $Ni^{2+}$ ions to AuNPs solution. This change could be due to $Ni^{2+}$ ion interaction with the functional groups present on the surface of AuNPs. The existing functional groups led to the aggregation of AuNPs. The pink color of the solution turned to purple at lower concentrations of $Ni^{2+}$ and then changed to violet at higher $Ni^{2+}$ concentrations. Also, the color of AuNPs solution turned to pale pink after addition of $Co^{2+}$ ions due to the formation of precipitation of particles (Annadhasan et al. 2015). The biologically synthesized AgNPs were prepared from the leaf extract of *Amomum subulatum* and were used as a selective colorimetric sensor for detection of $Zn^{2+}$ in drinking water. In the sensing process of $Zn^{2+}$ ions, the yellowish-brown color of biosynthesized plasmonic AgNPs changed to colorless and the LSPR absorption band at 425 nm was decreased and slightly shifted to a higher wavelength. This peak was completely disappeared at high concentration of $Zn^{2+}$ (Ihsan et al. 2015). Fresh extracts of different parts of neem were used for green synthesis of AgNPs. The obtained AgNPs were employed as colorimetric sensors for detection of heavy metal ions. The synthesized AgNPs using fresh neem leaf extracts were used for selective colorimetric detection of $Hg^{2+}$ and fabricated AgNPs via sun-dried neem leaf extract exhibited selective detection of $Hg^{2+}$ and $Pb^{2+}$. The obtained AgNPs using neem bark extract detected $Hg^{2+}$ and $Zn^{2+}$. Similarly, synthesized AgNPs using fresh and sun-dried mango leaf and green tea extracts successfully demonstrated selective colorimetric detection of $Hg^{2+}$ and $Pb^{2+}$. Prepared AgNPs using pepper seed extracts also showed selective colorimetric detection towards $Hg^{2+}$, $Pb^{2+}$, and $Zn^{2+}$ (Karthiga and Anthony 2013). A colorimetric sensor was designed for selective naked-eye detection of $Pb^{2+}$ using AgNPs synthesized from leaf extract of *Aconitum violaceum*. The phytosynthesized AgNPs exhibited a color change from yellow to red in the presence of $Pb^{2+}$ ions. The color change was observed due to the interaction of metal ions with catechins found on the surface of AgNPs. The interaction of catechin and $Pb^{2+}$ ions led to aggregation of nanoparticles. Moreover, the LSPR spectrum of AgNPs change after addition of $Pb^{2+}$ ions. A new absorption band was appeared at 520 nm and the intensity of the plasmonic band at 404 nm was decreased (Khan et al. 2018b). The prepared AuNPs using the aqueous leaf extract of *Rosa indica – wichuraiana* hybrid *Francois Guillot* were functionalized with glutathione (GSH). The synthesized GSH-AuNPs were employed as a highly responsive sensor for colorimetric detection of $Cd^{2+}$. The proposed sensor showed color change from ruby red to purple due to the aggregation of GSH-AuNPs in the presence of $Cd^{2+}$ (Manjumeena et al. 2015). AuNPs were prepared using the extracellular culture filtrate of the fungus *Aspergillus candidus* IF1. The biogenic AuNPs exhibited rapid aggregation by addition of cerium (Ce) and were demonstrated as a fast, precise colorimetric sensor towards $Ce^{3+}$. It was revealed that the aggregation was due to the formation of a coordinate complex between $Ce^{3+}$ ions and –COOH and –NH functional groups present on the surface of AuNPs. The interaction between $Ce^{3+}$ ions and AuNPs led to color change from red to purple (Priyadarshini et al. 2015). Riboflavin-functionalized AgNPs were synthesized using the *Cucumis melo* juice via a green synthesis approach. The prepared AgNPs were used as a colorimetric sensing platform for the selective and sensitive detection of trace concentrations of $Cu^{2+}$ ions in ground and tap water. In this study, colorimetric detection of $Cu^{2+}$ ions was performed based on the accelerated etching of AgNPs which resulted in the fading of yellow color of AgNPs and decrease of the LSPR absorption band intensity (Basiri et al. 2017). In another study, AgNPs were obtained using an aqueous extract of *Ficus benjamina* leaves and were evaluated as a colorimetric sensor for $Zn^{2+}$ detection. Upon addition of $Zn^{2+}$ ion to AgNPs solution, the yellow color of the solution turned to orange due to the aggregation of nanoparticles (Puente et al. 2019). The synthesized AgNPs using *Acalypha hispida* leaf extract showed a colorimetric response to detect a very low concentration of $Mn^{2+}$ in industrial effluent. Aggregation of AgNPs was observed to due to chelation of $Mn^{2+}$ ions in the presence of capping agents of AgNPs and subsequently a change in color of the solution, from reddish brown color to colorless (Sithara et al. 2017). Some reported green-synthesized plasmonic metal nanoparticles as colorimetric sensors for detection of heavy metal ions and the related analytical parameters are summarized in Table 3.

**Table 3** Reported green-synthesized plasmonic nanoparticles as colorimetric sensors for detection of heavy metal ions

| Microorganism | Biomolecule | Metal nanoparticle | Size and shape | Metal ion | Linear range | Detection limit | Sample | Ref. |
|---|---|---|---|---|---|---|---|---|
| *Syzygium aromaticum* | Phenolic compounds | Ag | 5–10 nm | $Hg^{2+}$ | – | 2 μM | – | Sangar et al. (2019) |
| *Terminalia catappa* (leaf extract) | Phenolic compounds<br>Flavonoid<br>Antioxidants | Ag | 10.9 nm | $Hg^{2+}$ | – | 0.85 μg $mL^{-1}$ | – | Devadiga et al. (2017) |
| Soap-root plant and *Hedysarum* | – | Ag | – | $Hg^{2+}$ | 10–100 μM | 2.2 μM | Lake water | Farhadi et al. (2012) |
| *Allium sativum* (garlic extract) | Amino acids<br>Carbohydrate<br>Vitamin (vitamin B6) | Ag | 10 nm, spherical | $Hg^{2+}$<br>$Pb^{2+}$ | 2–200 μM<br>0.05–1 mM | 2 μM | – | Ghosh et al. (2018) |
| *Murraya koenigii* (leaf extract) | Tannin<br>Flavonoids | Ag | 8.6 nm, spherical | $Hg^{2+}$ | 50 nM–500 μM | – | – | Kumar et al. (2017b) |
| Gum kondagogu | – | Ag | 5 ± 2.8 nm, | $Hg^{2+}$ | 0.05–1 μM | 0.05 μM | Ground water | Rastogi et al. (2014) |
| *Citrus limon*<br>*Citrus limetta* (fruit juice) | Citric acid<br>Ascorbic acid | Ag | 25–35 nm | $Hg^{2+}$ | – | 100 μM<br>Visual detection | Tap water | Ravi et al. (2013) |
| *Dahlia pinnata* (leaf extract) | Phenolic compounds<br>Carbohydrates | Ag | 15 nm, spherical | $Hg^{2+}$ | – | 10 μM<br>Visual detection | – | Roy et al. (2015) |
| Tunisian date palm petioles | Carboxymethyl cellulose | Ag | – | $Hg^{2+}$ | 25–200 nM | 5 nM | Drinking water | Sakly et al. (2017) |

(continued)

**Table 3** (continued)

| Microorganism | Biomolecule | Metal nanoparticle | Size and shape | Metal ion | Linear range | Detection limit | Sample | Ref. |
|---|---|---|---|---|---|---|---|---|
| *Mangifera indica* Mango (leaf extract) | Phenolic compound Flavonoid | Ag | 20.7 nm, spherical | $Hg^{2+}$ | 1–517 μM | – | – | Samari et al. (2018) |
| *Euphorbia geniculata* (leaf extract) | Alkaloids Flavonoids Polyphenols | Ag | 17 nm | $Hg^{2+}$ | 10–120 ppm | 10 ppm | – | Santhosh et al. (2019) |
| *Panax ginseng* (root extract) | – | Ag | 4–20 nm, spherical | $Hg^{2+}$ | 10 μM–1 mM | 5 μM | – | Tagad et al. (2017) |
| *Trichoderma harzianum* (Fungi) | Protein Cysteine Enzyme | Au | 26–34 nm, spherical | $Hg^{2+}$ | 0.001–1 μM | 2.6 nM | Tap water Ground water | Tripathi et al. (2014) |
| *Matricaria recutita* (stem extract) | Terpenoids Flavonoids Coumarins Polysaccharides | Ag | 11 nm, overall quasi-spherical | $Hg^{2+}$ | 5–15 ppm | 5 ppm | – | Uddin et al. (2017) |
| *Mangifera indica* (leaf extract) | Flavonoids Terpenoids Thiamine | Au | Spherical | $As^{3+}$ | 1–11 ppb | 1.2 ppb | – | Boruah et al. (2019) |

| | | | | | | | | |
|---|---|---|---|---|---|---|---|---|
| – | L-tyrosine | Ag | 8.2–32 nm, spherical, ellipsoids and rods | $Hg^{2+}$<br>$Mn^{2+}$ | $Hg^{2+}$: 20–150 nM<br>$Mn^{2+}$: 20–200 nM | $Hg^{2+}$: 19 nM (drinking water) and 26 nM (tap water)<br>$Mn^{2+}$: 23 nM (drinking water) and 26 nM (tap water) | Drinking water<br>Tap water | Annadhasan et al. (2014) |
| | | Au | 17–36 nm, spherical | $Hg^{2+}$<br>$Pb^{2+}$ | $Hg^{2+}$: 60–200 nM<br>$Pb^{2+}$: 20–100 nM | $Hg^{2+}$: 60 nM (drinking water) and 72 nM (tap water)<br>$Pb^{2+}$: 23 nM (drinking water) and 26 nM (tap water) | | |
| – | *N*-cholyl-L-valine (NaValC) | Au | 8–40 nm, spherical | $Co^{2+}$<br>$Ni^{2+}$ | 5–40 nM | 10 nM | Drinking water<br>Tap water | Annadhasan et al. (2015) |
| *Amomum subulatum* (leaf extract) | Phenolic compounds | Ag | 20–50 nm, spherical | $Zn^{2+}$ | 10–80 μM | 3.5 μM | Drinking water | Ihsan et al. (2015) |
| Neem (leaf and bark extract)<br>Mango (leaf extract)<br>Green tea<br>Paper (seed extract) | Gallic acid<br>Catechin<br>Phenolic compounds<br>Flavonoids | Ag | Spherical | $Hg^{2+}$<br>$Pb^{2+}$<br>$Zn^{2+}$ | – | – | – | Karthiga and Anthony (2013) |

(continued)

**Table 3** (continued)

| Microorganism | Biomolecule | Metal nanoparticle | Size and shape | Metal ion | Linear range | Detection limit | Sample | Ref. |
|---|---|---|---|---|---|---|---|---|
| *Aconitum violaceum* (leaf extract) | Polyphenols (catechin) | Ag | 20 nm, spherical | $Pb^{2+}$ | 0.5–25 μM | 0.1 μM | River water | Khan et al. (2018b) |
| *Rosa indica-wichuriana hybrid Francois guillot* (leaf extract) | – | Au | 5–13 nm, spherical | $Cd^{2+}$ | – | 30 nM<br>70 nM | Water sample | Manjumeena et al. (2015) |
| *Aspergillus candidus* IF1 (Fungi extract) | Protein<br>Peptide<br>Carbohydrate | Au | 22–25 nm, spherical | $Ce^{3+}$ | 2–50 ppm | 15 ppm (DLS)<br>35 ppm (UV–Vis) | – | Priyadarshini et al. (2015) |
| *Cucumis melo* (juice) | Vitamin A<br>Vitamin C<br>Riboflavin | Ag | 20 nm | $Cu^{2+}$ | 5–100 nM | 1.12 nM | Ground water<br>Tap water | Basiri et al. (2017) |
| *Ficus benjamina* (leaf extract) | – | Ag | 11.5 nm, spherical | $Zn^{2+}$ | 0.2–2 mM | 0.2 mM | – | Puente et al. (2019) |
| *Acalypha hispida* (leaf extract) | Phenolic compounds | Ag | 20–50 nm | $Mn^{2+}$ | 50–250 μM | 50 μM | Industrial effluent | Sithara et al. (2017) |
| *Momordica charantia* (fruit extract from peel, seed, and seed coat) | Phenolic compounds | Au | – | $Cd^{2+}$ | 0.2–0.8 ppm (0.18–0.71 μM) | 0.186 μM | – | Singh et al. (2018a) |

The biologically obtained plasmonic nanoparticles were also evaluated for detection of other types of pollutants. In-situ green fabrication of AgNPs was reported using flexible and transparent bacterial cellulose nanopapers. In this biosynthesis method, first Ag ions were adsorbed on bacterial cellulose nanopaper and then were reduced to AgNPs by the hydroxyl groups of cellulose nanofibers. The resulted AgNPs were employed as a sensitive sensing probe for the detection of cyanide ion ($CN^-$) and 2-mercaptobenzothiazole in water samples (Pourreza et al. 2015). Epigallocatechin gallate isolated from green tea synthesized AgNPs. A sensitive colorimetric sensor was established using the as-prepared stable AgNPs for colorimetric detection of kanamycin and sulfide ions (Singh et al. 2018b). Stable AuNPs were obtained using *Momordica charantia* fruit extract (peel, seed, and seed coat) and were evaluated as a colorimetric sensor for detection of $Cd^{2+}$ and thiophenol. The mechanism of colorimetric detection for both analytes was based on the aggregation of AuNPs and color of AuNPs solution changed from red to violet (Singh et al. 2018a).

In several studies, the application of nontoxic and eco-friendly metal nanoparticles as a dual or multifunctional sensor has been demonstrated. Moreover, these green prepared metal nanoparticles have been employed in other sensing approaches like electrochemical, fluorescence, or surface enhanced Raman spectroscopy (SERS). The obtained AgNPs from *Agaricus bisporus* were used for optical and electrochemical sensing of $Hg^{2+}$ ions (Sebastian et al. 2018). A biogenic route was developed for the synthesis of AuNPs with tunable size using *Citrus paradisi* extract. It was observed that the biofabricated AuNPs were able to detect $Pb^{2+}$, $Cu^{2+}$, $Hg^{2+}$, $Zn^{2+}$, and $Ca^{2+}$ ions through both fluorescent and plasmonic sensing strategies (Silva-De Hoyos et al. 2020). A multifunctional sensing approach was investigated for detection of $Cu^{2+}$ ions via biosynthesized AgNPs. The AgNPs have been obtained using bark extract of *Moringa oleifera*. These AgNPs were employed as a sensing probe for colorimetric, fluorescence, and electrochemical sensing of $Cu^{2+}$ ions (Sebastian et al. 2019). A dual sensor was developed for the detection of Thiram (toxic dithiocarbamate fungicide) using both colorimetric and electrochemical sensing methods. The crystalline spherical AgNPs which have been synthesized biologically by the stem extract of *Coscinium fenestratum* were exploited as a sensing probe for detection of Thiram in tap, canal, and river water samples (Ragam and Mathew 2019). Eco-friendly stable monometallic Ag and AuNPs and also bimetallic Ag/Au alloy nanoparticles were fabricated using Indian curry leaf plant (*Murraya koenigii Spreng*) under sonochemical condition at room temperature and pressure. These nanoparticles were employed for fluorometric detection of hazardous dithiocarbamate pesticide like Mancozbe in an aqueous medium (Alam et al. 2016). AuNPs were synthesized using *Acacia Nilotica* twig bark extract at room temperature and was employed for trace level electrochemical detection of nitrobenzene which is known as a hazardous pollutant (Emmanuel et al. 2014).

### *4.2 SERS-Based Green Sensors for Detection of Environmental Pollutions*

SERS as an advanced powerful analytical technique enables simple and rapid detection of various analytes in a single molecular level. In some reported studies, green-synthesized silver or gold nanoparticles were considered as SERS substrates for detection of numerous environmental pollutions. In one work, prepared Au–Ag core–shell and alloy nanoparticles using xylan extracted from bagasse were evaluated as a probe for detection of a common food contaminant, Sudan I. Xylan-capped Au@Ag nanoparticles with the advantage of enhanced Raman performance were able to detect trace concentration of Sudan I with detection limit of 0.126 ppm (Cai et al. 2019). Green-synthesized silver nanoparticle-reduced graphene oxide nanocomposite using *Psidium guajava* exhibited remarkable surface enhanced Raman signal for detection of methylene blue with a limit of detection of 0.01 μM (Chettri et al. 2017). In-situ synthesized AuNPs using a common edible fungus, *Tremella fuciformis* (TF) were used as a SERS substrate for trace detection of cationic dyes like methylene blue, Congo red and crystal violet as water contaminations. The TF as a capping agent on AuNPs prevented the aggregation of nanoparticles. It also provided an effective surface for adsorption of dyes which caused significant Raman signal enhancement (Tang et al. 2018). Bimetallic Au@Ag nanostructures were obtained through green synthesis using epigallocatechin Gallate isolated from tea leaves. These green-synthesized nanoparticles exhibited strong SERS signal for detection of Rhodamine 6G (Wang et al. 2016b). In another study, biosynthesized AuNPs using IPE of bacterial strain *Staphylococcus warneri* exhibited enhanced SERS signal towards detection of Rhodamine 6G. According to the obtained results, it was revealed that the green-synthesized AuNPs could be a superior SERS substrate for sensitive detection of different toxic chemical compounds, organic pollutants like nitro aromatics and dyes at a single molecular level (Nag et al. 2018).

## 5 Conclusion

This chapter focuses on the green and biological synthesis of metal nanoparticles and their application in sensing and detection of heavy metal ions and organic pollutants. Green synthesis of metal nanoparticles has displayed several advantages over traditional chemical methods such as producing environmental friendly and biocompatible nanoparticles, preventing the use of chemical reagents, and decreasing the formation of hazardous by-products and their side effects. Therefore, green synthesis method has been evolved as an important and popular synthetic branch of nanotechnology due to its advantages. Different microorganisms like bacteria, yeasts, fungi, and algae, and plants and pure bioagents have been described as potential biological nanofactories for the generation of metal nanoparticles. The green-synthesized nanoparticles are successfully being used in numerous applications including biological and food analysis, drug delivery, environmental monitoring, and sensing approaches. By increasing the level of harmful pollutants, there is an environmental concern to sensing and removal of toxic compounds. By considering unique properties of plasmonic metal nanoparticles, different LSPR-based sensors have been designed employing biosynthesized metal nanoparticles for detection of several toxic heavy metal ions and organic pollutants. Colorimetric and SERS-based sensors as powerful and highly selective and sensitive detection methods using nontoxic nanoparticles can provide a promising opportunity for simple and rapid detection of pollutants in soil and water resources, as attractive sensing methods for environmental monitoring applications.

**Acknowledgements** We are grateful to Department of Chemistry, Faculty of Science, K.N. Toosi University of Technology and Department of Marine Living Science, Ocean Sciences Research Center, Iranian National Institute for Oceanography and Atmospheric Science, Tehran, Iran.

## References

Abdeen S, Geo S, Sukanya S, Praseetha P, Dhanya R (2014) Biosynthesis of silver nanoparticles from actinomycetes for therapeutic applications. Int J Nano Dimens 5(2):155–162. https://doi.org/10.7508/IJND.2014.02.008

Abdel-Raouf N, Al-Enazi NM, Ibraheem IB (2017) Green biosynthesis of gold nanoparticles using Galaxaura elongata and characterization of their antibacterial activity. Arab J Chem 10:S3029–S3039. https://doi.org/10.1016/j.arabjc.2013.11.044

Abdel-Raouf N, Al-Enazi NM, Ibraheem IBM, Alharbi RM, Alkhulaifi MM (2018) Biosynthesis of silver nanoparticles by using of the marine brown alga Padina pavonia and their characterization. Saudi J Biol Sci. https://doi.org/10.1016/j.sjbs.2018.01.007

Aboelfetoh EF, El-Shenody RA, Ghobara MM (2017) Eco-friendly synthesis of silver nanoparticles using green algae (*Caulerpa serrulata*): reaction optimization, catalytic and antibacterial activities. Environ Monit Assess 189(7):349. https://doi.org/10.1007/s10661-017-6033-0

Adil SF, Assal ME, Khan M, Al-Warthan A, Siddiqui MRH, Liz-Marzán LM (2015) Biogenic synthesis of metallic nanoparticles and prospects toward green chemistry. Dalton Trans 44(21):9709–9717. https://doi.org/10.1039/C4DT03222E

Agnihotri M, Joshi S, Kumar AR, Zinjarde S, Kulkarni S (2009) Biosynthesis of gold nanoparticles by the tropical marine yeast Yarrowia lipolytica NCIM 3589. Mater Lett 63(15):1231–1234. https://doi.org/10.1016/j.matlet.2009.02.042

Ahlawat J, Sehrawat AR (2017) Nano dimensional (1–20 nm) silver nanoparticles: stem extract of Capparis decidua (FORSK.) EDGEW mediated synthesis and its characterization—a lab to land approach. Int J Curr Microbiol Appl Sci 6(10):1874–1883. https://doi.org/10.20546/ijcmas.2017.610.226

Ahmad A, Mukherjee P, Senapati S, Mandal D, Khan MI, Kumar R, Sastry M (2003) Extracellular biosynthesis of silver nanoparticles using the fungus *Fusarium oxysporum*. Colloids Surf B Biointerfaces 28(4):313–318. https://doi.org/10.1016/S0927-7765(02)00174-1

Ahmad A, Senapati S, Khan MI, Kumar R, Sastry M (2005) Extra-/intracellular biosynthesis of gold nanoparticles by an alkalotolerant fungus, *Trichothecium* sp. J Biomed Nanotechnol 1(1):47–53. https://doi.org/10.1166/jbn.2005.012

Ahmad S, Munir S, Zeb N, Ullah A, Khan B, Ali J, Bilal M, Omer M, Alamzeb M, Salman SM (2019) Green nanotechnology: a review on green synthesis of silver nanoparticles—an eco-friendly approach. Int J Nanomed 14:5087–5107. https://doi.org/10.2147/IJN.S200254

Ahn E-Y, Jin H, Park Y (2019) Assessing the antioxidant, cytotoxic, apoptotic and wound healing properties of silver nanoparticles green-synthesized by plant extracts. Mater Sci Eng C 101:204–216. https://doi.org/10.1016/j.msec.2019.03.095

Akther T, Mathipi V, Kumar NS, Davoodbasha M, Srinivasan H (2019) Fungal-mediated synthesis of pharmaceutically active silver nanoparticles and anticancer property against A549 cells through apoptosis. Environ Sci Pollut Res 26(13):13649–13657. https://doi.org/10.1007/s11356-019-04718-w

Alam MN, Das S, Batuta S, Mandal D, Begum NA (2016) Green-nanochemistry for safe environment: bio-friendly synthesis of fluorescent monometallic (Ag and Au) and bimetallic (Ag/Au alloy) nanoparticles having pesticide sensing activity. J Nanostruct Chem 6(4):373–395. https://doi.org/10.1007/s40097-016-0209-y

Ali J, Ali N, Wang L, Waseem H, Pan G (2019) Revisiting the mechanistic pathways for bacterial mediated synthesis of noble metal nanoparticles. J Microbiol Methods 159:18–25. https://doi.org/10.1016/j.mimet.2019.02.010

Anandan M, Poorani G, Boomi P, Varunkumar K, Anand K, Chuturgoon AA, Saravanan M, Prabu HG (2019) Green synthesis of anisotropic silver nanoparticles from the aqueous leaf extract of *Dodonaea viscosa* with their antibacterial and anticancer activities. Process Biochem 80:80–88. https://doi.org/10.1016/j.procbio.2019.02.014

Annadhasan M, Muthukumarasamyvel T, Sankar Babu V, Rajendiran N (2014) Green synthesized silver and gold nanoparticles for colorimetric detection of $Hg^{2+}$, $Pb^{2+}$, and $Mn^{2+}$ in aqueous medium. ACS Sustain Chem Eng 2(4):887–896. https://doi.org/10.1021/sc400500z

Annadhasan M, Kasthuri J, Rajendiran N (2015) Green synthesis of gold nanoparticles under sunlight irradiation and their colorimetric detection of $Ni^{2+}$ and $Co^{2+}$ ions. RSC Adv 5(15):11458–11468. https://doi.org/10.1039/C4RA14034F

Annamalai J, Nallamuthu T (2016) Green synthesis of silver nanoparticles: characterization and determination of antibacterial potency. Appl Nanosci 6(2):259–265

Apte M, Sambre D, Gaikawad S, Joshi S, Bankar A, Kumar AR, Zinjarde S (2013) Psychrotrophic yeast *Yarrowia lipolytica* NCYC 789 mediates the synthesis of antimicrobial silver nanoparticles via cell-associated melanin. AMB Express 3(1):32–39. https://doi.org/10.1186/2191-0855-3-32

Arokiyaraj S, Vincent S, Saravanan M, Lee Y, Oh YK, Kim KH (2017) Green synthesis of silver nanoparticles using *Rheum palmatum* root extract and their antibacterial activity against *Staphylococcus aureus* and *Pseudomonas aeruginosa*. Artif Cells Nanomed Biotechnol 45(2):372–379. https://doi.org/10.3109/21691401.2016.1160403

Arya A, Mishra V, Chundawat TS (2019) Green synthesis of silver nanoparticles from green algae (*Botryococcus braunii*) and its catalytic behavior for the synthesis of benzimidazoles. Chem Data Coll 20:100190. https://doi.org/10.1016/j.cdc.2019.100190

Ashengroph M (2014) Synthesis of silver nanoparticle from silver nitrate using culture extract of *Rhodotorula* sp. strain GM5. Pajoohandeh J 18(6):310–319

Babu PJ, Sharma P, Saranya S, Tamuli R, Bora U (2013) Green synthesis and characterization of biocompatible gold nanoparticles using *Solanum indicum* fruits. Nanomater Nanotechnol 3:4–10. https://doi.org/10.5772/56608

Bao Z, Cao J, Kang G, Lan CQ (2019) Effects of reaction conditions on light-dependent silver nanoparticle biosynthesis mediated by cell extract of green alga *Neochloris oleoabundans*. Environ Sci Pollut Res 26(3):2873–2881. https://doi.org/10.1007/s11356-018-3843-8

Barizuddin S, Bok S, Gangopadhyay S (2016) Plasmonic sensors for disease detection—a review. J Nanomed Nanotechnol 7(3):1000373–1000382. https://doi.org/10.4172/2157-7439.1000373

Basiri S, Mehdinia A, Jabbari A (2017) Biologically green synthesized silver nanoparticles as a facile and rapid label-free colorimetric probe for determination of $Cu^{2+}$ in water samples. Spectrochim Acta A 171:297–304. https://doi.org/10.1016/j.saa.2016.08.032

Basiri S, Mehdinia A, Jabbari A (2018) Green synthesis of reduced graphene oxide-Ag nanoparticles as a dual-responsive colorimetric platform for detection of dopamine and $Cu^{2+}$. Sensors Actuators B Chem 262:499–507. https://doi.org/10.1016/j.snb.2018.02.011

Behravan M, Panahi AH, Naghizadeh A, Ziaee M, Mahdavi R, Mirzapour A (2019) Facile green synthesis of silver nanoparticles using *Berberis vulgaris* leaf and root aqueous extract and its antibacterial activity. Int J Biol Macromol 124:148–154. https://doi.org/10.1016/j.ijbiomac.2018.11.101

Bhagat M, Anand R, Datt R, Gupta V, Arya S (2019) Green synthesis of silver nanoparticles using aqueous extract of *Rosa brunonii* Lindl and their morphological, biological and photocatalytic characterizations. J Inorg Organomet Polym 29(3):1039–1047. https://doi.org/10.1007/s10904-018-0994-5

Bhainsa KC, D'souza S (2006) Extracellular biosynthesis of silver nanoparticles using the fungus *Aspergillus fumigatus*. Colloids Surf B Biointerfaces 47(2):160–164. https://doi.org/10.1016/j.colsurfb.2005.11.026

Binupriya A, Sathishkumar M, Vijayaraghavan K, Yun S-I (2010) Bioreduction of trivalent aurum to nano-crystalline gold particles by active and inactive cells and cell-free extract of *Aspergillus oryzae* var. viridis. J Hazard Mater 177(1–3):539–545. https://doi.org/10.1016/j.jhazmat.2009.12.066

Boruah BS, Daimari NK, Biswas R (2019) *Mangifera indica* leaf extract mediated gold nanoparticles: a novel platform for sensing of As(III). IEEE Sensors Lett 3(3):1–3. https://doi.org/10.1109/LSENS.2019.2894419

Botha TL, Elemike EE, Horn S, Onwudiwe DC, Giesy JP, Wepener V (2019) Cytotoxicity of Ag, Au and Ag-Au bimetallic nanoparticles prepared using golden rod (*Solidago canadensis*) plant extract. Sci Rep 9(1):4169–4176. https://doi.org/10.1038/s41598-019-40816-y

Cai J, Li Y-CE, Liu C, Wang X (2019) Green and controllable synthesis of Au-Ag bimetal nanoparticles by xylan for SERS. ACS Sustain Chem Eng. https://doi.org/10.1021/acssuschemeng.9b00260

Castro L, Blázquez ML, Muñoz JA, González F, Ballester A (2013) Biological synthesis of metallic nanoparticles using algae. IET Nanobiotechnol 7(3):109–116. https://doi.org/10.1049/iet-nbt.2012.0041

Chahardoli A, Karimi N, Fattahi A (2017) Biosynthesis, characterization, antimicrobial and cytotoxic effects of silver nanoparticles using *Nigella arvensis* seed extract. Iran J Pharm Res 16(3):1167–1175. https://doi.org/10.22037/IJPR.2017.2066

Chandran SP, Chaudhary M, Pasricha R, Ahmad A, Sastry M (2006) Synthesis of gold nanotriangles and silver nanoparticles using *Aloe vera* plant extract. Biotechnol Prog 22(2):577–583. https://doi.org/10.1021/bp0501423

Chellapandian C, Ramkumar B, Puja P, Shanmuganathan R, Pugazhendhi A, Kumar P (2019) Gold nanoparticles using red seaweed *Gracilaria verrucosa*: green synthesis, characterization and biocompatibility studies. Process Biochem 80:58–63. https://doi.org/10.1016/j.procbio.2019.02.009

Chettri P, Vendamani V, Tripathi A, Singh MK, Pathak AP, Tiwari A (2017) Green synthesis of silver nanoparticle-reduced graphene oxide using *Psidium guajava* and its application in SERS for the detection of methylene blue. Appl Surf Sci 406:312–318. https://doi.org/10.1016/j.apsusc.2017.02.073

Chums-ard W, Fawcett D, Fung CC, Poinern GEJ (2019) Biogenic synthesis of gold nanoparticles from waste watermelon and their antibacterial activity against *Escherichia coli* and *Staphylococcus epidermidis*. Int J Res Med Sci 7(7):2499–2505. https://doi.org/10.18203/2320-6012.ijrms20192874

Correa-Llantén DN, Muñoz-Ibacache SA, Castro ME, Muñoz PA, Blamey JM (2013) Gold nanoparticles synthesized by *Geobacillus* sp. strain ID17 a thermophilic bacterium isolated from Deception Island, Antarctica. Microb Cell Factories 12(1):75–80. https://doi.org/10.1186/1475-2859-12-7

Dahoumane SA, Mechouet M, Wijesekera K, Filipe CD, Sicard C, Bazylinski DA, Jeffryes C (2017) Algae-mediated biosynthesis of inorganic nanomaterials as a promising route in nanobiotechnology—a review. Green Chem 19(3):552–587. https://doi.org/10.1039/C6GC02346K

Dameron C, Reese R, Mehra R, Kortan A, Carroll P, Steigerwald M, Brus L, Winge D (1989) Biosynthesis of cadmium sulphide quantum semiconductor crystallites. Nature 338(6216):596. https://doi.org/10.1038/338596a0

Dang H, Fawcett D, Poinern GEJ (2019) Green synthesis of gold nanoparticles from waste macadamia nut shells and their antimicrobial activity against *Escherichia coli* and *Staphylococcus epidermis*. Int J Res Med Sci 7(4):2499–2505. https://doi.org/10.18203/2320-6012.ijrms20191320

Das J, Velusamy P (2014) Catalytic reduction of methylene blue using biogenic gold nanoparticles from *Sesbania grandiflora* L. J Taiwan Inst Chem Eng 45(5):2280–2285. https://doi.org/10.1016/j.jtice.2014.04.005

Das SK, Das AR, Guha AK (2009) Gold nanoparticles: microbial synthesis and application in water hygiene management. Langmuir 25(14):8192–8199. https://doi.org/10.1021/la900585p

Das S, Chakraborty J, Chatterjee S, Kumar H (2018) Prospects of biosynthesized nanomaterials for the remediation of organic and inorganic environmental contaminants. Environ Sci Nano 5(12):2784–2808. https://doi.org/10.1039/C8EN00799C

Devadiga A, Shetty KV, Saidutta M (2017) Highly stable silver nanoparticles synthesized using *Terminalia catappa* leaves as antibacterial agent and colorimetric mercury sensor. Mater Lett 207:66–71. https://doi.org/10.1016/j.matlet.2017.07.024

Devi LS, Joshi SR (2015) Ultrastructures of silver nanoparticles biosynthesized using endophytic fungi. J Microsc Ultrastruct 3(1):29–37. https://doi.org/10.1016/j.jmau.2014.10.004

Duan H, Wang D, Li Y (2015) Green chemistry for nanoparticle synthesis. Chem Soc Rev 44(16):5778–5792. https://doi.org/10.1039/C4CS00363B

Dudhane AA, Waghmode SR, Dama LB, Mhaindarkar VP, Sonawane A, Katariya S (2019) Synthesis and characterization of gold nanoparticles using plant extract of *Terminalia arjuna* with antibacterial activity. Int J Nanosci Nanotechnol 15(2):75–82

Elegbede JA, Lateef A, Azeez MA, Asafa TB, Yekeen TA, Oladipo IC, Hakeem AS, Beukes LS, Gueguim-Kana EB (2019) Silver-gold alloy nanoparticles biofabricated by fungal xylanases exhibited potent biomedical and catalytic activities. Biotechnol Prog:e2829. https://doi.org/10.1002/btpr.2829

Elemike EE, Onwudiwe DC, Nundkumar N, Singh M, Iyekowa O (2019) Green synthesis of Ag, Au and Ag-Au bimetallic nanoparticles using *Stigmaphyllon ovatum* leaf extract and their in vitro anticancer potential. Mater Lett 243:148–152. https://doi.org/10.1016/j.matlet.2019.02.049

Elsupikhe RF, Shameli K, Ahmad MB, Ibrahim NA, Zainudin N (2015) Green sonochemical synthesis of silver nanoparticles at varying concentrations of κ-carrageenan. Nanoscale Res Lett 10(1):302–309. https://doi.org/10.1186/s11671-015-0916-1

Emmanuel R, Karuppiah C, Chen S-M, Palanisamy S, Padmavathy S, Prakash P (2014) Green synthesis of gold nanoparticles for trace level detection of a hazardous pollutant (nitrobenzene) causing methemoglobinaemia. J Hazard Mater 279:117–124. https://doi.org/10.1016/j.jhazmat.2014.06.066

Farhadi K, Forough M, Molaei R, Hajizadeh S, Rafipour A (2012) Highly selective $Hg^{2+}$ colorimetric sensor using green synthesized and unmodified silver nanoparticles. Sensors Actuatators B Chem 161(1):880–885. https://doi.org/10.1016/j.snb.2011.11.052

Francis S, Joseph S, Koshy EP, Mathew B (2018) Microwave assisted green synthesis of silver nanoparticles using leaf extract of elephantopus scaber and its environmental and biological applications. Artif Cells Nanomed Biotechnol 46(4):795–804. https://doi.org/10.1080/21691401

Gahlawat G, Choudhury AR (2019) A review on the biosynthesis of metal and metal salt nanoparticles by microbes. RSC Adv 9(23):12944–12967. https://doi.org/10.1039/C8RA10483B

Gao Y, Huang Q, Su Q, Liu R (2014) Green synthesis of silver nanoparticles at room temperature using kiwifruit juice. Spectrosc Lett 47(10):790–795. https://doi.org/10.1080/00387010.2013.848898

Gayda GZ, Demkiv OM, Stasyuk NY, Serkiz RY, Lootsik MD, Errachid A, Gonchar MV, Nisnevitch M (2019) Metallic nanoparticles obtained via "green" synthesis as a platform for biosensor construction. Appl Sci 9(4):720–735. https://doi.org/10.3390/app9040720

Gericke M, Pinches A (2006) Microbial production of gold nanoparticles. Gold Bull 39(1):22–28. https://doi.org/10.1007/BF03215529

Ghosh S, Patil S, Chopade N, Luikham S, Kitture R, Gurav D, Patil A, Phadatare S, Sontakke V, Kale S (2016) *Gnidia glauca* leaf and stem extract mediated synthesis of gold nanocatalysts with free radical scavenging potential. J Nanomed Nanotechnol 7(358):2–11. https://doi.org/10.4172/2157-7439.1000358

Ghosh S, Maji S, Mondal A (2018) Study of selective sensing of $Hg^{2+}$ ions by green synthesized silver nanoparticles suppressing the effect of $Fe^{3+}$ ions. Colloids Surf A Physicochem Eng Asp 555:324–331. https://doi.org/10.1016/j.colsurfa.2018.07.012

Gonnelli C, Cacioppo F, Giordano C, Capozzoli L, Salvatici C, Salvatici MC, Colzi I, Del Bubba M, Ancillotti C, Ristori S (2015) *Cucurbita pepo* L. extracts as a versatile hydrotropic source for the synthesis of gold nanoparticles with different shapes. Green Chem Lett Rev 8(1):39–47. https://doi.org/10.1080/17518253.2015.1027288

Gonnelli C, Giordano C, Fontani U, Salvatici MC, Ristori S (2018) Green synthesis of gold nanoparticles from extracts of *Cucurbita pepo* L. leaves: insights on the role of plant ageing. In: Advances in bionanomaterials. Springer, pp 155–164. https://doi.org/10.1007/978-3-319-62027-5_14

González-Ballesteros N, Prado-López S, Rodríguez-González J, Lastra M, Rodríguez-Argüelles M (2017) Green synthesis of gold nanoparticles using brown algae *Cystoseira baccata*: its activity in colon cancer cells. Colloids Surf B Biointerfaces 153:190–198. https://doi.org/10.1016/j.colsurfb.2017.02.020

Gour A, Jain NK (2019) Advances in green synthesis of nanoparticles. Artif Cells Nanomed Biotechnol 47(1):844–851. https://doi.org/10.1080/21691401.2019.1577878

Govindaraju K, Kiruthiga V, Kumar VG, Singaravelu G (2009) Extracellular synthesis of silver nanoparticles by a marine alga, *Sargassum wightii* Grevilli and their antibacterial effects. J Nanosci Nanotechnol 9(9):5497–5501. https://doi.org/10.1166/jnn.2009.1199

Govindaraju K, Krishnamoorthy K, Alsagaby SA, Singaravelu G, Premanathan M (2015) Green synthesis of silver nanoparticles for selective toxicity towards cancer cells. IET Nanobiotechnol 9(6):325–330. https://doi.org/10.1049/iet-nbt.2015.0001

Gowramma B, Keerthi U, Rafi M, Rao DM (2015) Biogenic silver nanoparticles production and characterization from native stain of *Corynebacterium* species and its antimicrobial activity. 3 Biotech 5(2):195–201. https://doi.org/10.1007/s13205-014-0210-4

He S, Guo Z, Zhang Y, Zhang S, Wang J, Gu N (2007) Biosynthesis of gold nanoparticles using the bacteria Rhodopseudomonas capsulata. Mater Lett 61(18):3984–3987. https://doi.org/10.1016/j.matlet.2007.01.018

He S, Zhang Y, Guo Z, Gu N (2008) Biological synthesis of gold nanowires using extract of *Rhodopseudomonas capsulata*. Biotechnol Prog 24(2):476–480. https://doi.org/10.1021/bp0703174

He Y, Wei F, Ma Z, Zhang H, Yang Q, Yao B, Huang Z, Li J, Zeng C, Zhang Q (2017) Green synthesis of silver nanoparticles using seed extract of *Alpinia katsumadai*, and their antioxidant, cytotoxicity, and antibacterial activities. RSC Adv 7(63):39842–39851. https://doi.org/10.1039/C7RA05286C

Homola J (2004) Surface plasmon resonance biosensors for food safety. In: Optical sensors. Springer, pp 145–172. https://doi.org/10.1007/978-3-662-09111-1_7

Huang J, Li Q, Sun D, Lu Y, Su Y, Yang X, Wang H, Wang Y, Shao W, He N (2007) Biosynthesis of silver and gold nanoparticles by novel sundried *Cinnamomum camphora* leaf. Nanotechnology 18(10):105104. https://doi.org/10.1088/0957-4484/18/10/105104

Hulkoti NI, Taranath T (2014) Biosynthesis of nanoparticles using microbes—a review. Colloids Surf B Biointerfaces 121:474–483. https://doi.org/10.1016/j.colsurfb.2014.05.027

Ibraheem I, Abd-Elaziz B, Saad W, Fathy W (2016) Green biosynthesis of silver nanoparticles using marine red algae Acanthophora specifera and its antimicrobial activity. J Nanomed Nanotechnol 7(409):2–5. https://doi.org/10.4172/2157-7439.1000409

Ibrahim HM (2015) Green synthesis and characterization of silver nanoparticles using banana peel extract and their antimicrobial activity against representative microorganisms. J Radiat Res Appl Sci 8(3):265–275. https://doi.org/10.1016/j.jrras.2015.01.007

Ihsan M, Niaz A, Rahim A, Zaman MI, Arain MB, Sharif T, Najeeb M (2015) Biologically synthesized silver nanoparticle-based colorimetric sensor for the selective detection of $Zn^{2+}$. RSC Adv 5(111):91158–91165. https://doi.org/10.1039/C5RA17055A

Iravani S (2011) Green synthesis of metal nanoparticles using plants. Green Chem 13(10):2638–2650. https://doi.org/10.1039/C1GC15386B

Jeeva K, Thiyagarajan M, Elangovan V, Geetha N, Venkatachalam P (2014) *Caesalpinia coriaria* leaf extracts mediated biosynthesis of metallic silver nanoparticles and their antibacterial activity against clinically isolated pathogens. Ind Crop Prod 52:714–720. https://doi.org/10.1016/j.indcrop.2013.11.037

Jha AK, Prasad K (2008) Yeast mediated synthesis of silver nanoparticles. Int J Nanosci Nanotechnol 4(1):17–22

Jha AK, Prasad K, Prasad K, Kulkarni A (2009) Plant system: nature's nanofactory. Colloids Surf B Biointerfaces 73(2):219–223. https://doi.org/10.1016/j.colsurfb.2009.05.018

Kahrilas GA, Wally LM, Fredrick SJ, Hiskey M, Prieto AL, Owens JE (2013) Microwave-assisted green synthesis of silver nanoparticles using orange peel extract. ACS Sustain Chem Eng 2(3):367–376. https://doi.org/10.1021/sc4003664

Kaler A, Jain S, Banerjee UC (2013) Green and rapid synthesis of anticancerous silver nanoparticles by *Saccharomyces boulardii* and insight into mechanism of nanoparticle synthesis. Biomed Res Int 2013:872940–872947. https://doi.org/10.1155/2013/872940

Karthiga D, Anthony SP (2013) Selective colorimetric sensing of toxic metal cations by green synthesized silver nanoparticles over a wide pH range. RSC Adv 3(37):16765–16774. https://doi.org/10.1039/c3ra42308e

Kathiraven T, Sundaramanickam A, Shanmugam N, Balasubramanian T (2015) Green synthesis of silver nanoparticles using marine algae *Caulerpa racemosa* and their antibacterial activity against some human pathogens. Appl Nanosci 5(4):499–504. https://doi.org/10.1007/s13204-014-0341-2

Kato Y, Yoshimura E, Suzuki M (2019) Synthesis of gold nanoparticles by extracellular components of *Lactobacillus casei*. ChemistrySelect 4(24):7331–7337. https://doi.org/10.1002/slct.201901046

Keshavamurthy M, Srinath B, Rai VR (2018) Phytochemicals-mediated green synthesis of gold nanoparticles using *Pterocarpus santalinus* L. (Red Sanders) bark extract and their antimicrobial properties. Particul Sci Technol 36(7):785–790. https://doi.org/10.1080/02726351.2017.1302533

Khan M, Tareq F, Hossen M, Roki M (2018a) Green synthesis and characterization of silver nanoparticles using *Coriandrum sativum* leaf extract. J Eng Sci Technol 13(1):158–166

Khan NA, Niaz A, Zaman MI, Khan FA, Tariq M (2018b) Sensitive and selective colorimetric detection of $Pb^{2+}$ by silver nanoparticles synthesized from *Aconitum violaceum* plant leaf extract. Mater Res Bull 102:330–336. https://doi.org/10.1016/j.materresbull.2018.02.050

Klaus T, Joerger R, Olsson E, Granqvist C-G (1999) Silver-based crystalline nanoparticles, microbially fabricated. Proc Natl Acad Sci 96(24):13611–13614. https://doi.org/10.1073/pnas.96.24.13611

Klekotko M, Brach K, Olesiak-Banska J, Samoc M, Matczyszyn K (2019) Popcorn-shaped gold nanoparticles: plant extract-mediated synthesis, characterization and multiphoton-excited luminescence properties. Mater Chem Phys 229:56–60. https://doi.org/10.1016/j.matchemphys.2019.02.066

Koduru JR, Kailasa SK, Bhamore JR, Kim K-H, Dutta T, Vellingiri K (2018) Phytochemical-assisted synthetic approaches for silver nanoparticles antimicrobial applications: a review. Adv Colloid Interf Sci 256:326–339. https://doi.org/10.1016/j.cis.2018.03.001

Krishnaswamy K, Vali H, Orsat V (2014) Value-adding to grape waste: green synthesis of gold nanoparticles. J Food Eng 142:210–220. https://doi.org/10.1016/j.jfoodeng.2014.06.014

Kumar V, Yadav SC, Yadav SK (2010) *Syzygium cumini* leaf and seed extract mediated biosynthesis of silver nanoparticles and their characterization. J Chem Technol Biotechnol 85(10):1301–1309. https://doi.org/10.1002/jctb.2427

Kumar D, Karthik L, Kumar G, Roa K (2011) Biosynthesis of silver anoparticles from marine yeast and their antimicrobial activity against multidrug resistant pathogens. Pharmacologyonline 3:1100–1111

Kumar B, Smita K, Cumbal L, Debut A (2017a) Green synthesis of silver nanoparticles using Andean blackberry fruit extract. Saudi J Biol Sci 24(1):45–50. https://doi.org/10.1016/j.sjbs.2015.09.006

Kumar V, Singh DK, Mohan S, Bano D, Gundampati RK, Hasan SH (2017b) Green synthesis of silver nanoparticle for the selective and sensitive colorimetric detection of mercury(II) ion. J Photochem Photobiol 168:67–77. https://doi.org/10.1016/j.jphotobiol.2017.01.022

Kumar I, Mondal M, Meyappan V, Sakthivel N (2019) Green one-pot synthesis of gold nanoparticles using *Sansevieria roxburghiana* leaf extract for the catalytic degradation of toxic organic pollutants. Mater Res Bull 117:18–27. https://doi.org/10.1016/j.materresbull.2019.04.029

Lakshmanan G, Sathiyaseelan A, Kalaichelvan P, Murugesan K (2018) Plant-mediated synthesis of silver nanoparticles using fruit extract of *Cleome viscosa* L.: assessment of their antibacterial and anticancer activity. Karbala Int J Mod Sci 4(1):61–68. https://doi.org/10.1016/j.kijoms.2017.10.007

Leela A, Vivekanandan M (2008) Tapping the unexploited plant resources for the synthesis of silver nanoparticles. Afr J Biotechnol 7(17):3162–3165

Liu Y, Kim S, Kim YJ, Perumalsamy H, Lee S, Hwang E, Yi T-H (2019) Green synthesis of gold nanoparticles using *Euphrasia officinalis* leaf extract to inhibit lipopolysaccharide-induced inflammation through NF-κB and JAK/STAT pathways in RAW 264.7 macrophages. Int J Nanomed 14:2945–2959. https://doi.org/10.2147/IJN.S199781

Logaranjan K, Raiza AJ, Gopinath SC, Chen Y, Pandian K (2016) Shape-and size-controlled synthesis of silver nanoparticles using aloe vera plant extract and their antimicrobial activity. Nanoscale Res Lett 11(1):520–528. https://doi.org/10.1186/s11671-016-1725-x

Mageswari A, Subramanian P, Ravindran V, Yesodharan S, Bagavan A, Rahuman AA, Karthikeyan S, Gothandam KM (2015) Synthesis and larvicidal activity of low-temperature stable silver nanoparticles from psychrotolerant *Pseudomonas mandelii*. Environ Sci Pollut Res 22(7):5383–5394. https://doi.org/10.1007/s11356-014-3735-5

Mahajan A, Arya A, Chundawat TS (2019) Green synthesis of silver nanoparticles using green alga (*Chlorella vulgaris*) and its application for synthesis of quinolines derivatives. Synth Commun 49(15):1926–1937. https://doi.org/10.1080/00397911.2019.1610776

Makarov V, Love A, Sinitsyna O, Makarova S, Yaminsky I, Taliansky M, Kalinina N (2014) "Green" nanotechnologies: synthesis of metal nanoparticles using plants. Acta Nat 6(1):35–44

Manikandan R, Manikandan B, Raman T, Arunagirinathan K, Prabhu NM, Basu MJ, Perumal M, Palanisamy S, Munusamy A (2015) Biosynthesis of silver nanoparticles using ethanolic petals extract of *Rosa indica* and characterization of its antibacterial, anticancer and anti-inflammatory activities. Spectrochim Acta A 138:120–129. https://doi.org/10.1016/j.saa.2014.10.043

Manjari G, Saran S, Arun T, Devipriya SP, Rao AVB (2017) Facile *Aglaia elaeagnoidea* mediated synthesis of silver and gold nanoparticles: antioxidant and catalysis properties. J Clust Sci 28(4):2041–2056. https://doi.org/10.1007/s10876-017-1199-8

Manjumeena R, Duraibabu D, Rajamuthuramalingam T, Venkatesan R, Kalaichelvan PT (2015) Highly responsive glutathione functionalized green AuNP probe for precise colorimetric detection of $Cd^{2+}$ contamination in the environment. RSC Adv 5(85):69124–69133. https://doi.org/10.1039/C5RA12427A

Masciangioli T, Zhang W-X (2003) Peer reviewed: environmental technologies at the nanoscale. Environ Sci Technol 37(5):102A–108A. https://doi.org/10.1021/es0323998

Massironi A, Morelli A, Grassi L, Puppi D, Braccini S, Maisetta G, Esin S, Batoni G, Della Pina C, Chiellini F (2019) Ulvan as novel reducing and stabilizing agent from renewable algal biomass:

application to green synthesis of silver nanoparticles. Carbohydr Polym 203:310–321. https://doi.org/10.1016/j.carbpol.2018.09.066

Mayer KM, Hafner JH (2011) Localized surface plasmon resonance sensors. Chem Rev 111(6):3828–3857. https://doi.org/10.1021/cr100313v

Mohammadlou M, Maghsoudi H, Jafarizadeh-Malmiri H (2016) A review on green silver nanoparticles based on plants: synthesis, potential applications and eco-friendly approach. Int Food Res J 23(2):446–463

Momin B, Rahman S, Jha N, Annapure US (2019) Valorization of mutant *Bacillus licheniformis* M09 supernatant for green synthesis of silver nanoparticles: photocatalytic dye degradation, antibacterial activity, and cytotoxicity. Bioprocess Biosyst Eng 42(4):541–553. https://doi.org/10.1007/s00449-018-2057-2

Mukherjee P, Senapati S, Mandal D, Ahmad A, Khan MI, Kumar R, Sastry M (2002) Extracellular synthesis of gold nanoparticles by the fungus *Fusarium oxysporum*. Chembiochem 3(5):461–463. https://doi.org/10.1002/1439-7633(20020503)3:5<461::AID-CBIC461>3.0.CO;2-X

Muthu K, Rathika C (2016) *Casuarina equisetifolia* leaf extract mediated biosynthesis of silver nanoparticles. J Nanosci Nanotechnol 2:166–168

Nadagouda MN, Iyanna N, Lalley J, Han C, Dionysiou DD, Varma RS (2014) Synthesis of silver and gold nanoparticles using antioxidants from blackberry, blueberry, pomegranate, and turmeric extracts. ACS Sustain Chem Eng 2(7):1717–1723. https://doi.org/10.1021/sc500237k

Nag S, Pramanik A, Chattopadhyay D, Bhattacharyya M (2018) Green-fabrication of gold nanomaterials using *Staphylococcus warneri* from *Sundarbans estuary*: an effective recyclable nanocatalyst for degrading nitro aromatic pollutants. Environ Sci Pollut Res 25(3):2331–2349. https://doi.org/10.1007/s11356-017-0617-7

Nair B, Pradeep T (2002) Coalescence of nanoclusters and formation of submicron crystallites assisted by *Lactobacillus* strains. Cryst Growth Des 2(4):293–298. https://doi.org/10.1021/cg0255164

Namvar F, Azizi S, Ahmad MB, Shameli K, Mohamad R, Mahdavi M, Tahir PM (2015) Green synthesis and characterization of gold nanoparticles using the marine macroalgae *Sargassum muticum*. Res Chem Intermed 41(8):5723–5730. https://doi.org/10.1007/s11164-014-1696-4

Nangia Y, Wangoo N, Goyal N, Shekhawat G, Suri CR (2009) A novel bacterial isolate *Stenotrophomonas maltophilia* as living factory for synthesis of gold nanoparticles. Microb Cell Factories 8(1):39–45. https://doi.org/10.1186/1475-2859-8-39

Narsaiah K, Jha SN, Bhardwaj R, Sharma R, Kumar R (2012) Optical biosensors for food quality and safety assurance—a review. J Food Sci Technol 49(4):383–406. https://doi.org/10.1007/s13197-011-0437-6

Nisbet E, Weiss R (2010) Top-down versus bottom-up. Science 328(5983):1241–1243. https://doi.org/10.1126/science.1189936

Onitsuka S, Hamada T, Okamura H (2019) Preparation of antimicrobial gold and silver nanoparticles from tea leaf extracts. Colloids Surf B Biointerfaces 173:242–248

Palomo J, Filice M (2016) Biosynthesis of metal nanoparticles: novel efficient heterogeneous nanocatalysts. Nano 6(5):84–99. https://doi.org/10.3390/nano6050084

Parial D, Pal R (2015) Biosynthesis of monodisperse gold nanoparticles by green alga *Rhizoclonium* and associated biochemical changes. J Appl Phycol 27(2):975–984. https://doi.org/10.1007/s10811-014-0355-x

Patel V, Berthold D, Puranik P, Gantar M (2015) Screening of cyanobacteria and microalgae for their ability to synthesize silver nanoparticles with antibacterial activity. Biotechnol Rep 5:112–119. https://doi.org/10.1016/j.btre.2014.12.001

Patra JK, Baek K-H (2016) Green synthesis of silver chloride nanoparticles using *Prunus persica* L. outer peel extract and investigation of antibacterial, anticandidal, antioxidant potential. Green Chem Lett Rev 9(2):132–142. https://doi.org/10.1080/17518253.2016.1192692

Pimprikar P, Joshi S, Kumar A, Zinjarde S, Kulkarni S (2009) Influence of biomass and gold salt concentration on nanoparticle synthesis by the tropical marine yeast *Yarrowia lipo-*

*lytica* NCIM 3589. Colloids Surf B Biointerfaces 74(1):309–316. https://doi.org/10.1016/j.colsurfb.2009.07.040

Pourreza N, Golmohammadi H, Naghdi T, Yousefi H (2015) Green in-situ synthesized silver nanoparticles embedded in bacterial cellulose nanopaper as a bionanocomposite plasmonic sensor. Biosens Bioelectron 74:353–359. https://doi.org/10.1016/j.bios.2015.06.041

Prema P, Iniya P, Immanuel G (2016) Microbial mediated synthesis, characterization, antibacterial and synergistic effect of gold nanoparticles using *Klebsiella pneumoniae* (MTCC-4030). RSC Adv 6(6):4601–4607. https://doi.org/10.1039/C5RA23982F

Priyadarshini E, Pradhan N, Panda P, Mishra B (2015) Biogenic unmodified gold nanoparticles for selective and quantitative detection of cerium using UV–vis spectroscopy and photon correlation spectroscopy (DLS). Biosens Bioelectron 68:598–603. https://doi.org/10.1016/j.bios.2015.01.048

Puente C, Gómez I, Kharisov B, López I (2019) Selective colorimetric sensing of Zn(II) ions using green-synthesized silver nanoparticles: *Ficus benjamina* extract as reducing and stabilizing agent. Mater Res Bull 112:1–8. https://doi.org/10.1016/j.materresbull.2018.11.045

Pušo J, Jonkuvienė D, Mačionienė I, Šalomskienė J, Jasutienė I, Kondrotas R (2014) Biosynthesis of silver nanoparticles using lingonberry and cranberry juices and their antimicrobial activity. Colloids Surf B Biointerfaces 121:214–221. https://doi.org/10.1016/j.colsurfb.2014.05.001

Ragam P, Mathew B (2019) Unmodified silver nanoparticles for dual detection of dithiocarbamate fungicide and rapid degradation of water pollutants. Int J Environ Sci Technol:1–14. https://doi.org/10.1007/s13762-019-02454-9

Raj S, Mali SC, Trivedi R (2018) Green synthesis and characterization of silver nanoparticles using *Enicostemma axillare* (Lam.) leaf extract. Biochem Biophys Res Commun 503(4):2814–2819. https://doi.org/10.1016/j.bbrc.2018.08.045

Rajesh S, Raja DP, Rathi J, Sahayaraj K (2012) Biosynthesis of silver nanoparticles using *Ulva fasciata* (Delile) ethyl acetate extract and its activity against *Xanthomonas campestris* pv. malvacearum. J Biopest 5:119–128

Ramakrishna M, Babu DR, Gengan RM, Chandra S, Rao GN (2016) Green synthesis of gold nanoparticles using marine algae and evaluation of their catalytic activity. J Nanostruct Chem 6(1):1–13. https://doi.org/10.1007/s40097-015-0173-y

Ramakritinan C, Kaarunya E, Shankar S, Kumaraguru A (2013) Antibacterial effects of Ag, Au and bimetallic (Ag-Au) nanoparticles synthesized from red algae. Solid State Phenom 201:211–230. https://doi.org/10.4028/www.scientific.net/SSP.201.211

Ramamoorthy R, Vanitha S, Krishnadev P (2019) Green synthesis of silver nanoparticles using red seaweed *Portieria hornemannii* (Lyngbye) PC silva and its antifungal activity against silkworm (*Bombyx mori* L.) muscardine pathogens. J Pharmacogn Phytochem 8(3):3394–3398

Rao C, Kulkarni G, Thomas PJ, Edwards PP (2002) Size-dependent chemistry: properties of nanocrystals. Chem Eur J 8(1):28–35. https://doi.org/10.1002/1521-3765(20020104)8:1<28::AID-CHEM28>3.0.CO;2-B

Rastogi L, Sashidhar R, Karunasagar D, Arunachalam J (2014) Gum kondagogu reduced/stabilized silver nanoparticles as direct colorimetric sensor for the sensitive detection of $Hg^{2+}$ in aqueous system. Talanta 118:111–117. https://doi.org/10.1016/j.talanta.2013.10.012

Rathod D, Golinska P, Wypij M, Dahm H, Rai M (2016) A new report of *Nocardiopsis valliformis* strain OT1 from alkaline Lonar crater of India and its use in synthesis of silver nanoparticles with special reference to evaluation of antibacterial activity and cytotoxicity. Med Microbiol Immunol 205(5):435–447. https://doi.org/10.1007/s00430-016-0462-1

Ravi SS, Christena LR, SaiSubramanian N, Anthony SP (2013) Green synthesized silver nanoparticles for selective colorimetric sensing of $Hg^{2+}$ in aqueous solution at wide pH range. Analyst 138(15):4370–4377. https://doi.org/10.1039/C3AN00320E

Ravindra B, Rajasab A (2014) A comparative study on biosynthesis of silver nanoparticles using four different fungal species. Int J Pharm Pharm Sci 6(1):372–376

Rivera-Rangel RD, González-Muñoz MP, Avila-Rodriguez M, Razo-Lazcano TA, Solans C (2018) Green synthesis of silver nanoparticles in oil-in-water microemulsion and nano-emulsion using geranium leaf aqueous extract as a reducing agent. Colloids Surf A Physicochem Eng Asp 536:60–67. https://doi.org/10.1016/j.colsurfa.2017.07.051

Rolim WR, Pelegrino MT, de Araújo LB, Ferraz LS, Costa FN, Bernardes JS, Rodigues T, Brocchi M, Seabra AB (2019) Green tea extract mediated biogenic synthesis of silver nanoparticles: characterization, cytotoxicity evaluation and antibacterial activity. Appl Surf Sci 463:66–74. https://doi.org/10.1016/j.apsusc.2018.08.203

Roy K, Sarkar CK, Ghosh CK (2015) Rapid colorimetric detection of $Hg^{2+}$ ion by green silver nanoparticles synthesized using *Dahlia pinnata* leaf extract. Green Process Synth 4(6):455–461. https://doi.org/10.1515/gps-2015-0052

Sadeghi B, Rostami A, Momeni S (2015) Facile green synthesis of silver nanoparticles using seed aqueous extract of *Pistacia atlantica* and its antibacterial activity. Spectrochim Acta A 134:326–332. https://doi.org/10.1016/j.saa.2014.05.078

Sakly N, Marzouk W, Ouada HB, Majdoub H (2017) Enhancing performances of colorimetric response of carboxymethylcellulose-stabilized silver nanoparticles: a fully eco-friendly assay for $Hg^{2+}$ detection. Sensors Actuatators B Chem 253:918–927. https://doi.org/10.1016/j.snb.2017.07.035

Samari F, Salehipoor H, Eftekhar E, Yousefinejad S (2018) Low-temperature biosynthesis of silver nanoparticles using mango leaf extract: catalytic effect, antioxidant properties, anticancer activity and application for colorimetric sensing. New J Chem 42(19):15905–15916. https://doi.org/10.1039/C8NJ03156H

Sangar S, Sharma S, Vats VK, Mehta S, Singh K (2019) Biosynthesis of silver nanocrystals, their kinetic profile from nucleation to growth and optical sensing of mercuric ions. J Clean Prod 228:294–302. https://doi.org/10.1016/j.jclepro.2019.04.238

Santhosh A, Sandeep S, Swamy NK (2019) Green synthesis of nano silver from euphorbia geniculata leaf extract: investigations on catalytic degradation of methyl orange dye and optical sensing of $Hg^{2+}$. Surf Interface 14:50–54. https://doi.org/10.1016/j.surfin.2018.11.004

Saratale RG, Karuppusamy I, Saratale GD, Pugazhendhi A, Kumar G, Park Y, Ghodake GS, Bharagava RN, Banu JR, Shin HS (2018a) A comprehensive review on green nanomaterials using biological systems: recent perception and their future applications. Colloids Surf B Biointerfaces 170:20–35. https://doi.org/10.1016/j.colsurfb.2018.05.045

Saratale RG, Saratale GD, Shin HS, Jacob JM, Pugazhendhi A, Bhaisare M, Kumar G (2018b) New insights on the green synthesis of metallic nanoparticles using plant and waste biomaterials: current knowledge, their agricultural and environmental applications. Environ Sci Pollut Res 25(11):10164–10183. https://doi.org/10.1007/s11356-017-9912-6

Sebastian M, Aravind A, Mathew B (2018) Green silver-nanoparticle-based dual sensor for toxic Hg(II) ions. Nanotechnology 29(35):355502. https://doi.org/10.1088/1361-6528/aacb9a

Sebastian M, Aravind A, Mathew B (2019) Green silver nanoparticles based multi-technique sensor for environmental hazardous Cu(II) ion. BioNanoScience 9(2):373–385. https://doi.org/10.1007/s12668-019-0608-x

Selvakumar R, Jothi NA, Jayavignesh V, Karthikaiselvi K, Antony GI, Sharmila P, Kavitha S, Swaminathan K (2011) As(V) removal using carbonized yeast cells containing silver nanoparticles. Water Res 45(2):583–592. https://doi.org/10.1016/j.watres.2010.09.034

Senapati S, Ahmad A, Khan MI, Sastry M, Kumar R (2005) Extracellular biosynthesis of bimetallic Au–Ag alloy nanoparticles. Small 1(5):517–520. https://doi.org/10.1002/smll.200400053

Senthilkumar P, Surendran L, Sudhagar B, Kumar DRS (2019) Facile green synthesis of gold nanoparticles from marine algae *Gelidiella acerosa* and evaluation of its biological potential. SN Appl Sci 1(4):284–295. https://doi.org/10.1007/s42452-019-0284-z

Seralathan J, Stevenson P, Subramaniam S, Raghavan R, Pemaiah B, Sivasubramanian A, Veerappan A (2014) Spectroscopy investigation on chemo-catalytic, free radical scavenging and bactericidal properties of biogenic silver nanoparticles synthesized using Salicornia brachiata aqueous extract. Spectrochim Acta A 118:349–355. https://doi.org/10.1016/j.saa.2013.08.114

Shah M, Fawcett D, Sharma S, Tripathy S, Poinern G (2015) Green synthesis of metallic nanoparticles via biological entities. Materials 8(11):7278–7308. https://doi.org/10.3390/ma8115377

Shankar SS, Ahmad A, Sastry M (2003) Geranium leaf assisted biosynthesis of silver nanoparticles. Biotechnol Prog 19(6):1627–1631. https://doi.org/10.1021/bp034070w

Sharma M, Yadav S, Ganesh N, Srivastava MM, Srivastava S (2019) Biofabrication and characterization of flavonoid-loaded Ag, Au, Au–Ag bimetallic nanoparticles using seed extract of the plant *Madhuca longifolia* for the enhancement in wound healing bio-efficacy. Prog Biomater 8(1):51–63. https://doi.org/10.1007/s40204-019-0110-0

Shen W, Qu Y, Li X, Pei X, You S, Yin Q, Wang J, Ma Q (2018) Comparison of gold nanoparticles biosynthesized by cell-free extracts of *Labrys*, *Trichosporon montevideense*, and *Aspergillus*. Environ Sci Pollut Res 25(14):13626–13632

Sheny D, Mathew J, Philip D (2012) Synthesis characterization and catalytic action of hexagonal gold nanoparticles using essential oils extracted from *Anacardium occidentale*. Spectrochim Acta A 97:306–310. https://doi.org/10.1016/j.saa.2012.06.009

Shore A (2018) Retraction: shape-specific silver nanoparticles prepared by microwave-assisted green synthesis using pomegranate juice for bacterial inactivation and removal. RSC Adv 8(69):39785–39785. https://doi.org/10.1039/C8RA90094A

Silva-De Hoyos LE, Sánchez-Mendieta V, Camacho-López MA, Trujillo-Reyes J, Vilchis-Nestor AR (2020) Plasmonic and fluorescent sensors of metal ions in water based on biogenic gold nanoparticles. Arab J Chem 13(1):1975–1985. https://doi.org/10.1016/j.arabjc.2018.02.016

Singaravelu G, Arockiamary J, Kumar VG, Govindaraju K (2007) A novel extracellular synthesis of monodisperse gold nanoparticles using marine alga, *Sargassum wightii* Greville. Colloids Surf B Biointerfaces 57(1):97–101. https://doi.org/10.1016/j.colsurfb.2007.01.010

Singh AK, Rathod V, Singh D, Ninganagouda S, Kulkarni P, Mathew J, Haq MU (2015) Bioactive silver nanoparticles from endophytic fungus *Fusarium* sp. isolated from an ethanomedicinal plant *Withania somnifera* (Ashwagandha) and its antibacterial activity. Int J Nanomater Biostruct 5:15–19

Singh K, Kukkar D, Singh R, Kukkar P, Kim K-H (2018a) Exceptionally stable green-synthesized gold nanoparticles for highly sensitive and selective colorimetric detection of trace metal ions and volatile aromatic compounds. J Ind Eng Chem 68:33–41. https://doi.org/10.1016/j.jiec.2018.07.026

Singh RK, Panigrahi B, Mishra S, Das B, Jayabalan R, Parhi PK, Mandal D (2018b) pH triggered green synthesized silver nanoparticles toward selective colorimetric detection of kanamycin and hazardous sulfide ions. J Mol Liq 269:269–277. https://doi.org/10.1016/j.molliq.2018.08.056

Sithara R, Selvakumar P, Arun C, Anandan S, Sivashanmugam P (2017) Economical synthesis of silver nanoparticles using leaf extract of *Acalypha hispida* and its application in the detection of Mn(II) ions. J Adv Res 8(6):561–568. https://doi.org/10.1016/j.jare.2017.07.001

Soshnikova V, Kim YJ, Singh P, Huo Y, Markus J, Ahn S, Castro-Aceituno V, Kang J, Chokkalingam M, Mathiyalagan R (2018) Cardamom fruits as a green resource for facile synthesis of gold and silver nanoparticles and their biological applications. Artif Cells Nanomed Biotechnol 46(1):108–117. https://doi.org/10.1080/21691401.2017.1296849

Srivastava SK, Yamada R, Ogino C, Kondo A (2013) Biogenic synthesis and characterization of gold nanoparticles by *Escherichia coli* K12 and its heterogeneous catalysis in degradation of 4-nitrophenol. Nanoscale Res Lett 8(1):70–78. https://doi.org/10.1186/1556-276X-8-70

Suganya KU, Govindaraju K, Kumar VG, Dhas TS, Karthick V, Singaravelu G, Elanchezhiyan M (2015) Blue green alga mediated synthesis of gold nanoparticles and its antibacterial efficacy against Gram positive organisms. Mater Sci Eng C 47:351–356. https://doi.org/10.1016/j.msec.2014.11.043

Suman T, Rajasree SR, Ramkumar R, Rajthilak C, Perumal P (2014) The green synthesis of gold nanoparticles using an aqueous root extract of *Morinda citrifolia* L. Spectrochim Acta A 118:11–16. https://doi.org/10.1016/j.saa.2013.08.066

Tagad C, Seo HH, Tongaonkar R, Yu YW, Lee JH, Dingre M, Kulkarni A, Fouad H, Ansari SA, Moh SH (2017) Green synthesis of silver nanoparticles using *Panax ginseng* root extract for the detection of $Hg^{2+}$. Sens Mat 29(2):205–215. https://doi.org/10.18494/SAM.2017.1475

Tang B, Liu J, Fan L, Li D, Chen X, Zhou J, Li J (2018) Green preparation of gold nanoparticles with *Tremella fuciformis* for surface enhanced Raman scattering sensing. Appl Surf Sci 427:210–218. https://doi.org/10.1016/j.apsusc.2017.08.008

Tao AR, Habas S, Yang P (2008) Shape control of colloidal metal nanocrystals. Small 4(3):310–325. https://doi.org/10.1002/smll.200701295

Thangamani N, Bhuvaneshwari N (2019) Green synthesis of gold nanoparticles using *Simarouba glauca* leaf extract and their biological activity of micro-organism. Chem Phys Lett 732:136587. https://doi.org/10.1016/j.cplett.2019.07.015

Tidke PR, Gupta I, Gade AK, Rai M (2014) Fungus-mediated synthesis of gold nanoparticles and standardization of parameters for its biosynthesis. IEEE Trans Nanobiosci 13(4):397–402. https://doi.org/10.1109/TNB.2014.2347803

Tripathi R, Gupta RK, Singh P, Bhadwal AS, Shrivastav A, Kumar N, Shrivastav B (2014) Ultrasensitive detection of mercury(II) ions in water sample using gold nanoparticles synthesized by *Trichoderma harzianum* and their mechanistic approach. Sensors Actuatators B Chem 204:637–646. https://doi.org/10.1016/j.snb.2014.08.015

Uddin I, Ahmad K, Khan AA, Kazmi MA (2017) Synthesis of silver nanoparticles using *Matricaria recutita* (Babunah) plant extract and its study as mercury ions sensor. Sens Biosens Res 16:62–67. https://doi.org/10.1016/j.sbsr.2017.11.005

Umamaheswari C, Lakshmanan A, Nagarajan N (2018) Green synthesis, characterization and catalytic degradation studies of gold nanoparticles against Congo red and methyl orange. J Photochem Photobiol 178:33–39. https://doi.org/10.1016/j.jphotobiol.2017.10.017

Vardhana J, Kathiravan G (2015) Biosynthesis of silver nanoparticles by endophytic fungi Pestaloptiopsis pauciseta isolated from the leaves of *Psidium guajava* Linn. Int J Pharm Sci Rev Res 31(1):29–31

Varshney R, Bhadauria S, Gaur M (2010) Biogenic synthesis of silver nanocubes and nanorods using sundried *Stevia rebaudiana* leaves. Adv Mater Lett 1(3):232–237. https://doi.org/10.5185/amlett.2010.9155

Velayutham K, Rahuman AA, Rajakumar G, Roopan SM, Elango G, Kamaraj C, Marimuthu S, Santhoshkumar T, Iyappan M, Siva C (2013) Larvicidal activity of green synthesized silver nanoparticles using bark aqueous extract of *Ficus racemosa* against *Culex quinquefasciatus* and *Culex gelidus*. Asian Pac J Trop Med 6(2):95–101. https://doi.org/10.1016/S1995-7645(13)60002-4

Velmurugan P, Anbalagan K, Manosathyadevan M, Lee K-J, Cho M, Lee S-M, Park J-H, Oh S-G, Bang K-S, Oh B-T (2014) Green synthesis of silver and gold nanoparticles using *Zingiber officinale* root extract and antibacterial activity of silver nanoparticles against food pathogens. Bioprocess Biosyst Eng 37(10):1935–1943. https://doi.org/10.1007/s00449-014-1169-6

Vivekanandhan S, Misra M, Mohanty AK (2009) Biological synthesis of silver nanoparticles using *Glycine max* (soybean) leaf extract: an investigation on different soybean varieties. J Nanosci Nanotechnol 9(12):6828–6833. https://doi.org/10.1166/jnn.2009.2201

Wadhwani SA, Shedbalkar UU, Singh R, Vashisth P, Pruthi V, Chopade BA (2016) Kinetics of synthesis of gold nanoparticles by *Acinetobacter* sp. SW30 isolated from environment. Indian J Microbiol 56(4):439–444. https://doi.org/10.1007/s12088-016-0598-0

Waghmare SS, Deshmukh AM, Sadowski Z (2014) Biosynthesis, optimization, purification and characterization of gold nanoparticles. Afr J Microbiol Res 8(2):138–146. https://doi.org/10.5897/AJMR10.143

Wang C, Kim YJ, Singh P, Mathiyalagan R, Jin Y, Yang DC (2016a) Green synthesis of silver nanoparticles by *Bacillus methylotrophicus*, and their antimicrobial activity. Artif Cells Nanomed Biotechnol 44(4):1127–1132. https://doi.org/10.3109/21691401.2015.1011805

Wang R, Yao Y, Shen M, Wang X (2016b) Green synthesis of Au@ Ag nanostructures through a seed-mediated method and their application in SERS. Colloids Surf A Physicochem Eng Asp 492:263–272. https://doi.org/10.1016/j.colsurfa.2015.11.076

Wei H, Abtahi SMH, Vikesland PJ (2015) Plasmonic colorimetric and SERS sensors for environmental analysis. Environ Sci Nano 2(2):120–135. https://doi.org/10.1039/C4EN00211C

Willets KA, Van Duyne RP (2007) Localized surface plasmon resonance spectroscopy and sensing. Annu Rev Phys Chem 58:267–297. https://doi.org/10.1146/annurev.physchem.58.032806.104607

Xue B, He D, Gao S, Wang D, Yokoyama K, Wang L (2016) Biosynthesis of silver nanoparticles by the fungus *Arthroderma fulvum* and its antifungal activity against genera of *Candida*, *Aspergillus* and *Fusarium*. Int J Nanomed 11:1899–1906. https://doi.org/10.2147/IJN.S98339

Yang N, WeiHong L, Hao L (2014) Biosynthesis of Au nanoparticles using agricultural waste mango peel extract and its in vitro cytotoxic effect on two normal cells. Mater Lett 134:67–70. https://doi.org/10.1016/j.matlet.2014.07.025

Yugandhar P, Haribabu R, Savithramma N (2015) Synthesis, characterization and antimicrobial properties of green-synthesised silver nanoparticles from stem bark extract of *Syzygium alternifolium* (Wt.) Walp. 3 Biotech 5(6):1031–1039. https://doi.org/10.1007/s13205-015-0307-4

Zahir AA, Rahuman AA (2012) Evaluation of different extracts and synthesised silver nanoparticles from leaves of *Euphorbia prostrata* against *Haemaphysalis bispinosa* and *Hippobosca maculata*. Vet Parasitol 187(3–4):511–520. https://doi.org/10.1016/j.vetpar.2012.02.001

Zayed MF, Eisa WH (2014) *Phoenix dactylifera* L. leaf extract phytosynthesized gold nanoparticles; controlled synthesis and catalytic activity. Spectrochim Acta A 121:238–244. https://doi.org/10.1016/j.saa.2013.10.092

# Ionic Liquids Modified Sensors and Biosensors for Detection of Environmental Contaminants

**Amina Saleem, Abdur Rahim, Nawshad Muhammad, and Fatima Abbas**

## Contents



## 1 Introduction

Over the ongoing decades, the world has encountered the antagonistic results of uncontrolled advancement of different human activities in, for instance, industry, transportation, agribusiness, and urbanization. The expansion in expectations for

A. Saleem
Interdisciplinary Research Centre in Biomedical Materials (IRCBM), COMSATS University Islamabad, Lahore, Pakistan

Department of Physics, COMSATS University Islamabad, Lahore, Pakistan
e-mail: aminasaleem@cuilahore.edu.pk

A. Rahim (✉) · N. Muhammad
Interdisciplinary Research Centre in Biomedical Materials (IRCBM), COMSATS University Islamabad, Lahore, Pakistan
e-mail: abdurrahim@cuilahore.edu.pk; nawshadmuhammad@cuilahore.edu.pk

F. Abbas
Department of Chemical Engineering, COMSATS University Islamabad, Lahore, Pakistan



Inamuddin, A. M. Asiri (eds.), *Nanosensor Technologies for Environmental Monitoring*, Nanotechnology in the Life Sciences,
https://doi.org/10.1007/978-3-030-45116-5_9

everyday comforts and higher buyer request have intensified contamination of the air with, for instance, $CO_2$ and other ozone-harming substances, NOx, $SO_2$, and particulate matter, of water with an assortment of synthetic concoctions, supplements, leachates, oil slicks, among others, and of the dirt because of the transfer of dangerous squanders, spreading of pesticides, slime, just as the utilization of expendable merchandise or nonbiodegradable materials and the absence of appropriate facilities for waste (das et al. 2007).

Incredible advances have been made in the examination of following poisons during ongoing decades, because of the improvement and refinement of explicit systems (Mauriz et al. 2006). A wide exhibit of undetected contaminants of developing ecological concern should be distinguished and measured in different natural parts and organic tissues. These toxins might be versatile and industrious in air, water, soil, silt, and environmental receptors even at low concentrations (Pichon and Chapuis-Hugon 2008).

The detecting devices for utilizing organic, ecological checking, procedure control, and pharmaceutical has been a thrust research domain in the ongoing years (Boisen et al. 2000; Karimi-Maleh et al. 2014c; Zhao et al. 2015). In this regard, electrochemical-based sensors satisfy enormous quantities of the necessities for such endeavors, particularly inferable from their ease of arranging, high selectivity and affectability, and speedy response (Beitollahi et al. 2014a; Karimi-Maleh et al. 2014a; Karimi-Maleh et al. 2015; Kazemi et al. 2016; Tajik et al. 2013, 2014).

The ionic conductivity of ILs can be modified for electrochemical applications because of large possibilities to select cations and anions (Beitollah et al. 2012; Bijad et al. 2013; Mokhtari et al. 2012). Owing to their excessive ionic conductivity, large electrochemical aperture, ionic liquids have been used as electrolytes, solvents, and foils in the electrochemical arrays of modified electrodes (Ensafi et al. 2012; Kakhki et al. 2013; Najafi et al. 2014; Sun et al. 2008; Tavana et al. 2012). The ionic liquids based sensors have been described for the direct movement of electrons of different types of electroactive mixtures, such as catechol, resorcinol, 4-aminophenol, hydrazine, ascorbic acid, folic acid, catecholamines, and Sudan I (Aguilar-Arteaga et al. 2010; Elyasi et al. 2013; Jamali et al. 2014; Karimi-Maleh et al. 2014b; Najafi et al. 2014; Zhu et al. 2010). As a result, it has been suggested that the use of ionic liquids can increase the reliability of the electrochemical reaction and promote the efficient displacement of direct electrons from various natural, pharmaceutical, and environmental compounds (Karimi-Maleh et al. 2015; Safavi et al. 2006; Sanati et al. 2014; Sun et al. 2009).

## 2 What Are Sensors and Biosensors

Sensors: “Sensor” is a source of the Latin word “feeling” that essentially means “to make a difference.” The first thing that comes to our minds when we hear this word sensor is the idea of the five basic human detectors: taste perception, eye perception, smell perception, visual perception, and sound perception. The work system of these facilities is reduced to collecting the information signal by the touch cells in

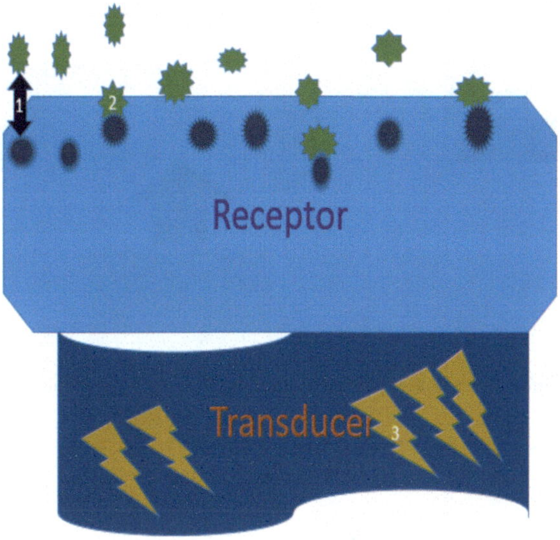


**Fig. 1** Graphical illustration of the biosensors process. (1) The analyte is enticed to the sites of the receptors. (2) Chemical reciprocity between the analyte generating an electrical signal. (3) The transducer converts the electrical signal into the processor

the light of external updates forming the interoperability herd. With this precise clarification of the meaning, a progressively efficient and specialized definition of the sensor could be elaborated as follows: It is a device that is stimulated and initiated from the ground and reacts to it (Schmid and Verger 1998) (Fig. 1).

Biosensor: Biosensors are systematic devices that coordinate a biocomponent/bioreceptor. As included chemicals, organelles, whole cells, tissues, immune systems, nucleic acids, aptamers, etc. with a suitable for distinguishing chemical compounds transduction structure. Normal transducers are mass-based electrochemical, optical transducers, such as piezoelectric transducers, surface acoustic transducers, and thermistors. Due to the special cooperation between the target particle and the biocomponent, an electrical signal is usually generated which can be estimated and recorded. Biosensors undoubtedly have favorable circumstances: selectivity to directly localize the analyte without the need for a pretreatment test or a negligible example of pretreatment, rapid examination with results in just a few minutes, low cost, prospects for reduction and versatility. Biosensors are far from difficult to operate and do not require well-prepared workers. In this way, the gadgets that are accessible to the industry can be promoted effectively in the advertising of the buyers (Turner et al. 1987) (Fig. 2).

## 3 What Are Ionic Liquids (ILs)

Recently, ionic liquids (IL) have become extensively popular for electrochemical and colorimetric studies in sensing applications (Barnes et al. 2008; Xiao et al. 2015; Zhao et al. 2010). ILs are salts that consist of cation and anion and melts up to 100 °C or below (Li et al. 2014; Liu et al. 2012; Papagni et al. 2011; Rogers and Seddon 2003; Ruiz et al. 2014).

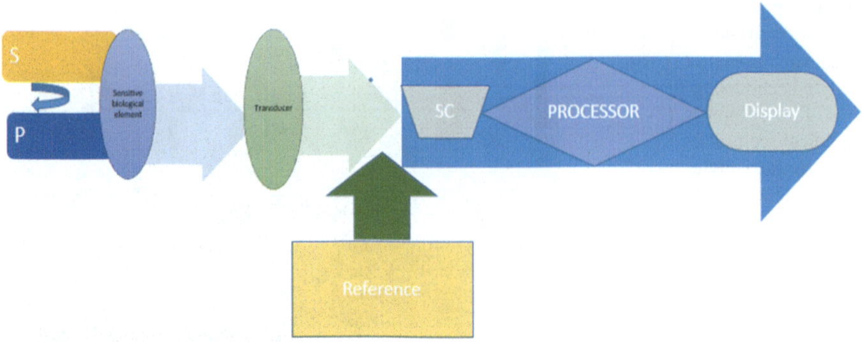


**Fig. 2** Graphical representation of the biosensor components

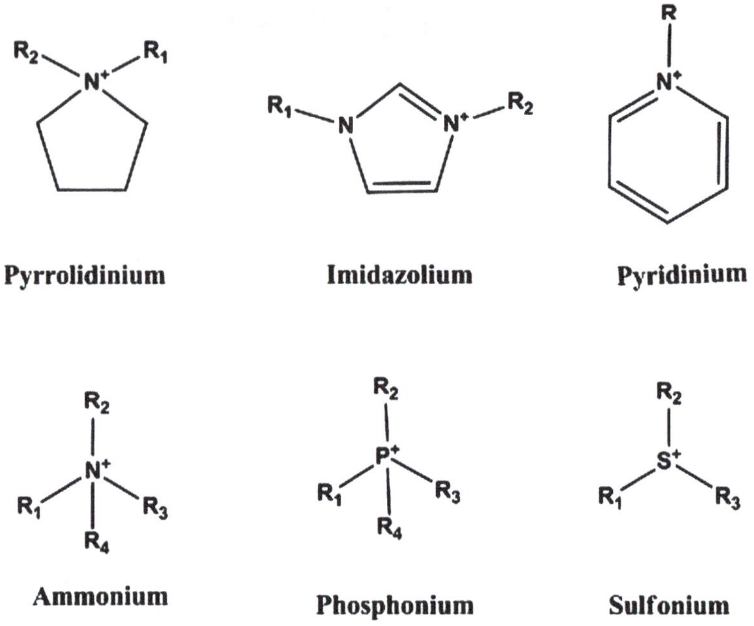


The properties of ILs that are particularly interested in scientists include large electrochemical windows, high conductivity, heat resistance, and low volatility (Ibrahim et al. 2016). It is possible to hypothetically combine $10^{18}$ ILs (Carmichael and Seddon 2000). In addition, the ILs are very customizable and can be structured according to a specific obligation. Therefore, the term "architectural solvent" is used to describe it (Cremer 2013; Ibrahim et al. 2016). It is conceivable to prepare liquid reagents at room temperature and below with a careful decision on the starting materials [39].

## *3.1 Structure of Ionic Liquids*

Ionic liquids include natural or synthetic cations and anions that may be natural or inorganic. Most used ionic liquids are prepared using the metathesis (Papagni et al. 2011) and acid-base neutralization (Kohagen et al. 2011).

### 3.1.1 Cations

The cation structure contains mainly charged nitrogen or phosphorus. The model's cations are imidazolium, pyridinium, pyrrolidinium, ammonium, sulfonium, or phosphonium. The properties of ionic liquids, including, consistency, solubility, and softening point in solvents, are subject to variations in structure as well as the regulation of the alkyl chain length on cation [22]. The structures of some of the major cations considered are shown in Scheme 1.

### 3.1.2 Anions

The anions of ionic liquids can be inorganic or organic and generally have a diffuse or certainly negative charge [72]. The structures of some of the most important cationic structures are shown in Scheme 2.

# 4 Applications of Ionic Liquids-Based Materials

Ionic liquid paying great attention in the field of sensors and biosensors due to their excellent properties, for example, negligible vapor pressure, high thermal, large liquidus range and chemical stability, low toxicity, good conductivity, wide electrochemical window, and dissolving capacity (Hu et al. 2012; Liu et al. 2010; Safavi et al. 2009). Ionic liquids (ILs) display complex behavior. Their simultaneous double nature as electrolyte and solvents carriers the existence of cations and anions with a structural agreement, which could form the basis of a new detection technology (Rehman and Zeng 2012). ILs have been extensively used as electrolytes in different electrochemical devices, e.g. photovoltaic cells, batteries, fuel cells, capacitors, and electroplating (Noda et al. 2003). Similarly, ionic liquids have many applications in the area of sensors and biosensors.

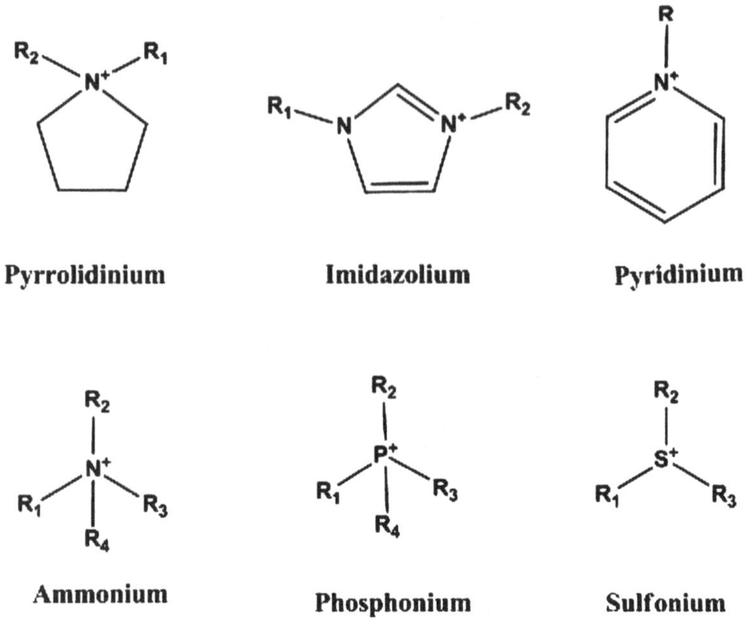


**Scheme 1** Cations in ionic liquids

# 5 Ionic Liquid Modified Materials for the Detection of Environmental Contaminants

## *5.1 Detection of Phenolic Compounds*

An extensive number of natural pollutants, generally circulated all through the earth, have an aromatic structure (Ortega et al. 1994). Phenol and substituted phenols, for example, chlorinated phenols and related sweet-smelling compounds are referred to be boundless as parts of modern waste (Barceló 1993). A lot of these phenolic compounds affect animal creatures and plants of which they effectively infiltrate the skin and layers, determining an expansive range of nontoxic, mutagenic, and hepatotoxic impacts, additionally influencing the bio catalyzed response rates in-breath and photosynthesis (Castillo et al. 1997; Yager et al. 1990). Simultaneous with the pattern in increasingly stringent contamination declining practices have come to the requirement for progressively solid strategies for determining phenolic-type compounds in the wastewater. These composites are available in the waste gushing waters from the plastics, natural chemical, steel, and oil businesses. They are viewed as toxins since they cause augmentation in oxygen demand and a terrible preference for consumable water supplies. Chlorophenols are delivered when phenol defiled water is chlorinated in the process of decontamination. These subordinates are offensive even in the low concentration in parts per billion (Mohler and Jacob 1957). Zhao et al. fabricated a phenol biosensor consist of polyaniline (PAN)/1-ethyl-3-methylimidazolium ethyl

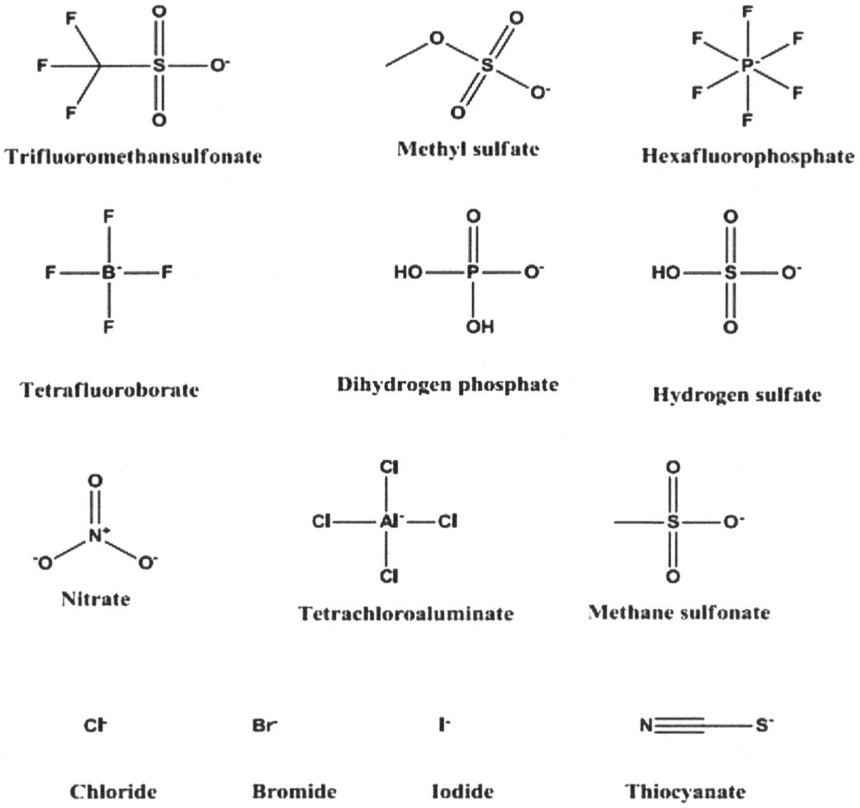


**Scheme 2** Anions in ionic liquids

sulfate ([Emim][ES])/carbon nanofiber (CNF) composite arranged by one-advance electropolymerization of aniline in the existence of [Emim][ES]. The PAN/[Emim][ES]/CNF composite film was electrochemically fabricated on the glassy carbon electrode (GCE) surface by applying a potential in the range of −0.2 to 0.9 V potential, in a mixture consisting of 2.0 M $H_2SO_4$ aniline, carbon nanofiber, and [Emim][ES] at scan rate of 60 $mVs^{-1}$. Then, the tyrosinase mixture was added to the outside surface of the PAN/[Emim][ES]/CNF composite film, followed by the addition of glutaraldehyde as a cross-linker. This anode displayed delicate amperometric response to the phenolic compounds such as p-cresol, catechol, phenol, and m-cresol. Compared with CNT, CNF has characteristic focal points, for example, lower cost, large scale manufacturing, better mechanical properties, and ease of surface functionalization (Zhang et al. 2009). Hong et al. used graphitic mesoporous carbon (GMC) which was fabricated by utilizing SBA-15 as a substrate and ionic liquid (IL) used as a source of carbon. A uniform and balanced GMC/BMIMPF6 composite film was developed on the electrode surface. For the simultaneous detection of catechol and hydroquinone, GMC/BMIMPF6 modified the glassy carbon (GC) electrode shows two redox couple peaks, both are well defined and independent of each other.

The redox peak current value in the case of GMC/BMIMPF$_6$/GC electrode was three folds higher than the GC/GMC electrode. It is evident from these results that GMC/IL composite film gave procedure better response for the determination of hydroquinone (HQ) and catechol (CC) (Hong et al. 2013). Gurban et al fabricated a modest and moderate biosensor using a screen-printed electrode for sensitive and selective determination of alkylphenols. The SPE electrode was modified with horseradish peroxidase (HRP) enzyme, ionic liquid (1-butyl-3-methylimidazolium hexafluorophosphate ([BMIM][PF6]), and single-walled carbon nanotubes (SWCNTs). The interaction between ionic liquid and SWCNTs was determined by FTIR, Raman spectroscopy, and cyclic voltammetry. The HRP-based biosensor shows high selectivity and great stability, permitting identification of the alkylphenols at potential of −0.2 V, 4-t-octylphenol detected in range between 5.5 and 97.7 μM and similarly, 4-n-nonylphenol was detected between 5.5 and 140 μM. The developed biosensors permitted a fast detection of alkylphenols and utilized further for real sample analysis of endocrine disrupters in food and water samples (Gurban et al. 2011). Li et al reported a method for the detection of bisphenol A in which CMK-3 modified nanocarbon ionic liquid paste was used as a working electrode. The surface morphology and nanostructure of CMK-3 and modified electrodes were determined using different characterization techniques such as SEM, XRD, TEM, and Raman. Electrochemical analysis of the modified electrode was studied through different electrochemical techniques i.e. cyclic voltammetry (CV) and electrochemical impedance spectroscopy (EIS). Owing to the great catalytic activity and high conductivity of ionic liquid modified electrode showed a very good electrochemical response toward bisphenol A (BPA), with increased oxidation peak current (Li et al. 2016). Bu et al developed a sensor for simultaneous detection of catechol and Hydroquinone utilizing an MWNTs-IL-Gel/GCE as a working electrode, Electrochemical analysis was conducted by using cyclic voltammetry and differential pulse voltammetry, modified electrode showed two independent well-defined oxidation–reduction peaks for both hydroquinone and catechol. High selectivity of $1.8 \times 10^{-7}$ M was attained for CC and HQ and detection limit of $6.0 \times 10^{-8}$ M and $6.7 \times 10^{-8}$, respectively. So, this method is applicable for the selective detection of phenolic compounds (Bu et al. 2011). Beitollahi et al reported a graphene-based electrode modified with benzoylferrocene. In addition, HMIMPF$_6$ (*n*-hexyl-3-methylimidazolium hexafluorophosphate) ionic liquid was utilized as a binder for the preparation of electrode modification. The redox phenomena of sulfite at the modified electrode surface were examined using electrochemical techniques. Compared to the bare electrode, the modified electrode shows good response and gave better sensitivity against phenol and sulfite. Owing to good selectivity and sensitivity, the graphene nano-sheets and ionic liquid (IL) modified electrode was successfully applied for the electrochemical detection of phenol and sulfite in real samples (Beitollahi et al. 2014b).

A very sensitive method was reported by Jie Ren for the detection of 4-nonylphenol at the glassy carbon electrode modified with a composite of gold nanoparticles/poly (ionic liquids) hollow nanospheres. The results demonstrated that there is a linear relationship between peak current and 4-NP concentrations in the range of 0.1–120 μM with a detection limit of $3.3 \times 10^{-8}$ M. This technique has a lower identification limit and more extensive linear range. The effects of interfering species on detection were studied and there were no changes in the results of 4-NP determination. Furthermore, the proposed assay has been used effectively for the detection of 4-NP in water samples (Ren et al. 2015).

## 5.2 Detection of Heavy Metals

The heavy metal pollution is the source of various diseases in the world, such as Minamata disease (organic mercury intoxication), iItai-Itai disease (cadmium disease), arsenic acid poisoning, and asthma-related to air pollution. Marine biological systems are very intricate, dynamic, and subject to numerous inside and outer connections that are liable to change after some time (Baby et al. 2010). Pollutants that enter the coastal waters and estuaries make significant issues making extensive harm the life and exercises of the living amphibian creatures and even to mass mortality. From pollutants, a gathering of heavy metals in marine environments is of worldwide significance. Important contributions have been made with reference to ocean and coastal spreading of different heavy metals (Förstner and Wittmann 2012).

A general overview of various ionic liquids modified sensors for detections of different analytes has been tabulated in Table 1. Some of the literature, like Bagheri et al., reported an electrochemical sensor for the detection of lead, mercury, and thallium. New composite materials which consist of ionic liquid 1-n-octyl pyridinium hexafluorophosphate (OPFP), graphene, and phosphorus ylide were used for the fabrication of electrode. The surface topography and electrochemical properties of the synthesized materials were confirmed using SEM, XRD, FT-IR, CV, and UV Vis. The fabricated sensor is of great significance due to selective and sensitive detection of Pb, Ti, and Hg simultaneously. Square wave voltammetry technique was employed to examine the analytical performance of the electrode. Hg, Pb, and Ti determined in the range from $1.25 \times 10^{-9}$ to $2.0\times 10^{-7}$ mol $L^{-1}$ (Bagheri et al. 2015). Li et al. have been fabricated a sensor for the determination of cadmium and lead in which hydroxyapatite-modified carbon ionic liquid electrode was. Hydroxyapatite in combination with ionic liquid shows the promising response for metals detections. Square-wave anodic stripping voltammetry (SWASV) was performed for analysis of certain heavy metals. Oxidation of two metals gave a separate and well-defined oxidation peak. Peak currents were measured at about −0.88 V for $Cd^{2+}$ and −0.34 V for $Pb^{2+}$. In this way, the proposed modified electrode was used potentially for the determination of metals and other toxins in water and medicinal products (Li et al. 2009). Bagheri et al. fabricated an electrode from a composite of triphenylphosphine, carbon nanotubes, and room temperature ionic liquid used as a binder and employed for the simultaneous and sensitive determination of $Cd^{2+}$, $Pb^{2+}$, and $Hg^{2+}$. The properties and surface morphology of the electrode have been characterized by scanning electron microscopy (SEM), and electrochemical impedance spectroscopy (EIS). Square-wave anodic stripping voltammetry (SWASV) was used to check the electrochemical response of modified electrode (Bagheri et al. 2013).

**Table 1** Ionic liquid modified sensors for various analyte detections

| Electrode | Method | Analyte | Detection limit | Linear range (M) | Ref. |
|---|---|---|---|---|---|
| GMC/BMIMPF6/GC | CV | Hydroquinone (HQ) and catechol(CC) | $5 \times 10^{-8}$ mol $L^{-1}$ (HQ) $6 \times 10^{-8}$ mol $L^{-1}$ (CC) | $1.0 \times 10^{-7}$ to $5.0 \times 10^{-5}$ (HQ) $1.0 \times 10^{-7}$ to $5.0 \times 10^{-5}$(CC) | Hong et al. (2013) |
| SWCNTs/([BMIM][PF6]/HRP | CV and EIS | Phenolic compounds | 1.1 μM for 4-t-octylphenol and 0.4 μM for 4-n-nonylphenol | 5.5 to 97.7 μM for 4-t-octylphenol and 5.5 and 140 μM for 4-n-nonylphenol | Gurban et al. (2011) |
| CMK-3/nano-CILPE | LSV | Bisphenol A | 0.05 μM | 0.2–150 μM | Li et al. (2016) |
| MWNTs-IL-Gel/GCE | DPV | Catechol (CC) and Hydroquinone (HQ) | $6.7 \times 10^{-8}$ M (HQ) and $6.0 \times 10^{-8}$ M (CC) | $1.8 \times 10^{-7}$ M | Bu et al. (2011) |
| BF/IL/GPE | SWV | Sulfite and phenol | 20.0 nM for sulfite | $5.0 \times 10^{-8}$ to $2.5 \times 10^{-4}$ M | Beitollahi et al. (2014b) |
| AuNPs/PILs/GCE | DPASV | 4-Nonylphenol | 0.033 μM | 0.1–120 μM | Ren et al. (2015) |
| Tyr/PANI–IL–CNF/GCE | SWASV | p-Cresol, m-Cresol, phenol and catechol | 0.1, 0.5, 0.1 and 0.1 nM respectively | $4.0 \times 10^{-10}$–$2.0 \times 10^{-6}$, $1.0 \times 10^{-9}$–$6.6 \times 10^{-6}$, $4.0 \times 10^{-10}$–$1.9 \times 10^{-6}$ & $4.0 \times 10^{-10}$–$2.1 \times 10^{-6}$ (M) respectively | Zhang et al. (2009) |
| IL/Gr/L/CPE | SWASV | $Tl^+$, $Pb^{2+}$, and $Hg^{2+}$ | $3.57 \times 10^{-10}$ mol $L^{-1}$ for $Tl^+$, $4.50 \times 10^{-10}$ mol $L^{-1}$ for $Pb^{2+}$ and $3.86 \times 10^{-10}$ mol $L^{-1}$ for $Hg^{2+}$ | $1.25 \times 10^{-9}$–$2.00 \times 10^{-7}$, $1.25 \times 10^{-9}$–$2.00 \times 10^{-7}$ and $1.25 \times 10^{-9}$–$2.00 \times 10^{-7}$ respectively | Bagheri et al. (2015) |
| HAP-CILE | SWASV | Cd and Pb | 1.57 μM for $Pb^{2+}$ and 2.93 μM for $Cd^{2+}$ | 0.001–0.1 μM for Pb and 0.0005 μM for Cd | Li et al. (2009) |
| PPh3/MWCNTs/IL/CPE | SWASV | Pb, Hg, and Cd | $6.0 \times 10^{-5}$ μM for Pb, $9.2 \times 10^{-5}$ μM for Hg and $7.4 \times 10^{-5}$ μM for Cd | $1 \times 10^{-4}$ to 0.15 μM for Pb, $1 \times 10^{-4}$ to 0.15 μM for Hg and $1 \times 10^{-4}$ to 0.15 μM for Cd | Bagheri et al. (2013) |
| Bi/GR/IL-SPE | SWASV | Pb, Cd | 0.08 μg $L^{-1}$for Cd 0.10 μg $L^{-1}$for Pb | 1–80 μg $L^{-1}$ | Liu et al. (2010) |

Wang et al. reported a new Bi/GR/IL modified SPE (screen printed electrode) which was fabricated using electrochemical and physicochemical methods and then used for the sensitive detection of Cd(II) and Pb(II) in rice samples. Because of the combined unique properties of IL, GR, and bismuth film, the fabricated electrode showed a few favorable benefits over conventional SPE. In addition, the simple and environmentally friendly procedure of preparation greatly extends the scope of research to mass production of "mercury-free" disposable sensors for heavy metal analysis, which is very promising for its global application, in biological, environmental, and food examination (Wang et al. 2014).

## 5.3 *Determination of Pesticides*

In recent years, thepesticides determination has become vital due to the extensive use of these compounds in agricultural fields. Owing to the widespread applications of pesticides, various enzyme-based sensors such as choline oxidase, peroxidase, tyrosinase, LAC, acetylcholinesterase, and other oxidoreductase have been developed for their determination, (De Castro and Herrera 2003; Marques and Yamanaka 2008; Rodriguez-Mozaz et al. 2004). Methomyl ($C_5H_{10}N_2O_2S$), S-methyl-1-N-[(methylcarbamoyl) oxythioacetimidate is an insecticide of the carbamate pesticide group were applied on agricultural crops protection. These have a high toxic effect because of its ready solubility in aqueous medium and are extremely dangerous to the environment. (Tomašević et al. 2010). The integration of an ILs with solid support has been shown to be a good system for the natural and synthetic enzyme immobilization. This new technology, referred to as the supporting ionic liquids phase (SILP) catalyst, amalgamate the benefits of ionic liquids with those of a heterogeneous material carried in the solid phase, resulting in low toxicity materials that are environmentally friendly (Dupont et al. 2002; Mehnert et al. 2002; Riisagera et al. 2006). In these materials preparation, ILs are covalently attached to a support surface or simply by depositing the ionic liquid on active surface. These supports are generally silicates, polymer, or clay minerals (Castillo et al. 2007; Gelesky et al. 2009; Kim et al. 2006). Platinum-based SILP nanoparticles and ionic liquids (1-butyl-3-methylimidazolium tetrafluoroborate) (Pt-BMI · $BF_4$) were deposited on MMT clay and used as a matrix for the immobilization of LAC (*Aspergillus oryzae*). This sensor was tested for the detection of Methomylus. Furthermore, the optimized biosensor was used to determine methomyl in carrot and tomato samples. The results attained with the proposed method were consistent with those obtained by the HPLC method.

# 6 Conclusions

In conclusion, the electrode modified with ionic liquids is employed preciously as sensors and biosensors for the determination of environmental contaminants i.e. phenolic compounds, heavy metals, and nitro compounds and hydrazine.

# References

Aguilar-Arteaga K, Rodriguez J, Barrado E (2010) Magnetic solids in analytical chemistry: a review. Anal Chim Acta 674(2):157–165

Baby J, Raj JS, Biby ET, Sankarganesh P, Jeevitha M, Ajisha S, Rajan SS (2010) Toxic effect of heavy metals on aquatic environment. Int J Biol Chem Sci 4(4)

Bagheri H, Afkhami A, Khoshsafar H, Rezaei M, Shirzadmehr A (2013) Simultaneous electrochemical determination of heavy metals using a triphenylphosphine/MWCNTs composite carbon ionic liquid electrode. Sensors Actuators B Chem 186:451–460

Bagheri H, Afkhami A, Khoshsafar H, Rezaei M, Sabounchei SJ, Sarlakifar M (2015) Simultaneous electrochemical sensing of thallium, lead and mercury using a novel ionic liquid/graphene modified electrode. Anal Chim Acta 870:56–66

Barceló D (1993) Environmental analysis: techniques, applications, and quality assurance. Elsevier Science; 1 Edition, p. 646

Barnes AS, Rogers EI, Streeter I, Aldous L, Hardacre C, Wildgoose GG, Compton RG (2008) Unusual voltammetry of the reduction of $O_2$ in [C4dmim][N(Tf)$_2$] reveals a strong interaction of $O_2^{\bullet-}$ with the [C4dmim] + cation. J Phys Chem C 112(35):13709–13715

Beitollah H, Goodarzian M, Khalilzadeh MA, Karimi-Maleh H, Hassanzadeh M, Tajbakhsh M (2012) Electrochemical behaviors and determination of carbidopa on carbon nanotubes ionic liquid paste electrode. J Mol Liq 173:137–143

Beitollahi H, Taher MA, Ahmadipour M, Hosseinzadeh R (2014a) Electrocatalytic determination of captopril using a modified carbon nanotube paste electrode: application to determination of captopril in pharmaceutical and biological samples. Measurement 47:770–776

Beitollahi H, Tajik S, Biparva P (2014b) Electrochemical determination of sulfite and phenol using a carbon paste electrode modified with ionic liquids and graphene nanosheets: application to determination of sulfite and phenol in real samples. Measurement 56:170–177

Bijad M, Karimi-Maleh H, Khalilzadeh MA (2013) Application of ZnO/CNTs nanocomposite ionic liquid paste electrode as a sensitive voltammetric sensor for determination of ascorbic acid in food samples. Food Anal Methods 6(6):1639–1647

Boisen A, Thaysen J, Jensenius H, Hansen O (2000) Environmental sensors based on micromachined cantilevers with integrated read-out. Ultramicroscopy 82(1–4):11–16

Bu C, Liu X, Zhang Y, Li L, Zhou X, Lu X (2011) A sensor based on the carbon nanotubes-ionic liquid composite for simultaneous determination of hydroquinone and catechol. Colloids Surf B Biointerfaces 88(1):292–296

Carmichael AJ, Seddon KR (2000) Polarity study of some 1-alkyl-3-methylimidazolium ambient-temperature ionic liquids with the solvatochromic dye, Nile Red. J Phys Org Chem 13(10):591–595

Castillo M, Domingues R, Alpendurada M, Barcelo D (1997) Persistence of selected pesticides and their phenolic transformation products in natural waters using off-line liquid solid extraction followed by liquid chromatographic techniques. Anal Chim Acta 353(1):133–142

Castillo MR, Fousse L, Fraile JM, García JI, Mayoral JA (2007) Supported ionic-liquid films (SILF) as two-dimensional nanoreactors for enantioselective reactions: surface-mediated selectivity modulation (SMSM). Chem Eur J 13(1):287–291

Cremer T (2013) Ionic liquid bulk and interface properties: electronic interaction, molecular orientation and growth characteristics. Springer Science & Business Media

das Neves M, Kimura T, Shimizu N, Nakajima M (2007) Dynamic biochemistry, process biotechnology and molecular biology. Global Sci Books 1:1–14

De Castro ML, Herrera M (2003) Enzyme inhibition-based biosensors and biosensing systems: questionable analytical devices. Biosens Bioelectron 18(2–3):279–294

Dupont J, Fonseca GS, Umpierre AP, Fichtner PF, Teixeira SR (2002) Transition-metal nanoparticles in imidazolium ionic liquids: recyclable catalysts for biphasic hydrogenation reactions. J Am Chem Soc 124(16):4228–4229

Elyasi M, Khalilzadeh MA, Karimi-Maleh H (2013) High sensitive voltammetric sensor based on Pt/CNTs nanocomposite modified ionic liquid carbon paste electrode for determination of Sudan I in food samples. Food Chem 141(4):4311–4317

Ensafi AA, Izadi M, Rezaei B, Karimi-Maleh H (2012) N-hexyl-3-methylimidazolium hexafluoro phosphate/multiwall carbon nanotubes paste electrode as a biosensor for voltammetric detection of morphine. J Mol Liq 174:42–47

Förstner U, Wittmann, GT. (2012) Metal pollution in the aquatic environment Springer Science & Business Media, 2 Edition, p. XVIII, 488, Springer-Verlag Berlin Heidelberg

Gelesky MA, Scheeren CW, Foppa L, Pavan FA, Dias SL, Dupont J (2009) Metal nanoparticle/ionic liquid/cellulose: new catalytically active membrane materials for hydrogenation reactions. Biomacromolecules 10(7):1888–1893

Gurban A-M, Rotariu L, Baibarac M, Baltog I, Bala C (2011) Sensitive detection of endocrine disrupters using ionic liquid—single walled carbon nanotubes modified screen-printed based biosensors. Talanta 85(4):2007–2013

Hong Z, Zhou L, Li J, Tang J (2013) A sensor based on graphitic mesoporous carbon/ionic liquids composite film for simultaneous determination of hydroquinone and catechol. Electrochim Acta 109:671–677

Hu C, Bai X, Wang Y, Jin W, Zhang X, Hu S (2012) Inkjet printing of nanoporous gold electrode arrays on cellulose membranes for high-sensitive paper-like electrochemical oxygen sensors using ionic liquid electrolytes. Anal Chem 84(8):3745–3750

Ibrahim MH, Hayyan M, Hashim MA, Hayyan A, Hadj-Kali MK (2016) Physicochemical properties of piperidinium, ammonium, pyrrolidinium and morpholinium cations based ionic liquids paired with bis (trifluoromethylsulfonyl) imide anion. Fluid Phase Equilib 427:18–26

Jamali T, Karimi-Maleh H, Khalilzadeh MA (2014) A novel nanosensor based on Pt: Co nanoalloy ionic liquid carbon paste electrode for voltammetric determination of vitamin B9 in food samples. LWT Food Sci Technol 57(2):679–685

Kakhki S, Shams E, Barsan MM (2013) Fabrication of carbon paste electrode containing a new inorganic–organic hybrid based on $[SiW_{12}O_{40}]^{4-}$ polyoxoanion and Nile blue and its electrocatalytic activity toward nitrite reduction. J Electroanal Chem 704:80–85

Karimi-Maleh H, Moazampour M, Ahmar H, Beitollahi H, Ensafi AA (2014a) A sensitive nanocomposite-based electrochemical sensor for voltammetric simultaneous determination of isoproterenol, acetaminophen and tryptophan. Measurement 51:91–99

Karimi-Maleh H, Sanati AL, Gupta VK, Yoosefian M, Asif M, Bahari A (2014b) A voltammetric biosensor based on ionic liquid/NiO nanoparticle modified carbon paste electrode for the determination of nicotinamide adenine dinucleotide (NADH). Sensors Actuators B Chem 204:647–654

Karimi-Maleh H, Tahernejad-Javazmi F, Ensafi AA, Moradi R, Mallakpour S, Beitollahi H (2014c) A high sensitive biosensor based on FePt/CNTs nanocomposite/N-(4-hydroxyphenyl)-3,5-dinitrobenzamide modified carbon paste electrode for simultaneous determination of glutathione and piroxicam. Biosens Bioelectron 60:1–7

Karimi-Maleh H, Rostami S, Gupta VK, Fouladgar M (2015) Evaluation of ZnO nanoparticle ionic liquid composite as a voltammetric sensing of isoprenaline in the presence of aspirin for liquid phase determination. J Mol Liq 201:102–107

Kazemi SH, Ghodsi E, Abdollahi S, Nadri S (2016) Porous graphene oxide nanostructure as an excellent scaffold for label-free electrochemical biosensor: detection of cardiac troponin I. Mater Sci Eng C 69:447–452

Kim NH, Malhotra SV, Xanthos M (2006) Modification of cationic nanoclays with ionic liquids. Micropor Mesopor Mater 96(1–3):29–35

Kohagen M, Brehm M, Lingscheid Y, Giernoth R, Sangoro J, Kremer F et al (2011) How hydrogen bonds influence the mobility of imidazolium-based ionic liquids. A combined theoretical and experimental study of 1-n-butyl-3-methylimidazolium bromide. J Phys Chem B 115(51):15280–15288

Li Y, Liu X, Zeng X, Liu Y, Liu X, Wei W, Luo S (2009) Simultaneous determination of ultra-trace lead and cadmium at a hydroxyapatite-modified carbon ionic liquid electrode by square-wave stripping voltammetry. Sensors Actuators B Chem 139(2):604–610

Li J, Peng X, Luo M, Zhao C-J, Gu C-B, Zu Y-G, Fu Y-J (2014) Biodiesel production from *Camptotheca acuminata* seed oil catalyzed by novel Brönsted–Lewis acidic ionic liquid. Appl Energy 115:438–444

Li Y, Zhai X, Liu X, Wang L, Liu H, Wang H (2016) Electrochemical determination of bisphenol A at ordered mesoporous carbon modified nano-carbon ionic liquid paste electrode. Talanta 148:362–369

Liu H, Liu Y, Li J (2010) Ionic liquids in surface electrochemistry. Phys Chem Chem Phys 12(8):1685–1697

Liu C-Z, Wang F, Stiles AR, Guo C (2012) Ionic liquids for biofuel production: opportunities and challenges. Appl Energy 92:406–414

Marques P, Yamanaka H (2008) Biossensores baseados no processo de inibição enzimática. Quím Nova 31(7):1791–1799

Mauriz E, Calle A, Montoya A, Lechuga LM (2006) Determination of environmental organic pollutants with a portable optical immunosensor. Talanta 69(2):359–364

Mehnert CP, Cook RA, Dispenziere NC, Afeworki M (2002) Supported ionic liquid catalysis—a new concept for homogeneous hydroformylation catalysis. J Am Chem Soc 124(44):12932–12933

Mohler E, Jacob LN (1957) Determination of phenolic-type compounds in water and industrial waste waters comparison of analytical methods. Anal Chem 29(9):1369–1374

Mokhtari A, Karimi-Maleh H, Ensafi AA, Beitollahi H (2012) Application of modified multiwall carbon nanotubes paste electrode for simultaneous voltammetric determination of morphine and diclofenac in biological and pharmaceutical samples. Sensors Actuators B Chem 169:96–105

Najafi M, Khalilzadeh MA, Karimi-Maleh H (2014) A new strategy for determination of bisphenol A in the presence of Sudan I using a ZnO/CNTs/ionic liquid paste electrode in food samples. Food Chem 158:125–131

Noda A, Susan MABH, Kudo K, Mitsushima S, Hayamizu K, Watanabe M (2003) Brønsted acid–base ionic liquids as proton-conducting nonaqueous electrolytes. J Phys Chem B 107(17):4024–4033

Ortega F, Domínguez E, Burestedt E, Emnéus J, Gorton L, Marko-Varga G (1994) Phenol oxidase-based biosensors as selective detection units in column liquid chromatography for the determination of phenolic compounds. J Chromatogr A 675(1–2):65–78

Papagni A, Trombini C, Lombardo M, Bergantin S, Chams A, Chiarucci M et al (2011) Cross-coupling of 5,11-dibromotetracene catalyzed by a triethylammonium ion tagged diphenylphosphine palladium complex in ionic liquids. Organometallics 30(16):4325–4329

Pichon V, Chapuis-Hugon F (2008) Role of molecularly imprinted polymers for selective determination of environmental pollutants—a review. Anal Chim Acta 622(1–2):48–61

Rehman A, Zeng X (2012) Ionic liquids as green solvents and electrolytes for robust chemical sensor development. Acc Chem Res 45(10):1667–1677

Ren J, Gu J, Tao L, Yao M, Yang X, Yang W (2015) A novel electrochemical sensor of 4-nonylphenol based on a poly(ionic liquid) hollow nanosphere/gold nanoparticle composite modified glassy carbon electrode. Anal Methods 7(19):8094–8099

Riisagera A, Fehrmanna R, Haumannb M, Wasserscheidb P (2006) Supported ionic liquids: versatile reaction and separation media. Top Catal 40(1–4):91–102

Rodriguez-Mozaz S, Marco M-P, de Alda MJL, Barceló D (2004) Biosensors for environmental monitoring of endocrine disruptors: a review article. Anal Bioanal Chem 378(3):588–598

Rogers RD, Seddon KR (2003) Ionic liquids—solvents of the future? Science 302(5646):792–793

Ruiz E, Ferro V, De Riva J, Moreno D, Palomar J (2014) Evaluation of ionic liquids as absorbents for ammonia absorption refrigeration cycles using COSMO-based process simulations. Appl Energy 123:281–291

Safavi A, Maleki N, Moradlou O, Tajabadi F (2006) Simultaneous determination of dopamine, ascorbic acid, and uric acid using carbon ionic liquid electrode. Anal Biochem 359(2):224–229

Safavi A, Maleki N, Farjami E, Mahyari FA (2009) Simultaneous electrochemical determination of glutathione and glutathione disulfide at a nanoscale copper hydroxide composite carbon ionic liquid electrode. Anal Chem 81(18):7538–7543

Sanati AL, Karimi-Maleh H, Badiei A, Biparva P, Ensafi AA (2014) A voltammetric sensor based on NiO/CNTs ionic liquid carbon paste electrode for determination of morphine in the presence of diclofenac. Mater Sci Eng C 35:379–385

Schmid RD, Verger R (1998) Lipases: interfacial enzymes with attractive applications. Angew Chem Int Ed 37(12):1608–1633

Sun W, Yang M, Li Y, Jiang Q, Liu S, Jiao K (2008) Electrochemical behavior and determination of rutin on a pyridinium-based ionic liquid modified carbon paste electrode. J Pharm Biomed Anal 48(5):1326–1331

Sun W, Duan Y, Li Y, Gao H, Jiao K (2009) Electrochemical behaviors of guanosine on carbon ionic liquid electrode and its determination. Talanta 78(3):695–699

Tajik S, Taher MA, Beitollahi H (2013) First report for simultaneous determination of methyldopa and hydrochlorothiazide using a nanostructured based electrochemical sensor. J Electroanal Chem 704:137–144

Tajik S, Taher MA, Beitollahi H (2014) Mangiferin DNA biosensor using double-stranded DNA modified pencil graphite electrode based on guanine and adenine signals. J Electroanal Chem 720:134–138

Tavana T, Khalilzadeh MA, Karimi-Maleh H, Ensafi AA, Beitollahi H, Zareyee D (2012) Sensitive voltammetric determination of epinephrine in the presence of acetaminophen at a novel ionic liquid modified carbon nanotubes paste electrode. J Mol Liq 168:69–74

Tomašević A, Kiss E, Petrović S, Mijin D (2010) Study on the photocatalytic degradation of insecticide methomyl in water. Desalination 262(1–3):228–234

Turner A, Karube I, Wilson GS (1987) Biosensors: fundamentals and applications: Oxford, New York: Oxford University Press, 1, p. 770

Wang Z, Wang H, Zhang Z, Liu G (2014) Electrochemical determination of lead and cadmium in rice by a disposable bismuth/electrochemically reduced graphene/ionic liquid composite modified screen-printed electrode. Sensors Actuators B Chem 199:7–14

Xiao C, Rehman A, Zeng X (2015) Evaluation of the dynamic electrochemical stability of ionic liquid–metal interfaces against reactive oxygen species using an in situ electrochemical quartz crystal microbalance. RSC Adv 5(40):31826–31836

Yager JW, Eastmond DA, Robertson ML, Paradisin WM, Smith MT (1990) Characterization of micronuclei induced in human lymphocytes by benzene metabolites. Cancer Res 50(2):393–399

Zhang J, Lei J, Liu Y, Zhao J, Ju H (2009) Highly sensitive amperometric biosensors for phenols based on polyaniline–ionic liquid–carbon nanofiber composite. Biosens Bioelectron 24(7):1858–1863

Zhao C, Bond AM, Compton RG, O'Mahony AM, Rogers EI (2010) Modification and implications of changes in electrochemical responses encountered when undertaking deoxygenation in ionic liquids. Anal Chem 82(9):3856–3861

Zhao H, Zang L, Zhao H, Zhang Y, Zheng Y, Zhang Z, Cao W (2015) Oxygen sensing properties of gadolinium labeled hematoporphyrin monomethyl ether based on filter paper. Sensors Actuators B Chem 206:351–356

Zhu Z, Li X, Zeng Y, Sun W (2010) Ordered mesoporous carbon modified carbon ionic liquid electrode for the electrochemical detection of double-stranded DNA. Biosens Bioelectron 25(10):2313–2317

# Nanobiosensors for Detection of Phenolic Compounds

**Fethi Achi, Amira Bensana, Abdallah Bouguettoucha, and Derradji Chebli**

## Contents



## Abbreviations

| | |
|---|---|
| @p-63 | Thiolated DNA sequence |
| 1-FP | 1-Formylpyrene |
| 3-DIPC | Sodium citrate-derived three-dimensional interconnected porous carbon |

F. Achi (✉)
Laboratoire Valorisation et Promotion des Ressources Sahariennes (VPRS), Kasdi Merbah University, Ouargla, Algeria

A. Bensana · A. Bouguettoucha · D. Chebli
Faculty of Technology, Department of Process Engineering, Ferhat Abbas University Sétif-1-, Setif, Algeria



Inamuddin, A. M. Asiri (eds.), *Nanosensor Technologies for Environmental Monitoring*, Nanotechnology in the Life Sciences,
https://doi.org/10.1007/978-3-030-45116-5_10

| | |
|---|---|
| ABPE | Acetylene black paste electrode |
| AC | Activated carbon |
| Acc | Acid-treated carbon cloth |
| ACF | Acriflavine |
| AChE | Acetylcholinesterase |
| AEP | Acetone-extracted propolis |
| AO | Acridine orange |
| AOX | Ascorbate oxidase |
| AP | Aminophenol |
| Apta | Aptamer |
| APTES | 3-Aminopropyltriethoxysilane |
| AWP | Azide-unit pendant water-soluble photopolymer |
| BC | Bacterial cellulose |
| BCA | Butyl carbitol acetate |
| BDD | Boron-doped diamond |
| BDND | Boron-doped nanocrystalline diamond |
| BiOx | Bismuth oxide |
| BMIM | 1-Butyl-3-methyimidazolium hexafluorophosphate |
| BOMC | Boron-doped ordered mesoporous carbon |
| BPA | Bisphenol A |
| BPHR | Peroxidase from *Brassica napus* hairy roots |
| BSA | Bovine serum albumin |
| CA | Caffeic acid |
| CB | Carbon black |
| CBP | Carbon black powder |
| CBPE | Carbon black paste electrode |
| CC | Catechol |
| CdTe | Cadmium telluride |
| CE | Carbon electrode |
| Ce | Cerium |
| CF | Carbon felt |
| CFG | Carboxyl-functionalized graphene |
| CFP | Carbon fiber paper |
| ChO | Choline oxidase |
| CMC | Carboxylmethyl cellulose |
| CNFs | Carbon nanofibers |
| CNH | Carbon nanohorn |
| CNTs | Carbon nanotubes |
| Coll | Colloid |
| CoPC | Cobalt phthalocyanine |
| $coVFc_{15}$ | Co-vinyl ferrocene |
| CP | Chlorophenol |
| CPE | Carbon paste electrode |
| CPO | Chloroperoxidase |

| | |
|---|---|
| CRG24H | Partially reduced graphene oxide |
| Cs | Chitosan |
| CTAB | Cetyltrimethylammonium bromide |
| CV | Cyclic voltammetry |
| Cys | Cysteine |
| CYST | Cysteamine |
| CZUF | Cross-linked zein ultrafine fibers |
| DAAO | D-amino acid oxidase |
| DCIL5 | Dicationic ionic liquid |
| DCP | Dichlorophenol |
| Den | Dendrimer |
| DGS | Diglycerylsilane |
| DHP | Dihexadecylphosphate |
| DMcT | 2,5-Dimercapto-1,3,4-thiadiazole |
| DMS | Disordered mesoporous silica |
| DNA | Deoxyribonucleic acid |
| DPNS | Dendritic platinum nanoparticles |
| DPV | Differential pulse voltammetry |
| DTSP | Dithiobis-*N*-succinimidyl propionate |
| DTTPS | Dithienotetraphenylsilane |
| EAPC | Enzyme adsorption, precipitation and cross-linking |
| EB cells | *Escherichia coli* bioreporter |
| EDC | 1-Ethyl-3-(3-dimethyl-aminopropyl)carbodiimide |
| EG | Exfoliated graphene |
| ELDH | Exfoliated layered double hydroxide |
| E-matrix | Enzymatic matrix |
| ESM | Eggshell membrane |
| F108 | Polyethylene oxide-polyoxypropylene-polyethylene oxide |
| FAM | 6-Fluorescein amidite |
| Fc | Ferrocene |
| FeOx | Hydroxy iron |
| FePc | Iron phthalocyanine |
| FESEM | Field emission scanning electron microscopy |
| FeTsPc | Iron tetrasulfonated phthalocyanine |
| FSM7.0 | Mesoporous silica powder (7.0 nm) |
| FTO | F-doped tin oxide |
| FYSSns | Flower-shaped yolk–shell $SiO_2$ |
| G | Graphene |
| GA | Glutaraldehyde |
| GCE | Glassy carbon electrode |
| GCHs | Ground cherry husks |
| GCs | Glycol chitosan |
| GE | Graphite electrode |
| GMA | Glycidyl methacrylate |

| | |
|---|---|
| Go | Graphene oxide |
| GOX | Glucose oxidase |
| g-PGE | Poly(ethylene glycol) |
| GR | Graphite |
| Gs | Graphene sheets |
| HA | Hydroxyapatite |
| Hb | Hemoglobin |
| HB82 | 2-Hydroxyethyl methacrylate (80.2%), butyl acrylate |
| HB91 | 2-Hydroxyethyl methacrylate (90.1%), butyl acrylate |
| HF | Hollow fiber |
| His | Histidine |
| HP | 2-Hydroypropyl |
| HQ | Hydroquinone |
| HRP | Horseradish peroxidase |
| HSA | Human serum albumin |
| h-$SiO_2$ | Helical silica |
| HT | Hydrogen-terminated |
| HTLc | Hydrotalcite-like compound |
| IL | Ionic liquid |
| ITO | Indium tin oxide |
| Km | Michaelis–Menten constant |
| L.O.D. | Limit of detection |
| L.R | Linear range |
| Lac | Laccase |
| L-Arg | L-arginine |
| LbL | Layer-by-layer |
| LDHs | Layered double hydroxides |
| L-dopa | Levo-Dopa |
| LSG | Laser scribed graphene |
| LSV | Linear sweep voltammetry |
| MAM | Melamine |
| MAPLE | Matrix assisted pulsed laser evaporation |
| MB | Methylene blue |
| MBA | Graphite microband arrays |
| MCF | Cellulose microfibers |
| MCH | 6-Mercapto-1-hexanol |
| MCM-41 | Mesoporous silica sieve |
| MEAs | Microelectrode arrays |
| MI | Molecular imprinting |
| MIPs | Molecular imprinting polymers |
| MnPc | Manganese phthalocyanine |
| MNPs | Magnetic nanoparticles |
| MO | Mineral oil |
| MOFs | Metal-organic frameworks |

| | |
|---|---|
| $MOS_2$ | Molybdenum disulfide |
| MTM | 3-Methyl thienyl methacrylate |
| MTPS | (3-Mercaptopropyl)-trimethoxy silane |
| MWCNTs | Multiwalled carbon nanotubes |
| Mxene | Two-dimensional transition metal carbides |
| N.R. | Not reported |
| NAC | *N*-acetyl-L-cysteine |
| NC | Nanocomposite |
| NCD | Nanocrystalline diamond |
| Ncl | Nitrocellulose |
| NCs | Nanocrystals |
| ND | Nanodiamond |
| NDs | Nanoneedles |
| NEs | Nanoellipsoids |
| $NEt^{+}_{4}$-pyrrole | [12-(Pyrrole-1-Yl)dodecyl] triethylammonium tertafluoroborate |
| NG | Hydrophilic nanographene |
| N-Gs | Nitrogen-doped graphene sheets |
| NHS | *N*-hydroxysulfosuccinimide sodium salt |
| NiTPPS | Ni(II) tetra Kis(4-sulfonatophenyl)porphyrin |
| NMCS | Nitrogen-doped mesoporous carbon nanosheet |
| NPBimBr | 1-[3-(*N*-pyrrolyl)propyl]-3-butyl midazolium bromide |
| NPG | Nanoporous gold |
| NPGF | Nanoporous gold film |
| NPGL | Nanoporous gold leaf |
| NPs | Nanoparticles |
| NPt | Nanoplatelets |
| NQ | Naphthoquinone |
| NR | Nanoribbon |
| NRs | Nanorods |
| Ns | Nanosheets |
| NT | Nitrophenol |
| NTAs | Nanotube arrays |
| NWs | Nanowires |
| NyM | Nylon membrane |
| OMC | Ordered mesoporous carbon |
| Osi | Organosilica |
| P(Gly) | Poly(glycine) |
| P(L-Arg) | Poly(L-arginine) |
| P2AE | Poly(2-anilinoethanol) |
| P4VP | Poly(4-vinyl pyridine) |
| PA | Phytic acid |
| Pal | Palygorskite |
| PAMAM | Poly(amidoamine) |
| PANI | Poly(aniline) |

| | |
|---|---|
| PASE | 1-Pyrenebutanoic acid succinimidyl ester |
| PATT | Poly(4′-pyrazine-2,2′,5′,2″-terthiophene |
| PB | Prussian blue |
| PC | Phosphatidylcholine |
| Pc | Phthalocyanine |
| PCA | Poly(citric acid) |
| PDA | Poly(dopamine) |
| PDm | Poly(L-dopa) |
| PDNPH | Poly(2,4-dinitrophenylhydrazine) |
| PE | Plastic electrode |
| PEDOT | Poly(3,4-ethylenedioxy-thiophene) |
| PEG | Polyethylene glycol |
| PEI | Poly(ethyleneimine) |
| PF6 | Hexafluorophosphate |
| PGA | Poly(glutaraldehyde) |
| PGE | Pencil graphite electrode |
| PGLA | Poly(glutamic acid) |
| PGluA | Poly(glutamate acid) |
| PGMA | Poly(glycidylmethacrylate) |
| PHEMA | Poly(2-hydroxyethyl methacrylate) |
| PhOSubPc | Hexa-phenoxy boron subphthalocyanine |
| PLT | Poly(L-tyrosine) |
| PMO | Poly(methyl orange) |
| p-NPDS | Para-nitrophenyl diazonium salt |
| POM | Polyoxometalate |
| POMA | Poly(O-methoxyaniline) |
| $PoX_{1B}$ | Peroxidase enzyme |
| PP | Polyphenol |
| PPD | Poly(O-phenylenediamine) |
| PPDA | *N*′-phenyl-P-phenylenediamine |
| p-PDA | Para-phenylene diamine |
| PPEGA | Poly(poly(ethylene glycol)acrylate) |
| pPhR | Poly(phenol red) |
| PPy | Poly(pyrrole) |
| PSS | Poly(syrene-4-sulfonate) |
| PTCA | 3,4,9,10-Perylene-tetracarboxylic acid |
| Pth | Poly(thionine) |
| PTn | Poly(thiophene) |
| PU | Poly(urethane) |
| PVA | Polyvinyl alcohol |
| PVC | Poly(vinyl chloride) |
| $PVF^+$ | Poly(vinylferrocenium) perchlorate |
| PVIM | Poly(vinylimidazolium) |

| | |
|---|---|
| PVS | Poly(vinyl sulfonate) |
| PVSA | Polyvinyl sulfonic acid |
| QDs | Quantum dots |
| rGo | Reduced graphene oxide |
| RhB | Rhodamine |
| Ru | Ruthenium |
| S- | Thiol-functionalized |
| SAMs | Self-assembled monolayers |
| SBA-15 | Mesoporous silica hybrid |
| SbQ | Pyridinium methyl sulfate |
| SC | Sinusoidal currents |
| SCs | Single crystals |
| SDBS | Sodium dodecylbenzene sulfonate |
| SEM | Scanning electron microscopy |
| SF | Silk fibroin |
| SISG | Silica sol-gel |
| SiTiMPs | Magnetic silica/titania xerogel |
| SNGC | Sonogel carbon electrode |
| SP | Silk peptide |
| SPE | Screen-printed electrode |
| SSDDNA | Single-stranded DNA |
| SV | Sinusoidal voltages |
| VIETLm$^{+}$lBr$^{-}$ | 1-Vinyl-3-ethyl imidazolium |
| WXX | Amino-modified carboxycellulose |
| SWCNTs | Single-walled carbon nanotubes |
| SWV | Square wave voltammetry |
| TCNQ | Tetracyanoquinodimethane |
| TCP | Trichlorophenol |
| TEM | Transmission electron microscopy |
| TESBA | Triethoxysilyl butyraldehyde |
| TLA | Thiolactic acid |
| TMO | Ternary metal oxide |
| T-$NH_2$ | Thiolactic acid amide |
| TNT | Titanium oxide nanotubes |
| TPC | Total phenolic content |
| TTF | Tetrathiafulvalene |
| Tyr | Tyrosinase |
| VFc | Vinylferrocene |
| W-lac D | Mutated bacterial laccase |
| XOD | Xanthine oxidase |
| β-CD | β-Cyclodextrin |

# 1 Introduction

Biosensor is an analytical tool able to convert biochemical reaction into a measurable signal. Electrochemical detection is one of the most sensing methods applied in monitoring analytes (Ronkainen et al. 2010; Thévenot et al. 1999, 2001). Electrochemical biosensors have been largely applied in many areas, including food analysis, medical applications, cancer diagnostics (Wang 2006) or for the determination of pesticides (Trojanowicz and Hitchman 1996) and heavy metals (Mehta et al. 2016; Achi et al. 2015). The electrochemical measurement of these analytes is mainly carried out using amperometry technique (Amine and Mohammadi 2018; Chaubey and Malhotra 2002).

Over the last two decades, considerable effort has been devoted to the construction of nanobiosensors for monitoring phenolic compounds. The use of nanomaterials to prepare biosensing platforms has main advantages including suitable microenvironment for immobilizing biorecognition elements at the surface of working electrodes. Accordingly, different types of electrodes were modified by variety of new nanomaterials such as nanoparticles (Wang 2005a), carbon nanotubes (Wang 2005b), conducting polymers (Teles and Fonseca 2008), graphene (Zhou et al. 2009; Pumera et al. 2010; Zhao et al. 2011), and gold nanoparticles (Yáñez-Sedeño and Pingarrón 2005; Pingarrón et al. 2008).

However, controlling the thickness of the electrodeposited biocomposite material remains a major challenge. Then, there has been much interest of studying the mass transport of species to provide fast diffusion rate (van Stroe-Biezen et al. 1993; Stewart 2003). Since the diffusion rate has a potential effect on the response time (Bartlett and Pratt 1995). Modeling of diffusion process provides an optimization of parameters affecting the analytical performance of biosensors such as thickness of enzymatic layer, and physicochemical parameters that indicate the limiting step of reaction mechanism. Here we briefly discuss the configuration of nanobiosensor for the detection of phenolic compounds, and its corresponding diffusion–reaction mechanism is reviewed.

This chapter assesses the relative importance of using nanomaterials for the construction of biosensors to detect phenolic compounds. The most common analytical performance parameters of phenolic compounds-based biosensor are described, including limit of detection, stability, linear range, affinity, sensitivity, and different strategies of enzyme immobilization are outlined in which a biosensing platform during measurements is more suitable.

# 2 Toxicity of Phenolic Compounds

The toxicity of phenolic compounds to human health is among the most widely studied of aquatic toxicology and environmental monitoring. For instance, bisphenol A (BPA) is an important one to be considered because of its potential risks to human health including its high ability to interact with different kinds of receptors causing mutagenic changes and inhibit DNA methylation (Erler and Novak 2010; Michałowicz 2014). Accordingly, several review papers have focused on defining the dangerous impact of BPA exposure (Rochester 2013; Singh and Li 2012). Interestingly, Rezg et al. (2014) have presented a general scheme for the harmful impact of BPA on human development and chronic human disease. Moreover, a further study of the shows the carcinogenic properties of BPA reported that it may be reasonably anticipated to be a human carcinogen in the breast and prostate (Seachrist et al. 2016; Prins et al. 2008).

The toxic effect and degradability of nitrophenol have been attracting the attention of many researchers (Arora et al. 2014; Megharaj et al. 1991). Indeed, nitrophenols can affect anaerobic treatment systems (Haghighi-Podeh and Bhattacharya 1999; Uberoi and Bhattacharya 1997) or metabolic activities of *Nostoc linckia* where toxicity increases by substitution of nitro groups at ortho and meta position of phenol (Megharaj et al. 1991). Toxicity effect of other various phenolic compounds has been reported in literature such as the generation of free radicals through auto-oxidation of catechol contributes to the toxic and oxidative effects of catechol on DNA damage or neurodegenerative diseases (Barreto et al. 2009; Cavalieri et al. 2002). Also, it has been reported that free p-cresol affects the functional capacity of white blood cells (De Smet 2003). Among benzene's phenolic metabolites, hydroquinone compound, which can be used not only as a depigmenting agent (Arndt and Fitzpatrick 1965; Fitzpatrick et al. 1966) but also to improve urea-nitrate recovery from soil and consequently decreases soil nitrate content and $NO_2$ emission (Chen et al. 2006). However, it has been reported that hydroquinone increases the genotoxic and mutagenic potential in mice (Jagetia et al. 2001).

Chlorinated phenols that commonly present in aqueous wastes are very weak acids containing an -OH group attached to a hydrocarbon (Jim 2014). These compounds have low biodegradability and high solubility in many organic solvents causing serious environmental problems (Olaniran and Igbinosa 2011), including important toxic effects in fish (Ge et al. 2017; Kishino and Kobayshi 1996) and human cells (Vlastos et al. 2016). Furthermore, Pera-Titus et al. (2004) have reviewed the Half-life times and kinetic constants for chlorophenol degradation and the different mechanistic degradation pathways of chlorophenols. Interestingly, the use of microalgae (e.g., *Pseudokirchneriella subcapitata*) to assess the toxic substances level in freshwater environments is very sensitive to chlorophenols compared to other aquatic organisms (Chen and Lin 2006). Toxicity of chlorophenols is dependent on their lipophilicity (van Gestel and Ma 1988) and their impact on the environment increases when burning of waste materials containing chlorinated phenols (Ahlborg et al. 1980). Thus, the structure of compound molecule including the number of chlorine atoms and their position in molecule are factors increasing chlorophenol toxicity (Czaplicka 2004). A further study on the analysis of cancer mortality of 21,863 workers exposed to phenoxy herbicides and chlorophenols revealed that workers exposed to higher chlorinated dioxins had an increased risk for all neoplasms compared with workers exposed to phenoxy herbicides (Kogevinas et al. 1997).

# 3 Theoretical Study of Diffusion Process

Electrochemical biosensors are composed of an electrode and matrix contains a mixture of nanomaterials and biorecognition element. The diffusion of analytes to the surface of the electrode forms a liquid film that separates the enzyme matrix from the solution (Scheme 1). As previously described, the diffusion rate of analytes in this film as well as in the matrix is an important factor to improve the sensitivity and stability of the biosensor.

The mathematical description of diffusion process according to the second Fick's law is as follows:

$$\frac{\partial C}{\partial T} = \frac{\partial^2 C}{\partial X^2} \pm V(S,M) \tag{1}$$

where *C* designed *S* (substrate) and *P* (product).

*V*(*S*, *M*) is the kinetic term of enzyme reaction in the case of ping-pong mechanism.

$$V(S,M) = \frac{V_{\max}}{1 + \left(\frac{K_S}{S}\right) + \left(\frac{K_M}{M}\right)} \tag{2}$$

where $V_{\max}$ is the maximum catalytic rate.

In the case of mediated biosensors, the oxidized $E_{(OX)}$ or reduced forms $E_{(Red)}$ of enzymes such as peroxidase (HRP) use hydrogen peroxide as mediator (M) to convert a wide variety of substrates (S) such as phenolic compounds to electroactive species (P) (Ozkan et al. 2015). The peroxidase reaction is as follows:

$$2H_2O_2 + E_{(OX)} \rightarrow 2H_2O + O_2 + E_{(Red)}$$

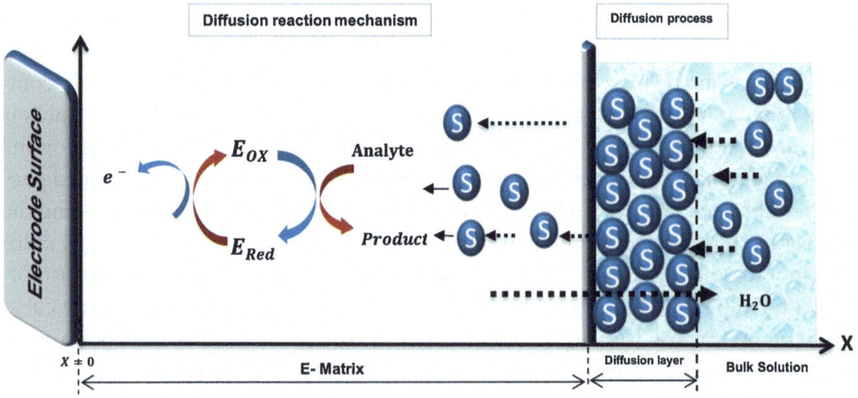


**Scheme 1** Scheme of biosensor configuration with immobilized enzyme (E-matrix), and formation of diffusion layer by mass transport of substrate through electrode surface

The diffusion–reaction system of substrate (S), product (P), and mediator (M) is as follows:

$$\begin{aligned} \partial S / \partial T &= \partial^2 S / \partial X^2 - V(S) \\ \{\partial P / \partial T &= \bar{D}_{pm}\left(\partial^2 P / \partial X^2\right) + V(S) \\ \partial M / \partial T &= \partial^2 M / \partial X^2 - V(S,M) \end{aligned} \quad (3)$$

## 3.1 *Reaction Mechanism*

Tyrosinase enzyme acts as biorecognition element by two distinct reaction mechanisms, the first one is cresolase activity in which tyrosinase uses $Cu^{2+}$ cations as cofactors for the oxygen insertion in ortho position of hydroxyl group of monophenols to obtain diphenols (Mayer and Harel 1979), which, in turn, transformed to quinones by dehydrogenation (Durán and Esposito 2000). The second type of reaction named catecholase activity, in which the oxidation of o-diphenols to benzoquinones occurs with hydrogen abstraction (Mayer and Harel 1979) (Scheme 2).

As well known, laccase enzyme has four copper atoms and oxidizes phenolic compounds by a radical catalysis (Scheme 3) (Claus 2004).

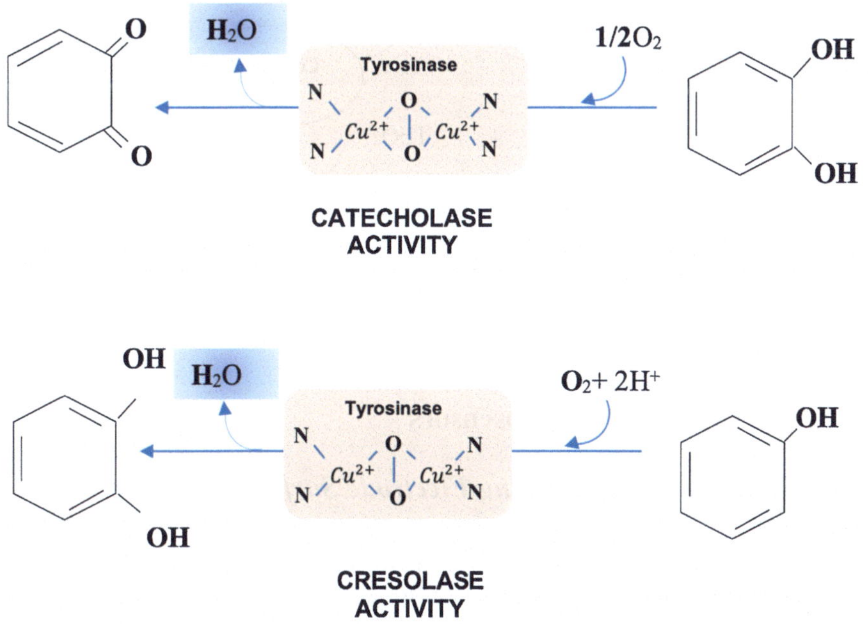


**Scheme 2** Types of tyrosinase activity for oxidation mechanism of phenol

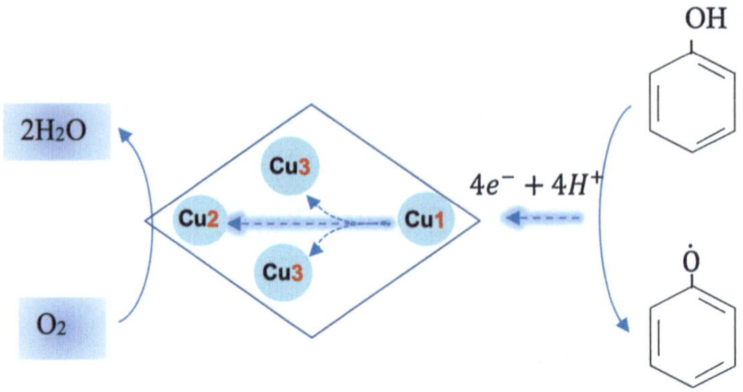


**Scheme 3** Oxidation mechanism of phenol by laccase enzyme

**Scheme 4** Oxidation mechanism of phenol with horseradish peroxidase

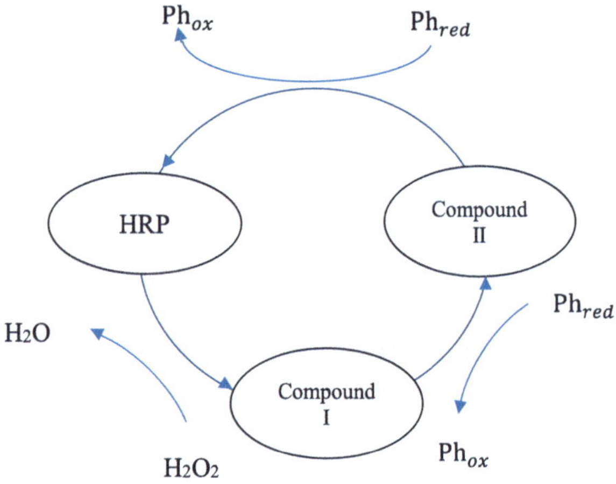


Differently, phenol oxidation as catalyzed by HRP enzyme and hydrogen peroxide is like the three substrate ping-pong mechanism, if considering phenols as two separate substrates and hydrogen peroxide as the third substrate (Danner et al. 1973) (Scheme 4).

## 4 Preparation of Nanobiosensors

### *4.1 Pretreatment of Working Electrode Surface*

As described in literature, amperometric measurements were performed in a stirred cell containing buffer solution with an optimized value of (pH) and at appropriate oxidation or reduction potential. A conventional three-electrode system was used comprising an auxiliary electrode, Ag/AgCl as reference electrode, and a working

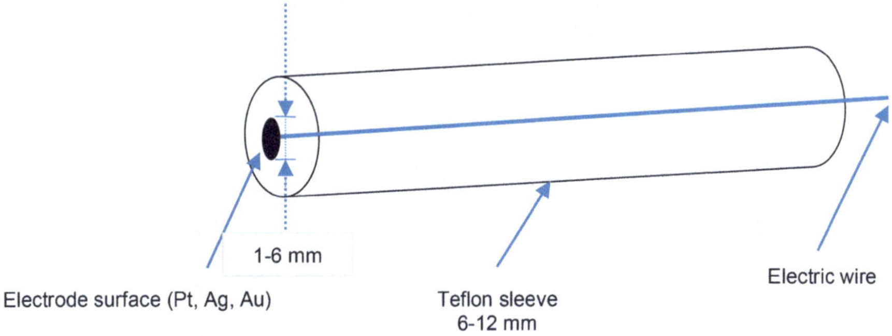


**Scheme 5** Schematic diagram showing the working electrode

electrode for fast electron transfer of electroactive species. Hence, the surface of working electrode should be polished using alumina slurries before immobilization of biorecognition element. The surface was then washed with ethanol or double distilled water and dried at ambient conditions. Afterward, the enzyme was dissolved in buffer solution and then dropped onto the surface of electrode. Finally, the modified electrode was kept under dry conditions at room temperature for 1 h or overnight and stored in a buffer solution at 4 °C under dry conditions for further use.

The electrode surface of the working electrode can be made of inert materials such as platinum, silver, or gold covered with Teflon sleeve 6–12 mm of diameter; the material was connected with an electric wire (Scheme 5).

## *4.2 Immobilization of Biorecognition Element*

Enzymes can be immobilized at the surface of the working electrode using a variety of techniques to keep their catalytic activity. The most important methods are, respectively, cross-linking, entrapment, and covalent binding.

### 4.2.1 Cross-Linking

In this method, biorecognition elements are fixed with agents such as glutaraldehyde, Nafion, or chitosan. Enzymes immobilized by cross-linking method usually suffer from significant loss of activity. In addition, the rate of overall mechanism is often limited by the diffusion process.

### 4.2.2 Covalent Binding

This method was based on pre-activation of the solid support surface by multifunctional reagents before enzyme deposition. A variety of reactive groups such as hydroxyl, amide, amino, and carboxyl could be used to attach enzyme molecules.

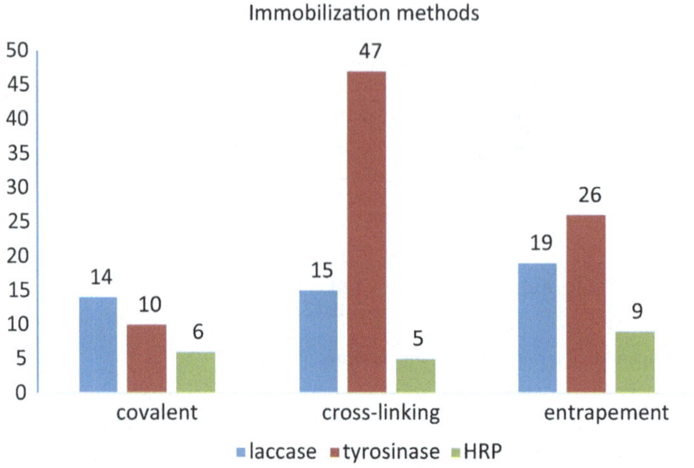


**Fig. 1** Immobilization techniques of biorecognition elements used to construct biosensors for phenolic compounds worldwide (2008–2018)

#### 4.2.3 Entrapment

This method based on cross-linking polymers to prepare lattice structure that prevents enzyme molecules to diffuse out, allowing for mass diffusion of substrate and product molecules. The method could be applied only to limited number of enzymes with no chemical modification (Dwevedi 2016).

Figure 1 shows a comparative study of the enzyme immobilization methods to design biosensing platforms for sensing phenolic compounds. Therefore, the mainly used technique to immobilize tyrosinase is cross-linking followed by entrapment and covalent linking. Furthermore, it can be seen that entrapment method followed by cross-linking and covalent linking, respectively, are the mostly used techniques for the immobilization of laccase, while entrapment, covalent binding, and cross-linking are, respectively, the most widely used methods to immobilize horseradish peroxidase.

# 5 Nanobiosensors for the Detection of Phenolic Compounds

## *5.1 Enzyme-Based Biosensors*

#### 5.1.1 Tyrosinase Biosensors

As shown in Table 1, the use of tyrosinaseenzyme as biorecognition element for phenol or p-cresol detection provides high affinity toward these analytes with good sensitivity (Oriero et al. 2015; Wang et al. 2016; Li et al. 2017; Shan et al. 2009;

**Table 1** Analytical characteristics of tyrosinase biosensors for detection of phenolic compounds

| | E-matrix/electrode | Sensitivity | L.R. | L.O.D. | pH | Km | Stability (%) | References |
|---|---|---|---|---|---|---|---|---|
| Phenol | Tyr/Pal/GCE | 1.897 A/M | 0.05–100 μM | N.R. | 7.0 | 49.3 μM | 80% 2 months | Chen and Jin (2010) |
| | Tyr/Titania sol–gel/CE | 1.4 A/M | 0.44–11 μM | 0.13 μM | 6.0 | N.R. | 55% 2 weeks | Kochana et al. (2008a, b) |
| | Tyr/PVA/Silica/ITO | 4.2 nA/μM | 10–150 μM | 10 μM | 6.8 | 21 nM | Life time 50% 4 days | Oriero et al. (2015) |
| | Tyr/SWNTs/LDHs/GCE | 1532 mA/M $cm^2$ | 7.95 nM–30.8 μM | 7.95 nM | 6.0 | 6.5 μM | 53% 30 days | Wang et al. (2016) |
| | Tyr/AuNPs/SPE | 15.7 μA/ppm | 47 ppb–15 ppm | 47 ppb | 7.0 | N.R. | 86% 4 days | Nurul Karim and Lee (2013) |
| | Tyr/rGo-AuNPs/ITO | 416 nA/μM | 0.084–0.38 μM | 72 nM | 7.0 | 2.27 μM | 50% (20 uses) | Penu et al. (2015) |
| | | 363 nA/μM | 0.741–19.5 μM | 0.318 μM | | | | |
| | Tyr-Nafion/CE | 330 nA/μM | 0–2 μM | 0.4 μM | 7.0 | 12.6 μM | 70% 35 days | Boukoureshtlieva et al. (2013) |
| | BiNPs/Tyr/GA/SPE | 44.056 nA/μM | 0–100 μM | 26 nM | 6.5 | 83 μM | N.R. | Mayorga-Martinez et al. (2013) |
| | Tyr/PVA/Silica/ITO | 2.8 nA/μM | 10–100 μM | 10 μM | 6.8 | 16 nM | Life time 50% 4 days | Oriero et al. (2015) |

(continued)

**Table 1** (continued)

| | E-matrix/electrode | Sensitivity | L.R. | L.O.D. | pH | Km | Stability (%) | References |
|---|---|---|---|---|---|---|---|---|
| p-Cresol | Tyr/SWNTs/LDHs/GCE | 4827 mA/M $cm^2$ | 3.98 nM–1.17 μM | 3.98 nM | 6.0 | 2.5 μM | 53% 30 days | Wang et al. (2016) |
| | Nafion/$PD_M$–Tyr/Au | 926 μA/mM | 0.1–5.6 μM | 20 nM | 7.0 | 14.8 μM | 78% 4 weeks | Guan et al. (2016) |
| | QDs/Cs/Tyr/GCE | 514 mA/M | 1 nM–20 μM | 0.5 nM | 6.5 | 4.75 μM | 80% 50 days | Han et al. (2015) |
| | Tyr/CuO/SBA-15/Au/GCE | 7.7 A/M $cm^2$ | 4 nM–11 μM | 1 nM | 6.0 | 17 μM | 86% 20 days | Liu et al. (2010) |
| | Tyr/PANI-IL-CNFs/GCE | 262 A/M.$cm^2$ | 0.4 nM – 2 μM | 0.1 nM | 7.0 | 1.33 μM | 94% 40 days | Zhang et al. (2009) |
| | Tyr/BiOx/GA/GCE | 12.3 A/M $cm^2$ | 1.3 nM–4 μM | 0.4 nM | 6.0 | 36 μM | 72% 47 days | Shan et al. (2009) |
| | Tyr/$Fe_3O_4$-Cs/GCE | 0.360 A/M | 0.083–78 μM | 25 nM | 6.5 | 130 μM | 53% 2 months | Wang et al. (2008) |
| | Tyr/PPy-NTs/GCE | 11.392 mA/mM $cm^2$ | 0.1–50 μM | 0.23 nM | 6.0 | 10 μM | 88% 30 days | Li et al. (2017) |
| | NAC-AuNPs//Tyr/Cs/GCE | 0.356 A/M | 0.15–50 μM | 80 nM | 7.0 | N.R. | 71% 1 month | Dong et al. (2017) |

Zhang et al. 2009); it can be immobilized with glutaraldehyde, nafion or using sol–gel method (Guan et al. 2016; Kochana et al. 2008a).

Tyrosinase biosensors constructed using gold nanoparticles (AuNPs) have fast electron transfer between electrode surface and enzyme with a sensitivity of 15.7 μA/ppm (Nurul Karim and Lee 2013; Penu et al. 2015). The use of polymers such as Nafion enhances considerably the stability of a biosensor; it could be mixed directly with the enzyme with a stability of 70% after 35 days or combined with poly(L-DOPA)-tyrosinase composite with a stability of 78% after 4 weeks (Boukoureshtlieva et al. 2013). Cross-linking tyrosinase with bismuth nanoparticles (BiNPs) by glutaraldehyde onto screen-printed electrode enables detecting phenol at 26 nM (Mayorga-Martinez et al. 2013). Interestingly, Chen and Jin (2010) have developed a method to retain tyrosinase bioactivity using palygorskite as natural clay with porous morphology with a stability of 80% after 2 months.

All tyrosinase biosensors constructed to detect p-cresol compound were found to be able to monitor phenols. The lowest detection limit for p-cresol detection was obtained using quantum dots mixed with chitosan (Dong et al. 2017; Wang et al. 2008) or with the combination of copper oxide and mesoporous silica material (Han 2015).

### 5.1.2 Laccase Biosensors

Table 2 shows laccase sensing platforms constructed using several nanomaterials, such as magnetic nanoparticles mixed with polymers (Xu 2009; Sezgintürk 2010), or using yolk–shell $SiO_2$ nanospheres (FYSSns) as an immobilized laccase supporter that provides good analytical performance with a sensitivity of 0.1477 μA/μM (Zheng et al. 2018). Laccase biosensor was successfully applied to determine hydroquinone and other phenolic compounds (P.Ibarra-Escutia et al., 2010; Gonzalez-Anton et al., 2017; Upan et al., 2016). The presence of metal oxides in polymeric composite based on $Fe_3O_4$ magnetic nanoparticles enhances stability of biosensor (Li et al., 2012; Yang et al., 2016). A mixture of gold nanoparticules and nafion increases considerably the stability of laccase biosensor for the detection of hydroquinone (96 % during 3 months) (Li et al., 2016).

Several immobilization techniques have been applied to improve the stability of a laccase biosensor such as incorporation of laccase in ZnO sol–gel using chitosan (Qu et al. 2015; Shimomura et al. 2011) or using electrophoretic deposition technique (96% after 50 days) (Verrastro et al. 2016). Immobilizing laccase enzyme by laser printing technology provides better analytical performance for catechol determination than using other immobilization methods (Touloupakis et al. 2014). Although covalent binding of laccase onto the surface of a silanized gold electrode provides a good analytical performance compared with the use of modified electrodes by organothiol layers cross-linked with glutaraldehyde (Sarika 2015).

**Table 2** Analytical characteristics of laccase biosensors for detection of phenolic compounds

| | E-matrix/electrode | Sensitivity | L.R. | L.O.D. | pH | Km | Stability (%) | References |
|---|---|---|---|---|---|---|---|---|
| Catechol | Lac MAPLE/SPCE | 13.4 nA/μM cm$^2$ | 1–60 μ M | 1 μM | 5.5 | N.R. | 80% 30 uses | Verrastro (2016) |
| | Lac/GA/BSA/Au | 0.2988 nA/μM | 1–100 μM | 7 μM | 6.8 | 57.5 μM | 50% (550 analysis) | Sarika et al. (2015) |
| | Lac/Gelatin/GA/SPE | 0.0119 mA/μM | 1–10 μM | 5 μM | 4.5 | N.R. | N.R. | Sezgintürk (2010) |
| | MB-MCM-41/PVA/Lac/Au | N.R. | 4–87.98 μM | 0.331 μM | 4.95 | 0.256 mM | N.R. | Xu (2009) |
| | Cs/Lac/MWCNTs/MCPE | 0.279 μA/μM | 0.1–165 μM | 33.4 nM | 5.6 | N.R. | 90% 45 days | Pang (2011) |
| Hydroquinone | PDA/Lac/$Fe_3O_4$/Au | 374 μA/mM.cm$^2$ | 0.2–95 μM | 30 nM | 4.0 | 0.34 mM | 85% 1 month | Yunyong Li et al. (2012) |
| | Lac/PVA-AWP/GR SPE | 9.44 nA/μM | 1.1–130 μM | 1.071 μM | 4.7 | N.R. | Life time > 6 months | P. Ibarra-Escutia et al. (2010) |
| | Lac/Silicagel/GA/CNTs/SPCE | 0.3523 mV/μM | 1–50 μM | 0.1 μM | 4.5 | N.R. | 80% 30 days | Upan et al. (2016) |
| | Nafion/$TiO_2$/CuCNFs-Lac/GCE | 24.6 μA/mM | 1–89.8 μM | 3.65 μM | 6.0 | N.R. | 93.45% 1 month | Yang et al. (2016) |
| | Lac/Bc-AuNPs/Nafion/GCE | 0.3029 μA/nM | 30–100 nM | 5.71 nM | 5.0 | N.R. | 96% 3 months | Li et al. (2016) |
| | ClSubPc-Lac/ITO<br>PhOSubPc-Lac/ITO<br>t-BuSubPc-Lac/ITO | 0.6 μA/μM<br>0.4 μA/μM<br>0.4 μA/μM | 0.5–20 μM<br>0.5–20 μM<br>5–20 μM | 0.169 μM<br>0.155 μM<br>0.485 μM | 7.0 | 1.73 mM<br>0.126 mM<br>2.06 mM | 89% 50 cycles | Gonzalez-Anton et al. (2017) |

### 5.1.3 Horseradish Peroxidase Biosensors

Platforms based on horseradish peroxidase are distinguished by their low affinity toward phenolic compounds (Santos et al. (2015); Ozoner et al. 2010); Kazemi and Khajeh 2011). However, HRP biosensor is the most able to detect some phenolic compounds at trace level such as bisphenol A, chlorophenolic derivatives (Wu et al. 2016), and nitrophenols. In (Table 3) was summarized the analytical performance of HRP biosensors for monitoring monophenolic compounds, which were prepared using conductive copolymers film (Ozoner et al. 2011) or using poly(citric acid)-block-poly(ethylene glycol) that enhances the analytical performance of HRP biosensor compared to the use of polyaniline nanofibers combined with chitosan and glutaraldehyde (Shamloo et al. 2013). Accordingly, the structure of polymeric films is determinant in the amperometric signal of the HRP biosensor (Korkut et al. 2015).

In addition, the determination of p-cresol by HRP biosensor was carried out simultaneously with other phenolic compounds using different types of electrodes and nanomaterials, such as sonogel–carbon electrode (ElKaoutit et al. 2008) and titania nanotube arrays (Kafi and Chen 2009), or using conducting polymers (Korkut et al. 2008). Interestingly, HRP biosensor stability for p-cresol detection can be improved using poly(allylamine hydrochloride)-wrapped multiwall carbon nanotubes (84.7% after 90 days).

As shown in (Table 3), dopamine biosensors were prepared using a variety of electrochemical electrodes on which HRP was covalently immobilized such as platinum electrode modified with functionalized multiwall carbon nanotubes (de Souza Ribeiro et al. 2013). As previously described, the lack of affinity of HRP-based platforms for phenolic compounds is often a factor limiting their development for detecting polyphenols. Accordingly, polymer platforms based on HRP enzyme exhibit low affinity of 0.112 mM to catechol detection (Mossanha et al. 2017). Furthermore, using ultramicroelectrode arrays with gold nanoparticles (AuNPs) and dithiobis-N-succinimidyl propionate (DTSP) enhances the sensitivity of HRP biosensor for monitoring catechol (228.6 μA $cm^{-2}$ $mM^{-1}$) (Orozco et al. 2009).

## 5.2 *Multienzyme and enzymeless Biosensors*

There are limited papers discussing the development of bi-enzyme biosensors for the detection of phenolic compounds (Barberis et al., 2017; Montereali et al., 2010; Liu et al., 2018; Monti et al., 2017; Jolanta Kochana et al. 2008). There have been other biological elements successfully used in biosensing platforms to improve electrochemical sensing performance of phenolic compounds biosensors. The analytical characteristics of this type of biosensors are summarized in (Table 4). It has been reported that laccase enzyme was the best biological element for the construction of biosensor to polyphenols determination (sensitivity= 89.066 nA/μM), hence, laccase was cross-linked with glutarldehyde and nafion at the surface of sonogel carbon elec-

**Table 3** Performance parameters of HRP biosensors for the detection of monophenols and chlorophenols

| | E-matrix/electrode | Sensitivity | Linear range | L.O.D. | pH | Stability (%) | References |
|---|---|---|---|---|---|---|---|
| Phenol | HRP/MWCNTs/CPE<br>HRP/MWCNTs/GR/CPE<br>HRP/Hydroxylated fullerene/GR/CPE | 27 nA/μM cm$^2$<br>33 nA/μM cm$^2$<br>15 nA/μM cm$^2$ | 5–85 μM<br>5–85 μM<br>5–200 μM | N.R. | 6.5 | 95% (200 analysis) | Santos et al. (2015) |
| | HRP/PANI NFs/Cs/GA/GCE<br>HRP/PCA-PEG-PCA/GCE | N.R. | 2.5–25 μM<br>2.5–40 μM | 2.5 μM<br>1.5 μM | 7.0 | 80% 6 weeks<br>77% 6 weeks | Shamloo et al. (2013) |
| | HRP/Nano-CuO/PSS film/PGE | N.R. | 0.01–0.5 mM | 0.1 μM | 7.0 | 94% 2 weeks | Kazemi and Khajeh (2011) |
| | HRP-SWCNTs-[BMIM][$PF_6$]/SPE | 134 mA/M cm$^2$ | 6.2–282 μM | 2.8 μM | 7.4 | N.R. | Gurban et al. (2011) |
| 4-CP | Poly(GMA-co MTM)/PPy/HRP/GCE | 60 nA/μM | 1–34 μM | 0.55 μM | 7.0 | 80% 40 days | Ozoner et al. (2010) |
| | (HRP/MWCNTs PAH)n /PSS/PAH/MPS/Au | 21.2 nA/μM | 0.8–16 μM | 0.2 μM | 6.5 | 84.7%<br>90 days | Liu et al. (2008) |
| 4-AP | (HRP/MWCNTs-PAH)n /PSS/PAH/MPS/Au | 205 nA/μM | 0.4–8 μM | 50 nM | 6.5 | 84.7%<br>90 days | Liu et al. (2008) |
| | HRP/NPG/GCE | 187.3 μA/mM cm$^2$ | 5–60 μM | 0.11 μM | 7.0 | 98.2%<br>32 days | Wu et al. (2016) |
| 2-AP | PPy/CNTs/HRP/Au | 40 nA/μM | 8–60.8 μM | 1.53 μM | 7.0 | 70% 1 month (700 uses) | Korkut et al. (2008) |
| 2,4-DCP | HRP/G/GCE | 5.95 μA/μM cm$^2$ | 0.01–13 μM | 5 nM | 7.0 | Lifetime<br>1 month | Q. Xu et al. (2012) |
| N.P. | MB-HRP-Cs/au/Ti/$TiO_2$ | 0.35 μA/μM | 0.3–120 μM | 90 nM | 7.0 | 92% 45 days | Kafi and Chen (2009) |
| Dopamine | HRP/EDC/MWNTs/Pt. | 1.98249 μA/μM | 32–44 μM | 2 μM | 6.0 | N.R. | de Souza Ribeiro et al. (2013) |

**Table 4** Analytical parameters of bienzyme and non-enzyme biosensors

| Analyte | E-matrix/electrode | Sensitivity | L.R. | L.O.D. | pH | Km | Stability (%) | References |
|---|---|---|---|---|---|---|---|---|
| Dopamine | DAAO-Hb/$MnO_2$NPs/PTn/GCE | 12.801 μA/μM | 0.04–9 μM | 41 nM | 6.0 | 0.98 μM | 60% 42 days | Shoja et al. (2017) |
| BPA | XOD-BSA-GA/GCE | 24.5 nA/ nM $cm^2$ | 1–41 nM | 1 nM | 7.5 | N.R. | 75% 15 days | Ben Messaoud et al. (2018) |
| 2,6-DCP | Cs/$PoX_{1B}$/Au | 1.5 μA/μM | 0.01–20 μM | 10 nM | 7.4 | 0.46 μM | Lifetime 3 months | El Ichi et al. (2009) |
| 4-CP | Cs/$PoX_{1B}$/Au | 1900 μA/μM | 1 pM–20 μM | 1 pM | 7.4 | 0.42 μM | | |
| BPA *(Aptamer biosensors)* | Apta-$NH_2$-$Fe_3O_4$/AuNPs/CNTs/ GA/GCE | N.R. | 1–600 nM | 300 pM | 7.4 | _ | 92% 2 weeks | Beiranvand (2017) |
| | ctDNA/SWNTs/Nafion/GCE | 0.722 A/M | 10 nM–20 μM | 5 nM | 8.5 | _ | N.R. | Jiang et al. (2013) |
| | Apta-Cs/SPGE | N.R. | 0.08–15 nM | 15 pM | 7.4 | _ | N.R. | Abnous (2018) |
| | MCH/Apta/Au@$Fe_2O_4$/MWCNTs/ GCE | 86,430 μA/ μM $cm^2$ | 0.1–8 nM | 0.03 nM | 7.6 | _ | 95% 7 days | Baghayeri (2018a) |
| | MCH/Apta/Au/Cu $Fe_2O_4$/ MWCNTs/GCE | – | 0.05–9 nM | 25.2 pM | 7.6 | _ | 96.1% 7 days | Baghayeri (2018b) |
| | PPy/@p-63/AuNPs/GCE | N.R. | 0.5 fM– 5 pM | 80 aM | 7.0 | _ | 93% 30 days | Ensafi et al. (2018) |
| Catechol HQ | Hb/$Cu_2S$ NRs/Nafion/GCE | 0.23 nA/μM 5.23 nA/μM | 7–110 μM 0.6–10 μM | 0.5 μM 30 nM | 7.8 | N.R. | 85% 30 days | M. Xu et al. (2012) |
| Phenol | EB cells-PVA/PAPG/Pt. | N.R. | 1–100 μM | 30 nM | 7.4 | N.R. | 80% 2 weeks | Shin and Lim (2016) |

(continued)

**Table 4** (continued)

| Analyte | | E-matrix/electrode | Sensitivity | L.R. | L.O.D. | pH | Km | Stability (%) | References |
|---|---|---|---|---|---|---|---|---|---|
| Catechol<br>Phenol<br>P-cresol | | Silica sol-gel/Hb/GCE | 3.55 nA/μM<br>3.65 nA/μM<br>3.55 nA/μM | 5–40 μM<br>5–50 μM<br>10–40 μM | 1.1 μM<br>0.8 μM<br>3.5 μM | 7.0 | N.R. | 80%<br>30 days | Kafi et al. (2008) |
| Phenol<br>Catechol<br>HQ | | [Cu($\mu_2$-hep)(hep-H)]$_2$. 2PF$_6$/CPE | 1.2293 μA/μM<br>0.9921 μA/μM<br>0.5135 μA/μM | 33.4 nM – 5.13 μM<br>28.3 nM–26.23 μM<br>54.1 nM–1.81 μM | 10.4 nM<br>8.6 nM<br>15.4 nM | 7.0 | 8.25 μM<br>3.89 μM<br>4.82 μM | 95%<br>200 days | Mobin et al. (2010) |
| Bienzyme | Phenol | Tyr-Lac/DGS/GR SPE | 11.067 μA/mM | N.R. | 2 μM | 6.0 | N.R. | N.R. | Montereali et al. (2010) |
| | | (C/(BSA-Lac/Tyr)$_2$/PU)/Pt. | 0.049 nA/μM | 0.1–140 mM | 1.59 μM | 7.4 | 2.572 mM | N.R. | Barberis et al. (2017) |
| | | PLT-Tyr-GOX/Au | 0.815 μA/μM | 0.05–10 μM | 17 nM | 7.0 | N.R. | N.R. | Liu et al. (2018) |
| | | AChE/ChO/PEI/Pt. Ir. | 0.336 nA/μM | 0–500 μM | N.R. | 12.85 | 635 μM | N.R. | Monti et al. (2017) |
| | **3-C.P.** | Tyr-Lac/Titania gel/CE | 817.7 mA/M | 0.98–7.9 μM | N.R. | 6.0 | N.R. | 52%<br>40 days | Jolanta Kochana et al. (2008) |
| | **C.A.** | Lac-Tyr-Cs–MWCNTs/ITO | 138 μA/mM | 0.4–11 μM | 24.9 nM | 5.5 | 16.6 μM | 90% 12 uses | Diaconu et al. (2011) |
| | | Tyr-Lac/DGS/GR SPE | 1.218 μA/mM | N.R. | 24 μM | 6.0 | N.R. | N.R. | Montereali et al. (2010) |
| | **C.C.** | Tyr-Lac/Titania gel/CE | 5380 mA/M | 0.20–23 μM | 0.13 μM | 6.0 | N.R. | 52%<br>40 days | Jolanta Kochana et al. (2008) |

| | | | | | | | | |
|---|---|---|---|---|---|---|---|---|
| Catechin based biosensors | GA-Nafion-Lac/SNGC/CE | 89.066 nA/μM | 0.04–3 μM | 0.10 μM | 5.0 | 38.02 μM | 84.6%<br>3 weeks | ElKaoutit et al. (2008) |
| | GA-Nafion-Tyr/SNGC/CE | 1.57 nA/μM | 0.6–10.3 μM | 0.33 μM | 6.5 | 72.55 μM | 75%<br>3 weeks | |
| | GA-Nafion-HRP/SNGC/CE | 15.925 nA/μM | 0.2–2 μM | 0.14 μM | 7.0 | 4.21 μM | 83.5%<br>3 weeks | |
| | Tyr-Lac/DGS/GR SPE | 0.435 μA/mM | N.R. | 40 μM | 6.0 | _ | N.R. | Montereali et al. (2010) |
| | BSA-Lac/AuNPs/$NH_2$-CNTs/GCE | 0.142 μA/μM | 1.7–30 μM | 0.20 μM | 5.0 | _ | N.R. | Amatatongchai et al. (2013) |
| | GA/Tyr/SPCE | 217 nA/μM | 0.05–23 μM | 30 nM | 7.0 | _ | 70%<br>53 days | Nadifiyine et al. (2013) |
| | Lac/Den(AuNPs)/PDATT/GCE | 9.6 μA/μM | 0.1–10 μM | 50 nM | 7.0 | _ | 92%<br>2 months | Rahman and Noh (2008) |
| | GA/Tyr/FePc/SWCNTs/SPE | 0.937 μA/μM | 5–250 μM | 0.89 μM | 7.0 | _ | N.R. | Apetrei et al. (2012) |

trode (ElKaoutit et al., 2008) or covalently immobilized into the surface of glassy carbon electrode in order to construct biosensor based on nanocomposites of $NH_2$-functionalized carbon nanotubes (CNT-$NH_2$) using gold nanoparticles (AuNPS) and bovine serum albumin (BSA) for total phenolic content (TPC) evaluation (Amatatongchai et al., 2013). Moreover, a highly stable response of catechin biosensor (92 % during 2 months) was obtained by immobilizing laccase with the formation of amide bonds between carboxylic acid groups of the dendrimer and the amine groups of laccase (Rahman and Noh 2008). However, using tyrosinase enzyme to construct biosensor for the determination of catechin derivatives in black and green teas provides a low detection limit of (LOD = 30 nM) (Nadifiyine et al., 2013). Similarly, (Apetrei et al., 2012) have prepared biosensor applied for the determination of catechin in vegetables samples using tyrosinase enzyme immobilized onto single wall carbon nanotubes screen-printed electrodes modified with iron(II) phthalocyanine.

Monitoring chlorophenol derivatives has been also realized with other biorecognition elements such as new protein ($POX_{1B}$) purified from garlic and immobilized at the surface of gold electrode using chitosan microspheres. The platform used to detect para-chlorophenol at picomolar level and provides ultrahigh sensitivity of 1900 μA/μM (El Ichi et al. 2009). Moreover, Shin and Lim (2016) have immobilized Escherichia coli with polyvinyl alcohol (PVA) at platinum electrode to prepare a microbial bioreporter for monitoring phenols at 30 nM, while Akyilmaz et al. (2017) have used candida tropicalis yeast cells to detect dopamine at 8 μM. Recently, vxanthine oxidase (XOD) was used as the biorecognition element to construct a bisphenol A biosensor combined with BSA and GA (Ben Messaoud et al. 2018).

## *5.3 Bienzyme-Based Biosensors*

There have been limited papers devoted to the development of bienzyme biosensors for the detection of phenol. Immobilizing tyrosinase with glucose oxidase simultaneously in polymer matrix provides a sensitivity of 0.815 μA/μM. Laccase and tyrosinase were also combined using chitosan matrix in order to demonstrate the effect of Tween 20 as a non-ionic surfactant (Diaconu et al. 2011), or using titania sol–gel matrix with a sensitivity of 5380 mA/M (Kochana et al. 2008b).

## *5.4 Hemoglobin, Hemin, and Biomimetic-Based Sensors*

Hemoglobin (Hb) has been used as a biorecognition element to provide high sensitivity of electrochemical biosensors. It could be fixed at the surface of a glassy carbon electrode with the sol–gel method (Kafi et al. 2008) or using Nafion to form a biocomposite film of cuprous sulfide ($Cu_2S$) nanorods that displays a stability of

85% after 30 days for polyphenol detection (Xu et al. 2012). It can be also combined with D-amino acid oxidase (DAAO) with a sensitivity of 12.801 μA/μM (Shoja et al. 2017). In order to enhance the biorecognition sites of biomimetic sensor, Mobin et al. (2010) have synthesized a new dinuclear copper(II) complex which contains an alcoholic OH group able to mimic the active sites of enzyme.

### *5.5 Aptamer/DNA-Based Biosensors*

Aptamer-based electrochemical biosensors were especially applied to monitoring bisphenol A component (Baghayeri et al. 2018a, b; Beiranvand et al. 2017). Most of apta-biosensors were constructed using glassy carbon as the electrochemical transducer. The surface of electrode was modified with a nanocomposite film of $NH_2$ functionalized by $Fe_3O_4$ and gold nanoparticles decorated with carbon nanotubes (LOD = 300 pM) (Beiranvand and Azadbakht 2017) (SWNTs) and Nafion (Jiang et al. 2013) or with a nanocomposite of gold nanoparticles (AuNPs) and graphene immobilized through the formation of thiol–gold (S–Au) bonds (Zhou et al. 2014). Interestingly, the use of a nanoporous gold film (NPGF) attached onto a glassy carbon electrode leads to detection of BPA at a low concentration of 0.056 nM (Zhu et al. 2015).

Recently, Abnous et al. (2018) have developed a new strategy for selective detection of BPA based on the use of terminal deoxynucleotidyl transferase as enzyme to form bridge on the surface of electrode in the absence of bisphenol A with a limit of detection equal to 15 pM. Interestingly, bisphenol A was successfully detected at an ultralow level (LOD = 80 attomolar) using highly selective electrochemical aptasensor based on molecularly imprinted pyrrole, and electrodeposition of gold nanoparticles (AuNPs) with the thiolated DNA sequence (p-63) and free bisphenol A complex (Ensafi et al. 2018).

## 6 List of Abbreviations of Nanomaterials

In order to avoid anomalies and to correct shortcomings in the abbreviation of nanomaterials used for the preparation of biosensors, we have constructed an abbreviated standard form of some words such as chitosan (e.g., abbreviated as "Cs." or "Chit.") or graphene (e.g., abbreviated as "G." or "Gr."). Furthermore, we have added a new abbreviation to investigate the possibility that no abbreviation words are confused with others, emphasizing differences between similar abbreviations. In this chapter, all abbreviated forms of materials and methods as well as techniques used in the field of phenolic compound biosensing are summarized in the list of abbreviations. The list is also extended to cover abbreviations used in enzymeless detection of phenolic compounds.

## 7 Conclusion

The development of biosensors for the detection of phenolic compounds with rapid response, high sensitivity, and long-term stability is now an area of significant research activity. This book chapter is aimed to show new strategies in developing biosensors for sensitive detection of phenolic compounds with the use of a wide range of nanomaterials and conducting polymers as well as biorecognition elements. Recent developments show that the functionalization and dispersion of nanomaterials are the main issues affecting the performance of biosensing platforms and limiting their application. The use of hydrophilic materials could effectively prevent aggregation and therefore aid in the formation of nanostructures for enzyme stabilization. A special focus is laid on studying theoretical approaches in order to predict an accurate estimation of physical and chemical parameters such as the thickness of liquid film, enzyme layer, and diffusion rate of analytes. Biosensors for the detection of phenolic compounds still suffer from some disadvantages such as lack of affinity, diffusion limitations, and weak stability.

## References

Abnous K, Danesh NM, Ramezani M, Alibolandi M, Taghdisi SM (2018) A novel electrochemical sensor for bisphenol A detection based on nontarget-induced extension of aptamer length and formation of a physical barrier. Biosens Bioelectron 119:204–208. https://doi.org/10.1016/j.bios.2018.08.024

Achi F, Bourouina-Bacha S, Bourouina M, Amine A (2015) Mathematical model and numerical simulation of inhibition based biosensor for the detection of Hg(II). Sensors Actuators B Chem 207:413–423. https://doi.org/10.1016/j.snb.2014.10.033

Ahlborg UG, Thunberg TM, Spencer HC (1980) Chlorinated phenols: occurrence, toxicity, metabolism, and environmental impact. CRC Crit Rev Toxicol 7:1–35. https://doi.org/10.3109/10408448009017934

Akyilmaz E, Canbay E, Dinçkaya E, Güvenç C, Yaşa İ, Bayram E (2017) Simultaneous determination of epinephrine and dopamine by using *Candida tropicalis* yeast cells immobilized in a carbon paste electrode modified with Single Wall carbon nanotube. Electroanalysis 29:1976–1984. https://doi.org/10.1002/elan.201700125

Amatatongchai M, Sroysee W, Laosing S, Chairam S (2013) Rapid Screening Method for Assessing Total Phenolic Content Using Simple Flow Injection System with Laccase based-biosensor. Int J Electrochem Sci 8:14

Amine A, Mohammadi H (2018) Amperometry. In: Reference module in chemistry, molecular sciences and chemical engineering. Elsevier, p B9780124095472142000

Arndt KA, Fitzpatrick TB (1965) Topical use of hydroquinone as a depigmenting agent. JAMA 194:965–967

Arora PK, Srivastava A, Singh VP (2014) Bacterial degradation of nitrophenols and their derivatives. J Hazard Mater 266:42–59. https://doi.org/10.1016/j.jhazmat.2013.12.011

Apetrei IM, Tutunaru D, Claudia-Veronica Popa, Apetrei C (2012) P1.1.16 Development of Amperometric Biosensor Based on Tyrosinase Immobilized in Phosphate-Doped Polypyrrole Film for Detection of Biogenic Amines. 4 Pages, 797 KB. https://doi.org/10.5162/IMCS2012/P1.1.16

Baghayeri M, Ansari R, Nodehi M, Razavipanah I, Veisi H (2018a) Voltammetric aptasensor for bisphenol A based on the use of a MWCNT/$Fe_3O_4$@gold nanocomposite. Microchim Acta 185. https://doi.org/10.1007/s00604-018-2838-y

Baghayeri M, Ansari R, Nodehi M, Razavipanah I, Veisi H (2018b) Label-free electrochemical Bisphenol A aptasensor based on designing and fabrication of a magnetic gold nanocomposite. Electroanalysis 30:2160–2166. https://doi.org/10.1002/elan.201800158

Barberis A, Garbetta A, Cardinali A, Bazzu G, D'Antuono I, Rocchitta G, Fadda A, Linsalata V, D'Hallewin G, Serra PA, Minervini F (2017) Real-time monitoring of glucose and phenols intestinal absorption through an integrated Caco-2TC7 cells/biosensors telemetric device: hypoglycemic effect of fruit phytochemicals. Biosens Bioelectron 88:159–166. https://doi.org/10.1016/j.bios.2016.08.007

Barreto G, Madureira D, Capani F, Aon-Bertolino L, Saraceno E, Alvarez-Giraldez LD (2009) The role of catechols and free radicals in benzene toxicity: an oxidative DNA damage pathway. Environ Mol Mutagen 50:771–780. https://doi.org/10.1002/em.20500

Bartlett PN, Pratt KFE (1995) Theoretical treatment of diffusion and kinetics in amperometric immobilized enzyme electrodes part I: redox mediator entrapped within the film. J Electroanal Chem 397:61–78. https://doi.org/10.1016/0022-0728(95)04236-7

Beiranvand S, Azadbakht A (2017) Electrochemical switching with a DNA aptamer-based electrochemical sensor. Mater Sci Eng C 76:925–933. https://doi.org/10.1016/j.msec.2017.03.028

Beiranvand ZS, Abbasi AR, Dehdashtian S, Karimi Z, Azadbakht A (2017) Aptamer-based electrochemical biosensor by using Au-Pt nanoparticles, carbon nanotubes and acriflavine platform. Anal Biochem 518:35–45. https://doi.org/10.1016/j.ab.2016.10.001

Ben Messaoud N, Ghica ME, Dridi C, Ben Ali M, Brett CMA (2018) A novel amperometric enzyme inhibition biosensor based on xanthine oxidase immobilised onto glassy carbon electrodes for bisphenol A determination. Talanta 184:388–393. https://doi.org/10.1016/j.talanta.2018.03.031

Boukoureshtlieva R, Yankova S, Beschkov V, Milusheva J, Naydenova G, Popova L, Yotov G, Hristov S (2013) Monitoring of the phenol biodegradation process with an electrochemical biosensor. Bulg Chem Commun 45:6

Cavalieri EL, Li K-M, Balu N, Saeed M, Devanesan P, Higginbotham S, Zhao J, Gross ML, Rogan EG (2002) Catechol ortho-quinones: the electrophilic compounds that form depurinating DNA adducts and could initiate cancer and other diseases. Carcinogenesis 23:1071–1077. https://doi.org/10.1093/carcin/23.6.1071

Chaubey A, Malhotra BD (2002) Mediated biosensors. Biosens Bioelectron 17:441–456. https://doi.org/10.1016/S0956-5663(01)00313-X

Chen J, Jin Y (2010) Sensitive phenol determination based on co-modifying tyrosinase and palygorskite on glassy carbon electrode. Microchim Acta 169:249–254. https://doi.org/10.1007/s00604-010-0320-6

Chen C-Y, Lin J-H (2006) Toxicity of chlorophenols to Pseudokirchneriella subcapitata under air-tight test environment. Chemosphere 62:503–509. https://doi.org/10.1016/j.chemosphere.2005.06.060

Chen L, Boeckx P, Zhou L, Cleemput O, Li R (2006) Effect of hydroquinone, dicyandiamide and encapsulated calcium carbide on urea N uptake by spring wheat, soil mineral N content and $N_2O$ emission. Soil Use Manag 14:230–233. https://doi.org/10.1111/j.1475-2743.1998.tb00156.x

Claus H (2004) Laccases: structure, reactions, distribution. Micron 35:93–96. https://doi.org/10.1016/j.micron.2003.10.029

Czaplicka M (2004) Sources and transformations of chlorophenols in the natural environment. Sci Total Environ 322:21–39. https://doi.org/10.1016/j.scitotenv.2003.09.015

Dong W, Han J, Shi J, Liang W, Zhang Y, Dong C (2017) Amperometric Biosensor for Detection of Phenolic Compounds Based on Tyrosinase, -Acetyl- -cysteine-capped Gold Nanoparticles and Chitosan Nanocomposite. Chinese Journal of Chemistry 35(8):1305–1310

Danner DJ, Brignac PJ, Arceneaux D, Patel V (1973) The oxidation of phenol and its reaction product by horseradish peroxidase and hydrogen peroxide. Arch Biochem Biophys 156:759–763. https://doi.org/10.1016/0003-9861(73)90329-9

De Smet R (2003) Toxicity of free p-cresol: a prospective and cross-sectional analysis. Clin Chem 49:470–478. https://doi.org/10.1373/49.3.470

Diaconu M, Litescu SC, Radu GL (2011) Bienzymatic sensor based on the use of redox enzymes and chitosan–MWCNT nanocomposite. Evaluation of total phenolic content in plant extracts. Microchim Acta 172:177–184. https://doi.org/10.1007/s00604-010-0486-y

Durán N, Esposito E (2000) Potential applications of oxidative enzymes and phenoloxidase-like compounds in wastewater and soil treatment: a review. Appl Catal B Environ 28:83–99. https://doi.org/10.1016/S0926-3373(00)00168-5

Dwevedi A (2016) Basics of enzyme immobilization. In: Enzyme immobilization. Springer International Publishing, Cham, pp 21–44

El Ichi S, Marzouki MN, Korri-Youssoufi H (2009) Direct monitoring of pollutants based on an electrochemical biosensor with novel peroxidase (POX1B). Biosens Bioelectron 24:3084–3090. https://doi.org/10.1016/j.bios.2009.03.036

ElKaoutit M, Naranjo-Rodriguez I, Temsamani KR, Hernández-Artiga MP, Bellido-Milla D, de Cisneros JLH-H (2008) A comparison of three amperometric phenoloxidase–Sonogel–carbon based biosensors for determination of polyphenols in beers. Food Chem 110:1019–1024. https://doi.org/10.1016/j.foodchem.2008.03.006

Ensafi AA, Amini M, Rezaei B (2018) Molecularly imprinted electrochemical aptasensor for the attomolar detection of bisphenol A Microchimica Acta 185. https://doi.org/10.1007/s00604-018-2810-x

Erler C, Novak J (2010) Bisphenol A exposure: human risk and health policy. J Pediatr Nurs 25:400–407. https://doi.org/10.1016/j.pedn.2009.05.006

Fitzpatrick TB, Arndt KA, El Mofty AM, Pathak MA (1966) Hydroquinone and psoralens in the therapy of hypermelanosis and vitiligo. Arch Dermatol 93:589–600. https://doi.org/10.1001/archderm.1966.01600230093025

Ge T, Han J, Qi Y, Gu X, Ma L, Zhang C, Naeem S, Huang D (2017) The toxic effects of chlorophenols and associated mechanisms in fish. Aquat Toxicol 184:78–93. https://doi.org/10.1016/j.aquatox.2017.01.005

van Gestel CAM, Ma W-C (1988) Toxicity and bioaccumulation of chlorophenols in earthworms, in relation to bioavailability in soil. Ecotoxicol Environ Saf 15:289–297. https://doi.org/10.1016/0147-6513(88)90084-X

Gonzalez-Anton R, Osipova MM, Garcia-Hernandez C, Dubinina TV, Tomilova LG, Garcia-Cabezon C, Rodriguez-Mendez ML (2017) Subphthalocyanines as electron mediators in biosensors based on phenol oxidases: application to the analysis of red wines. Electrochim Acta 255:239–247. https://doi.org/10.1016/j.electacta.2017.09.168

Guan Y, Liu L, Chen C, Kang X, Xie Q (2016) Effective immobilization of tyrosinase via enzyme catalytic polymerization of l -DOPA for highly sensitive phenol and atrazine sensing. Talanta 160:125–132. https://doi.org/10.1016/j.talanta.2016.07.003

Haghighi-Podeh MR, Bhattacharya SK (1999) Fate and toxic effects of nitrophenols on anaerobic treatment systems. Water Sci Technol 34:345–350. https://doi.org/10.1016/0273-1223(96)00664-6

Han E, Yang Y, He Z, Cai J, Zhang X, Dong X (2015) Development of tyrosinase biosensor based on quantum dots/chitosan nanocomposite for detection of phenolic compounds. Anal Biochem 486:102–106. https://doi.org/10.1016/j.ab.2015.07.001

Ibarra-Escutia P, Gómez JJ, Calas-Blanchard C, Marty JL, Ramírez-Silva MT (2010) Amperometric biosensor based on a high resolution photopolymer deposited onto a screen-printed electrode for phenolic compounds monitoring in tea infusions. Talanta 811(4–5):1636–1642

Jagetia GC, Menon KSL, Jain V (2001) Genotoxic effect of hydroquinone on the cultured mouse spleenocytes. Toxicol Lett 121:15–20. https://doi.org/10.1016/S0378-4274(00)00290-3

Jiang X, Ding W, Luan C, Ma Q, Guo Z (2013) Biosensor for bisphenol A leaching from baby bottles using a glassy carbon electrode modified with DNA and single walled carbon nanotubes. Microchim Acta 180:1021–1028. https://doi.org/10.1007/s00604-013-1025-4

Jim C (2014) Acidity of phenols—chemistry LibreTexts. https://chem.libretexts.org/Bookshelves/Organic_Chemistry/Supplemental_Modules_(Organic_Chemistry)/Phenols/Properties_of_Phenols/Acidity_of_Phenols. Accessed 3 Jan 2019

Kafi AKM, Chen A (2009) A novel amperometric biosensor for the detection of nitrophenol. Talanta 79:97–102. https://doi.org/10.1016/j.talanta.2009.03.015

Kafi AKM, Lee D-Y, Park S-H, Kwon Y-S (2008) Potential application of hemoglobin as an alternative to peroxidase in a phenol biosensor. Thin Solid Films 516:2816–2821. https://doi.org/10.1016/j.tsf.2007.04.123

Kazemi SH, Khajeh K (2011) Electrochemical studies of a novel biosensor based on the CuO nanoparticles coated with horseradish peroxidase to determine the concentration of phenolic compounds. Journal of the Iranian Chemical Society 8(S1):S152–S160

Kishino T, Kobayshi K (1996) Studies on the mechanism of toxicity of chlorophenols found in fish through quantitative structure-activity relationships. Water Res 30:393–399. https://doi.org/10.1016/0043-1354(95)00152-2

Kochana J, Nowak P, Jarosz-Wilkołazka A, Bieroń M (2008) Tyrosinase/laccase bienzyme biosensor for amperometric determination of phenolic compounds. Microchemical Journal 89(2):171–174

Kochana J, Gala A, Parczewski A, Adamski J (2008a) Titania sol–gel-derived tyrosinase-based amperometric biosensor for determination of phenolic compounds in water samples. Examination of interference effects. Anal Bioanal Chem 391:1275–1281. https://doi.org/10.1007/s00216-007-1798-6

Kochana J, Nowak P, Jarosz-Wilkołazka A, Bieroń M (2008b) Tyrosinase/laccase bienzyme biosensor for amperometric determination of phenolic compounds. Microchem J 89:171–174. https://doi.org/10.1016/j.microc.2008.02.004

Kogevinas M, Becher H, Benn T, Bertazzi PA, Boffetta P, Bueno-de-Mesqurta HB, Coggon D, Colin D, Flesch-Janys D, Fingerhut M, Green L, Kauppinen T, Littorin M, Lynge E, Mathews JD, Neuberger M, Pearce N, Saracci R (1997) Cancer mortality in workers exposed to phenoxy herbicides, chlorophenols, and dioxins an expanded and updated international cohort study. Am J Epidemiol 145:1061–1075. https://doi.org/10.1093/oxfordjournals.aje.a009069

Korkut S, Keskinler B, Erhan E (2008) An amperometric biosensor based on multiwalled carbon nanotube-poly(pyrrole)-horseradish peroxidase nanobiocomposite film for determination of phenol derivatives. Talanta 76:1147–1152. https://doi.org/10.1016/j.talanta.2008.05.016

Korkut S, Kilic MS, Erhan E (2015) Modified poly(pyrrole) film based biosensors for phenol detection. 9:4

Li G, Sun K, Li D, Lv P, Wang Q, Huang F, Wei Q (2016) Biosensor based on bacterial cellulose-Au nanoparticles electrode modified with laccase for hydroquinone detection. Colloids and Surfaces A: Physicochemical and Engineering Aspects 509:408–414

Li H, Hu X, Zhu H, Zang Y, Xue H (2017) Amperometric phenol biosensor based on a new immobilization matrix: polypyrrole nanotubes derived from methyl orange as dopant. Int J Electrochem Sci 12:6714–6728. https://doi.org/10.20964/2017.07.80

Liu L, Zhang F, Xi F, Lin X (2008) Highly sensitive biosensor based on bionanomultilayer with water-soluble multiwall carbon nanotubes for determination of phenolics. Biosens Bioelectron 24:306–312. https://doi.org/10.1016/j.bios.2008.04.003

Liu Y, Lei J, Ju H (2010) CuO-doped Mesoporous silica hybrid for rapid and sensitive Amperometric detection of phenolic compounds. Electroanalysis 22:2407–2412. https://doi.org/10.1002/elan.201000200

Liu L, Kang X, Chen C, Zhang H, Chen C, Xie Q (2018) L-tyrosine polymerization-based ultrasensitive multi-analyte enzymatic biosensor. Talanta 179:803–809. https://doi.org/10.1016/j.talanta.2017.12.014

Mayer AM, Harel E (1979) Polyphenol oxidases in plants. Phytochemistry 18:193–215. https://doi.org/10.1016/0031-9422(79)80057-6

Mayorga-Martinez CC, Cadevall M, Guix M, Ros J, Merkoçi A (2013) Bismuth nanoparticles for phenolic compounds biosensing application. Biosens Bioelectron 40:57–62. https://doi.org/10.1016/j.bios.2012.06.010

Megharaj M, Pearson HW, Venkateswarlu K (1991) Toxicity of phenol and three nitrophenols towards growth and metabolic activities of *Nostoc linckia*, isolated from soil. Arch Environ Contam Toxicol 21:578–584. https://doi.org/10.1007/BF01183881

Mehta J, Bhardwaj SK, Bhardwaj N, Paul AK, Kumar P, Kim K-H, Deep A (2016) Progress in the biosensing techniques for trace-level heavy metals. Biotechnol Adv 34:47–60. https://doi.org/10.1016/j.biotechadv.2015.12.001

Michałowicz J (2014) Bisphenol A—sources, toxicity and biotransformation. Environ Toxicol Pharmacol 37:738–758. https://doi.org/10.1016/j.etap.2014.02.003

Mobin SM, Sanghavi BJ, Srivastava AK, Mathur P, Lahiri GK (2010) Biomimetic sensor for certain phenols employing a copper(II) complex. Anal Chem 82:5983–5992. https://doi.org/10.1021/ac1004037

Montereali MR, Seta LD, Vastarella W, Pilloton R (2010) A disposable Laccase–Tyrosinase based biosensor for amperometric detection of phenolic compounds in must and wine. Journal of Molecular Catalysis B: Enzymatic 64(3–4):189–194

Monti P, Calia G, Marceddu S, Dettori MA, Fabbri D, Jaoua S, O'Neill RD, Migheli Q, Delogu G, Serra PA (2017) Low electro-synthesis potentials improve permselectivity of polymerized natural phenols in biosensor applications. Talanta 162:151–158. https://doi.org/10.1016/j.talanta.2016.10.019

Mossanha R, Erdmann CA, Santos CS, Wohnrath K, Fujiwara ST, Pessoa CA (2017) Construction of a biosensor based on SAM of thiolactic acid on gold nanoparticles stabilized by silsesquioxane polyelectrolyte for cathecol determination. Sensors Actuators B Chem 252:747–756 https://doi.org/10.1016/j.snb.2017.06.001

Nadifiyine S, Haddam M, Mandli J, Chadel S, Blanchard CC, Marty JL, Amine A (2013) Amperometric Biosensor Based on Tyrosinase Immobilized on to a Carbon Black Paste Electrode for Phenol Determination in Olive Oil. Analytical Letters 46(17):2705–2726

Nurul Karim M, Lee HJ (2013) Amperometric phenol biosensor based on covalent immobilization of tyrosinase on au nanoparticle modified screen printed carbon electrodes. Talanta 116:991–996. https://doi.org/10.1016/j.talanta.2013.08.003

Olaniran AO, Igbinosa EO (2011) Chlorophenols and other related derivatives of environmental concern: properties, distribution and microbial degradation processes. Chemosphere 83:1297–1306. https://doi.org/10.1016/j.chemosphere.2011.04.009

Oriero DA, Gyan IO, Bolshaw BW, Cheng IF, Aston DE (2015) Electrospun biocatalytic hybrid silica–PVA-tyrosinase fiber mats for electrochemical detection of phenols. Microchem J 118:166–175. https://doi.org/10.1016/j.microc.2014.09.005

Orozco J, Jiménez-Jorquera C, Fernández-Sánchez C (2009) Gold nanoparticle-modified ultramicroelectrode arrays for biosensing: a comparative assessment. Bioelectrochemistry 75:176–181. https://doi.org/10.1016/j.bioelechem.2009.03.013

Ozoner SK, Yalvac M, Erhan E (2010) Flow injection determination of catechol based on polypyrrole–carbon nanotube–tyrosinase biocomposite detector. Current Applied Physics 10(1):323–328

Ozkan SA, Kauffmann J-M, Zuman P (2015) Electroanalysis in biomedical and pharmaceutical sciences. Springer, Berlin

Ozoner SK, Yilmaz F, Celik A, Keskinler B, Erhan E (2011) A novel poly(glycine methacrylate-co-3-thienylmethyl methacrylate)-polypyrrole-carbon nanotube-horseradish peroxidase composite film electrode for the detection of phenolic compounds. Curr Appl Phys 11:402–408 https://doi.org/10.1016/j.cap.2010.08.010

Pang Y, Zeng G, Tang L, Zhang Y, Li Z, Chen L (2011) Laccase biosensor using magnetic multi-walled carbon nanotubes and chitosan/silica hybrid membrane modified magnetic carbon paste electrode. J Cent S Univ Technol 18:1849–1856. https://doi.org/10.1007/s11771-011-0913-1

Penu R, Obreja AC, Patroi D, Diaconu M, Radu GL (2015) Graphene and gold nanoparticles based reagentless biodevice for phenolic endocrine disruptors monitoring. Microchem J 121:130–135. https://doi.org/10.1016/j.microc.2015.03.002

Pera-Titus M, García-Molina V, Baños MA, Giménez J, Esplugas S (2004) Degradation of chlorophenols by means of advanced oxidation processes: a general review. Appl Catal B Environ 47:219–256. https://doi.org/10.1016/j.apcatb.2003.09.010

Pingarrón JM, Yáñez-Sedeño P, González-Cortés A (2008) Gold nanoparticle-based electrochemical biosensors. Electrochim Acta 53:5848–5866. https://doi.org/10.1016/j.electacta.2008.03.005

Prins GS, Tang W-Y, Belmonte J, Ho S-M (2008) Perinatal exposure to oestradiol and bisphenol A alters the prostate epigenome and increases susceptibility to carcinogenesis: bisphenol A and prostate carcinogenesis. Basic Clin Pharmacol Toxicol 102:134–138. https://doi.org/10.1111/j.1742-7843.2007.00166.x

Pumera M, Ambrosi A, Bonanni A, Chng ELK, Poh HL (2010) Graphene for electrochemical sensing and biosensing. TrAC Trends Anal Chem 29:954–965. https://doi.org/10.1016/j.trac.2010.05.011

Qu J, Lou T, Wang Y, Dong Y, Xing H (2015) Determination of catechol by a novel Laccase biosensor based on zinc-oxide sol-gel. Anal Lett 48:1842–1853. https://doi.org/10.1080/00032719.2014.1003427

Rahman MA, Noh H-B (2008) Direct Electrochemistry of Laccase Immobilized on Au Nanoparticles Encapsulated-Dendrimer Bonded Conducting Polymer: Application for a Catechin Sensor. Analytical Chemistry 80(21):8020–8027

Rezg R, El-Fazaa S, Gharbi N, Mornagui B (2014) Bisphenol A and human chronic diseases: current evidences, possible mechanisms, and future perspectives. Environ Int 64:83–90. https://doi.org/10.1016/j.envint.2013.12.007

Rochester JR (2013) Bisphenol A and human health: a review of the literature. Reprod Toxicol 42:132–155. https://doi.org/10.1016/j.reprotox.2013.08.008

Ronkainen NJ, Halsall HB, Heineman WR (2010) Electrochemical biosensors. Chem Soc Rev 39:1747. https://doi.org/10.1039/b714449k

Santos AS, Costa VS, Felício RC (2015) Comparative Study of Nanostructured Matrices Employed in the Development of Biosensors Based on HRP Enzyme for Determination of Phenolic Compounds. Electroanalysis 27(7):1572–1578

Sarika C, Rekha K, Narasimha Murthy B (2015) Studies on enhancing operational stability of a reusable laccase-based biosensor probe for detection of ortho-substituted phenolic derivatives. 3 Biotech 5:911–924. https://doi.org/10.1007/s13205-015-0292-7

Seachrist DD, Bonk KW, Ho S-M, Prins GS, Soto AM, Keri RA (2016) A review of the carcinogenic potential of bisphenol A. Reprod Toxicol 59:167–182. https://doi.org/10.1016/j.reprotox.2015.09.006

Sezgintürk MK, Odaci D, Pazarlioğlu N, Pilloton R, Dinçkaya E, Telefoncu A, Timur S (2010) Construction and Comparison of Laccase Biosensors Capable of Detecting Xenobiotics. Artificial Cells, Blood Substitutes, and Biotechnology 38(4):192–199

Shamloo A, Vossoughi M, Alemzadeh I, Naeini AT, Darvish M (2013) Two nanostructured polymers: Polyaniline Nanofibers and new linear-dendritic matrix of poly(citric acid)-block-poly(ethylene glycol) copolymers for environmental monitoring in novel biosensors. Int J Polym Mater 62:377–383. https://doi.org/10.1080/00914037.2012.710861

Shan D, Zhang J, Xue H-G, Zhang Y-C, Cosnier S, Ding S-N (2009) Polycrystalline bismuth oxide films for development of amperometric biosensor for phenolic compounds. Biosens Bioelectron 24:3671–3676. https://doi.org/10.1016/j.bios.2009.05.038

Shimomura T, Itoh T, Sumiya T, Hanaoka T, Mizukami F, Ono M (2011) Amperometric detection of phenolic compounds with enzyme immobilized in mesoporous silica prepared by electrophoretic deposition. Sensors Actuators B Chem 153:361–368. https://doi.org/10.1016/j.snb.2010.10.048

Shin HJ, Lim WK (2016) Comparative evaluation of an electrochemical bioreporter for detecting phenolic compounds. Prep Biochem Biotechnol 46:71–77. https://doi.org/10.1080/10826068.2014.979207

Shoja Y, Rafati AA, Ghodsi J (2017) Polythiophene supported $MnO_2$ nanoparticles as nano-stabilizer for simultaneously electrostatically immobilization of d-amino acid oxidase and hemoglobin as efficient bio-nanocomposite in fabrication of dopamine bi-enzyme biosensor. Mater Sci Eng C 76:637–645. https://doi.org/10.1016/j.msec.2017.03.155

Singh S, Li SS-L (2012) Epigenetic effects of environmental chemicals bisphenol A and phthalates. Int J Mol Sci 13:10143–10153. https://doi.org/10.3390/ijms130810143

de Souza Ribeiro FA, Tarley CRT, Borges KB, Pereira AC (2013) Development of a square wave voltammetric method for dopamine determination using a biosensor based on multiwall carbon nanotubes paste and crude extract of *Cucurbita pepo* L. Sensors Actuators B Chem 185:743–754. https://doi.org/10.1016/j.snb.2013.05.072

Stewart PS (2003) Diffusion in biofilms. J Bacteriol 185:1485–1491. https://doi.org/10.1128/JB.185.5.1485-1491.2003

van Stroe-Biezen SAM, Everaerts FM, Janssen LJJ, Tacken RA (1993) Diffusion coefficients of oxygen, hydrogen peroxide and glucose in a hydrogel. Anal Chim Acta 273:553–560. https://doi.org/10.1016/0003-2670(93)80202-V

Teles FRR, Fonseca LP (2008) Applications of polymers for biomolecule immobilization in electrochemical biosensors. Mater Sci Eng C 28:1530–1543. https://doi.org/10.1016/j.msec.2008.04.010

Thévenot DR, Toth K, Durst RA, Wilson GS (1999) Electrochemical biosensors: recommended definitions and classification. IUPAC, Great Britain

Thévenot DR, Toth K, Durst RA, Wilson GS (2001) Electrochemical biosensors: recommended definitions and classification. Anal Lett 34:635–659. https://doi.org/10.1081/AL-100103209

Touloupakis E, Chatzipetrou M, Boutopoulos C, Gkouzou A, Zergioti I (2014) A polyphenol biosensor realized by laser printing technology. Sensors Actuators B Chem 193:301–305. https://doi.org/10.1016/j.snb.2013.11.110

Trojanowicz M, Hitchman ML (1996) Determination of pesticides using electrochemical biosensors. TrAC Trends Anal Chem 15:38–45. https://doi.org/10.1016/0165-9936(96)88036-8

Uberoi V, Bhattacharya SK (1997) Toxicity and degradability of nitrophenols in anaerobic systems. Water Environ Res 69:146–156. https://doi.org/10.2175/106143097X125290

Upan J, Reanpang P, Chailapakul O, Jakmunee J (2016) Flow injection amperometric sensor with a carbon nanotube modified screen printed electrode for determination of hydroquinone. Talanta 146:766–771

Verrastro M, Cicco N, Crispo F, Morone A, Dinescu M, Dumitru M, Favati F, Centonze D (2016) Amperometric biosensor based on Laccase immobilized onto a screen-printed electrode by matrix assisted pulsed laser evaporation. Talanta 154:438–445. https://doi.org/10.1016/j.talanta.2016.03.072

Vlastos D, Antonopoulou M, Konstantinou I (2016) Evaluation of toxicity and genotoxicity of 2-chlorophenol on bacteria, fish and human cells. Sci Total Environ 551–552:649–655. https://doi.org/10.1016/j.scitotenv.2016.02.043

Wang J (2005a) Nanomaterial-based electrochemical biosensors. Analyst 130:421. https://doi.org/10.1039/b414248a

Wang J (2005b) Carbon-nanotube based electrochemical biosensors: a review. Electroanalysis 17:7–14. https://doi.org/10.1002/elan.200403113

Wang J (2006) Electrochemical biosensors: towards point-of-care cancer diagnostics. Biosens Bioelectron 21:1887–1892. https://doi.org/10.1016/j.bios.2005.10.027

Wang S, Tan Y, Zhao D, Liu G (2008) Amperometric tyrosinase biosensor based on Fe3O4 nanoparticles–chitosan nanocomposite. Biosensors and Bioelectronics 23(12):1781–1787

Wang H, Qin Y, Chen K, Xue H (2016) The phenol biosensor based on LDHs/SWNTs hybrid materials. Int J Electrochem Sci 11:15

Wu C, Liu Z, Sun H, Wang X, Xu P (2016) Selective determination of phenols and aromatic amines based on horseradish peroxidase-nanoporous gold co-catalytic strategy. Biosensors and Bioelectronics 79:843–849

Xinhua Xu, Ping Lu, Yumei Zhou, Zhenzhen Zhao, Meiqing Guo, (2009) Laccase immobilized on methylene blue modified mesoporous silica MCM-41/PVA. Materials Science and Engineering: C 29(7):2160–2164

Xu M, Cui L, Han R, Ai S (2012) Amperometric biosensor based on hemoglobin immobilized on $Cu_2S$ nanorods/nafion nanocomposite film for the determination of polyphenols. J Solid State Electrochem 16:2547–2554. https://doi.org/10.1007/s10008-012-1673-z

Yang J, Li D, Fu J, Huang F, Wei Q (2016) $TiO_2$-CuCNFs based laccase biosensor for enhanced electrocatalysis in hydroquinone detection. Journal of Electroanalytical Chemistry 766:16–23

Yáñez-Sedeño P, Pingarrón JM (2005) Gold nanoparticle-based electrochemical biosensors. Anal Bioanal Chem 382:884–886. https://doi.org/10.1007/s00216-005-3221-5

Yunyong Li, Qin C, Chen C, Fu Y, Ma M, Xie Q (2012) Highly sensitive phenolic biosensor based on magnetic polydopamine-laccase-$Fe_3O_4$ bionanocomposite. Sensors Actuators B Chem 168:46–53. https://doi.org/10.1016/j.snb.2012.01.013

Zhang J, Lei J, Liu Y, Zhao J, Ju H (2009) Highly sensitive amperometric biosensors for phenols based on polyaniline–ionic liquid–carbon nanofiber composite. Biosens Bioelectron 24:1858–1863. https://doi.org/10.1016/j.bios.2008.09.012

Zhao J, Chen G, Zhu L, Li G (2011) Graphene quantum dots-based platform for the fabrication of electrochemical biosensors. Electrochem Commun 13:31–33. https://doi.org/10.1016/j.elecom.2010.11.005

Zheng Y, Wang D, Li Z, Sun X, Gao T, Zhou G (2018) Laccase biosensor fabricated on flower–shaped yolk–shell $SiO_2$ nanospheres for catechol detection. Colloids Surf A Physicochem Eng Asp 538:202–209. https://doi.org/10.1016/j.colsurfa.2017.10.086

Zhou M, Zhai Y, Dong S (2009) Electrochemical sensing and biosensing platform based on chemically reduced Graphene oxide. Anal Chem 81:5603–5613. https://doi.org/10.1021/ac900136z

Zhou L, Wang J, Li D, Li Y (2014) An electrochemical aptasensor based on gold nanoparticles dotted graphene modified glassy carbon electrode for label-free detection of bisphenol A in milk samples. Food Chem 162:34–40. https://doi.org/10.1016/j.foodchem.2014.04.058

Zhu Y, Zhou C, Yan X, Yan Y, Wang Q (2015) Aptamer-functionalized nanoporous gold film for high-performance direct electrochemical detection of bisphenol A in human serum. Anal Chim Acta 883:81–89. https://doi.org/10.1016/j.aca.2015.05.002

# Noble Metal-Metal Oxide Hybrid Nanoparticles for Surface-Enhanced Raman Spectroscopy-Based Sensors

**Bramhaiah Kommula and Neena S. John**

## Contents



## Abbreviations

| | |
|---|---|
| 2-NT | 2-Naphthalenethiol |
| 4-ABT | 4-Aminobenzenethiol |
| 4-MBA | 4-Mercaptobenzoicacid |
| 4-MP | 4-Mercaptophenol |

B. Kommula · N. S. John (✉)
Centre for Nano and Soft Matter Sciences, Jalahalli, Bengaluru, India
e-mail: jsneena@cens.res.in



Inamuddin, A. M. Asiri (eds.), *Nanosensor Technologies for Environmental Monitoring*, Nanotechnology in the Life Sciences,
https://doi.org/10.1007/978-3-030-45116-5_11

| | |
|---|---|
| 4-MPY | 4-Mercaptopyridine |
| 4-NP | 4-Nitrophenol |
| ANTA | 5-Amino-3-nitro-1 H-1,2,4-triazole |
| ATP | Aminothiophenol |
| BT | Benzenethiol |
| CB | Conduction band |
| CL-20 | 2,4,6,8,10,12-Hexanitro-2,4,6,8,10,12-hexaazaisowurtzitane |
| CT | Charge transfer |
| CV | Crystal violet |
| DTNB | 5–5′-Dithio-bis (2-nitrobenzoic acid) |
| EF | Enhancement factor |
| EM | Electromagnetic |
| FESEM | Field emission scanning electron microscope |
| FIB | Focused ion beam |
| FOX-7 | 1,1-Diamino-2,2dinitroethene |
| HOMO | Highest occupied molecular orbitals |
| HRTEM | High-resolution transmission electron microscope |
| LUMO | Lowest unoccupied molecular orbitals |
| MB | Methylene blue |
| MBA | Mercaptobenzoic acid |
| MEPL | Metal-enhanced photoluminescence |
| MG | Malachite green |
| MO | Methyl orange |
| MPH | Mercaptophenol |
| NIR | Near-infrared |
| NPs | Nanoparticles |
| NRs | Nanorods |
| PATP | p-Aminothiophenol |
| PNTP | p-Nitrothiophenol |
| PSA | Prostate-specific antigen |
| PVP | Poly vinyl pyrrolidone |
| Py | Pyridine |
| R6G | Rhodamine 6G |
| RhB | Rhodamine B |
| SERS | Surface-enhanced Raman spectroscopy |
| SPR | Surface Plasmon resonance |
| TP | Thiophenol |
| VB | Valance band |

# 1 Introduction

Raman spectroscopy is an attractive and utmost research area for the surface scientists. It can be used for the analysis of adsorbed molecules on the chemical, biological and physical surfaces and interfaces at the molecular scale. However, the fundamental disadvantage of conventional Raman spectroscopy is its low sensitivity due to the low scattering process (Moskovits 2005; Schlücker 2014). To overcome this disadvantage, a potential tactic is surface-enhanced Raman spectroscopy (SERS), which is a surface phenomenon that intensifies the inherently weak Raman scattering signal of adsorbed analyte molecules on metal or semiconductor surfaces by many orders of magnitude (Betz et al. 2014; Sharma et al. 2012). SERS has many advantages such as significant signal enhancement, fingerprint chemical identification and inherent suitability for aqueous media. SERS has arisen as a most important analytic tool to find trace amounts of analytes in many fields such as chemical, biology, environment and food owing to its ultra-high sensitivity, potential rapidity and non-destructive nature (Schlücker 2014; Betz et al. 2014).

## *1.1 History of SERS*

The SERS effect was first reported by Fleischmann et al. in 1974 (Fleischmann et al. 1974) and observed a significant enhancement of the Raman band intensity of single monolayers of pyridine adsorbed on the electrochemically roughened surface of the Ag electrode from aqueous solution. Their approach was mainly to roughen the electrode surface to increase its surface area, and therefore the number of adsorbed molecules. Later, in 1977, Van Duyne (Jeanmaire and Van Duyne 1977) and Creighton (Albrecht and Creighton et al. 1977) first recognized that the enhancement in the Raman signal of analyte molecules was due to the excited localized surface plasmon resonance on the surface of the electrode, more than the contribution from increased surface area factor. Since then, considerable interest has been focused on the development of spectroscopic instrumentation, fabrication methods, theoretical modelling and novel detection schemes towards the realization of SERS for practical applications. In this chapter, we primarily aim at the introduction of recent research progress in the noble metal-metal oxide hybrids as SERS sensors, particularly, the detection of various molecules in different fields such as chemical identification, biological sensing and environmental analysis is highlighted.

## *1.2 Noble Metal Nanostructures*

Noble metal nanoparticles, primarily Ag, Au and Cu, are popularly employed as SERS substrates for the detection of analyte molecules. Other than Ag, Au and Cu, metals such as alkali metals (Li, Na, K, Rb and Cs), Al, Ga, In, Pt, Rh, Os, Ir and their alloys have also been explored as SERS substrates for the detection of molecules (Van Duyne et al. 1993; Sharma et al. 2012). Al and Cu nanostructures are cheaper than other metals. The foremost disadvantages of some of these metals are that they can easily get oxidized and exhibit relatively low enhancement factors. Nonetheless, if the new approaches can develop to overcome this drawback, novel

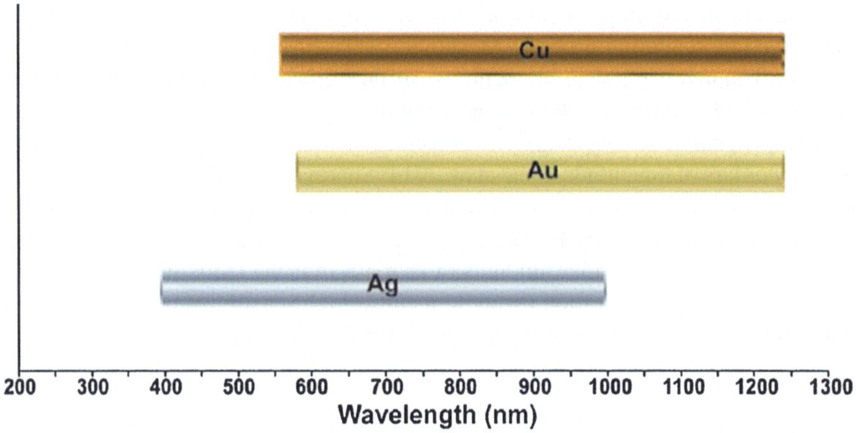


**Fig. 1** Approximate wavelength ranges of Ag, Au and Cu for SERS applications (Sharma et al. 2012)

avenues could be opened for the development of metallic SERS substrates. So far, plasmonic nanostructures of Ag, Au and Cu are the most commonly used SERS substrates due to their higher enhancement factors and availability of plasmonic resonance in the visible and near-infrared (NIR) regions (Sharma et al. 2012; Jahn et al. 2016). Furthermore, the tunability of plasmonics in the range of 300–1200 nm for Ag and 500–1200 nm for Au, which in turn allows a wide option to choose the excitation wavelength, also renders them as attractive materials (Fig. 1). It is well known that Ag nanoparticles (NPs) show much higher enhancement factors and improvement in SERS signal several times than similar Au NPs in the visible light region. The optimization of the enhancement of Raman scattering by plasmonic effects in noble metals is achieved by tuning the morphology (anisotropy) and their relative distribution (creates intrinsic hotspots) within the substrate (Reguera et al. 2017).

The enhancement of the Raman signal of the probed molecules is explained essentially by two theoretical mechanisms, including long-range electromagnetic (EM) enhancement and short-range chemical enhancement (Guerrini and Graham 2012; Schlücker 2014; Sharma et al. 2012; Jahn et al. 2016). The electromagnetic mechanism is mainly responsible for the improvement of Raman scattering. The surface plasmon polaritons can be excited at the metal nanoparticle surface if the wavelength of the incident light matches the plasmon resonance (Schlücker 2014; Jahn et al. 2016; Guerrini and Graham 2012). As a result, a strong electromagnetic field is induced in the metallic nanoparticle surface and the Raman modes of the adsorbed molecules on the surface of the metal nanoparticles are enhanced with the Raman intensity being proportional to the square of the incident electromagnetic

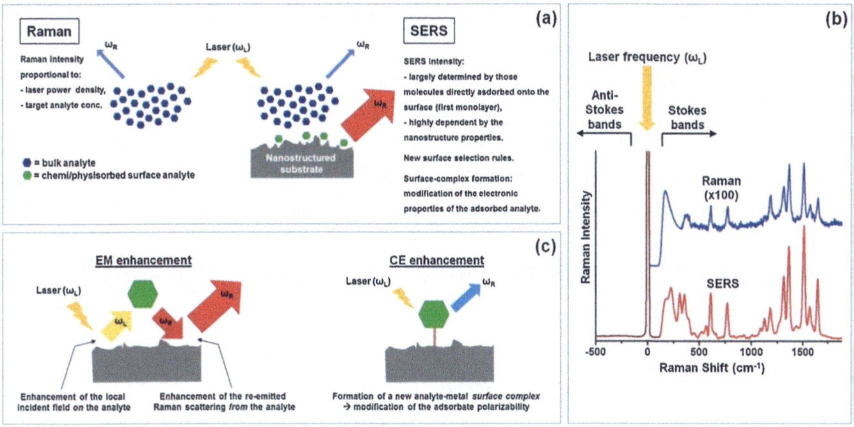


**Fig. 2** (**a**) Schematic representation of Raman and SERS phenomena (**b**) SERS spectra of R6G $10^{-7}$ M (red line) in silver hydroxylamine colloid and Raman spectra of R6G $10^{-3}$ M (blue line) in milli-Q water (**c**) pictorial representation of the electromagnetic and chemical enhancements in SERS (Guerrini and Graham 2012)

field intensity (Fig. 2). Depending on the structure of the plasmonic material, EM enhancement for SERS is theoretically calculated to reach factors of ~$10^{10}$–$10^{11}$ (Guerrini and Graham 2012; Schlücker 2014; Le et al. 2007).

Most of the reports have demonstrated that plasmonic resonances play an essential role in the observed Raman enhancements. However, the plasmonic theory alone could not explain all the SERS regarding varieties of molecules and substrates. Chemical enhancement mechanism associated with the direct interaction of the analyte molecules of the first adsorbed layer with the SERS substrate gives the enhancement factor up to ~$10^3$ (Schlücker 2014; Jahn et al. 2016; Wang and Kong 2015) and its magnitude depends on the chemical structure of the molecule. The chemical enhancement is the sum of various contributions as follows: (1) chemical interaction between the analyte and metal nanoparticle at ground state, (2) resonant excitation of charge transfer process between the metal nanoparticle and analyte molecule and (3) resonance Raman enhancement via excitation of an electronic transfer within the analyte molecule (McNay et al. 2011). Spectroscopically forbidden Raman modes can be observable under SERS conditions (Le et al. 2007). The total SERS enhancement factor is the product of both electromagnetic and chemical enhancement mechanism. For greatly optimized surfaces, the SERS may reach up to ~$10^{10}$–$10^{11}$ (Schlücker 2014; Le et al. 2007).

Metal nanoparticles have a few drawbacks such as aggregation due to ageing and strong fluorescence background upon laser illumination. Notably, in the case of fluorescent analytes, the fluorescence from the analyte molecules can mask the Raman bands. Therefore, to overcome these problems, graphene derivatives have been used as anchoring bases. The anchoring can prevent nanoparticle aggregation while graphene can act as electron reservoirs resulting in quenching of the fluorescence upon laser illumination. The combination of metal nanoparticles with the graphene derivatives gives rise to a host of hybrid materials that provide enhanced Raman signals of analytes by a synergic effect. Recently, Kavitha et al. prepared rGO-Ag (reduced graphene oxide) and rGO-Os nanoparticles hybrid films using a simple liquid–liquid interface method and demonstrated their potential as SERS substrates for the detection of dye molecules. The underlying concept is to use charge transfer resulting in chemical enhancement from the graphene and EM from the Ag nanoparticles along with fluorescence suppression (Kavitha et al. 2015, 2017).

## *1.3 Metal Oxide Nanostructures*

Metal oxides as SERS-active substrates have also been explored due to their extensive applications in biological interface. But, it is noticed that metal oxides generate weak SERS signals because of its low availability of the free electrons compared with noble metals. The observation of Raman enhancement on metal oxide substrates was reported in the 1980s. Firstly, Yamada et al. observed an enhanced Raman signal of pyridine over a cleaved surface of a NiO(1 1 0) single crystal (Yamada et al. 1982), and a year later, a similar Raman enhancement for pyridine was observed by the same group on $TiO_2$ (0 0 1) surface (Yamada and Yamamota 1983). Ueba in 1983 reported a theoretical discussion about the Raman polarizability of the adsorbed molecules on ZnO and $TiO_2$. The enhanced Raman signal of the adsorbed molecular species is due to the electronic excitation in the molecular site interacting with the excitons present in the semiconductor through non-radiative energy transfer (Ueba 1983). A similar effect was identified experimentally in 2006, the chemical bonding between 'N' group of pyridine and atomic sites of metal oxide causing the enhancement of the Raman signal via charge transfer mechanism, which is similar to the resonance Raman scattering effect. Later, the SERS phenomenon by charge transfer mechanism is observed by PérezLeón et al. in 2006 on mesoporous anatase $TiO_2$ films. The results demonstrated the formation of a bidentate or bridging linkage between the Ru-bpy dye and anatase $TiO_2$ (Pérez León et al. 2006).

The Raman signal enhancement employing metal oxides as SERS substrates can be due to the charge transfer between the metal oxide and analyte molecules. Concisely, the charge transfer in metal oxide-analyte molecules can occur via the following pathways (Han et al. 2017): (1) excitation of electrons from analyte molecule HOMO level to the conduction band (CB) of a metal oxide followed by transit back to the HOMO of analyte molecule via the release of Raman photon (Fig. 3a), (2) chemical bonding between the analyte molecule and metal oxides (charge transfer (CT) complex) resulting in enhancement of the polarizability and Raman signal of the molecules (Fig. 3b), (3) excitation of valence band (VB) electrons to the LUMO of the analyte molecule by light irradiation followed by transit back to

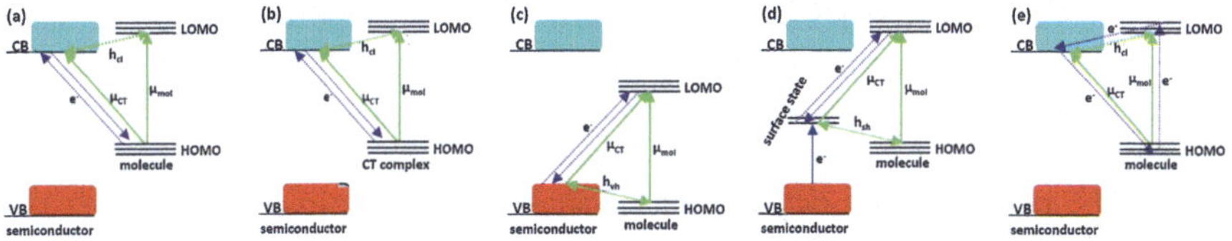

**Fig. 3** The charge transfer (CT) pathways in semiconductor-molecule systems: (**a**) molecule HOMO to CB, (**b**) CT complex to CB, (**c**) VB to molecule LUMO, (**d**) surface state to molecule LUMO and (**e**) CB to molecule HOMO (Han et al. 2017)

the valence band of metal oxide via the release of Raman photon (Fig. 3c), (4) excitation of valence electrons to the surface defect of metal oxide followed by further excitation to the LUMO of analyte molecule. The release of Raman photon via transit back to the surface state of metal oxide (Fig. 3d) (5) electrons of analyte (dye) molecules can be easily excited to the LUMO level by the visible light irradiation. These excited electrons are then injected into the conduction band of metal oxide via resonance tunnelling and can transit back to the ground state of analyte molecule via the release of Raman photon (Fig. 3e)(Han et al. 2017).

Lately, it was found that the semiconductor nanostructures can also exhibit higher SERS enhancement similar to noble metal nanostructures by creating oxygen vacancies or incorporating oxygen in the lattice of semiconductors. These oxygen defects will be responsible for the generation of local surface plasmon resonance modes and various intermittent energy levels in the bandgap (defect energy levels) of the semiconductor. Wu et al. reported a general route for the transformation of SERS inactive metal oxides like $Nb_2O_5$, $V_2O_5$ and $MoO_3$ nanostructures to the active SERS substrate by annealing in vacuum at various temperatures (Wu et al. 2017). Zheng et al. reported oxygen incorporation into the $MoS_2$ nanostructures to enhance the SERS performance of the $MoS_2$ substrates. The more exciting thing is that the enhancement factor of $MoS_2$ is continuously increasing to its maximum value with the increased oxygen incorporation as long as its phase structure remains undisturbed. Once the phase structure is changed, the EF value quickly drops to a minimal value (Zheng et al. 2017). Wang et al. indicated that the amorphous ZnO nanocages exhibited higher SERS activity compared with the crystalline structures, which is due to the high-efficiency interfacial charge transfer and availability of metastable electronic states in the amorphous ZnO nanocages (Wang et al. 2017a). Recently, R. Prabhu et al. fabricated various morphologies of $MoO_3$ nanostructures, including nanorod (NR) and sea urchin morphologies by microwave and chemical bath deposition approaches. The $MoO_3$nanostructures exhibited a morphology-dependent SERS activity. The sea urchin morphology exhibited higher EF values than the nanorods due to the availability of more oxygen vacancies and surface defects. It is well known that the low-temperature synthesis routes introduce nonstoichiometry and defects in the nanostructures (Prabhu et al. 2019).

The regeneration of SERS substrates is also a desired factor for the reuse of these SERS substrates. Generally, metal oxides such as ZnO, $TiO_2$, CuO and $SnO_2$ are considered as potential candidates for photocatalytic degradation of analyte molecules and also active substrates for SERS applications (Han et al. 2017; Ray and Pal 2017). The photocatalytic degradation of analytes over the metal oxides usually involves the following steps: (1) adsorption of reactant molecules on the surface of the metal oxides, (2) absorption of incident photons and formation of electron-hole pairs and (3) photocatalytic degradation of analyte molecules into simpler non-toxic molecules (Ibhadon and Fitzpatrick 2013; Yemmireddy and Hung 2017). After the degradation of analyte molecules, the substrates can be cleaned and reused as new substrates.

## *1.4 Noble Metal-Metal Oxide Hybrids*

Metal oxide nanoparticles, one of the most significant photocatalytic and photovoltaic materials, have been extensively investigated. However, single-component metal oxides typically show weak absorption in the visible light and strong absorption in UV light (Ibhadon and Fitzpatrick 2013; Yemmireddy and Hung 2017; Chan et al. 2011). Because of this reason, various attempts have been made to improve the light-harvesting properties of the metal oxides such as doping, dye sensitization and coupling with noble metal (Ag, Au and Cu) nanoparticles (Ray and Pal 2017). When combining the noble metal nanoparticle with metal oxides, the light absorption range can be extended, and the recombination rates of photoexcited electron-hole pairs can be significantly reduced.

The noble metal-metal oxide displays the following advantages:

1. Noble metals are most effective plasmonic materials and show excellent Raman enhancement.
2. The Fermi level of the noble metal NPs is doped into the bandgap of the metal oxides served as a doping level facilitating the charge transfer process.
3. Metal oxides can provide the chemical stability and inhibit the aggregation of noble metal nanoparticles and also tune the localized surface plasmon resonance of metal nanoparticles.
4. The hybrids show better Raman enhancement compared with their individual counterparts.
5. The better photocatalytic activity of the hybrids helps in degradation of analyte molecules adsorbed on the surface to get a clean surface for further use.

The extremely high SERS enhancement at the noble metal-metal oxide nanostructures is due to the electromagnetic enhancement and charge transfer mechanism. The EM mechanism has been explained briefly in the above section, and the additional enhancement was due to the charge transfer mechanism.

## *1.5 Charge Transfer Pathways in Metal Oxide-Noble Metal Nanostructures*

The enormous Raman signal enhancement of analyte molecules in metal oxide-noble metal nanostructures depends on the nature of the noble metal, probe molecule and assembly. The assembly of the SERS substrates affects the charge transfer direction and, as a result, additional EM field can be generated. The charge transfer pathways have been summarized as follows in metal oxide-noble metal nanostructures (Han et al. 2017).

### 1.5.1 Metal Oxide to Analyte Molecule to Noble Metal

The sandwich structure of $TiO_2$–4-mercaptobenzoic acid (MBA)-Ag NPs has been generated by a self-assembly technique. A higher SERS enhancement is obtained for MBA molecule with $TiO_2$-MBA-Ag NPs structure compared with that of $TiO_2$-MBA. The introduction of Ag offers additional charge transfer and EM through surface Plasmon resonance (SPR) effect in addition to the intrinsic $TiO_2$-MBA charge transfer. The high electronegativity nature of the Ag NPs aids to act as electron acceptors. The process of electron transfer from $TiO_2$ to Ag NPs via analyte molecules is shown in Fig. 4a (Jiang et al. 2012).

### 1.5.2 Noble Metal to Metal Oxide to Analyte Molecule

The direct interface of noble metals with metal oxides would lead to the charge transfer from metal to analyte molecule via metal oxides. In the case of Cu-ZnO-*p*-aminothiophenol (Han et al. 2017; Mao et al. 2012) and Ag-$TiO_2$-MBA (Yang et al. 2009) self-assembly, the as-deposited noble metal nanoparticles over metal oxides can inject the visible light-excited electrons (plasmon resonance absorption of noble

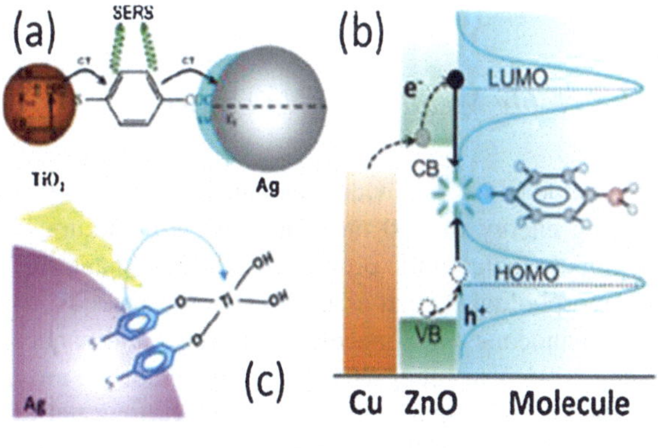


**Fig. 4** The charge transfer pathways in metal oxide-noble metal nanostructure: (**a**) $TiO_2$ metal oxide to analyte molecule to noble metal (Ag), (**b**) metal (Cu) to metal oxide (ZnO) to analyte molecule, (**c**) noble metal (Ag) to the molecule to metal oxide ($TiO_2$) (Han et al. 2017)

metal under visible light) into analyte molecule adsorbed on the metal oxide surface through the conduction band of metal oxide (Fig. 4b).

### 1.5.3 Noble Metal to Analyte Molecule to Metal Oxide

The charge transfer from noble metals to semiconductor via analyte molecules was observed by fabricating self-assembled noble metal-analyte-metal oxide nanostructures. In the case of Ag-mercaptophenol (MPH)-$TiO_2$ nanostructures (Fig. 4c), the enhancement of $b_2$ modes of MPH molecules is associated with the charge transfer between Ag NPs and MPH-$TiO_2$ complex and also depends on the incident laser with sufficient energy to excite the charge transfer electronic transitions from MPH molecule to $TiO_2$ (Han et al. 2017; Ji et al. 2011, 2012).

# 2 Types of Noble Metal-Metal Oxide Nanoparticles

As we know, most of the standard SERS substrates are of single-time use only and are mostly valuable, high-cost noble metal substrates (Jahn et al. 2016; Fateixa et al. 2015). Hence, these SERS substrates cannot be explored for routine analytical technique. Thus, the research community has been focused on the development of reusable SERS substrates based on metal-metal oxide hybrids. The association of noble metal NPs with the metal oxides can enhance the Raman scattering while the metal oxides are well-known photocatalysts for the decomposition of organic pollutants under light illumination (Chan et al. 2011; Liu et al. 2016b). The enhancement could be attributed to the synergic effect that occurs at the interface of the noble metal and metal oxide domains in the hybrid, which means that the transfer of electrons from the metal oxide to the Fermi level of the noble metal nanoparticles through the intimate contact between them (Jiang et al. 2014; Liu et al. 2017b). The noble metal nanoparticles can act as nano reservoirs for the electrons, which drastically changes the properties of the hybrids. The utility of various noble metal-metal oxide hybrid materials has been discussed briefly in the following sections.

## *2.1 Noble Metal-ZnO Nanoparticles*

Zinc oxide (ZnO) is a wide bandgap semiconductor with ~3.3 eV bandgap, exciton binding energy of ~60 meV at room temperature, wurtzite crystal structure and exhibits optoelectronic properties. The ZnO nanomaterials can be grown into various morphologies such as nanospheres, nanorods, nanoflowers, nanoshells, nanobelts, nanocones, nanoneedles, nanorings and nanocages (Sirelkhatim et al. 2015; Wang 2004; Yang et al. 2017). ZnO is one of the most common and versatile semiconductors with many applications in various fields such as photocatalysis, dye-sensitized solar cells, sensors, functional ceramics and light-emitting diodes

(Sirelkhatim et al. 2015; Wang 2004). However, doping or hybridization with noble metal nanoparticles has been followed for obtaining enhanced Raman scattering towards detection of analyte molecules. The activity towards SERS of noble metal-ZnO nanostructures is due to the following reasons: (1) high refractive index ZnO has the ability to confine the light to enhance the SERS effect and (2) abundantly achievable morphologies of the ZnO are in favour for the combination of noble metals to improve SERS activity. Furthermore, ZnO has numerous advantages including high chemical stability, superhydrophobicity, biocompatibility, tunable optoelectronic properties and photocatalytic self-cleaning effect that can be synchronized with the SERS effect of the ZnO nanostructures to achieve versatility and multifunctionality (Sirelkhatim et al. 2015; Yang et al. 2017). Therefore, considerable effort has been devoted to the fabrication of high-performance SERS substrates for practical use as analytical tool. The SERS performance of the substrates mainly depends on the hybrid material morphology. 'Thus, a variety of synthesis methods have been reported towards realizing efficient SERS hybrid subsrates. The methods include solution phase, sol-gel, hydrothermal, microwave, photochemical, chemical vapour deposition, sputtering, e-beam, lithography techniques, and so on (Liu et al. 2016b; Li et al. 2017).

Several studies have been focused on enhancing the properties of ZnO by decorating with noble metal nanoparticles. Flower-shaped Au-ZnO nanostructures were prepared by Liu and coworkers using seeding growth followed by wet-chemical etching of Au-ZnO core-shell nanostructures. This hybrid structure was used as a SERS substrate for the detection of *p*-aminothiophenol (PATP) molecules. They observed enhanced SERS of the PATP molecules over hybrid compared with that of bare Au and ZnO NPs and explained it as due to the charge transfer contribution (Liu et al. 2015). Sivashanmugan et al. reported the use of Ag nanoclusters deposited on a focused ion beam (FIB) made ZnO nano dome as SERS substrate for high selectivity for single molecular detection (Sivashanmugan et al. 2015). Sun et al. fabricated ZnO-Au hybrid nanostructures employing hydrothermal and electric beam evaporation deposition technique. The thickness of the Au is controlled by the deposition time (Sun et al. 2012).

Huang et al. prepared a unique Au- or Ag-decorated 3D hierarchical ZnO/Si nanomace SERS substrates employing two different approaches (Fig. 5a). Initially, Si nanoneedles were grafted on to ZnO nanorods by vapour–liquid–solid growth followed by the decoration of noble metals such as Ag or Au NPs over these structures employing galvanic displacement reaction and the fabricated substrates were used for the detection of R6G. The results show the detection limit as low as $10^{-16}$ M (Fig. 5a) with $8.7 \times 10^{7}$ enhancement factor (Huang et al. (2015a). Cheng et al. fabricated 3D ordered arrays of Si/ZnO nanotrees decorated with Ag NPs. The 3D structures exhibited high sensitivity and good reproducibility compared with 2D nanostructures due to the additional contribution from the enhanced light trapping because of multi-scattering (Cheng et al. 2010). Park et al. fabricated Au-decorated 3D ZnO nanostructures using lithography and atomic layer deposition techniques. A high density of Au NPs separated by nanoscale gaps was achieved and these helped in realizing high SERS enhancement (Park et al. 2016). The fabrication of

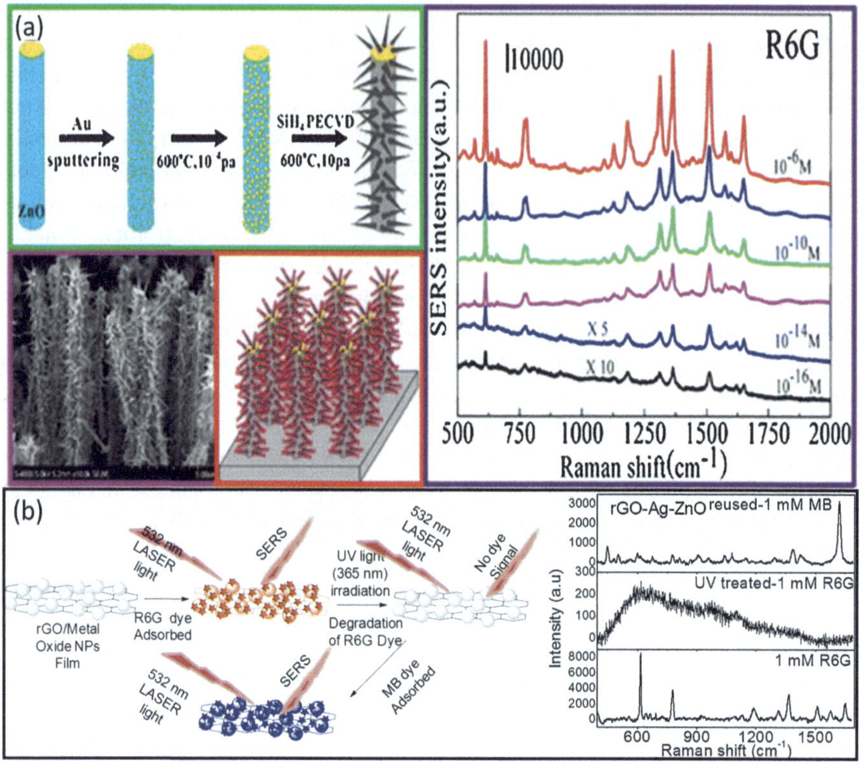


**Fig. 5** (**a**) Schematic illustration and synthesis of Au NPs deposited 3D hierarchical ZnO/Si nanomace SERS substrates and SERS spectra of R6G with various concentrations (Huang et al. (2015a) and (**b**) Schematic of the renewable process along with SERS spectra of 1 mM R6G before and after UV treatment and that of 1 mM MB dye on the UV-treated rGO-ZnO-Ag hybrid film (Bramhaiah et al. 2017)

effective hotspots on the SERS substrate becomes essential to obtain high sensitivity. He et al. fabricated multi-hotspots urchin-like Ag NPs-ZnO hollow nanosphere arrays using simple wafer-scale methods. Due to the multi-hotspots, the enhancement for R6G is as high as $10^8$ and it shows excellent stability and reproducibility of the SERS signal (He et al. 2014). Bramhaiah et al. reported hybrid films of rGO-ZnO-Ag nanoparticles for the SERS detection of fluorescent dyes. It is well known that the assembly of multiple nanomaterials with various functions on reduced graphene oxide surface is an advanced step for the preparation of multifunctional materials. The synergy of graphene and nanoparticles has already been utilized in multiple applications like catalysis, pollutant removal and SERS. They showed that by a minimal amount of Ag doping in the rGO-ZnO system (Fig. 5b), the EF can be increased to higher orders from $10^2$ to $10^4$ (Bramhaiah et al. 2017). These substrates can be regenerated for multiple uses by simple UV light irradiation to degrade the dyes in the wet condition. The presence of rGO and Ag NPs also assists in faster photodegradation kinetics (Bramhaiah et al. 2016). From the above reports, the 3D large area nanocomposites are advantageous for use as substrates in practical applications involving biosensing and environmental pollutants. Table 1 displays the various reports on noble metal-based ZnO nanostructures for SERS applications.

## *2.2 Noble Metal–$TiO_2$ Nanoparticles*

Titanium dioxide ($TiO_2$) is commercially produced from the early twentieth century, for their wide use in pigments, sunscreens, paints, ointments, toothpaste, etc. Fujishima and Honda reported a phenomenon of photocatalytic water splitting over a $TiO_2$ electrode under UV light irradiation (Fujishima and Honda 1972). Since then, massive efforts have been devoted to the research of $TiO_2$ material. $TiO_2$ is also a wide bandgap semiconductor and is one of the most prominent oxide materials owing to excellent physical and chemical properties (Nakata and Fujishima 2012). $TiO_2$ has many applications in photocatalysis, energy conversion, pollutant degradation and sensors. $TiO_2$ is considered very close to an ideal semiconductor for photocatalysis because of its high stability, low cost and safe material for both humans and environment (Schneider et al. 2014; Rahimi et al. 2016).

Noble metal-based $TiO_2$composites attracted significant attention as SERS substrates due to the good enhancement in Raman signals from the charge transfer ($TiO_2$-adsorbed molecule) and synergic EM contribution of noble metals coated on $TiO_2$. Huang et al. prepared anatase and rutile $TiO_2$/Ag composites employing the facile and green photochemical method. The nanocomposites show excellent sensitivity for R6G and CV dyes. The reusability of SERS substrates is obtained by UV light irradiation. In comparison with the non-porous materials, porous supporting materials can have more surface area and lower mass density that form an added benefit for supported SERS substrates to obtain higher SERS enhancement. Zou et al. fabricated porous $TiO_2$-Ag core-shell nanocomposites, and it exhibited high SERS enhancement factor along with self-regenerating property (Zou et al. 2013). Au-$TiO_2$-Au nanosheets have been prepared by Jiang et al., and when compared to $TiO_2$-Au substrates, Au-$TiO_2$-Au nanosheets exhibited much stronger signal and reproducibility (Jiang et al. 2015a). Xu et al. made a large area Ag-decorated $TiO_2$ nanograss employing solvothermal approach (Fig. 6). The Ag protrusion and gaps between them act as 3D SERS hotspots. The Ag-$TiO_2$ shows good homogeneity with high detection for R6G (Fig. 6), and it can be self-cleaned by visible light irradiation by photodegradation of analyte molecules (Xu et al. 2013).

Various crystal planes of $TiO_2$ have different surface energies whose order is as follows: $\gamma\{1\,1\,0\}$ (1.09 J $m^{-2}$) > $\gamma\{0\,0\,1\}$ (0.90 J $m^{-2}$) > $\gamma\{1\,0\,0\}$ (0.53 J $m^{-2}$) > $\gamma\{1\,0\,1\}$ (0.44 J $m^{-2}$). Therefore, the materials having more {1 1 0} and {0 0 1} facets can be responsible for the high activity. Ag nanocrystals were photochemically deposited over the surface of brookite and rutile by Guo et al. The increasing of $AgNO_3$ concentration in the solution leads to the morphology changes from smaller nanoparticles to nanoplates and polyhedrons. A proper amount of polyvinyl pyrrolidone (PVP) was required for the large area nanoplate formation with better SERS signal

**Table 1** Noble metal-based ZnO nanostructures for SERS applications

| System | Synthesis method | Analyte | EF | Reference |
|---|---|---|---|---|
| Ag nanoclusters on ZnO nanodome | Focused ion beam for ZnO and Ag clusters by e-beam deposition | MG<br>CV | ~$10^6$ | Sivashanmugan et al. (2015) |
| Ag NPs on ZnO Nano arrays | Hydrothermal method | R6G | ~$1.24 \times 10^5$ | Li et al. (2017) |
| Flower-shaped Au-ZnO hybrid nanoparticles | Seeding growth and subsequent wet-chemical etching | PATP | – | Liu et al. (2015) |
| Si/ZnO/Ag hybrid nanotrees | Hydrothermal growth of ZnO nanorods<br>Ag NPs-photochemical reduction and deposition | R6G | ~$1 \times 10^6$ | Cheng et al. (2010) |
| Au-decorated 3D ZnO nanostructures | Prism holographic Lithography and atomic layer deposition techniques | BT | ~$1.4 \times 10^5$–3 mm Au layer<br>~$9.1 \times 10^3$–5 mm Au layer | Park et al. (2016) |
| Zinc oxide (ZnO)/silver (Ag) composite microspheres | Solution-based | 4-MPY | ~$9.0 \times 10^4$ | Song et al. (2007) |
| Urchin-like Ag nanoparticle (NP)/ZnO hollow nanosphere (HNS) arrays | Nanosphere lithography (NSL) and solution processes | R6G | ~$10^8$ | He et al. (2013) |
| Colloidal ZnO-Ag and Au | Solution phase | ZnO | Ag-$10^4$<br>Au-$10^2$ | Rumyantseva et al. (2013) |
| Silver nanoparticles (AgNPs) on the side walls of ZnO nanowires | Wet chemical method | PNTP | ~$2.8 \times 10^7$ | Satheeshkumar and Yang (2014) |
| ZnO nanowires/Ag NPS | ZnO-low-pressure Chemical vapour transport and deposition system<br>Ag-solution phase-380-nm UV Light | PNTP | ~$10^6$ | Chen et al. (2011a) |
| Vertically aligned Au-coated ZnO NRs | Hydrothermal route | MB | $1 \times 10^{-12}$ M | Sinha et al. (2011) |
| Ag/ZnO nanoparticles | Heating reflux approach | RhB | – | Li et al. (2016b) |

(continued)

**Table 1** (continued)

| System | Synthesis method | Analyte | EF | Reference |
|---|---|---|---|---|
| Ag nano-islands on ZnO nanosheets | UV light induced-photochemical reaction | | ~$10^7$ | Xu et al. (2017a, b) |
| Porous ZnO nanosheets decorated with Ag nanoparticles | Solvent method | R6G | ~$10^8$ | Zhao et al. (2015a, b) |
| ZnO/Au composite nanoarrays | | R6G | ~$1.2 \times 10^7$ | Chen et al. (2009) |
| ZnO nanowires/Ag NPs | ZnO-hydrothermal Ag NPs-solution phase (UV light) | Glucose | ~$6.36 \times 10^{11}$ | Kang et al. (2015). |
| ZnO/Ag 3D nanostructures | Rapid thermal oxidation of metallic Zn films at 500 °C Ag-ion beam sputter Deposition | Anta, FOX-7, and CL-20 | ~$10^7$ ~$10^7$ ~$10^4$ | Shaik et al. (2016) |
| ZnO nanorod/Ag NPs | ZnO-vapour-phase transport process Ag-ionic Beam sputtering | R6G | ~$1.2 \times 10^{10}$ | Lu et al. (2016) |
| ZnO NRs/Ag NPs | ZnO-hydrothermal Ag NPs-photochemical | RhB | ~$5.04 \times 10^7$ | Huang et al. (2015a, b) |
| ZnO NRs/Ag NPs | ZnO-hydrothermal (microwave) Ag NPs-thermal evaporation | R6G | ~$7 \times 10^5$ | Pimentel et al. (2017) |
| 3D hierarchical ZnO–Ag hybrids | Organic-chemical-assisted hydrothermal method | R6G | ~$1.17 \times 10^7$ | He et al. (2014) |

(Guo et al. 2014a). The morphologies of the $TiO_2$ can influence the SERS spectra of analyte molecules. Han et al. fabricated nanotube, nanolace and nanopore $TiO_2$ morphologies by adjusting the anodization parameters and decorated them by Au NPs for SERS applications. The nanopore $TiO_2$/Au structure exhibited a high SERS activity with good reproducibility and reusability (Han et al. 2015). A thin coating of $TiO_2$ can effectively protect the noble metal NPs from unwanted oxidation. For this, Bao et al. prepared $TiO_2$-coated Ag nanowires employing electrochemical deposition technique. The bifunctional $TiO_2$-coated Ag nanowires have been employed as efficient self-cleaning and recyclable SERS substrates (Bao et al. 2014; Zhou et al. 2012). An ordered array of Au semishells over hollow $TiO_2$ spheres has been prepared by Li et al. using facile atomic layer deposition technique. The substrates exhibited high sensitivity and reusability for R6G and brilliant cresyl blue dyes (Li et al. 2012). Various noble metal NPs/$TiO_2$ nanocomposites have been

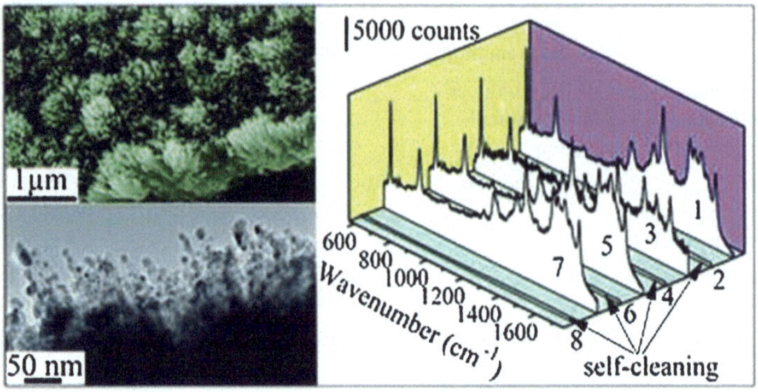


**Fig. 6** FESEM and HRTEM images of large area Ag-$TiO_2$nanograss along with reusability for R6G dye (Xu et al. 2013)

prepared by different methods and have shown good sensitivity, reproducibility and reusability for various analytes, whose details have been listed in Table 2.

## 2.3 *Noble Metal–CuO and $Cu_2O$ Nanoparticles*

The oxides of copper ($Cu_xO$) are fascinating materials because of their remarkable properties including optical, electrical, thermal and magnetic properties. Generally, the oxides of copper are in two crystal forms such as a copper oxide (CuO) or cupric (II) oxide and cuprous oxide ($Cu_2O$) (Zoolfakar et al. 2014; Zhang et al. 2014; Sun and Yang 2014). Copper oxide (CuO) is an antiferromagnetic *p*-type semiconductor with an indirect bandgap of 1–2 eV. It is non-toxic, chemically stable, electrochemically active, abundant, low cost and is easy to prepare various morphologies. It has been employed in multiple applications including catalysis, batteries, magnetic storage media, solar energy conversion, gas sensing and field emission. Cuprous oxide ($Cu_2O$) is a typical semiconductor with a direct bandgap of 2.17 eV. $Cu_2O$ is also environmentally friendly and naturally abundant but less stable (Zoolfakar et al. 2014; Zhang et al. 2014; Sun and Yang 2014). These two copper-based oxides have dissimilar properties. Both oxides are significant in many applications such as catalysis, batteries, magnetic storage media, solar energy conversion, gas sensing and field emission. However, CuO is more stable than $Cu_2O$ due to the stability of Cu(II) ions in ambient conditions (Zoolfakar et al. 2014; Zhang et al. 2014; Sun and Yang 2014; Kudelski et al. 1998). The first report of $Cu_2O$-based SERS is by Kudelski and coworkers in 1998. The $Cu_2O$ surface shows very weak SERS peaks of pyridine with a small enhancement (Kudelski et al. 1998). Later in 2007, Wang et al. reported CuO as SERS substrate for 4-mercaptopyridine and observed a ~ $10^2$ enhancement factor (Wang et al. 2007).Till date, there are only a few reports available on CuO and $Cu_2O$ and also their hybrids with noble metal nanostructures.

**Table 2** Noble metal-based $TiO_2$ nanostructures for SERS applications

| Material | Synthesis method | Analyte | EF/Conc. | Reference |
|---|---|---|---|---|
| Bluewing of butterfly E. mulciber template-3D Ag/ $TiO_2$ | Chemical reduction method | R6G | ~$1.1 \times 10^5$ | Chen et al. (2014a) |
| Ag-decorated $TiO_2$ nanograss | Solvothermal | R6G and 4-ATP | ~$1.2 \times 10^7$ ~$2.9 \times 10^6$ | Xu et al. (2013) |
| Porous $TiO_2$–Ag core-shell nanocomposite | Solution phase | 4-MPy | ~$6.5 \times 10^5$ | Zou et al. (2013) |
| Ag nanocrystals on TiO2 (Brookite and rutile) | Photochemical method | R6G | ~$3.0 \times 10^5$ | Guo et al., (2014a) |
| $TiO_2$/4-mercaptopyridine (4-Mpy)/Ag sandwich | Hydrothermal | 4-MPy | – | Xue et al. (2015) |
| Au-$TiO_2$ | Electrochemical | R6G | ~$5 \times 10^4$ | Han et al. (2015) |
| Urchin-like$TiO_2$@ Ag | Hydrothermal | R6G | ~$7.6 \times 10^6$ | Zhou et al. (2017) |
| $TiO_2$ nanowires (NWs)/Ag NPs | Wafer scale methods | R6G | ~$1 \times 10^8$ | Shan et al. (2015) |
| $TiO_2$-coated Ag nanowire arrays | Electro deposition-Ag Chemical hydrolysis growth –$TiO_2$ | 2,4-Dichlorophenoxyacetic acid | – | Bao et al. (2014) |
| Au-coated $TiO_2$ nanotube arrays | ZnO template | R6G | $1 \times 10^{-9}$ M | Li et al. (2010) |
| $TiO_2$-coated Ag nanowire arrays | Electro deposition-Ag Chemical hydrolysis growth –$TiO_2$ | R6G and methyl parathion | $10^{-6}$ M | Zhou et al. (2012). |
| Au semishells on $TiO_2$ Spheres | Self-assembly and atomic layer deposition | R6G | ~$1.4 \times 10^5$ | Li et al. (2012) |
| Au nanoclusters/ $TiO_2$ | – | 4-NP | – | Qi et al. (2015) |
| Porous Ag/$TiO_2$ composite | Spin coating-$TiO_2$ Photochemical-Ag | CV | $10^{-10}$ M | Zhang et al. (2015) |
| $TiO_2$ nanotubes/Ag or Cu | Sputter deposition of Ag or Cu | Py | 0.05 M | Roguska et al. (2009) |

(continued)

**Table 2** (continued)

| Material | Synthesis method | Analyte | EF/Conc. | Reference |
|---|---|---|---|---|
| Ag/$TiO_2$ | Hydrothermal | 4-MBA | ~$7.8 \times 10^5$ | Xie et al. (2014) |
| Ag film on $TiO_2$-catalysed Ag nanoparticles | $TiO_2$-sol-gel method<br>Ag NPs-photochemical<br>Ag film-magnetron sputtering | R6G | ~$1.2 \times 10^7$ | Li et al. (2016a) |
| $TiO_2$ nanotubes/Ag or Au or Cu | Ag or Au or Cu—Sputter deposition | Py | 0.05 M | Roguska et al. (2011) |
| Rutile and anatase $TiO_2$/Ag nanocomposites | Photochemical method | R6G<br>CV | ~$6.46 \times 10^6$ | Huang et al. (2017b) |
| Ag–$TiO_2$ nanocomposites | Sol-gel technique | MO | – | Prakash et al. (2016) |
| $TiO_2$/Ag nanoparticles | Hydrothermal | 4-ATPH<br>MB<br>CV | $10^{-7}$ M | Dai et al. (2017) |
| Ag-loaded $TiO_2$ nanotube film | $TiO_2$-electrochemical anodization<br>Ag-sputtered | R6G | $10^{-6}$ M | Ling et al. (2016). |

Yang et al. reported a facile in situ solution-phase method for the synthesis of homogeneous Ag-$Cu_2O$ composite microstructure with tunable Ag NPs concentration over truncated octahedral $Cu_2O$ structures. The results showed that the SERS enhancement factor for RhB is around ~$7.8 \times 10^4$ (Yang et al. 2017). The same group reported the one-pot synthesis of Ag-$Cu_2O$ nano frames (Yang et al. 2014a, b) and their SERS applications. The presence of eight hexagonal (1 1 1) faces with well-defined interior voids and the large surface area made them good SERS substrates. The (1 1 1) facets of $Cu_2O$ having surface Cu atoms with dangling bonds exhibit higher energy and superior properties (Ho and Huang 2009). The fabricated Ag-$Cu_2O$ nano frames display highly sensitive SERS signals with excellent reproducibility and stability. The foremost problem for the copper oxides is its oxidizing state, which affects the stability of the CuO. To overcome this problem, Jayram et al. prepared superhydrophobic Ag@CuO nanoflowers, and they got an EF factor around ~$2.0 \times 10^7$ (Jayram et al. 2016). Pal et al. developed Au-CuO by redox transformation reaction. Very sharp tips and nanometer-scale gaps between tips of the Au-CuO nanoflowers were anticipated for the observed EF around ~$10^8$ (Pal et al. 2014). Hsieh et al. reported the comparison results of the solution phase Ag-CuO and solid-phase Ag-CuO composites. The solution phase composites exhibited tenfold higher Raman enhancement, which is due to the large active surface area and higher density of dynamic hotspots resulting from solution-phase approach (Hsieh et al. 2014). The wet chemical method used by Wang et al. for the preparation of Ag-$Cu_2O$ and Au-$Cu_2O$ nanocomposites and the Raman line mapping exhibits excellent uniformity of the composites. The same substrates have been applied for the detection of plasticizers such as dimethyl phthalate and dibutyl phthalate (Wang et al. 2014a).

Octahedral $Cu_2O$-Au composite microstructures are obtained employing a facile solution-phase method and show an enhancement factor for MBA as ~$7.2 \times 10^5$ (Chen et al. 2016). The density of Au NPs over the $Cu_2O$ octahedral can be controlled by tuning the concentration of gold precursor. Figure 7 shows the Fermi level equilibrium and effect of the Au content over the $Cu_2O$ on SERS spectra of MBA. They found that the overloading of Au NPs can decrease the efficiency of the $Cu_2O$-Au composite by blocking the active sites. Various noble metal-coated CuO/$Cu_2O$ composites as SERS substrates are given in Table 3.

## 2.4 *Noble Metal-Iron Oxide Nanoparticles*

Iron oxides are the most common metal oxides widely used as magnetic pigments in recording and storage media, catalysis and magneto-optical devices (Wu et al. 2015). Among various iron oxide nanoparticle systems, magnetite ($Fe_3O_4$) and maghemite ($\gamma Fe_2O_3$) are the only Food and Drug Administration (FDA) approved magnetic nanosystems for biological applications (Wu et al. 2015; Hola et al. 2015). These have been widely explored due to their nontoxicity, biocompatibility and high magnetization. As we know that the usage of noble metal NPs as SERS substrates not only increases the cost but also tend to aggregate, and it is challenging to separate and recover them (Cai et al. 2014; Tang et al. 2015). To overcome this problem,

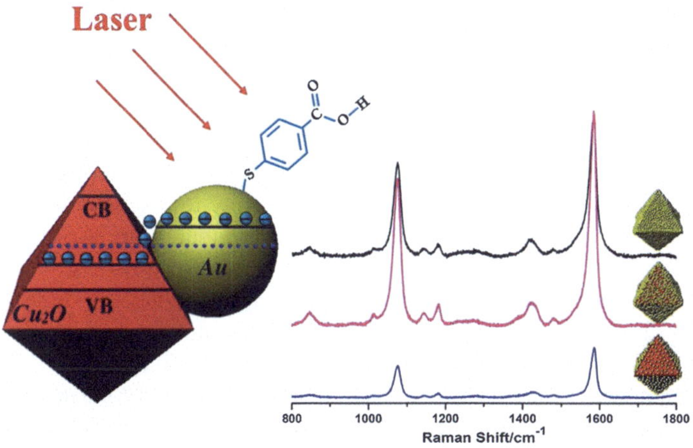


**Fig. 7** Fermi level equilibration in $Cu_2O$-Au composite microstructures along with SERS spectra of MBA with increasing the Au content over the $Cu_2O$ octahedra (Chen et al. 2016)

**Table 3** Noble metal-based $CuO/Cu_2O$ nanostructures for SERS applications

| Material | Method | Analyte | EF/Conc. | Reference |
|---|---|---|---|---|
| Ag NPs/truncated octahedral $Cu_2O$ | Solution phase | RhB | ~$7.8 \times 10^4$ | Yang et al. (2014a) |
| Ag@CuO | SILAR method | R6G | ~$2.0 \times 10^7$ | Jayram et al. (2016) |
| Octahedral $Cu_2O$-Au composite microstructures | Solution phase | 4-MBA | ~$7.2 \times 10^5$ | Chen et al. (2016) |
| Ag NP-decorated $Cu_2O$ nanoframes | Solution phase | RhB | ~$10^5$ | Yang et al. (2014b) |
| Hierarchical Au–CuO nanocomposite | Solution phase and hydrothermal | 4-ATP | ~$10^8$ | Pal et al. (2014) |
| Ag/CuO nanocomposite | Plasma treatment | 4-ATP | ~$10^5$ | Hsieh et al. (2014) |
| $Cu_2O$-Au and $Cu_2O$-Ag nanocomposites | Solution phase | R6G | $1 \times 10^{-10}$ M | Wang et al. (2014a, d) |
| Ag/$Cu_2O$ film | Hydrothermal | p-ATP | $1 \times 10^{-7}$ M | Ji et al. (2014) |
| Au@$Cu_2O$ nanostructures | Solution phase | – | – | Zhang et al. (2016) |
| Ag@$Cu_2O$ core-shell NPs | Solution phase | 4-MBA | ~$3.21 \times 10^5$ | Chen et al. (2017a) |
| $Cu_2O$ polyhedrons / Ag NPs | Solution phase | 4-MBA | ~$3.21 \times 10^6$ | Luo et al. (2016) |
| Ag island films over CuO thin films | CuO-sol-gel Ag-thermally vapour deposition | 4-MPy | – | Wang et al. (2009) |
| Au@$Cu_2O$ core-shell nanoparticles | Solution phase | 4-MBA | – | Chen et al. (2018) |

the combination of the iron oxide and noble metal nanoparticles in composites can result in simultaneous magnetic activity and optical response and also these can be easily separated and concentrated by applying external magnet. Furthermore, the optical properties of the noble metals can be controlled by monitoring the magnitude of the external magnetic field (Yang et al. 2012).

Yang et al. reported a facile solid-phase synthesis for the Ag-coated $Fe_3O_4$ microspheres under an argon atmosphere and employed for the clean, reproducible SERS substrates, which is achieved by externally subjecting to magnetic field. Figure 8 shows the schematic of the reversible SERS behaviour using Ag-coated $Fe_3O_4$ microspheres (Yang et al. 2012). The analyte molecules can be easily captured or detached using an external magnet and also get analysed by SERS. These reproducible SERS substrates were used to detect the analytes such as methyl parathion and 4-mercaptopyridine and detection limit is achieved as low as $10^{-12}$ M for 4-aminothiophenol (Yang et al. 2012). Cai et al. also reported Au nanocube-coated $Fe_3O_4$ NPs and Tang et al. Au NR-coated $Fe_3O_4$ microspheres for SERS detection of pesticides and reusability was done employing external magnet (Cai et al. 2014; Tang et al. 2015). Sun et al. prepared Ag@$Fe_3O_4$core-shell nanospheres employing

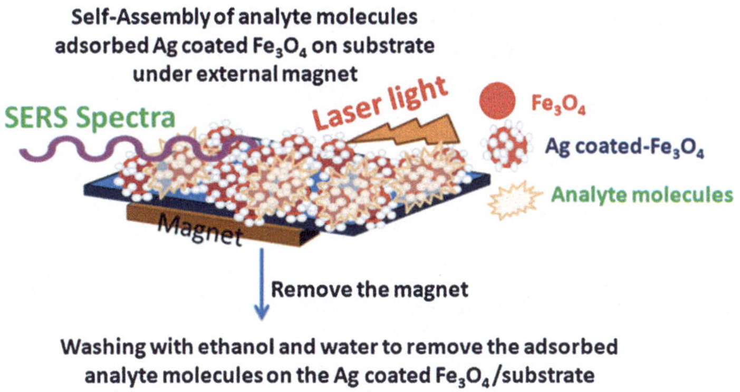


**Fig. 8** Schematic illustration of reversible SERS behaviour of noble metal-coated $Fe_3O_4$ nanohybrids

in situ reduction of $AgNO_3$ and $Fe(NO_3)_3$ with ethylene glycol as a reducing agent. They used these nanostructures as SERS substrates for 4-aminothiophenol (4-ATP) and R6G. The substrates showed high reproducibility, stability and reusability. The reusability is achieved by repeated washing, followed by magnetic separation (Sun et al. 2013).

Numerous approaches have been explored for the fabrication of adequate, stable noble metal-iron oxide nanoparticle systems. Zhu et al. reported a one-pot method for the preparation of Ag-$Fe_3O_4$ nanocomposite by a redox reaction between $Ag_2O$ and $Fe(OH)_2$ in the absence of further reductant at moderate temperature and atmospheric conditions (Zhu et al. 2015) and used for SERS detection of 4-mercaptobenzoic acid (4-MBA). The results showed that the EF of 4-MBA is up to~$5.2 \times 10^{-6}$ and the detection limit is down to ~$10^{-10}$ M. Ding et al. prepared 1D $Fe_3O_4$-Au nanochains by the magnetic field-induced assembly followed by Ag decoration by in situ reduction and produced 3D $Fe_3O_4$@Au@Ag nanoflowers. These structures possess a large number of hotspots, significantly enhancing the Raman signal with an EF of $2.2 \times 10^9$ for R6G (Ding et al. 2016). The following Table 4 provides the various noble metal-based iron oxide nanocomposites for SERS applications.

## 2.5 *Noble Metal-$SiO_2$ Nanoparticles*

Silica ($SiO_2$) particles have a foremost role in the nanotechnology due to their significant features including tunable size from 5 nm to 1000 nm, distinctive optical properties, high specific surface area, low density, adsorption capacity, encapsulation capacity, biocompatibility and low toxicity (Bitar et al. 2012). These features make $SiO_2$ nanoparticles being extensively employed as an inert solid support and entrapping matrix for the noble metal nanoparticles (Lin et al. 2014). Noble metal

**Table 4** Noble metal-based iron oxide nanostructures for SERS applications

| Material | Method | Analyte | EF/Conc. | Reference |
|---|---|---|---|---|
| 3D $Fe_3O_4$@Au@Ag nanoflowers | External static magnetic field. | R6G | ~$2.2 \times 10^9$ | Ding et al. (2016). |
| Au-nanorod@$Fe_3O_4$ | Solvothermal | 4-ATP | – | Tian et al. (2012) |
| Ag-coated $Fe_3O_4$ microsphere | Solid-phase synthesis | 4-ATP | $1.0 \times 10^{-12}$ M | Yang et al. (2012) |
| Au nanocube-coated $Fe_3O_4$ | Solution phase | CV | ~$2.08 \times 10^7$ | Cai et al. (2014) |
| Ag–$Fe_3O_4$nanohybrids | Solution phase | 2-NT | ~$1.14 \times 10^3$ | Huang et al. (2011) |
| Au nanorod-coated $Fe_3O_4$ microspheres | Hydrothermal | 4-ATP | ~$2.13 \times 10^5$ | Tang et al. (2015) |
| Ag@$Fe_3O_4$core-shell nanospheres | Solvothermal | 4-ATP R6G | $1.0 \times 10^{-11}$ M | Sun et al. (2013) |
| $Fe_3O_4$/Ag composites | Solvothermal | p-ATP R6G | $1.0 \times 10^{-7}$ M | Guo et al. (2015) |
| $Fe_3O_4$-Au core-shell nanostructures | Hydrothermal | R6G | ~$10^6$ | Wheeler et al. (2012) |
| Ag-$Fe_3O_4$ nanocomposites | Solvothermal | R6G | $1.09 \times 10^{-7}$ M | Guo et al. (2014b) |
| Ag-$Fe_3O_4$ nanocomposites | Solution phase | 4-ATP | $1 \times 10^{-7}$ M | Joshi et al. (2014) |
| Ag-$Fe_3O_4$ nanocomposites | Solution phase | 4-MBA | ~$5.2 \times 10^6$ | Zhu et al. (2015) |
| $Fe_2O_3$-Ag hollow microspheres | Solvothermal | 4-ABT | ~$9.6 \times 10^4$ | Weng et al. (2016) |
| Spindle-shaped $Fe_2O_3$@Au nanoparticle | Solution phase | Py | ~$3 \times 10^6$ | Shen et al. (2015) |
| Chitosan capped $\gamma$-$Fe_2O_3$ NPs coated with Ag NPs | Solution phase | 4-ATP | $10\times 10^{-12}$ M | Kaloti and Kumar (2016) |
| Ag-$Fe_3O_4$ nanocomposites | Solution phase | 4-MBA R6G CV | ~$4.1 \times 10^6$ | Fan et al. (2017) |
| $Fe_2O_3$/Au/Ag nanostructures | Co-precipitation | TP and ATP | $1 \times 10^{-6}$ M | Han et al. (2012) |
| $Fe_2O_3$@Ag | Self-assembly and seeding | TP, ATP and R6G | ~$10^7$ | Chen et al. (2011b) |

NPs embedded in fumed $SiO_2$-based substrates offer the advantage of high roughness and prevents the aggregation of noble metal NPs, as a result, stabilizing the SERS effect (Jayram et al. 2015). Numerous reports are present on the noble metal-$SiO_2$ nanocomposites for ultrasensitive detection of analytes with enhanced effects. Moreover, attractive methods have been reported to synthesize nanocomposites by chemical processes and physical techniques including sol-gel, solution phase, vacuum coating, sputtering and plasma etching, etc. (Wu et al. 2013). For the synthesis of noble metal-$SiO_2$ nanocomposites, Stöber-based approaches are appropriate.

Generally, coupling agents with special functional groups including poly(vinylpyrrolidone) (PVP), 3-mercaptopropyltrimethoxysilane and 3-aminopropyltrimethoxysilane are employed to functionalize the surface of noble metal nanoparticles for coating of silica shell. Kha et al. reported $SiO_2$-coated Ag nanocubes with enhanced SERS and metal-enhanced photoluminescence (MEPL). The SERS and MEPL intensity of R6G can be manipulated by tuning the thickness of the $SiO_2$ shell over Ag nanocubes (Kha et al. 2015). Zhang et al. reported Ag NPs coated silica nanolayers on to a large scale and uniform silicon nanowire array employing simple chemical etching and metal reduction method for the SERS application. The detection limit for R6G is as low as $10^{-16}$ M and enhancement factor as large as $10^{14}$ (Zhang et al. 2013). The higher sensitivity, stability and lower detection limit is attributed to the presence of $SiO_2$ monolayer around the Ag NPs that prevents the unwanted surface oxidation of Ag NPs. Lin et al. prepared Au NP-embedded mesoporous $SiO_2$ substrate for the detection of 3,3′-diethylthiodicarbocyanineiodide (DTDC), MBA, thiophenol (TP), 2-naphthalene diol (NT), 5,5′-dithiobis-(2-nitrobenzoic acid) (DTNB), etc. and obtained enhancement factor to the order of $10^8$–$10^{13}$ (Lin et al. 2014). Jayram et al. demonstrated that the fabrication of monodispersed Ag-embedded $SiO_2$-nanostructured thin films aimed at high enhancement factor for R6G, with a detection limit as low as $10^{-18}$ mol $L^{-1}$ (Jayram et al. 2015). Wang et al. prepared pillar-shaped arrays of Ag/$SiO_2$ employing the sputtering technique followed by annealing under Ar atmosphere. Some of the Ag NPs got squeezed into the $SiO_2$ sublayer through pinholes, forming hotspots for the SERS (Wang et al. 2016). Multi-Au NP-embedded mesoporous $SiO_2$ microspheres were prepared by Chen et al. and used as a SERS substrate for the detection of biomolecules. Figure 9 shows the schematic illustration of SERS and reusability of the prepared substrate. The self-filtering and thermal stability made them as useful reusable SERS substrates for biosensing (Chen et al. 2017b).

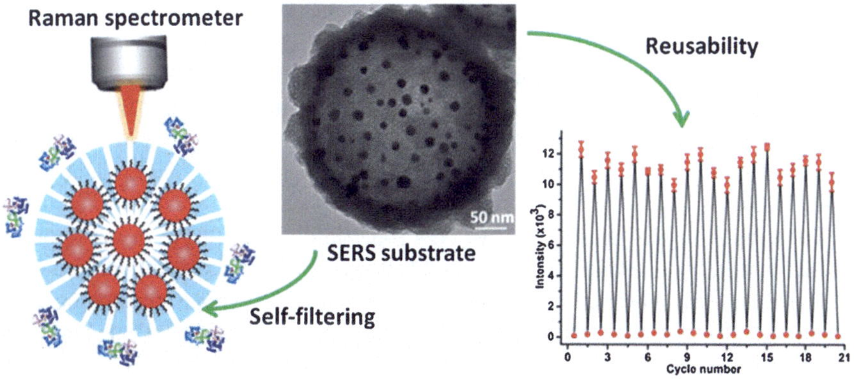


**Fig. 9** Schematic of the Au NPs/ mesoporous $SiO_2$ along with TEM image and reusability of SERS substrate for MBA (Chen et al. 2017b)

Lai et al. prepared silica-coated Ag nanorod aggregate with MBA molecules and used for SERS detection of prostate-specific antigen (PSA). The limit of detection of PSA was achieved as low as 0.3 fg $mL^{-1}$ (Lai et al. 2015). Paper-based $Ag@SiO_2$ nanocubes were fabricated by Mekonnen et al. and applied for the detection of melamine. They found good linearity up to 1 mg $L^{-1}$ with a limit of detection of 0.06 mg $L^{-1}$ (Mekonnen et al. 2017). The following Table 5 represents the various $SiO_2$-based noble metal NPs composites reported in literature for SERS applications.

## 2.6 *Noble Metal-Alumina Nanoparticles*

Alumina ($Al_2O_3$) is widely used as a supporting material due to their chemical inertness, resistant to oxidation, high dielectric constant, thermal conductivity, low magnetic conductivity and stability. The preparation of the substrate is similar to the fumed silica (Toccafondi et al. 2015; Xu et al., 2017b). The main advantage of this material over Teflon and latex microspheres is its low cost, large porosity and high surface area. Alumina-based substrates, due to their high efficiency, low-cost and simplicity of the preparation is made to use extensively in many practical applications (Toccafondi et al. 2015; Xu et al. 2017b). Nanocomposites formed with alumina and noble metal NPs have great importance. It is well known that the noble metal NPs morphology could be damaged at a moderately low temperature around 120 °C and also their surfaces can be readily oxidized or sulfurized at normal atmospheres, thus losing their sensitivity. Therefore, they are not suitable for high-temperature sensing and thermal cleaning. Henceforth, alumina was used as a supporting matrix or material for the noble metal NPs (Xu et al. 2017b). Ma et al. reported that wrapping of Ag nanorods with ultrathin $Al_2O_3$ layers employing atomic layer deposition technique could help in stabilizing the Ag morphology up to 400 °C (Ma et al. 2015a; Ma et al. 2016). Ag NRs are coated with subnanometer thick pinhole containing $Al_2O_3$ layer. Due to the strong chemisorption between $Al_2O_3$ and –COOH groups, $Al_2O_3$-coated Ag NRs showed superior detection towards a variety of carboxylic acids with high efficiency. The schematic detection and chemisorption process are shown in Fig. 10a. Figure 10b shows the SERS spectra of dipicolinic acid adsorbed on Ag $NRs@Al_2O_3$ substrate with a concentration range from $10^{-4}$ to $10^{-8}$ M. The advantage of these structures is that they display superior sensitivity and ability to work for a long time in corrosive and harsh environments (Ma et al. 2016).

The preparation of anisotropic Au NPs (boat like morphology) in mesoporous $\gamma Al_2O_3$ matrix by a simple sol-gel method followed by heat treatment was reported by Dandapat et al. The films show good SERS sensitivity owing to the unique morphology of the Au NPs. The limit of detection obtained for methylene blue was around $10^{-8}$ M with an EF $2.5 \times 10^5$ (Dandapat et al. 2013). Alumina is well known as an adsorbent in chromatographic separations and has a high affinity towards binding molecules with strong polarity. Xu et al. fabricated $Au/Al_2O_3$ nanocomposites via electrostatic interactions and demonstrated them as SERS substrates for the

**Table 5** Noble metal-based $SiO_2$ nanostructures for SERS applications

| Material | Method | Analyte | EF/Conc. | Reference |
|---|---|---|---|---|
| Ag@$SiO_2$core-shell nanoparticles-Si nanowires | Chemical etching | R6G | ~$10^{14}$ | Zhang et al. (2013) |
| Ag-$SiO_2$ Janus Particles | Pickering emulsion method | RhB | ~$5 \times 10^4$ | Panwar et al. (2017) |
| Au NPs@meso$SiO_2$ composites | Hydrothermal | DTNB | ~$10^{-12}$ M. | Lin et al. (2014) |
| Ag-embedded $SiO_2$-nanostructured thin films | Solution phase | R6G | ~$7.79 \times 10^8$ | Jayram et al. (2015) |
| Ag-$SiO_2$ nanocubes | Solution phase | R6G | ~$1.28 \times 10^6$ | Nguyen et al. (2017). |
| Ag@$SiO_2$ nanocubes | Ag-Polyol method | R6G | ~$1.2 \times 10^6$ | Kha et al. (2015) |
| Au NP-decorated $SiO_2$ mask | Sol-gel | R6G | ~$6.5 \times 10^7$ | Chen et al. (2014c) |
| Pillar-cap shaped arrays of Ag/$SiO_2$ | Solution phase | 4-MBA | $10^{-3}$ M | Wang et al., (2016) |
| Au/$SiO_2$ Nanocomposites | Sono-electrochemical pulse deposition | R6G | ~$5.4 \times 10^8$ | Chang et al. (2012) |
| $SiO_2$@Au core-shell | Solution phase | 1,2-Bis(4-pyridyl) ethylene | $10^{-11}$ M | Saini et al. (2015) |
| Au NPs/m$SiO_2$ | Solution phase | 4-MBA | ~$2.01 \times 10^7$ | Chen et al. (2017b) |
| Au–$SiO_2$ composite | Solution phase | R6G | – | Wang et al. (2014d) |
| Silica-coated Ag nanorods | Sol-gel | 4-MBA | ~$1.58 \times 10^8$ | Lai et al. (2015) |
| Ag nanocubes @ $SiO_2$ | Ag-Polyol method | Melamine | ~$1.93 \times 10^7$ | Mekonnen et al. (2017) |
| Ag Nanorod-$SiO_2$ core–shell array nanostructure | Ag-oblique angle deposition | 4-MP | ~$10^6$ | Song et al. (2011) |
| Core-shell $SiO_2$@ Ag | Solution phase | R6G | ~$1.09 \times 10^6$ | Tzounis et al. (2014) |
| Ag@$SiO_2$ nanoparticles | Solution phase | Pyrene | ~$10^{10}$ | Shanthil et al. (2017) |
| Ag@$SiO_2$/Ag core-shell nanosphere arrays | $SiO_2$-CVD Ag-thermal evaporation | R6G | ~$3 \times 10^7$ | Liu et al. (2016a, b) |
| Au@$SiO_2$ | Solution phase | TP and MG | $5 \times 10^{-7}$ M<br>$5 \times 10^{-9}$ M | Wang et al. (2014c) |
| $SiO_2$-isolated Ag Islands | Sputtering | 4-ATP | ~$4.41 \times 10^5$ | Wang et al. (2014b) |
| $SiO_2$–Ag | Glancing angle deposition technique | R6G | ~$2.38 \times 10^9$ | Wu and Cunningham (2011) |

(continued)

**Table 5** (continued)

| Material | Method | Analyte | EF/Conc. | Reference |
|---|---|---|---|---|
| $SiO_2$@Ag nanocomposite rods | Solution phase | Penicillin G sodium<br>Chloramphenicol | ~3.17 × $10^8$<br>~2.50 × $10^8$ | Zhao et al. (2015a, b) |
| Ag@$SiO_2$ core-shell | Ag-sputtering<br>$SiO_2$-inductively coupled plasma-enhanced chemical vapour deposition | CV | ~6.6 × $10^6$ | Zhao et al. (2016) |
| $SiO_2$/Ag | $SiO_2$-sol-gel<br>Ag-electron-irradiation | 4-ATP | ~1.694 × $10^5$ | Phatangare et al. (2016) |
| Ag-coated monolayer array of $SiO_2$ spheres | Solution phase | Brilliant green | ~4 × $10^4$ | Wu et al., 2014 |
| $SiO_2$@Ag | Solution phase | Bisphenol A | 1.46 × $10^{-11}$ M | Yin et al. (2018) |
| β-CD dimer@Ag@ $SiO_2$ NPs | Sol gel | Perylene | $10^{-8}$ M | Hahm et al. (2016) |

detection of adenosine triphosphate. The results showed the lowest detection of 5 × $10^{-9}$ M with excellent linearity (Xu et al. 2017b). Table 6 displays the various reports on SERS employing noble metal-based alumina substrates.

## 2.7 *Metal–Mn Oxide Nanoparticles*

Manganese oxides are emerging functional metal oxides due to their potential applications in various fields such as catalysis, ion exchange, electrochemical, molecular adsorption and SERS applications (Remucal and Ginder-Vogel 2014; Liu et al. 2013; Jana et al. 2009). Most of the reports are devoted to the engineering of manganese oxides for energy storage and conversion applications. The great interest in the field of battery industry is due to its theoretical capacitance (308 mAh $g^{-1}$). The natural abundance, low cost and low-toxicity nature of the Mn oxides render them as valuable functional materials among transition metal oxides (Liu et al. 2013). In the literature, only a handful of reports are available for SERS application of Mn oxide hybrid with noble metal NPs. Jana et al. reported a flower-like Ag-doped $MnO_2$ 3D nanostructures for SERS application. The composites were prepared using the wet chemical method. The SERS effect was demonstrated using R6G and aminothiophenol as probe molecules with an EF of $10^4$ and $10^6$, respectively (Jana et al. 2009). So far, various morphologies have been developed, and among them one-dimensional structures such as nanowires, nanorods and nanotubes exhibit distinctive physical and chemical properties. The nanorod morphology can provide a smooth surface and unique sharp edges for uniform distribution of Au NPs and

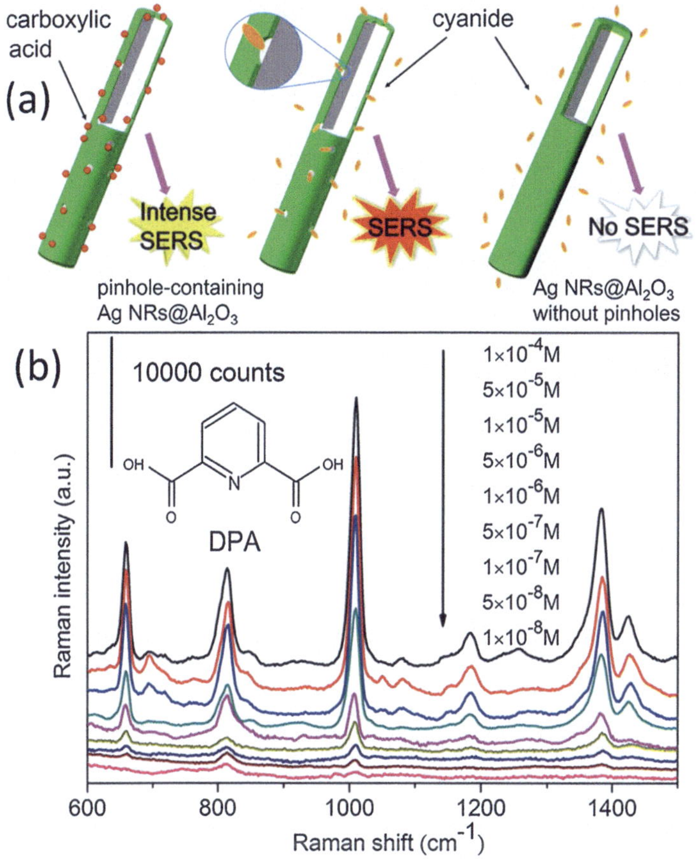


**Fig. 10** (**a**) The schematic displays SERS measurement of analytes over Ag NRs@$Al_2O_3$ shells. (**b**) Raman spectra of dipicolinic acid of various concentrations adsorbed on Ag NRs coated with $Al_2O_3$ (Ma et al. 2016)

hotspots. Hence, one can expect stable SERS signal (Jiang et al., 2015b). Jiang et al. prepared Au-decorated single-crystal tetragonal α-$MnO_2$nanorods employing hydrothermal and sputtering deposition technique (Fig. 11). The results demonstrate that the composite shows stable and intense SERS signal for methylene blue (Fig. 11). Furthermore, it has been used for the detection of methyl parathion pesticide, a broad concentration range from 1 mM to 100 ppm with a detection limit of 0.001 ppm (Jiang et al. 2015b).

Pradhan et al. fabricated Au-MnOOH nanocomposite flowers through redox reaction between the Mn(II) acetate and $HAuCl_4$ in aqueous solution without any surfactant and template. These were employed as a SERS substrate with thiophenol and aminothiophenol as probe molecules. A giant SERS signal enhancement in comparison to the individual components is noticed and gives the best single molecular detection level with an EF of $10^{15}$. The high sensitivity is due to the charge

**Table 6** Noble metal-based alumina ($Al_2O_3$) nanostructures for SERS applications

| Material | Method | Analyte | EF/Conc. | Reference |
|---|---|---|---|---|
| Au/nanoporous alumina layer | Electrochemical | CV<br>4-MPy | ~$3.4 \times 10^7$<br>~$4.6 \times 10^6$ | Yu et al. (2017) |
| $Al_2O_3$/Au NPs | Sonoelectrochemical deposition-dissolution cycles | R6G | ~$4.4 \times 10^8$ | Chang et al. (2013) |
| Boat like Au NPs/ $\gamma Al_2O_3$ | Solution phase | MB | ~$2.59 \times 10^6$ | Dandapat et al. (2013) |
| Ag/anodic aluminium oxide | Electrochemical | 4-MPy | ~$2.7 \times 10^5$ | Ji et al. (2009) |
| Au/$\gamma Al_2O_3$ Nanocomposite | Solution phase | ATP | $5 \times 10^{-9}$ M | Xu et al. (2017b) |
| Ag@$Al_2O_3$nanobowl arrays | $Al_2O_3$-anodization<br>Ag-thermal evaporation | p-thiocresol | ~$7.4 \times 10^6$ | Kang et al., 2016. |
| Ag/$Al_2O_3$ films | Solution phase | R6G | $2 \times 10^{-5}$ M | Mai et al., 2012. |
| Ag-nanoparticles/ $Al_2O_3$/Au-nanograting | $Al_2O_3$-atomic layer deposition<br>Au-nanoimprint lithography and metal deposition | p-thiocresol | ~$5.2 \times 10^7$ | Wang et al. (2017b) |
| $Al_2O_3$ Shell-coated Ag nanorods | Ag-oblique angle deposition<br>$Al_2O_3$-atomic layer deposition | 4-MBA | ~$4 \times 10^7$ | Ma et al. (2016). |
| Au/$Al_2O_3$ colloids | Sonoelectrochemical | R6G | $2 \times 10^{-10}$ M | Mai et al. (2011). |
| Ag/$Al_2O_3$ | $Al_2O_3$-Anodization<br>Ag-sputtering | RhB | $1 \times 10^{-6}$ M | Malek et al. (2014). |
| Ag Nanorods/ ultra-thin $Al_2O_3$ layers | $Al_2O_3$-atomic layer deposition<br>Ag NRs-oblique angle deposition | MB and Py | $5 \times 10^{-6}$ M<br>$1 \times 10^{-2}$ M | Ma et al. (2015a) |
| Ag NPs/porous alumina | Solution phase | CV | $10^{-10}$ M | Liu et al., (2017a) |
| Au/$Al_2O_3$@Au<br>Au/$Al_2O_3$@Ag | Au-rapid thermal annealing<br>$Al_2O_3$-atomic layer deposition | 4-NBT | ~$10^7$<br>~$10^9$ | Hu et al. (2014) |

transfer and electromagnetic effects (Pradhan et al. 2014a). The same group reported the fabrication of various Mn oxides by changing the amount of Ag content in the reaction. The composites, Ag-MnOOH and Ag-$MnO_2$ with 8% Ag content show an EF of $10^{10}$ with excellent thermal stability for thiophenol (Pradhan et al., 2014b). $MnO_2$/Au nanowall film has been fabricated by Zhou et al. using hydrothermal method followed by a sputtering technique. The substrates show high sensitivity, stability and reproducibility towards detection of crystal violet dye with an EF of $2.31 \times 10^8$ at the concentration $10^{-8}$ M (Zhou et al. 2015). The following Table 7 presents the various noble metal-based Mn oxide SERS substrates reported in literature.

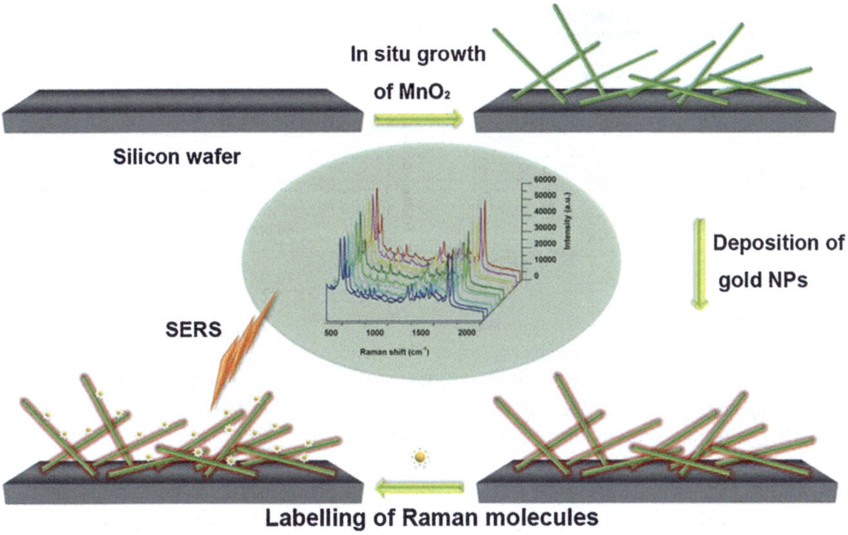


**Fig. 11** Fabrication of Au NP-decorated single-crystal tetragonal $\alpha$-$MnO_2$nanorods on Si wafer along with stable and intense SERS spectra of methylene blue (20 measured sites) (Jiang et al., 2015b)

**Table 7** Noble metal-based Mn oxide nanostructures for SERS applications

| Material | Method | Analyte | EF/Conc. | Reference |
|---|---|---|---|---|
| Ag-doped $MnO_2$ | $MnO_2$-wet chemical<br>Ag-photochemical | R6G and ATP | ~$3.7 \times 10^4$<br>~$2.23 \times 10^6$ | Jana et al. (2009) |
| Tetragonal $\alpha$-$MnO_2$nanorods/Au NPs | $MnO_2$-hydrothermal<br>Au-sputter deposition | MB | ~$1.61 \times 10^6$ | Jiang et al. (2015b) |
| Ag-$MnO_2$<br>Ag-MnOOH | Wet chemical method | TP | ~$10^{10}$ | Pradhan et al. (2014b) |
| $MnO_2$/Au hybrid nanowall film | $MnO_2$-hydrothermal<br>Au-thermal evaporation | CV | ~$1.38 \times 10^8$ | Zhou et al. (2015) |
| Hierarchical Hollow Au–MnOOH flowers | Wet chemical method | ATP | ~$10^{15}$ | Pradhan et al. (2014a) |

## 2.8 *Noble Metal-Other Metal Oxide Nanoparticles*

The cleaning strategies for SERS substrates are mostly intensive and time-consuming, which desires expensive apparatus and specific operations such as ultraviolet irradiation, magnet separation and plasma treatment, and so on. These methods consume time from several minutes to hours. Hence, it is essential to clean and regenerate SERS substrates in a more convenient and efficient manner. Thermal annealing is a quick and convenient tactic to detach adsorbed analyte molecules on

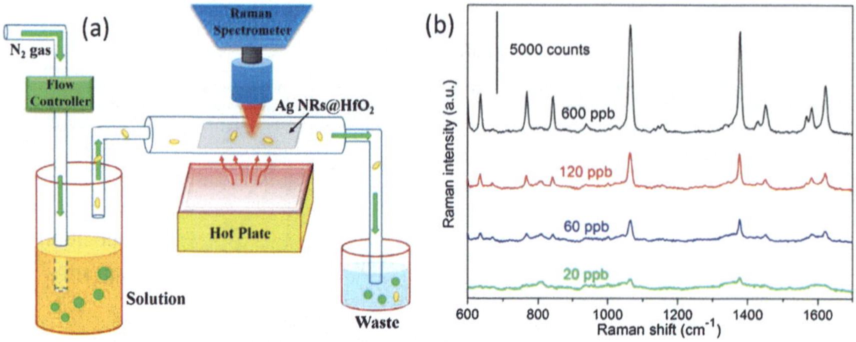


**Fig. 12** (**a**) Pictorial presentation of the SERS gas sensing device (**b**) Raman spectra of 2-naphthalenethiol (2-NAT)probed at various concentrations, 600, 120, 60 and 20 ppb on Ag NRs@$HfO_2$ substrate after 40 min of gas flow (Ma et al. 2016)

the surface of the SERS substrates. Most of the metal oxides are not stable in alkaline or acid environments limiting the application of the noble metal-metal oxides substrates (Ma et al. 2016). For high SERS with much stronger thermal robustness and excellent chemical stability, Ma et al. choose a hafnia ($HfO_2$) material, which possesses high melting point around ~2800 °C and chemical inertness to protect Ag nanorods. They found that the substrate is thermally stable and SERS active (Fig. 12). They reused the SERS substrate for detection of methylene blue and crystal violet dyes. The reusability is achieved by thermal treatment for several seconds at 50–700 °C (Ma et al. 2016).

Chen et al. reported a well-aligned zinc gallate ($ZnGa_2O_4$) nanorod array on Si substrate employing chemical vapour deposition and used as a template for the fabrication of SERS substrate by deposition of Ag NPs on the surface of zinc gallate nanorods. The SERS substrates display good sensitivity and reproducibility for melamine detection with a low concentration of $10^{-7}$ M (EF ~ $10^7$) (Chen et al., 2014b). In the later year, the same group reported urchin-like $LaVO_4$ microspheres as a template for fabrication of Au-$LaVO_4$ composite and showed good SERS with EF for MBA around ~$10^6$. They also show melamine detection with a low concentration of $10^{-9}$ M, which is better than an earlier report (Chen et al. 2015a). Hollow $BiVO_4$ microspheres preparation in the presence of ethylenediaminetetraacetic acid employing the hydrothermal method were decorated by Ag NPs and showed a higher photocatalytic activity under visible light and higher SERS activity for R6G (~$10^7$), which is much higher than the ZnO-Au and $LaVO_4$-Au systems (Chen et al., 2015b). To enhance the SERS signal, the surface of the SERS substrates should be extremely clean. For this Huang et al. prepared a clean 3D chestnut-like Ag/$WO_{3-x}$ nanostructures employing the hydrothermal method followed by an in situ redox reaction on the $WO_{2.72}$ that are inherently weak reducing species. Because of high-quality clean surfaces, the limit of detection for methylene blue dye concentration could be achieved down to $2.9 \times 10^{-13}$ M (Huang et al. 2017a,). Pisarek et al. reported a nanoporous zirconia ($ZrO_2$) loaded with Ag nanoparticles for use as

**Table 8** Noble metal-based other metal oxide nanostructures for SERS applications

| Materials | Method | Analyte | EF/Conc. | Reference |
|---|---|---|---|---|
| 3D chestnut like $Ag/WO_{3-x}$ | Solution phase | 4-ATP | ~$1.4 \times 10^7$ | Huang et al. (2017a) |
| $CeO_2$-on-Ag particles | Hydrothermal | PVP | – | Chang et al. (2014) |
| Ag Nanorods @ $HfO_2$ | Ag-glancing angle deposition $HfO_2$-atomic layer deposition | MB | ~$6.1 \times 10^7$ | Ma et al. (2016) |
| Hollow $BiVO_4$/Ag composite | Hydrothermal | R6G | ~$10^7$ | Chen et al. (2015b) |
| Urchin-like $LaVO_4$/ Au composite | Hydrothermal | 4-MBA | ~$10^6$ | Chen et al. (2015a) |
| $ZnGa_2O_4$Nanorod/ Ag NPs | $ZnGa_2O_4$—chemical vapour transport and condensation Ag-sputtering | R6G Melamine | ~$10^7$ | Chen et al. (2014b) |
| Ag/$ZrO_2$-nanotube/ Zr composite | $ZrO_2$-electrochemical Ag-sputtering | Py and MBA | ~$5 \times 10^5$ | Pisarek et al. (2014) |
| Ag films on $ZrO_2$ nanopore arrays | $ZrO_2$-electrochemical Ag-physical vapour deposition | Py and Sodium 2-mercaptoethanesulfonate | ~$10^5$ | Pisarek et al. (2017) |
| Au/porous $ZrO_2$ | Sol-gel | 4-MBA | $10^{-11}$ M | Chen et al., (2017c) |
| Ag /$SnO_2$ | Bio-template | R6G | ~$10^6$ | Liu et al. (2012) |
| $MoO_3$Nanowires@ Au NPs | Hydrothermal | Melamine | ~$8 \times 10^5$ | Liang et al. (2016) |
| Ag NPs-Ag molybdate nanowires | Photochemical | p-ATP | $1.0 \times 10^{-11}$ M | Bao et al. (2013) |
| Ag/$Ag_2MoO_4$ | Hydrothermal | p-Phenylenediamine | $3 \times 10^{-12}$ M | Fodjo et al. (2013) |

SERS substrate. The results display that the composite is active and stable for the detection of mercaptobenzoic acid, pyridine and rhodamine dye derivatives (Pisarek et al. 2014). Tin oxide ($SnO_2$) is also a suitable material for the optical coatings because of its transparency in the visible region. Furthermore, it is electroconductive. Lee et al. prepared a uniform monodisperse $SnO_2$-coated Au nanoparticles and studied their colloidal stability with respect to pH. Remarkably, the $SnO_2$-coated Au nanoparticles were stable up to a pH of 12.5 (Lee et al. 2013). Liu et al. prepared 3D Ag biomorphic $SnO_2$ nanostructures employing bio templates (Euploea mulciber—fore wings scales) and showed an enhancement factor around ~$10^6$ for rhodamine 6G dye (Liu et al. 2012). Table 8 presents the various other metal oxide-based noble metals reported for SERS applications.

## 3 Summary

SERS is a promising technique with superior advantage for the rapid and sensitive detection of molecules via identification of their vibrational signatures. Prominent attention has been shown towards the analysis of food additives and chemical contaminants. The progress in the manufacturing technology has made available, miniature and affordable Raman spectrophotometers that commercially unlock the doors to the possibility of portable SERS detection. Consequently, the development of low-cost and reusable SERS substrates holds great significance. This chapter gives an overview of the enhancement of Raman scattering employing various metal/metal oxide nanostructures utilizing their electromagnetic and charge transfer contributions to the enhancement and detection sensitivity. The analytes ranging from dyes and food adulterants to pesticides have been detected with good sensitivity proving their potential for on-field applications. The current trend is to use metal oxide supported or protected noble metal nanoparticles for robust SERS substrates that are renewable with less input energy.

## References

Albrecht MG, Creighton JA (1977) Anomalously intense Raman spectra of pyridine at a silver electrode. J Am Chem Soc 99:5215–5217. https://doi.org/10.1021/ja00457a071

Bao ZY, Lei DY, Dai J, Wu Y (2013) In situ and room-temperature synthesis of ultra-long Ag nanoparticles-decorated Ag molybdate nanowires as high-sensitivity SERS substrates. Appl Surf Sci 287:404–410. https://doi.org/10.1016/j.apsusc.2013.09.167

Bao ZY, Liu X, Dai J, Wu Y, Tsang YH, Lei DY (2014) In situ SERS monitoring of photocatalytic organic decomposition using recyclable $TiO_2$-coated Ag nanowire arrays. Appl Surf Sci 301:351–357. https://doi.org/10.1016/j.apsusc.2014.02.078

Betz JF, Wei WY, Cheng Y, White IM, Rubloff GW (2014) Simple SERS substrates: powerful, portable, and full of potential. Phys Chem Chem Phys 16:2224–2239. https://doi.org/10.1039/C3CP53560F

Bitar A, Ahmad NM, Fessi H, Elaissari A (2012) Silica-based nanoparticles for biomedical applications. Drug Discov Today 17:1147–1154. https://doi.org/10.1016/j.drudis.2012.06.014

Bramhaiah K, Singh VN, John NS (2016) Hybrid materials of ZnO nanostructures with reduced graphene oxide and gold nanoparticles: enhanced photodegradation rates in relation to their composition and morphology. Phys Chem Chem Phys 18:1478–1486. https://doi.org/10.1039/C5CP05081B

Bramhaiah K, Singh VN, Kavitha C, John NS (2017) Films of reduced graphene oxide with metal oxide nanoparticles formed at a liquid/liquid interface as reusable surface enhanced Raman scattering substrates for dyes. J Nanosci Nanotechnol 17:2711–2719. https://doi.org/10.1166/jnn.2017.13431

Cai W, Wang X, Yan Y (2014) Controllable fabrication and sensitive detection based on SERS substrates with au nanocubes coated $Fe_3O_4$. Mater Res Bull 52:1–5. https://doi.org/10.1016/j.materresbull.2013.12.046

Chan SHS, Yeong Wu T, Juan JC, Teh CY (2011) Recent developments of metal oxide semiconductors as photocatalysts in advanced oxidation processes (AOPs) for treatment of dye wastewater. J Chem Technol Biotechnol 86:1130–1158. https://doi.org/10.1002/jctb.2636

Chang CC, Yang KH, Liu YC, Hsu TC, Mai FD (2012) Surface-enhanced Raman scattering-active au/$SiO_2$ nanocomposites prepared using sonoelectrochemical pulse deposition methods. ACS Appl Mater Interfaces 4:4700–4707. https://doi.org/10.1021/am3017366

Chang CC, Yu CC, Liu YC, Yang KH (2013) $Al_2O_3$-modified surface-enhanced Raman scattering-active gold nanoparticles on substrates by using sonoelectrochemical pulse deposition. J Electroanal Chem 696:38–44. https://doi.org/10.1016/j.jelechem.2013.03.005

Chang S, Ruan S, Wu E, Huang W (2014) $CeO_2$ thickness-dependent SERS and catalytic properties of $CeO_2$-on-Ag particles synthesized by $O_2$-assisted hydrothermal method. J Phys Chem C 118:19238–19245. https://doi.org/10.1021/jp506187d

Chen L, Luo L, Chen Z, Zhang M, Zapien JA, Lee CS, Lee ST (2009) ZnO/au composite nanoarrays as substrates for surface-enhanced Raman scattering detection. J Phys Chem C 114:93–100. https://doi.org/10.1021/jp908423v

Chen IC, Liou YCM, Yang J, Shieh TY (2011a) Preparation of silver nanoparticles on zinc oxide nanowires by photocatalytic reduction for use in surface-enhanced Raman scattering measurements. J Raman Spectrosc 42:339–344. https://doi.org/10.1002/jrs.2727

Chen L, Seo HK, Mao Z, Jung YM, Zhao B (2011b) Tunable plasmon properties of $Fe_2O_3$@Ag substrate for surface-enhanced Raman scattering. Anal Methods 3:1622–1627. https://doi.org/10.1039/C0AY00729C

Chen J, Su H, You X, Gao J, Lau WM, Zhang D (2014a) 3D $TiO_2$ sub microstructures decorated by silver nanoparticles as SERS substrate for organic pollutants detection and degradation. Mater Res Bull 49:560–565. https://doi.org/10.1016/j.materresbull.2013.09.040

Chen L, Jiang D, Liu X, Qiu G (2014b) $ZnGa_2O_4$Nanorod arrays decorated with Ag nanoparticles as surface-enhanced Raman-scattering substrates for melamine detection. ChemPhysChem 15:1624–1631. https://doi.org/10.1002/cphc.201400050

Chen LY, Yang KH, Chen HC, Liu YC, Chen CH, Chen QY (2014c) Innovative fabrication of a au nanoparticle-decorated $SiO_2$ mask and its activity on surface-enhanced Raman scattering. Analyst 139:1929–1937. https://doi.org/10.1039/C3AN02089D

Chen L, Wu M, Xiao C, Yu Y, Liu X, Qiu G (2015a) Urchin-like $LaVO_4$/au composite microspheres for surface-enhanced Raman scattering detection. J Colloid Interface Sci 443:80–87. https://doi.org/10.1016/j.jcis.2014.12.011

Chen L, Yu Y, Wu M, Huang J, Liu Y, Liu X, Qiu G (2015b) Synthesis of hollow $BiVO_4$/Ag composite microspheres and their Photocatalytic and surface-enhanced Raman scattering properties. ChemPlusChem 80:871–877. https://doi.org/10.1002/cplu.201402434

Chen L, Zhao Y, Zhang Y, Liu M, Wang Y, Qu X, Liu Y, Li J, Liu X, Yang J (2016) Design of $Cu_2O$-au composite microstructures for surface-enhanced Raman scattering study. Colloids Surf A Physicochem Eng Asp 507:96–102. https://doi.org/10.1016/j.colsurfa.2016.07.053

Chen L, Sun H, Zhao Y, Zhang Y, Wang Y, Liu Y, Zhang X, Jiang Y, Hua Z, Yang J (2017a) Plasmonic-induced SERS enhancement of shell-dependent Ag@$Cu_2O$ core-shell nanoparticles. RSC Adv 7:16553–16560. https://doi.org/10.1039/C7RA01187C

Chen M, Luo W, Zhang Z, Wang R, Zhu Y, Yang H, Chen X (2017b) Synthesis of multi-au-nanoparticle-embedded Mesoporous silica microspheres as self-filtering and reusable substrates for SERS detection. ACS Appl Mater Interfaces 9:42156–42166. https://doi.org/10.1021/acsami.7b16618

Chen Y, Shen J, Huang Z, Zhu P, Xiong X, Ouyang F (2017c) One-step synthesis of au/porous $ZrO_2$ SERS-active nanocomposite: fabrication and tunable optical properties. Jo. Alloys Compd 721:118–125. https://doi.org/10.1016/j.jallcom.2017.05.104

Chen L, Zhang F, Deng XY, Xue X, Wang L, Sun Y, Feng JD, Zhang Y, Wang Y, Jung YM (2018) SERS study of surface plasmon resonance induced carrier movement in au@ $Cu_2O$ core-shell nanoparticles. Spectrochim. Acta A 189:608–612. https://doi.org/10.1016/j.saa.2017.08.065

Cheng C, Yan B, Wong SM, Li X, Zhou W, Yu T, Shen Z, Yu H, Fan HJ (2010) Fabrication and SERS performance of silver-nanoparticle-decorated Si/ZnO nanotrees in ordered arrays. ACS Appl Mater Interfaces 2:1824–1828. https://doi.org/10.1021/am100270b

Dai H, Sun Y, Ni P, Lu W, Jiang S, Wang Y, Li Z, Li Z (2017) Three-dimensional $TiO_2$ supported silver nanoparticles as sensitive and UV-cleanable substrate for surface enhanced Raman scattering. Sensors Actuators B 242:260–268. https://doi.org/10.1016/j.snb.2016.10.085

Dandapat A, Pramanik S, Bysakh S, De G (2013) Boat-like au nanoparticles embedded mesoporous $\gamma$-$Al_2O_3$ films: an efficient SERS substrate. J Nanopart Res 15:1804. https://doi.org/10.1007/s11051-013-1804-1

Ding Q, Zhou H, Zhang H, Zhang Y, Wang G, Zhao H (2016) 3D $Fe_3O_4$@au@Ag nanoflowers assembled magnetoplasmonic chains for in situ SERS monitoring of plasmon-assisted catalytic reactions. J Mater Chem A 4:8866–8874. https://doi.org/10.1039/C6TA02264B

Fan C, Zhu S, Xin H, Tian Y, Liang E (2017) Tunable and enhanced SERS activity of magneto-plasmonic Ag-$Fe_3O_4$ nanocomposites with one pot synthesize method. J Opt 19:015401. https://doi.org/10.1088/2040-8986/19/1/015401

Fateixa S, Nogueira HI, Trindade T (2015) Hybrid nanostructures for SERS: materials development and chemical detection. Phys Chem Chem Phys 17:21046–21071. https://doi.org/10.1039/C5CP01032B

Fleischmann M, Hendra PJ, McQuillan AJ (1974) Raman spectra of pyridine adsorbed at a silver electrode. Chem Phys Lett 26:163–166. https://doi.org/10.1016/0009-2614(74)85388-1

Fodjo EK, Li DW, Marius NP, Albert T, Long YT (2013) Low temperature synthesis and SERS application of silver molybdenum oxides. J Mater Chem A 1:2558–2566. https://doi.org/10.1039/C2TA01018F

Fujishima A, Honda K (1972) Electrochemical photolysis of water at a semiconductor electrode. Nature 238:37–38. https://doi.org/10.1038/238037a0

Guerrini L, Graham D (2012) Molecularly-mediated assemblies of plasmonic nanoparticles for surface-enhanced Raman spectroscopy applications. Chem Soc Rev 41:7085–7107. https://doi.org/10.1039/C2CS35118H

Guo TL, Li JG, Ping DH, Sun X, Sakka Y (2014a) Controlled photocatalytic growth of Ag nanocrystals on Brookite and rutile and their SERS performance. ACS Appl Mater Interfaces 6:236–243. https://doi.org/10.1021/am404027m

Guo H, Zhao A, Gao Q, Li D, Zhang M, Gan Z, Wang D, Tao W, Chen X (2014b) One-step synthesis of Ag-$Fe_3O_4$ nanocomposites and their SERS properties. J Nanopart Res 16:2538. https://doi.org/10.1007/s11051-014-2538-4

Guo H, Zhao A, Wang R, Wang D, Wang L, Gao Q, Sun H, Li L, He Q (2015) Generalized green synthesis of $Fe_3O_4$/Ag composites with excellent SERS activity and their application in fungicide detection. J Nanopart Res 17:494. https://doi.org/10.1007/s11051-015-3286-9

Hahm E, Jeong D, Cha MG, Choi JM, Pham XH, Kim HM, Kim H, Lee YS, Jeong DH, Jung S, Jun BH (2016) β-CD dimer-immobilized Ag assembly embedded silica nanoparticles for sensitive detection of polycyclic aromatic hydrocarbons. Sci Rep 6:26082. https://doi.org/10.1038/srep26082

Han SY, Guo QH, Xu MM, Yuan YX, Shen LM, Yao JL, Liu W, Gu RA (2012) Tunable fabrication on iron oxide/Au/Ag nanostructures for surface enhanced Raman spectroscopy and magnetic enrichment. J Colloid Interface Sci 378:51–57. https://doi.org/10.1016/j.jcis.2012.04.047

Han D, Huang H, Du D, Lang X, Long K, Hao Q, Qiu T (2015) Facile synthesis of gold-capped $TiO_2$ nanocomposites for surface-enhanced Raman scattering. Mater Chem Phys 153:88–92. https://doi.org/10.1016/j.matchemphys.2014.12.038

Han XX, Ji W, Zhao B, Ozaki Y (2017) Semiconductor-enhanced Raman scattering: active nanomaterials and applications. Nanoscale 9:4847–4861. https://doi.org/10.1039/C6NR08693D

He X, Yue C, Zang Y, Yin J, Sun S, Li J, Kang J (2013) Multi-hot spot configuration on urchin-like Ag nanoparticle/ZnO hollow nanosphere arrays for highly sensitive SERS. J Mater Chem A 1:15010–15015. https://doi.org/10.1039/C3TA13450D

He X, Wang H, Zhang Q, Li Z, Wang X (2014) Exotic 3D hierarchical ZnO–Ag hybrids as recyclable surface-enhanced Raman scattering substrates for multifold organic pollutant detection. Eur J Inorg Chem 2014(14):2432–2439. https://doi.org/10.1002/ejic.201402014

Ho JY, Huang MH (2009) Synthesis of submicrometer-sized $Cu_2O$ crystals with morphological evolution from cubic to hexapod structures and their comparative photocatalytic activity. J Phys Chem C 113:14159–14164. https://doi.org/10.1021/jp903928p

Hola K, Markova Z, Zoppellaro G, Tucek J, Zboril R (2015) Tailored functionalization of iron oxide nanoparticles for MRI, drug delivery, magnetic separation and immobilization of biosubstances. Biotechnol Adv 33:1162–1176. https://doi.org/10.1016/j.biotechadv.2015.02.003

Hsieh S, Lin PY, Chu LY (2014) Improved performance of solution-phase surface-enhanced Raman scattering at Ag/CuO nanocomposite surfaces. J Phys Chem C 118:12500–12505. https://doi.org/10.1021/jp503202f

Hu Z, Liu Z, Li L, Quan B, Li Y, Li J, Gu C (2014) Wafer-scale double-layer stacked Au/$Al_2O_3$@Au nanosphere structure with tunable nanospacing for surface-enhanced Raman scattering. Small 10:3933–3942. https://doi.org/10.1002/smll.201400509

Huang J, Sun Y, Huang S, Yu K, Zhao Q, Peng F, Yu H, Wang H, Yang J (2011) Crystal engineering and SERS properties of Ag–$Fe_3O_4$ nanohybrids: from heterodimer to core-shell nanostructures. J Mater Chem 21:17930–17937. https://doi.org/10.1039/C1JM13045E

Huang J, Chen F, Zhang Q, Zhan Y, Ma D, Xu K, Zhao Y (2015a) 3D silver nanoparticles decorated zinc oxide/silicon heterostructured nanomace arrays as high-performance surface-enhanced Raman scattering substrates. ACS Appl Mater Interfaces 7:5725–5735. https://doi.org/10.1021/am507857x

Huang Q, Liu S, Wei W, Yan Q, Wu C (2015b) Selective synthesis of different ZnO/Ag nanocomposites as surface enhanced Raman scattering substrates and highly efficient photocatalytic catalysts. RSC Adv 5:27075–27081. https://doi.org/10.1039/C5RA01068C

Huang J, Ma D, Chen F, Chen D, Bai M, Xu K, Zhao Y (2017a) Green in situ synthesis of clean 3D chestnut like Ag/$WO_{3-x}$ nanostructures for highly efficient, recyclable and sensitive SERS sensing. ACS Appl Mater Interfaces 9:7436–7446. https://doi.org/10.1021/acsami.6b14571

Huang Q, Li J, Wei W, Wu Y, Li T (2017b) Synthesis, characterization and application of $TiO_2$/Ag recyclable SERS substrates. RSC Adv 7:26704–26709. https://doi.org/10.1039/C7RA03112B

Ibhadon AO, Fitzpatrick P (2013) Heterogeneous photocatalysis: recent advances and applications. Catalysts 3:189–218. https://doi.org/10.3390/catal3010189

Jahn M, Patze S, Hidi IJ, Knipper R, Radu AI, Mühlig A, Yüksel S, Peksa V, Weber K, Mayerhöfer T, Cialla-May D (2016) Plasmonic nanostructures for surface enhanced spectroscopic methods. Analyst 141:756–793. https://doi.org/10.1039/C5AN02057C

Jana S, Pande S, Sinha AK, Sarkar S, Pradhan M, Basu M, Saha S, Pal T (2009) A green chemistry approach for the synthesis of flower-like Ag-doped $MnO_2$ nanostructures probed by surface-enhanced Raman spectroscopy. J Phys Chem C 113:1386–1392. https://doi.org/10.1021/jp809561p

Jayram ND, Sonia S, Kumar PS, Marimuthu L, Masuda Y, Mangalaraj D, Ponpandian N, Viswanathan C, Ramakrishna S (2015) Highly monodispersed Ag embedded $SiO_2$ nanostructured thin film for sensitive SERS substrate: growth, characterization and detection of dye molecules. RSC Adv 5:46229–46239. https://doi.org/10.1039/C5RA04355G

Jayram ND, Aishwarya D, Sonia S, Mangalaraj D, Kumar PS, Rao GM (2016) Analysis on superhydrophobic silver decorated copper oxide nanostructured thin films for SERS studies. J Colloid Interface Sci 477:209–219. https://doi.org/10.1016/j.jcis.2016.05.051

Jeanmaire DL, Van Duyne RP (1977) Surface Raman spectroelectrochemistry: part I. heterocyclic, aromatic, and aliphatic amines adsorbed on the anodized silver electrode. J Electroanal Chem Interfacial Electrochem 84:1–20. https://doi.org/10.1016/S0022-0728(77)80224-6

Ji N, Ruan W, Wang C, Lu Z, Zhao B (2009) Fabrication of silver decorated anodic aluminum oxide substrate and its optical properties on surface-enhanced Raman scattering and thin film interference. Langmuir 25:11869–11873. https://doi.org/10.1021/la901521j

Ji W, Xue X, Ruan W, Wang C, Ji N, Chen L, Li Z, Song W, Zhao B, Lombardi JR (2011) Scanned chemical enhancement of surface-enhanced Raman scattering using a charge-transfer complex. ChemCommun 47:2426–2428. https://doi.org/10.1039/C0CC03697H

Ji W, Kitahama Y, Xue X, Zhao B, Ozaki Y (2012) Generation of pronounced resonance profile of charge-transfer contributions to surface-enhanced Raman scattering. J Phys Chem C 116:2515–2520. https://doi.org/10.1016/S0022-0728(77)80224-6

Ji R, Sun W, Chu Y (2014) One-step hydrothermal synthesis of Ag/$Cu_2O$ heterogeneous nanostructures over cu foil and their SERS applications. RSC Adv 4:6055–6059. https://doi.org/10.1039/C3RA44281K

Jiang X, Li X, Jia X, Li G, Wang X, Wang G, Li Z, Yang L, Zhao B (2012) Surface-enhanced Raman scattering from synergistic contribution of metal and semiconductor in $TiO_2$/MBA/Ag (Au) and Ag (Au)/MBA/$TiO_2$ assemblies. J Phys Chem C 116:14650–14655. https://doi.org/10.1021/jp302139e

Jiang R, Li B, Fang C, Wang J (2014) Metal/semiconductor hybrid nanostructures for Plasmon-enhanced applications. Adv Mater 26:5274–5309. https://doi.org/10.1002/adma.201400203

Jiang L, Liang X, You T, Yin P, Wang H, Guo L, Yang S (2015a) A sensitive SERS substrate based on Au/$TiO_2$/Au nanosheets. Spectrochim Acta A 142:50–54. https://doi.org/10.1016/j.saa.2015.01.040

Jiang T, Zhang L, Jin H, Wang X, Zhou J (2015b) In situ controlled sputtering deposition of gold nanoparticles on $MnO_2$ nanorods as surface-enhanced Raman scattering substrates for molecular detection. Dalton Trans 44:7606–7612. https://doi.org/10.1039/C4DT03774J

Joshi P, Zhou Y, Ahmadov TO, Zhang P (2014) Quantitative SERS-based detection using Ag-$Fe_3O_4$ nanocomposites with an internal reference. J Mater Chem C 2:9964–9968. https://doi.org/10.1039/C4TC01550A

Kaloti M, Kumar A (2016) Synthesis of chitosan-mediated silver coated $\gamma$-$Fe_2O_3$ (Ag–$\gamma$-$Fe_2O_3$@Cs) superparamagnetic binary nanohybrids for multifunctional applications. J Phys Chem C 120:17627–17644. https://doi.org/10.1021/acs.jpcc.6b05851

Kang HW, Leem J, Sung HJ (2015) Photoinduced synthesis of Ag nanoparticles on ZnO nanowires for real-time SERS systems. RSC Adv 5:51–57. https://doi.org/10.1039/C4RA11296B

Kang M, Zhang X, Liu L, Zhou Q, Jin M, Zhou G, Gao X, Lu X, Zhang Z, Liu J (2016) High-density ordered Ag@ $Al_2O_3$ nanobowl arrays in applications of surface-enhanced Raman spectroscopy. Nanotechnology 27:165304. https://doi.org/10.1088/0957-4484/27/16/165304

Kavitha C, Bramhaiah K, John NS, Ramachandran BE (2015) Low cost, ultra-thin films of reduced graphene oxide–Ag nanoparticle hybrids as SERS based excellent dye sensors. Chem Phys Lett 629:81–86. https://doi.org/10.1016/j.cplett.2015.04.026

Kavitha C, Bramhaiah K, John NS, Aggarwal S (2017) Improved surface-enhanced Raman and catalytic activities of reduced graphene oxide–osmium hybrid nano thin films. R Soc Open Sci 4:170353. https://doi.org/10.1098/rsos.170353

Kha NM, Chen CH, Su WN, Rick J, Hwang BJ (2015) Improved Raman and photoluminescence sensitivity achieved using bifunctional Ag@$SiO_2$nanocubes. Phys Chem Chem Phys 17:21226–21235. https://doi.org/10.1039/C4CP05217J

Kudelski A, Grochala W, Janik-Czachor M, Bukowska J, Szummer A, Dolata M (1998) Surface-enhanced Raman scattering at (SERS) oxide copper (I). J Raman Spectrosc 29:431–435. https://doi.org/10.1002/(SICI)1097-4555(199805)29:5<431::AID-JRS260>3.0.CO;2-S

Lai W, Zhou J, Liu Y, Jia Z, Xie S, Petti L, Mormile P (2015) 4MBA-labeled Ag-nanorod aggregates coated with $SiO_2$: synthesis, SERS activity, and biosensing applications. Anal Methods 7:8832–8838. https://doi.org/10.1039/C5AY01886B

Le Ru EC, Blackie E, Meyer M, Etchegoin PG (2007) Surface enhanced Raman scattering enhancement factors: a comprehensive study. J Phys Chem C 111:13794–13803. https://doi.org/10.1021/jp0687908

Lee SH, Rusakova I, Hoffman DM, Jacobson AJ, Lee TR (2013) Monodisperse $SnO_2$-coated gold nanoparticles are markedly more stable than analogous $SiO_2$-coated gold nanoparticles. ACS Appl Mater Interfaces 5:2479–2484. https://doi.org/10.1021/am302740z

Li X, Chen G, Yang L, Jin Z, Liu J (2010) Multifunctional au-coated $TiO_2$ nanotube arrays as recyclable SERS substrates for multifold organic pollutants detection. Adv Funct Mater 20:2815–2824. https://doi.org/10.1002/adfm.201000792

Li X, Hu H, Li D, Shen Z, Xiong Q, Li S, Fan HJ (2012) Ordered array of gold semi-shells on $TiO_2$ spheres: an ultrasensitive and recyclable SERS substrate. ACS Appl Mater Interfaces 4:2180–2185. https://doi.org/10.1021/am300189n

Li S, Li D, Zhang QY, Tang X (2016a) Surface enhanced Raman scattering substrate with high-density hotspots fabricated by depositing Ag film on $TiO_2$-catalyzed Ag nanoparticles. J Alloys Compd 689:439–445. https://doi.org/10.1016/j.jallcom.2016.07.303

Li Z, Zhu K, Zhao Q, Meng A (2016b) The enhanced SERS effect of Ag/ZnO nanoparticles through surface hydrophobic modification. Appl Surf Sci 377:23–29. https://doi.org/10.1016/j.apsusc.2016.03.084

Li F, Wu S, Zhang L, Li Z (2017) Ag nanoparticles decorated ZnO nanoarrays with enhanced surface-enhanced Raman scattering and field emission property. J Mater Sci 28:16233–16238. https://doi.org/10.1007/s10854-017-7526-x

Liang X, Zhang XJ, You TT, Wang GS, Yin PG, Guo L (2016) Controlled assembly of one-dimensional $MoO_3$@ au hybrid nanostructures as SERS substrates for sensitive melamine detection. CrystEngComm 18:7805–7813. https://doi.org/10.1039/C6CE01678B

Lin CC, Chang CW (2014) AuNPs@ meso$SiO_2$ composites for SERS detection of DTNB molecule. Biosens Bioelectron 51:297–303. https://doi.org/10.1016/j.bios.2013.07.065

Lin CC, Yang YM, Liao PH, Chen DW, Lin HP, Chang HC (2014) A filter-like AuNPs@MS SERS substrate for *Staphylococcus aureus* detection. Biosens Bioelectron 53:519–527. https://doi.org/10.1016/j.bios.2013.10.017

Ling Y, Zhuo Y, Huang L, Mao D (2016) Using Ag-embedded $TiO_2$ nanotubes array as recyclable SERS substrate. Appl Surf Sci 388:169–173. https://doi.org/10.1016/j.apsusc.2016.01.257

Liu B, Zhang W, Lv H, Zhang D, Gong X (2012) Novel Ag decorated biomorphic $SnO_2$ inspired by natural 3D nanostructures as SERS substrates. Mater Lett 74:43–45. https://doi.org/10.1016/j.matlet.2011.12.086

Liu X, Chen C, Zhao Y, Jia B (2013) A review on the synthesis of manganese oxide nanomaterials and their applications on lithium-ion batteries. J Nanomater 2013. https://doi.org/10.1155/2013/736375

Liu L, Yang H, Ren X, Tang J, Li Y, Zhang X, Cheng Z (2015) Au–ZnO hybrid nanoparticles exhibiting strong charge-transfer-induced SERS for recyclable SERS-active substrates. Nanoscale 7:5147–5151. https://doi.org/10.1039/C5NR00491H

Liu LW, Zhou QW, Zeng ZQ, Jin ML, Zhou GF, Zhan RZ, Chen HJ, Gao XS, Lu XB, Senz S, Zhang Z (2016a) Induced SERS activity in Ag@ $SiO_2$/Ag core-shell nanosphere arrays with tunable interior insulator. J Raman Spectrosc 47:1200–1206. https://doi.org/10.1002/jrs.4941

Liu S, Regulacio MD, Tee SY, Khin YW, Teng CP, Koh LD, Guan G, Han MY (2016b) Preparation, functionality, and application of metal oxide-coated noble metal nanoparticles. Chem Rec 16:1965–1990. https://doi.org/10.1002/tcr.201600028

Liu S, Yu J, Wang T, Li F (2017a) A multifunctional Ag/PAOCG reusable substrate for p-nitrophenol reduction and SERS applications. J Mater Sci 52:13748–13763. https://doi.org/10.1007/s10853-017-1461-3

Liu X, Iocozzia J, Wang Y, Cui X, Chen Y, Zhao S, Li Z, Lin Z (2017b) Noble metal-metal oxide nanohybrids with tailored nanostructures for efficient solar energy conversion, photocatalysis and environmental remediation. Energy Environ Sci 10:402–434. https://doi.org/10.1039/C6EE02265K

Lu J, Xu C, Nan H, Zhu Q, Qin F, Manohari AG, Wei M, Zhu Z, Shi Z, Ni Z (2016) SERS-active ZnO/Ag hybrid WGM microcavity for ultrasensitive dopamine detection. Appl Phys Lett 109:073701. https://doi.org/10.1063/1.4961116

Luo H, Zhou J, Zhong H, Zhou L, Jia Z, Tan X (2016) Polyhedron $Cu_2O$@ Ag composite microstructures: synthesis, mechanism analysis and structure-dependent SERS properties. RSC Adv 6:99105–99113. https://doi.org/10.1039/C6RA20856H

Ma L, Huang Y, Hou M, Xie Z, Zhang Z (2015a) Silver nanorods wrapped with ultrathin $Al_2O_3$ layers exhibiting excellent SERS sensitivity and outstanding SERS stability. Sci Rep 5:12890. https://doi.org/10.1038/srep12890

Ma L, Huang Y, Hou M, Li J, Xie Z, Zhang Z (2015b) Pinhole-containing, subnanometer-thick $Al_2O_3$ shell-coated Ag nanorods as practical substrates for quantitative surface-enhanced Raman scattering. J Phys Chem C 120:606–615. https://doi.org/10.1021/acs.jpcc.5b11043

Ma L, Wu H, Huang Y, Zou S, Li J, Zhang Z (2016) High-performance real-time SERS detection with recyclable Ag Nanorods@$HfO_2$ substrates. ACS Appl Mater Interfaces 8:27162–27168. https://doi.org/10.1021/acsami.6b10818

Mai FD, Yu CC, Liu YC, Yang KH, Juang MY (2011) Preparation of surface-enhanced Raman scattering-active au/$Al_2O_3$ colloids by Sonoelectrochemical methods. J Phys Chem C 115:13660–13666. https://doi.org/10.1021/jp203931r

Mai FD, Yang KH, Liu YC, Hsu TC (2012) Improved stabilities on surface-enhanced Raman scattering-active Ag/$Al_2O_3$ films on substrates. Analyst 137:5906–5912. https://doi.org/10.1039/C2AN35829H

Malek K, Brzózka A, Rygula A, Sulka GD (2014) SERS imaging of silver coated nanostructured Al and $Al_2O_3$ substrates. The effect of nanostructure. J Raman Spectrosc 45:281–291. https://doi.org/10.1002/jrs.4452

Mao Z, Song W, Xue X, Ji W, Li Z, Chen L, Mao H, Lv H, Wang X, Lombardi JR, Zhao B (2012) Interfacial charge-transfer effects in semiconductor–molecule–metal structures: influence of contact variation. J Phys Chem C 116:14701–14710. https://doi.org/10.1021/jp304051r

McNay G, Eustace D, Smith WE, Faulds K, Graham D (2011) Surface-enhanced Raman scattering (SERS) and surface-enhanced resonance Raman scattering (SERRS): a review of applications. Appl Spectrosc 65:825–837. https://doi.org/10.1366/11-06365

Mekonnen ML, Su WN, Chen CH, Hwang BJ (2017) Ag@$SiO_2$nanocube loaded miniaturized filter paper as a hybrid flexible plasmonic SERS substrate for trace melamine detection. Anal Methods 9:6823–6829. https://doi.org/10.1039/C7AY02192E

Moskovits M (2005) Surface-enhanced Raman spectroscopy: a brief retrospective. J Raman Spectrosc 36:485–496. https://doi.org/10.1002/jrs.1362

Nakata K, Fujishima A (2012) $TiO_2$ photocatalysis: design and applications. J Photochem Photobiol 13:169–189. https://doi.org/10.1016/j.jphotochemrev.2012.06.001

Nguyen MK, Su WN, Chen CH, Rick J, Hwang BJ (2017) Highly sensitive and stable Ag@$SiO_2$ nanocubes for label-free SERS-photoluminescence detection of biomolecules. Spectrochim Acta A 175:239–245. https://doi.org/10.1016/j.saa.2016.12.024

Pal J, Ganguly M, Dutta S, Mondal C, Negishi Y, Pal T (2014) Hierarchical Au-CuO nanocomposite from redox transformation reaction for surface-enhanced Raman scattering and clock reaction. CrystEngComm 16:883–893. https://doi.org/10.1039/C3CE41766B

Panwar K, Jassal M, Agrawal AK (2017) Ag-$SiO_2$ Janus particles based highly active SERS macroscopic substrates. Appl Surf Sci 411:368–373. https://doi.org/10.1016/j.apsusc.2017.03.105

Park SG, Kwon JD, Mun CW, Cho B, Kim CS, Song M, Kim DH, Jeon TY, Jeon HC (2016) Fabrication of Au-decorated 3D ZnO nanostructures as recyclable SERS substrates. IEEE Sensors J 16:1387–1390. https://doi.org/10.1109/JSEN.2015.2418787

Pérez León C, Kador L, Peng B, Thelakkat M (2006) Characterization of the adsorption of Ru-bpy dyes on mesoporous $TiO_2$ films with UV– Vis, Raman, and FTIR spectroscopies. J Phys Chem B 110:8723–8730. https://doi.org/10.1021/jp0561827

Phatangare AB, Dhole SD, Dahiwale SS, Mathe VL, Bhoraskar SV, Late DJ, Bhoraskar VN (2016) Surface chemical bonds, surface-enhanced Raman scattering, and dielectric constant of $SiO_2$ nanospheres in-situ decorated with Ag-nanoparticles by electron-irradiation. J Appl Phys 120:234901. https://doi.org/10.1063/1.4971866

Pimentel A, Araújo A, Coelho B, Nunes D, Oliveira M, Mendes M, Águas H, Martins R, Fortunato E (2017) 3D ZnO/Ag surface-enhanced Raman scattering on disposable and flexible cardboard platforms. Materials 10:1351. https://doi.org/10.3390/ma10121351

Pisarek M, Roguska A, Kudelski A, Holdynski M, Janik-Czachor M, Hnida K, Sulka GD (2014) Ag/$ZrO_2$-NT/Zr hybrid material: a new platform for SERS measurements. Vib Spectrosc 71:85–90. https://doi.org/10.1016/j.vibspec.2014.01.005

Pisarek M, Krajczewski J, Wierzbicka E, Hołdyński M, Sulka GD, Nowakowski R, Kudelski A, Janik-Czachor M (2017) Influence of the silver deposition method on the activity of platforms for chemometric surface-enhanced Raman scattering measurements: silver films on $ZrO_2$ nanopore arrays. Spectrochim Acta Part A 182:124–129. https://doi.org/10.1016/j.saa.2017.04.005

Prabhu R, Bramhaiah K, Singh KK, John NS (2019) Single Sea urchin–$MoO_3$ nanostructure for surface enhanced Raman spectroscopy of dyes. Nanoscale Adv 1:2426–2434. https://doi.org/10.1039/C9NA00115H

Pradhan M, Sinha AK, Pal T (2014a) Redox transformation reaction for hierarchical hollow au–MnOOH flowers for high SERS activity. RSC Adv 4:30315–30324. https://doi.org/10.1039/C4RA03544E

Pradhan M, Sinha AK, Pal T (2014b) Mn oxide-silver composite nanowires for improved thermal stability, SERS and electrical conductivity. Chem Eur J 20:9111–9119. https://doi.org/10.1002/chem.201304518

Prakash J, Kumar P, Harris RA, Swart C, Neethling JH, van Vuuren AJ, Swart HC (2016) Synthesis, characterization and multifunctional properties of plasmonic Ag–$TiO_2$ nanocomposites. Nanotechnology 27:355707. https://doi.org/10.1088/0957-4484/27/35/355707

Qi D, Yan X, Wang L, Zhang J (2015) Plasmon-free SERS self-monitoring of catalysis reaction on au nanoclusters/$TiO_2$ photonic microarray. Chem Commun 51:8813–8816. https://doi.org/10.1039/C5CC02468D

Rahimi N, Pax RA, Gray EM (2016) Review of functional titanium oxides. I: $TiO_2$ and its modifications. Prog Solid State Chem 44:86–105. https://doi.org/10.1016/j.progsolidstchem.2016.07.002

Ray C, Pal T (2017) Recent advances of metal–metal oxide nanocomposites and their tailored nanostructures in numerous catalytic applications. J Mater Chem A 5:9465–9487. https://doi.org/10.1039/C7TA02116J

Reguera J, Langer J, de Aberasturi DJ, Liz-Marzán LM (2017) Anisotropic metal nanoparticles for surface enhanced Raman scattering. Chem Soc Rev 46:3866–3885. https://doi.org/10.1039/C7CS00158D

Remucal CK, Ginder-Vogel M (2014) A critical review of the reactivity of manganese oxides with organic contaminants. Environ Sci Process Impact 16:1247–1266. https://doi.org/10.1039/C3EM00703K

Roguska A, Kudelski A, Pisarek M, Lewandowska M, Dolata M, Janik-Czachor M (2009) Raman investigations of $TiO_2$ nanotube substrates covered with thin Ag or Cu deposits. J Raman Spectrosc 40:1652–1656. https://doi.org/10.1002/jrs.2314

Roguska A, Kudelski A, Pisarek M, Opara M, Janik-Czachor M (2011) Surface-enhanced Raman scattering (SERS) activity of Ag, au and cu nanoclusters on $TiO_2$-nanotubes/Ti substrate. Appl Surf Sci 257:8182–8189. https://doi.org/10.1016/j.apsusc.2010.12.048

Rumyantseva A, Kostcheev S, Adam PM, Gaponenko SV, Vaschenko SV, Kulakovich OS, Ramanenka AA, Guzatov DV, Korbutyak D, Dzhagan V, Stroyuk A (2013) Nonresonant surface-enhanced Raman scattering of ZnO quantum dots with au and Ag nanoparticles. ACS Nano 7:3420–3426. https://doi.org/10.1021/nn400307a

Saini A, Maurer T, Lorenzo II, Santos AR, Beal J, Goffard J, Gérard D, Vial A, Plain J (2015) Synthesis and SERS application of $SiO_2$@Au nanoparticles. Plasmonics 10:791–796. https://doi.org/10.1007/s11468-014-9866-1

Satheeshkumar E, Yang J (2014) Photochemical decoration of silver nanoparticles on ZnO nanowires as a three-dimensional substrate for surface-enhanced Raman scattering measurement. J Raman Spectrosc 45:407–413. https://doi.org/10.1002/jrs.4477

Schlücker S (2014) Surface-enhanced Raman spectroscopy: concepts and chemical applications. Angew Chem Int Ed 53:4756–4795. https://doi.org/10.1002/anie.201205748

Schneider J, Matsuoka M, Takeuchi M, Zhang J, Horiuchi Y, Anpo M, Bahnemann DW (2014) Understanding $TiO_2$ photocatalysis: mechanisms and materials. Chem Rev 114:9919–9986. https://doi.org/10.1021/cr5001892

Shaik UP, Hamad S, Ahamad Mohiddon M, Soma VR, Ghanashyam Krishna M (2016) Morphologically manipulated Ag/ZnO nanostructures as surface enhanced Raman scattering probes for explosives detection. J Appl Phys 119:093103. https://doi.org/10.1063/1.4943034

Shan Y, Yang Y, Cao Y, Yin H, Long NV, Huang Z (2015) Hydrogenated black $TiO_2$ nanowires decorated with Ag nanoparticles as sensitive and reusable surface-enhanced Raman scattering substrates. RSC Adv 5:34737–34743. https://doi.org/10.1039/C5RA04352B

Shanthil M, Fathima H, George Thomas K (2017) Cost-effective plasmonic platforms: glass capillaries decorated with Ag@ $SiO_2$ nanoparticles on inner walls as SERS substrates. ACS Appl Mater Interfaces 9:19470–19477. https://doi.org/10.1021/acsami.6b12478

Sharma B, Frontiera RR, Henry AI, Ringe E, Van Duyne RP (2012) SERS: materials, applications, and the future. Mater Today 15:16–25. https://doi.org/10.1016/S1369-7021(12)70017-2

Shen HX, Zou WJ, Yang ZL, Yuan YX, Xu MM, Yao JL, Gu RA (2015) Surface-enhanced Raman spectroscopy on single $Fe_2O_3$@Au spindle nanoparticle: polarization dependence and FDTD simulation. J Opt 17:114014. https://doi.org/10.1088/2040-8978/17/11/114014

Sinha G, Depero LE, Alessandri I (2011) Recyclable SERS substrates based on au-coated ZnO nanorods. ACS Appl Mater Interfaces 3:2557–2563. https://doi.org/10.1021/am200396n

Sirelkhatim A, Mahmud S, Seeni A, Kaus NHM, Ann LC, Bakhori SKM, Hasan H, Mohamad D (2015) Review on zinc oxide nanoparticles: antibacterial activity and toxicity mechanism. Nano-Micro Lett 7:219–242. https://doi.org/10.1007/s40820-015-0040-x

Sivashanmugan K, Liao JD, Liu BH, Yao CK, Luo SC (2015) Ag nanoclusters on ZnO nanodome array as a hybrid SERS-active substrate for trace detection of malachite green. Sensors Actuators B Chem 207:430–436. https://doi.org/10.1016/j.snb.2014.10.088

Song W, Wang Y, Hu H, Zhao B (2007) Fabrication of surface-enhanced Raman scattering-active ZnO/Ag composite microspheres. J Raman Spectrosc 38:1320–1325. https://doi.org/10.1002/jrs.1769

Song C, Chen J, Abell JL, Cui Y, Zhao Y (2011) Ag–$SiO_2$ core–shell nanorod arrays: morphological, optical, SERS, and wetting properties. Langmuir 28:1488–1495. https://doi.org/10.1021/la203772u

Sun S, Yang Z (2014) Recent advances in tuning crystal facets of polyhedral cuprous oxide architectures. RSC Adv 4:3804–3822. https://doi.org/10.1039/C3RA45445B

Sun L, Zhao D, Ding M, Zhao H, Zhang Z, Li B, Shen D (2012) A white-emitting ZnO–Au nanocomposite and its SERS applications. Appl Surf Sci 258:7813–7819. https://doi.org/10.1016/j.apsusc.2012.04.039

Sun L, He J, An S, Zhang J, Ren D (2013) Facile one-step synthesis of Ag@$Fe_3O_4$ core-shell nanospheres for reproducible SERS substrates. J Mol Struct 1046:74–81. https://doi.org/10.1016/j.molstruc.2013.04.048

Tang X, Dong R, Yang L, Liu J (2015) Fabrication of au nanorod-coated $Fe_3O_4$ microspheres as SERS substrate for pesticide analysis by near-infrared excitation. J Raman Spectrosc 46:470–475. https://doi.org/10.1002/jrs.4658

Tian Y, Chen L, Zhang J, Ma Z, Song C (2012) Bifunctional au-nanorod@ $Fe_3O_4$ nanocomposites: synthesis, characterization, and their use as bioprobes. J Nanopart Res 14:998. https://doi.org/10.1007/s11051-012-0998-y

Toccafondi C, La Rocca R, Scarpellini A, Salerno M, Das G, Dante S (2015) Thin nanoporous alumina-based SERS platform for single cell sensing. Appl Surf Sci 351:738–745. https://doi.org/10.1016/j.apsusc.2015.05.169

Tzounis L, Contreras-Caceres R, Schellkopf L, Jehnichen D, Fischer D, Cai C, Uhlmann P, Stamm M (2014) Controlled growth of Ag nanoparticles decorated onto the surface of $SiO_2$ spheres: a nanohybrid system with combined SERS and catalytic properties. RSC Adv 4:17846–17855. https://doi.org/10.1039/C4RA00121D

Ueba H (1983) Theory of Raman scattering from molecules adsorbed at semiconductor surfaces. Surf Sci 131:328–346. https://doi.org/10.1016/0039-6028(83)90282-0

Van Duyne RP, Hulteen JC, Treichel DA (1993) Atomic force microscopy and surface-enhanced Raman spectroscopy. I. Ag island films and Ag film over polymer nanosphere surfaces supported on glass. J Chem Phys 99:2101–2115. https://doi.org/10.1063/1.465276

Wang ZL (2004) Zinc oxide nanostructures: growth, properties and applications. J Phys Condens Matter 16:R829–R858. https://doi.org/10.1088/0953-8984/16/25/R01

Wang AX, Kong X (2015) Review of recent progress of plasmonic materials and nano-structures for surface-enhanced Raman scattering. Materials 8:3024–3052. https://doi.org/10.3390/ma8063024

Wang Y, Hu H, Jing S, Wang Y, Sun Z, Zhao B, Zhao C, Lombardi JR (2007) Enhanced Raman scattering as a probe for 4-mercaptopyridine surface-modified copper oxide nanocrystals. Anal Sci 23:787–791. https://doi.org/10.2116/analsci.23.787

Wang Y, Song W, Ruan W, Yang J, Zhao B, Lombardi JR (2009) SERS spectroscopy used to study an adsorbate on a nanoscale thin film of CuO coated with Ag. J Phys Chem C 113:8065–8069. https://doi.org/10.1021/jp900052q

Wang J, Cui F, Chu S, Jin X, Pu J, Wang Z (2014a) In situ growth of Noble-metal nanoparticles on $Cu_2O$ Nanocubes for surface-enhanced Raman scattering detection. ChemPlusChem 79:684–689. https://doi.org/10.1002/cplu.201400028

Wang Y, Zhao X, Chen L, Chen S, Wei M, Gao M, Zhao Y, Wang C, Qu X, Zhang Y, Yang J (2014b) Ordered nanocap array composed of $SiO_2$-isolated Ag islands as SERS platform. Langmuir 30:15285–15291. https://doi.org/10.1021/la5032834

Wang W, Guo Q, Xu M, Yuan Y, Gu R, Yao J (2014c) On-line surface enhanced Raman spectroscopic detection in a recyclable Au@ $SiO_2$ modified glass capillary. J Raman Spectrosc 45:736–744. https://doi.org/10.1002/jrs.4553

Wang W, Meng Z, Zhang Q, Jia X, Xi K (2014d) Synthesis of stable au–$SiO_2$ composite nanospheres with good catalytic activity and SERS effect. J Colloid Interface Sci 418:1–7. https://doi.org/10.1016/j.jcis.2013.11.043

Wang Y, Zhang M, Yan C, Chen L, Liu Y, Li J, Zhang Y, Yang J (2016) Pillar-cap shaped arrays of Ag/$SiO_2$ multilayers after annealing treatment as a SERS-active substrate. Colloids Surf A 506:96–103. https://doi.org/10.1016/j.colsurfa.2016.05.100

Wang X, Shi W, Jin Z, Huang W, Lin J, Ma G, Li S, Guo L (2017a) Remarkable SERS activity observed from amorphous ZnO nanocages. Angew Chem Int Ed 56:9851–9855. https://doi.org/10.1002/anie.201705187

Wang Y, Jin A, Quan B, Liu Z, Li Y, Xia X, Li W, Yang H, Gu C, Li J (2017b) Large-scale Ag-nanoparticles/$Al_2O_3$/Au-nanograting hybrid nanostructure for surface-enhanced Raman scattering. Microelectron Eng 172:1–7. https://doi.org/10.1016/j.mee.2017.01.024

Weng X, Feng Z, Guo Y, Feng JJ, Hudson SP, Zheng J, Ruan Y, Laffir F, Pita I (2016) Recyclable SERS substrates based on $Fe_2O_3$-Ag hybrid hollow microspheres with crumpled surfaces. New J Chem 40:5238–5244. https://doi.org/10.1039/C6NJ00473C

Wheeler DA, Adams SA, López-Luke T, Torres-Castro A, Zhang JZ (2012) Magnetic $Fe_3O_4$-au core-shell nanostructures for surface enhanced Raman scattering. Ann Phys 524:670–679. https://doi.org/10.1002/andp.201200161

Wu HY, Cunningham BT (2011) Plasmonic coupling of $SiO_2$–Ag "post-cap" nanostructures and silver film for surface enhanced Raman scattering. Appl Phys Lett 98:153103. https://doi.org/10.1063/1.3555342

Wu SH, Mou CY, Lin HP (2013) Synthesis of mesoporous silica nanoparticles. Chem Soc Rev 42:3862–3875. https://doi.org/10.1039/C3CS35405A

Wu MC, Lin MP, Chen SW, Lee PH, Li JH, Su WF (2014) Surface-enhanced Raman scattering substrate based on a Ag coated monolayer array of $SiO_2$ spheres for organic dye detection. RSC Adv 4:10043–10050. https://doi.org/10.1039/C3RA45255G

Wu W, Wu Z, Yu T, Jiang C, Kim WS (2015) Recent progress on magnetic iron oxide nanoparticles: synthesis, surface functional strategies and biomedical applications. Sci Technol Adv Mater 16:023501. https://doi.org/10.1088/1468-6996/16/2/023501

Wu H, Wang H, Li G (2017) Metal oxide semiconductor SERS-active substrates by defect engineering. Analyst 142:326–335. https://doi.org/10.1039/C6AN01959E

Xie Y, Jin Y, Zhou Y, Wang Y (2014) SERS activity of self-cleaning silver/titania nanoarray. Appl Surf Sci 313:549–557. https://doi.org/10.1016/j.apsusc.2014.06.020

Xu SC, Zhang YX, Luo YY, Wang S, Ding HL, Xu JM, Li GH (2013) Ag-decorated $TiO_2$nanograss for 3D SERS-active substrate with visible light self-cleaning and reactivation. Analyst 138:4519–4525. https://doi.org/10.1039/C3AN00750B

Xu L, Li S, Li F, Zhang H, Wang D, Chen M, Chen F (2017a) Ultraviolet light-induced photochemical reaction for controlled fabrication of Ag nano-islands on ZnO nanosheets: an advanced inexpensive substrate for ultrasensitive surface-enhanced Raman scattering analysis. Opt Mater Express 7:3137–3146. https://doi.org/10.1364/OME.7.003137

Xu L, Xu Q, Guo X, Ying Y, Wu Y, Wen Y, Yang H (2017b) Facile synthesis of $Au/Al_2O_3$ nanocomposites for improving the detection sensitivity of adenosine triphosphate. RSC Adv 7:25746–25752. https://doi.org/10.1039/C7RA03683C

Xue X, Xu D, Ruan W, Chen L, Chang L, Zhao B (2015) Enhanced Raman scattering when scatterer molecules located in $TiO_2$/Ag nanojunctions. RSC Adv 5:64235–64239. https://doi.org/10.1039/C5RA11667H

Yamada H, Yamamoto Y (1983) Surface-enhanced Raman scattering (SERS) of chemisorbed species on various kinds of metals and semiconductors. Surf Sci 134:71–90. https://doi.org/10.1016/0039-6028(83)90312-6

Yamada H, Yamamoto Y, Tani N (1982) Surface-enhanced Raman scattering (SERS) of adsorbed molecules on smooth surfaces of metals and a metal oxide. Chem Phys Lett 86:397–400. https://doi.org/10.1016/0009-2614(82)83531-8

Yang L, Jiang X, Ruan W, Yang J, Zhao B, Xu W, Lombardi JR (2009) Charge-transfer-induced surface-enhanced Raman scattering on Ag-$TiO_2$ Nano composites. J Phys Chem C 113:16226–16231. https://doi.org/10.1021/jp903600r

Yang L, Bao Z, Wu Y, Liu J (2012) Clean and reproducible SERS substrates for high sensitive detection by solid phase synthesis and fabrication of Ag-coated $Fe_3O_4$ microspheres. J Raman Spectrosc 43:848–856. https://doi.org/10.1002/jrs.3106

Yang L, Lv J, Sui Y, Fu W, Zhou X, Ma J, Li Q, Sun M, Mu Y, Chen Y, Wang J (2014a) Ag-$Cu_2O$ composite microstructures with tunable Ag contents: synthesis and surface-enhanced (resonance) Raman scattering (SE (R) RS) properties. RSC Adv 4:17249–17254. https://doi.org/10.1039/C4RA00675E

Yang L, Lv J, Sui Y, Fu W, Zhou X, Ma J, Su S, Zhang W, Lv P, Wu D, Mu Y (2014b) Fabrication of $Cu_2O$/Ag composite nanoframes as surface-enhanced Raman scattering substrates in a successive one-pot procedure. CrystEngComm 16:2298–2304. https://doi.org/10.1039/C3CE42052C

Yang L, Yang Y, Ma Y, Li S, Wei Y, Huang Z, Long NV (2017) Fabrication of semiconductor ZnO nanostructures for versatile SERS application. Nano 7:398. https://doi.org/10.3390/nano7110398

Yemmireddy VK, Hung YC (2017) Using Photocatalyst metal oxides as antimicrobial surface coatings to ensure food safety-opportunities and challenges. Comp Rev Food Sci Food Saf 16:617–631. https://doi.org/10.1111/1541-4337.12267

Yin W, Wu L, Ding F, Li Q, Wang P, Li J, Lu Z, Han H (2018) Surface-imprinted $SiO_2$@Ag nanoparticles for the selective detection of BPA using surface enhanced Raman scattering. Sensors Actuators B 258:566–573. https://doi.org/10.1016/j.snb.2017.11.141

Yu J, Shen M, Liu S, Li F, Sun D, Wang T (2017) A simple technique for direct growth of au into a nanoporous alumina layer on conductive glass as a reusable SERS substrate. Appl Surf Sci 406:285–293. https://doi.org/10.1016/j.apsusc.2017.02.103

Zhang CX, Su L, Chan YF, Wu ZL, Zhao YM, Xu HJ, Sun XM (2013) Ag@ $SiO_2$core-shell nanoparticles on silicon nanowire arrays as ultrasensitive and ultrastable substrates for surface-enhanced Raman scattering. Nanotechnology 24:335501. https://doi.org/10.1088/0957-4484/24/33/335501

Zhang Q, Zhang K, Xu D, Yang G, Huang H, Nie F, Liu C, Yang S (2014) CuO nanostructures: synthesis, characterization, growth mechanisms, fundamental properties, and applications. Prog Mater Sci 60:208–337. https://doi.org/10.1016/j.pmatsci.2013.09.003

Zhang Z, Yu J, Yang J, Lv X, Wang T (2015) Preparation of sensitive and recyclable porous Ag/$TiO_2$ composite films for SERS detection. Appl Surf Sci 359:853–859. https://doi.org/10.1016/j.apsusc.2015.10.197

Zhang S, Jiang R, Guo Y, Yang B, Chen XL, Wang J, Zhao Y (2016) Plasmon modes induced by anisotropic gap opening in Au@$Cu_2O$ nanorods. Small 12:4264–4276. https://doi.org/10.1002/smll.201600065

Zhao K, Lin J, Guo L (2015a) ZnO/Ag porous nanosheets used as substrate for surface-enhanced Raman scattering to detect organic pollutant. RSC Adv 5:53524–53528. https://doi.org/10.1039/C5RA06735A

Zhao K, Wu C, Deng Z, Guo Y, Peng B (2015b) Preparation of silver decorated silica nano-composite rods for catalytic and surface-enhanced Raman scattering applications. RSC Adv 5:52726–52736. https://doi.org/10.1039/C5RA08076B

Zhao M, Guo H, Liu W, Tang J, Wang L, Zhang B, Xue C, Liu J, Zhang W (2016) Silica cladding of Ag nanoparticles for high stability and surface-enhanced Raman spectroscopy performance. Nanoscale Res Lett 11:403. https://doi.org/10.1186/s11671-016-1604-5

Zheng Z, Cong S, Gong W, Xuan J, Li G, Lu W, Geng F, Zhao Z (2017) Semiconductor SERS enhancement enabled by oxygen incorporation. Nat Commun 8:1993. https://doi.org/10.1038/s41467-017-02166-z

Zhou Y, Chen J, Zhang L, Yang L (2012) Multifunctional $TiO_2$-coated Ag nanowire arrays as recyclable SERS substrates for the detection of organic pollutants. Eur J Inorg Chem 2012:3176–3182. https://doi.org/10.1002/ejic.201200009

Zhou M, Liu X, Yu B, Cai J, Liao C, Ni Z, Zhang Z, Ren Z, Bai J, Fan H (2015) $MnO_2$/au hybrid nanowall film for high-performance surface-enhanced Raman scattering substrate. Appl Surf Sci 333:78–85. https://doi.org/10.1016/j.apsusc.2015.02.014

Zhou W, Yin BC, Ye BC (2017) Highly sensitive surface-enhanced Raman scattering detection of hexavalent chromium based on hollow sea urchin-like $TiO_2$@ Ag nanoparticle substrate. Biosens Bioelectron 87:187–194. https://doi.org/10.1016/j.bios.2016.08.036

Zhu S, Fan C, Wang J, He J, Liang E, Chao M (2015) Realization of high sensitive SERS substrates with one-pot fabrication of Ag-$Fe_3O_4$ nanocomposites. J Colloid Interface Sci 438:116–121. https://doi.org/10.1016/j.jcis.2014.09.015

Zoolfakar AS, Rani RA, Morfa AJ, O'Mullane AP, Kalantar-Zadeh K (2014) Nanostructured copper oxide semiconductors: a perspective on materials, synthesis methods and applications. J Mater Chem C 2:5247–5270. https://doi.org/10.1039/C4TC00345D

Zou X, Silva R, Huang X, Al-Sharab JF, Asefa T (2013) A self-cleaning porous $TiO_2$–Ag core-shell nanocomposite material for surface-enhanced Raman scattering. Chem Commun 49:382–384. https://doi.org/10.1039/C2CC35917K

# Molecularly Imprinted Nanosensors for Microbial Contaminants

**Neslihan Idil, Monireh Bakhshpour, Işık Perçin, and Adil Denizli**

## Contents



## Abbreviations

| | |
|---|---|
| ADH | Alcohol dehydrogenase |
| AgNPs | Silver nanoparticles |
| AuNPs | Gold nanoparticles |
| BSA | Bovine serum albumin |
| CFU | Colony forming unit |
| CNTs | Carbon nanotubes |
| CPE | Carbon paste electrode |
| CV | Cyclic voltammetry |
| EHEC | Enterohemorrhagic *E. coli* |
| EIS | Electrochemical impedance spectroscopy |
| EPEC | Enteropathogenic *E. coli* |

N. Idil · I. Perçin
Department of Biology, Hacettepe University, Ankara, Turkey
e-mail: nsurucu@hacettepe.edu.tr; ipercin@hacettepe.edu.tr

M. Bakhshpour · A. Denizli (✉)
Department of Chemistry, Hacettepe University, Ankara, Turkey
e-mail: denizli@hacettepe.edu.tr



Inamuddin, A. M. Asiri (eds.), *Nanosensor Technologies for Environmental Monitoring*, Nanotechnology in the Life Sciences,
https://doi.org/10.1007/978-3-030-45116-5_12

| | |
|---|---|
| GCE | Glassy carbon electrode |
| GE | Gold electrode |
| GNPs | Gold nanoparticles |
| GO | Graphene oxide |
| HAU | Hemagglutinin unit |
| LOD | Limit of detection |
| LOQ | Limit of quantification |
| MAH | *N*-methacryloyl-L-histidine methyl ester |
| MGCE | Magnetic carbon paste electrode |
| MIP | Molecular imprinting |
| MIPs | Molecularly imprinted polymers |
| MWCNTs | Multi-walled carbon nanotubes |
| N-GQDs | Nitrogen-doped graphene quantum dots |
| NPs | Nanoparticles |
| PDA-SIP | Polydopamine surface imprinted polymer |
| PEG | Polyethylene glycol |
| PFU | Plaque forming units |
| PGE | Pencil graphite electrode |
| PPy | Polypyrrole |
| QCM | Quartz crystal microbalance |
| QDs | Quantum dots |
| RGSs-CS | Chitosan (CS) doped with reduced graphene sheets (RGSs) |
| SA | Streptavidin |
| SEM | Scanning electron microscopy |
| SIP | Spore-imprinted polymer |
| SPCE | Screen-printed carbon electrode |
| SPE | Carbon screen-printed electrode |
| SPR | Surface plasmon resonance |
| SWCNTs | Single-walled carbon nanotubes |
| ZIKV | Zika virus |

## 1 Introduction

The importance of the identification of microorganisms is rapidly increasing. Therefore, it is necessary to detect these cells in complex environments. In particular, the development of selective sensors for the detection of pathogenic microorganisms is of particular interest due to the widespread use of the developed systems and their significant environmental impact (Kaittanis et al. 2010; Ait Lahcen et al. 2018; Saylan et al. 2019).

The incidence of microbial infections has become an important health problem. Therefore, real-time, quantitative, and qualitative detection of pathogenic microorganisms in water and food products is very important. The rapid detection of these microorganisms has a major role in preventing the emergence of infections and taking control measures. Culturing, biochemical tests, serological tests, antibody-based methods, DNA/RNA-based analyzes, and immunological methods are used for the isolation and identification of pathogenic microorganisms from clinical specimens. Many of these conventional methods for identifying microorganisms are time-consuming, labor-intensive, and costly. In recent years, many attempts have been made to develop alternative methods for microorganism detection. These methods include the use of various components such as silica nanoparticles, micro-liquids, liquid crystals, and carbon nanotubes. These components are integrated into sensor systems for microorganism detection. Sensor-based technologies are more advanced techniques, and they promise developments because they deliver fast, reliable, and precise results (Nguyen et al. 2014; Wang and Salazar 2016; Zhao et al. 2016). Biosensors have been promising candidates providing inexpensive monitoring and screening of related molecule with high selectivity and specificity in many environmental and clinical applications (Idil et al. 2017).

Optical, piezoelectric, and electrochemical sensors for detecting microbial cells have been developed and reported in the literature. Optical sensors are the most widely used sensors for the detection of pathogenic microorganisms because of their advantages such as fast response time, portability, reliability, high selectivity, and sensitivity. Optical sensors are mostly based on fluorescence and surface plasmon resonance (SPR). Optical measurements can be achieved by monitoring the change of the optical signal that takes place between a modified nanomaterial and a target molecule. On the other hand, electrochemical sensors have the ability of measuring the change in electrical current, conductance, impedance, or potential occurring between the interface of the electrode and target sample. In addition, electrochemical sensors have some advantages over optical sensors such as low cost, small size, and simple operation (Pérez-López and Merkoçi 2011; Munawar et al. 2019).

Recent developments in nanomaterial-based sensor systems offer many advantages for detecting microbial contaminants and their related products. Unique properties of nanomaterials make them useful for designing effective sensor systems showing high sensitivity, multiplexing capability, and reduced detection time. Nano-sized materials (1–100 nm) have a higher surface area, surface reactivity, electrical conductivity, and magnetic properties when compared with bulk polymers. Furthermore, the properties of nanomaterials can be modified by changing the composition, size, shape, and surface functionalization. Thus, electronic, conductive, and spectroscopic properties of nanomaterials can be designed by changing the structural features of nanomaterials such as size, composition, and self-assembling. In this regard, gold, magnetic, silver, silica nanoparticles, gold nanorods, and quantum dots have been used for the detection of microbial cells and their contaminants in the literature (Inbaraj and Chen 2016; Wang and Duncan 2017).

Antibodies are used as recognition elements by means of the specific interaction between the target cell and the binding region of the antibody for cell recognition and detection in many nanosensor systems. Although native antibodies exhibit high selectivity and sensitivity, they have disadvantages such as short-term stability and tendency to denature under different physiological conditions. It is among different approaches to take advantage of the superior properties of biomimetic materials (mimicking the natural order) in the design of artificial systems. In this context, molecular imprinting is a promising and effective method for developing sensitive and portable sensor systems with high selectivity. Molecularly imprinted polymers (MIPs) are of particular interest due to their high selectivity, low cost and easy synthesis, high stability in different chemical and physical conditions, and reusability. These advantages make the materials prepared with molecular imprinting technology superior to natural recognition agents. MIPs have been established in due course as alternative choices to natural antibodies for nanosensor applications revealing similar affinity, specificity, and even sensitivity in analytical measurements (Idil and Mattiasson 2017; Perçin et al. 2017).

In this chapter, various nanomaterial-based sensors using molecular imprinting technology reported in the literature for the detection of microbial contaminants have been summarized. The unique characteristics of MIPs as biorecognition elements within designing of promising sensing tools and their biomimetic nanosensor applications in detection of microbial contaminants have been highlighted. In addition, principles, limitations, and advantages of recent strategies applied for microbial detection have been summarized.

## 2 Molecular Imprinting

Molecular imprinting (MIP) is an effective and feasible method for creating selective recognition sites for the target molecule/cell on the polymer matrix (Bereli et al. 2008). This approach basically stands on the molecular detection, identification, capturing, and/or recognition events taking place around the target molecule. Molecular imprinting technology is used to synthesize highly crosslinked polymers capable of selective recognition. Functional monomers and crosslinkers are used for co-polymerization with the target molecule/cell. Thus, three-dimensional recognition sites are formed which are complementary in shape and size with the target molecule/cell. In this process, the target molecule, functional monomers, crosslinker(s), initiator, and progenic solvent are used. Monomer polymerization occurs in the presence of the target molecule incorporated into the polymer matrix. The process begins with the dissolution of the target molecule, crosslinker, functional monomer, and initiator in a progenic solvent. Functional monomers having the ability to interact with the target molecule are preferred. A successful molecular imprinting and recognition require the formation of a stable target molecule-monomer complex. After the monomers are placed around the target molecule, the position is fixed by co-polymerization of the crosslinked monomers. The resulting polymer is a porous matrix with microcavities containing three-dimensional structure complementary to the target molecule. Thus, by removing the target molecules from the polymer by washing them with solvent, binding sites are formed which recognize the target molecule shape. As a result, polymers with high selectivity and affinity for the target molecule selectively recognize

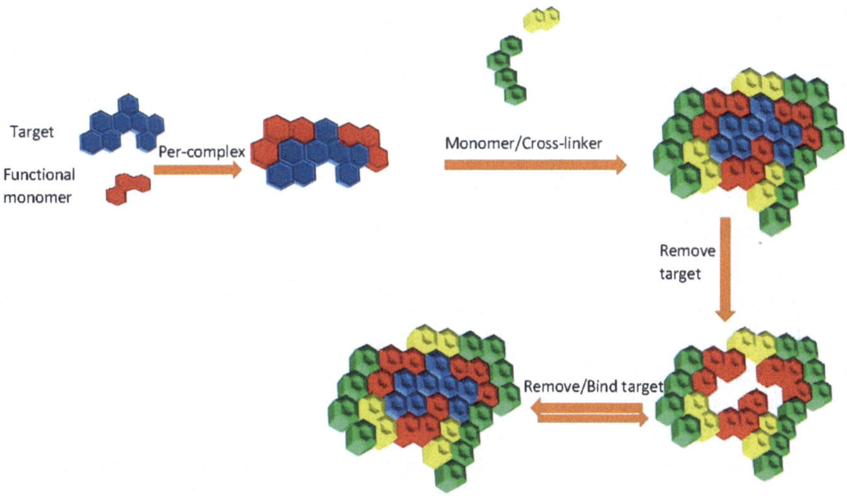


**Fig. 1** Schematic presentation of the preparation of MIPs

and bind the molecule (Idil et al. 2017; Saylan et al. 2017a, b; Safran et al. 2019). In Fig. 1, schematic presentation of the preparation of MIPs was illustrated.

The most important point in molecular imprinting is the maintenance of stability of interactions between monomers and target molecule during polymerization. Therefore, the solvent used in the molecular imprinting method has a significant effect on stability and water-like organic solvents are used to improve stability in many applications. Another aspect that affects the success of molecular imprinting is that rigidity (mechanically stable) of imprinted polymers. The quantity of cross-linker added to polymerization medium must be in an appropriate ratio to obtain a stable polymer structure. It is also important to use suitable solvents to control the formation of pores in the structure during polymerization. In addition, types and concentrations of monomers and initiators, and experimental conditions have significant effects on the formation of MIPs with superior quality (Wulff 1995). It is noteworthy to mention that chemical and physical structures of the target molecules have to be taken into consideration while preferring functional and assistant monomers since these properties are as notable as other factors. Therefore, effects of different ratios of monomer/crosslinker have to be well-examined to optimize polymerization reaction and obtain stable binding competency between target molecule and imprinted cavities (Saylan et al. 2017a). In this concept, many molecules are successfully imprinted using different types of monomers, crosslinkers. Drugs, hormones, proteins, amino acids, carbohydrates, dyes, pesticides, nucleotides, coenzymes, food additives, and steroids such as cholesterol are some examples of molecularly imprinted molecules reported in the literature (Osman et al. 2013; Shaikh et al. 2015; Bakhshpour et al. 2017; Perçin et al. 2019). MIPs could be efficiently prepared by different types of molecular imprinting strategies (1) bulk imprinting, (2) surface imprinting, (3) microcontact imprinting, and (4) epitope imprinting. They have been exploited as artificial receptors and extensively applied in the fields of purification, separation, and isolation processes, catalysis, drug delivery (Bakhshpour et al. 2018), and sensor technology (Ertürk and Mattiasson 2017).

## 2.1 Molecular Imprinting Techniques

MIPs referred as biomimetic polymeric materials have been prodigiously organized in order to create selective recognition regions for analytes of interest. Herewith, composing intended sensitivity serving for specific purposes in anticipated applications could be accomplished by stabilizing polymers and proper organization of polymeric components during imprinting. In this respect, appropriate imprinting strategies have been designated for obtaining recognition sites showing high affinity towards template molecule keeping its own identical geometry (Frasco et al. 2017). In Fig. 2, general progress of molecular imprinting was shown.

In bulk imprinting, template molecule is added to pre-polymerization mixture. Polymerization was allowed to perform and then, entrapped template molecule was removed from the polymeric matrix. By the way, readily accessible recognition regions having the shape memory cavity of template molecule were produced within the bulk material and able to selectively capture the target molecule. However, this imprinting strategy is more available for relatively small molecules. Size and dimension of large macromolecules, restricted mass transfer, slow binding kinetics, limited structural stability, and transportation of target molecule throughout thick bulk material can be respected as the disadvantages in the imprinting of large molecules including microbial cells. Low accessibility during removal step eventually resulted in low selectivity. All these drawbacks have been tackled with alternative

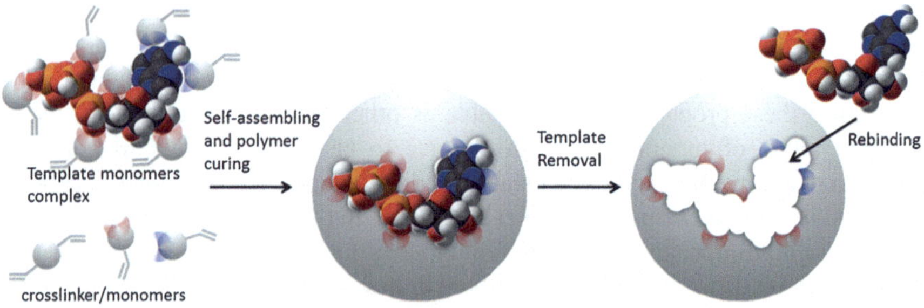


**Fig. 2** General progress of molecular imprinting: First, template molecule, crosslinker, and functional monomers are dissolved in the respective solvent leading to self-assembly of the compounds. Then, polymerization takes place. When the template is removed, cavities remain which resemble the template in size, shape, and surface chemistry. Since it fits into the cavity, the template will be incorporated preferentially (Reprinted with permission of Wackerlig and Schirhagl 2015)

approaches such as surface and epitope imprinting to extend application fields (Ertürk and Mattiasson 2017).

Surface imprinting has received great interest in recent years. In surface imprinting, larger molecules could easily be imprinted and template removal/rebinding occurs without mass transfer by enabled diffusion. In the process, thin polymeric films were synthesized with stamping the template molecule so that imprinted binding sites were placed near the surface of polymeric matrix. Stamping of the template molecule provides the creation of more robust tools applicable for analytical measurements. Especially, facile accessible binding regions enable to solve the major problem in the transportation of cells to and from the cavities (Ertürk and Mattiasson 2017). In Fig. 3, fabrication of surface imprinting of proteins on sensors and its proposed working mechanism was shown.

Microcontact imprinting is another alternative approach to create well-characterized polymeric matrices for imprinting of large, unsteady, and complex macromolecules. In this strategy, a single efficient monolayer was prepared using stamped template molecule during the imprinting process. The most outstanding point is that print is never completely entrapped through the polymer; in this manner, it will be effortless to remove template and rebind target molecules even particulate cells (Idil and Mattiasson 2017). In Fig. 4, schematic representation of the microcontact imprinting on Quartz Crystal Microbalance (QCM) chip was indicated.

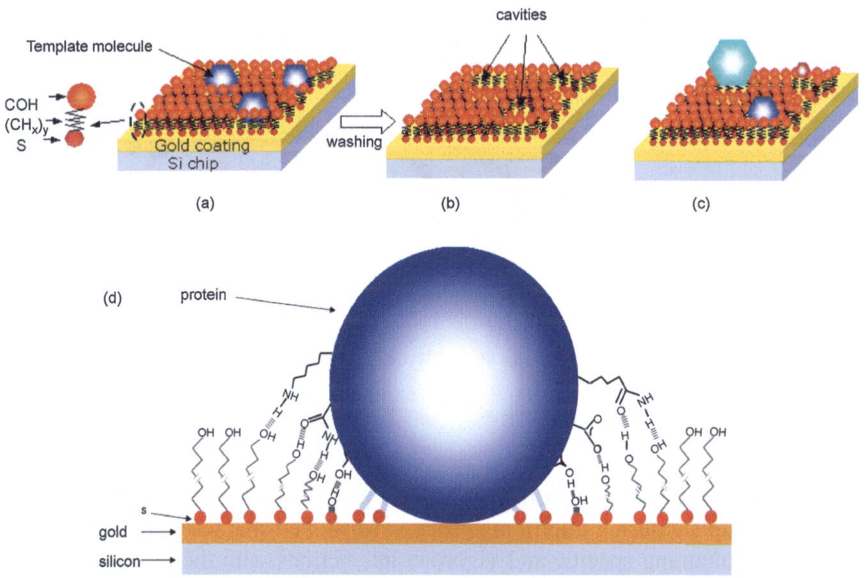


**Fig. 3** Fabrication of protein-imprinted SAMs sensor and its proposed working mechanism. (**a**) Co-adsorption of template protein molecules and thiol SAMs onto the gold surface. (**b**) Cavities created after washing off the templates. (**c**) Selective adsorption of the template protein molecules against other molecules. (**d**) Hypothetical binding mechanism showing the hydrogen bonds between protein and –OH end groups of the thiol (Reprinted with permission of Wang et al. 2008)

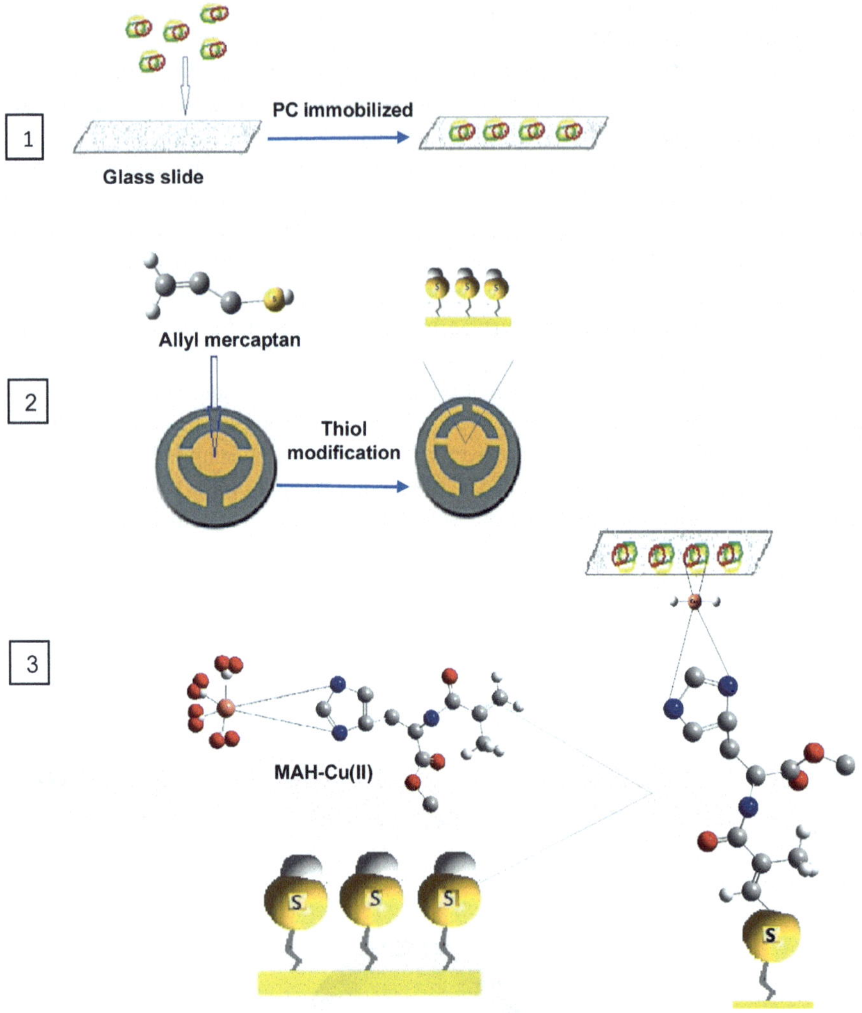


**Fig. 4** Schematic representation of the PC-μCIP/QCM sensor (Reprinted with permission of Bakhshpour et al. 2017)

In epitope imprinting, a specific part, component, and domain of the molecule is used as a template during imprinting process. Most well-marked advantage of this approach is attaining specific and vigorous interactions with the usage of certain fragment of molecules instead of entire molecule. Exactly like this, nonspecific binding can be reduced while the affinity can be advanced (Ertürk and Mattiasson 2017). In Fig. 5, schematic diagram representation of epitope imprinting were demonstrated.

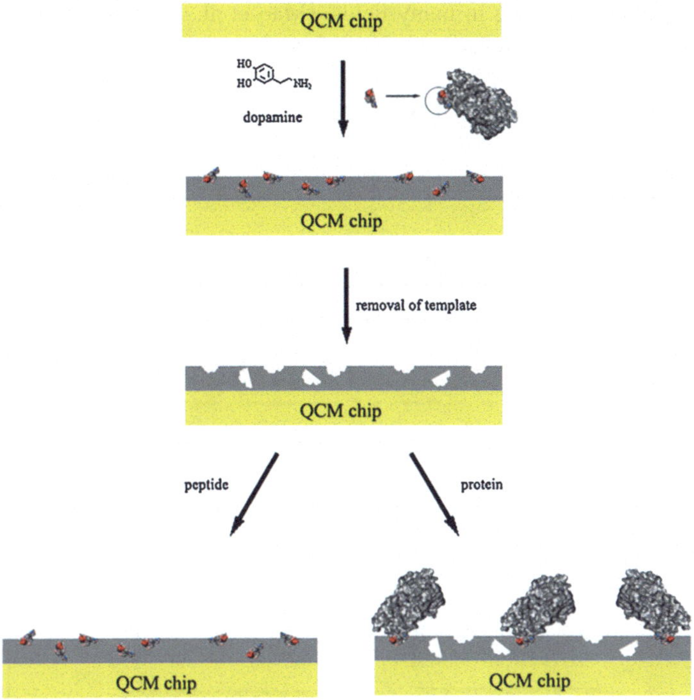


**Fig. 5** Schematic diagram representation of epitope imprinting (Reprinted with permission of Lu et al. 2012)

# 3 Sensors

## *3.1 Optical Nanosensors*

Optical nanosensors can be used for the detection of target molecules with the measurement of a change in the optical characteristics of the transducer surface (Wolfbeis 2016; Bharadwaj et al. 2016; Saylan et al. 2017b). In the literature and the market, there are multiple optical nanosensors, including the resonant mirror optical sensor, optrode-based fiber optical sensors, time-resolved fluorescence, evanescent wave fiber optical sensors, interferometric sensors, and SPR sensors (Eissa et al. 2015; Hu et al. 2016; do et al. 2017). Nanosensors are used as an original application in many fields including food safety, diagnosis, laboratory medicine, and especially clinical examinations (Giamblanco et al. 2015; Zhang et al. 2015a, b; Kabessa et al. 2016). SPR nanosensors optical sensing tools which provide real-time monitoring of interactions of a various template and chemical analytes. These nanosensors depend on measuring any shift in the resonance angle related to the mass of the bound template at the SPR surfaces. Nowadays, the advantages of SPR nanosensors are attributed to using in many fields. (Mao et al. 2009; Dudak and Boyaci 2014; Anik et al. 2016; Cunha et al. 2018; Cheeveewattanagul et al. 2017; Erdem et al. 2019). In Fig. 6, schematic illustration of antibody-based sensing strategies in SPR biosensors used for *E. coli* detection were shown.

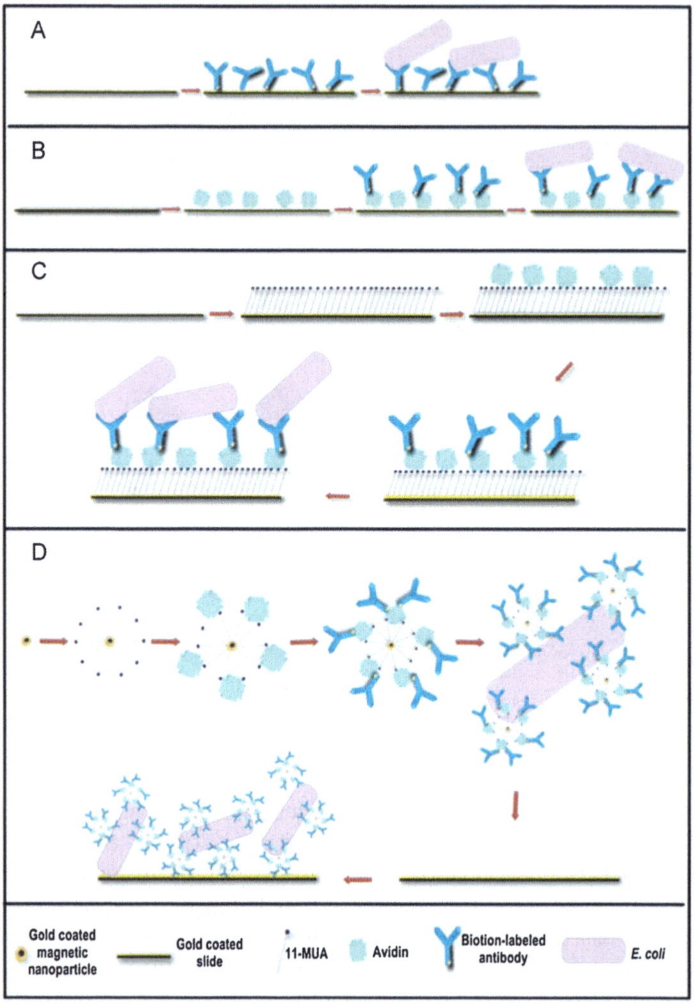


**Fig. 6** Schematic illustration of sensing strategies in *SPR*biosensor (**a**) nonspecific adsorption of antibody on to the SPR surface and *E. coli* detection, (**b**) oriented immobilization of antibody on to the SPR surface using (via) avidin–biotin interaction and *E. coli* detection, (**c**) covalent immobilization of antibody onto the SPR surface and *E. coli* detection, (**d**) magnetic gold nanoparticle-based assay protocol comprised of the steps, immunomagnetic separation of *E. coli,* and SPR measurements for *E. coli* enumeration (Torun et al. 2012)

## 3.2 Applications of SPR Nanosensors

Among the other research fields, clinical research is an interesting research area in recent years. Microorganisms, viruses, mycotoxins, and microbial metabolites have been effectively detected using SPR sensors. In literature, there are a few studies aimed to detect viruses. Among these, Uzun et al. (2009) prepared hepatitis antibody-imprinted surface plasmon resonance sensor for the diagnosis of hepatitis in human serum. This study has an importance since Hepatitis B virus is one of the major infection problems in healthcare settings. The surface antigen arises in the serum of patients before biochemical evidence occurs during the hepatitis B virus infection. Hepatitis B surface antigen persists during the disease until the surface antibody on the antigen becomes detectable. They showed heterogeneity surface and obeyed the Langmuir adsorption isotherm model. In this study, they reported 208.2mIU/mL limit of detection value.

The rapid and real-time detection of pathogenic microorganisms causing infectious diseases is important for health and safety areas. In spite of antibiotics, new and multi-drug-resistant pathogens are incessantly appearing. Nowadays, SPR sensor is easily used to portable, rapid, and cost-effective diagnosis. SPR system has been used for the detection of whole cells such as *Escherichia coli* (*E. coli*), *Salmonella* sp., and *Listeria* sp. These systems were successfully used for pathogenic bacteria detection. In a previous study performed in our research group, bacteria-imprinted nanofilm was visualized by Yılmaz et al. for the detection of *E. coli* (Yilmaz et al. 2015). They used *N*-methacryloyl-L-histidine methyl ester (MAH) as a functional monomer for sensitive and selective detection. As seen in Fig. 7, Scanning Electron Microscopy (SEM) photographs were shown for the characterization of SPR nanosensors. In this study, limit of detection (LOD) was found as $1.54 \times 10^6$ CFU/mL for *E. coli*. In our previous study, label-free, microcontact-imprinted SPR nanosensor was fabricated for rapid detection of *Salmonella paratyphi* (Perçin et al. 2017). Characterization of SPR chips were performed with ellipsometry and SEM. Real-time detection was examined within the bacterial concentrations of 2.5–15 $\times$ $10^6$ CFU/mL, and LOD was reported as $1.4 \times 10^6$ CFU/mL.

Erdem et al. (2019) developed label-free molecularly imprinted nanoparticles-based SPR sensors for detection of *Enterococcus faecalis* (*E. faecalis*) from aqueous and seawater samples. They reported detection range with a low limit of detection (down to ~100 bacteria/mL) and a high correlation coefficient value (>0.99). They demonstrated a selectivity and specificity while other bacteria such as *E. coli*, *Bacillus subtilis* (*B. subtilis*), and *Staphylococcus aureus* (*S. aureus*). The preparation of *E. faecalis*-imprinted plasmonic sensor was shown in Fig. 8.

Gur et al. (2019) combined surface imprinted technology with Au nanoparticles for detection of *E. coli*. They used Cu(II) ions to provide specific and selective interaction with *E. coli*. They reported a comparatively low limit of detection value (1 CFU/mL) to the SPR nanosensor system. They developed an ultrasensitive detection method by a combination of the signal enhancing properties of Au nanoparticles

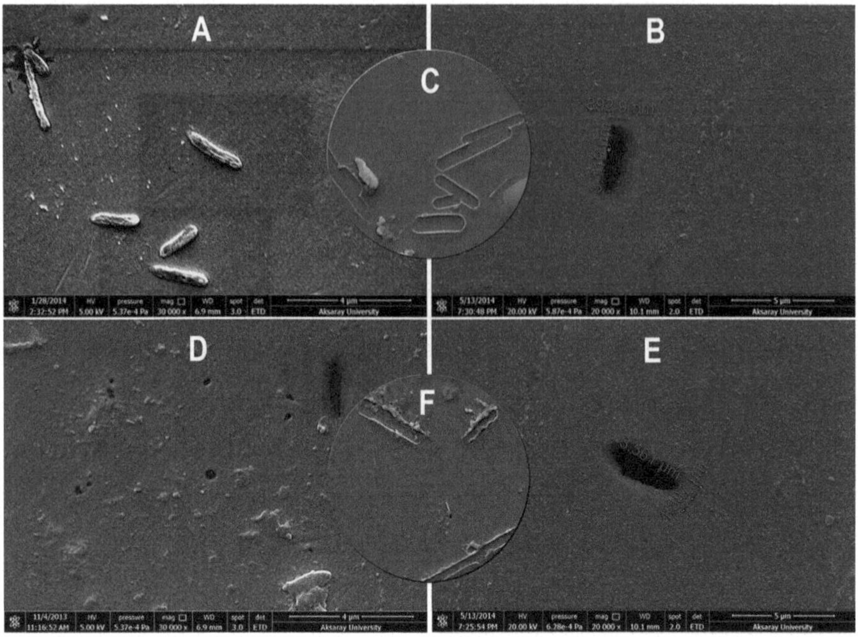


**Fig. 7** SEM photograph of *E-coli*-imprinted SPR sensor (Reprinted with permission of Yilmaz et al. 2015)

and surface imprinting technique. They used *S. aureus*, *Klebsiella pneumoniae* (*K. pneumoniae*), and *Pseudomonas aeruginosa* (*P. aeruginosa*) for verifying the selectivity of the SPR system. They reported the highest response for *E. coli*, as expected. Also, they demonstrated the detection capacity of the SPR system for *E. coli* in an artificial urine sample.

In other methods for detection of microorganisms by SPR, Ashiba et al. (2017) investigated the detection of norovirus (virus-like particles) using an SPR sensor. Norovirus, a human enteric pathogen, causes major disease during health care. Norovirus contamination occurs from contaminated water and food. Then, this infection easily spread through fecal or oral ways. They used a V-shaped trench chip equipped in their sensor experiments. They reported 0.01 ng/mL limit of detection concentration, which corresponds to 100 virus-like particles included in the detection region of the V-trench.

Bai et al. (2012) designed a portable sensor for the detection of avian influenza virus H5N1. They used aptamer-based SPR sensor in poultry swab samples and successfully detected avian influenza virus H5N1. Avian influenza, as known as "bird flu" is A-type influenza virus causing huge economic losses. They biotinylated and immobilized aptamer on the surface of sensor chip that coated with streptavidin. According to the results, the refraction index was increased with binding the avian influenza virus H5N1 onto the sensor surface. Also, they demonstrated a linear relationship between increasing the refraction and virus concentration in the range of

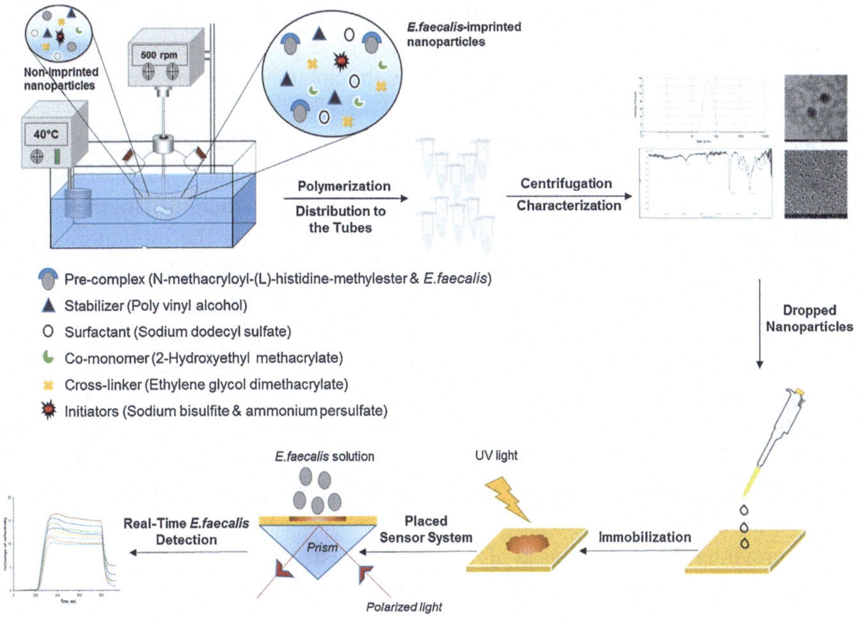


**Fig. 8** The schematic preparation via SPR sensor (Reprinted with permission of Erdem et al. 2019)

0.128–1.28 hemagglutinin unit (HAU). They reported 1.5 h detection time for the same concentration of avian H5N1 influenza virus in poultry swab samples.

Prabowo et al. (2017) introduced a portable SPR sensor system for rapid detection of human enterovirus 71. They reduced the experimental time required for the human enterovirus 71 via quantification from 6 days to several minutes. LOD was reported as 67 virus particles per milliliter.

Riedel et al. (2014) presented the diagnosis of different stages of Epstein–Barr virus infections in clinical serum samples using an SPR sensor system. This was achieved by simultaneous detection of the antibodies against three different antigens present in the virus. Sensor platform was constructed via hybridization of complementary oligonucleotides. Proposed sensor system was efficiently applied for the same sensing surface repeatedly.

Torrance et al. (2006) obtained an SPR sensor for the detection of the virus with oriented immobilization of functionalized single chain (scFv) antibody molecules. They cloned scFv into a genetic vector that expressed with the light chain constant domain of human immunoglobulin with a C-terminal cysteine residue combination. They also demonstrated over more than 25 cycles of specific binding and dissociation curves. They did not show any specific binding in the scFv molecules without C-terminal cysteine. The resonance was increased with increasing of Cowpea mosaic virus from 12.5 to 50 μg/mL. Successive exposure of the SPR chip to both of the molecules (cowpea mosaic virus and BSA as a negative control) verifies a selective response to cowpea mosaic virus.

Lautner et al. (2010) prepared aptamer SPR sensor for sensitive detection of apple stem pitting virus coat proteins. Different experimental parameters were examined for obtaining best binding, such as the aptamer flanking, surface coverage, and type of spacer molecules.

Vaisocherová et al. (2007) prepared SPR sensor for specific and selective detection of anti-Epstein–Barr virus attacking towards the immune system of human. They investigated three immobilization chemistries for attaching Bovine serum albumin (BSA) and a respective synthetic peptide that using the immunoreaction between the anti-Epstein–Barr virus and respective synthetic peptide. They reported 0.2 ng/mL detection limit of the anti-Epstein–Barr virus by functionalized SPR sensor. Also, they showed the repeatability of the developed SPR system.

In another study, Baac et al. (2006) developed antibody-based SPR for the detection of an intact viral pathogen. This technique was achieved by the detection of intact viral pathogen with three different functional layers: amine reactive cross linker, protein A, and mouse IgG antibody. The selectivity of antibody-based SPR was showed via tobacco mosaic virus and the SPR sensor verified a selective response for intact viral pathogen.

In a previous study, Altintas et al. (2015) synthesized MIPs for targeting *E. coli* bacteriophage MS2 in order to detect waterborne viruses using SPR sensor. The bacteriophage selected as template molecule was immobilized to silica microbeads and therefore, MIPs having high affinity binding cavities were created and LOD was found as $5 \times 10^6$ plaque-forming units (PFU) per milliliter. Principles of the high affinity molecularly imprinted polymer production using a novel solid-phase synthesis method applied in this study were shown in Fig. 9. Summary of MIP-based and other detection technologies used for the detection of bacteria, viruses, fungi, and mycotoxins were given in Table 1.

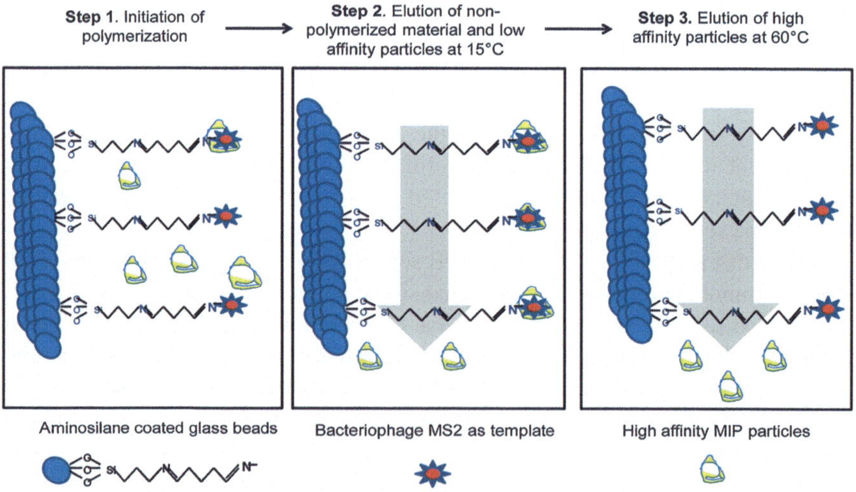


**Fig. 9** Principles of the high affinity molecularly imprinted polymer production using a novel solid-phase synthesis method (Reprinted with permission of Altintas et al. 2015)

**Table 1** Summary of MIP-based and other detection technologies used for detection of bacteria, viruses, fungi, and mycotoxins

| Target | Technology | Ligand/Functional monomer | Limit of detection value | Reference |
|---|---|---|---|---|
| *Cowpea mosaic virus* | Molecular imprinting | – | 12.5 μL | Torrance et al. (2006) |
| *Bacteriophage MS2 for E.coli* | Molecular imprinting | *N*-isopropylacrylamide | $5 \times 10^6$ pfu/ mL | Altintas et al. (2015) |
| *Fusarium culmorum* | Molecular imprinting | – | 0.06 pg | Zezza et al. (2006) |
| *Puccinia striiformis* | Molecular imprinting | – | $10^4$ CFU/mL | Skottrup et al. (2007) |
| *Enterococcus faecalis* | Molecular imprinting | *N*-methacryloyl-(L)-histidine-methylester | Down to ~100 bacteria/mL | Erdem et al. (2019) |
| *E. coli* | Molecular imprinting | *N*-methacryloyl-(L)-histidine-methylester | $1.54 \times 10^6$ CFU/mL | Yilmaz et al. (2015) |
| *Salmonella paratyphi* | Molecular imprinting | *N*-methacryloyl-(L)-histidine-methylester | $1.4 \times 10^6$ CFU/mL | Perçin et al. (2017) |
| *Salmonella typhimurium* | SPR immunosensor | Polyclonal antibody | $1.25 \times 10^5$ cells/mL | Mazumdar et al. (2007) |
| *Salmonella typhimurium* | SPR sensor | O-specific anti-*Salmonella* antibodies | $2.50 \times 10^5$ cells/mL | Barlen et al. (2007) |
| *Salmonella enteritidis* | SPR sensor | O-specific anti-*Salmonella* antibodies | $2.50 \times 10^8$ cells/mL | Barlen et al. (2007) |
| *E. coli* | Plastic optical fiber | – | $10^4$–$10^8$ CFU/mL | Wandemur et al. (2014) |
| *Ochratoxin A* | Gold hollow balls (AuHBs) with dendritic surface sensor | Anti-Ochratoxin A monoclonal antibody | 0.01 ng/mL | Fu (2007) |
| *Ochratoxin A* | Prototype capable sensor | Anti-Ochratoxin A antibody | 0.5 μg/kg | Todescato et al. (2014)[a] |
| *Ochratoxin A* | AuNPs-enhanced sensor | Anti-Ochratoxin A antibody | 0.19 ng/mL | Karczmarczyk et al. (2016b) |
| *Aflatoxin M1* | AuNPs-enhanced sensor | Primary antibody | 18 pg/mL | Karczmarczyk et al. (2016a) |

## 3.3 *Piezoelectric Nanosensors*

Piezoelectric crystals are well sensitive to any mass change or pressure on their surfaces. Therefore, piezoelectric crystals are used in trace quantitative analysis and microbalances for thin-film technology. The adsorption of target molecules on the surface of piezoelectric crystals the oscillating frequency decreases. A combination of a simple converter receptor can be used for detection of a target in the sensors system. QCM sensor as piezoelectric crystals-based sensor is a simple and convenient sensor platform. This approach is well-suited as transducer elements for chemical sensors. QCM sensor is an extremely sensitive mass-measuring device such as nano-grams. Dynamic monitoring of biochemical interactions is shown in the QCM sensor systems. The advantages of these systems are quick-response time, cost-effective, mass-sensitive, and high-resolution (Yilmaz et al. 2015). The mass of the QCM sensor is increased with the binding of the template but the oscillating frequency is decreased.

## *3.4 Applications of QCM Nanosensors*

QCM sensor is used in many areas such as diagnosing of diseases (Bakhshpour et al. 2019), environmental pollutants detection, and food contaminants. Li et al. (2011) prepared magnetic nanobeads amplification QCM immunosensor for the detection of AI H5N1 virus. They used polyclonal antibodies for the detection of the virus. They showed a change in the frequency when the H5N1 viruses were captured by immobilized polyclonal antibodies. They reported 0.0128 HAU for detection of the H5N1 virus. In another study, Wang and Li (2013) developed a novel QCM aptasensor for the detection of AIV (avian influenza virus) H5N1. They used ssDNA crosslinked polymeric hydrogel. They used aptamer against the surface protein of AIV H5N1 for selective and highly sensitive detection. Also, they prepared the crosslinked structure in the polymer by hybridization between the aptamer and ssDNA. A selected aptamer with high affinity and specificity against AIV H5N1 surface protein was used and hybridization was performed between the aptamer and ssDNA. They showed 0.0128 HAU for AIV H5N1 in 30 min with high sensitivity. Su et al. (2003) developed a QCM sensor for the detection of dengue virus. They used two monoclonal antibodies for selective detection of dengue virus specifically for envelope protein (E-protein) and non-structural 1 protein (NS-1 protein). They used three immobilization method including glutaraldehyde, protein A, and carbodiimide methods. They demonstrated the best results with the protein A method. They reported 100-fold greater sensitive than the conventional sandwich ELISA method. Karczmarczyk et al. (2017) prepared a QCM sensor with dissipation monitoring (QCM-D). They modified the gold surface of sensor with a mixed thiol self-assembled monolayer (SAM). They used antibodies for sensitive detection of Ochratoxin A in red wine. They reported 0.16 ng/mL limit of detection value for Ochratoxin A in the detection range between 0.2 and 40 ng/mL. Tang et al. (2018) designed a sensitive immunosensing QCM for the detection of Aflatoxin B1 in foodstuffs. Firstly, they immobilized phenoxy-derived dextran molecule on the gold surface of QCM. Then, they assembled BSA that was conjugated with Concanavalin A. They used monoclonal anti-Aflatoxin B1 antibody for amplification of the sensor signal. They reported 0.83 ng/kg limit of detection value.

Furthermore, there are many studies for the detection of microorganisms using different methods by MIP-based QCM sensor. Summary of MIP-based QCM detection technologies used for bacteria and yeasts was given in Table 2.

**Table 2** Summary of MIP-based QCM detection technologies used for bacteria and yeasts

| Microorganisms | Technology | Monomer | Limit of detection value | Reference |
|---|---|---|---|---|
| *E. coli* | Molecular imprinted | MAH | $3.72 \times 10^5$ CFU/mL | Yilmaz et al. (2015) |
| *Campylobacter fetus* | Molecular imprinted | – | $10^3$ CFU/mL | Brooks et al. (2004) |
| *Saccharomyces cerevisiae* | Surface imprinting | Polyurethane | $10^4$–$10^6$ CFU/mL | Dickert and Hayden (2002) |
| *Saccharomyces cerevisiae* | Surface imprinting | Titanium ethylate | $10^4$–$10^6$ CFU/mL | Dickert and Hayden (2002) |
| *E. coli* | Surface imprinting | Polyureathane | 0.1 mg/mL | Findeisen et al. (2012) |
| *E. coli* | Electro polymerization imprinted | Pyrrole | $10^3$–$10^9$ CFU/mL | Tokonami et al. (2012) |
| *Pseudomonas aeruginosa* | Electro polymerization imprinted | Pyrrole | $10^3$–$10^9$ CFU/mL | Tokonami et al. (2012) |
| *Acinetobacter calcoaceticus* | Electro polymerization imprinted | Pyrrole | $10^3$–$10^9$ CFU/mL | Tokonami et al. (2012) |
| *Serratia marcescens* | Electro polymerization imprinted | Pyrrole | $10^3$–$10^9$ CFU/mL | Tokonami et al. (2012) |
| *E. coli* | Bulk imprinted | Tetraethoxysilane | $10^2$ CFU/mL | Starosvetsky et al. (2012) |

## 3.5 Electrochemical Nanosensors

Electrochemical sensors have been one of the most common applied sensing tools for the detection of wide range of target molecules such as glucose (Meng et al. 2018), heavy metal ions (Anjuran and Chinnapiyan 2019), phenolic compounds (Maallah et al. 2019), tumor markers (Chikkaveeraiah et al. 2012), nucleic acids (Ferapontova 2018), cells (Ruiyi et al. 2018), microbial cells (Idil et al. 2017), and viruses (Babamiri et al. 2018) with the advantage of high sensing accuracy. This biosensing principle could further be manipulated in order to recognize other molecules by newly improved sensing approaches. In addition, electrochemical nanosensors could be easily integrated into point-of-care analyzing instruments and offers the creation of on-site detection and easy clinical diagnosis systems (Kuss et al. 2018). For these reasons, the applications of electrochemical nanosensors in the field of microbial detection and quantification have gained considerable attention from many researchers and industries (Amiri et al. 2018).

Electrochemical sensors have been introduced as attractive approaches to examine the content of a target analyte. In this approach, a process constantly converted to an electronic signal and an electrochemical element presents as transduction constituent. In general, events whether induce signal variation in conductance/impedance and detectable current, or charge change. Electrochemical sensors can be grouped depending on the mechanism applied to detect the signal from the transducer: (1) conductometric, (2) amperometric, (3) potentiometric, and (4) voltammetric sensors.

Electrochemical sensors have some components such as working electrode, a transducer element, and recording equipment. Reference electrodes have been typically produced by Ag/AgCl and located at a place to keep a stable potential. Sensing electrodes have served as a transduction layer in the formation of reaction. Counter (auxiliary) electrodes have a role on providing a contact tool of applying potential to the working electrode by implementing the charge transmission between two points. The working mechanism of electrochemical sensors can be explained by the formation of electrochemical reaction originated from signal transduction. The signal which could be raised from various types of transducers is recorded.

Conductometric sensors have the ability to detect target analytes by the determination of the change in electrolytic conductivity of the analyte solution. Sensing mechanism of these sensors are based on the differentiation of ions by transducers. They present the opportunity of rapid and sensitive detection of the target analytes with the advantage of easy miniaturization due to any requirement of a reference electrode. On the other hand, electrical charges could affect the conductivity change which resulted in the decrease in selectivity. Amperometric sensors are capable of measuring the current signal for the detection of target analyte concentration by recording the current signal through a sample with prominent sensitivity. Potentiometric sensors have applied for the detection of the difference in potential by working with three electrodes (two working electrodes and a reference electrode) offering outstanding sensitivity and selectivity.

Nowadays, nanomaterial-based technologies have become popular and combined with electrochemical detection techniques for the development of microbial sensors. The major superiorities of these integrated sensor systems are miniaturization and portability, high-throughput screening, enhanced sensitivity and selectivity, and simple and rapid immobilization. On the other hand, they offer sufficient opportunity to improve inexpensive, available, and applicable sensing platforms (Amiri et al. 2018). Furthermore, these systems achieve this unique success through some recognition elements such as enzymes, antibodies, nucleic acids, aptamers, and related molecules/cells. In addition, it is necessary to emphasize that low LOD values can be obtained by incorporation of nano-sized materials including gold and metallic nanoparticles, conductive polymeric materials, carbon nanotubes, and graphene oxide. These materials have also a crucial role in the development of well-performance analytical sensing platforms with the usage of cutting-edge elements offering conductivity.

One of the most frequently used working electrodes in electrochemical detection are glassy carbon electrodes (GCEs), and they could easily be applied for creation of functional MIP-based sensing systems. The other major approach to advance analytical power is the application of screen-printed electrodes (SPEs) which enables functionalization of their surfaces with nanomaterials. Their facilities in the arrangement of shape, dimension, and conductivity make significant contribution to improve analytical properties of biosensing. The small size of the SPEs offers the advantage of reduced amount of volume requirement, which enables the integration of these elements into the portable sensing tools (Lopes et al. 2017). In addition, surfaces of SPEs could be modified by MIPs to produce efficient sensing platforms (Silva et al. 2016). Graphene is another applicable alternative because of high surface to volume ratio. It also provides high conductivity and enables rapid electron movement. Besides, it could be integrated in sensing systems with the advantage of enabling polymer stability. Carbon nanotubes (CNTs) and multi-walled carbon nanotubes (MWCNTs) have extensively been preferred to compose MIP-based sensing platforms due to their ability in the improvement of electrical signal production. On the other sites, gold nanoparticles have also been applied in conjunction with MWCNTs for extending conductivity and sensitivity of sensing devices (Crapnell et al. 2019). Quantum dots (QDs) are nano-sized crystal structures with semiconductor properties and enable the design of high-speed diagnostic systems with optical and electrochemical properties. The most impressive feature of QDs is their adjustable sizes. The color of the fluorescence they made changes according to the size of the QDs. Following the introduction of new approaches in the direction of increasing demands in the clinical diagnosis, the polymers prepared by molecular imprinting having selective recognition property for the target molecule/cell have been appeared as promising candidates (Pedrero et al. 2017).

For ages, diagnosis of the infections from which millions of people suffering have been a focus point due to sexually transmission and spreading of diseases, emergence of infections caused by contaminated water, food, and body fluids. Many microorganisms contributing to these infections remain one of the most common causative agents of mortality. Therefore, rapid and accurate detection and quantification of these strains is essential in order to determine the most efficient treatment procedure applied on patients (Kuss et al. 2018).

In this context, analytical measurements based on electrodes contribute to make sustainable achievements for generating desirable sensing systems. In recent years, there have been many scientific publications reporting the applications of electrochemical sensors aimed to detect various whole cells of microorganisms and their by-products (Kuss et al. 2018). Among these microorganisms, *E. coli* is known as a non-pathogenic strain and a member of normal microbial flora. However, some serotypes of this strain such as EPEC (Enteropathogenic *E. coli*) and EHEC (Enterohemorrhagic *E. coli*) were pathogenic strains. EPEC particularly affects particularly infants and young children and leads to diarrhea (Ochoa and Contreras 2011). EHEC is the causative agent of bloody diarrhea and life-threatening diseases such as hemolytic uremic syndrome (Nguyen and Sperandio 2012). These bacterial strains are likely to be transmitted by drinking contaminated water or food supplies.

*S. aureus* is one of the most important causes of hospital infections showing high resistance to commonly used antimicrobial agents. *S. aureus* causes serious local and systemic infections such as bacteremia, endocarditis, pneumonia, and food poisoning sepsis. *S. aureus* enterotoxin leads to food poisoning and is one of the most common reasons of bacterial food poisoning in the worldwide. *Vibrio cholerae*

(*V. cholerae*) is the etiological agent of human cholera characterized with profound vomiting and diarrhea which lead to dehydration eventually mortality. The source of spread of these microorganism is drinking contaminated water or fecal-orally through contaminated water and ingesting contaminated food. Accordingly, *V. cholerae* is regarded as one of the most common agents of epidemic and pandemic diseases. *Bacillus cereus* (*B. cereus*) and *Clostridium perfringens* (*C. perfringens*) are two agents of food poisoning.

*Salmonella*sp. has appeared as one of the most common bacterial strains founds in food product for a long while. It is the causative agent of Salmonellosis related to the consumption of meat, milk, and eggs leads to abdominal cramps, fever, vomiting, and diarrhea. However, spreading of these strains could be attributed to the contaminated fruits and vegetables recently (Cinti et al. 2017).

Considering all these citations, recent advances in sensor technology provide the opportunity to detect a single cell which has a vital importance for identifying microorganisms having especially low infective doses. This approach also ensures to eliminate pre-treatment step as much as possible. In this regard, it appears obvious that highly sensitive and selective methods have been still required to quantify microbial cells presenting a valid choice (Cinti et al. 2017). In Fig. 10, schematic representation of whole cell imprinting was indicated.

Molecular imprinting technology has been incorporated in sensing systems as recognition elements for the detection, identification, and quantification of many target molecules/cells including microorganisms. In contrast to small/macromolecules, there is major ongoing challenges and limitations in imprinting of huge and complex featured microorganisms. Surface imprinting is an alternative approach enabling diffusion of microorganisms and provides easy removing them from the imprinted regions (Hayden and Dickert 2001)(Dickert and Hayden 2002). It is important to state that various functional monomers could be preferred to produce microorganism-imprinted polymers such as non-conducting polymers, sol-gels, and pyrrole conducting polymers (Tokonami et al. 2012) SAMs (Idil et al. 2017). Whole microorganism cell-imprinted polymers were successfully generated by designing selective recognition regions both on polymeric beads and in polymeric films (Idil et al. 2017). In Fig. 11, a schematic diagram of fabrication procedure of bacteria-mediated bioimprinted films for bacterial detection was illustrated.

In literature, many publications reported that MIPs were used for imprinting of bacteria and viruses. Apart from the other polymerization methods, electropolymerization method provides the feasibility to customize the layer thickness. This property directly affects the integration of target molecule/cell into the polymeric structure having specific recognition sites. In addition, OH-groups of lipopolisaccaride layer found in Gram-negative cell walls facilitate the polymerization around them (Crapnell et al. 2019).

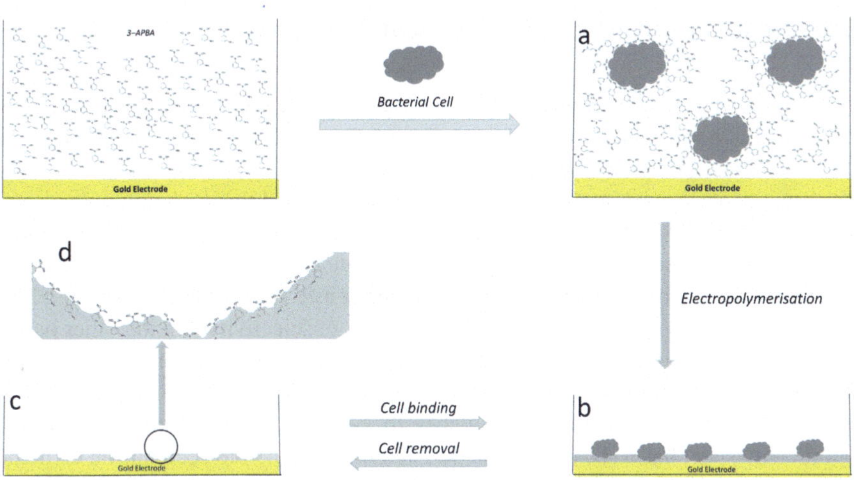


**Fig. 10** Schematic representation of whole cell imprinting. (**a**) Boronic acid groups attach to the surface of bacteria. (**b**) Electropolymerization preserves the structure of the cell at the polymer network. (**c**) Bacterial cells are removed and the complementary cavities remain. (**d**) Cavities containing boronic acid groups increase affinity of the preserved complementary shape of target bacteria (Reprinted with permission of Golabi et al. 2017)

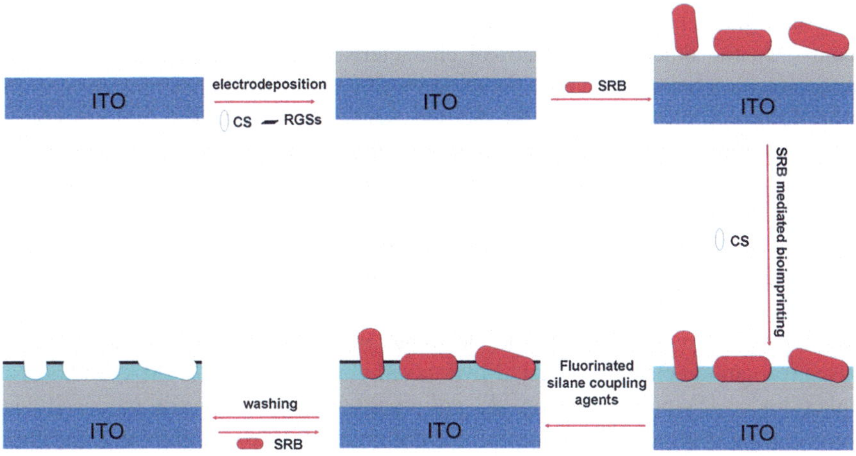


**Fig. 11** Schematic diagram of fabrication procedure of bacteria-mediated bioimprinted films for bacterial detection (Reprinted with permission of Qi et al. 2013)

## 3.6 Applications of Electrochemical Nanosensors

In the last decades, technological developments have been made widely in the fields of food safety and quality, agriculture, environmental monitoring, pharmaceutical and biomedical processes even clinical diagnosis. From the point of sensor technology, analytical devices impact on real-time analyzing and this technology has confronted with MIPs quite lately. These incorporated platforms have been frequently based on electrochemical detection, and they state desirable properties. It is supposed that electrochemical analysis enables high sensitivity, rapid response, and low limit of detection values, whereas MIPs ensure high selectivity. On the other hand, choice of compounds in polymeric matrix and appropriate imprinting strategy according to the template molecule play a crucial role in the formation of MIP-based electrochemical sensors particularly to get low detection limits. The other significant aspect has to be stressed is that determining the bio-recognition element used in imprinting phase to create sensing systems with high altitude.

In literature, there is an influential focus on electrochemical sensors able to detect microorganisms using antibodies, DNA, bacteriophage, aptamers called as immunosensors, genosensors, phagosensors (Liébana et al. 2013), aptasensors (Bagheryan et al. 2016), respectively. These electrochemical sensors have great potential to detect target microbial cells in real sample analysis including food. Nanomaterial-based sensing platforms could be constructed by various nano-structured elements along with synthetic receptors as molecular recognition elements. On the other hand, electrical characteristics of carbon-based nano-sized structures have been selected for designing well-defined electrochemical sensors in order to detect microbial cells. By the way, sensitive and selective methodologies have been improved for rapid detection of pathogenic microorganisms. In Table 3, representative examples of electrochemical detection using nano-sized materials for microbial detection was given.

In our previous study of a research collaboration, whole cell-imprinted capacitive sensor was developed to detect a model microorganism, *E. coli*. In the first step, bacterial stamps were prepared by immobilization of them onto the glass slides. In the second step, *E. coli*-imprinted gold electrodes were designed with an amino acid-based recognition element, MAH, 2-hydroxyethyl methacrylate (HEMA) and cross-linker under UV-polymerization. *E. coli*-imprinted surfaces were produced via microcontact imprinting by putting *E. coli* immobilized surface of glass slide together with the modified surface of gold electrode, like a sandwich (Fig. 12). In this study, real-time *E. coli* detection was carried out via measuring capacitance between $1.0 \times 10^2$–$1.0 \times 10^7$ CFU/mL bacterial concentration and LOD was found to be 70 CFU/mL with a recovery of 81–97%. In addition, selectivity of proposed capacitive sensor was clarified when target microorganism, *E. coli* presented along with competitive microorganisms which are known to have similar/different shape and size or cell wall structure. Eventually, this MIP-based sensing system was capable of detecting *E. coli* in river water and apple juice samples. It could be concluded that described sensor has been exploited as a promising tool for monitoring *E. coli* in contaminated water or food supplies (Idil et al. 2017).

**Table 3** Representative examples of electrochemical detection using nano-sized materials for microbial detection

| Target | Platform | Detection technique | Limit of Detection /Real sample | References |
|---|---|---|---|---|
| *E. coli CECT 675* | Aptasensor SWCNT | Potentiometric | 6 CFU/mL, milk 26 CFU/mL, apple juice | Zelada-Guillén et al. (2010) |
| *E. coli* | L-cysteine functionalized $Fe_3O_4$ NPs | CV | $10^1$ CFU/mL | Panhwar et al. (2019) |
| *E. coli* | Genosensor DNA nanopyramid | Amperometric | 1.20 CFU/mL | Giovanni et al. (2015) |
| *E. coli O157:H7* | Immunosensor BSA-conjugated 3D ag nanoflowers | Electrochemical | $10^2$ CFU/mL | Huang et al. (2016) |
| *Vibrio cholerae* | Genosensor gold nanoparticles/latex microsphere (AuNPs–PSA) | Electrochemical | PCR products, 1 fM | Liew et al. (2015) |
| *Vibrio cholerae* | $CeO_2$ nanowire-modified microelectrode | Impedimetric | Linear range: $1.0 \times 10^2$–$1.0 \times 10^4$ CFU/mL | Tam and Thang (2016) |
| *Vibrio cholerae* *Salmonella* sp., *Shigella* sp. | Multiplex DNA sensor QDs (PbS, CdS, ZnS) | Square-wave anodic stripping voltammetry | Multiplex detection limit in the attomolar range | Vijian et al. (2016) |
| *Bacillus cereus* | Genosensor AuNPs modified PGE | Electrochemical | 10 CFU/mL Milk, infant formula | Izadi et al. (2016) |
| *Staphylococcus aureus* | Aptasensor AgNPs | Electrochemical/ stripping voltammetry | 1 CFU/mL Real water | Abbaspour et al. (2015) |
| *Staphylococcus aureus* | AuNPs-GO | Impedimetric | 10 CFU/mL Water, fish | Jia et al. (2014) |
| *Salmonella enterica Serovar typhimurium* | Aptasensor GCE GNPs-GO | EIS | 3 CFU/mL Pork meat | Ma et al. (2014) |
| *Salmonella pullorum Salmonella gallinarum* | Immunosensor SPE GNPs | CV | $3 \times 10^3$ CFU/mL Eggs, chicken meat | Fei et al. (2015) |
| *Salmonella typhimurium* | Aptasensor GE PPy | Impedimetric | 3 CFU/mL Apple juice | Sheikhzadeh et al. (2016) |

(continued)

**Table 3** (continued)

| Target | Platform | Detection technique | Limit of Detection /Real sample | References |
|---|---|---|---|---|
| *Salmonella pullorum* | Immunosensor SPE CNTs | Amperometric | Chicken meat 60–100 CFU/mL | Chunyan Chai (2013) |
| *Salmonella typhimurium* | Immunosensor | Impedimetric | $5 \times 10^2$ CFU/mL Milk | Dong et al. (2013) |
| DNA of *C. perfringens* | SA/ADH/$Fe_3O_4$ nanocomposites | Amperometric | Range: $10^{-12}$–$10^{-6}$ M | Jiang et al. (2014) |

*GCE* glassy carbon electrode, *GNPs* gold nanoparticles, *GO* graphene oxide, *EIS* electrochemical impedance spectroscopy, *CV* cyclic voltammetry, *GE* gold electrode, *PPy* polypyrrole, *CNTs* carbon nanotubes, *SPE* carbon screen-printed electrode, *SWCNT* single-walled carbon nanotubes, *AgNPs* silver nanoparticles, *PGE* pencil graphite electrode, *SA* streptavidin, *ADH* alcohol dehydrogenase, *QDs* quantum dots, *NPs* nanoparticles, *AuNPs* gold nanoparticles, *CFU* colony forming unit

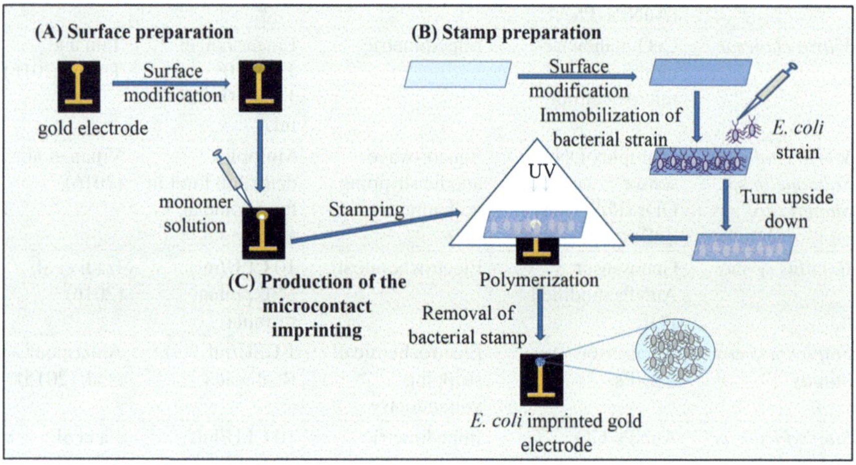


**Fig. 12** Schematic representation of microcontact imprinting of *E. coli* onto the polymer modified surfaces. (**a**) preparation of electrode surface, (**b**) preparation of bacteria stamps, (**c**) production of the microcontact imprinting (Reprinted with permission of Idil et al. 2017)

In literature, scientific publications relied on microbial detection have been reported some imprinting strategies using microbial specific moieties, signaling molecules, spores, harmful toxins, proteins of cell wall instead of whole microbial cell.

In a study conducted to detect *Aeromonas hydrophila* and *P. aeruginosa*, 2,5-dimethyl-4-hydroxy-3(2*H*)-furanone was imprinted because of its chemical and size similarity to N-acyl-homoserine-lactones (AHLs) referred as quorum signaling small organic molecules. Magnetic $Fe_3O_4$ was selected to create MIP-based electrochemical sensor. Differential pulse voltammetry (DPV) was applied to measure AHLs and LOD was found as $8 \times 10^{-10}$ mol/L for AHLs. The created sensor was able to quantify microbial quorum signaling molecules. In the future developments, obtained findings would be the basis to form analyzing assays in many potential clinical and food applications with confirmed stability and specificity (Jiang et al. 2016).

In another study, flagella filaments of *Proteus mirabilis* found on the surface of bacterial cell were detected by molecularly imprinted artificial receptors. Many bacterial strains have flagella that could provide movement behavior of cell and be used as biomarkers in environment applications. They used both commercial single-walled carbon nanotubes screen-printed electrodes (SWCNTs-SPEs) and a home-made paper-based carbon-printed electrodes. The fabricated detection platform has an LOD of 0.6 ng/mL provides label-free, rapid, cost-effective analysis with high selectivity (Khan et al. 2017).

Apart from these studies, a research group aimed to detect protein A with the fabrication of polymers using bulk imprinting. In this study, protein A, a major component of bacterial cell wall, was used as a template molecule and 3-aminophenol was selected as monomer for imprinting process. Imprinted polymer was attached onto the SWCNTs and SPEs were used for electrochemical detection. The obtained results showed that LOD was found as 0.6 nM (Khan et al. 2016). In Fig. 13, schematic representation of the synthetic process used in this study was demonstrated.

In a previous study aimed to detect Zika virus (ZIKV), electrochemical biosensor based on surface imprinted polymers and graphene oxide composites was fabricated. In Fig. 14, scheme depicting SIPs-graphene oxide composites preparation on a gold surface for making the ZIKV biosensor was shown.

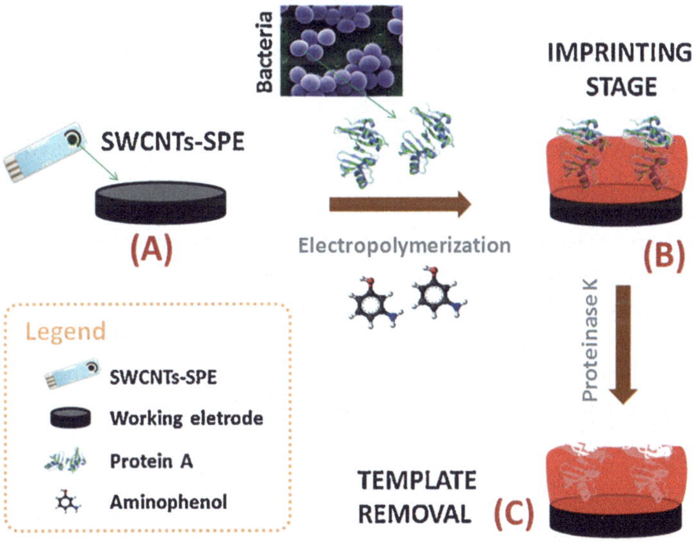


**Fig. 13** Schematic representation of the synthetic process. (**a**) Working area of SWCNTs-SPE; (**b**) Imprinting stage after electropolymerization of *AP* along with PA and (**c**) Binding site formation after template removal by proteinase K (Reprinted with permission of Khan et al. 2016)

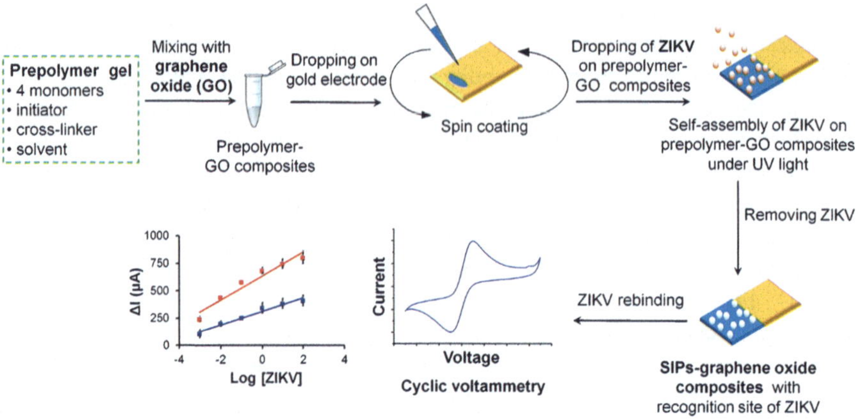


**Fig. 14** Scheme depicting SIPs-graphene oxide composites preparation on a gold surface for making the ZIKV biosensor. When Zika virus was added to the composites, some particles would have chances to be on the copolymers sites. At that time, the copolymers were partially polymerized and the self-assembly process by some prepolymer around the virus could occur. After the polymerization process had finished, the virus was removed and the recognition sites became cavities (Reprinted with permission of Tancharoen et al. 2019)

In a previous study, fungal toxin detection was performed using MIP-based electrochemical sensor. Aflatoxin B1 is an aflatoxin produced by certain fungal strains such as *Aspergillus flavus* and *Aspergillus parasiticus*. The main source of this harmful toxin is environmental conditions, by the way contaminated foods. It has genotoxic and carcinogenic effect; therefore, it leads to public health problems. In the study, *p*-aminothiophenol was preferred as monomer for the preparation of self-assembled monolayer in order to functionalize gold nanoparticles. Electropolymerization was performed in the presence of Aflatoxin B1 (AFB1). After the removal of target molecule, AFB1, specific recognition cavities which were capable of capturing were produced. The detection was carried out via linear sweep voltammetry. It was observed that binding of AFB1 resulted in an increase in conductivity. The applicability concentration range was examined between 3.2 fM and 3.2 μM and LOQ (limit of quantification) was found as 3 fM. In addition, the selectivity of designed sensing system was verified with obtained results from other aflatoxins and Ochratoxin A (Jiang et al. 2015). Representative examples of MIP-based electrochemical sensing platforms for detection of bacteria, viruses, and fungal toxins were summarized in Table 4.

**Table 4** Representative examples of MIP-based electrochemical sensing platforms for detection of bacteria, viruses, and fungal toxins

| Target | Methodology | Monomer/Polymer<br>Signal compound | Linear Range Limit of Detection | References |
|---|---|---|---|---|
| HIV-1 gene | MIP-based electrochemiluminescence | *o*-Phenylenediamine<br>Europium sulfide nanocrystals | $3 \times 10^{-15} - 0.3 \times 10^{-9}$ M<br>$0.3 \times 10^{-15}$ M | Babamiri et al. (2018) |
| HIV-p24 | MIP-based electrochemical sensor | Acrylamide<br>MWCNTs modified GCE | $1.0 \times 10^{-4}$–2 ng $cm^{-3}$<br>0.083 pg $cm^{-3}$ | Ma et al. (2017) |
| Zika virus | Electrochemical sensor | Surface imprinted polymers<br>Graphene oxide composites | $10^{-3}$–$10^{2}$ PFU/mL | Tancharoen et al. (2019) |
| Bacteriophage | MIP-based capacitive biosensor | *N*-hydroxymethyl acrylamide and PEG-400 DMA | $10^{1}$–$10^{5}$ PFU/mL<br>10 PFU/mL | Ertürk and Lood (2018) |
| *E. coli* | MIP-based capacitive biosensor | *N*-hydroxymethyl acrylamide and PEG-400 DMA | $10^{2}$–$10^{7}$ CFU/mL<br>100 CFU/mL | Ertürk and Lood (2018) |
| *E. coli* O157:H7 | Electrochemiluminescence sensor | PDA-SIP<br>N-GQDs | $10^{1}$–$10^{7}$ CFU/mL<br>8 CFU/mL | Chen et al. (2017) |
| *Bacillus subtilis* endospore | Electrochemical sensor | Endospore imprinted polypyrrole/ poly(3-methylthiophene) composite films<br>GCE | $10^{4}$–$10^{7}$ CFU/mL<br>$10^{2}$ CFU/mL | Namvar and Warriner (2007) |
| *S. aureus* | Electrochemical disinfection | Cell-imprinted artificial antibody | –<br>500 CFU/mL | Zhang et al. (Zhang et al. 2015a, b) |
| Sulfate-reducing bacteria | Electrochemical impedance spectroscopy | RGSs-CS hybrid film | $10^{4}$–$10^{8}$ CFU/mL | Qi et al. (2013) |
| *E. coli* | Microcontact-imprinted capacitive biosensor | MAH-$Cu^{2}$ | 70 CFU/mL | Idil et al. (2017) |
| *E. coli* | Cell-imprinted capacitive biosensor | Boronic acid-doped polyaniline methacrylic acid-based MIPs<br>Carbon nanotube conductive film | 36 CFU/mL | Lu et al. (2018) |

(continued)

**Table 4** (continued)

| Target | Methodology | Monomer/Polymer Signal compound | Linear Range Limit of Detection | References |
|---|---|---|---|---|
| *Staphylococcus epidermitis* | Electrochemical impedance spectroscopy | Poly(3-aminophenylboronic acid)-based imprinted polymer | $10^3$–$10^7$ CFU/mL | Golabi et al. (2017) |
| *Staphylococcal* exotoxin | MIP-based potentiometric sensor | Methacrylic acid | $10^{-3}$ M | Ahari et al. (2017) |
| *Aeromonas hydrophila and P. aeruginosa* | Magnetic MIP-based electrochemical sensor | Methacrylic acid MGCE | N-acyl-homoserine-lactones, $8 \times 10^{-10}$ Mol/L | Jiang et al. (2016) |
| Staphylococcus aureus-protein A | Electrochemical impedance spectroscopy | 3-aminophenol<br>SPCE | Protein A, 0.6 nM | Khan et al. (2016) |
| *Proteus mirabilis-* Flagella filaments | Molecularly imprinted artificial receptors | Phenol<br>SWCNTs-SPEs | Flagellar filaments, 0.6 ng/mL | Khan et al. (2017) |
| Ochratoxin A | MIP-based electrochemical sensor | Polypyrrole<br>MWCNTs modified GCE | 0.050 and 1.0 μM<br>LOD: 0.0041 μM (1.7 μg/L)<br>LOQ: 0.014 μM (5.7 μg/L) | Pacheco et al. (2015) |

*SWCNTs* single-walled carbon nanotubes, *MWCNTs* multi-walled carbon nanotubes, *GCE* glassy carbon electrode, *N-GQDs* nitrogen-doped graphene quantum dots, *PDA-SIP* polydopamine surface imprinted polymer, *RGSs-CS* chitosan (CS) doped with reduced graphene sheets (RGSs), *MAH N*-methacryloyl-L-histidine methylester, *CPE* carbon paste electrode, *SIP* spore-imprinted polymer, *MGCE* magnetic carbon paste electrode, *SPCE* screen-printed carbon electrode, *PEG* polyethylene glycol

## 4 Conclusion

In the last decades, biosensor technology has gained great attention for the detection of microorganisms. Several attempts were made to make innovative methods. Many nanostructures could be applied in conjunction with sensing systems; therefore, important advances have been implemented for developing competences and functionality of sensing platforms. It has been proven that following platforms steadily and quietly achieved microbial recognition, capturing, detection, identification, and quantification. On the other hand, microbial metabolites, by-products, and toxins have been detected using these nanosensors. Selectivity, stability, and comparability of constructed sensing systems should be examined for interpreting their potential in routine analysis and large-scale applications. On the whole, verification of the applicability of electrochemical nanosensors in complex real samples has a great importance in microbial detection from the standpoint of moving this tremendous strategy into the market place.

## References

Abbaspour A, Norouz-Sarvestani F, Noori A, Soltani N (2015) Aptamer-conjugated silver nanoparticles for electrochemical dual-aptamer-based sandwich detection of *Staphylococcus aureus*. Biosens Bioelectron 15(68):149–155. https://doi.org/10.1016/j.bios.2014.12.040

Ahari H, Hedayati M, Akbari-adergani B, Kakoolaki S, Hosseini H, Anvar A (2017) *Staphylococcus aureus* exotoxin detection using potentiometric nanobiosensor for microbial electrode approach with the effects of pH and temperature. Int J Food Prop 20(2):1578–1587. https://doi.org/10.1080/10942912.2017.1347944

Ait Lahcen A, Arduini F, Lista F, Amine A (2018) Label-free electrochemical sensor based on spore-imprinted polymer for *Bacillus cereus* spore detection. Sensors Actuators B Chem 276(10):114–120. https://doi.org/10.1016/j.snb.2018.08.031

Altintas Z, Gittens M, Guerreiro A, Thompson KA, Walker J, Piletsky S, Tothill IE (2015) Detection of waterborne viruses using high affinity molecularly imprinted polymers. Anal Chem 87(13):6801–6807. https://doi.org/10.1021/acs.analchem.5b00989

Amiri M, Bezaatpour A, Jafari H, Boukherroub R, Szunerits S (2018) Electrochemical methodologies for the detection of pathogens. ACS Sensors 3(6):1069–1086. https://doi.org/10.1021/acssensors.8b00239

Anik Ü, Tepeli Y, Diouani MF (2016) Fabrication of electrochemical model influenza a virus biosensor based on the measurements of neuroaminidase enzyme activity. Anal Chem 88(12):6151–6153. https://doi.org/10.1021/acs.analchem.6b01720

Anjuran V, Chinnapiyan V (2019) Electrochemical sensors for heavy metals detection in Gracilaria corticata using multiwalled carbon nanotubes modified glassy carbon electrode. J Anal Chem 74(3):276–285. https://doi.org/10.1134/s106193481903002x

Ashiba H, Sugiyama Y, Wang X, Shirato H, Higo-Moriguchi K, Taniguchi K, Ohki Y, Fujimaki M (2017) Detection of norovirus virus-like particles using a surface plasmon resonance-assisted fluoroimmunosensor optimized for quantum dot fluorescent labels. Biosens Bioelectron 93:260–266. https://doi.org/10.1016/j.bios.2016.08.099

Baac H, Hajós JP, Lee J, Kim D, Kim SJ, Shuler ML (2006) Antibody-based surface plasmon resonance detection of intact viral pathogen. Biotechnol Bioeng 94(4):815–819. https://doi.org/10.1002/bit.20882

Babamiri B, Salimi A, Hallaj R (2018) A molecularly imprinted electrochemiluminescence sensor for ultrasensitive HIV-1 gene detection using EuS nanocrystals as luminophore. Biosens Bioelectron 117:332–339. https://doi.org/10.1016/j.bios.2018.06.003

Bagheryan Z, Raoof JB, Golabi M, Turner APF, Beni V (2016) Diazonium-based impedimetric aptasensor for the rapid label-free detection of *Salmonella typhimurium* in food sample. Biosens Bioelectron 80:566–573. https://doi.org/10.1016/j.bios.2016.02.024

Bai H, Wang R, Hargis B, Lu H, Li Y (2012) A SPR aptasensor for detection of avian influenza virus H5N1. Sensors 12(9):12506–12518. https://doi.org/10.3390/s120912506

Bakhshpour M, Özgür E, Bereli N, Denizli A (2017) Microcontact imprinted quartz crystal microbalance nanosensor for protein C recognition. Colloids Surf B Biointerfaces 151:264–270. https://doi.org/10.1016/j.colsurfb.2016.12.022

Bakhshpour M, Yavuz H, Denizli A (2018) Controlled release of mitomycin C from PHEMAH–Cu(II) cryogel membranes. Artif Cells Nanomed Biotechnol 46(1):946–954. https://doi.org/10.1080/21691401.2018.1439840

Bakhshpour M, Piskin AK, Yavuz H, Denizli A (2019) Quartz crystal microbalance biosensor for label-free MDA MB 231 cancer cell detection via notch-4 receptor. Talanta 204:840–845. https://doi.org/10.1016/j.talanta.2019.06.060

Barlen B, Mazumdar SD, Lezrich O, Kämpfer P, Keusgen M (2007) Detection of *Salmonella* by surface Plasmon resonance. Sensors 7:1427–1446. https://doi.org/10.3390/s7081427

Bereli N, Andaç M, Baydemir G, Say R, Galaev IY, Denizli A (2008) Protein recognition via ion-coordinated molecularly imprinted supermacroporous cryogels. J Chromatogr A 1190(1–2):18–26. https://doi.org/10.1016/j.chroma.2008.02.110

Bharadwaj R, Mukherji S, Mukherji S (2016) Probing the localized surface plasmon field of a gold nanoparticle-based fibre optic biosensor. Plasmonics 11(3):753–761. https://doi.org/10.1007/s11468-015-0106-0

Brooks BW, Devenish J, Lutze-Wallace CL, Milnes D, Robertson RH, Berlie-Surujballi G (2004) Evaluation of a monoclonal antibody-based enzyme-linked immunosorbent assay for detection of *Campylobacter fetus* in bovine preputial washing and vaginal mucus samples. Vet Microbiol 103(1–2):77–84. https://doi.org/10.1016/j.vetmic.2004.07.008

Cheeveewattanagul N, Morales-Narváez E, Hassan ARHA, Bergua JF, Surareungchai W, Somasundrum M, Merkoçi A (2017) Straightforward immunosensing platform based on graphene oxide-decorated nanopaper: a highly sensitive and fast biosensing approach. Adv Funct Mater 27:1702741–1702749. https://doi.org/10.1002/adfm.201702741

Chen S, Chen X, Zhang L, Gao J, Ma Q (2017) Electrochemiluminescence detection of *Escherichia coli* O157:H7 based on a novel polydopamine surface imprinted polymer biosensor. ACS Appl Mater Interfaces 9(6):5430–5436. https://doi.org/10.1021/acsami.6b12455

Chikkaveeraiah BV, Bhirde AA, Morgan NY, Eden HS, Chen X (2012) Electrochemical immunosensors for detection of cancer protein biomarkers. ACS Nano 6(8):6546–6561. https://doi.org/10.1021/nn3023969

Chunyan Chai GL (2013) Rapid evaluation of *Salmonella pullorum* contamination in chicken based on a portable amperometric sensor. J Biosens Bioelectron 4:4. https://doi.org/10.4172/2155-6210.1000137

Cinti S, Volpe G, Piermarini S, Delibato E, Palleschi G (2017) Electrochemical biosensors for rapid detection of foodborne *Salmonella*: a critical overview. Sensors 17(8):1910–1932. https://doi.org/10.3390/s17081910

Crapnell RD, Hudson A, Foster CW, Eersels K, van Grinsven B, Cleij TJ, Banks CE, Peeters M (2019) Recent advances in electrosynthesized molecularly imprinted polymer sensing platforms for bioanalyte detection. Sensors 19(5):1204–1232. https://doi.org/10.3390/s19051204

Cunha I, Biltes R, Sales MGF, Vasconcelos V (2018) Aptamer-based biosensors to detect aquatic phycotoxins and cyanotoxins. Sensors 18(7):2367–1401. https://doi.org/10.3390/s18072367

Dickert FL, Hayden O (2002) Bioimprinting of polymers and sol-gel phases. Selective detection of yeasts with imprinted polymers. Anal Chem 74:1302–1306. https://doi.org/10.1021/ac010642k

do Nascimento NM, Juste-Dolz A, Grau-García E, Román-Ivorra JA, Puchades R, Maquieira A, Morais S, Gimenez-Romero D (2017) Label-free piezoelectric biosensor for prognosis and diagnosis of systemic lupus erythematosus. Biosens Bioelectron 90:166–173. https://doi.org/10.1016/j.bios.2016.11.004

Dong J, Zhao H, Xu M, Maa Q, Ai S (2013) A label-free electrochemical impedance immunosensor based on AuNPs/PAMAM-MWCNT-chi nanocomposite modified glassy carbon electrode for detection of Salmonella typhimurium in milk. Food Chem 141(3):1980–1986. https://doi.org/10.1016/j.foodchem.2013.04.098

Dudak FC, Boyaci IH (2014) Peptide-based surface plasmon resonance biosensor for detection of staphylococcal enterotoxin B. Food Anal Methods 7(2):506–511. https://doi.org/10.1007/s12161-013-9739-9

Eissa S, Siaj M, Zourob M (2015) Aptamer-based competitive electrochemical biosensor for brevetoxin-2. Biosens Bioelectron 69:148–154. https://doi.org/10.1016/j.bios.2015.01.055

Erdem Ö, Saylan Y, Cihangir N, Denizli A (2019) Molecularly imprinted nanoparticles based plasmonic sensors for real-time *Enterococcus faecalis* detection. Biosens Bioelectron 126:608–614. https://doi.org/10.1016/j.bios.2018.11.030

Ertürk G, Lood R (2018) Bacteriophages as biorecognition elements in capacitive biosensors: phage and host bacteria detection. Sensors Actuators B Chem 258:535–543. https://doi.org/10.1016/j.snb.2017.11.117

Ertürk G, Mattiasson B (2017) Molecular imprinting techniques used for the preparation of biosensors. Sensors 17(2):288–305. https://doi.org/10.3390/s17020288

Fei J, Dou W, Zhao G (2015) A sandwich electrochemical immunosensor for *Salmonella pullorum* and *Salmonella gallinarum* based on a screen-printed carbon electrode modified with an ionic liquid and electrodeposited gold nanoparticles. Microchim Acta 182(13–14):2267–2275. https://doi.org/10.1007/s00604-015-1573-x

Ferapontova EE (2018) DNA electrochemistry and electrochemical sensors for nucleic acids. Annu Rev Anal Chem 11:197–218. https://doi.org/10.1146/annurev-anchem-061417-125811

Findeisen A, Wackerlig J, Samardzic R, Pitkänen J, Anttalainen O, Dickert FL, Lieberzeit PA (2012) Artificial receptor layers for detecting chemical and biological agent mimics. Sensors Actuators B Chem 170:196–200. https://doi.org/10.1016/j.snb.2011.08.025

Frasco MF, Truta LAANA, Sales MGF, Moreira FTC (2017) Imprinting technology in electrochemical biomimetic sensors. Sensors 17(3):523–551. https://doi.org/10.3390/s17030523

Fu XH (2007) Surface plasmon resonance immunoassay for Ochratoxin a based on nanogold hollow balls with dendritic surface. Anal Lett 40(14):2641–2652. https://doi.org/10.1080/00032710701588366

Giamblanco N, Conoci S, Russo D, Marletta G (2015) Single-step label-free hepatitis B virus detection by a piezoelectric biosensor. RSC Adv 5:38152–38158. https://doi.org/10.1039/c5ra03467a

Giovanni M, Setyawati MI, Tay CY, Qian H, Kuan WS, Leong DT (2015) Electrochemical quantification of *Escherichia coli* with DNA nanostructure. Adv Funct Mater 25(25):3840–3846. https://doi.org/10.1002/adfm.201500940

Golabi M, Kuralay F, Jager EWH, Beni V, Turner APF (2017) Electrochemical bacterial detection using poly(3-aminophenylboronic acid)-based imprinted polymer. Biosens Bioelectron 93:87–93. https://doi.org/10.1016/j.bios.2016.09.088

Gür SD, Bakhshpour M, Denizli A (2019) Selective detection of *Escherichia coli* caused UTIs with surface imprinted plasmonic nanoscale sensor. Mater Sci Eng C 104:109869–109876. https://doi.org/10.1016/j.msec.2019.109869

Hayden O, Dickert FL (2001) Selective microorganism detection with cell surface imprinted polymers. Adv Mater 13(19):1480–1483. https://doi.org/10.1002/1521-4095(200110)13:19<1480::AID-ADMA1480>3.0.CO;2-V

Hu T, Zhang L, Wen W, Zhang X, Wang S (2016) Enzyme catalytic amplification of miRNA-155 detection with graphene quantum dot-based electrochemical biosensor. Biosens Bioelectron 77:451–456. https://doi.org/10.1016/j.bios.2015.09.068

Huang H, Liu M, Wang X, Zhang W, Yang DP, Cui L, Wang X (2016) Label-free 3D Ag nanoflower-based electrochemical immunosensor for the detection of *Escherichia coli* O157:H7 pathogens. Nanoscale Res Lett 11:507. https://doi.org/10.1186/s11671-016-1711-3

Idil N, Mattiasson B (2017) Imprinting of microorganisms for biosensor applications. Sensors 17(4):708–715. https://doi.org/10.3390/s17040708

Idil N, Hedström M, Denizli A, Mattiasson B (2017) Whole cell based microcontact imprinted capacitive biosensor for the detection of *Escherichia coli*. Biosens Bioelectron 87:807–815. https://doi.org/10.1016/j.bios.2016.08.096

Inbaraj B, Chen BH (2016) Nanomaterial-based sensors for detection of foodborne bacterial pathogens and toxins as well as pork adulteration in meat products. J Food Drug Anal 24(1):15–28. https://doi.org/10.1016/j.jfda.2015.05.001

Izadi Z, Sheikh-Zeinoddin M, Ensafi AA, Soleimanian-Zad S (2016) Fabrication of an electrochemical DNA-based biosensor for *Bacillus cereus* detection in milk and infant formula. Biosens Bioelectron 80:582–589. https://doi.org/10.1016/j.bios.2016.02.032

Jia F, Duan N, Wu S, Ma X, Xia Y, Wang Z, Wei X (2014) Impedimetric aptasensor for Staphylococcus aureus based on nanocomposite prepared from reduced graphene oxide and gold nanoparticles. Microchim Acta 181(9–10):967–974. https://doi.org/10.1007/s00604-014-1195-8

Jiang D, Liu F, Zhang L, Liu L, Liu C, Pu X (2014) An electrochemical strategy with molecular beacon and hemin/G-quadruplex for the detection of *Clostridium perfringens* DNA on screen-printed electrodes. RSC Adv 4:57064–57070. https://doi.org/10.1039/c4ra09834j

Jiang M, Braiek M, Florea A, Chrouda A, Farre C, Bonhomme A, Bessueille F, Vocanson F, Zhang A, Jaffrezic-Renault N (2015) Aflatoxin B1 detection using a highly-sensitive molecularly-imprinted electrochemical sensor based on an electropolymerized metal organic framework. Toxins (Basel) 7(9):3540–3553. https://doi.org/10.3390/toxins7093540

Jiang H, Jiang D, Shao J, Sun X (2016) Magnetic molecularly imprinted polymer nanoparticles based electrochemical sensor for the measurement of Gram-negative bacterial quorum signaling molecules (N-acyl-homoserine-lactones). Biosens Bioelectron 75:411–419. https://doi.org/10.1016/j.bios.2015.07.045

Kabessa Y, Eyal O, Bar-On O, Korouma V, Yagur-Kroll S, Belkin S, Agranat AJ (2016) Standoff detection of explosives and buried landmines using fluorescent bacterial sensor cells. Biosens Bioelectron 79:784–788. https://doi.org/10.1016/j.bios.2016.01.011

Kaittanis C, Santra S, Perez JM (2010) Emerging nanotechnology-based strategies for the identification of microbial pathogenesis. Adv Drug Deliv Rev 62(4–5):408–423. https://doi.org/10.1016/j.addr.2009.11.013

Karczmarczyk A, Dubiak-Szepietowska M, Vorobii M, Rodriguez-Emmenegger C, Dostálek J, Feller KH (2016a) Sensitive and rapid detection of aflatoxin M1 in milk utilizing enhanced SPR and p(HEMA) brushes. Biosens Bioelectron 81:159–165. https://doi.org/10.1016/j.bios.2016.02.061

Karczmarczyk A, Reiner-Rozman C, Hageneder S, Dubiak-Szepietowska M, Dostálek J, Feller KH (2016b) Fast and sensitive detection of ochratoxin a in red wine by nanoparticle-enhanced SPR. Anal Chim Acta 937:143–150. https://doi.org/10.1016/j.aca.2016.07.034

Karczmarczyk A, Haupt K, Feller KH (2017) Development of a QCM-D biosensor for Ochratoxin a detection in red wine. Talanta 166:193–197. https://doi.org/10.1016/j.talanta.2017.01.054

Khan MAR, Khan MAR, Moreira FTC, Riu JF, Sales MG (2016) Plastic antibody for the electrochemical detection of bacterial surface proteins. Sensors Actuators B Chem 233:697–704. https://doi.org/10.1016/j.snb.2016.04.075

Khan MAR, Aires Cardoso AR, Sales MGF, Merino S, Tomás JM, Rius FX, Riu J (2017) Artificial receptors for the electrochemical detection of bacterial flagellar filaments from Proteus mirabilis. Sensors Actuators B Chem 244:732–741. https://doi.org/10.1016/j.snb.2017.01.018

Kuss S, Amin HMA, Compton RG (2018) Electrochemical detection of pathogenic bacteria-recent strategies, advances and challenges. Chem Asian J 13(19):2758–2769. https://doi.org/10.1002/asia.201800798

Lautner G, Balogh Z, Bardóczy V, Mészáros T, Gyurcsányi RE (2010) Aptamer-based biochips for label-free detection of plant virus coat proteins by SPR imaging. Analyst 35(5):918–926. https://doi.org/10.1039/b922829b

Li D, Wang J, Wang R, Li Y, Abi-Ghanem D, Berghman L et al (2011) A nanobeads amplified QCM immunosensor for the detection of avian influenza virus H5N1. Biosens Bioelectron 26(10):4146–4154. https://doi.org/10.1016/j.bios.2011.04.010

Liébana S, Spricigo DA, Cortés MP, Barbé J, Llagostera M, Alegret S, Pividori MI (2013) Phagomagnetic separation and electrochemical magneto-genosensing of pathogenic bacteria. Anal Chem 85(6):3079–3086. https://doi.org/10.1021/ac3024944

Liew PS, Lertanantawong B, Lee SY, Manickam R, Lee YH, Surareungchai W (2015) Electrochemical genosensor assay using lyophilized gold nanoparticles/latex microsphere label for detection of Vibrio cholerae. Talanta 1(139):167–73. https://doi.org/10.1016/j.talanta.2015.02.054

Lopes F, Pacheco JG, Rebelo P, Delerue-Matos C (2017) Molecularly imprinted electrochemical sensor prepared on a screen printed carbon electrode for naloxone detection. Sensors Actuators B Chem 243:745. https://doi.org/10.1016/j.snb.2016.12.031

Lu CH, Zhang Y, Tang SF, Fang ZB, Yang HH, Chen X, Chen GN (2012) Sensing HIV related protein using epitope imprinted hydrophilic polymer coated quartz crystal microbalance. Biosens Bioelectron 31(1):439–444. https://doi.org/10.1016/j.bios.2011.11.008

Lu W, Mugo S (2018) A Cell-Imprinted Polymer Capacitive Biosensor for the Detection of Escherichia coli, ECS Meet. Abstr. MA2018-01 2444. https://doi.org/10.1149/MA2018-01/42/2444

Ma X, Jiang Y, Jia F, Yu Y, Chen J, Wang Z (2014) An aptamer-based electrochemical biosensor for the detection of *Salmonella*. J Microbiol Methods 98:94–98. https://doi.org/10.1016/j.mimet.2014.01.003

Ma Y, Shen XL, Zeng Q, Wang HS, Wang LS (2017) A multi-walled carbon nanotubes based molecularly imprinted polymers electrochemical sensor for the sensitive determination of HIV-p24. Talanta 164:121–127. https://doi.org/10.1016/j.talanta.2016.11.043

Maallah R, Moutcine A, Laghlimi C, Smaini MA, Chtaini A (2019) Electrochemical bio-sensor for degradation of phenol in the environment. Sens Biosensing Res 24:100279. https://doi.org/10.1016/j.sbsr.2019.100279

Mao C, Liu A, Cao B (2009) Virus-based chemical and biological sensing. Angew Chem Int Ed 48(37):6790–6810. https://doi.org/10.1002/anie.200900231

Mazumdar SD, Hartmann M, Kämpfer P, Keusgen M (2007) Rapid method for detection of *Salmonella* in milk by surface plasmon resonance (SPR). Biosens Bioelectron 22(9–10):2040–2046. https://doi.org/10.1016/j.bios.2006.09.004

Meng W, Wen Y, Dai L, He Z, Wang L (2018) A novel electrochemical sensor for glucose detection based on ag@ZIF-67 nanocomposite. Sensors Actuators B Chem 260:852–860. https://doi.org/10.1016/j.snb.2018.01.109

Munawar A, Ong Y, Schirhagl R, Tahir MA, Khan WS, Bajwa SZ (2019) Nanosensors for diagnosis with optical, electric and mechanical transducers. RSC Adv 9:6793–6803. https://doi.org/10.1039/c8ra10144b

Namvar A, Warriner K (2007) Microbial imprinted polypyrrole/poly(3-methylthiophene) composite films for the detection of *Bacillus* endospores. Biosens Bioelectron 22(9–10):2018–2024. https://doi.org/10.1016/j.bios.2006.08.039

Nguyen Y, Sperandio V (2012) Enterohemorrhagic *E. coli* (EHEC) pathogenesis. Frontiers in cellular and infection. Microbiology 2:90. https://doi.org/10.3389/fcimb.2012.00090

Nguyen PD, Tran TB, Nguyen DTX, Min J (2014) Magnetic silica nanotube-assisted impedimetric immunosensor for the separation and label-free detection of *Salmonella typhimurium*. Sensors Actuators B Chem 197:314–320. https://doi.org/10.1016/j.snb.2014.02.089

Ochoa TJ, Contreras CA (2011) Enteropathogenic *E. coli* (EPEC) infection in children. Curr Opin Infect Dis 24(5):478–483. https://doi.org/10.1097/QCO.0b013e32834a8b8b.Enteropathogenic

Osman B, Uzun L, Beşirli N, Denizli A (2013) Microcontact imprinted surface plasmon resonance sensor for myoglobin detection. Mater Sci Eng C 33(7):3609–3614. https://doi.org/10.1016/j.msec.2013.04.041

Pacheco JG, Castro M, Machado S, Barroso MF, Nouws HPA, Delerue-Matos C (2015) Molecularly imprinted electrochemical sensor for Ochratoxin a detection in food samples. Sensors Actuators B Chem 215:107–112. https://doi.org/10.1016/j.snb.2015.03.046

Panhwar S, Hassan SS, Mahar RB, Carlson K, Rajput MH, Talpur MY (2019) Highly Sensitive and Selective Electrochemical Sensor for Detection of Escherichia coli by Using L-Cysteine Functionalized Iron Nanoparticle, Journal of the Electrochemical Society, 166(4):B227–B235. https://doi.org/10.1149/2.0691904jes

Pedrero M, Campuzano S, Pingarrón JM (2017) Quantum dots as components of electrochemical sensing platforms for the detection of environmental and food pollutants: a review. J AOAC Int 100(4):950–961. https://doi.org/10.5740/jaoacint.17-0169

Perçin I, Idil N, Bakhshpour M, Yılmaz E, Mattiasson B, Denizli A (2017) Microcontact imprinted plasmonic nanosensors: powerful tools in the detection of *Salmonella paratyphi*. Sensors (Switzerland) 17(6):1375. https://doi.org/10.3390/s17061375

Perçin I, Idil N, Denizli A (2019) Molecularly imprinted poly(N-isopropylacrylamide) thermosensitive based cryogel for immunoglobulin G purification. Process Biochem 80:181–189. https://doi.org/10.1016/j.procbio.2019.02.001

Pérez-López B, Merkoçi A (2011) Nanomaterials based biosensors for food analysis applications. Trends Food Sci Tech 22:625–639. https://doi.org/10.1016/j.tifs.2011.04.001

Prabowo BA, Wang RYL, Secario MK, Ou PT, Alom A, Liu JJ, Liu KC (2017) Rapid detection and quantification of enterovirus 71 by a portable surface plasmon resonance biosensor. Biosens Bioelectron 92:186–191. https://doi.org/10.1016/j.bios.2017.01.043

Qi P, Wan Y, Zhang D (2013) Impedimetric biosensor based on cell-mediated bioimprinted films for bacterial detection. Biosens Bioelectron 39(1):282–288. https://doi.org/10.1016/j.bios.2012.07.078

Riedel T, Rodriguez-Emmenegger C, de los Santos Pereira A, Bědajánková A, Jinoch P, Boltovets PM, Brynda E (2014) Diagnosis of Epstein-Barr virus infection in clinical serum samples by an SPR biosensor assay. Biosens Bioelectron 55:278–284. https://doi.org/10.1016/j.bios.2013.12.011

Ruiyi L, Fangchao C, Haiyan Z, Xiulan S, Zaijun L (2018) Electrochemical sensor for detection of cancer cell based on folic acid and octadecylamine-functionalized graphene aerogel microspheres. Biosens Bioelectron 119:156–162. https://doi.org/10.1016/j.bios.2018.07.060

Safran V, Göktürk I, Derazshamshir A, Yılmaz F, Sağlam N, Denizli A (2019) Rapid sensing of Cu+2 in water and biological samples by sensitive molecularly imprinted based plasmonic biosensor. Microchem J 148:141–150. https://doi.org/10.1016/j.microc.2019.04.069

Saylan Y, Yilmaz F, Özgür E, Derazshamshir A, Yavuz H, Denizli A (2017a) Molecular imprinting of macromolecules for sensor applications. Sensors (Switzerland) 17(4):898. https://doi.org/10.3390/s17040898

Saylan Y, Akgönüllü S, Duygu Ç, Derazshamshir A, Bereli N, Yılmaz F, Denizli A (2017b) Development of surface plasmon resonance sensors based on molecularly imprinted nanofilms for sensitive and selective detection of pesticides. Sensors Actuators B Chem 241:446–454. https://doi.org/10.1016/j.snb.2016.10.017

Saylan Y, Erdem Ö, Cihangir N, Denizli A (2019) Detecting fingerprints of waterborne bacteria on a sensor. Chemosensors 7(3):33. https://doi.org/10.3390/chemosensors7030033

Shaikh H, Sener G, Memon N, Bhanger MI, Nizamani SM, Üzek R, Denizli A (2015) Molecularly imprinted surface plasmon resonance (SPR) based sensing of bisphenol a for its selective detection in aqueous systems. Anal Methods 7:4661–4670. https://doi.org/10.1039/c5ay00541h

Sheikhzadeh E, Chamsaz M, Turner APF, Jager EWH, Beni V (2016) Label-free impedimetric biosensor for *Salmonella typhimurium* detection based on poly [pyrrole-co-3-carboxyl-pyrrole] copolymer supported aptamer. Biosens Bioelectron 80:194–200. https://doi.org/10.1016/j.bios.2016.01.057

Silva BVM, Rodríguez BAG, Sales GF, Sotomayor MDPT, Dutra RF (2016) An ultrasensitive human cardiac troponin T graphene screen-printed electrode based on electropolymerized-molecularly imprinted conducting polymer. Biosens Bioelectron 77:978–985. https://doi.org/10.1016/j.bios.2015.10.068

Skottrup P, Hearty S, Frøkiær H, Leonard P, Hejgaard J, O'Kennedy R et al (2007) Detection of fungal spores using a generic surface plasmon resonance immunoassay. Biosens Bioelectron 22(11):2724–2729. https://doi.org/10.1016/j.bios.2006.11.017

Starosvetsky J, Cohen T, Cheruti U, Bilanovi D, Armon R (2012) Effects of physical parameters on bacterial cell adsorption onto pre-imprinted sol-gel films. J Biomater Nanobiotechnol:499–507. https://doi.org/10.4236/jbnb.2012.324051

Su CC, Wu TZ, Chen LK, Yang HH, Tai DF (2003) Development of immunochips for the detection of dengue viral antigens. Anal Chim Acta 479(2):117–123. https://doi.org/10.1016/S0003-2670(02)01529-5

Tam PD, Thang CX (2016) Label-free electrochemical immunosensor based on cerium oxide nanowires for *Vibrio cholerae* O1 detection. Mater Sci Eng C 58:953–959. https://doi.org/10.1016/j.msec.2015.09.027

Tancharoen C, Sukjee W, Thepparit C, Jaimipuk T, Auewarakul P, Thitithanyanont A, Sangma C (2019) Electrochemical biosensor based on surface imprinting for Zika virus detection in serum. ACS Sensors 4(1):69–75. https://doi.org/10.1021/acssensors.8b00885

Tang Y, Tang D, Zhang J, Tang D (2018) Novel quartz crystal microbalance immunodetection of aflatoxin B1 coupling cargo-encapsulated liposome with indicator-triggered displacement assay. Anal Chim Acta 1031:161–168. https://doi.org/10.1016/j.aca.2018.05.027

Todescato F, Antognoli A, Meneghello A, Cretaio E, Signorini R, Bozio R (2014) Sensitive detection of Ochratoxin a in food and drinks using metal-enhanced fluorescence. Biosens Bioelectron 57:125–132. https://doi.org/10.1016/j.bios.2014.01.060

Tokonami S, Saimatsu K, Nakadoi Y, Furuta M, Shiigi H, Nagaoka T (2012) Vertical immobilization of viable bacilliform bacteria into polypyrrole films. Anal Sci J Jpn Soc Anal Chem 28(4):319–321. https://doi.org/10.2116/analsci.28.319

Torrance L, Ziegler A, Pittman H, Paterson M, Toth R, Eggleston I (2006) Oriented immobilisation of engineered single-chain antibodies to develop biosensors for virus detection. J Virol Methods 134(1–2):164–170. https://doi.org/10.1016/j.jviromet.2005.12.012

Torun O, Boyacı IH, Temür E, Tamer U (2012) Comparison of sensing strategies in SPR biosensor for rapid and sensitive enumeration of bacteria. Biosens Bioelectron 37:53–60

Uzun L, Say R, Unal S, Denizli A (2009) Production of surface plasmon resonance based assay kit for hepatitis diagnosis. Biosens Bioelectron 24(9):2878–2884. https://doi.org/10.1016/j.bios.2009.02.021

Vaisocherová H, Mrkvová K, Piliarik M, Jinoch P, Šteinbachová M, Homola J (2007) Surface plasmon resonance biosensor for direct detection of antibody against Epstein-Barr virus. Biosens Bioelectron 22(6):1020–1026. https://doi.org/10.1016/j.bios.2006.04.021

Vijian D, Chinni SV, Yin LS, Lertanantawong B, Surareungchai W (2016) Non-protein coding RNA-based genosensor with quantum dots as electrochemical labels for attomolar detection of multiple pathogens. Biosens Bioelectron 77:805–811. https://doi.org/10.1016/j.bios.2015.10.057

Wackerlig J, Schirhagl R (2015) Applications of molecularly imprinted polymer nanoparticles and their advances toward industrial use: a review anal. Chem 88:250–261. https://doi.org/10.1021/acs.analchem.5b03804

Wandemur G, Rodrigues D, Allil R, Queiroz V, Peixoto R, Werneck M, Miguel M (2014) Plastic optical fiber-based biosensor platform for rapid cell detection. Biosens Bioelectron 54:661–666. https://doi.org/10.1016/j.bios.2013.11.030

Wang Y, Duncan TV (2017) Nanoscale sensors for assuring the safety of food products. Curr Opin Biotechnol 44:74–86. https://doi.org/10.1016/j.copbio.2016.10.005

Wang R, Li Y (2013) Hydrogel based QCM aptasensor for detection of avian influenza virus. Biosens Bioelectron 42:148–155. https://doi.org/10.1016/j.bios.2012.10.038

Wang Y, Salazar JK (2016) Culture-independent rapid detection methods for bacterial pathogens and toxins in food matrices. Compr Rev Food Sci F 15(1):183–205. https://doi.org/10.1111/1541-4337.12175

Wang Y, Zhou Y, Sokolov J, Rigas B, Levon K, Rafailovich M (2008) A potentiometric protein sensor built with surface molecular imprinting method. Biosens Bioelectron 24(1):162–166. https://doi.org/10.1016/j.bios.2008.04.010

Wolfbeis OS (2016) Fiber-optic chemical sensors and biosensors. Anal Chem 78:3859–3874. https://doi.org/10.1021/acs.analchem.5b04298

Wulff G (1995) Molecular imprinting in cross-linked materials with the aid of molecular templates—a way towards artificial antibodies. Angew Chem Int Ed Engl 34(17):1812–1832. https://doi.org/10.1002/anie.199518121

Yilmaz E, Majidi D, Ozgur E, Denizli A (2015) Whole cell imprinting based *Escherichia coli* sensors: a study for SPR and QCM. Sens Actuators B Chem 209:714–721. https://doi.org/10.1016/j.snb.2014.12.032

Zelada-Guillén GA, Bhosale SV, Riu J, Rius FX (2010) Real-time potentiometric detection of bacteria in complex samples. Anal Chem 82(22):9254–9260. https://doi.org/10.1021/ac101739b

Zezza F, Pascale M, Mulè G, Visconti A (2006) Detection of Fusarium culmorum in wheat by a surface plasmon resonance-based DNA sensor. J Microbiol Methods 66(3):529–537. https://doi.org/10.1016/j.mimet.2006.02.003

Zhang Q, Zhang D, Lu Y, Yao Y, Li S, Liu Q (2015a) Graphene oxide-based optical biosensor functionalized with peptides for explosive detection. Biosens Bioelectron 68:494–499. https://doi.org/10.1016/j.bios.2015.01.040

Zhang Z, Guan Y, Li M, Zhao A, Ren J, Qu X (2015b) Highly stable and reusable imprinted artificial antibody used for in situ detection and disinfection of pathogens. Chemical 6:2822. https://doi.org/10.1039/c5sc00489f

Zhao Y, Wang H, Zhang P, Sun C, Wang X et al (2016) Rapid multiplex detection of 10 foodborne pathogens with an up-converting phosphor technology-based 10-channel lateral flow assay. Sci Rep 6:21342. https://doi.org/10.1038/srep21342

# Nanomaterials as Toxic Gas Sensors and Biosensors

**Jaison Jeevanandam, Abirami Kaliyaperumal, Mohanarangan Sundararam, and Michael K. Danquah**

## Contents



J. Jeevanandam
Faculty of Engineering and Science, Department of Chemical Engineering, Curtin University, Miri, Sarawak, Malaysia

A. Kaliyaperumal
Center for Environmental Studies, Anna University, Chennai, Tamil Nadu, India

M. Sundararam
Crystal Growth Centre, Anna University, Chennai, Tamil Nadu, India

M. K. Danquah (✉)
Chemical Engineering Department, University of Tennessee, Chattanooga, TN, USA
e-mail: michael-danquah@utc.edu



Inamuddin, A. M. Asiri (eds.), *Nanosensor Technologies for Environmental Monitoring*, Nanotechnology in the Life Sciences,
https://doi.org/10.1007/978-3-030-45116-5_13

## Abbreviations

| | |
|---|---|
| $Al_2O_3$ | Aluminium oxide |
| AuNPs | Gold nanoparticles |
| BRET | Bioluminescence resonance energy transfer |
| CdSe | Cadmium selenide |
| CEA | Carcinoembryonic antigen |
| $CeO_2$ | Cerium dioxide |
| CHS | Chondroitin sulfate |
| CNMs | Carbon nanomaterials |
| CNTs | Carbon nanotubes |
| CO | Carbon monoxide |
| CPE | Carbon paste electrode |
| CRET | Chemiluminescence resonance energy transfer |
| CTCs | Circulating tumor cells |
| CV | Cyclic voltammetry |
| DNA | Deoxyribonucleic acid |
| EIS | Electrochemical impedance spectroscopy |
| EMF | Electromotive force |
| FETs | Field effect transistors |
| Gly | Glyphosate |
| GO | Graphene oxide |
| GOx | Glucose oxidase enzyme |
| GR | Graphene |
| $H_2O_2$ | Hydrogen peroxide |
| $H_2S$ | Hydrogen sulfide |
| Hb | Hemoglobin |
| hCG | Human chorionic gonadotropin |
| HOPG | Highly oriented pyrolytic graphite |
| IL-6 | Interleukin-6 |
| LbL | Layer-by-layer |
| MgO | Magnesium oxide |
| $MnO_2$ NPs | Manganese dioxide NPs |
| MNPs | Metal nanoparticles |
| MWNTs | Multi-walled carbon nanotubes |
| NADH | Nicotinamide–adenine dinucleotide |
| $NH_3$ | Ammonia |
| NiOx | Nickel oxide |
| NIR | Near infrared |
| NO | Nitrogen oxide |
| NPs | Nanoparticles |

| | |
|---|---|
| PAH | Poly (allylamine hydrochloride) |
| PAni | Polyaniline |
| PEDOT | Poly (3,4-ethylenedioxythiophene) |
| PEG | Polyethylene glycol |
| PEI | Polyethyleneimine |
| PNA | Peptide nucleic acid |
| POC | Point-of-care |
| PPy | Polypyrrole |
| PSA | Prostate-specific antigen |
| PSS | Poly (styrene sulfonate) |
| PSS | Polystyrene sulfonate |
| Pt | Platinum |
| PTh | Polythiophene |
| PVP | Polyvinylpyrrolidone |
| QCM | Quartz crystal microbalance |
| QDs | Quantum dots |
| RNAs | Ribonucleic acid |
| SAW | Surface acoustic wave |
| $SnO_2$ | Tin dioxide |
| SPANI | Poly (anilinesulfonic acid) |
| ssDNA | Single-stranded DNA |
| STW | Surface transverse wave |
| SWNTs | Single-walled carbon nanotubes |
| $TiO_2$ | Titanium dioxide |
| TMF-α | Tumor necrosis factor-α |
| USD | United States Dollars |
| UV | Ultraviolet |
| VOCs | Volatile organic compounds |
| $WO_3$ | Tungsten oxide |
| ZnO | Zinc oxide |
| ZnS | Zinc sulfide |

# 1 Introduction

Sensors are the general term that is used for the materials that are used to detect and sense a physical parameter and converts them to electrical current (Pallas-Areny and Webster 2012). Sensors consist of four main parts, namely analyte or an element to be detected, receptor, transducer, and signal processing unit (Töppel et al. 2018). The analyte or the molecule to be sensed is the key for the fabrication of sensors, based on which the type of sensors to be used for detection will be decided (Ajay Piriay et al. 2017). The transducer is an essential component of a sensor, which converts the physical quantities into electrical signals (Kocakulak and Butun 2017). These converted electrical signals are amplified and converted into readable signals via signal processors, which can be displayed in a digital device (Erden et al. 2016). There are numerous types of sensors that are under extensive research and have been used in several applications (Khaydukova et al. 2017). Among these sensors, gas sensors are the widely utilized sensor type in several industries as well as in commercial markets (Dey 2018). In 2016, the overall market of gas sensors throughout the world is about USD 812.3 million, which is expected to be about USD 1297.6 million in 2023 (Bogue 2014). The expected growth in the gas sensor market between 2017 and 2023 is about 6.83%, which is based on the gas type, technology, geography, and end-use application (Markets 2017). Companies such as Amphenol Corporation (United States) and Figaro Engineering Inc. (Japan) are the significant producers of gas sensors for end-use applications including aerospace, medical, transportation, and residential safety (Kato et al. 2000; Shan et al. 2018).

Generally, micro-sized particles are utilized for gas sensor fabrication to monitor distinct gas molecules that are toxic to humans above its threshold limit value (Lupan et al. 2016). These micro-sized particles exhibited enhanced potential in sensing various toxic gases and help to avoid their exposure toward humans (Acosta et al. 2009). However, the lack of precision in the detection, disability in detecting multiple gases, and formation of impurities during the detection of gases are the limitations of using micro-sized particles for detection of gases (Carregal-Romero et al. 2013; Kim and Kim 2014). Thus, nano-sized particles and materials are introduced as novel materials to fabricate gas sensors (Llobet 2013). These nanomaterials possess exclusive size-dependent entities with an elevated ratio of surface and volume as well as effects due to confinement in the quantum regime (Eranna 2016). Additionally, the size, morphology, and surface charge of the nanosized materials can be manipulated based on the desired sensor applications (Lyson-Sypien et al. 2015). These advantages of nanomaterials pave the way for a separate market for nanomaterial-based toxic gas sensors which are now gaining attention among industries and researchers (Šutka and Gross 2016).

The emergence of nanotechnology has led to the fabrication of biosensors, similar to toxic gas sensors (Pandey et al. 2008). These biosensors are the latest trend in biomedical sciences which is beneficial in detecting any form of biological analytes (Rai et al. 2012). The advantages of nanomaterials in toxic gas sensors have made the biomedical scientists to try out nano-sized particles for the fabrication of biosensors (Mishra and Rajakumari 2019). The incorporation of nanomaterials in biosensors has led to various wearable and in-situ biochip advancement that can help in the detection of biological analytes and in the diagnosis of diseases (Chandra and Segal 2016). Currently, the global market of biosensors is about 21.2 billion USD, which is expected to reach about 31.5 billion USD in 2024 (Markets 2019). Due to the high demand for novel toxic gas and biosensors, numerous nanomaterials are introduced to fabricate sensors with high detection efficiency (Mehrotra 2016).

Metal, metal oxide, carbon and polymer nanoparticles, either in nanoformulated or nanocomposite forms are the nanomaterials, that are usually utilized for sensor fabrication in recent times. The diverse variety of nanomaterials used in the sensor fabrication has led to the development of semiconductor (Dey 2018), sensitive (Wang et al. 2016a), hybrid (Chatterjee et al. 2015), fluorescent (Zhou et al. 2016), room temperature (Shafiei et al. 2015), and impedimetric (Vignesh et al. 2015) sensors to detect toxic gases. Likewise, microbial (Ponamoreva et al. 2019), electrochemical (Rotariu et al. 2016), surface plasmon resonance (Olaru et al. 2015), whole-cell based (Saini et al. 2019), and lab-on-a-chip (Jamshaid et al. 2016) are the distinct biosensors that are fabricated using nanomaterials. Thus, the aim of the present chapter is to discuss about nanomaterials that are used to fabricate sensors with the ability to detect toxic gases and biological molecules. In addition, the drawbacks of nanomaterials as sensors, and the efficiency and limitations of green synthesized nanomaterials for sustainable, toxic gas detection, and biosensing applications are also discussed.

# 2 Nanomaterials as Toxic Gas Sensors

Nanomaterials are widely utilized to fabricate swift-responding nanosensors with electrochemical, optical, thermal, piezoelectric, and other properties for low concentration chemical compound detection (Tallury et al. 2010). Recently, nanomaterial-enhanced sensors such as carbon nanotubes, nanosized metal particles (particularly gold and silver), graphene, semiconductors, quantum dots, and silicon-based nanomaterials were tailored for detecting and measuring contaminants such as ions of heavy metals, hydrogen peroxide ($H_2O_2$), organophosphate pesticides, toxic gases, and industrial wastewaters (Su et al. 2012). Metals, oxides of metals, metal complexes, polymers, and carbon-based nanomaterials are extensively under research for the fabrication of toxic gas sensors.

## *2.1 Metal and Metal Oxide Nanomaterials*

### 2.1.1 Metal Oxide Nanomaterials

Numerous metal oxide nanomaterials are reported to be appropriate for reducing, combustible, or oxidizing gas detection via measurements of modifications in conductivity. It has been proven that the oxide nanomaterials show a sensitive conductivity response upon detecting a gas molecule (Kanazawa et al. 2001). It is noteworthy that the choice of metal oxides for fabricating a gas sensor is based on their electronic structure. The oxides of metals possess an inclusive electronic structure range that is classified into oxides of transition metals and non-transition metals, which are further subclassified into oxides of pre-transition metals and post-transition metals. Oxides of pre-transition metals, such as MgO, are naturally inert, due to their hefty gaps in their electronic bands, which makes it tedious to form electrons or holes. These materials are infrequently designated to fabricate gas sensors as it is hard to measure their electrical conductivity (Kanazawa et al. 2001). The oxide materials of post-transition metals can alter their properties to perform as a better toxic gas sensors. Thus, these materials are reported to be highly sensitive than oxides of pre-transition metals toward the environment. Nevertheless, instability in structures and difficulties in optimizing parameters of these nanomaterials limits their usage in conductometric gas sensor applications. Only oxides of transition metals with electronic configurations of d0 and d10 are employed in real gas sensor application. The configuration of d0 is present in binary oxides of transition metals, whereas the configuration of d10 exists in the oxides of post-transition metals. Even though, several semiconductor oxides of metals that are delicate to noxious gas molecules are reported to be n-type semiconductors. Also, it was demonstrated that certain p-type semiconductors are also utilized as materials to fabricate gas sensors. It is reported that about 10% content of NiOx is required to alter n-type into p-type conductivity. When temperature elevates, an increase in the sensitivity of n-type film toward reducing gases can be noted, which is vice versa in p-type NiOx film (Wisitsoraat et al. 2009). Thus, semiconductors of p-type nature can operate in a comparatively lesser temperature than n-type, for toxic gas sensor applications.

Sensors that are fabricated with semiconductors of metal oxides are primarily applied to sense-gas molecules via target gas and surface oxygen mediated redox reactions (Yamazoe and Shimanoe 2008). This process requires two initial steps, namely (1) redox reactions, where scattered oxygen ions over material surface will react with target gas molecules to create electronic variations in the surface of oxides; and later (2) the disparity is transferred via transduction into a noticeable change in the electrical resistance of sensors. The variations in resistance can be noticed by determining the capacitance alterations, function of work function, optical characteristics, mass, or reaction energy (Kanan et al. 2009).

Tin dioxide is a noteworthy metal oxide nanoparticles that are used for toxic gas detection. It is a granular n-type material with density-dependent electrical conductivity of surface pre-adsorbed oxygen ions. It is reported that the tin dioxide resistance alters, depending on the variations in the gas concentration (Batzill and Diebold 2005; Wang et al. 2010), with nonlinear association between concentration and resistance of target gas. Also, other semiconducting oxides of metals such as tungsten trioxide are generally utilized for noxious gas detecting applications. Oxides of tungsten nanomaterials as anode that are fabricated via electrochemical tungsten etching illustrate outstanding hydrogen and nitrogen oxide detection responses (Endres et al. 1996). However, the pure $WO_3$ respond poorly toward $NH_3$ with declined selectivity, due to the interference of NOx. Thus, vanadium and copper decorated $WO_3$ are utilized as additives of catalysts to recover their response, eliminate anomalous sensor performance and to utilize $WO_3$ in gas sensors (Hoefer et al. 1997). Other nanosized metal oxides ($TiO_2$) are highly beneficial as layers of sensitivity to elevate dielectric permittivity of sensors to adsorb gas molecules

(Fraiwan et al. 2011). Likewise, distinct ZnO nanostructures including nanowires, nanobelts, and tetrapods are fabricated as monitoring devices for ultrasensitive $H_2S$ and NO gas detection (Gupta et al. 2010).

Moreover, semiconducting sensors of metallic oxides that are fabricated using nanomaterials are gaining importance in large-scale applications. However, elevated operational temperature demand in certain sensor involves high budget and intricate arrangements than other room temperature sensors, which limits their large-scale application. Researchers have developed a unique method to overcome these limitations, which includes silicon-based integrated circuit fabricated microsensor mediated micro-heater technology (Lee et al. 1996). Additionally, pulsating operating temperature mode to heat at short intervals (Jaegle et al. 1999) was also introduced, which simplifies the sensor operation and reduces the consumption of power. Further, studies about nanosized oxides of metal structures have revealed that semiconductor nanowires can recover the response time and sensitivity of gas sensors (Comini 2006). Another challenge is the requirement of extended recovery period after exposure of each gas, which restricts their employment in certain e-nose like sensor devices, as the concentration of gas is rapidly and regularly under modifications. Moreover, the defects and instability in the structure of indicators also reduce the applications in the field of sensors. In recent times, there are numerous reports to prove the efficiency of metal oxide nanomaterials in detecting toxic gas molecules with high sensitivity (Dey 2018). However, novel solutions are required to overcome the limitations and defects of semiconducting oxides of metals, which is achievable via extensive sensor research in the future.

### 2.1.2 Metal Nanoparticles

Nanosized metal particles are deposited over substrate surface for elevating the ratio between area and volume and also to favor the adsorption of gas molecules. When these nanomaterials are bound with analyte, the molecules of analytes alter their substrate entities as gas molecules are adsorbed to the metal (Jimenez-Cadena et al. 2007). Mostly, the preparation for deposition is performed via metal precursor vaporization, to afford consequent annealing of the nanofilms or particles. For instance, an electromotive force (emf) measuring electrode in a cell with high concentration of oxygen was fabricated using a zirconium and yttrium metal component along with 5–10 nm-sized iridium nanoparticles. In the existence of oxygen at 650 °C, the sensor conductivity was altered within 10–20 s of response time. The sensor response was explained via oxidization reaction, where the nanosized iridium particles permit a better interface between the substrate and the molecules to provide an enhanced contact area for the analyte (Kimura and Goto 2005). Li et al. (2002) fabricated an electrode of platinum with surface embedded nanosized gold particles along with a silver–silver chloride composite as a reference electrode. The nanosized metal particles are reported to be beneficial as a layer of sensors to permit the detection of hydrogen and sulfate analytes with ultrahigh sensitivity and selectivity at low temperature (Li et al. 2002). Moreover, gold and platinum nanoclusters are widely accepted as a catalyst to upsurge sensor sensitivity. Generally, nanoclusters are integrated or fabricated into alloys with distinct nanomaterials to enhance their sensitivity, selectivity, and to increase their kinetic oxygen reduction limitation (Hallil et al. 2010). Further, Wang et al. (2010) prepared an extremely sensitive hydrogen gas nanosensor using platinum nanocluster-decorated graphene oxide (GO) between a prepatterned pair of titanium–gold electrodes with microgap (Xiang et al. 2010). Currently, there are several nanomaterials, especially biosynthesized nanoparticles, are employed in the fabrication of toxic gas nanosensors (Xu et al. 2018b). It is noteworthy that most of the studies reported the utilization of metal oxide nanomaterials, rather than a nano-metals, to fabricate substrates of nanosensor to detect toxic sensors. However, the utilization of nanosized metal particles on the substrate with inert molecules to fabricate noxious gas nanosensors should be studied extensively in the future.

### 2.1.3 Nanosensors with Metal Complexes

Transition metal nanomaterials are proficient in modifying interactions of atoms which makes them to be extensively beneficial as receptors to detect and analyze various gases. This behavior is subjugated to prepare an analyte with complex supra-molecules for selectivity enhancement of numerous toxic gas sensors. Elosua et al. (2006) stated that optical fiber sensor fabricated by coating gold and silver complex was used to detect volatile alcoholic compounds. Nanoscale Fizeau interferometer with complex alcoholic vapor dopants is proved to possess vapochromic property to act as recognition layer of the nanosensor. The solvents coordinate with centers of silver metal and break the silver and gold–silver bonds to provide initial orange to red color to the complex. The results revealed claim the nanosensor with metal nanoparticles can be used for at least 3 months without degradation for the effective detection of toxic gases (Elosúa et al. 2006).

Brousseau et al. (1997) revealed carbon dioxide detection using reactions of carbon dioxide with hydroxysilanes, alcohols, and amines under optimum pressure and temperature. The study utilized three bifunctional nano-receptors along with a group of amines that coordinates with ions of metal in the copper octane di-yl bis (phosphonate) thin film and are coated over a microbalance made up of quartz crystal. The sensor and the analyte reaction at optimum temperature were reversible which is highly dependent on the receptor and the concentration of carbon dioxide (Brousseau et al. 1997). The primary benefit of metallic sensing layer complexes as sensing layer is the interactions that are reversible between the devices and the analytes. The formation of coordination bonds in gas detection are fragmented by temperature increment or chemical alterations in the sensor. Besides, specific receptors are intended to cooperate with precise analytes to increase their selectivity. Thus, fabrication of sensors with these complexes is expensive and is used only to functionalize specific receptors that can be beneficial as nanostructured transducer materials (Jimenez-Cadena et al. 2007).

## 2.2 *Polymer Nanomaterials*

Polymers are used in gas sensor fabrication as they possess exceptional physicochemical properties and are classified into polymers with conducting and nonconducting properties.

### 2.2.1 Conducting Polymers

Polymers with electric conductive property are broadly considered as toxic gas sensors due to their sensitive electrical conductivity alterations via assorted inorganic and organic gas exposure. This property has led to critical material examination for fabricating layers that can detect gas molecules in sensors (McNaghten et al. 2009; Shrivas et al. 2007). Polymers with conducting property such as polyaniline (PAni), polypyrrole (PPy), polythiophene (PTh), and their byproducts are widely employed as noxious gas determining nanosized materials. It can be noted that pure polymers with low conductivity to perform as an individual gas determining material. Hence, extended research to select specific polymers for nanocomposite fabrication with metallic nanoparticles is highly recommended in the future. Earlier reports demonstrated that polymer conductivity is upgraded via exclusive doping approach by reactions of protonation or redox. Later, polymers are converted into conductors or semiconductors after the reversible doping process. It is significant to note that the level of doping can be altered via chemical-based target gas and polymer reactions, making the analyte detection to be a more practical with conducting polymers. Several polymers are doped through redox reactions and specific polymers are utilized for gases, which are inactive at room temperature to redox reactions. For instance, redox reactions will not happen in CO at room temperature, however, PAni can lead to a change in their redox potential (Hatfield et al. 1994). Thus, conducting polymer-based nanomaterials are directly used as transducers in toxic gas detecting nanosensors to reflect electrical property modifications.

### 2.2.2 Nonconducting Polymers

Polymers without conducting property are broadly employed as absorptive coatings on diverse sensor devices, where general polymeric transducers with distinct physisorption are used as sensor fabricating material. For example, layers of polymer that lead to resonance frequency modifications, enthalpy, and dielectric constant upon analyte desorption or absorption are coated on the surface of surface transverse wave (STW), quartz crystal microbalance (QCM), calorimetric sensor devices, capacitive and surface acoustic wave (SAW) for sensitive toxic gas detection. Later, these sensor devices are transformed into an output electrical signal via monitored polymer properties (Bai and Shi 2007). Even though the basic polymer with non-conducting principles are logical in gas sensors, their recital is highly intricate, even after coating them on the sensor devices. For example, STW devices with resonance property are coated with sensitive thin nanosized layers of polymer to feature extensive rewards of relative sensitive gas probing, inclusive electrical property, and low sensor oscillator noise than bulk counterparts of SAW (Hagleitner et al. 2002). Furthermore, membrane of polymer with nonconducting properties was also proved to be utilized in semiconducting oxides of metal-based gas sensors as sieves of molecules, to augment their inclusive selectivity for sensitive layers of polymer introduction (Avramov et al. 2000).

Numerous approaches have been reviewed for polymer nanocomposite fabrication with exclusive properties for sensor applications (Kaushik et al. 2015). Among several approaches based on emulsion polymerization such as conventional emulsion, soapless emulsion, microemulsion, suspension, dispersion, and precipitation polymerization approaches (Wong et al. 1995), mini-emulsion polymerization is considered as a potential approach to fabricate functional and flexible nanomaterials (Anderson and Daniels 2003). Nohria et al. (2006) utilized a thin film of poly (anilinesulfonic acid) (SPANI) to construct a sensor to evaluate humidity. The 90 nm sized thin films were deposited via the spin coating method or by the layer-by-layer (LbL) nanoparticle assembly approach, which added negative charges on the coupled layer polycation substrate and a polyanion named poly (allylamine hydrochloride) (PAH), and poly (styrene sulfonate) (PSS), respectively. After the deposition of layers, the PSS is substituted by SPANI to yield a film with 26 nm of thickness. When these nanofilm sensors are under atmospheric exposure, its resistance declines with an increase in relative humidity along with 15–30 s of response time (Nohria et al. 2006).

Nanosized gas sensors with polymers have merits such as enhanced sensitivities at reduced response times, compared to bulk polymers. Furthermore, sensors with polymers function at room temperature, which make them superior than metal oxide nanoparticles as they detect toxic gases at elevated temperatures. Therefore, the lower consumption of energy by nano polymers permits their utilization in toxic gas detection units that are operated by batteries. Additionally, advantages including the low fabrication cost, simple structures, and portability, as well as reproducibility (Landfester 2006) make these nanostructures to gain focus of researchers to use them in nanosensor fabrication. Gas sensors with polymers also possess hindrances such as instability, poor selectivity, and irreversibility. Further, the performance can be altered due to the working environment. Besides, there are only few evidences to explain the actual working principle of toxic gas-sensing polymers. Despite these demerits, polymer-based nanosensors can be used as a lower power-consuming toxic gas sensors in the future.

## 2.3 *Carbon-Based Nanomaterials*

Carbon nanotubes (CNTs) and graphene show a high potential for miniaturized chemical sensor development, due to their excellent large surface-mediated adsorptive capacity, better electrical property modulation upon exposure to analytes via greater cross-sectional interaction zone, capability to fine-tune electrical nanostructure entities by tailoring their composition or size and the configuration ease into distinct geometries (Goustouridis et al. 2005; Munoz and Steinthal 1999). SWNTs are cylinders with individual diametric layers of 1–3 nm and micron length of rolled graphite. Similarly, multi-walled carbon nanotubes (MWNTs) are fabricated via concentric SWNTs with various nanosized diameters. Both SWNTs and MWNTs are employed to fabricate various sensors, due to their exclusive physico-electronic properties (Ellis and Star 2016). The adsorption processes are highly favored due to their unique atomic arrangement on SWNTs surface and their enhanced ratio between area and volume, which increases their sensitivity to various gas molecules in the atmosphere (Zaporotskova et al. 2016). Nanotubes can exhibit both metallic and semiconductor properties with their extraordinary electrical conductivity which depends on their structural chirality (Kavitha and Kalpana 2017). Semiconducting nanotubes are highly beneficial in device construction including field-effect transistors (FETs). These nanotube devices are utilized for analyte determination via electrical surface conductivity modifications of CNTs (Tran and Mulchandani 2016). This can lead to consequences such as, transfer of charge from analyte to the nanotube, or may elevate the scattering potential of analyte (Park et al. 2017a). These exclusive effects of CNTs are utilized for electrode of third gate-mediated FET device modulation (Jang et al. 2016). Alternatively, the current will decline without altering the CNTs characteristics, if the center of scattering is an analyte. Even though CNTs possess sensitivity toward the surrounding chemical environment, they lack selectivity which is a major limitation to use them in large-scale sensor application. Thus, several functionalization processes have been implemented for sensitivity and selectivity enhancement of CNTs as gas sensors. Among functionalization processes, distinct polymeric material coating and doping or oxides and metal particle deposition are proven to be significant in improving the gas detection properties of CNTs at room temperature (Lee et al. 2018).

Gas sensors with graphene and pristine CNTs have firm restrictions, such as low analyte sensitivity, declined selectivity, irreversibility, and long recovery time (Bekyarova et al. 2004). Graphene and CNTs functionalized with diverse materials are essential to overcome the limitations of free CNTs via chemical property modifications and sensing performance enhancements. Functionalization is significant in altering the chemistry of carbon nanotubes and graphene and manipulating their chemical properties is highly essential for utilizing them in potential applications including sensor fabrication (Dong et al. 2004). There are numerous CNTs and graphene functionalization approaches that are reported in literatures (Wright 2017) such as defect method, covalent sidewall functionalization approach, noncovalent exohedral, and endohedral functionalization. CNTs functionalized with metal, metal oxides, and polymer nanoparticles can result in improved electronic properties, selectivity, and response to specific gases through target molecule interface with functional additives is distinct, compared to free and bulk carbon materials (Li et al. 2011).

### 2.3.1 Carbon Nanotubes Functionalized with Metal, Metal Oxide, and Polymer Nanomaterials

The conductivity of nanotubes can be altered significantly with sidewall functionalization of chemical reactants, which is essential to specific gas sensor application. Various approaches have been reported for covalent carbon nanotube functionalization such as creation of defect site and functionalization via defects, embedding acids of carboxyl group on caps of carbon nanotube end and following acid derivatization. Polymer surfactant nanotube wrapping for noncovalent functionalization is proven to reserve the physical entity and solubilization of nanotubes. Pure carbon nanotubes do not possess appreciable sensitivity toward certain gases, whereas CNTs decorated with nanoparticles comparatively improves gas-sensing performance. Nanosized metal particles including platinum, lead, aluminium, and tin are widely utilized to decorate CNTs, allowing determination of selective gases, such as $H_2$, $NH_3$, $NO_2$, and CO (Fowler et al. 2009; Li et al. 2010; Stankovich et al. 2006; Yan et al. 2010). The oxide of metal NPs modified with CNT as hybrid was developed as sensing films that displayed advantages such as high catalytic activity, efficient charge transfer, and capacity to adsorb. Numerous studies have demonstrated the brilliant sensing ability of CNTs/$SnO_2$ and CNTs/ZnO hybrid sensors for carbon monoxide, nitrogen dioxide, and ammonia gas detected at low temperatures (Ma et al. 2012). Polymer-functionalized CNTs also considered as an essential nanomaterial to recover the CNTs and polymers of organic nature's compatibility in sensor fabrication and provide ultrahigh sensitivity and selectivity for the determination of gases. The unique characteristics of CNTs combined with polymer properties such as delocalized bonds, high permeability, and low density have established the possibility of detecting diverse gas molecules with swift response, excellent sensitivity, and great reproducibility (Alshammari et al. 2017). In addition, various other CNT modifications via doping or coating are recommended to further augment the CNT's gas-detecting properties (Gardner and Bartlett 1993; Penza et al. 2008).

### 2.3.2 Graphene Functionalized with Metal, Metal Oxide, and Polymer Nanomaterials

In several gas-sensing applications, metal nanoparticles are embedded with graphene to upsurge the sensitivity, detection limit, selectivity, or a mixture of all these properties (Kim 2009). In several cases, graphene nanomaterials are modified via electrochemical metal salt reduction by using graphene flakes that are attained from oxide of graphene. In specific scenarios, the deposition of nanosized particles over graphene can be achieved via chemical metal salt reduction by adding reducing agent, followed by that are yielded after adsorption of nanosized particles in solution (Liu et al. 2011). Noble metals that are active as catalysts are extensively utilized to elevate graphene-based chemical sensor sensitivity toward a wide range of gases. Lately, the advancement of room temperature-operating sensors with oxides of metal along with enhanced sensitivity and truncated fabrication cost has fascinated much consideration. When molecules of gases are adsorbed on the surface of functionalized oxides of metal with graphene sensor film, nanosized oxide of metal particles act as sensing and element of transduction ability, whereas oxide of graphene act as a mesh with high conductivity. These nanomaterial-based sensor intensifies their transducing property resulting in larger alterations in conductance, compared to previous results demonstrated for chemical synthesized graphene-based sensors (Bekyarova et al. 2004). Numerous literatures also reported that the graphene is incorporated into matrices of polymer to yield novel nanosized composites. It was recommended that graphene nanosheets can provide highly active polyaniline nucleation sites and exclusive electron transfer pathways (Yuan and Shi 2013).

## *2.4 Nanocomposites*

Generally, gas sensors are fabricated by two types of materials such as organic polymers with conductivity and inorganic oxides of metals. Gas sensors using coveted functional and conductive organic conducting polymers are proven to improve gas-detecting performance (Castro et al. 2011). Although these nanocomposites are unstable and exhibit comparatively meager sensitivity (Janata and Josowicz 2003) due to excellent polymers with conducting and affinity property toward environmental moisture and volatile organic compounds (VOCs). Gas detectors fabricated via oxides of inorganic metals exhibited upsurge detecting qualities due to oxygen-mediated stoichiometric alterations and surface-active electrical charges (Capone et al. 2000). The function of these devices at higher temperatures led to gradual modifications in metal oxide nanostructure properties. The high-temperature function can lead to grain boundary fusion, which can alter nanostructure stability and shorten the sensor lifetime. In addition, high-temperature operating nanocomposite sensor needs a discrete assembly of temperature-controlled heating complex and consumes extra heating power. The shortcomings of organic materials such as low stability and conductivity, and inorganic materials such as functioning at high temperatures and sophisticated processability act as a hurdle to be employed in large-scale gas sensor fabrication. Thus, the nanocomposites are utilized to promote effective and peculiar gas-sensing ability and allow them to operate at low temperature. A promising approach for conductometric gas sensor improvement is to utilize composite nanomaterials that are fabricated via semiconducting metal oxide or metal or carbon nanoparticles along with organic polymer or inorganic matrix (Barsan et al. 2007).

Suri et al. (2002) described that nanosized magnetic composites possess a significant part in noxious gas sensor applications. Nanosized composites of iron oxide and poly pyrrole that are prepared via instantaneous polymerization and gelation process were utilized for the fabrication of sensors to determine humidity. The results revealed that these sensors are highly beneficial in toxic gas determination such as carbon dioxide, nitrogen, and methane. Further, the study also stated that the sensitivity of nanosensor was in the order of carbon dioxide > nitrogen > methane, due to their kinetic diameter variation of gas molecules (Suri et al. 2002). Moreover, various methods are utilized to combine CNMs with polymers to yield functional nanocomposites with specific extraordinary properties for technological purposes (Yu et al. 2017). Polyaniline (PANI), polypyrrole (PPy), polythiophene (PTh), and poly(3,4-ethylenedioxythiophene) (PEDOT) conducting polymers are exploited as matrices to integrate various CNMs (Xue et al. 2017). The polymer matrices embedded with carbon nanomaterials are an attractive approach for combining their electrical and mechanical entities (Liu and Kumar 2014). These novel nanosized composites created novel chances, not only in sensor applications, but also in electrochemical capacitor, solar cells, transistors, and molecular electronic devices (Christ et al. 2017). In recent times, nanosized composite with carbon polymers, and nanosized metal particles (MNPs) with unique alignments and proportions are broadly examined (Kondratiev et al. 2016; Monerris et al. 2016). Further, Trakhtenberg et al. (2012) reported numerous metal oxide nanocomposites that are utilized to fabricate efficient nanocomposites for the sensitive and selective determination of noxious gases in the environment at room temperature (Trakhtenberg et al. 2012). All these studies revealed that nanocomposites can be an excellent nanosensor fabrication material for gas sensor applications, compared to free nanoparticles.

# 3 Nanomaterials as Biosensors

Similar to toxic gas sensors, biosensors are also fabricated using metal, metal oxide, carbon, polymer, and composite nanomaterials.

## *3.1 Metal and Metal Oxide Nanomaterials*

The noble metals including gold, silver, and platinum (Pt) are commonly utilized in the field of biosensor application. These metal nanoparticles (MNP) are significant in detecting microorganisms such as viruses, bacteria, and pathogens with enhanced sensitivity and specificity. These MNPs incorporated biosensors with biological recognition receptors namely antibody, enzyme, nucleic acid, and cells are also utilized for formidable disease monitoring applications such as cardiovascular and cancer diseases. The MNPs act as mediators to transfer the signals that are obtained via modification on their surface to transducers in the form of electrochemical, piezoelectric, and optical signals. The gold NPs are in conjugation with primary or secondary antibodies through gold–antibody ionic interaction, hydrophobic interaction, and dative binding phenomena. Conjugation of nanoparticles with antibodies is achieved via other chemical interactions such as thiol derivative chemisorption, bifunctional linkers, and via adapter molecules namely Streptavidin and biotin (Ijeh 2011; Ljungblad 2009).

### 3.1.1 Glucose Biosensors

The MNPs are conjugated with a glucose oxidase enzyme (GOx) and ferrocyanide for the glucose sensor signal enhancement. The most popular techniques for conjugation are amperometry, electrochemical impedance spectroscopy (EIS), and cyclic voltammetry (CV). The nanosized gold particles are attached covalently to GOx and CV measurement is used to obtain the quantitative biological glucose level in complex samples. Similarly, the ferrocyanide acts as a mediator for transferring electrons during GOx reaction for glucose detecting nanosensors. A typical example of the fabrication of the glucose sensor with nanosized gold electrode particles was demonstrated by Zhang et al. (2005) to detect highly concentrated with enhanced sensitivity and 8.2 μM limit of detection (Zhang et al. 2005). Likewise, nanosized platinum particles are incorporated with GOx and Nafion on the highly oriented pyrolytic graphite (HOPG) surface to fabricate sensors with stability, controllability, and reproducibility. The results showed that these nanosensors exhibited a linear response (25 μM) with 15 mM limit for determining glucose (Liu and Huang 2012). Similarly, manganese dioxide NPs ($MnO_2$ NPs) were also conjugated with Nafion and GOx on the graphene nanoribbon surface under the influence of 0.1 M phosphate buffer at pH 7.4. The amperometric glucose measurement for the nanosensor prepared with $MnO_2$ NPs at an operating potential of +0.50 was reported to be as high as 56.32 μA/mmol $cm^2$ (Vukojević et al. 2018). However, the detachment of GOx from the Nafion is common in the preparation, which is the major limitation for glucose biosensor fabrication. Thus, these challenges can be avoided by treating the initial electrode with positively charged polyethyleneimine (PEI) polymer, which was followed by adding 5% (w/v) of negatively charged Nafion on the PEI surface. Later, GOx has to be attached to the PEI/Nafion layer to initiate strong bond formation between them (Tsiafoulis et al. 2005). Furthermore, selenium nanoparticles are conjugated with mesoporous silica composite (MCM-41) matrix. In this study, the carbon paste electrode (CPE) was bound together with mesoporous silica composite material which was later attached to selenium nanoparticles to embed on the MCM-41 surface. Since GOx reaction liberates the electron, the amperometric measurement will yield the quantity of glucose present in the sample by slight modifications in the surface of MCM-41 that are initiated by selenium metal (Yusan et al. 2018).

### 3.1.2 DNA Biosensors

Nanosized gold and silver particles are widely utilized in DNA biosensor fabrication to detect the target DNA molecules. Metallic nanoparticles possess the ability to bind with target DNA molecule at 0.1 M of sodium chloride concentration, due to the salt aggregation effect. Further, an increment in the salt concentration to the 2 M of sodium chloride concentration will lead to aggregation of ssDNA molecules with NPs (Jamdagni et al. 2016). On the other hand, the peptide nucleic acid (PNA) and PNA–DNA bound with NPs are measured at 600 nm and 520 nm, respectively, in a UV–visible spectroscopy, which indicates the colorimetric DNA sensing ability of gold–silver nanoparticles (Kanjanawarut and Su 2009). The alterations in the DNA motif via gold NP are based on the pH variation in fluorescent quenching method (Liu et al. 2006). In this scenario, the DNA hybridization determination is possible via gold nanoparticles encapsulated by streptavidin and biotinylated oligomer target by stripping potentiometric method (Madhurantakam et al. 2018). Additionally, gold NPs linked with cysteamine-modified electrode that are encapsulated with oligonucleotide and a group of mercaptohexyl at the 5′-phosphate of DNA end and chitosan layer onto SAM-modified gold NPs are highly significant in electrochemical biosensor fabrication to detect target DNA (Mazloum-Ardakani et al. 2015).

#### 3.1.3 Blood Pressure Sensors

A simple and common method for monitoring blood pressure is to use piezoelectric sensors that detect the pressure of blood flow in vessels, based on piezo resistivity, capacitance, and piezo electricity mechanism. Nanosized metal particles including gold, aluminium, silver, and copper are considered as the best materials for the fabrication of integrated circuits to be included in piezoelectric sensors (Xu et al. 2018a). Conversely, nanosized gold–copper alloy particles are used for uric acid determination, depending on an enzymatic reaction through electrochemical measurement (Wang et al. 2001). Moreover, the electro-catalytic effect of silver with platinum nanoparticles on reduced oxide of graphene is utilized for tumor necrosis factor-α (TNF-α) detection through electrochemical measurements (Pingarrón et al. 2008). It is noteworthy that TNF-α is an essential part of proinflammatory cytokines which is associated with salt-sensitive hypertension and is related to renal injury (Mehaffey and Majid 2017).

### *3.2 Carbon-Based Nanomaterials*

The hollow cavity in carbon nanotubes (CNTs) attracts several researchers to use them in the fabrication of nanosized biosensors (Adhikari and Chowdhury 2010), as hollow cavities provide a chemically inert environment (Khlobystov et al. 2005; Manzetti 2013). Further, these hollow cavities also provide a potential electromagnetic or a magnetic response site for biosensor and nanoreactor development via electric or electromagnetic impulses (Khlobystov 2011). The CNT structures are classified according to the molecule of interest such as single-walled carbon nanotubes (SWCNTs) and multi-walled carbon nanotubes (MWCNTs) via binding to their biosensor surfaces. However, SWCNTs with one carbon layer can easily transfer an electric signal after an analyte or mediator molecule attachment to their modified surfaces (Hirata et al. 2008).

CNTs are accepted as a potential material for electrochemical biosensor fabrication, owing to their enhanced electron transferability to the electrodes. Thus, CNTs are reported to be working on both second and third electrode systems using the electrochemical cell, which converts or transfers a biological element detection into electrochemical signal. However, the hydrophobic nature of CNTs and strong intermolecular p–p interactions are the major limitations, while developing CNT-based biosensors (Thirumalraj et al. 2017; Wang and Dai 2015). For instance, cationic surfactants effectively bind with negatively charged DNA for designing functionalized CNTs. In addition, cationic functionalities are introduced with amine groups to allow further attachment of other targeting molecules and fluorophore markers to track cells. Several studies demonstrated that RNAs, short double-stranded and single-stranded DNAs, possess the ability to dissolve into SWCNTs, whereas the macromolecules are attached to the open MWCNTs cavity in a nonspecific fashion (Sanz et al. 2011). Furthermore, water-soluble, PEGylated phospholipid functionalized SWCNTs were developed for cancer drug delivery and as a noninvasive diagnostic tool (Liu et al. 2008). In addition, MWCNTs conjugated with poly ethylenimine to enhance binding properties of DNA and these conjugated nanoparticles are proved to be highly sensitive and low toxicity for efficient detection and targeted delivery of genes (Liu et al. 2005).

Graphene sheets with high surface area are fabricated into semi-conductive films and conductive interfaces to be useful in biosensor development. Graphene-dependent biological sensors are broadly utilized in medical diagnostic applications. These biological sensors are utilized in the determination of significant biomolecules including cytochromes, reduced nicotinamide–adenine dinucleotide (NADH) form, hemoglobin (Hb), and individual amino acids (Kuila et al. 2011; Song et al. 2012). Further, these biosensors are involved in trace element detection that are present in urine samples (Tahernejad-Javazmi et al. 2018). Graphene-based biosensors also possess ability to detect heavy metals and larger biomolecules including DNA (Huang et al. 2015a; Mishra et al. 2017).

### *3.3 Polymer Nanomaterials*

The polymeric nanoparticle-based biosensors are highly beneficial compared to metal and ceramic materials due to their mild synthetic conditions, scalable downstream processing, flexibility, low operating temperature, nontoxic, and biocompatibility. It can be noted that the conducting polymers are proven to possess high sensitivity, precise reproducibility, and eliminates nonspecific binding with the analytes in biosensors (Park et al. 2014). The macroscale hydrogel-based materials are durable, elastic, transparent, and biocompatible for the prosthetic device fabrication such as wearable or even implantable biosensor devices. Poly (3,4-ethylenedioxythiophene) (PEDOT), polyaniline (PANI), polypyrrole (PPy), and poly indole are the general conducting polymers that are employed in biosensor fabrication. These nanosized polymers are conjugated with CNTs, MNPs, metal oxides, and other nanomaterials to enhance their sensitivity and selectivity while detection biomolecules (Annabi et al. 2014; Oliva et al. 2017). The polylactic acid (PLA) and PLA–alginate nanoparticles are efficiently encapsulated platinum–porphyrin complex for oxygen detection based on fluorescent measurement. A linear fluorescence response of oxygen concentrations (0–6 mM) was recorded with sensitivity toward oxygen indirectly measures and represents the presence of 0–10 mM of glucose within 4 seconds via catalytic reaction. These biocompatible and implantable glucose biosensors are proposed to be highly beneficial as in-situ glucose sensors for diabetic patients (Pandey et al. 2019). Moreover, the glyphosate (Gly) herbicide was detected via p-aminothiophenol-functionalized AuNPs with Gly as a template molecule that is fabricated using the electro-polymerization technique. The study emphasized that a linear sweep voltammetry with $[Fe(CN)^{6}]^{4-}$and $[Fe(CN)^{6}]^{3-}$ solution can be utilized for the quantitative detection of 1 pM and 1 μM concentration of Gly, respectively (Do et al. 2015).

The template-oriented nanosized polymer particles are utilized in innumerable forms including porous membranes, vesicles, micelles, and macromolecules for biosensor applications. The mesoporous conducting nanocomposite polymer of micro-sized flowers of graphene-nanosized polyaniline fibers (PANInf-GMF) was reported to be beneficial in cholesterol biosensor fabrication. This polymer material was placed over the glass substrate that is coated with Indium–tin oxide via drop-casting approach and this novel biosensor electrochemically sensed cholesterol with a 1.93 mg/dL as limit of detection (Lakshmi et al. 2016). Similarly, PPy was used for the electrochemical enhancement of skeletal muscle cell proliferation, whereas PEDOT and polystyrene sulfonate (PSS) were utilized for the detection of dextran sulfate. The conjugation of PPy, PEDOT and PSS as a nanomaterial was used in the fabrication of biosensors with enhanced conductivity that can increase skeletal cell differentiation and helps in monitoring in vitro dextran sulfate (Harman et al. 2015). Likewise, PANIs are formed via polymerization with a chiral monomer chondroitin sulfate (CHS) for chondroitin sulfate detection (Yuan and Kuramoto 2004). The incorporation of bio-dopants namely chondroitin sulfate, dextran sulfate, and alginate, into polymers of PEDOT are further used to detect biomolecules such as proteins, fibronectin, and collagen (Molino et al. 2014). Other studies with PPy and PANI showcased that the polymeric nanoparticles can be utilized to detect cell proliferation and dextran sulfate, respectively (Gilmore et al. 2009; Yuan and Kuramoto 2003). In addition, most biosensors fabricated via polymer are utilized for peptide and protein determination. For instance, a potentiometric biosensor for urea determination was developed via a copolymer conductive thiophene, and poly (3-hexylthiophene-co-3-thiopheneacetic acid). The urease-immobilized polymer film electrode was covalently surface linked with carboxyl group of polymer film to detect urea with a detection range of about 1–5 mM (Lai et al. 2017).

### 3.4 Nanocomposites

The construction of biosensors with nanocomposites provides additional reinforcement for the signal transaction and to use them in diverse applications. The nanocomposites are broadly classified into four types such as metal, ceramic, polymer, and magnetic composites. The metallic composites of rare earth elements are further divided into bimetallic colloids. Metallic oxides, including $Al_2O_3$ and MgO in combination with other metal oxides, CNT and platinum, iron, magnesium, and nickel bimetallic colloids in combination with other metals including ruthenium, copper, palladium, and silver (Janas and Liszka 2018; Khalil et al. 2018). The bimetallic nanocomposites are exclusively used in implantable biosensor preparation as they possess specific biological entities such as diffusion coefficient, biocompatibility, and biodegradation rate. It also helps in eliminating infected cells with the influence of external magnetic field (Pankhurst et al. 2016).

The graphene (GR)- and graphene oxide (GO)-based nanocomposites show high conductivity of electricity, enhanced area in the surface, defective site access, and better activity as electrocatalysts. These properties of GR and GO make them as a highly efficient material to fabricate nanocomposite-based biosensors. The functional groups of oxygen present in the GO are hydrophilic and hence, have high binding efficiency with MNPs, oxides of metals, nanosized semiconducting particles, polymers, and quantum-sized dots to recover their biosensing electrochemical ability. The nanocomposite-based biosensors are broadly classified into three types are enzymatic, nonenzymatic, and immunosensors, which is based on the type of biomarkers to be detected and the desired applications. The biomarkers such as hydrogen peroxides ($H_2O_2$) and DNA are detected using GO-based nanocomposite biosensors. The P-L-His–reduced GO and $CeO_2$–reduced GO is used for $H_2O_2$ detection, whereas, chitosan–GO, GR–reduced GO and gold NPs–reduced GO are beneficial for DNA detection. It can be noted that GR and reduced GO-based nanocomposites are extensively utilized for glucose biosensor fabrication. Most of the enzymatic reactions in humans are demonstrated to be based on NADH-dependent reactions. These NADH molecules are detected through nanocomposites that contain gold and silver NPs with reduced GO such as gold NPs–reduce GO–PAH (poly (allylamine hydrochloride)), and gold–silver NPs–P(L-Cys)–ErGO. Further, the cholesterol molecules are detected with nanocomposites of GR and reduced GO, apart from gold, palladium, and platinum, the additional molecules in blood serum are detected by other nanocomposites that consist of chitosan, $TiO_2$ nanowires, PPy (polypyrrole), and PVP (polyvinylpyrrolidone) (Istrate et al. 2016; Komathi et al. 2016; Pakapongpan and Poo-Arporn 2017; Pramanik et al. 2018; Radhakrishnan and Kim 2015; Tığ 2017; Vilian and Chen 2014; Wu et al. 2017; Zou et al. 2019).

Besides, graphene-based nanocomposite biosensors are employed for dopamine, bilirubin, ascorbic acid (AA), and uric acid (UA) determination via nonenzyme electrodes that are fabricated using multilayer graphene nanoflake films (MGNFs) (Shang et al. 2008; Thangamuthu et al. 2018). It is reported in various literatures that nanocomposites of GR and reduced GO with metal or metal oxides are highly beneficial in glucose, $H_2O_2$, and cholesterol sensors (Dhara et al. 2015; Gao et al. 2014; Lakshmi et al. 2016; Zor et al. 2014). Further, the graphene-based nanocomposites are used to develop immunosensors for biomarker determination such as human chorionic gonadotropin (hCG), carcinoembryonic antigen (CEA), interleukin-6 (IL-6), and prostate-specific antigen (PSA). The early detection of breast cancer markers such as ERBB2c and CD24c is quantified through gold–graphene oxide nanocomposites. Moreover, the tumor cells were identified by bio-probe that is fabricated via GO–polyaniline (PANI) nanocomposites along with CdSe quantum dot-functionalized polystyrene microspheres (Saeed et al. 2017; Wang et al. 2018). Similarly, graphene-coated gold and silver NPs as nanocomposites are used to determine CEA antigen (Huang et al. 2015b). Likewise, electrochemical label-free immunosensor was fabricated to sense PSA via nanosized chitosan–graphene–methylene blue composite (Yáñez-Sedeño et al. 2017). Also, circulating tumor cells (CTCs) were captured from the blood via graphene oxide conjugated with Neutravidin to restrain and detect the biotinylated epithelial cell that is adhesive to antibody (Yoon et al. 2013). Recently, Saeed et al. (2017) produced nanosized graphene oxide–gold composites modified with DNA aptamer (ERBB2c and CD24c) for early breast cancer detection. A sandwich-type strategy to build sensor was utilized for fabrication which has led to sensitive detection of ERBB2 and CD24 (Saeed et al. 2017). In addition, Wang et al. (2018) demonstrated that CdSe quantum dot-functionalized polystyrene microspheres can be used as a bio-probe along with GO–polyaniline (PANI) nanocomposites for ultrasensitive tumor cell detection. The results revealed that the detection limit was about 3 cells/mL, which was attributed to the high rate of electron transfer and enhanced loading of tumor cell on the surface of nanosized composites (Wang et al. 2018).

## 3.5 *Other Novel Nanomaterials*

The nanomaterial-based biosensors are showing promises to determine molecules and are in high demand in the current pharmaceutical market. The current scenario seems to be challenging to enhance the electrochemical sensors for real-time measurement, point-of-care (POC) analysis, portability, and direct analysis of multiple target analytes. Nanomaterials have been employed as immobilized bioreceptor, electrode modifiers, and electroactive labels, to increase the direct or indirect signals for the detection and quantify biomolecules (García-Mendiola et al. 2018). Since nanomaterials possess an enhanced surface area, the biomolecules are attached to the electrodes via physical adsorption through van der Waals forces (Yáñez-Sedeño et al. 2017). Thus, the biomolecules are immobilized to either supramolecular or coordinative binding biological species surfaces. For instance, SWCNT are coated with poly (adamantane–pyrrole), which is anchored to glucose oxidase (GOX) and gold nanoparticles are bound to β-cyclodextrin. The adamantane-tagged GOX was immobilized over the novel nanomaterials for the glucose biosensor fabrication and the glucose was measured by the potentiostatic method at 0.7 V (Holzinger et al. 2009).

The luminescent semiconducting nanocrystals are called as quantum dots (QDs), which are predominant nanomaterial that are used for the biosensor fabrication. The in vitro immunoassays-based biosensors are fabricated via quantum dots for nucleic acid determination via fluorescence resonance energy transfer (FRET) method. The core-shell CdSe–ZnS QDs with size-tunable fluorescence entities are used for live tracking of cell, in vivo imaging, drug discovery, and other biological diagnostic purposes (Michalet et al. 2005). The bioluminescence resonance energy transfer (BRET) principle for QD655-Luc8 determination is reported to be useful for in vivo imaging of endogenous chromophores (So et al. 2006). The effective sensing of biomolecules using FRET and BRET is based on graphene oxide (GO) with quantum dots (QDs) quenching approach (Dong et al. 2010). The DNA and oligonucleotide are detected through FRET quenching, which reveals the optical transduction of QD (Freeman et al. 2013). The principles of FRET and BRET are used along with quenching approach to transfer charges and chemiluminescence resonance energy transfer (CRET) as light-mediated transducers for certain biosensing purposes with QDs (Algar et al. 2010). Further, streptavidin-conjugated QDs are utilized in the imaging of prostate-specific antigen (PSA) and ssDNA via near-infrared (NIR) light using surface plasmon resonance phenomena (Malic et al. 2011). Moreover, the ferromagnetic iron oxide nanomaterial is mostly used in the determination of numerous bio-analytes such as DNA, mRNA, proteins, enzymes, drugs, pathogens, and tumor cells (Haun et al. 2010).

## 4 Nanomaterials as Sustainable Gas and Biosensors

It is noteworthy from the previous sections that the nanomaterials are extensively utilized in the fabrication of enhanced toxic gas and biosensors. However, the toxicity and stability of nanomaterials toward humans and the environment are a major concern, while utilizing them in sensors (Jeevanandam et al. 2018). Several literatures reported that chemically synthesized nanomaterials are toxic to human cells and organs, irrespective of the type, concentration, or dose (Khan et al. 2017). Chemical-based nanomaterial synthesis approaches utilized toxic precursors, stabilizing and reducing agents, which encapsulate over nanomaterials to exhibit toxicity toward human cells (Andra et al. 2019). These toxic nanomaterials are impossible to be used as implantable biochips and other biomedical applications (Darwish et al. 2019). Even in toxic gas sensors, it is better to avoid toxic nanomaterials as they may lead to hazard effects toward the environment after its lifetime (Valsami-Jones and Lynch 2015). Thus, green or biosynthesis approaches are introduced for the fabrication of nanomaterials to be used as sensors (Mandal et al. 2018). These biosynthesized nanomaterials are formed by using biomolecules and hence they are less or nontoxic to human cells, compared to chemically synthesized nanomaterials (Yola et al. 2014). The recent trends in sensor applications are to utilize biosynthesized nontoxic nanomaterials for the fabrication of sustainable sensors that are not harmful to both humans and the environment (Omobayo Adio et al. 2016).

Additionally, the energy used to fabricate sensors with green synthesized nanomaterial is much lesser compared to sensors derived from a chemical synthesis approach, which makes them as a significant method to develop sustainable sensors (Sharma et al. 2017).

## 4.1 Metal-Based Nanomaterials

Numerous metal, metal oxide nanoparticles, and nanocomposites were synthesized via green synthesis approach for the fabrication of efficient and sustainable toxic gas sensors. Pandey et al. (2013) fabricated an eco-friendly nanosized gold particle via reducing agent from the extracts of guar gum. These green-synthesized nanosized particles were employed as an optical sensor to detect aqueous ammonia via surface plasmon resonance. The obtained results emphasized that the green-synthesized gold nanoparticles possess properties of good reproducibility, excellent sensitivity with less than 10 s response time, and detection limit of 1 parts-per-billion (Pandey et al. 2013). Likewise, Zhao et al. (2015) synthesized spherical shaped, 6–10 nm-sized zinc oxide nanoparticles coated with silver using leaf extract from *Tribulus terrestris* to be used as a nontoxic and economic gas sensor at room temperature. The study revealed that the silver-coated zinc oxide nanoparticles possess enhanced and sustainable ethanol gas-sensing properties than pure zinc oxide nanoparticles (Zhao et al. 2015). Moreover, Li et al. (2008) fabricated porous tin dioxide nanoparticles by using an innovative ionic liquid template at room temperature named 1-hexadecyl-3-methylimidazolium bromide using a biogenic sol–gel approach. The obtained micro and mesopore nanoparticles were employed to detect gases such as carbon monoxide and hydrogen. The result revealed that the tin dioxide nanoparticles have the potential to be a sensitive and swift responding sustainable gas sensor (Li et al. 2008). Silver nanoparticles synthesized using ultraviolet radiation (Dubas and Pimpan 2008), *Cyamopsis tetragonaloba* (Pandey et al. 2012) and gold nanoparticles via locus bean gum (Tagad et al. 2014) are the other green synthesized nanomaterials that possess the ability to be fabricated as sustainable gas sensors.

Recently, Kalanur et al. (2015) fabricated palladium–tungsten oxide nanocomposites via polyvinyl pyrrolidone as a template using ultraviolet radiation-assisted photochemical method, without using any toxic chemicals. These green-synthesized metal nanocomposites exhibited excellent gas-sensing ability against hydrogen with excellent sensitivity (Kalanur et al. 2015). Likewise, Manoj et al. (2018) introduced novel green method to synthesized monodispersed copper nanoparticles using carboxymethyl cellulose. These green synthesized copper nanoparticles were combined with multi-walled carbon nanotubes and glassy carbon to detect nitrite oxidation. The result revealed that the novel nanocomposite possesses excellent nitrite oxidation detection property with great sensitivity (Manoj et al. 2018). Moreover, Tomer et al. (2019) reported that the silver nanoparticles fabricated via cyanobacterial extracts possess enhanced ammonia sensing ability (Tomer et al. 2019). Similarly, silver nanoparticles synthesized via *Duranta erecta* extract (Ismail et al. 2018), silver nanoparticle decorated carbon nanotube (Banihashemian et al. 2019), and palladium-deposited multi-walled nanotubes (Yoo et al. 2019) are the other significant green synthesized nanomaterials that are employed for sustainable toxic gas sensor fabrication.

Apart from toxic gas sensors, green-synthesized metal nanoparticles are also used to fabricate biosensors. Zhang et al. (2013) synthesized hybrid membrane structures of nanosized gold-reduced graphene oxide particles using uric acid, glucose, ascorbic acid, and dopamine as reducing agent to be beneficial as a hydrogen peroxide sensor. The result emphasized that the self-assembled hybrid membranes possess enhanced hydrogen peroxide detection potential with high selectivity, stability, and low detection limit (Zhang et al. 2013). Likewise, Liu et al. (2012a, b) reported a novel biogenic nanosized gold particles–graphene sheet nanohybrids using polyoxometalate as a reducing and encapsulating agent. The result demonstrated that the nanohybrids possess elevated catalytic activity with better sensitivity, stability, swift response, varied range of linearity, and low limit of detection (Liu et al. 2012a). Meanwhile, silver–graphene nanocomposites were synthesized via reducing tannic acid. The electrochemical and the Raman scattering results emphasized that the nanocomposites possess surface-enhanced Raman-scattering properties with enhanced hydrogen peroxide detection ability. The study confirmed that the nanocomposites can be used as an enzyme-less, amperometric hydrogen peroxide and glucose sensor with rapid response time of less than 2 s (Zhang et al. 2012b). Moreover, silver nanoparticles as optical fiber-based hydrogen peroxide sensors (Tagad et al. 2013), palladium-decorated poly(3,4-ethylenedioxythiophene) as a nonenzymatic biosensor to detect hydrogen peroxide (Jiang et al. 2013), biotemplated synthesized gold nanoparticles—bacteria cellulose nanofiber nanocomposites (Zhang et al. 2010), starch-stabilized silver nanoparticles (Vasileva et al. 2011), *Sargassum alga* synthesized palladium nanoparticles (Momeni and Nabipour 2015) are greener nanomaterials that are used in sustainable biosensor fabrication.

In recent times, Gayda et al. (2019) listed the metallic nanoparticles that are attained via biogenic method and used as a biosensor construction platform (Gayda et al. 2019). Bollella et al. (2017) fabricated nanosized gold and silver particles via reducing quercetin for biosensor applications. These spherical shaped, 5–8 nm sized, green-synthesized nanoparticles showed enhanced efficiencies as third-generation lactose biosensors (Bollella et al. 2017). Meanwhile, Su et al. (2016) fabricated nanostructure of cobalt oxide hydroxide using a photochemical approach without template, surfactant, or toxic solvents of organic nature at optimum temperature. The result showed that these nanostructures are highly beneficial in the sunlight mediated bifunctional detection of hydrogen peroxide and can be used to fabricate novel biosensors (Su et al. 2016). Silver-reduced graphene oxide–carbon nanotube nanocomposites were fabricated recently via a novel green synthesis approach for the biosensor applications. The result demonstrated that the nanocomposite with unique sensing properties exhibited excellent ability to detect dopamine with a response time of 12 s and 0.033 μM of detection limit (Khan et al. 2016). Similarly, green synthesized zinc oxide nanoparticles (Muthuchamy et al. 2018), graphene nanoribbon–silver nanoparticle–polyphenol oxidase composite

(Sandeep et al. 2018), gold nanoparticles (Raj and Sudarsanakumar 2018), cyclodextrin-capped gold nanoparticles (Zhang et al. 2019), and molybdenum disulfide–gold nanoparticles (Wu et al. 2018) are the other recent nanomaterials that are used to develop sustainable biosensors.

## 4.2 *Carbon-Based Nanomaterials*

Similar to metals, carbon-based nanomaterials were also used to fabricate toxic gas and biosensors. It is noteworthy that most of the carbon-based nanomaterials are nanocomposites as carbon materials are highly reactive and cause toxic reactions toward humans and microbes. Nanosized composites of tin dioxide with reduced graphene oxide to detect nitrogen dioxide at low temperatures (Zhang et al. 2014), graphene–zinc oxide nanoparticle hybrid (Kavitha et al. 2012) and conjugated polymer–carbon nanotube (Dai et al. 2002) are widely used in the gas sensor applications. However, they are synthesized via chemical approaches that use toxic chemicals and are not suitable for biosensor applications. Nanosized sheets of graphene decorated with nanosized zirconia particle hybrid for the enzyme-less detection of methyl parathion (Gong et al. 2012), indium-doped tin dioxide nanoparticle-graphene nanohybrids as nitrogen dioxide sensors (Cui et al. 2013), tin dioxide nanoparticle-decorated graphene sheets as cataluminescence gas sensors (Song et al. 2011), and mesoporous spherical tin dioxide–graphene nanocomposites as highly sensitive formaldehyde sensor (Chen et al. 2016) are the carbon-based nanomaterials that are used as sustainable gas sensors. In recent times, zinc oxide loaded with nanosized silver particles along with reduced graphene oxide are designed as hybrid for low-temperature detection of acetylene gas (Iftekhar Uddin et al. 2015), *Justicia glauca* leaf extract synthesized graphene oxide reduced with nanosized silver particle decoration as nitrobenzene sensor (Karuppiah et al. 2015), graphene-based sensors (Wang et al. 2016c), and nanosized hybrid composite of iron oxide-reduced graphene oxide to detect nitrogen dioxide at room temperature (Zhang et al. 2017) are the latest graphene-based nanomaterials that are used in sustainable toxic gas sensor fabrication. Likewise, carbon nanotubes (Davis et al. 2003; Wanna et al. 2006) and nanodots (Sun and Lei 2017; Wang et al. 2016b) were also synthesized via green approaches and are used to fabricate sustainable gas sensors.

Several green synthesized carbon-based nanomaterials were also employed for biosensor fabrication. Nayak et al. (2013) demonstrated a novel green synthesis method for the fabrication of nanosized hybrid of graphene-carbon nanotube decorated with zinc oxide particles as composites using solar energy. These novel nanohybrids were subjected to perform as a transducer in an organophosphorus biosensor and the result revealed that the nanohybrids exhibited a linear Paraoxon detection response of 1–26 nM with a detection limit of 1 pM (Nayak et al. 2013). Likewise, a biogenic nanosized composite film that is fabricated with glucose oxidase, gold particles, polyvinyl alcohol, and carbon nanotubes by Zhang et al. (2011). The result from the study reported that the nanocomposite exhibited a linear amperometric response toward concentration of glucose (0.5–8 mM) with a sensitivity of 16.6 μA $mM^{-1}$ $cm^{-2}$ (Zhang et al. 2011). Similarly, nanosized composites of gold reduced graphene oxide were synthesized via aqueous extract of rose flower as reducing agent. These novel green synthesized nanocomposites exhibited enhanced efficiency to be an effective glucose sensor with a low (10 μM) limit of detection (Amouzadeh Tabrizi and Varkani 2014). Recently, carbon dots (Liu et al. 2015), nickel–cobalt oxide–graphene nanohybrids (Ko et al. 2017), photoluminescent self-doped carbon dots (Wang et al. 2017), carbon dots-manganese dioxide nanosheets (Qu et al. 2017), and composites made up of nanosized silver particle, reduced oxide of graphene and carbon nanotube (Lorestani et al. 2015) are synthesized using green methods and are used to fabricate sustainable biosensors.

### 4.3 Polymer-Based Nanomaterials

Polymer nanomaterials and individual nanomaterials functionalized with polymers were widely used to fabricate toxic gas and biosensors. Venditti et al. (2013) introduced an innovative osmosis-based method for polyaniline–gold nanoparticle fabrication. Further, these nanocomposites are doped with hydrogen sulfide to detect ammonia gas and the result demonstrated that these nanomaterials possess enhanced sensitivity against ammonia up to 10 parts-per-million, compared to other vapor organic compounds (Venditti et al. 2013). Moreover, Nia et al. (2015) reported that assembly of silver nanoparticle decorated graphene oxide over glassy carbon electrode via an amperometry method. Polypyrrole nanofibers were attached to the electrodes and are employed as nonenzymatic hydrogen peroxide sensor. The result showed that these nanocomposites possess enhanced ability to detect gases with a detection limit of 1.099 (Moozarm Nia et al. 2015). Likewise, Wang et al. (2008) produced antimony-doped tin dioxide nanoparticles via a new block copolymer with amphiphilic property named poly(ethylene-*co*-butylene)-*block*-poly (ethylene oxide). These doped nanoparticles were employed to detect formaldehyde gas in the atmosphere and the nanoparticles exhibited good and swift responses in detecting the gas with good recovery (Wang et al. 2008). In recent times, conducting polymer–nanoparticle composite based chemo-electrical gas sensors (Park et al. 2017b), zinc oxide thin film prepared with inorganic green sodium–carboxymethyl cellulose polymer as acetone sensors (Muthukrishnan et al. 2016), uniformly decorated silver nanoparticles on polypyrrole as hydrogen peroxide sensor (Nia et al. 2015) and polyaniline–samarium-doped titanium dioxide nanocomposites (Ramesan and Sampreeth 2018) are the green synthesized polymer nanomaterials that are used to fabricate sustainable toxic gas sensors.

Huang et al. (2011) fabricated a bio-electro-chemically active infinite coordination polymer nanoparticles and reported that they possess enhanced glucose detection property (Huang et al. 2011). Likewise, Zhang et al. (2010) synthesized nanocomposites via nanosized gold particles and bacterial cellulose nanofiber for hydrogen peroxide detection with 1 μM of detection limit (Zhang et al. 2010).

Similarly, Asati et al. (2009) fabricated a polymer-coated nanosized cerium oxide particles for oxidation monitoring in cells and antioxidants (Asati et al. 2009). Moreover, Liu et al. (2012a, b) synthesized a nanodot polymer with doped nitrogen that is rich in carbon and with photoluminescent property for biosensor applications, especially for the label-free copper ion detection in body fluids (Liu et al. 2012b). Furthermore, Ruan et al. (2013) described a green synthesis approach for polydopamine–graphene composite film fabrication that is modified with enzyme electrode to be useful as a glucose sensor (Ruan et al. 2013). In recent times, novel carbon nanotube–poly (brilliant green), carbon nanotube–poly (3,4-ethylenedioxythiophene) as enzyme-based biosensors with electrochemical property (Barsan et al. 2016), gold nanoparticles stabilized by alcohol oxidase protein that is encapsulated with polyaniline as amperometric alcohol biosensor (Chinnadayyala et al. 2015) and silver nanoparticle embedded in bacterial cellulose nano-paper as plasmonic sensors (Pourreza et al. 2015) are the novel green synthesized polymeric nanomaterials that help in sustainable biosensor fabrication.

## *4.4 Novel Nanocomposites*

Nanocomposites are the latest addition to the broad set of nanomaterials that are effective for sensor applications. Punicalagin green functionalized copper–copper oxide–zinc oxide nanocomposite as potential electrochemical transducer (Fuku et al. 2016), nanosized composites of graphene oxide reduced by zinc oxide for nitrogen dioxide gas-sensing application (Kumar et al. 2015a) and tin dioxide-based hierarchical nano-microstructures (Jiang et al. 2009) are the novel nanocomposites that are extensively under research as sustainable toxic gas sensors. Meanwhile, mesoporous spherical tin dioxide–graphene nanoparticles as formaldehyde gas sensors (Chen et al. 2016), copper oxide-y@ zinc oxide-α nanosized composites for improved room temperature nitrogen dioxide detection applications (Geng et al. 2017) and silver–iron oxide core–shell magnetic nanocomposite (Mirzaei et al. 2016) are the recent novel nano-sized composites that are used to fabricate sustainable toxic gas sensors. Similarly, photoluminescent carbon dots are synthesized with willow bark to fabricate gold nanoparticles-reduced graphene oxide nanocomposites, which is useful as a glucose sensor (Qin et al. 2013). Likewise, gold–chitosan nanocomposites as caffeic acid sensors (Di Carlo et al. 2012), silver nanoparticle-graphene oxide nanocomposite as tryptophan sensor (Li et al. 2013), palladium–gold–carbon dot nanocomposite to sense colitoxin DNA in human serum (Huang et al. 2017), graphene-nanosized gold particles hybrid as surface-enhanced Raman-scattering biosensor (Khalil et al. 2016) and nanosized flower-shaped reduced graphene oxide–iron oxide hybrid composites for sensing riboflavin (Madhuvilakku et al. 2017) are the novel nanocomposite-based sustainable biosensors that are under extensive research. Even though there are several nanomaterials that are synthesized using green and biosynthesis approaches, there exist certain limitations while using nanomaterials as sensors which have to be addressed before employing them for large-scale sensor applications.

## 5 Limitations and Future Perspective

The nanomaterial sensors are becoming an integral part of any sensor research due to their enormous applications (Luo et al. 2006). It is noteworthy that nanosensors are the hot topic in recent nanotechnology researches to develop enhanced sensors with rapid sensitivity and excellent detection limit (Ding et al. 2010). However, there exist several limitations to use nanomaterials in the fabrication of large-scale and commercial toxic gas and biosensors (Wang et al. 2013). Most of the commercial nanosized toxic gas and biosensors are fabricated via chemical approaches as shown in previous sections. These chemically synthesized nanomaterial-based sensors may lead to toxicity toward humans and other organisms (El-Safty et al. 2007). Moreover, it may accumulate as wastes over its lifetime and the toxic chemicals upon degradation may react with the environment (soil, air, or water) and lead to hazardous effects (Khin et al. 2012). This has led to the emergence of green and biosynthesis approaches to fabricate nanomaterials for sensor applications. These synthesis approaches have provided less or nontoxic nanomaterials for sensor fabrications (Zhang et al. 2012a). However, nanomaterials from green or biosynthesis are unstable most of the time and agglomerate to become micro-sized particles which alter their significant sensor properties (Wahab et al. 2018). Thus, it is highly essential to blend chemical, green, and biosynthesis approaches in the right proportion as a hybrid approach to fabricate sensors to avoid toxicity and improve stability with enhanced sensing properties. Moreover, the latest nanomaterials are focused on detecting a specific molecule, either gas or biomolecules, with enhanced detection limit and sensitivity (Yang et al. 2015). The future research of nanomaterial-based sensors must focus on multi-molecule detection, which can detect several toxic gases and biomolecules. This is highly possible by fabricating complex nanocomposites that can be tailored based on the requirement of sensors for desired applications.

The future of toxic gas and biosensors is based on their incorporation with wearable technologies and in biochips that can be implanted in the patients (Shafiee et al. 2018). Generally, nanomaterials are designed to either sense toxic gases or biomolecules (Smith et al. 2017). In future, multitasking nanomaterials will be designed to detect both toxic gas and biomolecules (Gu and Zhang 2018). These multitasking nanomaterial-based sensors will reduce the cost as well as the time and effort needed to produce such sensors (Casanova-Chafer and Llobet 2019). Moreover, these novel sensors will be incorporated into fabrics to sense toxic gases in the environment, inhalation or exposure level of toxic gas toward humans and detect biomolecules related to disease in patients (Hu et al. 2018; Jang et al. 2018). Recently, fabrics with sensors are available which possess sensors and can change color upon sensing toxic gas in the environment (He et al. 2019). Likewise, multitasking nanomaterial-based sensors can be incorporated in smartwatches and mobile phones to detect toxic gases as well as disease-related biomolecules (Tiwari et al. 2019). These smart technologies will serve as a display to know the level of detecting toxic gas or biomolecules and can be monitored by physicians (Tabatabaee et al. 2019). Such sensors will be the future of biomedical sciences which will reduce the burden of

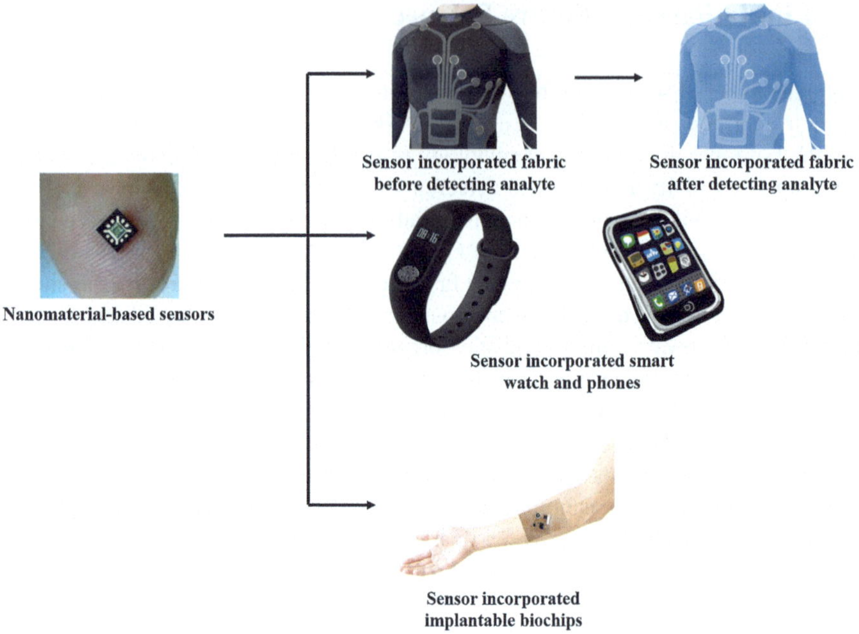


**Fig. 1** Summary of futuristic nanomaterial-based sensors for toxic gas and biosensor applications

patients and physicians in continuous monitoring of analytes. It will also help in the fabrication of fabrics that can monitor both toxic gases and biomolecules, which can help firefighters, workers in mines, and the people who work in extreme conditions. It is noteworthy that nanomaterials can be used to fabricate sensor substrate (Kumar et al. 2015b), bind with the analyte for enhancing the detection (Cardinal et al. 2017), transducer (Zaretski et al. 2016), signal processors (Lu et al. 2011), and as an individual sensing material (Wang et al. 2003). Thus, nanomaterials will replace the current sensor market and enhance its ability to be present in any tools that are used by consumers for sensing toxic gases and will be used as implantable biochips in the future that can sense disease and deliver drugs in the target site. Figure 1 is the summary of futuristic nanomaterial-based sensors in different toxic gas and biosensor applications.

## 6 Conclusion

This chapter is a summary of distinct nanomaterials that are used for the fabrication of toxic gas and biosensors. Metal, carbon, polymer-based nanomaterials, and nanocomposites are widely used in sensor fabrication applications. However, it was noteworthy that chemically synthesized nanomaterials are proved to be toxic toward humans as well as environment and are not suitable for implantable biosensors. Thus, green and biosynthesis are introduced to fabricate sustainable, nontoxic, and stable sensors to detect toxic gases and biomolecules. Biosynthesis using extracts from plants and bacteria as well as green synthesis using sunlight, ultraviolet, visible, and infrared radiation are used as catalyst and reducing agent to fabricate nanomaterials for sensor applications. However, there are several limitations to utilize green and biosynthesized nanomaterials for large-scale and commercial sensor applications. In future, it is possible to overcome these limitations with the advantages and swift progresses in nanotechnology field for the emergence of futuristic sensors. These futuristic sensors can be incorporated into fabrics, smartwatch, phones, paints, and several other tools that a consumer will use in their day-to-day life for toxic gases and biomolecules detection.

**Acknowledgments** The authors would like to acknowledge their respective departments for their support while drafting this chapter.

## References

Acosta MA, Ymele-Leki P, Kostov YV, Leach JB (2009) Fluorescent microparticles for sensing cell microenvironment oxygen levels within 3D scaffolds. Biomaterials 30:3068–3074

Adhikari S, Chowdhury R (2010) The calibration of carbon nanotube based bionanosensors. J Appl Phys 107:124322

Ajay Piriya VS, Joseph P, Daniel SCGK, Lakshmanan S, Kinoshita T, Muthusamy S (2017) Colorimetric sensors for rapid detection of various analytes. Mater Sci Eng C 78:1231–1245

Algar WR, Tavares AJ, Krull UJ (2010) Beyond labels: a review of the application of quantum dots as integrated components of assays, bioprobes, and biosensors utilizing optical transduction. Anal Chim Acta 673:1–25

Alshammari AS, Alenezi MR, Lai KT, SRP S (2017) Inkjet printing of polymer functionalized CNT gas sensor with enhanced sensing properties. Mater Lett 189:299–302

Amouzadeh Tabrizi M, Varkani JN (2014) Green synthesis of reduced graphene oxide decorated with gold nanoparticles and its glucose sensing application. Sensors Actuators B 202:475–482. https://doi.org/10.1016/j.snb.2014.05.099

Anderson CD, Daniels ES (2003) Emulsion polymerisation and latex applications, vol 14. iSmithers Rapra Publishing.

Andra S, Balu SK, Jeevanandham J, Muthalagu M, Vidyavathy M, San Chan Y, Danquah MK (2019) Phytosynthesized metal oxide nanoparticles for pharmaceutical applications. Naunyn Schmiedebergs Arch Pharmacol 392(7):755–771

Annabi N et al (2014) 25th anniversary article: rational design and applications of hydrogels in regenerative medicine. Adv Mater 26:85–124

Asati A, Santra S, Kaittanis C, Nath S, Perez JM (2009) Oxidase-like activity of polymer-coated cerium oxide nanoparticles. Angew Chem Int Ed 48:2308–2312

Avramov ID, Rapp M, Voigt A, Stahl U, Dirschka M (2000) In: Comparative studies on polymer coated SAW and STW resonators for chemical gas sensor applications. IEEE, pp 58–65

Bai H, Shi G (2007) Gas sensors based on conducting polymers. Sensors 7:267–307

Banihashemian SM, Hajghassem H, Nikfarjam A, Azizi Jarmoshti J, Abdul Rahman S, Boon Tong G (2019) Room temperature ethanol sensing by green synthesized silver nanoparticle decorated SWCNT. Mater Res Express 6:065602. https://doi.org/10.1088/2053-1591/ab07de

Barsan N, Koziej D, Weimar U (2007) Metal oxide-based gas sensor research: how to? Sensors Actuators B 121:18–35

Barsan MM, Pifferi V, Falciola L, Brett CMA (2016) New CNT/poly (brilliant green) and CNT/poly (3,4-ethylenedioxythiophene) based electrochemical enzyme biosensors. Anal Chim Acta 927:35–45

Batzill M, Diebold U (2005) The surface and materials science of tin oxide. Prog Surf Sci 79:47–154

Bekyarova E, Davis M, Burch T, Itkis ME, Zhao B, Sunshine S, Haddon RC (2004) Chemically functionalized single-walled carbon nanotubes as ammonia sensors. J Phys Chem B 108:19717–19720

Bogue R (2014) Towards the trillion sensors market. Sens Rev 34:137–142

Bollella P, Schulz C, Favero G, Mazzei F, Ludwig R, Gorton L, Antiochia R (2017) Green synthesis and characterization of gold and silver nanoparticles and their application for development of a third generation lactose biosensor. Electroanalysis 29:77–86. https://doi.org/10.1002/elan.201600476

Brousseau LC, Aurentz DJ, Benesi AJ, Mallouk TE (1997) Molecular design of intercalation-based sensors. 2. Sensing of carbon dioxide in functionalized thin films of copper octanediylbis (phosphonate). Anal Chem 69:688–694

Capone S, Siciliano P, Quaranta F, Rella R, Epifani M, Vasanelli L (2000) Analysis of vapours and foods by means of an electronic nose based on a sol–gel metal oxide sensors array. Sens Actuators B 69:230–235

Cardinal MF, Vander Ende E, Hackler RA, McAnally MO, Stair PC, Schatz GC, Van Duyne RP (2017) Expanding applications of SERS through versatile nanomaterials engineering. Chem Soc Rev 46:3886–3903

Carregal-Romero S et al (2013) Multiplexed sensing and imaging with colloidal nano-and microparticles. Annu Rev Anal Chem 6:53–81

Casanova-Chafer J, Llobet E (2019) Carbon nanomaterials integrated in rugged and inexpensive sensing nanoscale materials for warfare agent detection. Nanoscience for Security 13

Castro M, Kumar B, Feller J-F, Haddi Z, Amari A, Bouchikhi B (2011) Novel e-nose for the discrimination of volatile organic biomarkers with an array of carbon nanotubes (CNT) conductive polymer nanocomposites (CPC) sensors. Sensors Actuators B 159:213–219

Chandra P, Segal E (2016) Nanobiosensors for personalized and onsite biomedical diagnosis. The Institution of Engineering and Technology

Chatterjee SG, Chatterjee S, Ray AK, Chakraborty AK (2015) Graphene–metal oxide nanohybrids for toxic gas sensor: a review. Sensors Actuators B 221:1170–1181

Chen S et al (2016) One-pot synthesis of mesoporous spherical $SnO_2$@graphene for high-sensitivity formaldehyde gas sensors. RSC Adv 6:25198–25202

Chinnadayyala SR, Santhosh M, Singh NK, Goswami P (2015) Alcohol oxidase protein mediated in-situ synthesized and stabilized gold nanoparticles for developing amperometric alcohol biosensor. Biosens Bioelectron 69:155–161

Christ JF, Aliheidari N, Ameli A, Pötschke P (2017) 3D printed highly elastic strain sensors of multiwalled carbon nanotube/thermoplastic polyurethane nanocomposites. Mater Des 131:394–401

Comini E (2006) Metal oxide nano-crystals for gas sensing. Anal Chim Acta 568:28–40

Cui S et al (2013) Indium-doped $SnO_2$ nanoparticle–graphene nanohybrids: simple one-pot synthesis and their selective detection of $NO_2$. J Mater Chem A 1:4462–4467

Dai L, Soundarrajan P, Kim T (2002) Sensors and sensor arrays based on conjugated polymers and carbon nanotubes. Pure Appl Chem 74:1753–1772

Darwish A, Sayed GI, Hassanien AE (2019) The impact of implantable sensors in biomedical technology on the future of healthcare systems. In: Intelligent pervasive computing systems for smarter healthcare

Davis JJ, Coleman KS, Azamian BR, Bagshaw CB, MLH G (2003) Chemical and biochemical sensing with modified single walled carbon nanotubes. Chem A Eur J 9:3732–3739

Dey A (2018) Semiconductor metal oxide gas sensors: a review. Mater Sci Eng B 229:206–217
Dhara K, Ramachandran T, Nair BG, TGS B (2015) Single step synthesis of Au–CuO nanoparticles decorated reduced graphene oxide for high performance disposable nonenzymatic glucose sensor. J Electroanal Chem 743:1–9
Di Carlo G et al (2012) Green synthesis of gold–chitosan nanocomposites for caffeic acid sensing. Langmuir 28:5471–5479
Ding B, Wang M, Wang X, Yu J, Sun G (2010) Electrospun nanomaterials for ultrasensitive sensors. Mater Today 13:16–27
Do MH et al (2015) Molecularly imprinted polymer-based electrochemical sensor for the sensitive detection of glyphosate herbicide. Int J Environ Anal Chem 95:1489–1501
Dong XM, Fu RW, Zhang MQ, Zhang B, Rong MZ (2004) Electrical resistance response of carbon black filled amorphous polymer composite sensors to organic vapors at low vapor concentrations. Carbon 42:2551–2559
Dong H, Gao W, Yan F, Ji H, Ju H (2010) Fluorescence resonance energy transfer between quantum dots and graphene oxide for sensing biomolecules. Anal Chem 82:5511–5517
Dubas ST, Pimpan V (2008) Green synthesis of silver nanoparticles for ammonia sensing. Talanta 76:29–33. https://doi.org/10.1016/j.talanta.2008.01.062
Ellis JE, Star A (2016) Carbon nanotube based gas sensors toward breath analysis. ChemPlusChem 81:1248–1265
Elosúa C, Bariáin C, Matías IR, Arregui FJ, Luquin A, Laguna M (2006) Volatile alcoholic compounds fibre optic nanosensor. Sensors Actuators B 115:444–449
El-Safty SA, Prabhakaran D, Ismail AA, Matsunaga H, Mizukami F (2007) Nanosensor design packages: a smart and compact development for metal ions sensing responses. Adv Funct Mater 17:3731–3745
Endres HE et al (1996) A thin-film $SnO_2$ sensor system for simultaneous detection of CO and $NO_2$ with neural signal evaluation. Sensors Actuators B 36:353–357
Eranna G (2016) Metal oxide nanostructures as gas sensing devices. CRC Press, Boca Raton
Erden F, Velipasalar S, Alkar AZ, Cetin AE (2016) Sensors in assisted living: a survey of signal and image processing methods. IEEE Signal Process Mag 33:36–44
Fowler JD, Allen MJ, Tung VC, Yang Y, Kaner RB, Weiller BH (2009) Practical chemical sensors from chemically derived graphene. ACS Nano 3:301–306
Fraiwan L, Lweesy K, Bani-Salma A, Mani N (2011) In: A wireless home safety gas leakage detection system. IEEE, pp 11–14
Freeman R, Girsh J, Willner I (2013) Nucleic acid/quantum dots (QDs) hybrid systems for optical and photoelectrochemical sensing. ACS Appl Mater Interfaces 5:2815–2834
Fuku X, Kaviyarasu K, Matinise N, Maaza M (2016) Punicalagin green functionalized Cu/$Cu_2O$/ZnO/CuO nanocomposite for potential electrochemical transducer and catalyst. Nanoscale Res Lett 11:386
Gao W, Tjiu WW, Wei J, Liu T (2014) Highly sensitive nonenzymatic glucose and $H_2O_2$ sensor based on $Ni(OH)_2$/electroreduced graphene oxide—multiwalled carbon nanotube film modified glass carbon electrode. Talanta 120:484–490
García-Mendiola T et al (2018) Carbon nanodots based biosensors for gene mutation detection. Sensors Actuators B 256:226–233
Gardner JW, Bartlett PN (1993) Design of conducting polymer gas sensors: modelling and experiment. Synth Met 57:3665–3670
Gayda ZG et al (2019) Metallic nanoparticles obtained via "green" synthesis as a platform for biosensor construction. Appl Sci 9. https://doi.org/10.3390/app9040720
Geng X, Zhang C, Luo Y, Debliquy M (2017) Preparation and characterization of CuxO1-y@ZnO1-α nanocomposites for enhanced room-temperature $NO_2$ sensing applications. Appl Surf Sci 401:248–255
Gilmore KJ et al (2009) Skeletal muscle cell proliferation and differentiation on polypyrrole substrates doped with extracellular matrix components. Biomaterials 30:5292–5304

Gong J, Miao X, Wan H, Song D (2012) Facile synthesis of zirconia nanoparticles-decorated graphene hybrid nanosheets for an enzymeless methyl parathion sensor. Sensors Actuators B 162:341–347. https://doi.org/10.1016/j.snb.2011.12.094

Goustouridis D, Manoli K, Chatzandroulis S, Sanopoulou M, Raptis I (2005) Characterization of polymer layers for silicon micromachined bilayer chemical sensors using white light interferometry. Sensors Actuators B 111:549–554

Gu B, Zhang Q (2018) Recent advances on functionalized upconversion nanoparticles for detection of small molecules and ions in biosystems. Adv Sci 5:1700609

Gupta SK, Joshi A, Kaur M (2010) Development of gas sensors using ZnO nanostructures. J Chem Sci 122:57–62

Hagleitner C, Lange D, Hierlemann A, Brand O, Baltes H (2002) CMOS single-chip gas detection system comprising capacitive, calorimetric and mass-sensitive microsensors. IEEE J Solid State Circuits 37:1867–1878

Hallil H, Chebila F, Menini P, Pons P, Aubert H (2010) In: Feasibility of wireless gas detection with an FMCW RADAR interrogation of passive RF gas sensor. IEEE, pp 759–762

Harman DG et al (2015) Poly (3,4-ethylenedioxythiophene): dextran sulfate (PEDOT: DS)—a highly processable conductive organic biopolymer. Acta Biomater 14:33–42

Hatfield JV, Neaves P, Hicks PJ, Persaud K, Travers P (1994) Towards an integrated electronic nose using conducting polymer sensors. Sensors Actuators B 18:221–228

Haun JB, Yoon TJ, Lee H, Weissleder R (2010) Magnetic nanoparticle biosensors. Wiley Interdisciplinary Rev Nanomed Nanobiotechnol 2:291–304

He T, Shi Q, Wang H, Wen F, Chen T, Ouyang J, Lee C (2019) Beyond energy harvesting-multi-functional triboelectric nanosensors on a textile. Nano Energy 57:338–352

Hirata T et al (2008) Chemical modification of carbon nanotube based bio-nanosensor by plasma activation. Japanese J Appl Phys 47:2068

Hoefer U, Böttner H, Felske A, Kühner G, Steiner K, Sulz G (1997) Thin-film $SnO_2$ sensor arrays controlled by variation of contact potential—a suitable tool for chemometric gas mixture analysis in the TLV range. Sensors Actuators B 44:429–433

Holzinger M, Bouffier L, Villalonga R, Cosnier S (2009) Adamantane/β-cyclodextrin affinity biosensors based on single-walled carbon nanotubes Adamantane/β-cyclodextrin affinity biosensors based on single-walled carbon nanotubes. Biosens Bioelectron 24:1128–1134

Hu H, Sun Z, Jornet JM (2018) In: Through-the-body localization of implanted biochip in wearable nano-biosensing networks. IEEE, pp 278–283

Huang P, Mao J, Yang L, Yu P, Mao L (2011) Bioelectrochemically active infinite coordination polymer nanoparticles: one-pot synthesis and biosensing property. Chem A Eur J 17:11390–11393. https://doi.org/10.1002/chem.201101634

Huang H, Bai W, Dong C, Guo R, Liu Z (2015a) An ultrasensitive electrochemical DNA biosensor based on graphene/Au nanorod/polythionine for human papillomavirus DNA detection. Biosens Bioelectron 68:442–446

Huang J, Tian J, Zhao Y, Zhao S (2015b) Ag/Au nanoparticles coated graphene electrochemical sensor for ultrasensitive analysis of carcinoembryonic antigen in clinical immunoassay. Sensors Actuators B 206:570–576

Huang Q, Lin X, Zhu J-J, Tong Q-X (2017) Pd-Au@ carbon dots nanocomposite: Facile synthesis and application as an ultrasensitive electrochemical biosensor for determination of colitoxin DNA in human serum. Biosens Bioelectron 94:507–512

Iftekhar Uddin ASM, Phan D-T, Chung G-S (2015) Low temperature acetylene gas sensor based on Ag nanoparticles-loaded ZnO-reduced graphene oxide hybrid. Sensors Actuators B 207:362–369. https://doi.org/10.1016/j.snb.2014.10.091

Ijeh MO (2011) Covalent gold nanoparticle-antibody conjugates for sensivity improvement in LFIA

Ismail M, Khan MI, Akhtar K, Khan MA, Asiri AM, Khan SB (2018) Biosynthesis of silver nanoparticles: a colorimetric optical sensor for detection of hexavalent chromium and ammonia in aqueous solution. Phys E Low-dimens Syst Nanostruct 103:367–376. https://doi.org/10.1016/j.physe.2018.06.015

Istrate O-M, Rotariu L, Marinescu VE, Bala C (2016) NADH sensing platform based on electrochemically generated reduced graphene oxide–gold nanoparticles composite stabilized with poly (allylamine hydrochloride). Sensors Actuators B 223:697–704

Jaegle M, Wöllenstein J, Meisinger T, Böttner H, Müller G, Becker T, Braunmühl CB-V (1999) Micromachined thin film $SnO_2$ gas sensors in temperature-pulsed operation mode. Sensors Actuators B 57:130–134

Jamdagni P, Khatri P, Rana JS (2016) Nanoparticles based DNA conjugates for detection of pathogenic microorganisms. Int Nano Lett 6:139–146

Jamshaid T, Neto ETT, Eissa MM, Zine N, Kunita MH, El-Salhi AE, Elaissari A (2016) Magnetic particles: from preparation to lab-on-a-chip, biosensors, microsystems and microfluidics applications. TrAC Trends Anal Chem 79:344–362

Janas D, Liszka B (2018) Copper matrix nanocomposites based on carbon nanotubes or graphene. Mater Chem Front 2:22–35

Janata J, Josowicz M (2003) Conducting polymers in electronic chemical sensors. Nat Mater 2:19

Jang S, Kim S, Geier ML, Hersam MC, Dodabalapur A (2016) Inkjet printed carbon nanotubes in short channel field effect transistors: influence of nanotube distortion and gate insulator interface modification. Flexible Printed Electron 1:035001

Jang JS et al (2018) Flexible gas sensors: glass-fabric reinforced Ag nanowire/Siloxane composite heater substrate: sub-10 nm metal@ metal oxide nanosheet for sensitive flexible sensing platform (Small 44/2018). Small 14:1870201

Jeevanandam J, Barhoum A, Chan YS, Dufresne A, Danquah MK (2018) Review on nanoparticles and nanostructured materials: history, sources, toxicity and regulations. Beilstein J Nanotechnol 9:1050–1074

Jiang L-Y, Wu X-L, Guo Y-G, Wan L-J (2009) $SnO_2$-based hierarchical nanomicrostructures: facile synthesis and their applications in gas sensors and lithium-ion batteries. J Phys Chem C 113:14213–14219

Jiang F, Yue R, Du Y, Xu J, Yang P (2013) A one-pot 'green'synthesis of Pd-decorated PEDOT nanospheres for nonenzymatic hydrogen peroxide sensing. Biosens Bioelectron 44:127–131

Jimenez-Cadena G, Riu J, Rius FX (2007) Gas sensors based on nanostructured materials. Analyst 132:1083–1099

Kalanur SS, Yoo I-H, Lee Y-A, Seo H (2015) Green deposition of Pd nanoparticles on $WO_3$ for optical, electronic and gasochromic hydrogen sensing applications. Sensors Actuators B 221:411–417. https://doi.org/10.1016/j.snb.2015.06.086

Kanan S, El-Kadri O, Abu-Yousef I, Kanan M (2009) Semiconducting metal oxide based sensors for selective gas pollutant detection. Sensors 9:8158–8196

Kanazawa E, Sakai G, Shimanoe K, Kanmura Y, Teraoka Y, Miura N, Yamazoe N (2001) Metal oxide semiconductor $N_2O$ sensor for medical use. Sensors Actuators B 77:72–77

Kanjanawarut R, Su X (2009) Colorimetric detection of DNA using unmodified metallic nanoparticles and peptide nucleic acid probes. Anal Chem 81:6122–6129

Karuppiah C et al (2015) Green synthesized silver nanoparticles decorated on reduced graphene oxide for enhanced electrochemical sensing of nitrobenzene in waste water samples. RSC Adv 5:31139–31146. https://doi.org/10.1039/C5RA00992H

Kato K, Kato Y, Takamatsu K, Udaka T, Nakahara T, Matsuura Y, Yoshikawa K (2000) Toward the realization of an intelligent gas sensing system utilizing a non-linear dynamic response. Sensors Actuators B 71:192–196

Kaushik A, Kumar R, Arya SK, Nair M, Malhotra BD, Bhansali S (2015) Organic–inorganic hybrid nanocomposite-based gas sensors for environmental monitoring. Chem Rev 115:4571–4606

Kavitha M, Kalpana AM (2017) Carbon nanotubes: properties and applications—a brief review i-Manager's. J Electron Eng 7:1

Kavitha T, Gopalan AI, Lee K-P, Park S-Y (2012) Glucose sensing, photocatalytic and antibacterial properties of graphene–ZnO nanoparticle hybrids. Carbon 50:2994–3000

Khalil I, Julkapli N, Yehye W, Basirun W, Bhargava S (2016) Graphene–gold nanoparticles hybrid—synthesis, functionalization, and application in a electrochemical and surface-enhanced Raman scattering biosensor. Materials (Basel) 9:406

Khalil I, Rahmati S, Julkapli NM, Yehye WA (2018) Graphene metal nanocomposites—recent progress in electrochemical biosensing applications. J Ind Eng Chem 59:425–439

Khan A, Khan AAP, Asiri AM, Abu-Zied BM (2016) Green synthesis of thermally stable Ag-rGO-CNT nano composite with high sensing activity. Compos Part B Eng 86:27–35. https://doi.org/10.1016/j.compositesb.2015.09.018

Khan I, Saeed K, Khan I (2017) Nanoparticles: properties, applications and toxicities. Arab J Chem 12(7):908–931

Khaydukova M, Medina-Plaza C, Rodriguez-Mendez ML, Panchuk V, Kirsanov D, Legin A (2017) Multivariate calibration transfer between two different types of multisensor systems. Sensors Actuators B 246:994–1000

Khin MM, Nair AS, Babu VJ, Murugan R, Ramakrishna S (2012) A review on nanomaterials for environmental remediation. Energ Environ Sci 5:8075–8109

Khlobystov AN (2011) Carbon nanotubes: from nano test tube to nano-reactor. ACS Nano 5:9306–9312

Khlobystov AN, Britz DA, GAD B (2005) Molecules in carbon nanotubes. Acc Chem Res 38:901–909

Kim SJ (2009) The effect on the gas selectivity of CNT-based gas sensors by binder in SWNT/Silane sol solution. IEEE Sens J 10:173–177

Kim JS, Kim G-W (2014) New non-contacting torque sensor based on the mechanoluminescence of ZnS: Cu microparticles. Sensors Actuators A Phys 218:125–131

Kimura T, Goto T (2005) Ir–YSZ nano-composite electrodes for oxygen sensors. Surf Coat Technol 198:36–39

Ko T-H, Radhakrishnan S, Seo M-K, Khil M-S, Kim H-Y, Kim B-S (2017) A green and scalable dry synthesis of $NiCo_2O_4$/graphene nanohybrids for high-performance supercapacitor and enzymeless glucose biosensor applications. J Alloys Compd 696:193–200

Kocakulak M, Butun I (2017) In: An overview of wireless sensor networks towards internet of things. IEEE, pp 1–6

Komathi S, Muthuchamy N, Lee KP, Gopalan AI (2016) Fabrication of a novel dual mode cholesterol biosensor using titanium dioxide nanowire bridged 3D graphene nanostacks. Biosens Bioelectron 84:64–71

Kondratiev VV, Malev VV, Eliseeva SN (2016) Composite electrode materials based on conducting polymers loaded with metal nanostructures. Russ Chem Rev 85:14

Kuila T, Bose S, Khanra P, Mishra AK, Kim NH, Lee JH (2011) Recent advances in graphene-based biosensors. Biosens Bioelectron 26:4637–4648

Kumar N, Srivastava AK, Patel HS, Gupta BK, Varma GD (2015a) Facile synthesis of ZnO-reduced graphene oxide nanocomposites for $NO_2$ gas sensing applications. Eur J Inorg Chem 2015:1912–1923

Kumar R, Al-Dossary O, Kumar G, Umar A (2015b) Zinc oxide nanostructures for $NO_2$ gas–sensor applications: a review. Nano-Micro Lett 7:97–120

Lai C-Y, Foot P, Brown J, Spearman P (2017) A urea potentiometric biosensor based on a thiophene copolymer. Biosensors 7:13

Lakshmi G, Sharma A, Solanki PR, Avasthi DK (2016) Mesoporous polyaniline nanofiber decorated graphene micro-flowers for enzyme-less cholesterol biosensors. Nanotechnology 27:345101

Landfester K (2006) Synthesis of colloidal particles in miniemulsions. Annu Rev Mat Res 36:231–279

Lee D-D, Chung W-Y, Choi M-S, Baek J-M (1996) Low-power micro gas sensor. Sensors Actuators B 33:147–150

Lee Y, Trocchia SM, Warren SB, Young EF, Vernick S, Shepard KL (2018) Electrically controllable single-point covalent functionalization of spin-cast carbon-nanotube field-effect transistor arrays. ACS Nano 12:9922–9930

Li H, Wang Q, Xu J, Zhang W, Jin L (2002) A novel nano-Au-assembled amperometric $SO_2$ gas sensor: preparation, characterization and sensing behavior. Sens Actuators B 87:18–24

Li L-L et al (2008) Room temperature ionic liquids assisted green synthesis of nanocrystalline porous $SnO_2$ and their gas sensor behaviors. Cryst Growth Des 8:4165–4172. https://doi.org/10.1021/cg800686w

Li Y, Fan X, Qi J, Ji J, Wang S, Zhang G, Zhang F (2010) Palladium nanoparticle-graphene hybrids as active catalysts for the Suzuki reaction. Nano Res 3:429–437

Li Z et al (2011) Graphene buffered galvanic synthesis of graphene–metal hybrids. J Mater Chem 21:13241–13246

Li J, Kuang D, Feng Y, Zhang F, Xu Z, Liu M, Wang D (2013) Green synthesis of silver nanoparticles–graphene oxide nanocomposite and its application in electrochemical sensing oftryptophan. Biosens Bioelectron 42:198–206

Liu A, Huang S (2012) A glucose biosensor based on direct electrochemistry of glucose oxidase immobilized onto platinum nanoparticles modified graphene electrode. Sci China Phys Mech Astron 55:1163–1167

Liu Y, Kumar S (2014) Polymer/carbon nanotube nano composite fibers—a review. ACS Appl Mater Interfaces 6:6069–6087

Liu Y et al (2005) Polyethylenimine-grafted multiwalled carbon nanotubes for secure noncovalent immobilization and efficient delivery of DNA. Angew Chem Int Ed 44:4782–4785

Liu D, Bruckbauer A, Abell C, Balasubramanian S, Kang D-J, Klenerman D, Zhou D (2006) A reversible pH-driven DNA nanoswitch array. J Am Chem Soc 128:2067–2071

Liu Z, Robinson JT, Sun X, Dai H (2008) PEGylated nanographene oxide for delivery of water-insoluble cancer drugs. J Am Chem Soc 130:10876–10877

Liu Y, Chen M, Mohebbi M, Wang ML, Dokmeci MR (2011) In: The effect of sequence length on DNA decorated CNT gas sensors. IEEE, pp 2156–2159

Liu R et al (2012a) Facile synthesis of au-nanoparticle/polyoxometalate/graphene tricomponent nanohybrids: an enzyme-free electrochemical biosensor for hydrogen peroxide. Small 8:1398–1406. https://doi.org/10.1002/smll.201102298

Liu S et al (2012b) Hydrothermal treatment of grass: a low-cost, green route to nitrogen-doped, carbon-rich, photoluminescent polymer Nanodots as an effective fluorescent sensing platform for label-free detection of Cu(II) ions. Adv Mater 24:2037–2041. https://doi.org/10.1002/adma.201200164

Liu S, Zhao N, Cheng Z, Liu H (2015) Amino-functionalized green fluorescent carbon dots as surface energy transfer biosensors for hyaluronidase. Nanoscale 7:6836–6842

Ljungblad J (2009) Antibody-conjugated gold nanoparticles integrated in a fluorescence based biochip

Llobet E (2013) Gas sensors using carbon nanomaterials: a review. Sensors Actuators B 179:32–45

Lorestani F, Shahnavaz Z, Mn P, Alias Y, NSA M (2015) One-step hydrothermal green synthesis of silver nanoparticle-carbon nanotube reduced-graphene oxide composite and its application as hydrogen peroxide sensor. Sensors Actuators B 208:389–398

Lu G, Park S, Yu K, Ruoff RS, Ocola LE, Rosenmann D, Chen J (2011) Toward practical gas sensing with highly reduced graphene oxide: a new signal processing method to circumvent run-to-run and device-to-device variations. ACS Nano 5:1154–1164

Luo X, Morrin A, Killard AJ, Smyth MR (2006) Application of nanoparticles in electrochemical sensors and biosensors. Electroanal Int J Devoted to Fundam Pract Asp Electroanal 18:319–326

Lupan O, Postica V, Cretu V, Wolff N, Duppel V, Kienle L, Adelung R (2016) Single and networked CuO nanowires for highly sensitive p-type semiconductor gas sensor applications. Phys Status Solidi Rapid Res Lett 10:260–266

Lyson-Sypien B, Radecka M, Rekas M, Swierczek K, Michalow-Mauke K, Graule T, Zakrzewska K (2015) Grain-size-dependent gas-sensing properties of $TiO_2$ nanomaterials. Sensors Actuators B 211:67–76

Ma H, Zhou W, Yuan W, Jie Z, Liu H, Li X (2012) The gas sensing mechanism of the low-dimension carbon composites with metal oxide quantum dots. Phys Proc 32:31–38

Madhurantakam S, Babu KJ, Rayappan JBB, Krishnan UM (2018) Nanotechnology-based electrochemical detection strategies for hypertension markers. Biosens Bioelectron 116:67–80

Madhuvilakku R, Alagar S, Mariappan R, Piraman S (2017) Green one-pot synthesis of flowers-like $Fe_3O_4$/rGO hybrid nanocomposites for effective electrochemical detection of riboflavin and low-cost supercapacitor applications. Sens Actuators B 253:879–892

Malic L, Sandros MG, Tabrizian M (2011) Designed biointerface using near-infrared quantum dots for ultrasensitive surface plasmon resonance imaging biosensors. Anal Chem 83:5222–5229

Mandal D, Mishra S, Singh RK (2018) Green synthesized nanoparticles as potential nanosensors. In: Environmental, chemical and medical sensors. Springer, pp 137–164

Manoj D, Saravanan R, Santhanalakshmi J, Agarwal S, Gupta VK, Boukherroub R (2018) Towards green synthesis of monodisperse Cu nanoparticles: an efficient and high sensitive electrochemical nitrite sensor. Sensors Actuators B 266:873–882. https://doi.org/10.1016/j.snb.2018.03.141

Manzetti S (2013) Molecular and crystal assembly inside the carbon nanotube: encapsulation and manufacturing approaches. Adv Manuf 1:198–210

Markets Ma (2017) Gas sensors market by gas type (oxygen, carbon monoxide, carbon dioxide, ammonia, chlorine, hydrogen sulfide, nitrogen oxide, volatile organic compounds, hydrocarbons), technology, end-use application, geography—global forecast 2023. www.marketsandmarkets.com. https://www.marketsandmarkets.com/Market-Reports/gas-sensor-market-245141093.html. 2019

Markets Ma (2019) Biosensors market by type (sensor patch and embedded device), product (wearable and nonwearable), technology (electrochemical and optical), application (POC, home diagnostics, research lab, food & beverages), and geography—global forecast to 2024. www.marketsandmarkets.com. https://www.marketsandmarkets.com/PressReleases/biosensors.asp

Mazloum-Ardakani M, Hosseinzadeh L, Taleat Z (2015) Synthesis and electrocatalytic effect of Ag@ Pt core–shell nanoparticles supported on reduced graphene oxide for sensitive and simple label-free electrochemical aptasensor. Biosens Bioelectron 74:30–36

McNaghten ED, Parkes AM, Griffiths BC, Whitehouse AI, Palanco S (2009) Detection of trace concentrations of helium and argon in gas mixtures by laser-induced breakdown spectroscopy. Spectrochim Acta Part B Atom Spectrosc 64:1111–1118

Mehaffey E, Majid DSA (2017) Tumor necrosis factor-α, kidney function, and hypertension. Am J Physiol Renal Physiol 313:F1005–F1008

Mehrotra P (2016) Biosensors and their applications—a review. J Oral Biol Craniofac Res 6:153–159. https://doi.org/10.1016/j.jobcr.2015.12.002

Michalet X et al (2005) Quantum dots for live cells, in vivo imaging, and diagnostics. Science 307:538–544

Mirzaei A, Janghorban K, Hashemi B, Bonyani M, Leonardi SG, Neri G (2016) A novel gas sensor based on Ag/$Fe_2O_3$ core-shell nanocomposites. Ceram Int 42:18974–18982

Mishra RK, Rajakumari R (2019) Nanobiosensors for biomedical application: present and future prospects. In: Characterization and biology of nanomaterials for drug delivery. Elsevier, pp 1–23

Mishra RK, Nawaz MH, Hayat A, Nawaz MAH, Sharma V, Marty J-L (2017) Electrospinning of graphene-oxide onto screen printed electrodes for heavy metal biosensor. Sensors Actuators B 247:366–373

Molino PJ et al (2014) Influence of biodopants on PEDOT biomaterial polymers: using QCM-D to characterize polymer interactions with proteins and living cells. Adv Mater Interfaces 1:1300122

Momeni S, Nabipour I (2015) A simple green synthesis of palladium nanoparticles with *Sargassum alga* and their electrocatalytic activities towards hydrogen peroxide. Appl Biochem Biotechnol 176:1937–1949

Monerris MJ, Arévalo FJ, Fernández H, Zon MA, Molina PG (2016) Electrochemical immunosensor based on gold nanoparticles deposited on a conductive polymer to determine estrone in water samples. Microchem J 129:71–77

Moozarm Nia P, Lorestani F, Meng WP, Alias Y (2015) A novel non-enzymatic $H_2O_2$ sensor based on polypyrrole nanofibers–silver nanoparticles decorated reduced graphene oxide nano composites. Appl Surf Sci 332:648–656. https://doi.org/10.1016/j.apsusc.2015.01.189

Munoz BC, Steinthal G, Sunshine S (1999) Conductive polymer-carbon black composites-based sensor arrays for use in an electronic nose. Sensor Rev 19:300–305

Muthuchamy N, Atchudan R, Edison TNJI, Perumal S, Lee YR (2018) High-performance glucose biosensor based on green synthesized zinc oxide nanoparticle embedded nitrogen-doped carbon sheet. J Electroanal Chem 816:195–204. https://doi.org/10.1016/j.jelechem.2018.03.059

Muthukrishnan K, Vanaraja M, Boomadevi S, Karn RK, Singh V, Singh PK, Pandiyan K (2016) Studies on acetone sensing characteristics of ZnO thin film prepared by sol–gel dip coating. J Alloys Compounds 673:138–143

Nayak P, Anbarasan B, Ramaprabhu S (2013) Fabrication of organophosphorus biosensor using ZnO nanoparticle-decorated carbon nanotube–graphene hybrid composite prepared by a novel green technique. J Phys Chem C 117:13202–13209. https://doi.org/10.1021/jp312824b

Nia PM, Meng WP, Alias Y (2015) Hydrogen peroxide sensor: uniformly decorated silver nanoparticles on polypyrrole for wide detection range. Appl Surf Sci 357:1565–1572

Nohria R, Khillan RK, Su Y, Dikshit R, Lvov Y, Varahramyan K (2006) Humidity sensor based on ultrathin polyaniline film deposited using layer-by-layer nano-assembly. Sensors Actuators B 114:218–222

Olaru A, Bala C, Jaffrezic-Renault N, Aboul-Enein HY (2015) Surface plasmon resonance (SPR) biosensors in pharmaceutical analysis. Crit Rev Anal Chem 45:97–105

Oliva N, Jo C, Wang K, Artzi N (2017) Designing hydrogels for on-demand therapy. Acc Chem Res 50:669–679

Omobayo Adio S, Basheer C, Zafarullah K, Alsharaa A, Siddiqui Z (2016) Biogenic synthesis of silver nanoparticles; study of the effect of physicochemical parameters and application as nanosensor in the colorimetric detection of $Hg^{2+}$ in water. Int J Environ Anal Chem 96:776–788

Pakapongpan S, Poo-Arporn RP (2017) Self-assembly of glucose oxidase on reduced graphene oxide-magnetic nanoparticles nanocomposite-based direct electrochemistry for reagentless glucose biosensor. Mater Sci Eng C 76:398–405

Pallas-Areny R, Webster JG (2012) Sensors and signal conditioning. Wiley, New York

Pandey P, Datta M, Malhotra BD (2008) Prospects of nanomaterials in biosensors. Anal Lett 41:159–209

Pandey S, Goswami GK, Nanda KK (2012) Green synthesis of biopolymer–silver nanoparticle nanocomposite: an optical sensor for ammonia detection. Int J Biol Macromol 51:583–589. https://doi.org/10.1016/j.ijbiomac.2012.06.033

Pandey S, Goswami GK, Nanda KK (2013) Green synthesis of polysaccharide/gold nanoparticle nanocomposite: an efficient ammonia sensor. Carbohydr Polym 94:229–234. https://doi.org/10.1016/j.carbpol.2013.01.009

Pandey G, Chaudhari R, Joshi B, Choudhary S, Kaur J, Joshi A (2019) Fluorescent Biocompatible platinum-porphyrin–doped polymeric hybrid particles for oxygen and glucose biosensing. Sci Rep 9:5029

Pankhurst Q, Jones S, Dobson J (2016) Applications of magnetic nanoparticles in biomedicine: the story so far. J Phys D Appl Phys 49:R167–R181

Park S, Kwon O, Lee J, Jang J, Yoon H (2014) Conducting polymer-based nanohybrid transducers: a potential route to high sensitivity and selectivity sensors. Sensors 14:3604–3630

Park RS, Hills G, Sohn J, Mitra S, Shulaker MM, HSP W (2017a) Hysteresis-free carbon nanotube field-effect transistors. ACS Nano 11:4785–4791

Park S, Park C, Yoon H (2017b) Chemo-electrical gas sensors based on conducting polymer hybrids. Polymers 9:155

Penza M, Rossi R, Alvisi M, Cassano G, Signore MA, Serra E, Giorgi R (2008) Surface modification of carbon nanotube networked films with Au nanoclusters for enhanced $NO_2$ gas sensing applications. J Sensors 2008:1–8

Pingarrón JM, Yañez-Sedeño P, González-Cortés A (2008) Gold nanoparticle-based electrochemical biosensors. Electrochim Acta 53:5848–5866

Ponamoreva ON, Rossinskaya IV, Alferov VA, Vlasova YA, Esikova TZ, Reshetilov AN (2019) Caprolactam detection using two types of microbial biosensors. Water Chem Ecol 118:30–35

Pourreza N, Golmohammadi H, Naghdi T, Yousefi H (2015) Green in-situ synthesized silver nanoparticles embedded in bacterial cellulose nanopaper as a bionanocomposite plasmonic sensor. Biosens Bioelectron 74:353–359

Pramanik K, Sarkar P, Bhattacharyay D, Majumdar P (2018) One step electrode fabrication for direct electron transfer cholesterol biosensor based on composite of polypyrrole, green reduced graphene oxide and cholesterol oxidase. Electroanalysis 30:2719–2730

Qin X, Lu W, Asiri AM, Al-Youbi AO, Sun X (2013) Green, low-cost synthesis of photoluminescent carbon dots by hydrothermal treatment of willow bark and their application as an effective photocatalyst for fabricating Au nanoparticles–reduced graphene oxide nanocomposites for glucose detection. Cat Sci Technol 3:1027–1035

Qu F, Pei H, Kong R, Zhu S, Xia L (2017) Novel turn-on fluorescent detection of alkaline phosphatase based on green synthesized carbon dots and $MnO_2$ nanosheets. Talanta 165:136–142

Radhakrishnan S, Kim SJ (2015) An enzymatic biosensor for hydrogen peroxide based on one-pot preparation of $CeO_2$-reduced graphene oxide nanocomposite. RSC Adv 5:12937–12943

Rai M, Gade A, Gaikwad S, Marcato PD, Durán N (2012) Biomedical applications of nanobiosensors: the state-of-the-art. J Braz Chem Soc 23:14–24

Raj DR, Sudarsanakumar C (2018) Colorimetric and fiber optic sensing of cysteine using green synthesized gold nanoparticles. Plasmonics 13:327–334

Ramesan MT, Sampreeth T (2018) In situ synthesis of polyaniline/Sm-doped $TiO_2$ nanocomposites: evaluation of structural, morphological, conductivity studies and gas sensing applications. J Mater Sci Mater Electron 29:4301–4311

Rotariu L, Lagarde F, Jaffrezic-Renault N, Bala C (2016) Electrochemical biosensors for fast detection of food contaminants—trends and perspective. TrAC Trends Anal Chem 79:80–87

Ruan C et al (2013) One-pot preparation of glucose biosensor based on polydopamine–graphene composite film modified enzyme electrode. Senors Actuators B 177:826–832

Saeed AA, Sánchez JLA, O'Sullivan CK, Abbas MN (2017) DNA biosensors based on gold nanoparticles-modified graphene oxide for the detection of breast cancer biomarkers for early diagnosis. Bioelectrochemistry 118:91–99

Saini R, Hegde K, Brar SK, Verma M (2019) Advances in whole cell-based biosensors in environmental monitoring. In: Tools, techniques and protocols for monitoring environmental contaminants. Elsevier, pp 263–284

Sandeep S, Santhosh AS, Swamy NK, Suresh GS, Melo JS, Chamaraja NA (2018) A biosensor based on a graphene nanoribbon/silver nanoparticle/polyphenol oxidase composite matrix on a graphite electrode: application in the analysis of catechol in green tea samples. New J Chem 42:16620–16629

Sanz V et al (2011) Optimising DNA binding to carbon nanotubes by non-covalent methods. Carbon 49:1775–1781

Shafiee A, Ghadiri E, Kassis J, Pourhabibi Zarandi N, Atala A (2018) Biosensing technologies for medical applications, manufacturing, and regenerative medicine. Curr Stem Cell Rep 4:105–115

Shafiei M, Hoshyargar F, Lipton-Duffin J, Piloto C, Motta N, O'Mullane AP (2015) Conversion of n-type CuTCNQ into p-type nitrogen-doped CuO and the implication for room-temperature gas sensing. J Phys Chem C 119:22208–22216. https://doi.org/10.1021/acs.jpcc.5b06894

Shan L, Freidman D, Kennedy C, Persak W, Lau K (2018) In: Backward compatible connectors for next generation PCIe electrical I/O. IEEE, pp 1798–1804

Shang NG et al (2008) Catalyst-free efficient growth, orientation and biosensing properties of multilayer graphene nanoflake films with sharp edge planes. Adv Funct Mater 18:3506–3514

Sharma V, Tiwari P, Mobin SM (2017) Sustainable carbon-dots: recent advances in green carbon dots for sensing and bioimaging. J Mater Chem B 5:8904–8924

Shrivas AG, Bavane RG, Mahajan AM (2007) In: Electronic nose: a toxic gas sensor by polyaniline thin film conducting polymer. IEEE, pp 621–623

Smith MK, Martin-Peralta DG, Pivak PA, Mirica KA (2017) Fabrication of solid-state gas sensors by drawing: an undergraduate and high school introduction to functional nanomaterials and chemical detection. J Chem Educ 94:1933–1938

So M-K, Xu C, Loening AM, Gambhir SS, Rao J (2006) Self-illuminating quantum dot conjugates for in vivo imaging. Nat Biotechnol 24:339

Song H, Zhang L, He C, Qu Y, Tian Y, Lv Y (2011) Graphene sheets decorated with $SnO_2$ nanoparticles: in situ synthesis and highly efficient materials for cataluminescence gas sensors. J Mater Chem 21:5972–5977

Song Y, He Z, Hou H, Wang X, Wang L (2012) Architecture of $Fe_3O_4$-graphene oxide nanocomposite and its application as a platform for amino acid biosensing. Electrochim Acta 71:58–65

Stankovich S et al (2006) Graphene-based composite materials. Nature 442:282

Su S, Wu W, Gao J, Lu J, Fan C (2012) Nanomaterials-based sensors for applications in environmental monitoring. J Mater Chem 22:18101–18110

Su C-Y, Lan W-J, Chu C-Y, Liu X-J, Kao W-Y, Chen C-H (2016) Photochemical green synthesis of nanostructured cobalt oxides as hydrogen peroxide redox for bifunctional sensing application. Electrochim Acta 190:588–595. https://doi.org/10.1016/j.electacta.2015.12.092

Sun X, Lei Y (2017) Fluorescent carbon dots and their sensing applications. TrAC Trends Anal Chem 89:163–180

Suri K, Annapoorni S, Sarkar AK, Tandon RP (2002) Gas and humidity sensors based on iron oxide–polypyrrole nanocomposites. Sensors Actuators B 81:277–282

Šutka A, Gross KA (2016) Spinel ferrite oxide semiconductor gas sensors. Sensors Actuators B 222:95–105

Tabatabaee RS, Golmohammadi H, Ahmadi SH (2019) Easy diagnosis of jaundice: a smartphone-based nanosensor bioplatform using photoluminescent bacterial nanopaper for point-of-care diagnosis of hyperbilirubinemia. ACS Sensors 4(4):1063–1071

Tagad CK, Dugasani SR, Aiyer R, Park S, Kulkarni A, Sabharwal S (2013) Green synthesis of silver nanoparticles and their application for the development of optical fiber based hydrogen peroxide sensor. Sens Actuators B 183:144–149. https://doi.org/10.1016/j.snb.2013.03.106

Tagad CK, Rajdeo KS, Kulkarni A, More P, Aiyer RC, Sabharwal S (2014) Green synthesis of polysaccharide stabilized gold nanoparticles: chemo catalytic and room temperature operable vapor sensing application. RSC Adv 4:24014–24019. https://doi.org/10.1039/C4RA02972K

Tahernejad-Javazmi F, Shabani-Nooshabadi M, Karimi-Maleh H (2018) Gold nanoparticles and reduced graphene oxide-amplified label-free DNA biosensor for dasatinib detection. New J Chem 42:16378–16383

Tallury P, Malhotra A, Byrne LM, Santra S (2010) Nanobioimaging and sensing of infectious diseases. Adv Drug Deliv Rev 62:424–437

Thangamuthu M, Gabriel W, Santschi C, Martin O (2018) Electrochemical sensor for bilirubin detection using screen printed electrodes functionalized with carbon nanotubes and graphene. Sensors 18:800

Thirumalraj B, Kubendhiran S, Chen S-M, Lin K-Y (2017) Highly sensitive electrochemical detection of palmatine using a biocompatible multiwalled carbon nanotube/poly-l-lysine composite. J Colloid Interface Sci 498:144–152

Tığ GA (2017) Highly sensitive amperometric biosensor for determination of NADH and ethanol based on Au-Ag nanoparticles/poly (L-cysteine)/reduced graphene oxide nanocomposite. Talanta 175:382–389

Tiwari S, Sharma V, Mujawar M, Mishra YK, Kaushik A, Ghosal A (2019) Biosensors for epilepsy management: state-of-art and future aspects. Sensors 19:1525

Tomer AK, Rahi T, Neelam DK, Dadheech PK (2019) Cyanobacterial extract-mediated synthesis of silver nanoparticles and their application in ammonia sensing. Int Microbiol 22:49–58. https://doi.org/10.1007/s10123-018-0026-x

Töppel T, Lausch H, Brand M, Hensel E, Arnold M, Rotsch C (2018) Structural integration of sensors/actuators by laser beam melting for tailored smart components. JOM 70:321–327

Trakhtenberg LI, Gerasimov GN, Gromov VF, Belysheva TV, Ilegbusi OJ (2012) Gas semiconducting sensors based on metal oxide nanocomposites. J Mater Sci Res 1:56

Tran T-T, Mulchandani A (2016) Carbon nanotubes and graphene nano field-effect transistor-based biosensors. TrAC Trends Anal Chem 79:222–232

Tsiafoulis CG, Trikalitis PN, Prodromidis MI (2005) Synthesis, characterization and performance of vanadium hexacyanoferrate as electrocatalyst of $H_2O_2$. Electrochem Commun 7:1398–1404

Valsami-Jones E, Lynch I (2015) How safe are nanomaterials? Science 350:388–389

Vasileva P, Donkova B, Karadjova I, Dushkin C (2011) Synthesis of starch-stabilized silver nanoparticles and their application as a surface plasmon resonance-based sensor of hydrogen peroxide. Colloids Surf A Physicochem Eng Asp 382:203–210

Venditti I, Fratoddi I, Russo MV, Bearzotti A (2013) A nanostructured composite based on polyaniline and gold nanoparticles: synthesis and gas sensing properties. Nanotechnology 24:155503. https://doi.org/10.1088/0957-4484/24/15/155503

Vignesh RH, Sankar KV, Amaresh S, Lee YS, Selvan RK (2015) Synthesis and characterization of $MnFe_2O_4$ nanoparticles for impedometric ammonia gas sensor. Sensors Actuators B 220:50–58

Vilian ATE, Chen S-M (2014) Simple approach for the immobilization of horseradish peroxidase on poly-l-histidine modified reduced graphene oxide for amperometric determination of dopamine and $H_2O_2$. RSC Adv 4:55867–55876

Vukojević V, Djurdjić S, Ognjanović M, Fabian M, Samphao A, Kalcher K, Stanković DM (2018) Enzymatic glucose biosensor based on manganese dioxide nanoparticles decorated on graphene nanoribbons. J Electroanal Chem 823:610–616

Wahab AW, Karim A, La Nafie N, Sutapa IW (2018) Synthesis of silver nanoparticles using *Muntingia calabura* L. leaf extract as bioreductor and applied as glucose nanosensor. Oriental J Chem 34:3088–3094

Wang Z, Dai Z (2015) Carbon nanomaterial-based electrochemical biosensors: an overview. Nanoscale 7:6420–6431

Wang J, Xu D, Kawde A-N, Polsky R (2001) Metal nanoparticle-based electrochemical stripping potentiometric detection of DNA hybridization. Anal Chem 73:5576–5581

Wang Y, Jiang X, Xia Y (2003) A solution-phase, precursor route to polycrystalline $SnO_2$ nanowires that can be used for gas sensing under ambient conditions. J Am Chem Soc 125:16176–16177

Wang YD, Djerdj I, Antonietti M, Smarsly B (2008) Polymer-assisted generation of antimony-doped $SnO_2$ nanoparticles with high crystallinity for application in gas sensors. Small 4:1656–1660. https://doi.org/10.1002/smll.200800644

Wang C, Yin L, Zhang L, Xiang D, Gao R (2010) Metal oxide gas sensors: sensitivity and influencing factors. Sensors 10:2088–2106

Wang P, Jornet JM, Malik MGA, Akkari N, Akyildiz IF (2013) Energy and spectrum-aware MAC protocol for perpetual wireless nanosensor networks in the Terahertz Band. Ad Hoc Networks 11:2541–2555

Wang F et al (2016a) A highly sensitive gas sensor based on CuO nanoparticles synthesized via a sol–gel method. RSC Adv 6:79343–79349

Wang L et al (2016b) Facile, green and clean one-step synthesis of carbon dots from wool: application as a sensor for glyphosate detection based on the inner filter effect. Talanta 160:268–275

Wang T et al (2016c) A review on graphene-based gas/vapor sensors with unique properties and potential applications. Nano-Micro Lett 8:95–119. https://doi.org/10.1007/s40820-015-0073-1

Wang R, Wang X, Sun Y (2017) One-step synthesis of self-doped carbon dots with highly photoluminescence as multifunctional biosensors for detection of iron ions and pH. Sensors Actuators B 241:73–79

Wang J, Wang X, Tang H, Gao Z, He S, Li J, Han S (2018) Ultrasensitive electrochemical detection of tumor cells based on multiple layer CdS quantum dots-functionalized polystyrene microspheres and graphene oxide–polyaniline composite. Biosens Bioelectron 100:1–7

Wanna Y, Srisukhumbowornchai N, Tuantranont A, Wisitsoraat A, Thavarungkul N, Singjai P (2006) The effect of carbon nanotube dispersion on CO gas sensing characteristics of polyaniline gas sensor. J Nanosci Nanotechnol 6:3893–3896

Wisitsoraat A, Tuantranont A, Comini E, Sberveglieri G, Wlodarski W (2009) Characterization of n-type and p-type semiconductor gas sensors based on NiOx doped $TiO_2$ thin films. Thin Solid Films 517:2775–2780

Wong KKL, Tang Z, Sin JKO, Chan PCH, Cheung PW, Hiraoka H (1995) In: Study on selectivity enhancement of tin dioxide gas sensor using non-conducting polymer membrane. IEEE, pp 42–45

Wright KD (2017) Functionalization of carbon materials with metals

Wu S et al (2017) Development of glucose biosensors based on plasma polymerization-assisted nanocomposites of polyaniline, tin oxide, and three-dimensional reduced graphene oxide. Appl Surf Sci 401:262–270

Wu J et al (2018) Two-dimensional molybdenum disulfide (MoS2) with gold nanoparticles for biosensing of explosives by optical spectroscopy. Sens Actuators B 261:279–287

Xiang Q, Meng G, Zhang Y, Xu J, Xu P, Pan Q, Yu W (2010) Ag nanoparticle embedded-ZnO nanorods synthesized via a photochemical method and its gas-sensing properties. Sens Actuators B 143:635–640

Xu F, Li X, Shi Y, Li L, Wang W, He L, Liu R (2018a) Recent developments for flexible pressure sensors: a review. Micromachines 9:580

Xu K, Fu C, Gao Z, Wei F, Ying Y, Xu C, Fu G (2018b) Nanomaterial-based gas sensors: a review. Instrum Sci Technol 46:115–145

Xue L, Wang W, Guo Y, Liu G, Wan P (2017) Flexible polyaniline/carbon nanotube nanocomposite film-based electronic gas sensors. Sensors Actuators B 244:47–53

Yamazoe N, Shimanoe K (2008) Theory of power laws for semiconductor gas sensors. Sensors Actuators B 128:566–573

Yan J, Wei T, Shao B, Fan Z, Qian W, Zhang M, Wei F (2010) Preparation of a graphene nanosheet/polyaniline composite with high specific capacitance. Carbon 48:487–493

Yáñez-Sedeño P, Campuzano S, Pingarrón J (2017) Carbon nanostructures for tagging in electrochemical biosensing: a review. J Carbon Res 3:3

Yang C, Denno ME, Pyakurel P, Venton BJ (2015) Recent trends in carbon nanomaterial-based electrochemical sensors for biomolecules: a review. Anal Chim Acta 887:17–37

Yola ML, Gupta VK, Eren T, Şen AE, Atar N (2014) A novel electro analytical nanosensor based on graphene oxide/silver nanoparticles for simultaneous determination of quercetin and morin. Electrochim Acta 120:204–211

Yoo I-H, Kalanur SS, Seo H (2019) Deposition of Pd nanoparticles on MWCNTs: Green approach and application to hydrogen sensing. J Alloys Compd 788:936–943. https://doi.org/10.1016/j.jallcom.2019.02.298

Yoon HJ et al (2013) Sensitive capture of circulating tumour cells by functionalized graphene oxide nanosheets. Nat Nanotechnol 8:735

Yu X, Zhang W, Zhang P, Su Z (2017) Fabrication technologies and sensing applications of graphene-based composite films: advances and challenges. Biosens Bioelectron 89:72–84

Yuan G-L, Kuramoto N (2003) Synthesis and chiroptical properties of optically active poly (N-alkylanilines) doped and intertwined with dextran sulfate in aqueous solution. Macromolecules 36:7939–7945

Yuan GL, Kuramoto N (2004) Synthesis of helical polyanilines using chondroitin sulfate as a molecular template. Macromol Chem Phys 205:1744–1751

Yuan W, Shi G (2013) Graphene-based gas sensors. J Mater Chem A 1:10078–10091
Yusan S, Rahman MM, Mohamad N, Arrif TM, Latif AZA, Ma MA, Nik W (2018) Development of an amperometric glucose biosensor based on the immobilization of glucose oxidase on the Se-MCM-41 mesoporous composite. J Anal Methods Chem 2018:1–8
Zaporotskova IV, Boroznina NP, Parkhomenko YN, Kozhitov LV (2016) Carbon nanotubes: sensor properties. A review. Mod Electron Mater 2:95–105
Zaretski AV et al (2016) Metallic nanoislands on graphene as highly sensitive transducers of mechanical, biological, and optical signals. Nano Lett 16:1375–1380
Zhang S, Wang N, Yu H, Niu Y, Sun C (2005) Covalent attachment of glucose oxidase to an Au electrode modified with gold nanoparticles for use as glucose biosensor. Bioelectrochemistry 67:15–22
Zhang T, Wang W, Zhang D, Zhang X, Ma Y, Zhou Y, Qi L (2010) Biotemplated synthesis of gold nanoparticle–bacteria cellulose nanofiber nanocomposites and their application in biosensing. Adv Func Mater 20:1152–1160
Zhang H, Meng Z, Wang Q, Zheng J (2011) A novel glucose biosensor based on direct electrochemistry of glucose oxidase incorporated in biomediated gold nanoparticles–carbon nanotubes composite film. Sensors Actuators B Chem 158:23–27. https://doi.org/10.1016/j.snb.2011.04.057
Zhang Y, Gao G, Qian Q, Cui D (2012a) Chloroplasts-mediated biosynthesis of nanoscale Au-Ag alloy for 2-butanone assay based on electrochemical sensor. Nanoscale Res Lett 7:475
Zhang Y et al (2012b) One-pot green synthesis of Ag nanoparticles-graphene nanocomposites and their applications in SERS, $H_2O_2$, and glucose sensing. RSC Adv 2:538–545. https://doi.org/10.1039/C1RA00641J
Zhang P, Zhang X, Zhang S, Lu X, Li Q, Su Z, Wei G (2013) One-pot green synthesis, characterizations, and biosensor application of self-assembled reduced graphene oxide–gold nanoparticle hybrid membranes. J Mater Chem B 1:6525–6531. https://doi.org/10.1039/C3TB21270J
Zhang H, Feng J, Fei T, Liu S, Zhang T (2014) $SnO_2$ nanoparticles-reduced graphene oxide nanocomposites for $NO_2$ sensing at low operating temperature. Sensors Actuators B Chem 190:472–478
Zhang H, Yu L, Li Q, Du Y, Ruan S (2017) Reduced graphene oxide/α-$Fe_2O_3$ hybrid nanocomposites for room temperature $NO_2$ sensing. Sensors Actuators B Chem 241:109–115. https://doi.org/10.1016/j.snb.2016.10.059
Zhang NMY et al (2019) One-step synthesis of cyclodextrin-capped gold nanoparticles for ultra-sensitive and highly-integrated plasmonic biosensors. Sensors Actuators B Chem 286:429–436
Zhao Z-Y, Wang M-H, Liu T-T (2015) *Tribulus terrestris* leaf extract assisted green synthesis and gas sensing properties of Ag-coated ZnO nanoparticles. Mater Lett 158:274–277. https://doi.org/10.1016/j.matlet.2015.05.155
Zhou X, Zeng Y, Liyan C, Wu X, Yoon J (2016) A fluorescent sensor for dual-channel discrimination between phosgene and a nerve-gas mimic. Angew Chem Int Ed 55:4729–4733
Zor E, Saglam ME, Akin I, Saf AO, Bingol H, Ersoz M (2014) Green synthesis of reduced graphene oxide/nanopolypyrrole composite: characterization and $H_2O_2$ determination in urine. RSC Adv 4:12457–12466
Zou N, Wei X, Zong Z, Li X, Wang Z, Wang X (2019) A novel enzymatic biosensor for detection of intracellular hydrogen peroxide based on 1-aminopyrene and reduced graphene oxides. J Chem Sci 131:28

# Flexible Substrate-Based Sensors in Health Care and Biosensing Applications

**Paramita Karfa, Kartick Chandra Majhi, and Rashmi Madhuri**

## Contents



## Abbreviations

| | |
|---|---|
| CNT | Carbon nanotube |
| CP | Conducting polymers |
| CVD | Chemical vapor deposition |
| ECGs | Electrocardiograms |
| EEG | Electroencephalography |
| EMG | Electromyography |
| FET | Field-effect transistor |
| ITO | Indium tin oxide |
| LOx | Lactate oxidase |
| MWCNT | Multi-walled carbon nanotube |
| NTC | Negative temperature coefficient |
| PANI | Polyaniline |
| PC | Polycarbonate |

P. Karfa · K. C. Majhi · R. Madhuri (✉)
Department of Applied Chemistry, Indian Institute of Technology (Indian School of Mines), Dhanbad, Jharkhand, India



Inamuddin, A. M. Asiri (eds.), *Nanosensor Technologies for Environmental Monitoring*, Nanotechnology in the Life Sciences,
https://doi.org/10.1007/978-3-030-45116-5_14

| | |
|---|---|
| PCO | Polymers of cyclic olefin |
| PDMS | Poly(dimethylsiloxane) |
| PEDOT:PSS | (3,4-Ethylenedioxythiophene) poly(styrene sulfonate) |
| PEN | Poly(ethylene naphthalate) |
| PEO | Polyethelene oxide |
| PET | Polyethylene terephthalate |
| PI | Polyimide |
| PMMA | Poly(methyl methacrylate) |
| POC | Point of care diagnostic |
| POE | Polyolefin elastomer |
| PP | Polypropylene |
| PPY | Polypyrrole |
| PS | Polystyrene |
| PTC | Positive temperature coefficient |
| PVDF | Polyvinylidene fluoride |
| Rgo | Reduced graphene oxide |
| SAALD | Spatial atmospheric atomic layer deposition |
| SWCNT | Single-walled carbon nanotube |
| TENG | Triboelectric nanogenerator |
| TTF | Tetrathiafulvalene |

## 1 Introduction

Sensor word can be defined as a device, which can produce measurable signal because of the biological, physical, and/or chemical stimuli responses. On a similar basis, they may be categorized as biological, physical, or chemical sensor depending on the analyte they are detecting or quantifying (Akyildiz et al. 2002). Sensor is actually a combination of receptor and transducer (Chong and Kumar 2003). In a sensor, the analyte gets engrossed to the receptor site, which produces an electrical energy as a result of interaction between them. After that, the transducer does the next work by transducing the electrical signal to the readable format, which will further analyze. Sensors are highly useful nowadays in a variety of applications like environmental monitoring, healthcare application, food adulterant determination, water examination, fabrication of drugs, forensic examination, and many more. However, to fulfill the recent demand and match with recent scenario, upgradation in sensor technology is happening in a day-to-day manner. Researchers are working in this field to develop the sensors with increased specificity, better sensitivity toward the analyte, low-cost fabrication process, and easy availability (Guth et al. 2009; Windmiller et al. 2013).

Recently, a new term i.e. flexible sensor came in limelight. Flexible sensor is made up of malleable material, which can be molded in any kind of design or modified with any kind of substrate (Nambiar and Yeow 2011). Before the flexible sensors, non-flexible or rigid sensors are more popular, which are mainly synthesized of silicon compounds.

Rigid sensors are not appropriate for monitoring human physiological factors because of its stiffness, which causes damage to the sensor with little stress (Castillejo et al. 2013). Therefore, flexible substrate-based sensor has gained interest of the researchers, owing to their numerous properties like lightweight, easy to stretch/fold/bend, effortlessly portable, biocompatible, high chemical stability and transparency. Other than monitoring human physiological factors, this type of sensors is also very popular in batteries, display in electronics, soft robotics, wearable electronics, mobile phones, solar cells, light-emitting diodes, robotics, and aerospace, owing to their high economical design/manufacture, biocompatibility, and multifunctional nature (Xu et al. 2003). On the commercial front, flexible sensors are designed or classified as chemical sensor, optical sensor, biosensor, strain and pressure sensor, temperature sensor, pH sensor, electronic sensor, etc. Flexible and wearable sensors have various biomedical applications also such as in point of care diagnosis of human health, electronic skin, smart medical prosthetics, etc. as shown in Fig. 1 (Xu et al. 2018a).

There are several reasons or advantages of increasing popularity of flexible sensors over rigid sensors, some of them are briefed below (Yang 2006):

1. Sensibilities of soft flexible sensor are comparable to that of rigid substrate sensor, even at very low temperatures.
2. Selectivity of the flexible sensor toward the analyte is more as compared to that of the rigid sensor. They can easily capture the analyte molecule, which is hindered in the conventional sensor due to its rigid surface.

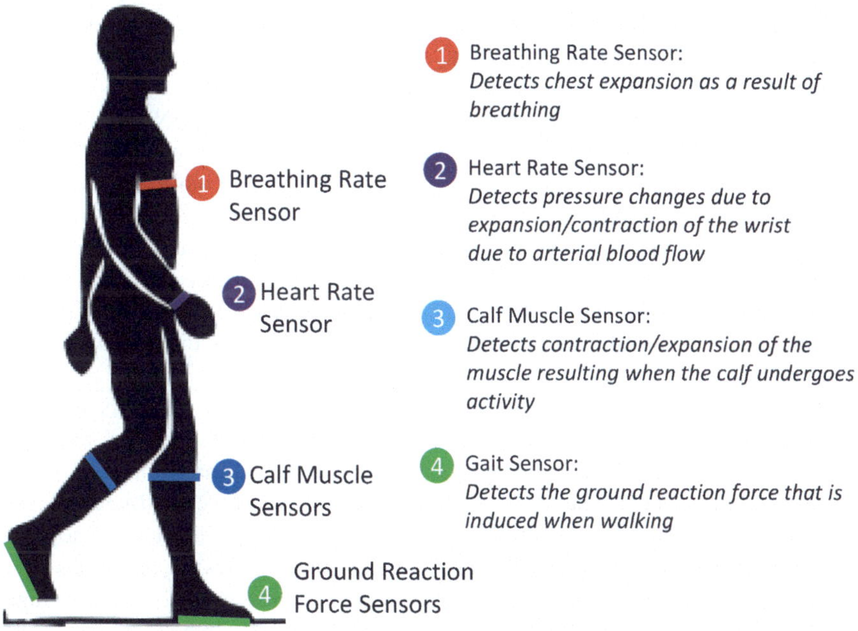


**Fig. 1** Schematic diagram showing different types of wearable sensor used for human health monitoring (Taken with permission from (Xu et al. 2018a)

3. Flexible sensor possesses good stability and repeatability because of its stretchable/bendable nature.
4. With the use of flexible sensor, different devices can be designed, which can be folded, stretched, bended without any loss of performances, whereas, on the other hand, rigid substrate-based sensor, when thinned gets deformed and lose their activity.
5. Signal transduction of high quality is generated by the flexible sensor than the conventional rigid sensors.

## 1.1 Materials Used for Manufacturing of Flexible Sensors

Designing of flexible sensors totally depends upon the expenses of manufacturing and availability of materials used for their preparation. Most commonly, flexible synthetic polymers like polystyrene (PS), poly(dimethylsiloxane) (PDMS), polypropylene (PP), poly(ethylene naphthalate) (PEN), polycarbonate (PC), poly(methyl methacrylate) (PMMA), polymers of cyclic olefin (PCO), poly(3,4-ethylenedioxythiophene) polyimide (PI), polyethylene terephthalate (PET) were used for preparation of base of the sensor that can be converted into thin films also, if needed (Han et al. 2017). Among them, the widely used soft polymer material, as a substrate, is polydimethylsiloxane (PDMS), which is a silicone-based elastomer and commercially popular with the name of Ecoflex, Dragon skin, or Silbione. PDMS has several advantageous properties like low elastic modulus, good stretching ability, durability, flexibility, optical transparency and can be synthesized by simple steps. Soft elastomers-based flexible sensors like PDMS-based sensor is more convenient to synthesize than the other thermoplastics like PEN because of the less easy stretching and conformability of the later. PDMS can be synthesized through formation of resin mixture and curing in a respective mold. Pu et al. have synthesized flexible substrate-based glucose sensor with five polydimethylsiloxane (PDMS) layers and were fabricated using the technique of micro-molding (Pu et al. 2016). Luo et al. have synthesized resistive electronic skin for pressure sensing using graphene platelet and multiwalled carbon nanotube on PDMS substrate. The prepared sensor showed high-pressure sensitivities and outstanding stability (Luo et al. 2019).

PET is another popular polymer, which is a synthetic polymer of esters and used as an electric insulator, has mechanical properties like inertness, optical transparency, and thermal stability. Similarly, Teonex the commercial name of PEN is also popular polymer, which is better than PET in their intrinsic properties like durability and chemical resistance. In addition, PEN also has better optical transparency and very popular in optical devices. Chun et al. have synthesized flexible pressure sensor using PEN as the flexible substrate and successfully transferred CNT film of high electrical conductivity on it through chemical-free process (Chun et al. 2018). The prepared sensor shows piezoresistive responses which can be used for various healthcare application like to detect the motion of hands, wrist, etc. Another polymer, which is popular in flexible sensor, is PI with commercial name Kepton, which can be easily stretched, molded, and folded, have low cost, high biocompatibility, easy to manipulate, and therefore used in fabrication of biosensors, bioelectronics, fuel cells, etc.

Metals and ceramics are nowadays avoided for preparation of flexible substrate material in spite of their outstanding electrical properties, which is good for efficient signal transduction but due to their rigidness, low elasticity they are vulnerable to mechanical damage. Textile, paper, silk are ecofriendly flexible materials, which are easily available and very much economical. Decreasing the size of the paper from micrometer to nanometer increases the optical transparency of the paper, which can be further used in different sensing modalities. Different nanocelluloses are also used as a substrate for biosensors nowdays such as bacterial nanocellulose, cellulose nanocrystal, nano-fibrillated cellulose (Lee et al. 2015).

Flexible sensor has two parts stretchable and flexible substrate and the active sensing material, which can be in solid or liquid forms. To prepare the flexible sensor device, the active materials were coated or attached to the flexible materials, which can be able to convert the sensing stimuli in information transducing or sense the stimuli, which later on send to the signal generating readable electronic devices. The popular active sensing material in solid form, which can be used in flexible sensor, can be conducting polymers, metals, metal nanowires, semiconductor material, carbon-based material, nanoparticles like graphene, CNT, or nanofibers of polymer. The popular liquid form of active sensing material is ionic liquids (Barlow et al. 2002).

Conducting polymers (CP) which are included in the flexible material are (3,4-ethylenedioxythiophene) poly(styrene sulfonate) (PEDOT:PSS), polyaniline (PANI), polypyrrole (PPY), which have conjugated $\pi$ systems with organic backbone. CP has a great affinity toward biological analytes through functional group modification for biosensing application. Nanoparticles of metal, carbon-based nanomaterial like carbon nanotube (CNT), graphene-like 2D nanomaterial having good electronic properties, mechanical properties, flexibility, wear-resistance are used for the formation of electrode material for sensors and are also sometimes used to form the core of the sensor (Yamada et al. 2011).

## *1.2 Fabrication of Flexible Sensor*

The different method for the synthesis of flexible sensors is always a point of research because they demand innovation in material and new cost-effective synthesis process for better commercialization. Till date the process through, which the flexible sensors can be fabricated are as follows:

- Screen printing.
- Inkjet printing.
- Gravure printing.
- Photolithography.

- Direct writing.
- Roll to roll printing.
- Spin coating.
- In-situ polymerization.
- Drop casting.
- Weaving method.

Among them, ink-printing process and weaving method are very popular for fabrication of flexible sensors. In ink-printing process, conductive ink is prepared from the metallic or inorganic particles suspension, which is deposited on to the polymer substrate with curing at high temperature. Whereas, in weaving method, conductive fibers are weaved on the textile to form fresh wearable known as e-textile. They can sustain high pressure, have high robustness, versatility and can be used in biomedical engineering and healthcare application.

Some of the synthesis processes commonly used for fabrication of flexible sensor are discussed below. For example, Dubourg et al. synthesized flexible humidity sensor using laser ablation method on PET (poly-ethylene terephthalate) and used screen printed $TiO_2$ as the active material which is shown in Fig. 2 (WDubourg et al. 2017).

Lou et al. have synthesized electronic skin based on piezo-resistive pressure sensor using polyaniline hollow nanospheres as the active component. The sensor was fabricated using PDMS substrate through conventional photolithography method (Luo et al., 2017). Romeo et al. have synthesized nonenzymatic electrochemical

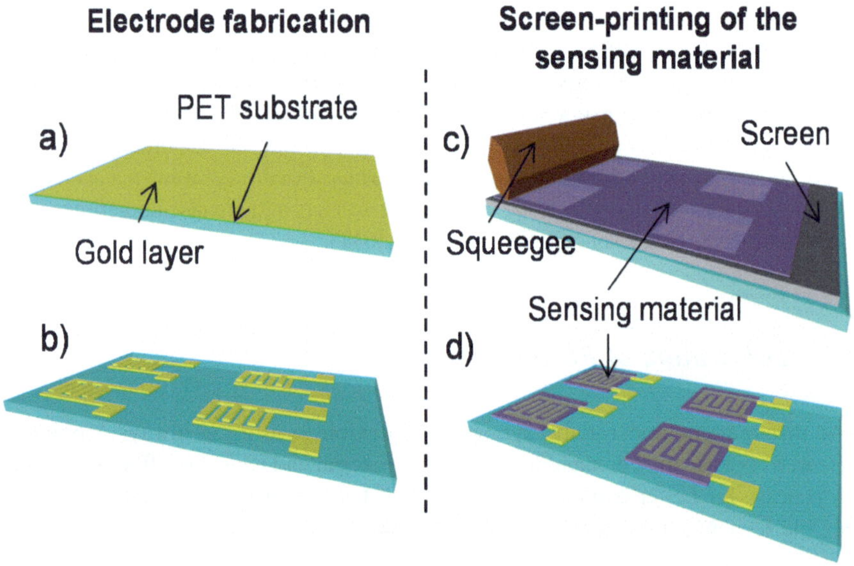


**Fig. 2** Schematic representation showing different steps for the synthesis of humidity sensor (**a**) substrate deposited with gold layer, (**b**) laser ablation of the deposited gold layer, (**c**) incorporation of the $TiO_2$ material through screen printing, and (**d**) the whole prepared sensor. Adapted from Ref. (WDubourg et al. 2017)

flexible glucose sensor using inkjet printing for tear fluid analysis. For the synthesis of the flexible sensor CuO microparticles were used as the active sensing material and polyethylene terephthalate as the substrate for electrode preparation (Romeo et al. 2018). Kim et al. have synthesized piezo-resistive electronic skin of high flexibility, low resistivity of bimodal nature that is both temperature and pressure sensor using inkjet process for the synthesis of core of the sensor. They have used organic conductive elastomer PEDOT:PSS/polyurethane dispersion as the substrate and silver nanoparticle as the active sensing material for flexible sensor fabrication (Kim et al. 2017).

### *1.3 Signal Transduction*

In the designing of sensor, role of transducers is very important. Depending on the sensing principle, the stress induced on the material causes electrical parameter variation, which can be detected and produced in the form of signal. The signal will be detected or read by the transducer and it will provide us the information in understandable format. The transducer can be of various types (Takei et al. 2015). The most commonly used signal transducers are as follows:

- Piezoelectricity.
- Resistivity.
- Capacitive.
- Triboelectricity.
- Field-effect transistor (FET).
- Optical.
- Electrical.
- Amperometry.

Among these transduction methods, the most famous is the piezoelectric transduction method, where voltage is generated in accordance to the applied pressure. In resistivity transduction method, variation of pressure and strain causes change in resistance, which is caused by structural deformation of the flexible sensor material. Capacitive transduction depends on the change in the dielectric constant between distances of the electrode caused by the stress induced on the material. Electrochemical sensing technique is also very popular, owing to their high sensitive and selective, which is generated as a result of electron transfer reaction between the analyte and the receptors.

### *1.4 Salient Features of Flexible or Wearable Sensor*

For large-scale commercialization of flexible sensor, some important features must be considered:

- Gauge factor or sensitivity.
- Linearity.
- Self-powering.
- Self-healing.
- Self-cleaning.
- Transparency.

More stretchable the conductor more is their use as an interconnecting material and electrode material, while high piezoresistive material shows potential application in electrochemical and mechanical sensor (Zhao et al. 2017). Linearity refers to the relative change of the signal with respect to the stress applied, which is represented in the form of straight line. Mostly the strain shows linearity in less strain conditions and nonlinearity in large strain conditions. Nonlinearity of sensor occurs due to morphology deformation, possible occurred due to stress. So, highly stretchable sensor must have high linearity with challenging sensing property.

Electronic devices may get damaged after working for several times, therefore, the self-healing property of the electronic devices both electrically and mechanically is a powerful restoration phenomenon. The incorporation of thermally sensitive ionic liquids, self-healing property of the sensors can be improved. Patchable pressure sensor with incorporated storage and power generation devices enhances the self-powering system in the sensor. Supercapacitor and triboelectric nanogenerator (TENG) is incorporated to the sensor to increase their self-powering property. Li et al. have synthesized a triboelectric transducer-based sensor using nanowire arrays deposited on Kapton substrate through ion etching. This sensor has contact angle of 152°, which allows easy removal of water along with dirt and other contaminants enabling their self-cleaning property (Li et al. 2015). In addition, to all these properties, the sensor must be optically transparent so that it must be invisible during daily activities. Roh et al. have synthesized highly transparent, 100% stretchable, with 62.3 gauge factor strain sensor using PU-PEDOT:PSS/SWCNT/PU-PEDOT:PSS hybrid (Roh et al. 2015).

## 2 Flexible Substrate for Healthcare Application

People nowadays are very much fretful about their longevity and lifestyle, which made them progressively more interested in healthcare and daily life care. With the increase in the population age, numerous unwanted diseases captured our life, which needs to be diagnosed as early as possible, so the facility of daily rapid diagnosis of the disease through detection of human bio-signals at every given location and time is very important at the present time. Personal health monitoring devices not only facilitates the elderly human beings but also help in real-time checkup to professional athletes and detection of early damages caused in various people suffering from chronic ailment (Christodouleas et al. 2018). The discovery of sensor has transformed human life to the next level by sensing or monitoring a disease in few seconds or minutes. Fast diagnosis has broad spectrum of function in various disciplines like food safety, clinical medicine, monitoring of environmental pollution, immunoassays, clinical medicine, etc. Development of quick, inexpensive, portable, stable, accurate, point-of-care diagnostic (POC)assay for timely detection of disease to solve health care problem is of great demand in current time, because of the absence of laboratory facilities, trained personnel, adequate financial support in the field of clinic and health care (Shafiee et al. 2015). POC diagnoses provide us with rapid and timely detection, monitoring, and counseling of disease for better clinical management. POC is better than conventional detection technique because they do not require heavy instruments, trained personnel, and the patient does not need to travel to the hospital, have high accuracy, and cost-effectiveness.

Flexible and wearable sensor with high-quality signal transduction is nowadays used for POC diagnosis based on the detection of various parameters in human beings. These sensors are used for healthcare applications of human beings in the form of several wearable devices like bands, watches, sunglasses, clothes, etc. (Nayak et al. 2016). However, these types of wearable sensors are still underdeveloped because of bulky circuit system, complicated power supply, poor sensitivity and little skin contact, and limited detection toward multiple bio-signals. The main advantages of flexible sensors with respect to conventional sensors are that they are thin, flexible, devoid of mechanical deformation, easy to fabricate, inexpensive, made up from easily available material, disposable, easy to use, have easy interaction with the analyte, have wearable design, multifunctional sensing application, and most importantly can be miniaturized through various nanoscale morphologies, which keep them closer contact to body (Ha et al. 2018). The flexible sensor can easily detect three types of human bio-signals:

1. Physical.
2. Biochemical.
3. Electrophysiological.

Physical health indicator comprises of blood pressure, pulse rate, motion of limb, walking, respiratory rate, temperature, humidity, etc. Biochemical indicator comprises of body fluids, metabolites, proteins, glucose, cholesterol, ascorbic acid, uric acid, pH, and blood oxygen. Electrophysiological signals include electromyography (EMG), electrocardiograms (ECGs), and electroencephalography (EEG). Based on detection of these bio-signals, different flexible sensors are designed for healthcare applications (Segev-Bar and Haick 2013). Some of the recent flexible sensors like pressure, temperature, pH-based sensors are discussed in the next sections:

## *2.1 Pressure or Strain Sensor*

One of the standardized applications of flexible sensor is pressure or strain sensor. These types of pressure sensor are mainly based on piezoelectric, piezocapacitive, and piezoresistive transducers, used for detection of various human motions for older disable long-suffering patient, monitoring sports and healthcare performances, rehabilitation, as well as physical therapy (Segev-Bar and Haick 2013). Pressure sensor mainly detects the physiological change which is caused by bending and stretching motion of the body like bending of the hand, legs, arms, motion of chest, face, movements, due to emotional expression, speaking, breathing, tremor caused due to Parkinson disease. From the pressure sensor, we can also detect various cardiovascular diseases through monitoring, blood pressure, pulse rate, etc. Using pressure and strain sensor of high sensitivity in the form of personal devices, which can regularly and periodically monitor heart rate and blood pressure, one can eliminate various diseases like arrhythmia and hypertension. Pressure miniaturized sensor can also detect health abnormalities like chronic lung diseases e.g. asthma, apnea, and respiration rate problem, with the measurement of movement caused in trachea, movements caused due to expiratory and inspiratory breathing, thoracic cavity expansion and contraction (Zang et al. 2015).

Yeh et al. have synthesized implantable wireless pressure sensor for long period monitoring of cardiovascular pressure. This type of sensor is inserted into the stent inside the blood vessel. They have used Parylene-C as the biocompatible polymer and PDMS as the substrate. The sensitivity of sensor was approximately $6.19 \times 10^{-2}$ $kPa^{-1}$and the linearity of the sensor over the ranges of 0–6 kPa and 6–33 kPa, which is 91.5% and 87.6%, respectively (Yeh et al. 2019). Xu et al. have synthesized flexible pressure sensor using silver-plated polyester (PET) and polyaniline (PANI) as the conductive polymers, which is polymerized in situ on the PET fabric for the detection of various body movements. The use of silver imposed the sensor with antibacterial and antimicrobial properties. The response time of the sensor is 0.2–0.3 s (Xu et al. 2019). Pignanelli et al. have synthesized PDMS substrate rooted flexible pressure sensor based on capacitive sensing for rehabilitation and health monitoring which is shown in Fig. 3. The pressure sensor was prepared through four different patterning methods i.e. photolithography, replica model, inverse mold, and nonuniform microstructure. The sensitivity of the pressure sensor is found to be 0.7298 $kPa^{-1}$ (Pignanelli et al. 2019). Kou et al. have synthesized micro-patterned sandwiched PDMS/graphene composite-based pressure sensor for wireless detection of movements of human facial and muscle movement shown in Fig. 4. Gold pattern was used as the active electrode material, the sensitivity of the sensor was found to be 0.24 $kPa^{-1}$ in the low-pressure regime and in the high-pressure regime it is of 0.0078 $kPa^{-1}$ (Kou et al. 2018). Jang and his coworkers have fabricated piezoresistive flexible pulse sensor which can detect the human pulse. In this sensor, the PEDOT:PSS was used as sensing flexible film and copper tape was used as the two counter electrode and on the other side the silver paste electrode was used at the edge of the sensor. The sensitivity of the sensor was measured to be 62.56 $kPa^{-1}$ in low-pressure range and 8.32 $kPa^{-1}$ in the high-pressure range (Jang et al. 2019).

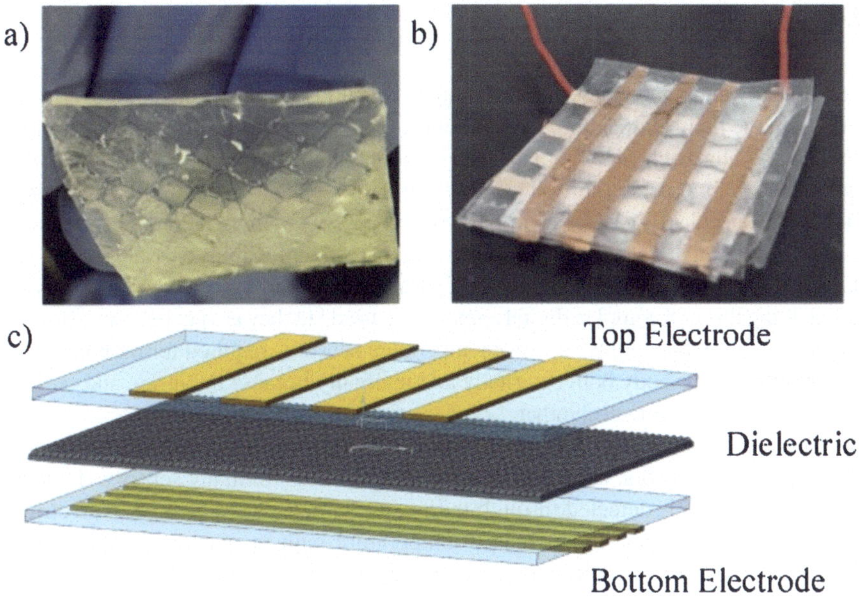


**Fig. 3** Illustration showing (**a**) different layers of the flexible electrode (**b**) camera image of the fabricated electrode (b) full set up of the sensor. Taken permission from (Pignanelli et al. 2019)

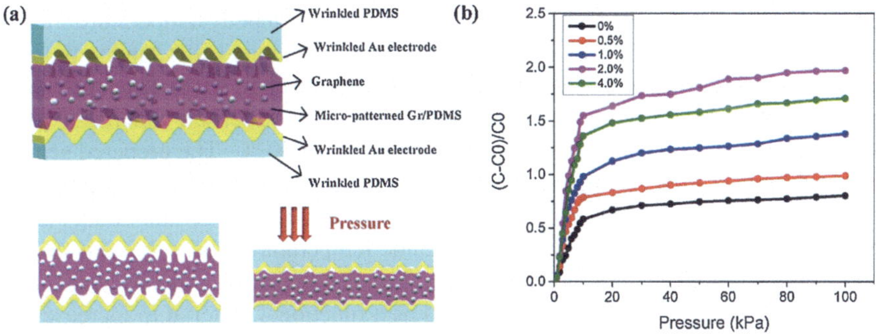


**Fig. 4** Diagrammatic representation of (**a**) the whole pressure sensor (**b**) change in capacitance with different concentration of graphene. Taken permission from (Kou et al. 2018)

## 2.2 Temperature Sensor

Human body temperature does not vary much, but certain activities or situation leads to change in the body temperature like emotions, illness, and certain activity. The normal body temperature of healthy human beings is found in the range of 35–37.5 °C (Moser and Gijis, 2007). A person suffering from several illnesses like hypothermia, hyperthermia, cardiovascular diseases, healing of wound causes certain variation in temperature and needs effective detection devices for better treatment (Sibinski et al. 2010). Till date, point temperature measurement of a particular area through paste on sensor, spatial imaging through complex digital cameras are used but they are not the point-of-care diagnosis, so flexible highly sensitive temperature sensor is required for monitoring of temperature of our body both externally and internally because the inside and outside temperature varies from each other.

Temperature sensors of flexible nature are based on thermoelectric effect, thermocouples, thermistors, and various optical approaches. In many sensors, the temperature variation is detected through thermistor, which causes change in electrical resistance in two types: for increased temperature with increasing resistance it is positive temperature coefficient (PTC) and for increasing temperature with decreased resistance it is negative temperature coefficient (NTC) (Shih et al. 2010). Aliane et al. have developed temperature sensor through screen printing technique following thermistor effect using PET and PEN as substrate. The prepared sensor showed good sensitivity of 0.06 V/1C (Aliane et al. 2014). Huang et al. have fabricated flexible temperature sensor with polyvinylidene fluoride (PVDF) and polyethelene oxide (PEO) filled with graphite for monitoring of body temperature shown in Fig. 5. The prepared sensor showed high accuracy of about 0.1 °C and with temperature sensing ranging from about 25 to 42 °C (Huang et al. 2018). Hao et al. have synthesized thermochromic ink-based flexible temperature sensor, which can visualize the temperature change ranging from about 26 to 50 °C. The prepared sensor was synthesized using xanthan gum and pectin gum as the substrate. The prepared sensor has sensitivity of 53–130%°C$^{-1}$ in above-mentioned temperature range (Hao et al. 2018). Kim et al. have fabricated organic temperature sensor through spatial atmospheric atomic layer deposition (SAALD). The sensor was encapsulated with thin film of $Al_2O_3$ over PVA, which is used as a functional material to increase the linearity, endurance, and repeatability, in the sensor polyethylene naphthalate was used as the substrate. The sensor can detect temperature variation of about 25 °C–90 °C with high stability and reproducibility (Kim et al. 2019).

## 2.3 pH Sensor

The chemical state of the body can be easily determined by measuring the pH of different fluids present in the body. Body pH value in different area can be varied due to various factors like sweat pH can be changed by some skin diseases like

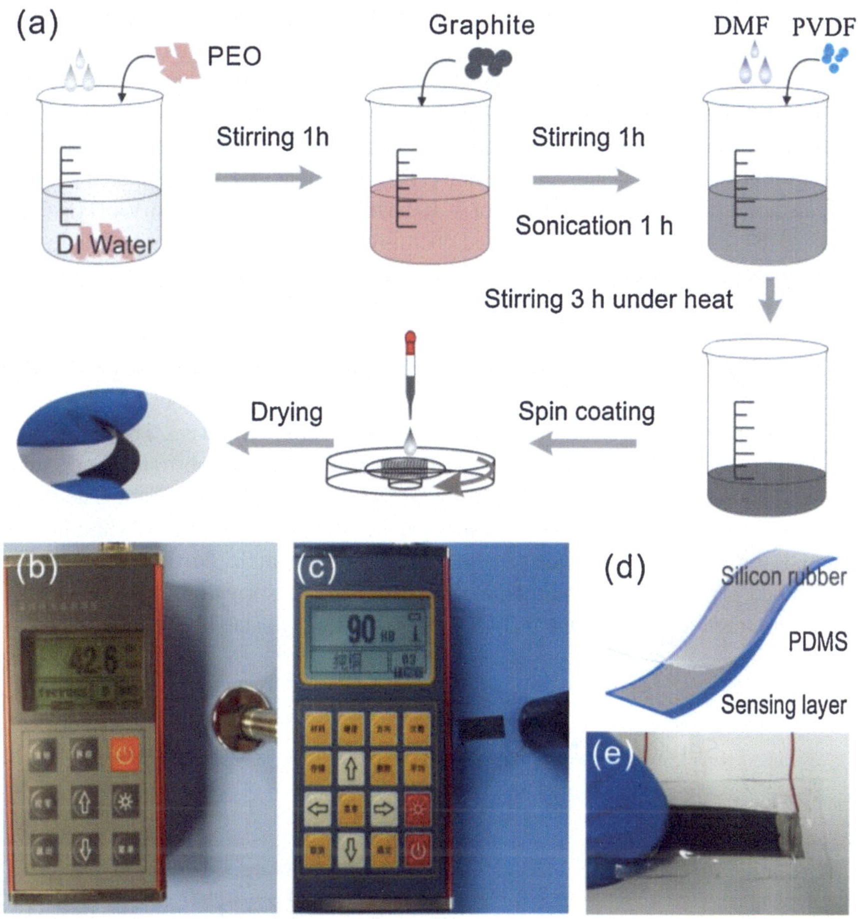


**Fig. 5** Schematic representation showing (**a**) synthesis procedure of the prepared temperature sensor; temperature sensing layer thickness and hardness (**b**, **c**) (**d**) graphical representation of the temperature sensor (**e**) camera image of the fabricated sensor. Taken permission from (Huang et al. 2018)

fungal infection, dermatitis. Similarly, area of wound changes the pH with the increase in the amount of bacterial colonization inside it, which increases the difficulty of the wound by slow drying (Rahimi et al. 2016). Increase in gastric juice, whose pH is 1.0–3.5 gives alert of any gastrointestinal diseases. The pH of heart, brain, liver, skeletal muscle always remains neutral but any change in the pH indicates abnormal health. Conventional pH monitoring system includes glass probes, which is very much expensive, required removal after dressing, required trained personnel, cannot be detected at home, in short, such kind of point-of-care diagnosis is not available (Guinovart et al. 2014).

Recent years provide us with several sensing technologies, which can detect the pH of the body through several transduction methods like electrical, optical, and chemo-mechanical transduction method. Nowadays flexible pH test strip sensors are used for the detection of pH, which is fitted on optical fibers have high accuracy, sensitivity, and flexibility. Some of the pH sensors used for healthcare application are discussed here. Rahimi et al. have fabricated transparent and flexible potentiometric pH sensor for wound pH detection through cost-effective laser scribing method. The flexible substrate used here is indium tin oxide (ITO) coated on PET. The prepared pH sensor can detect physiological pH value of 4–10 having sensitivity of −55 mV/pH (Rahimi et al. 2018). Liu et al. have synthesized chemi-resistive flexible pH for wearable biomedical electronic devices, water quality maintaining for better environmental condition, and food safety monitoring. The sensor was prepared through growing single-walled carbon nanotube (SWCNT) on PET substrate. The sensor has great linearity, good performances, less power consumption, and high sensitivity in the pH range of 5–9 with a response time of 22.6 s (Liu et al. 2016). Smith et al. have fabricated well conductive flexible pH sensor using cotton yarn for monitoring pH of body sweat. They have used the dipping and drying method for the synthesis of the flexible substrate using multi-walled carbon nanotube (MWCNT), PEDOT:PSS, PANI, and cotton warn. The sensitivity of the sensor is around $-61 \pm 2$ mV $pH^{-1}$ in the pH range of 2–12 (Smith et al. 2019). Rahimi et al. have fabricated economical, biocompatible, flexible pH sensor for wound bed pH determination as shown in Fig. 6. The sensor was prepared by PANI-coated commercial paper as the flexible substrate with high stability, repeatability, and

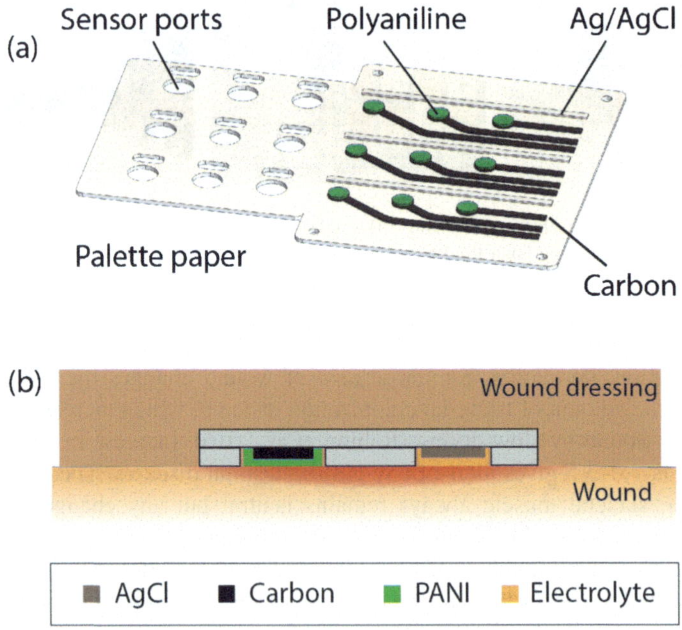


**Fig. 6** (**a**) Graphical representation of the pH sensor (**b**) working principle of the sensor on the wound bed. Adapted from (Rahimi et al. 2016)

good linearity of $R^2 = 0.9734$. The sensitivity of the electrode is −50 mV/pH with detection range of pH 4–10 in different buffer solutions (Rahimi et al. 2016).

## 2.4 E-skin

In spite of the advancement of the wearable sensor, they still have certain demerits like lack of flexibility, brittleness, easy deformation, low stability, bulkiness, requirement of adhesives to settle on skin renders them unfavorable for healthcare monitoring (Hammock et al. 2013). Therefore, user-friendly, entirely flexible, miniaturized device for health monitoring is required, which can withstand extreme level of external strain. Electronic skin is the sensor, which has multimodal sensing property and ability to detect various bio-signals of human body. They have high flexibility, high accuracy, and mimics human skin in various activity. It is one of the great achievements of flexible sensor also known as e-skin, which can adhere to the human skin, made up flexible material and can detect various physiological parameters of body. They are formed of flexible pixel arrays where various external stimuli like temperature, pressure, strain are converted simultaneously into electrical signals (Núñez et al. 2017). E-skin devices have same water vapors permeability, thickness, elastic modulus, thermal mass, temperature as that of our skin. These types of soft sensors, which have mechanically invisible electronic interfaces with the human skin, get laminated at the epidermis of the skin with van der Waals forces and detect various bio-signals for health monitoring. Through e-skin we can detect various cardiovascular information, brain activity, movement of eyes, body temperature, body oxygen level, electrolytes present in epidermis, as well as pH of sweat (Chortos and Bao 2014). For such kind of versatility, the electronic skin must have self-healing, self-powering, self-cleaning properties with high flexibility. Researchers are studying to discover new materials, which have all above properties that can be incorporated into electronic skin for best possible health monitoring.

Some of the articles reported for the fabrication of electronic skin used in different healthcare applications are discussed here. Shi et al. have synthesized electronic skin for detection of pulse rate and movement of thumb by mounting it in the human body illustrated in Fig. 7. They prepared nanostructured Cu electrode through electrode-less deposition method and transferred it into adhesive flexible film of PET to form the sensor. The prepared sensor shows high sensitivity of 2.22 $kPa^{-1}$ (Shi et al. 2019). Suen et al. have fabricated a flexible tactile electronic skin sensor with high flexibility, high force sensitivity for detection of multiple stimuli like bending, pressure, and torsion forces. The space between the two electrodes in the sensing platform is interlocked by zinc oxide nano rods, which increases the conductivity of the system. The substrate of the electrode is PDMS layer, which inputs great flexibility to the system. The proposed sensor has high sensitivity of −0.768 $kPa^{-1}$, high recovery time of 12 ms with great response time of 14 ms (Suen et al. 2018). He et al. have synthesized multifunctional electronic

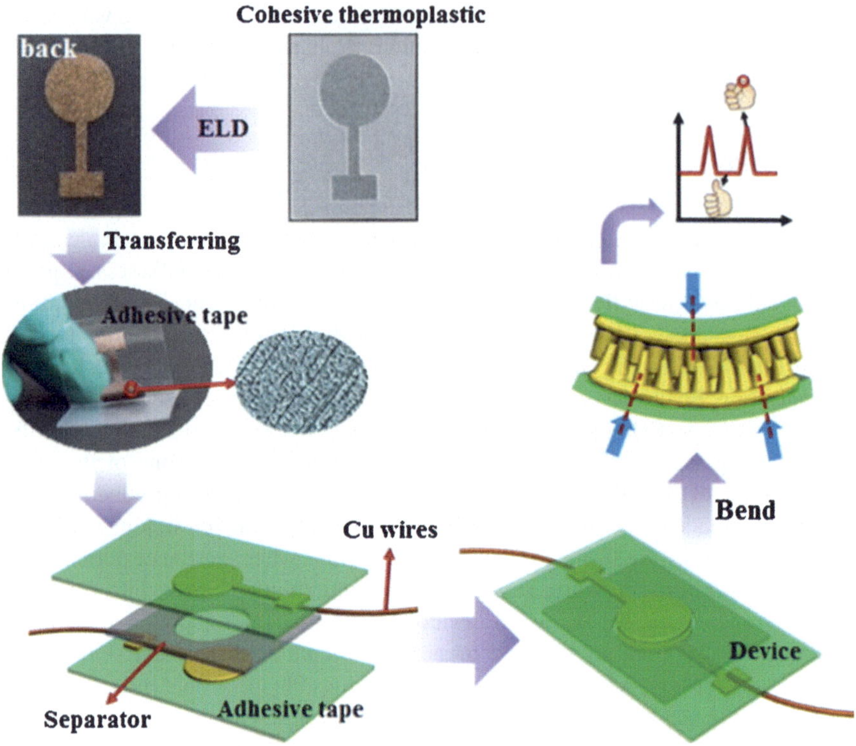


**Fig. 7** Graphical illustration showing the working principle and set up of electronic skin-based sensor for movement of human body. Taken permission from (Shi et al. 2019)

skin, which is hybridized with tetrapod ZnO (T-ZnO) nanostructures and piezoelectric polyvinylidene fluoride (PVDF) anchored on PET fabric screen. The prepared electrode has self-powering, self-cleaning, and gas-sensing properties, thus used in health monitoring as well as atmospheric gas detection (He et al. 2017). Li et al. have synthesized electronic skin sensor for detection of vibration of vocal cord, pulse rate, elbow and finger movement with high stability and recyclability. They have used silver nanoparticle paste hybridized hydrophobic polyolefin elastomer (POE) membrane as the flexible substrate. The prepared sensor shows fast response time of 10 ms with gauge factor of 3953 and withstand 30% of strength (Li et al. 2019).

## 2.5 *Flexible Substrate for Biosensing Application*

Devices that have biological component to perform the body healthcare and disease management are known as biosensor (Malhotra and Chaubey 2003). Choice of the analyte mainly depends on the application of the biosensor in different fields. Due to

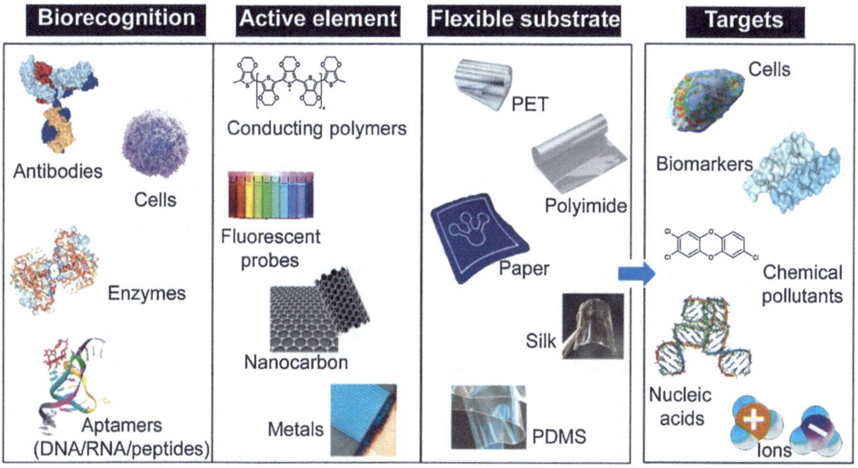


**Fig. 8** Schematic representation of the components needed to prepare flexible biosensor. Adapted from (Xu et al. 2018b)

plethora of utilization, wearable biosensors have achieved great responses in the field of healthcare and ailment diagnosis. This type of sensor detects various biological components like different biomarkers, biomolecules like dopamine, lactate, glucose, ascorbic acid, DNA, RNA, whole cells for fatal diseases at very lower detection. Flexible biosensor usually consists of several components shown in Fig. 8 (Xu et al. 2018b):

1. Substrate which provides flexible support to the sensor system mainly made up of various synthetic polymers, papers, or textiles.
2. Bio-recognition sites like antibodies, aptamers, enzymes, oligonucleotide, which is specific and selective for particular analyte used.
3. An active material, which increases the conductivity and helps in signal transduction depending on detection mechanism from bioreceptor and analyte.
4. Specific analyte molecule depending upon the application.

Flexible substrate-based biosensors have high transparency, strechability, stability, biocompatibility, good portability, have light weight, inexpensive, and disposable. These types of sensors have inherent power unit, data acquisition mechanism, and signal transmission through wireless modules. Biosensors have mainly electrochemical method for detection like voltammetry, amperometry, field-effect transistor (FET), etc. because electrochemical methods possess high selectivity and sensitivity. Biosensors flexible electrodes are mainly fabricated through sputtering, lithography, drop casting, inkjet printing, stamping, electrodeposition, or chemical vapor deposition (CVD) methods (Shah et al. 2015). As this book chapter is based on healthcare application, we have discussed the use of flexible biosensors in physiological environment. Flexible biosensors in healthcare application can be used as external to the human body, internal to the body, or can be applied for the detection of analytes/biomarker taken from biological fluids.

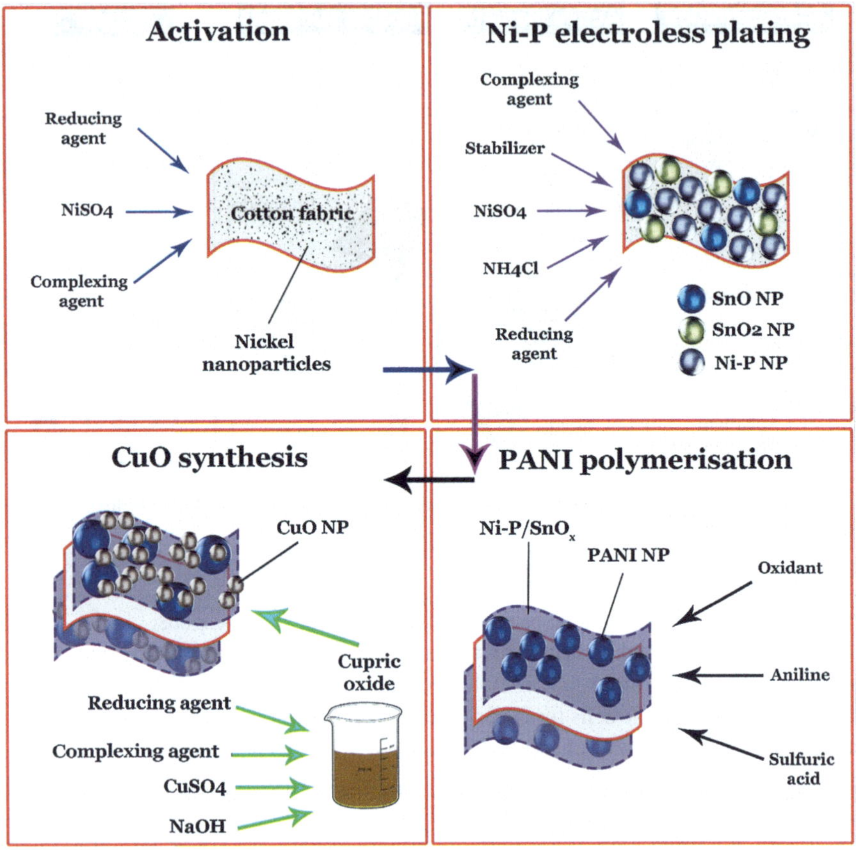


**Fig. 9** Illustration showing the fabrication process of the flexible biosensor made up of $NiP_{0.1}$-$SnO_x$/PANI/CuO (NTPC) nanocomposite. Adapted from (Sedighi et al. 2019)

Sha and his coworkers have synthesized flexible, low-cost, nonenzymatic uric acid sensor through electrochemical detection technique (Sha et al. 2019). They have hydrothermally synthesized $MoS_2$ grown on aluminum foil, which is then place on flexible cellophane tape. The prepared composite has sensitivity of 98.3 ± 1 nA $\mu M^{-1}$ and detection limit of 10–400 μM. Sedighi et al. have synthesized glucose sensor of nonenzymatic type for the detection of glucose level of human blood for diabetic patients which is shown in Fig. 9. They have synthesized CuO, $NiP_{0.1}$-$SnO_x$, and PANI nanoparticle on flexible carbon fabrics to increase the conductivity of the sensor. The prepared sensor has sensitivity of 1625 μA $mM^{-1}$ in the ranges of 0.001–1 mM and 1325 $\mu AmM^{-1}$ $cm^{-2}$ in the range of 1–10 mM (Sedighi et al. 2019). Munje et al. have synthesized flexible enzyme based biosensor for the detection of glucose in sweat through the electrochemical detection technique. Flexible polyamide substrate was used as the supporting material in which gold/zinc oxide thin film was grown as the electrode material. The sensor has detection limit of 0.01–200 mg/

dL with sensitivity of 19.5 μA $mM^{-1}$ $cm^{-2}$ (Munje et al. 2017). Xuan et al. have fabricated a wearable glucose sensor for the detection of glucose in human sweat through amperometry as detection technique. They have used flexible polyimide as the substrate where reduced graphene oxide (rGO) nanostructures are grown through microfabrication method and used as the electrode. The sensor has sensitivity of 48 μA $mM^{-1}$ $cm^{-2}$ with detection limit of 0–2.4 mM (Xuan et al. 2018). Kang et al. have synthesized wearable glucose biosensor for routine monitoring of glucose level in chronic diabetic suffering patients using thick polyimide (PI) flexible substrate. The substrate was grown with networks of single-walled carbon nanotube (SWCNT) films, which is further surface functionalized with glucose oxidase (GOD). The prepared sensor shows sensitivity of 41.397 $\mu M^{-1}$ within the detection range of 50 μM to 1 mM (Kang et al. 2019). Jia et al. have synthesized flexible enzymatic tattoo-based wearable biosensor for lactate sensing from human perspiration. They have used carbon nanotube (CNT) and tetrathiafulvalene (TTF) functionalized with lactate oxidase (LOx) as the electrode and flexible GORE-TEX textile as the substrate. The sensitivity of the sensor is found10.31 μA/mM $cm^2$ with high linearity and stability (Jia et al. 2013). Some of the flexible substrate based biosensors are summarized in Table 1 [66–75].

**Table 1** Summary of some of the flexible substrate-based biosensor for healthcare application

| S. N. | Analyte | Flexible substrate | Detection mode | Detection range | References |
|---|---|---|---|---|---|
| 1. | Lactate | PET | Potentiometric | 0.2–3 mM | Chou et al. (2018) |
| 2. | Glucose | R-go | Amperometric | 2 μM to 13 mM | Wang et al. (2017) |
| 3. | Urea | PANI | Amperometric | 0.04 mM | Meibodi and Haghjoo (2014) |
| 4. | Glucose | Bacterial cellulose | CV | 0–50 mM | Lv et al. (2018) |
| 5. | Glucose | CSF | Chronoamperometry | 0–5 mM | Chao et al. (2018) |
| 6. | Lactate | Tatoo-based paper | Amperometric | 1–20 | Jia et al. (2013) |
| 7. | DNA | Polyimide | Low-temperature solution-processed IZO TFTs | – | Jung et al. (2014) |
| 8. | Sodium | PET | Potentiometric | 0.1–100 | Bandodkar et al. (2014) |
| 9. | Glutamate | PDMS | Chronoamperometry | 1 μM to 1400 μM | Nguyen et al. (2019) |
| 10. | Glucose | Polyurethan | Amperometric | 0–20 mM | Fang et al. (2018) |

*PET* polyethylene terephthate, *PDMS* polydimethyl siloxane, *R-go* reduced graphene oxide, *CSF* carbonized silk fabric, *PANI* polyaniline, *CV* cyclic voltammetry, *TFT* thin-film transistors(s), *IZO* In-Zn-O

## 3 Future Aspects and Conclusion

In the last few years, there is rapid growth in the landscape of flexible sensor for various applications with increase in the intrinsic searchability, engineering of sensor geometry, inclusion of active conductive material, incorporation of multimodal sensing platform, and use of new efficient signal transduction technique. This chapter includes versatile use of flexible sensor in biomedical and healthcare application. We have discussed here various types of sensors under different categories like pressure, temperature, and pH sensors. Large emphasis is nowadays given in multifunctional wearable human e-skin sensors, which have self-healing, self-powering, self-cleaning properties with great biocompatibility, elasticity, and flexibility, which can be easily attached to human skin for vital biological signal detection. Despite the emerging popularity of the flexible sensor, there are certain tribulations, which need great attention like:

1. The study of the sensing capabilities of the flexible sensors under harsh conditions like high humidity, high and very low temperature is rarely studied; they need certain super-hydrophobic or temperature resistive coating material for better performances.
2. The devoted attention toward increasing the sensitivity of the flexible sensor has surpassed other problems like packaging, integration of power, signal, and data processing unit in a small area. Thus, the sensor needs better packaging to survive in unfriendly environment.
3. Awareness must be increased about the use of this type of electronic skin-like wearable biosensor for quick detection and recovery of the diseases. People in spite of the great development of the wearable flexible sensor rely on conventional detection techniques, therefore, flexible sensor needs enhanced public promotion to enhance their popularity among common people.
4. Battery-free, self-powering human-friendly multiple sensor devices are also lacking in the market, which need to be studied more.
5. The fabrication method of the sensor with ultrafast laser-induced chemical technology should be carefully considered and explored greatly.

With the settlement of the above issues and combining unique effort of the multidisciplinary science fields like material science, nanotechnology, microelectronics, chemistry, and physics can amplify the rapid development of the flexible substrate in near future.

**Acknowledgements** *Author declaration*: Ms. Karfa and Mr. Majhi have given the major contribution in writing this book chapter along with drawing the figures and tables, taking the copyright permission, etc.

## References


Akyildiz IF, Weilian S, Sankarasubramaniam Y, Cayirci E (2002) Wireless sensor networks: a survey. Comput Netw 38(4):393–422

Aliane A, Fischer V, Galliari M, Tournon L, Gwoziecki R, Serbutoviez C, Chartier I, Coppard R (2014) Enhanced printed temperature sensors on flexible substrate. Microelectron J 45(12):1621–1626

Bandodkar AJ, Molinnus D, Mirza O, Guinovart T, Windmiller JR, Valdés-Ramírez G, Andrade FJ, Schöning MJ, Wang J (2014) Epidermal tattoo potentiometric sodium sensors with wireless signal transduction for continuous non-invasive sweat monitoring. Biosens Bioelectron 54:603–609

Barlow F, Lostetter A, Elshabini A (2002) Low cost flex substrates for miniaturized electronic assemblies. Microelectron Reliab 42(7):1091–1099

Castillejo P, Martinez J-F, Rodriguez-Molina J, Cuerva A (2013) Integration of wearable devices in a wireless sensor network for an E-health application. IEEE Wirel Commun 20(4):38–49

Chao HC, Ran R, Yang Z, Lv R, Shen W, Kang F, Huang Z-H (2018) An efficient flexible electrochemical glucose sensor based on carbon nanotubes/carbonized silk fabrics decorated with Pt microspheres. Sensors Actuators B Chem 256:63–70

Chong C-Y, Kumar SP (2003) Sensor networks: evolution, opportunities, and challenges. Proc IEEE 91(8):1247–1256

Chortos A, Bao Z (2014) Skin-inspired electronic devices. Mater Today 17(7):321–331

Chou J-C, Yan S-J, Liao Y-H, Lai C-H, Chen J-S, Chen H-Y, Wu C-Y, You-Xiang W (2018) Reaction of NiO film on flexible substrates with buffer solutions and application to flexible arrayed lactate biosensor. Microelectron Reliab 83:249–253

Christodouleas DC, Kaur B, Chorti P (2018) From point-of-care testing to eHealth diagnostic devices (eDiagnostics). ACS Central Sci 4(12):1600–1616

Chun S, Son W, Choi C (2018) Flexible pressure sensors using highly-oriented and free-standing carbon nanotube sheets. Carbon 139:586–592

Fang Y, Wang S, Liu Y, Xu Z, Zhang K, Guo Y (2018) Development of cu nanoflowers modified the flexible needle-type microelectrode and its application in continuous monitoring glucose in vivo. Biosens Bioelectron 110:44–51

Guinovart T, Valdés-Ramírez G, Windmiller JR, Andrade FJ, Wang J (2014) Bandage-based wearable potentiometric sensor for monitoring wound pH. Electroanalysis 26(6):1345–1353

Guth U, Vonau W, Zosel J (2009) Recent developments in electrochemical sensor application and technology—a review. Meas Sci Technol 20(4):042002

Ha M, Lim S, Ko H (2018) Wearable and flexible sensors for user-interactive health-monitoring devices. J Mater Chem B 6(24):4043–4064

Hammock ML, Chortos A, Tee BC-K, Tok JB-H, Bao Z (2013) 25th anniversary article: the evolution of electronic skin (e-skin): a brief history, design considerations, and recent progress. Adv Mater 25(42):5997–6038

Han S-T, Peng H, Sun Q, Venkatesh S, Chung K-S, Lau SC, Zhou Y, Roy VAL (2017) An overview of the development of flexible sensors. Adv Mater 29(33):1700375

Hao L, Ding J, Yuan N, Xu J, Zhou X, Dai S, Chen B (2018) Visual and flexible temperature sensor based on a pectin-xanthan gum blend film. Org Electron 59:243–246

He H, Yongming F, Zang W, Wang Q, Xing L, Zhang Y, Xue X (2017) A flexible self-powered T-ZnO/PVDF/fabric electronic-skin with multi-functions of tactile-perception, atmosphere-detection and self-clean. Nano Energy 31:37–48

Huang Y, Zeng X, Wang W, Guo X, Hao C, Pan W, Liu P et al (2018) High-resolution flexible temperature sensor based graphite-filled polyethylene oxide and polyvinylidene fluoride composites for body temperature monitoring. Sensors Actuators A Phys 278:1–10

Jang H-H, Park J-S, Choi B (2019) Flexible piezoresistive pulse sensor using biomimetic PDMS mold replicated negatively from shark skin and PEDOT: PSS thin film. Sensors Actuators A Phys 286:107–114

Jia W, Bandodkar AJ, Valdés-Ramírez G, Windmiller JR, Yang Z, Ramírez J, Chan G, Wang J (2013) Electrochemical tattoo biosensors for real-time noninvasive lactate monitoring in human perspiration. Anal Chem 85(14):6553–6560

Jung J, Kim SJ, Lee KW, Yoon DH, Kim Y-g, Kwak HY, Dugasani SR, Park SH, Kim HJ (2014) Approaches to label-free flexible DNA biosensors using low-temperature solution-processed InZnO thin-film transistors. Biosens Bioelectron 55:99–105

Kang B-C, Park B-S, Ha T-J (2019) Highly sensitive wearable glucose sensor systems based on functionalized single-wall carbon nanotubes with glucose oxidase-nafion composites. Appl Surf Sci 470:13–18

Kim K, Jung M, Kim B, Kim J, Shin K, Kwon O-S, Jeon S (2017) Low-voltage, high-sensitivity and high-reliability bimodal sensor array with fully inkjet-printed flexible conducting electrode for low power consumption electronic skin. Nano Energy 41:301–307

Kim SW, Rehman MM, Sajid M, Rehman MMU, Gul J, Jo JD, Choi KH (2019) Encapsulation of polyvinyl alcohol based flexible temperature sensor through spatial atmospheric atomic layer deposition system to enhance its lifetime. Thin Solid Films 673:44–51

Kou H, Zhang L, Tan Q, Liu G, Lv W, Lu F, Dong H, Xiong J (2018) Wireless flexible pressure sensor based on micro-patterned Graphene/PDMS composite. Sensors Actuators A Phys 277:150–156

Lee J, Kwon H, Seo J, Shin S, Koo JH, Pang C, Son S et al (2015) Conductive fiber-based ultrasensitive textile pressure sensor for wearable electronics. Adv Mater 27(15):2433–2439

Li X, Yeh M-H, Lin Z-H, Guo H, Yang P-K, Wang J, Wang S, Yu R, Zhang T, Wang ZL (2015) Self-powered triboelectric nanosensor for microfluidics and cavity-confined solution chemistry. ACS Nano 9(11):11056–11063

Li M, Chang K, Zhong W, Xiang C, Wang W, Liu Q, Liu K, Wang Y, Lu Z, Dong W (2019) A highly stretchable, breathable and thermoregulatory electronic skin based on the polyolefin elastomer nanofiber membrane. Appl Surf Sci 486:249–256

Liu L, Shao J, Li X, Zhao Q, Nie B, Xu C, Ding H (2016) High performance flexible pH sensor based on carboxyl-functionalized and DEP aligned SWNTs. Appl Surf Sci 386:405–411

Lou Z, Chen S, Wang L, Shi R, Li L, Jiang K, Chen D, Shen G (2017) Ultrasensitive and ultraflexible e-skins with dual functionalities for wearable electronics. Nano Energy 38:28–35

Luo Y, Wu D, Yang Z, Chen Q, Yu X, Wang M, Lin L, Wang L, Sun D (2019) Direct write of a flexible high-sensitivity pressure sensor with fast response for electronic skins. Org Electron 67:10–18

Lv P, Zhou H, Mensah A, Feng Q, Wang D, Hu X, Cai Y, Lucia LA, Li D, Wei Q (2018) A highly flexible self-powered biosensor for glucose detection by epitaxial deposition of gold nanoparticles on conductive bacterial cellulose. Chem Eng J 351:177–188

Malhotra BD, Chaubey A (2003) Biosensors for clinical diagnostics industry. Sensors Actuators B Chem 91(1–3):117–127

Meibodi ASE, Haghjoo S (2014) Amperometric urea biosensor based on covalently immobilized urease on an electrochemically polymerized film of polyaniline containing MWCNTs. Synth Met 194:1–6

Moser Y, Gijs MAM (2007) Miniaturized flexible temperature sensor. J Microelectromech Syst 16(6):1349–1354

Munje RD, Muthukumar S, Prasad S (2017) Lancet-free and label-free diagnostics of glucose in sweat using zinc oxide based flexible bioelectronics. Sensors Actuators B Chem 238:482–490

Nambiar S, Yeow JTW (2011) Conductive polymer-based sensors for biomedical applications. Biosens Bioelectron 26(5):1825–1832

Nayak S, Blumenfeld NR, Laksanasopin T, Sia SK (2016) Point-of-care diagnostics: recent developments in a connected age. Anal Chem 89(1):102–123

Nguyen TNH, Nolan JK, Park H, Lam S, Fattah M, Page JC, Joe H-E et al (2019) Facile fabrication of flexible glutamate biosensor using direct writing of platinum nanoparticle-based nanocomposite ink. Biosens Bioelectron 131:257–266

Núñez CG, Navaraj WT, Polat EO, Dahiya R (2017) Energy-autonomous, flexible, and transparent tactile skin. Adv Funct Mater 27(18):1606287

Pignanelli J, Schlingman K, Carmichael TB, Rondeau-Gagné S, Ahamed MJ (2019) A comparative analysis of capacitive-based flexible PDMS pressure sensors. Sensors Actuators A Phys 285:427–436

Pu Z, Wang R, Wu J, Yu H, Xu K, Li D (2016) A flexible electrochemical glucose sensor with composite nanostructured surface of the working electrode. Sensors Actuators B Chem 230:801–809

Rahimi R, Ochoa M, Parupudi T, Zhao X, Yazdi IK, Dokmeci MR, Tamayol A, Khademhosseini A, Ziaie B (2016) A low-cost flexible pH sensor array for wound assessment. Sensors Actuators B Chem 229:609–617

Rahimi R, Brener U, Chittiboyina S, Soleimani T, Detwiler DA, Lelièvre SA, Ziaie B (2018) Laser-enabled fabrication of flexible and transparent pH sensor with near-field communication for in-situ monitoring of wound infection. Sensors Actuators B Chem 267:198–207

Roh E, Hwang B-U, Kim D, Kim B-Y, Lee N-E (2015) Stretchable, transparent, ultrasensitive, and patchable strain sensor for human–machine interfaces comprising a nanohybrid of carbon nanotubes and conductive elastomers. ACS Nano 9(6):6252–6261

Romeo A, Moya A, Leung TS, Gabriel G, Villa R, Sánchez S (2018) Inkjet printed flexible non-enzymatic glucose sensor for tear fluid analysis. Appl Mater Today 10:133–141

Sedighi A, Montazer M, Mazinani S (2019) Synthesis of wearable and flexible NiP0. 1-SnOx/PANI/CuO/cotton towards a non-enzymatic glucose sensor. Biosens Bioelectron 135:192–199

Segev-Bar M, Haick H (2013) Flexible sensors based on nanoparticles. ACS Nano 7(10):8366–8378

Sha R, Vishnu N, Badhulika S (2019) $MoS_2$ based ultra-low-cost, flexible, non-enzymatic and non-invasive electrochemical sensor for highly selective detection of uric acid in human urine samples. Sensors Actuators B Chem 279:53–60

Shafiee H, Asghar W, Inci F, Yuksekkaya M, Jahangir M, Zhang MH, Durmus NG, Gurkan UA, Kuritzkes DR, Demirci U (2015) Paper and flexible substrates as materials for biosensing platforms to detect multiple biotargets. Sci Rep 5:8719

Shah S, Smith J, Stowell J, Christen JB (2015) Biosensing platform on a flexible substrate. Sensors Actuators B Chem 210:197–203

Shi Z, Wu XS, Zhang H, Chai H, Li CM, Lu ZS, Ling Y (2019) Flexible electronic skin with nanostructured interfaces via flipping over electroless deposited metal electrodes. J Colloid Interface Sci 534:618–624

Shih W-P, Tsao L-C, Lee C-W, Cheng M-Y, Chang C, Yang Y-J, Fan K-C (2010) Flexible temperature sensor array based on a graphite-polydimethylsiloxane composite. Sensors 10(4):3597–3610

Sibinski M, Jakubowska M, Sloma M (2010) Flexible temperature sensors on fibers. Sensors 10(9):7934–7946

Smith RE, Totti S, Velliou E, Campagnolo P, Hingley-Wilson SM, Ward NI, Varcoe JR, Crean C (2019) Development of a novel highly conductive and flexible cotton yarn for wearable pH sensor technology. Sensors Actuators B Chem 287:338–345

Suen M-S, Lin Y-C, Chen R (2018) A flexible multifunctional tactile sensor using interlocked zinc oxide nanorod arrays for artificial electronic skin. Sensors Actuators A Phys 269:574–584

Takei K, Honda W, Harada S, Arie T, Akita S (2015) Toward flexible and wearable human-interactive health-monitoring devices. Adv Healthc Mater 4(4):487–500

Wang B, Wu Y, Chen Y, Weng B, Li C (2017) Flexible paper sensor fabricated via in situ growth of cu nanoflower on RGO sheets towards amperometrically non-enzymatic detection of glucose. Sensors Actuators B Chem 238:802–808

WDubourg G, Segkos A, Katona J, Radović M, Savić S, Niarchos G, Tsamis C, Crnojević-Bengin V (2017) Fabrication and characterization of flexible and miniaturized humidity sensors using screen-printed $TiO_2$ nanoparticles as sensitive layer. Sensors 17(8):1854

Windmiller J, Ray J, Wang J (2013) Wearable electrochemical sensors and biosensors: a review. Electroanalysis 25(1):29–46

Xu Y, Jiang F, Newbern S, Huang A, Ho C-M, Tai Y-C (2003) Flexible shear-stress sensor skin and its application to unmanned aerial vehicles. Sensors Actuators A Phys 105(3):321–329

Xu F, Li X, Shi Y, Li L, Wang W, Liang H, Liu R (2018a) Recent developments for flexible pressure sensors: a review. Micromachines 9(11):580

Xu M, Obodo D, Yadavalli VK (2018b) The design, fabrication, and applications of flexible biosensing devices—a review. Biosens Bioelectron 110:23–37

Xu R, Wang W, Sun J, Wang Y, Wang C, Ding X, Ma Z, Mao Y, Yu D (2019) A flexible, conductive and simple pressure sensor prepared by electroless silver plated polyester fabric. Colloids and Surfaces A: Physicochemical and Engineering Aspects 578:123554

Xuan X, Yoon HS, Park JY (2018) A wearable electrochemical glucose sensor based on simple and low-cost fabrication supported micro-patterned reduced graphene oxide nanocomposite electrode on flexible substrate. Biosens Bioelectron 109:75–82

Yamada T, Hayamizu Y, Yamamoto Y, Yomogida Y, Izadi-Najafabadi A, Futaba DN, Hata K (2011) A stretchable carbon nanotube strain sensor for human-motion detection. Nat Nanotechnol 6(5):296

Yang G (2006) In: Yang G-Z (ed) Body sensor networks, vol 1. Springer, London

Yeh C-C, Lo S-H, Xu M-X, Yang Y-J (2019) Fabrication of a flexible wireless pressure sensor for intravascular blood pressure monitoring. Microelectron Eng 213:55–61

Zang Y, Zhang F, Di C-a, Zhu D (2015) Advances of flexible pressure sensors toward artificial intelligence and health care applications. Mater Horizons 2(2):140–156

Zhao S, Li J, Cao D, Zhang G, Li J, Li K, Yang Y et al (2017) Recent advancements in flexible and stretchable electrodes for electromechanical sensors: strategies, materials, and features. ACS Appl Mater Interfaces 9(14):12147–12164

# Lab-on-a-Chip Devices for Water Quality Monitoring

**Ashish Kapoor, Sivasamy Balasubramanian, Ponnuchamy Muthamilselvi, Vijay Vaishampayan, and Sivaraman Prabhakar**

## Contents



## Abbreviations

| | |
|---|---|
| AAS | Atomic absorbance spectroscopy |
| BOD | Biochemical oxygen demand |
| BPA | Bisphenol A |
| COC | Cyclic olefin copolymer |
| DFR | Dry film resist |
| DNA | Deoxyribonucleic acid |
| DOC | Dissolved oxygen concentration |
| GC-MS | Gas chromatography-mass spectrometry |
| HLF | Humic-like fluorescence |
| ICP-OES | Inductively coupled plasma optical emission spectrometry |
| LC-MS | Liquid chromatography–mass spectrometry |
| LIF | Laser-induced fluorescence |
| LOC | Lab-on-a-chip |
| PDMS | Polydimethylsiloxane |
| PMMA | Polymethyl methacrylate |

A. Kapoor (✉) · S. Balasubramanian · P. Muthamilselvi · V. Vaishampayan · S. Prabhakar
Department of Chemical Engineering, SRM Institute of Science and Technology, Kattankulathur, India



Inamuddin, A. M. Asiri (eds.), *Nanosensor Technologies for Environmental Monitoring*, Nanotechnology in the Life Sciences,
https://doi.org/10.1007/978-3-030-45116-5_15

| | |
|---|---|
| PTFE | Polytetrafluoroethylene |
| RO | Reverse osmosis |
| TLF | Tryptophan-like fluorescence |
| μPAD | Microfluidic paper-based analytical device |
| μPESI | Micropillar array electrospray ionization |
| μ-TAS | Micro total analytical systems |

## 1 Introduction

Water is an essential ingredient for the sustenance of life. Access to clean water is of prime importance for the well-being of individuals as well as the entire society. Increasing industrialization and change in lifestyles have led to widespread contamination of water resources. The polluted water has immediate effect on the health of the entire ecosystem. Pollution prevention and control strategies are being rapidly developed and installed to check any adverse consequences. The success of any pollution prevention and control measure depends on the accurate and rapid analytical determination of various contaminants present in water (Ho et al. 2005).

Although many established techniques are in practice, most of such methods need access to sophisticated equipment. Special means of sample handling and preparation are needed, which can be performed only by trained personnel. A new class of lab-on-a-chip (LOC) technology has emerged for water quality monitoring in recent times to address these challenges and provide user-friendly platforms for on-the-spot detection of contaminants (Gardeniers and Van Den Berg 2004; Whitesides 2006). In this chapter, we cover the following topics: a discussion on conventional techniques for water quality monitoring, introduction to lab-on-a-chip technology, materials and fabrication techniques involved in developing LOC-based platforms, and summarize various platforms developed for the determination of chemical and biological contaminants and water quality parameters. Finally, we present challenges and future directions of development in this field.

## 2 Conventional Techniques for Water Quality Monitoring

The well-established analytical methods used for estimation of water contaminants include atomic absorbance spectroscopy (AAS) (Kumari et al. 2013), liquid chromatography–mass spectrometry (LC-MS), gas chromatography–mass spectrometry (GC-MS) (Botitsi et al. 2011) inductively coupled plasma optical emission spectrometry (ICP-OES) (Asfaram et al. 2016), fluorescence spectroscopy (Carstea et al. 2016). While majority of these techniques offer high sensitivity, accuracy, and low detection limits, most of the water quality monitoring approaches based on these techniques are constrained by one or more of the factors enlisted here, that limits their applicability for portable, on-site detection:

1. Sample collection and transportation to central labs with chances of deterioration.
2. Sophisticated sample preparation and testing.
3. Need for trained professionals for equipment operation.
4. High costs involved in equipment operation and maintenance.
5. Time-consuming.
6. Large sample and reagent volume requirements.

Increasing necessity and interest in monitoring of water quality parameters at point-of-use has led to exploration of new scientific and technical methodologies. Baker et al. used a portable luminescence spectrophotometer for analysis of river and wastewater samples and benchmarked the results with conventional spectrophotometer (Baker et al. 2004). Sorenson et al. evaluated the feasibility of using online fluorometers for real-time evaluation of the microbial quality of water for drinking purposes (Sorensen et al. 2018). Online fluorescence measurements, namely tryptophan-like fluorescence (TLF) and humic-like fluorescence (HLF), were found to correlate well to *Escherichia coli* content as well as total bacterial cell counts. Hence, these were better indicators for microbial quality in public water supply in comparison to turbidity, the conventionally used indicator for microbial content in water. Such promising developments have certainly broadened the scope of conventional techniques. However, many of these adaptations of benchtop equipments for portable applications still carry the inherent limitations as mentioned above. This has motivated the researchers to look further for path-breaking technologies to cater the need for affordable and effective alternatives.

# 3 Lab-on-a-Chip Technology

In broad terms, lab-on-a-chip (LOC) refers to miniaturization of macroscale laboratory techniques. LOC technology employs microfabrication techniques to develop devices that process or handle minute quantities of fluids. Such devices are also often referred to as micro total analytical systems (μ-TAS) or microfluidic systems. The advantages offered by lab-on-a-chip technology include:

1. Low volume requirements for samples and reagents.
2. Low cost of fabrication.
3. Ease of design iteration.
4. Less waste product generation.
5. Compactness.
6. Amenable for portable applications.

In addition to the merits offered by the virtue of miniaturization, LOC systems take advantage of unique physical phenomena observed at microscale. Fluid flows in microscale devices are primarily laminar (low Reynolds number) due to small dimensions of channels (Stone et al. 2004). Reynolds number signifies the ratio of inertial to viscous forces. Hence, viscous forces dominate when compared to inertial forces.

$$Re = \frac{\rho D v}{\mu}$$

where *Re* is Reynolds number (dimensionless), $\rho$ = density of fluid (kg/m$^3$), $D$ = diameter of the microchannel (m), $v$ = velocity of the fluid (m/s), and $\mu$ = viscosity of the fluid (Pa s). The laminar flow behavior has useful implications that have been utilized for various practical applications like fabrication of microelectrode arrays, composite fibers, preparation of crystals, and selective patterning of biomolecules and cells (Kenis et al. 2000; Frampton et al. 2011; Puigmartí-Luis 2014).

## 3.1 Materials of Construction for Lab-on-a-Chip (LOC) Devices

The substrate materials used in the fabrication of microfluidics system based on the evolution of materials are discussed here. The use of silicon and glass as a substrate material in fabrication of microfluidic device goes back to 1970s where the microscale gas chromatography and electrophoresis systems were designed for molecular separation processes. Over the years more advancement in micromachining techniques from microelectronics industries paved the way to the developments of silicon- and glass-based microchemical systems. Some of the applications of silicon-based microfluidic are chemical microreactors, analysis of DNA samples, drug delivery systems, environmental and forensic analysis (Ren et al. 2013). Daridon et al. employed a silicon microfluidic chip sandwiched between two Pyrex glass plates for the determination of ammonia in water using Berthelot reaction (Daridon et al. 2001). A fiber-optic system was used for absorbance detection of ammonia in aqueous solution. Sainiemi et al. fabricated a circular silicon-wafer based platform comprising of 60 identical micropillar array electrospray ionization (μPESI) chips (Sainiemi et al. 2011). The platform was fixed in front of a mass spectrophotometer on a digitally controlled rotating table and demonstrated for rapid analysis of verapamil drug samples. The sample volume requirement was merely 1.5 μL, substantially lesser than used in established commercial equipment. The flow was driven by capillary forces in microchannels and 60 samples could be measured in 80 min without any cross-contamination. Similarly, the glass-based microscale systems have also been used in analytical applications (Wang et al. 2002). Broyles et al. showed integrated functioning on a quartz microfluidic analytical device (Broyles et al. 2003). The operations including sample pretreatment, open channel electrochromatographic separation, and laser-induced fluorescence (LIF) detection were performed on a single chip. The device was successfully demonstrated for determination of four polycyclic aromatic hydrocarbons (anthracene, pyrene, 1,2-benzofluorene, and benzo[a]pyrene) in water. The standard technique used in the fabrication of silicon- and glass-based device is photolithography and etching. The channel size achieved using this technique is less than 100 nm. The advantages of silicon and glass as the substrate materials include: hard and multilayer channels with hydrophobic surface, good thermal stability, excellent resistance to oxidizing agents, very high solvent compatibility, very stable to surface charge, transparency of glass, good oxygen barrier. Although the silicon and glass substrates possess many advantages, some of the problems such as high cost of operation, use of toxic chemicals during fabrication, requirement of safe processing facilities, high temperature, pressure, and ultraclean environment limit its use. As continuous progress in research to overcome the above-mentioned disadvantages, the emphasis on substrates was changed from silicon and glass substrate to elastomers and plastic-related materials.

Elastomers are flexible and contain entangled cross-linked polymer chain. For instance, polydimethylsiloxane (PDMS) was adopted as a relatively low-cost material that facilitated easy fabrication of microfluidic devices. The merits of using PDMS as substrate are high optical transparency, high elasticity, multilayer channels, and the channel size less than 1 μm in the device can be achieved, gas permeability, and provides compatible surface for cell growth in bioresearch. Due to these advantages, PDMS-based microfluidic devices were used in bioresearch such as single-cell analysis, cell growth, cell screening, and biochemical tests. Despite these advantages of PDMS as substrates, it also has some limitations viz. incompatibility to certain organic solvents, less thermostability, moderate resistance to oxidizer, and unstable surface charges. Npota et al. employed Sandell–Kolthoff's reaction in a PDMS-based microfluidic device to estimate total inorganic iodine content in drinking water (Inpota et al. 2018). A fiber optic cable from the spectrofluorometer was installed above the detection zone of microfluidic channel to capture the fluorimetric signal. The system had a linear operating range 50–400 μg iodide/L and a limit of detection of 7.7 μg iodide/L without any necessity of heating of the reaction mixture.

In contrast to elastomers, plastics of thermoset and thermoplastics types also emerged as alternative materials that offered medium to high thermostability, moderate to good resistance to oxidizers and solvents. Polyvinyl chloride, polystyrene, polycarbonate, polymethyl methacrylate (PMMA), and polyethylene terephthalate are few examples of thermoplastics materials. These materials are fabricated using thermo-molding techniques (i.e., casting/polymerization and photopolymerization). Yan et al. developed a PMMA-based miniaturized colorimeter for water quality detection (Yan et al. 2019). The optimization schemes were used to design the structure and achieve high-precision and low-fluid consumption. An improvement of two orders in precision and reduction in volume consumption was reported for the optimized design. Hydrogels were developed as three-dimensional polymer chains of hydrophilic nature in aqueous medium in which about 99% of the content is water. The hydrogels are highly porous; allow very smaller molecules to diffuse through the pores and not the bulk fluid so these are biocompatible. Hence it is more suitable for encapsulation of cells in cell-culture based bioresearch experiments. When compared to the elastomers and plastic-based substrates, hydrogels are restricted to cell-culture tests particularly in tissue-engineering related areas.

The use of cellulosic-based substrate as another class of materials dated back to 1800s, centuries ago, even before the use of silicon and glass as substrates in microfluidics applications. The first scientific work reported in the literature was litmus paper, and in the late 1940s the use of paper in chromatography and electrophoresis was reported (Martin and Synge 1941; Kunkel and Tiselius 1951). In 1980s, the emergence of home pregnancy test kid made up of nitrocellulosic membrane played a vital role in the development of paper-based analytical devices. It was 2007, after the pioneering work by George Whitesides's group on microfluidic paper-based analytical device (μPAD), the research in the use of paper as substrate has exponentially increased (Martinez et al. 2007). The advantages such as abundant availability, low cost, biocompatibility, lightweight, flexibility, transport through capillary action without pump, amenable to chemical modification, fabrication, biodegradability, and easy disposal by incineration have attracted the paper as a well suitable substrate in the development of microfluidic analytical devices in chemical, biochemical, and environmental applications. Extensive information on the developments of paper-based microfluidic in diagnostic applications is published in literature (Gong and Sinton 2017). Barghouthi and Amereih developed a spot-testing platform for detection of fluoride in drinking groundwater using photometric measurements (Barghouthi and Amereih 2017). Macherey–Nagel chromatography paper (MN 827) was used as the substrate and impregnated with aluminium quinalizarin complex for use as fluoride sensor. The method allowed determination of fluoride in the range 0.0–2.0 mg/L with limit of detection as 0.1 mg/L. Recently composite materials, or combination of materials as hybrids, are also being developed (Blazej et al. 2006; Mu et al. 2009; Talaei et al. 2009; Zhu et al. 2012). For instance, combination of PDMS and glass, integration of metal electrodes with glass into polymeric materials, and doping of nanomaterial have attracted attention as hybrid materials (Ren et al. 2013). Gallardo-Gonzalez et al. developed a passive microfluidic platform to achieve in situ and real--time potentiometric measurements in flowing water for field applications (Gallardo-Gonzalez et al. 2019). The device had two parts: PDMS-based microfluidic component and a microelectrode system. Ammonium detection in flowing water was exhibited with limit of detection of $4 \times 10^{-5}$ M and response time around 10 s. Table 1 summarizes the advantages and the limitations of various substrates in fabrication of LOC devices.

### 3.2 Applications of Lab-on-a-Chip Devices in Water Quality Monitoring

Numerous miniaturized analytical devices, based on "lab-on-a-chip" approach, have been demonstrated with the objective of rapid and in situ monitoring of various water quality parameters (Marle and Greenway 2005). Figure 1a,b shows an illustration of typical lab-on-a-chip devices for water quality tests and analysis on

**Table 1** Merits and demerits of various substrate materials used for fabrication of LOC devices

| Substrate materials | Merits | Demerits |
|---|---|---|
| Silicon and glass | Fabrication of sub-100 nm channel dimensions are achievable, possible to fabricate multilayered structures, mechanically strong, thermally stable, chemically inert, compatible with wide range of solvents, hydrophobic surface, and optically transparent (glass) | Cost intensive fabrication, hazardous chemicals are involved in certain fabrication steps, ultraclean and safe process facilities are required |
| Elastomers | Flexible, transparent, elastic, fabrication of the devices with features of order of microns can be achieved, permeable to gas and usually biocompatible | Incompatible to some organic solvents, and may not withstand high temperatures |
| Plastics | Thermoset plastics are mechanically robust and offer good stiffness, good resistance to most of the solvents, thermoplastics can be remolded, possible surface modification for specific applications | Expensive, thermoset plastics cannot be remolded after curing, requires advanced fabrication facilities, and trained professionals are needed for fabrication |
| Paper | Cheap, widely available, compatible with most of the chemicals, lightweight, flexible, suitable for passive flows without peripheral pumping devices, suitable for surface modification, biodegradable, suitable for use and throw applications | Absorbs moisture, lacks in mechanical strength, not reusable, and sample loss occurs during fluid transport due to adsorption in the porous fibrous structure. |
| Hydrogel | High porosity, biocompatibility, high mass transfer rates, and can be shaped as per requirements | Difficulty in storage, one-time use, difficult in bonding with other substrates |
| Hybrid | Biocompatible, compatible for organic chemicals, reusable/disposable | Complex in fabrication, fluid flow manipulation, need of clean and controlled facilities for fabrication |

opaque substrates (paper, and silicon) and transparent substrates (glass, PDMS, and PMMA). The representation shown in the figure is for spot-test type LOC devices. Similar image acquisition and analysis strategies are adopted for flow best LOC test devices (lateral flow assays).

The development of any analytical device hinges upon several factors, such as availability of substrate material, its compatibility with reagents and fluid samples to be tested, selection of appropriate detection modality, and costs involved. Table 2 illustrates the features of various lab-on-a-chip devices for water quality monitoring.

## 3.3 *Role of Nanotechnology in Lab-on-a-Chip Devices*

Nanotechnology has numerous potential applications in combination with microfluidics particularly in contaminant sensing, water treatment, and water quality parameters monitoring (Su et al. 2012; Das et al. 2017). Lab-on-a-chip devices allow the

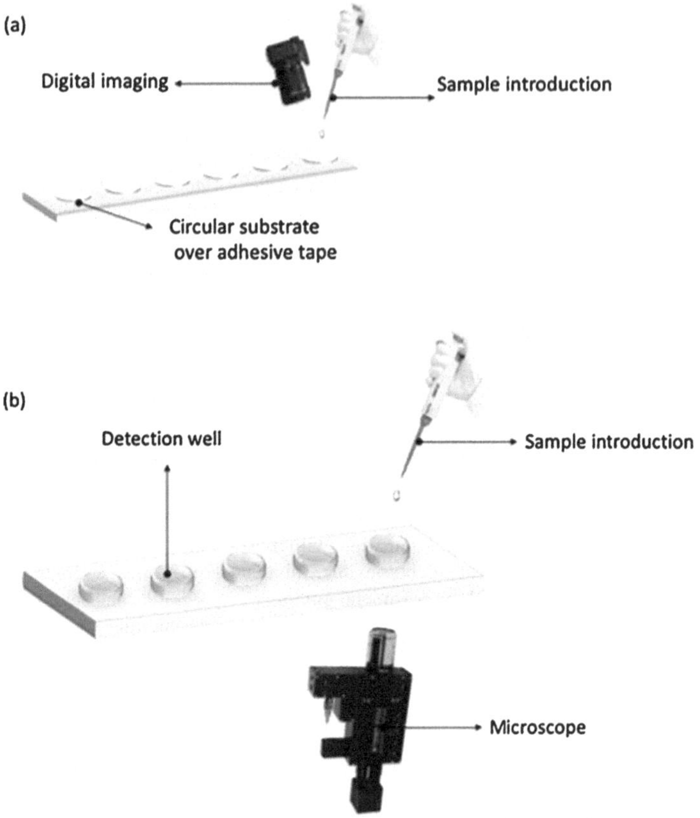


**Fig. 1** Representative schematic diagram for spot-test type LOC device using (**a**) opaque substrates and (**b**) transparent substrates

synthesis and screening of nanoparticles of various shapes and structures (Zhao et al. 2011). For purification and synthesis purpose, microreactors of various kinds are more essential than single channels. They offer more surface area and contact time for the synthesis reactions. In situ growth of nanoparticles or nanowires of photocatalysts like zinc oxide and titanium dioxide, creates more opportunities for lab-on-a-chip devices to use as a micro-photocatalytic reactor (Azzouz et al. 2018). These methods can minimize the interaction time and increases the efficiency and performance. Desalination in microchambers suggests new opportunities to understand the fundamentals of transport of ions on nano and microscale (Roelofs et al. 2015). This can lead to new technique of microfluidic ionic transport. By integrating nanoelectrodes and nanoparticles these devices can perform qualitative and quantitative analysis of the contaminants present in the water samples. The coating of nanomaterials can control the fluid delivery for particular channels. The fixation of nanomaterials in a single channel can avail the concept of microfluidic packed bed reactor. The fluid movement, as well as contact time, highly depends on the particle

**Table 2** Water quality monitoring using lab-on-a-chip devices

| Analyte(s)/parameters | Device substrate | Detection method (or modality) | Limit of detection (LOD) | References |
|---|---|---|---|---|
| Lead | Epitaxial graphene on Si-terminated face of SiC in 3D-printed resin chip | Conductivity | 95 nM | Santangelo et al. (2019) |
| Cocaine | Polydimethylsiloxane/ polytetrafluoroethylene (PDMS/PTFE) chip with integrated screen-printed electrode | Electrochemical | 0.15 ng/L | Abdelshafi et al. (2019) |
| Copper(II)sulfate, 1,3-dihydroxybenzene and 1,4-benzoquinone | Poly(methyl methacrylate) | Bioluminescence | 3 µM (copper(II)sulfate), 15 mM (1,3-dihydroxybenzene), and 2 µM (1,4-benzoquinone) | Denisov et al. (2018) |
| Chromium(VI) | Chromatography no. 1 paper | Colorimetry | 0.5 mg/L | Li et al. (2015) |
| Nickel(II) | Chromatography no. 1 paper | Colorimetry | 0.5 mg/L | Li et al. (2015) |
| Copper(II) | Chromatography no. 1 paper | Colorimetry | 0.8 mg/L | Li et al. (2015) |
| Iron(II) | Whatman filter paper grade 1 | Colorimetry | 1 ppm | Ghosh et al. (2019) |
| Ammonium | PDMS microfluidic device with silicon transducer | Electrochemical | $4 \times 10^{-5}$ M | Gallardo-Gonzalez et al. (2019) |
| Nitrite, nitrate | Whatman filter paper grade 1 and 4 | Colorimetry | 1 µM (nitrite) and 19 µM (nitrate) | Jayawardane et al. (2014) |
| 2,4-Dichlorophenol | Plastic microfluidic chip with incorporated electrodes | Potential difference measurements | 0.1 ppm | Ho et al. (2019) |
| Calcium ($Ca^{2+}$) | Advantec no. 5C filter paper modified with ionophore-doped ion-selective optode nanospheres | Distance-based colorimetry | 0.05 mmol/L | Shibata et al. (2019) |

(continued)

**Table 2** (continued)

| Analyte(s)/parameters | Device substrate | Detection method (or modality) | Limit of detection (LOD) | References |
|---|---|---|---|---|
| Phenolic compounds (phenol, bisphenol A (BPA), dopamine, catechol, and m-and p-cresol) | Filter paper | Colorimetry | 0.86 (±0.1) μg/L for each of the phenolic compounds | Alkasir et al. (2012) |
| Nitrite | Whatman filter paper grade 1 | Colorimetry | 0.52 mg/L | Lopez-Ruiz et al. (2014) |
| Mercury(II) | Glass fiber paper | Colorimetry | 0.01 μM | Chen et al. (2016) |
| Mercury(II) | Filter paper | Distance-based colorimetry | 0.93 μg/mL | Cai et al. (2017) |
| *Escherichia coli*, *Saccharomyces cerevisiae*, and *Aeromonas hydrophila* | Dry film resist (DFR)-based microfluidic chip bonded with multimode fiber pigtails | Absorbance measurements (optical) | $1.0 \times 10^5$ cells/mL for *A. hydrophila* and *E. coli*; $1.0 \times 10^6$ cells/mL for *S. cerevisiae* | Kamuri et al. (2019) |
| *E. coli* | Soda lime glass microfluidic chip (NS-12A, Perkin Elmer, USA) | Fluorescence detection | $10^4$ cfu/mL | Schwartz and Bercovici (2014) |
| *E. coli* | PDMS microfluidic chip | Dielectrophoretic impedance measurements | 300 cfu/mL | Kim et al. (2015) |
| pH | Whatman filter paper grade 1 | Colorimetry | pH detection range 2–12 | Lopez-Ruiz et al. (2014) |
| pH | Whatman filter paper grade 1 | Colorimetry | pH detection range 0–10 | Kapoor et al. (2017) |

size and shape of nanomaterials in these cases. These concepts can be useful for the development of water purifier sticks for remote applications. By Infusing the semi-permeable membrane between the channels and applying pressure can give opportunity to use microfluidic reverse osmosis (RO) systems for small-scale pretreatment purpose. In the case of paper-based microfluidic sensors, the paper strip coated with the nanomaterials offers an excellent potential for sensing purposes and several other diverse applications (Ngo et al. 2011; Chen et al. 2014).

## 4 Challenges

The impetus to develop portable, affordable, and simple devices capable of performing on-site measurements is the key driver for lab-on-a-chip technology for water quality monitoring (Heibati et al. 2017). LOC technology has witnessed a rapid and steady growth, originating from silicon and glass-based chips to polymeric and paper-based devices. Various materials, fabrication techniques, and detection methodologies are being pursued to make the technology more accessible and effective (Wu and Gu 2011). With global research efforts and successful reports, the time is ripe to stress upon commercialization of these devices for practical applications. However, several challenges need to be overcome before realization of the desired goal. Some of the key challenges include pretreatment steps, selectivity, reproducibility, shelf life of sensor, detection readout, signal amplification, and interfacing. The interdisciplinary nature of the technology (Fig. 2) calls for inputs from diverse disciplines including basic sciences (physics, chemistry, biology), as well as engineering sciences (fluid mechanics, optics, material science, and nanotechnology) (Figueredo et al. 2016; Syedmoradi et al. 2017; Xu et al. 2018; Hárendarčíková and Petr 2018). On one hand, the synergy of ideas originating from different disciplines leads to novel scientific outcomes and technological breakthroughs. At the same time, it brings a unique set of challenges with respect to seamless integration of apparently disconnected domains, for instance, collecting the continuous digital signal output from a contaminant originating from a biotic system in water bodies. Consortiums for interdisciplinary research are likely to facilitate overcoming the existing challenges.

## 5 Conclusions

Water is an essential lifeline for the human existence and it needs to be protected. There are many well-established conventional methods but still they are limited due to the cost of operation, requirement of large space, use of high volume of sample, and highly skilled man power. The LOC microfluidic devices offer promising features in diagnosing and detecting the contaminants present in the water. The advantages of using these devices include cost-effective, smaller size, low volume of

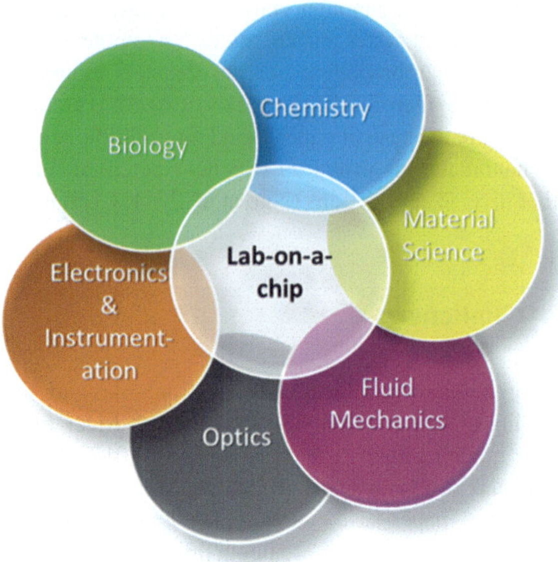


**Fig. 2** Interdisciplinary approach for the development of lab-on-a-chip devices for water quality assessment

samples, less-skilled manpower, reusability, and environmentally friendly. Due to these many advantages, LOC devices are promising in analyzing and monitoring the quality of water. Several interdisciplinary efforts are already being made to improve the versatility and performance of LOC devices, such as integration of detection procedure with smartphone-based imaging, mobile and internet applications for instant analysis and communication of results, and incorporation of unique features of nanotechnology in LOC devices. The future devices would not only integrate analytical features involved in detection of contaminants, on a single device, but also are likely to combine other functionalities, such as water purification on the same or an attached platform. Lab-on-a-chip devices hold immense promise to be viable solutions for point-of-use water quality monitoring applications.

## References

Abdelshafi NA, Bell J, Rurack K, Schneider RJ (2019) Microfluidic electrochemical immunosensor for the trace analysis of cocaine in water and body fluids. Drug Test Anal 11:492–500. https://doi.org/10.1002/dta.2515

Alkasir RSJ, Ornatska M, Andreescu S (2012) Colorimetric paper bioassay for the detection of phenolic compounds. Anal Chem 84:9729–9737. https://doi.org/10.1021/ac301110d

Anjum M, Shekhar H, Hyun SH, et al (2004) A disposable BOD microsensor using a polymer substrate. In: Proceedings of IEEE sensors. pp. 1202–1205

Asfaram A, Ghaedi M, Ghezelbash GR (2016) Biosorption of $Zn^{2+}$, $Ni^{2+}$ and $Co^{2+}$ from water samples onto *Yarrowia lipolytica* ISF7 using a response surface methodology, and analyzed by inductively coupled plasma optical emission spectrometry (ICP-OES). RSC Adv 6:23599–23610. https://doi.org/10.1039/c5ra27170c

Azzouz I, Habba YG, Capochichi-Gnambodoe M et al (2018) Zinc oxide nano-enabled microfluidic reactor for water purification and its applicability to volatile organic compounds. Microsyst Nanoeng:4. https://doi.org/10.1038/micronano.2017.93

Baker A, Ward D, Lieten SH et al (2004) Measurement of protein-like fluorescence in river and waste water using a handheld spectrophotometer. Water Res 38:2934–2938. https://doi.org/10.1016/j.watres.2004.04.023

Barghouthi Z, Amereih S (2017) Field method for estimation of fluoride in drinking groundwater by photometric measurement of spot on aluminium quinalizarin reagent paper. Arab J Chem 10:S2919–S2925. https://doi.org/10.1016/j.arabjc.2013.11.024

Blazej RG, Kumaresan P, Mathies RA (2006) Microfabricated bioprocessor for nanoliter-scale Sanger DNA sequencing. PNAS 103:1567–1569. https://doi.org/10.1073/pnas.0602476103

Botitsi HV, Garbis SD, Economou A, Tsipi DF (2011) Current mass spectrometry strategies for the analysis of pesticides and their metabolites in food and water matrices. Mass Spectrom Rev 30:907–939. https://doi.org/10.1002/mas.20307

Broyles BS, Jacobson SC, Ramsey JM (2003) Sample filtration, concentration, and separation integrated on microfluidic devices. Anal Chem 75:2761–2767. https://doi.org/10.1021/ac025503x

Cai L, Fang Y, Mo Y et al (2017) Visual quantification of Hg on a microfluidic paper-based analytical device using distance-based detection technique. AIP Adv 7:085214. https://doi.org/10.1063/1.4999784

Carstea EM, Bridgeman J, Baker A, Reynolds DM (2016) Fluorescence spectroscopy for wastewater monitoring: a review. Water Res 95:205–219. https://doi.org/10.1016/j.watres.2016.03.021

Chen GH, Chen WY, Yen YC et al (2014) Detection of mercury(II) ions using colorimetric gold nanoparticles on paper-based analytical devices. Anal Chem 86:6843–6849. https://doi.org/10.1021/ac5008688

Chen W, Fang X, Li H et al (2016) A simple paper-based colorimetric device for rapid mercury(II) assay. Sci Rep 6. https://doi.org/10.1038/srep31948

Daridon A, Sequeira M, Pennarun-Thomas G et al (2001) Chemical sensing using an integrated microfluidic system based on the Berthelot reaction. Sensors Actuators B Chem 76:235–243. https://doi.org/10.1016/S0925-4005(01)00573-1

Das R, Vecitis CD, Schulze A et al (2017) Recent advances in nanomaterials for water protection and monitoring. Chem Soc Rev 46:6946–7020. https://doi.org/10.1039/c6cs00921b

Denisov I, Lukyanenko K, Yakimov A et al (2018) Disposable luciferase-based microfluidic chip for rapid assay of water pollution. Luminescence 33:1054–1061. https://doi.org/10.1002/bio.3508

Figueredo F, Garcia PT, Cortón E, Coltro WKT (2016) Enhanced analytical performance of paper microfluidic devices by using $Fe_3O_4$ nanoparticles, MWCNT, and graphene oxide. ACS Appl Mater Interfaces 8:11–15. https://doi.org/10.1021/acsami.5b10027

Frampton JP, Lai D, Sriram H, Takayama S (2011) Precisely targeted delivery of cells and biomolecules within microchannels using aqueous two-phase systems. Biomed Microdevices 13:1043–1051. https://doi.org/10.1007/s10544-011-9574-y

Gallardo-Gonzalez J, Baraket A, Boudjaoui S et al (2019) A fully integrated passive microfluidic lab-on-a-chip for real-time electrochemical detection of ammonium: sewage applications. Sci Total Environ 653:1223–1230. https://doi.org/10.1016/j.scitotenv.2018.11.002

Gardeniers JGE, Van Den Berg A (2004) Lab-on-a-chip systems for biomedical and environmental monitoring. Anal Bioanal Chem 378:1700–1703. https://doi.org/10.1007/s00216-003-2435-7

Ghosh R, Vaishampayan V, Mahapatra A et al (2019) Enhancement of limit of detection by inducing coffee-ring effect in water quality monitoring microfluidic paper-based devices. Desalin Water Treat 156:316–322. https://doi.org/10.5004/dwt.2019.23715

Gong MM, Sinton D (2017) Turning the page: advancing paper-based microfluidics for broad diagnostic application. Chem Rev 117:8447–8480. https://doi.org/10.1021/acs.chemrev.7b00024

Hárendarčíková L, Petr J (2018) Smartphones and microfluidics: marriage for the future. Electrophoresis 39:1319–1328. https://doi.org/10.1002/elps.201700389

Heibati M, Stedmon CA, Stenroth K et al (2017) Assessment of drinking water quality at the tap using fluorescence spectroscopy. Water Res 125:1–10. https://doi.org/10.1016/j.watres.2017.08.020

Ho CK, Robinson A, Miller DR, Davis MJ (2005) Overview of sensors and needs for environmental monitoring. Sensors 5:4–37. https://doi.org/10.3390/s5010004

Ho WF, Nguyen LT, Yang KL (2019) A microfluidic sensor for detecting chlorophenols using cross-linked enzyme aggregates (CLEAs). Lab Chip 19:634–640. https://doi.org/10.1039/C8LC01065J

Inpota P, Strzelak K, Koncki R et al (2018) Microfluidic analysis with front-face fluorometric detection for the determination of total inorganic iodine in drinking water. Anal Sci 34:161–167. https://doi.org/10.2116/analsci.34.161

Jayawardane BM, Wei S, McKelvie ID, Kolev SD (2014) Microfluidic paper-based analytical device for the determination of nitrite and nitrate. Anal Chem 86:7274–7279. https://doi.org/10.1021/ac5013249

Kamuri MF, Abidin ZZ, Jun LH et al (2019) Performance evaluation of free-space fibre optic detection in a lab-on-chip for microorganism. J Sensors 2019. https://doi.org/10.1155/2019/1026905

Kapoor A, Balasubramanian S, Vaishampayan V, Ghosh R (2017) Lab-on-a-chip: a potential tool for enhancing teaching-learning in developing countries using paper microfluidics. In: 2017 international conference on transforming engineering education (ICTEE). IEEE, pp 1–7. https://doi.org/10.1109/icteed.2017.8586151

Kenis PJA, Ismagilov RF, Takayama S et al (2000) Fabrication inside microchannels using fluid flow. Acc Chem Res 33:841–847. https://doi.org/10.1021/ar000062u

Kim HD, Yi SJ, Kim KC (2013) Simultaneous measurement of dissolved oxygen concentration and velocity field in microfluidics using oxygen-sensitive particles. Microfluid Nanofluid 15:139–149. https://doi.org/10.1007/s10404-012-1130-4

Kim M, Jung T, Kim Y et al (2015) A microfluidic device for label-free detection of *Escherichia coli* in drinking water using positive dielectrophoretic focusing, capturing, and impedance measurement. Biosens Bioelectron 74:1011–1015. https://doi.org/10.1016/j.bios.2015.07.059

Kumari M, Tripathi S, Pathak V, Tripathi BD (2013) Chemometric characterization of river water quality. Environ Monit Assess 185:3081–3092. https://doi.org/10.1007/s10661-012-2774-y

Kunkel HG, Tiselius A (1951) Electrophoresis of proteins on filter paper. J Gen Physiol 35:89–118. https://doi.org/10.1085/jgp.35.1.89

Li M, Cao R, Nilghaz A et al (2015) "Periodic-table-style" paper device for monitoring heavy metals in water. Anal Chem 87:2555–2559. https://doi.org/10.1021/acs.analchem.5b00040

Lopez-Ruiz N, Curto VF, Erenas MM et al (2014) Smartphone-based simultaneous pH and nitrite colorimetric determination for paper microfluidic devices. Anal Chem 86:9554–9562. https://doi.org/10.1021/ac5019205

Marle L, Greenway GM (2005) Microfluidic devices for environmental monitoring. TrAC Trends Anal Chem 24:795–802. https://doi.org/10.1016/j.trac.2005.08.003

Martin AJP, Synge RLM (1941) A new form of chromatogram employing two liquid phases. Biochem J 35:1358–1368. https://doi.org/10.1042/bj0351358

Martinez AW, Phillips ST, Butte MJ, Whitesides GM (2007) Patterned paper as a platform for inexpensive, low-volume, portable bioassays. Angew Chem Int Ed 46:1318–1320. https://doi.org/10.1002/anie.200603817

Mu X, Liang Q, Hu P et al (2009) Laminar flow used as "liquid etch mask" in wet chemical etching to generate glass microstructures with an improved aspect ratio. Lab Chip 9:1994–1996. https://doi.org/10.1039/b904769g

Ngo YH, Li D, Simon GP, Garnier G (2011) Paper surfaces functionalized by nanoparticles. Adv Colloid Interf Sci 163:23–38. https://doi.org/10.1016/j.cis.2011.01.004

Puigmartí-Luis J (2014) Microfluidic platforms: a mainstream technology for the preparation of crystals. Chem Soc Rev 43:2253–2271. https://doi.org/10.1039/C3CS60372E

Ren K, Zhou J, Wu H (2013) Materials for microfluidic chip fabrication. Acc Chem Res 46:2396–2406. https://doi.org/10.1021/ar300314s

Roelofs SH, Van Den Berg A, Odijk M (2015) Microfluidic desalination techniques and their potential applications. Lab Chip 15:3428–3438. https://doi.org/10.1039/C5LC00481K

Sainiemi L, Nissilä T, Kostiainen R et al (2011) A microfabricated silicon platform with 60 microfluidic chips for rapid mass spectrometric analysis. Lab Chip 11:3011–3014. https://doi.org/10.1039/c1lc20275h

Santangelo MF, Shtepliuk I, Filippini D et al (2019) Real-time sensing of lead with epitaxial graphene-integrated microfluidic devices. Sensors Actuators B Chem:425–431. https://doi.org/10.1016/j.snb.2019.03.021

Schwartz O, Bercovici M (2014) Microfluidic assay for continuous bacteria detection using antimicrobial peptides and isotachophoresis. Anal Chem 86:10106–10113. https://doi.org/10.1021/ac5017776

Shibata H, Hiruta Y, Citterio D (2019) Fully inkjet-printed distance-based paper microfluidic devices for colorimetric calcium determination using ion-selective optodes. Analyst 144:1178–1186. https://doi.org/10.1039/c8an02146e

Sorensen JPR, Vivanco A, Ascott MJ et al (2018) Online fluorescence spectroscopy for the real-time evaluation of the microbial quality of drinking water. Water Res 137:301–309. https://doi.org/10.1016/j.watres.2018.03.001

Stone HA, Stroock AD, Ajdari A (2004) Engineering flows in small devices. Annu Rev Fluid Mech 36:381–411. https://doi.org/10.1146/annurev.fluid.36.050802.122124

Su S, Wu W, Gao J et al (2012) Nanomaterials-based sensors for applications in environmental monitoring. J Mater Chem 22:18101–18110. https://doi.org/10.1039/c2jm33284a

Syedmoradi L, Daneshpour M, Alvandipour M et al (2017) Point of care testing: the impact of nanotechnology. Biosens Bioelectron 87:373–387. https://doi.org/10.1016/j.bios.2016.08.084

Talaei S, Frey O, van der Wal PD et al (2009) Hybrid microfluidic cartridge formed by irreversible bonding of SU-8 and PDMS for multi-layer flow applications. Proc Chem 1:381–384. https://doi.org/10.1016/j.proche.2009.07.095

Wang J, Escarpa A, Pumera M, Feldman J (2002) Capillary electrophoresis-electrochemistry microfluidic system for the determination of organic peroxides. J Chromatogr A 952:249–254. https://doi.org/10.1016/S0021-9673(02)00075-4

Whitesides GM (2006) The origins and the future of microfluidics. Nature 442:368–373. https://doi.org/10.1038/nature05058

Wu J, Gu M (2011) Microfluidic sensing: state of the art fabrication and detection techniques. J Biomed Opt 16:080901. https://doi.org/10.1117/1.3607430

Xu D, Huang X, Guo J, Ma X (2018) Automatic smartphone-based microfluidic biosensor system at the point of care. Biosens Bioelectron 110:78–88

Yan JC, Ren J, Ren LL et al (2019) A novel structure design and fabrication method for low liquid consumption and high precision device of colorimeter in water quality detection. Sensors Actuators A Phys 289:1–10. https://doi.org/10.1016/j.sna.2019.02.016

Zhao CX, He L, Qiao SZ, Middelberg APJ (2011) Nanoparticle synthesis in microreactors. Chem Eng Sci 66:1463–1479. https://doi.org/10.1016/j.ces.2010.08.039

Zhu Z, Frey O, Ottoz DS et al (2012) Microfluidic single-cell cultivation chip with controllable immobilization and selective release of yeast cells. Lab Chip 12:906–915. https://doi.org/10.1039/c2lc20911j

# Advance Nanostructure-Based Electrochemical Sensors for Pharmaceutical Drugs Detection

**Razium Ali Soomro, Nazar Hussain Kalwar, Sana Jawaid, and Mawada Mohamed Tunesi**

## Contents



## Abbreviations

| | |
|---|---|
| 2D | Two dimensional |
| AgNPs | Silver nanoparticles |
| AS | Adipic acid |

R. A. Soomro (✉)
National Centre of Excellence in Analytical Chemistry, University of Sindh, Jamshoro, Pakistan

College of Material Science and Engineering, Beijing University of Chemical Technology, Beijing, China

N. H. Kalwar
Institute of Chemistry, Shah Abdul Latif University, Khairpur, Pakistan

Faculty of Engineering, Department of Mechanical Engineering, University of Selcuk Campus, Konya, Turkey

S. Jawaid
National Center of Excellence in Analytical Chemistry, University of Sindh, Jamshoro, Pakistan

M. M. Tunesi
Department of Chemistry, Istanbul University, Istanbul, Turkey



Inamuddin, A. M. Asiri (eds.), *Nanosensor Technologies for Environmental Monitoring*, Nanotechnology in the Life Sciences,
https://doi.org/10.1007/978-3-030-45116-5_16

| | |
|---|---|
| ASA | Acetylsalicylic acid |
| AuDE | Gold disk electrode |
| BRB | Briton Robinson buffer |
| CA | Citric acid |
| CAP | Captopril |
| CdS QDs | Cadmium sulfide quantum dots |
| CE | Capillary electrophoresis |
| CuO | Copper oxide |
| *D* | Diffusion coefficient |
| DBX | Doxorubicin |
| DNA | Deoxy ribose nucleic acid |
| DPV | Differential pulse voltammetry |
| DS | Double stranded |
| ELISA | Enzyme-linked immunosorbent assay |
| GC | Gas chromatography |
| GR-TH | Thionine-functionalized graphene composite |
| GS-Nf/TH/Pt | Nafion/thionine/Pt nanoparticles |
| HPLC | High-performance liquid chromatography |
| HR-SEM | High-resolution scanning electron microscope |
| IDA | Interdigitated array |
| ITO | Indium tin oxide |
| MIP | Imprinted polymer |
| MWCNTs | Multiwall carbon nanotubes |
| MXT | Methotrexate |
| NAC | *N*-acetyl-L-cysteine |
| NAL | Nalbuphine hydrochloride |
| N-GQDs | Nitrogen-doped graphene quantum dots |
| OTC | Oxytetracycline |
| PBS | Phosphate buffer solution |
| PEC | Photoelectrochemical |
| Pt–Cu | – |
| SEM | Scanning electron microscope |
| SERS | Surface-enhanced Raman scattering |
| TLC | Thin-layer chromatography |

# 1 Introduction

The recent accomplishments in nanotechnology particularly nanomaterial-based electrochemical systems have led to the development of an inimitable platform that has profoundly improved the sensory characteristics of the conventional electrochemical systems. The combination of an electrochemical method with nanomaterials of diverse nature and exceptional properties has significantly contributed to the fundamental biological research, healthcare diagnostics, food safety, pharmaceutical procedures, and environmental monitoring. The interdisciplinary nature of this synergic platform has not only widen the scope of sensor systems but has uncovered new pathways for the development of flexible and portable personal care and field applicable devices. The most crucial factor that has contributed to such an advancement is the improved electrode engineering met by the integration of intricate nanostructures of distinct characteristics (Wang et al. 2015).

The nanomaterials based on their higher electro-active surface area, greater charge isolation capacity, higher active site availability can significantly improve the electron-transfer kinetic and thus the sensitive of the sensor under consideration (Soomro et al. 2015a). Unlike the solution-based catalysis, the electro-catalytic reactions, facilitated by the nanomaterials, are largely influenced by the shape, geometry, structure, and morphology of the nanostructure. This reliance is a complex combination of various factors including active area, dimensions, surface free energy, surface functionality, and shape-depended charge density isolated at the apex of nanostructure (Soomro et al. 2016a).

Moreover, it is widely known that presence of any surface functionality can heavily alter the process at the electrode–solution interface. Thus, the synthesizing nanomaterials with morphology that can facilitate both the electron-transfer kinetics and support low-over potential values during electro-catalytic processes are highly demanding (Soomro et al. 2015b). When it comes to designing such nanostructures, structural reproducibility is an important issue to be considered.

This structural reproducibility not only governs the properties of the nanostructure but is a crucial factor responsible for the production of sensitive yet reliable electrochemical sensors. To overcome this issue of structural reproducibility, numerous approaches including the use of different molecules such as proteins, amino acids, and deoxyribose nucleic acid (DNA) have already been used as an effective template (Voet and Tame 2017). In addition, monosaccharide, polymers, and surfactants have also been reported to control and direct the growth of nanostructures (Abdelaal 2015; Banerjee et al. 2013; Kwak et al. 2013). The use of template not only ensures structural reproducibility in the grown nanostructure but the directional capability of the template can be tuned to produce desired morphology that would possess superior electrochemical characteristics compared to their zero-dimensional counterparts. Electrochemical sensors are currently used in various areas including food science, environmental monitoring, marine sector, and health care. The most growing use of these electrochemical sensors has been observed in the area of pharmaceutical drug detection (Soomro et al. 2017; Tunesi et al. 2016, 2017). Pharmaceutical drugs comprise of wide range of organic compounds that are active even at low concentrations. These drugs, aside from their benefiting use in disease diagnosis and treatment, are becoming a serious concern for public health due to their abusive usage and neglected presence in the environment (Sanvicens et al. 2011).

The uncontrolled use of antibiotic has been linked with increased microbial resistance and hardship in the treatment of common infections (Ray et al. 2018). One of the major routes by which these drugs end up in the environment is the hospital effluent which contains a mixture of administered or disposed drugs.

Similarly, the use of drugs for the treatment of animals (veterinary drugs) is also a major source by which such pharmaceutical drugs end up in the environment (Schröder et al. 2016). In regard to the detection and monitoring of such drugs, the electrochemical sensors particularly nanomaterial-based electrochemical sensors are gaining substantial scientific attention. The capability of electrochemical sensors to detect pharmaceutical drugs in real time with high sensitivity and low cost classifies them as an ideal analytical tool for diverse applications (McKeating et al. 2016).

This chapter provides a comprehensive understanding of electrochemical sensor systems particularly electrochemical nano-sensors which deal with the detection and quantification of pharmaceutical drug that falls in the category of antibiotic and therapeutic drugs. Each subsection discusses a variety of strategies offered for some representative drugs on the bases of the technique such as voltammetric, amperometric, and photo-electrochemical approaches. The discussion covers crucial aspects such as design, morphology of the nano-transducer, and the recognition elements in the case of biosensors.

To conclude, recent improvements in the electrochemical sensors including the design and nature of nanomaterials that are responsible for selective and sensitive detection of pharmaceutical drugs will be discussed.

## 2 The Need of Electrochemical Sensors for Pharmaceutical Drug Detection

Pharmaceuticals are diverse group of organic compounds that are significantly active even at low concentrations. The global rise in the use of pharmaceuticals aside from its benefiting applications in disease diagnosis, treatment, and prevention has led to significant increase in the concentration of these drugs. The uncontrolled usage and persistent nature of such drugs are a growing concern for public health (Sanvicens et al. 2011). For example, the abusive use of antibiotics is not only associated with increased microbial resistance but has also significantly reduced the treatment effectiveness against common infectious diseases, leading to frequent hospitalization and overall increase in the health care cost (Ray et al. 2018).

The common household pharmaceuticals which end up in water system after improper disposal and excretion are not subject to any pre-disposal treatment. The contraception hormones which are active even at nanogram level are usually detected in the wastewater system. This hormone has been associated with problems-related endocrine system (Schröder et al. 2016). To monitor and quantify such drugs, to date, majority of chemical analysis is carried using conventional techniques such as thin-layer chromatography (TLC), gas chromatography (GC), and high-performance liquid chromatography (HPLC) (Ahammad et al. 2018).

In addition, the capillary electrophoresis (CE) and enzyme-linked immunosorbent assay (ELISA) are also considered reliable for variety of analyses (Lan et al. 2017). Despite, the befitting and sensitive applications of these techniques, the associated drawbacks such as prolonged analysis time, complex pre-sample preparation, and expensive sample analysis mark them as an unsuitable candidate for diverse range of applications. One of the main constraints associated with these techniques is the need of a proper infrastructure setup. This might not be a viable option when it comes to underdeveloped areas and field-based operations.

Contrary to these conventional techniques, the electrochemical sensors with competitive sensitivities are well-positioned to analyze diverse range of samples from variety of matrices with minimum sample analysis complexities and without the prerequisite of an infrastructure. The convenience of an electrochemical sensor to be modified in reference to the type of analyte and nature of medium enables fast, easy, and reliable measurements with possibility of miniaturization for field applicability at minimum cost (McKeating et al. 2016). Depending on the type of the transducer, the electrochemical sensors can further be classified as voltammetric, amperometric, and photo-electrochemical sensors.

## 3 Voltammetric Sensor for Antibiotics

The voltammetric electrochemical sensors are based on measuring current variation between the working electrode and the analyte present in an active electrolytic solution. In case of nanobiosensors, the working electrode is modified with active transducer layer that is responsible for identifying and producing a specific response (peak) against the varying potential (Chillawar et al. 2015). This capability of recognizing a particular target species via simple voltammetric peak at a specific potential classifies the voltammetric approach highly sensitive and selective. The working electrode in case of voltmetric sensor is often modified using simple drop-casting layer of diverse nanomaterials in combination with biorecognition element (Soomro et al. 2016b).

In regard to drug detection, kanamycin is a common antibiotic widely explored using voltammetric electrochemical sensors. A recent study demonstrated the use of hierarchical nanoporous Pt–Cu nanocomposite in combination with thionine-functionalized graphene for the electrochemical aptamer-based detection of kanamycin (Qin et al. 2016). The hierarchical nanoporous Pt–Cu nanocomposite was synthesized using de-alloying method. The $Pt_4$ $Cu_{18}$ $Al_{78}$ ternary alloy was considered the primary source alloy which was treated with 1 M NaOH followed by annealing the sample at 200 °C for 20 min to produce de-alloyed material. The material was then treated with concentrated $HNO_3$ solution for 1 h to purify the synthesized hierarchical nanoporous Pt–Cu alloy.

The graphene was prepared using one-step reduction method from graphene oxide prepared from graphite powder using an improved Hummer's method. The graphene oxide was then treated with thionine followed by heat treatment of 100 °C for 12 h to produce thionine-functionalized graphene composite (GR-TH). The sensor was developed by layer-to-layer deposition of the as-synthesized nanostructures over pre-polished glassy carbon electrode. Initially, GR-TH was deposited over the electrode surface followed by deposition of specific volume of nanoporous Pt-Cu nanocomposite. The aptamer was assembled on the surface by subjecting the modified electrode to the aptamer solution overnight where the interaction between Pt and–$NH_2$ of aptamer was the responsible force for the adsorption of aptamer over the modified surface. Finally, the aptamer-assembled surface was treated with 1% BSA solution to block the remaining active site to restrict the nonspecific binding. The sensor used differential pulse voltammetry as primary mode of measurement.

The system used aptamter as the bio-recognition element where the graphene-based composite acted as an efficient charge-transfer facilitator. In this scenario, nanoporous Pt–Cu composite provides sufficient surface area to immobilize good density of aptameter which ultimately increases the sensitivity of the electrode. This composite, coupled with good charge facilitator (graphene) allowed the working electrode to detect kanamycin up to 0.42 pg $mL^{-1}$. The measured sensitivity in case of kanamycin nanoporous Pt–Cu composite is relatively greater than $Ag@Fe_3O_4$ which utilized the same charge transfer facilitator (Yu et al. 2013). This signifies the use of nanostructures as transducers. The nanoporous material which provides additional surface area (internal) can significantly improve the immobilization density compared to their nonporous competitors.Another aptasensor with cyclic voltammetry and differential pulse voltammetry (DPV) as basic modes for measurement has been proposed for sensitive detection of tetracycline (Zhou et al. 2012). The proposed approach includes the use of carbonyl functionalized multiwall carbon nanotubes (MWCNTs) to work simultaneously as electrode transducer and substrate for anti-tetracycline aptamer immobilization. The sensor relied on the inhibition of DPV signal observed for potassium hexa-ferrocyanide acting as redox probe. The MWCNTs with carbonyl functionalization provide both large surface area and favorable forces for immobilizing good density of aptamer for the target drug.

This ensured successful detection of tetracycline with signal sensitivity measurable up to 5 nM. The sensor also demonstrated extreme robustness when considered for the detection of tetracycline in milk samples. The aptamer-based voltammetric sensor in this case had superior analytical characteristics compared to colorimetric, bioluminescent biosensor system (Virolainen et al. 2008; Jeon et al. 2008). Despite, the increased use of high surface area possessing nanomaterials, the signal sensitivity problem in case of electrochemical biosensors still persists. Thus to overcome these issues, various strategies, including the use of combination of different nano-sized material, have also been considered. In this context, the use of porous carbon nanorods in conjugation with multifunctional graphene composite consisting of metal oxide and metal ($Fe_3O_4$ and Au) constituents have also been considered as an electrochemical biosensing substrate. This has been promising against the detection of streptomycin antibiotic from food samples (Yin et al. 2017).

The sensor considered the use of nonporous carbon for improving the immobilization efficiency where the oxide/metal composite provides a stable platform for signal amplification. The metal oxide ($Fe_3O_4$) and gold nanoparticles in this case were synthesized using simple hydrothermal and classical Frens method respectively, whereas the graphene was obtained using improved Hummers method.

The multifunctional graphene composite was prepared by continuous shaking of graphene–$Fe_3O_4$ and the as-prepared Au NPs for 12 h.

The electrode was then prepared by simply drop casting specific volume of the designed composite over pre-polished glassy carbon electrode followed by amptamer immobilization. The devised setup based on its unique signal amplification process, realized 0.028 ng/mL as limit of detection for streptomycin antibiotic using differential pulse voltammetry as the mode for quantification.

*N*-acetyl-L-cysteine (NAC) is a common mucolytic agent which is used in the treatment of pulmonary secretions and chronic respiratory diseases. This NAC drug is also known for its effectiveness against hepatotoxicity due to acetaminophen overdose and hepatitis C. Despite the electro-active nature of NAC, the conventional electrodes fail to produce significantly stable response which then hinders the sensitive quantification of the drug from biological matrices. In regard to the quantification of NAC, Tunesi et al. (2016) have demonstrated the use of copper oxide (CuO) nanostructures as an active transducing material for direct electrochemical oxidation of NAC.

The study provides a comprehensive view of how distinct morphologies of CuO can influence the signal response designated for NAC oxidation. The CuO nanostructures in this case were produced using simple hydrothermal method with the assistance of templates i.e. NAC itself, adipic acid (AS) and citric acid (CA). The nanostructures were obtained via growth of CuO nuclei in the presence of the templates that directed and controlled the morphology of the nanostructures.

The hydrothermal treatment of 85 °C for 8 h resulted in distinct structural features including interlaced disc-shaped structures, of individual flakes and large spheres consisting of layers composed of tiny nanoparticles as shown in Fig. 1.

The interesting aspect included the use of NAC, itself as a template. This not only highlighted the drug–CuO interaction but also the potential of pharmaceutical drugs to be considered as an effective template. The electrode modification in this case was achieved using slurry-based approach where the decided volume of the nano-dispersion was casted over a pre-polished electrode.

The modified layer was then allowed to dry under ambient air condition followed by a layer of Nafion® to prevent the surface erosion. The electrochemical assessment was carried in a competitive manner where the oxidation of NAC was profiled using cyclic voltammetry in potential range of −0.15 to 0.30 V with 0.1 M phosphate buffer solution (PBS) buffer as an active electrolytic solution. Figure 2 shows the observed variation in the electrocatalytic response of the representative electrodes modified with different morphologies of CuO nanostructures. The profile is evident of high current achieved using CuO nanostructures formed using citric acid as an active template. The high current with low-over potential value in this case was considered the result of high surface area, and favorable interaction that perceived between the surface-bound functionality (carbonyl) and hydroxyl moiety of NAC.

The combination of attractive morphological features and surface functionalization enabled sensitive detection of NAC with detection limit as low as 0.01 μM. The relatively poor performance of other CuO nanostructures in this case was credited

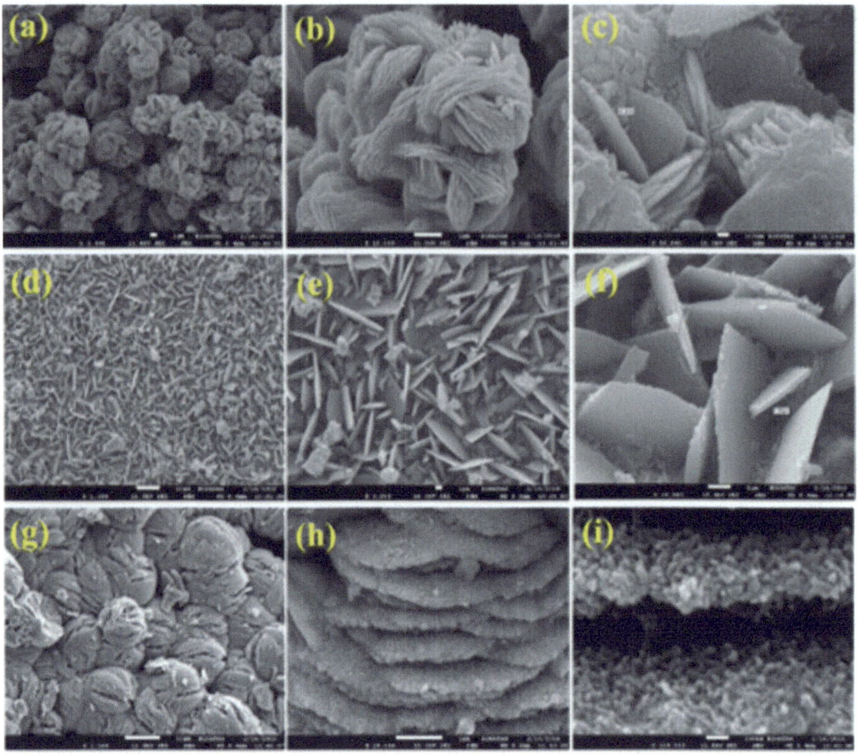


**Fig. 1** Different morphologies for CuO obtained using (**a–c**) NAC, (**d–f**) AS and (**g–i**) CA as a template (Tunesi et al. 2016)

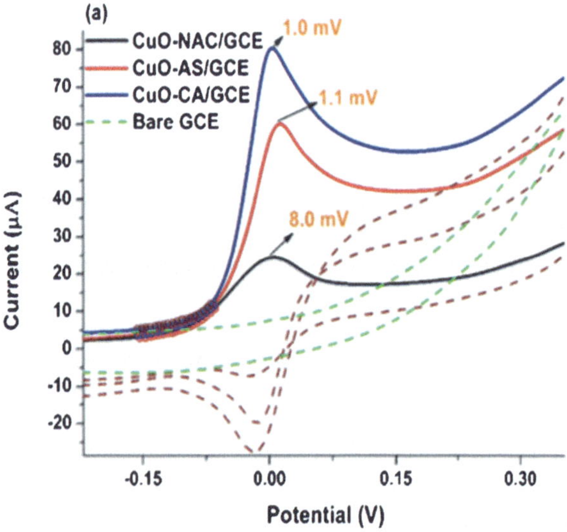


**Fig. 2** CV profile for different CuO nanostructures against electrooxidation of NAC (Tunesi et al. 2016)

to the weak interaction observed for less charged moieties associated with adipic acid and feeble interactions offered by the analogous NAC molecules bound to the surface of CuO nanostructures.

Other than NAC, the determination of nalbuphine hydrochloride (NAL), a phenanthrene derivative is highly considered based on its clinical and pharmacological importance (Elqudaby et al. 2016). The NAL due to its requirement of high overpotential and poor electrode response is a less explored drug. The electrode material that has proven effective against NAL determination includes pencil graphite, gold nanoparticles, and NiO/functional single-walled carbon nanotubes nanocomposite (Elqudaby et al. 2016; Shaikh et al. 2015; Fouladgar 2016). Similarly, the use of NiO, CuO, and $Co_3O_4$ nanostructures for direct electro-catalytic oxidation of NAL has also been proposed (Kalwar et al. 2017).

In this case, the nanostructures of NiO, CuO, and $Co_3O_4$ were prepared using hydrothermal method with nonsteroidal anti-inflammatory drug (acetyl salicylic acid) (ASA) used as a template that controls and directs the growth of nuclei to form different morphological features of nickel, copper, and cobalt oxides (Fig. 3). The study provides evidence of signal variation in accordance to the difference of morphological features associated with nickel, copper, and cobalt oxide nanostructures.

The electrochemical performance of the devised electrodes was studied using CV in the potential range of 0–0.8 V against 0.001 μM NAL. The modified electrodes in this case demonstrated significant electrocatalytic activity toward NAL; however, the CuO nanostructures exhibited the best response in terms of low-over potential value and high current density. The observed superior characteristics were considered the consequence of unique morphology of CuO nanostructures which consists of sharp spike-like structural features. The associated high surface area with greater number of active site supports faster electron transfer process during electrocatalytic oxidation of NAL as proved by relatively greater electron transfer coefficient of 0.69.

The high electron transfer rate can significantly reduce the conductive resistance of the electrode via encouragement of conduction centers subsequent to increased number of contact-points for NAL oxidation. The quantification of NAL was carried using DPV as mode of operation with the NAL concentration range of 0.001–2.25 μM within 0.1 M Briton Robinson Buffer BRB (pH 7.0). The CuO nanostructures-based electrode proved to be highly sensitive toward NAL with detection limit estimated to be $1 \times 10^{-4}$ (S/N = 3) μM. The robust nature of the devised electrode was evident when tested for practical application i.e. quantification of NAL from human urine and clinical wastewater samples. When it comes to electrochemical signal sensitivity and stability, one of the major constrains that limits the current enhancement is the technique primarily used for electrode modification i.e. drop casting.

Since the approach allows direct deposition of nanomaterials dispersion over the surface of the electrode, the distribution and coverage of the nanostructures are difficult to control. Thus, it is not only responsible for signal fluctuations during repetitive measurements but also responsible for the reduction of current from actual value due to stacking/overlapping of nanostructures (Soomro et al. 2017). Moreover,

**Fig. 3** HR-SEM images of CuO nanostructures in situ grown over ITO substrate directly use as working electrode in the determination of NAC (Tunesi et al. 2017)

this approach additionally requires the use of polymeric binders to inhibit the erosion of the casted nanomaterials. These polymeric binders not only decrease the performance of electrode but also contribute to the overall production cost of the electrode under construction. Contrary to this, the in situ growth of nanostructures directly over the surface of flexible electrode can serve as a suitable alternative (Kalwar et al. 2014).

The use of indium tin oxide (ITO)-based electrode with in situ grown nanostructures offer a suitable way-out to produce ordered assembly of nanostructures with high degree of structural similarity. This approach not only facilitates uniform surface–analyte interaction but also ensures complete coverage of the electrode surface which then contributes in overall current density intensification. Moreover, the low-background current associated with ITO substrate also increases the sensitivity of the sensor under construction (Von Weber et al. 2015; Dung et al. 2013).

The use of this technique has been observed for the detection of *N*-acetylcysteine (NAC). The study demonstrates the potential of an ordered morphological feature with high structural uniformity and good electrode surface coverage for the direct electrooxidation of NAC (Tunesi et al. 2017). The study indicated that the in situ grown nanostructures over ITO substrate acts as an active conduction center that can drastically enhance the electron-transfer process between the modified layer and collector. The in situ growth of CuO nanostructures over ITO had been achieved using low-temperature hydrothermal approach with pre-clean ITO substrate submerged within the precursor solution. To ensure the formation of 2D nanostructures, succinic acid was used as an effective growth template.

The scanning electron microscope (SEM) images for the in situ grown CuO layer are shown in Fig. 4. As observed, the grown nanostructures possess highly ordered assembly with structural features similar to flowers. The high-resolution image further shows that the flakes are composed via assembly of tiny nanoparticles.

The ITO-based electrode proved to be highly electro-active against the oxidation of NAC with experimental measurement evident of high current density and signal reproducibility. The quantification of NAC was carried using DPV based on its low background current where the sensor demonstrated signal sensitivity up to $1.2 \times 10^{-3}$ μM (S/N = 3) with charge transfer coefficient ($\alpha$) and diffusion coefficient ($D$) values of 0.65 and $1.62 \times 10^{-2}$ cm$^2$ s$^{-1}$, respectively. The observed analytical characteristics were better than those obtained for slurry-based electrode.

Similar platform has also been considered for the electrochemical determination of captopril (CAP), an antihypertensive drug (Soomro et al. 2017). However, in this case, malonic acid was considered as an effective growth template. The dicarboxylic acid nature of malonic acid enabled the formation of flake-like morphological features possessing excellent size homogeneity and structural similarity.

This electrode was then tested for its capability against the oxidation of CAP in aqueous buffer system. CV and DPV were utilized to study the electrode behavior in reference to conventional glassy carbon electrode. The charge transfer coefficient ($\alpha$) for ITO-based electrode was determined to be 1.13 time higher than conventional GCE. The DPV was used for the quantification, with signal sensitivity down recognized up to $2 \times 10^{-3}$ μM of CAP. The electrode also demonstrates excellent working capability when used for the determination of CAP from commercial tablets and human urine samples.

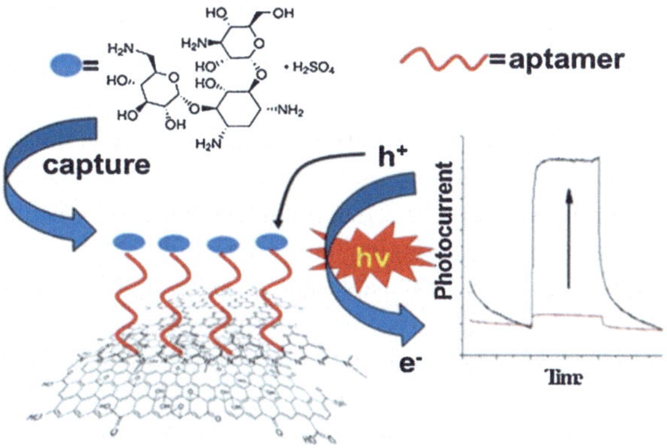


**Fig. 4** Electrochemical aptasensing of kanamycin using graphite-like carbon nitride with graphene (Li et al. 2014)

Despite the wide use of voltammetric sensors, the approach still needs lots of modification and advancement to reach practical maturity. Unlike CV, amperometric mode is relatively more sensitive and practical. The following section covers the use of amperometric sensor system designed for the detection and quantification of antibiotic drugs.

## 4 Amperometric Sensor for Antibiotics

Amperometric sensors are highly considered for the production of miniaturized system capable of real-time practical applications (Tomassetti et al. 2015). In an amperometric analysis, the amplitude of the redox reaction flow is measured against a fixed potential for a specific duration of time. The convectional amperometric sensor considers the use of label enzymes that are responsible for the production of an electroactive species whose current response is then proportional to the concentration of target analyte (Hayat et al. 2014). Whereas, in case of nanomaterial-based amperometric sensors, the incorporation of highly active nanomaterials has significantly increased the response time of the sensor with an enhancement in the electron flow rate.

This makes the amperometric sensor highly sensitive even when the concentration of target analyte is very low. An example includes the use of amperometric approach for the detection of chloramphenicol. An amperometric immunosensor system based on cadmium sulfide quantum dots (CdS QDs) modified with dendrimer attached to gold nanoparticles pre-functionalized with conducting polymer has been considered for chloramphenicol (Kim et al. 2010). The sensor relies on the electrode consisting of Au NPs formed via electrode deposition approach followed by formation of conducting polymeric layer using potential cyclic method.

The immuno-recognition site for chloramphenicol was introduced via amine-terminated dendrimer, which also allowed easy attachment of CdS QDs via covalent interaction between the amine-terminated dendrimer and carboxylic moiety bounded to CdS QDs. The amperometric response was based on the competitive immuno-interaction between the free and labeled chloramphenicol for active sites of the antibody. This enabled sensitive detection of chloramphenicol in concentration range of 50–950 pg $mL^{-1}$ with sensor sensitivity up to 45 pg $mL^{-1}$.

Another approach allows selective amperometric detection of drugs considering the use of aptamer as receptors. In this category, the use of DNA is most common. The detection of oxytetracycline (OTC) is very challenging based on its structural similarity to the other tetracyclines such as doxycycline or tetracycline. This issue of selectively recognizing and quantifying OTC from other counterparts can be resolved using ssDNA aptamer as indicated by Kim and coworkers (Kim et al. 2009). In this case, the sensor utilized thiol-modified aptamer immobilized over gold interdigitated array (IDA) electrode chip to specifically detect OTC. The use of highly specific aptamer with gold nanoparticles as transducer enabled detection of OTC in wide concentration range of 1–100 nM.

To improve the conductivity, specific surface area and potential activity and combination of different nanostructures have also been considered. A recent effort in this context has been observed in the detection of kanamycin (Wei et al. 2012). The system relied on anti-kanamycin, antibodies immobilized over the surface of electrode which had been modified with sandwich layers of Nafion/thionine/Pt nanoparticles (GS-Nf/TH/Pt). The use of graphene in combination with Pt nanoparticles provided a conductive platform capable of supporting fast electron transfer meditation thionine enabling sensitivity of 5.74 pg mL $^{-1}$ for kanamycin.

The discussed amperometric approach proved sensitive toward the detection of kanamycin from animal-derived foods. An aptamer-based amperometric biosensor based on the use of silver nanoparticles in combination with reduced graphene oxide has also been proposed for CAP (Liu et al. 2017). In this case, the interaction of CAP with aptamer previously immobilized over the nanocomposite (AgNPs/rGO) results in the decline of amperometric response for AgNPs/rGO. The signal inhibition was directly associated with the concentration of CAP molecules present in the system. This approach realized good signal sensitivity with 0.65 ng mL$^{-1}$ obtained as the limit of detection for CAP.

Despite the high sensitive and practical feasibility of the amperometric approach. The use of noble metal nanoparticles in combination with costly carbonaceous materials such as graphene and carbon nanotubes limits the scope of amperometric approach for the development of cheap, field-portable sensor systems. In this context, the recently advanced photo-electrochemical approach based on its cheap photoactive material is a suitable alternative. The following section discusses recently developed photo-electrochemical sensors for the detection of antibiotics.

## 5 Photo-electrochemical Sensors for Antibiotics

The photo-electrochemical (PEC) sensors are considered one of the most advanced versions of the electrochemical sensors family. The synergic coupling of photoactive material with electrochemical setup has established a platform capable of recognizing different molecules of biological and clinical importance (Shi et al. 2016). The use of photoactive nanomaterials with highly selective recognition sites has not only enhanced the sensitivity of photo-electrochemical sensor but has also widened the scope of PEC for low-level drug detection (Yan et al. 2015).The PEC aptasensor suggested by Li et al. (2014) uses graphite-like carbon nitride as photoactive material in conjugation with aptamer as bio-recognition element for the detection of kanamycin. The study discusses the inclusion of graphene with the photoactive material and how this inclusion can enhance the visible light photoactive response (Fig. 4). In this case, the engineered photoactive nanocomposite provides greater surface area for satisfactory immobilization of aptamer via $\pi$–$\pi$ stacking interactions. The PEC sensor for kanamycin provided sensitive signal up to 0.2 nM, which is relatively greater than its photo-free counterparts.

In PEC, the photoactivity of the material is the key parameter responsible for the sensitivity of the measured signal. Thus, the use of nanomaterial that is capable of harnessing more light besides larger surface area can drastically enhance the signal sensitivity and selectivity of the PEC sensor under investigation. In this context, Yan et al. (2015) demonstrated the capability of BiOI as a photoactive material for the sensing of oxytetracycline (OTC).

The PEC aptasensor was designed to possess larger surface area for greater immobilization of anti-OTC aptamer with tunable photo-response to achieve the desired signal. This was achieved by doping graphene within the BiOI under vigorous stirring condition. The PEC sensor demonstrated excellent working response toward OTC where inclusion of graphene facilitates the charge transfer between the electrode and transducer interface. This high current photo-response is high desirable when the sensing is based on signal inhibition strategy. In this case, the PEC biosensor was capable of detecting OTC up to 0.9 nM using signal inhibition approach.

The use of nitrogen-doped graphene quantum dots (N-GQDs) has also attracted PEC researchers. The combination of highly active N-GQDs with low-cost binding aptamer has been tested for the detection of chloramphenicol (CAP) (Liu et al. 2015). The large quantum size area of the N-GQDs enables dense immobilization of nucleobases aptamer. The affinity-based interaction of CAP with immobilized aptamer can significantly enhance the generation of photocurrent. The N-GQDs based PEC biosensor has reported detection limit of 3.1 nM for CAP which is comparable to its conventional counter-parts such as HPLC and capillary electrophoresis.

Despite the collective advantages of PEC biosensor, the choice of photoactive material capable of harnessing greater light intensity with sufficient large surface area for photoactivity and immobilization is still a challenging issue. However, the recent advancement in nano-engineering is indicative of achieving this in the near future.

## 6 Electrochemical Sensors for Therapeutic Drug Detection Conclusion and Future Perspective

Other than antibiotics, monitoring of the therapeutic drugs is highly desirable based on their pharmacokinetics assessment. The requirement of low dosage and associated toxicity of therapeutic drugs limit their analysis using conventional approaches. In this context, the electrochemical sensor in combination with nanomaterials has emerged as an ideal monitoring system for therapeutic drugs (Martinkova and Pohanka 2015). The speedy advancement in the development of practical electrochemical sensors for the therapeutic drug is the evidence of their promising approach for many common and new drugs which would be launched in the market (Masson and Pelletier 2015).

At present only few selected drugs have been tested and successfully detected using nanomaterial-based electrochemical sensors (McKeating et al. 2016). Since the development of electrochemical sensor for therapeutic drugs is relatively new area, this section discusses the electrochemical sensors without classifying them in accordance to the mode of investigation i.e., voltammetric or amperometric, etc.

Methotrexate (MXT) is one of the most widely administered drugs for different types of cancer-related diseases. The high toxicity of MXT has been directly associated with the duration of exposure and thus, the patients are subjected to higher doses of MXT for shorter interval to achieve maximum drug efficiency. This as a result, requires rapid and highly sensitive monitoring system to ensure elimination of MXT subsequent to the treatment (Ramsey et al. 2018). In this regard, the DNA-based electrochemical immune-sensors are highly promising. The recent effort of Chen et al. (2018) demonstrates the use of DNA immobilized graphene oxide–modified GCE electrode for electrochemical sensing of MTX. In this case, graphene oxide was utilized as a support to immobilize DNA and to facilitate the charge transfer process at the electrode surface.

The electrochemical probing was achieved using guanine as redox mediator where the exogenous action of MXT over the immobilized DNA-produced gradual increment in the current measured for the oxidation of guanine. This immunosensor-enabled sensitive detection of MXT, with limit of detection achieved as 7.6 $3 \times 10^{-9}$ mol $L^{-1}$.

In a similar approach, cisplatin which is a common drug used for the treatment of various solid tumors has been detected using DNA immobilized electrode. In a methodology discussed by Yardım et al. (2017) the successful use of DNA modified electrochemically reduced graphene oxide has been demonstrated for the sensitive detection of cisplatin. The study compared the performance of the devised electrode with conventional electrodes where high sensitivity was achieved as a consequence of the high electron sensitizing capability of graphene oxide. The oxidation of guanine and adenine from the DNA was recognized as the basic signal for the quantification of cisplatin.

The interaction of the target drug with DNA-modified electrode led to a decline in the current response noted for the oxidation of guanine and adenine. A linear inhibition was observed for in the concentration range of 10–100 mM and 1–100 mM for guanine and adenine peaks, respectively. The detection limits in this case were determined to be 0.3 and 0.2 mM using guanine and adenine peak currents, respectively. Another DNA-based approach has been considered for the detection of cyclophosphamide (Sigen et al. 2009). The study demonstrates the use of screen-printed electrodes modified with carbon nanotubes for DNA immobilization. In this case, the interaction of drug with the double-stranded (ds) calf thymus DNA results in the change of redox response for guanine within DNA.

The observed redox change is enhanced by the carbon nanotubes used previously for the modification of the electrode. The design and modification process of the screen-printed electrode is illustrated as Fig. 5. The improved sensitivity and selectivity in case of cyclophosphamide have been achieved via the use of molecular imprinted polymer (MIP) (Huang et al. 2017). The use of composite material based

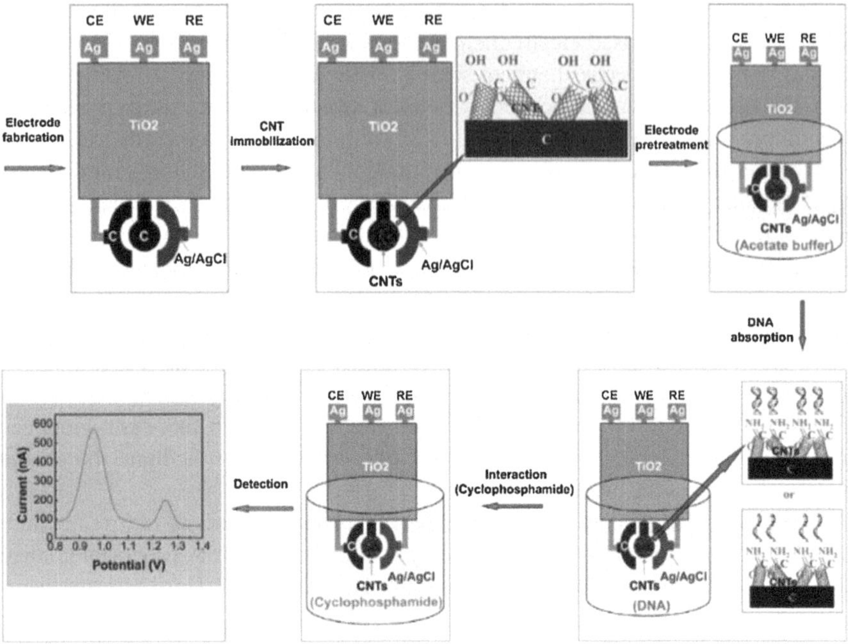


**Fig. 5** The DPV-based quantification of cyclophosphamide using CNT-modified DNA biosensor (Sigen et al. 2009)

on nitrogen and sulfur co-doped activated graphene can serve as a model tool for selective recognition of cyclophosphamide. The application of graphene not only provides sufficient surface area but also promotes the fast transfer of the electrons. In addition, the electropolymerized MIP layers allow selective capturing and quantification of cyclophosphasmide. In this approach, DPV in combination with $Fe(CN)_6^{3-/4-}$ has been considered a probe for signal measurement. In another approach, the detection of doxorubicin (DBX), cancer-related drugs, has been suggested by Bahner et al. (2018).

The electrochemical sensor relied on the use of aptamer which was co-immobilized with mercaptohexanol on gold. The drug–electrode interaction was studied using the electrochemical impedance spectroscopy (EIS), where the quantification was based on the signal inhibition of the probing agent (ferro/ferricyanide) as a result of the said interaction. EIS enabled sensitive detection of DBX in concentration range of 31–125 nM with sensitivity down to 28 nM. The observed sensitivity in this case is much greater than the non-aptamer-based electrochemical sensor designed for DBX.

In a much-advanced approach, use of surface-enhanced Raman scattering (SERS) spectroscopy in combination of electrochemical methods has been used to drug–DNA reactivity (Ilkhani et al. 2016). A self-assembled monolayer protected gold disk electrode (AuDE) coated with reduced graphene oxide and decorated with plasmonic gold coated with $Fe_2Ni@Au$ magnetic nanoparticles has been used for the functionalization of double-stranded DNA (dsDNA). This platform has been used for the detection of DBX. The coupled approach provides higher sensitivity with limit of detection for DBX estimated to be 8 μg/mL. The new platforms are highly sensitive to agents that can bind with DNA and thus can largely facilitate the sensory studies for various kinds of therapeutic drugs.

Despite all the reports discussed in this section for electrochemical sensor designed for monitoring therapeutic drugs, there still needs a lot of research to ensure their proximity to the patients and for real-time drug analysis.

# 7 Conclusion and Future Perspective

Electrochemical sensor systems based on nanomaterials are promising candidates for clinical diagnostics and environmental analysis. The discussed techniques and strategies in this chapter are evident of growing efforts toward the increased sensitivity, selectivity, speedy analysis with field portability of the sensors. However, there are still a lot of challenges to be met before such sensor systems could be driven into the domestic market and be considered for point of care diagnostics. One of the major challenges in this regard is the signal reproducibility of the electrochemical sensor in matrices other than standard. Besides this, the use of electrochemical sensor for continuous monitoring of analyte is still an under-explored area.

The electrochemical sensor with capability of real-time and on-site monitoring can significantly reduce the number of representative samples and analysis time. This would not only reduce the overall cost of the sensor but would provide new opportunities toward end-point detection applications. In addition to this, a challenging issue is to overcome the problem of mass-transfer restriction by facilitation of fluidic motion at the nanoscale. The papers discussed in this chapter signify the viable integration of nanomaterials and electrochemical system which has led to the unparalleled maturity to the electrode-based sensor devices. However, one major aspect that marks these sensor devices as proof of concept rather than to applicable as real-time devices is their limited integration within the miniaturized system which would enable their application in more relevant practical applications.

In this context, the limited work is directly related to the lesser studies available for the biocompatibility and toxicity of the nano-transducer-containing devices. In addition to this, studies related to morphological variation of the nano-material-based electrode before and after use would provide a complete profile for the sensor system that would enable better understanding of the sensors venerability and inherent problem of electrode stability. Thus, to drive the electrochemical sensor system into the domestic market, research trends must focus of key issues of stability, reproducibility, portability, and most important of all affordability.

The advancement of biosenors in commercial market could be achieved by simplifying the synthesis protocols for transducers, efficient and innovative approches toward electrode modification which would not be possible without considering the integration of nanotechnology, neurology, drug pharmacology, environmental science, electronics, and electrochemistry. Nevertheless, the continuous and determined efforts of scientific community are a clear evidence of the greater optimism that this objective would be achieved in the coming years.

**Acknowledgments** The authors thank, The Scientific and Technological Research Council of Turkey (TÜBİTAK) for providing financial support under 2221-Visiting Scientists Fellowship Programme for doing research at Department of Mechanical Engineering, Faculty of Engineering, University of Selcuk, Campus, 42079 Konya- Turkey. The authors also thank Higher Education Commission of Pakistan (HEC) for their support under International Research Support Initiative Programme (IRSIP) and appreciate the facilities provided by National Centre of Excellence in Analytical Chemistry, University of Sindh, Jamshoro, 76080, Pakistan. The authors acknowledge the support and encouragement provided by Prof. Dr. Tufail Hussain Sherazi and Prof. Dr. Sirajuddin affiliated with NCEAC, Jamshoro Pakistan.

***Conflict of Interest*:** The corresponding author states that there is no conflict of interest.

## References

Abdelaal HM (2015) ChemistryOpen 4:72–75
Ahammad AS, Islam T, Hasan MM, Mozumder MI, Karim R, Odhikari N et al (2018) J Electrochem Soc 165:B174–BB83
Bahner N, Reich P, Frense D, Menger M, Schieke K, Beckmann D (2018) Anal Bioanal Chem 410:1453–1462
Banerjee S, Maji T, Paira TK, Mandal TK (2013) Macromol Rapid Commun 34:1480–1486
Chen J, Fu B, Liu T, Yan Z, Li K (2018) Electroanalysis 30(2):288–295
Chillawar RR, Tadi KK, Motghare RV (2015) J Anal Chem 70:399–418
Dung NQ, Patil D, Jung H, Kim J, Kim D (2013) Sensors Actuatators B Chem 183:381–387
Elqudaby HM, Hendawy HA, Souaya ER, Mohamed GG, Eldin GM (2016) Int J Electrochem 8621234, 1–9
Fouladgar M (2016) Sensors Actuatators B Chem 230:456–462
Hayat A, Catanante G, Marty JL (2014) Sensors 14:23439–23461
Huang B, Xiao L, Dong H, Zhang X, Gan W, Mahboob S et al (2017) Talanta 164:601–607
Ilkhani H, Hughes T, Li J, Zhong CJ, Hepel M (2016) Biosens Bioelectron 80:257–264
Jeon M, Kim J, Paeng K-J, Park S-W, Paeng IR (2008) Microchem J 88:26–31
Kalwar NH, Soomro RA, Sherazi STH, Hallam KR, Khaskheli AR (2014) Int J Metals 126103, 1–11
Kalwar NH, Tunesi MM, Soomro RA, Amir M, Avci A, Hallam KR et al (2017) J Electroanal Chem 807:137–144
Kim YS, Niazi JH, Gu MB (2009) Anal Chim Acta 634:250–254
Kim D-M, Rahman MA, Do MH, Ban C, Shim Y-B (2010) Biosens Bioelectron 25:1781–1788
Kwak G, Hwang J, Cheon J-Y, Woo MH, Jun K-W, Lee J et al (2013) J Phys Chem C 117:1773–1779
Lan L, Yao Y, Ping J, Ying Y (2017) Biosens Bioelectron 91:504–514
Li R, Liu Y, Cheng L, Yang C, Zhang J (2014) Anal Chem 86:9372–9375
Liu Y, Yan K, Okoth OK, Zhang J (2015) Biosens Bioelectron 74:1016–1021
Liu S, Lai G, Zhang H, Yu A (2017) Microchim Acta 184:1445–1451
Martinkova P, Pohanka M (2015) Anal Lett 48:2509–2532
Masson J-F, Pelletier JN (2015) Nanomedicine 10:521–524
McKeating KS, Aube A, Masson J-F (2016) Analyst 141:429–449

Qin X, Yin Y, Yu H, Guo W, Pei M (2016) Biosens Bioelectron 77:752–758
Ramsey LB, Balis FM, O'Brien MM, Schmiegelow K, Pauley JL, Bleyer A (2018) et al. Oncologist 23:52–61
Ray MJ, Trick WE, Lin MY (2018) Infect Control Hosp Epidemiol 39:1–6
Sanvicens N, Mannelli I, Salvador J-P, Valera E, Marco M-P (2011) TrAC Trends Anal Chem 30:541–553
Schröder P, Helmreich B, Škrbić B, Carballa M, Papa M, Pastore C et al (2016) Environ Sci Pollut Res Int 23:12835–12866
Shaikh T, Nafady A, Talpur FN, Agheem MH, Shah MR, Sherazi STH et al (2015) Sensors Actuatators B Chem 211:359–369
Shi H, Zhao J, Wang Y, Zhao G (2016) Biosens Bioelectron 81:503–509
Sigen W, Ruili W, Paul JS, Sha C (2009) Curr Nanosci 5:312–317
Soomro RA, Ibupoto ZH, Abro MI, Willander M (2015a) Sensors Actuatators B Chem 209:966–974
Soomro RA, Hallam KR, Ibupoto ZH, Tahira A, Jawaid S, Hussain Sherazi ST et al (2015b) RSC Adv 5:105090–105097
Soomro RA, Hallam KR, Ibupoto ZH, Tahira A, Sherazi STH, Sirajjuddin et al (2016a) Electrochim Acta 190:972–979
Soomro RA, Nafady A, Hallam KR, Jawaid S, Al Enizi A, Sherazi STH et al (2016b) Anal Chim Acta 948:30–39
Soomro RA, Tunesi MM, Karakus S, Kalwar N (2017) RSC Adv 7:19353–19362
Tomassetti M, Serone M, Angeloni R, Campanella L, Mazzone E (2015) Sensors 15:3435–3452
Tunesi MM, Soomro RA, Ozturk R (2016) J Electroanal Chem 777:40–47
Tunesi MM, Soomro RA, Ozturk R (2017) J Mater Chem C 5:2708–2716
Virolainen NE, Pikkemaat MG, Elferink JWA, Karp MT (2008) J Agric Food Chem 56:11065–11070
Voet ARD, Tame JRH (2017) Curr Opin Biotechnol 46:14–19
Von Weber A, Baxter ET, White HS, Anderson SL (2015) J Phys Chem C 119:11160–11170
Wang S, Xu L-P, Zhang X (2015) Sci Adv Mater 7:2084–2102
Wei Q, Zhao Y, Du B, Wu D, Li H, Yang M (2012) Food Chem 134:1601–1606
Yan K, Liu Y, Yang Y, Zhang J (2015) Anal Chem 87:12215–12220
Yardım Y, Vandeput M, Çelebi M, Şentürk Z, Kauffmann J-M (2017) Electroanalysis 29: 1451–1458
Yin J, Guo W, Qin X, Zhao J, Pei M, Ding F (2017) Sensors Actuatators B Chem 241:151–159
Yu S, Wei Q, Du B, Wu D, Li H, Yan L et al (2013) Biosens Bioelectron 48:224–229
Zhou L, Li D-J, Gai L, Wang J-P, Li Y-B (2012) Sensors Actuatators B Chem 162:201–208

# Green Sensors for Environmental Contaminants

**Mahmoud El-Maghrabey, Rania El-Shaheny, Fathalla Belal, Naoya Kishikawa, and Naotaka Kuroda**

## Contents



M. El-Maghrabey (✉)
Department of Pharmaceutical Analytical Chemistry, Faculty of Pharmacy, Mansoura University, Mansoura, Egypt
e-mail: dr_m_hamed@mans.edu.eg

Department of Analytical Chemistry for Pharmaceuticals, Course of Pharmaceutical Sciences, Graduate School of Biomedical Sciences, Nagasaki University, Nagasaki, Japan

R. El-Shaheny
Department of Pharmaceutical Analytical Chemistry, Faculty of Pharmacy, Mansoura University, Mansoura, Egypt

Department of Hygienic Chemistry, Course of Pharmaceutical Sciences, Graduate School of Biomedical Sciences, Nagasaki University, Nagasaki, Japan

F. Belal, Professor
Department of Pharmaceutical Analytical Chemistry, Faculty of Pharmacy, Mansoura University, Mansoura, Egypt

N. Kishikawa · N. Kuroda, Professor
Department of Analytical Chemistry for Pharmaceuticals, Course of Pharmaceutical Sciences, Graduate School of Biomedical Sciences, Nagasaki University, Nagasaki, Japan



Inamuddin, A. M. Asiri (eds.), *Nanosensor Technologies for Environmental Monitoring*, Nanotechnology in the Life Sciences,
https://doi.org/10.1007/978-3-030-45116-5_17

## Abbreviations

| | |
|---|---|
| BPA | Bisphenol A |
| BPEI | Branched polyethyleneimine |
| CNOs | Onion-like carbon nanoparticles |
| CNT NCs | Carbon nanotube nanocomposites |
| CQDs | Carbon quantum dots |
| EDA | Ethylenediamine |
| EDTA | Ethylenediaminetetraacetic acid |
| GAC | Green analytical chemistry |
| GCE | Glassy carbon electrode |
| GQDs | Graphene quantum dots |
| HPI | 2-(4-Hydroxyphenyl)-4,5-di(2-pyridyl)imidazole |
| MNPs | Metal nanoparticles |
| N-CQDs | Nitrogen-doped carbon quantum dots |
| NPs | Nanoparticles |
| NPys | Nanopyramids |
| P-CQDs | Phosphorus-doped carbon quantum dots |
| QDs | Quantum dots |
| SPE | Screen-printed electrode |

## 1 Introduction

In the modern age, human activity is being multiplied in areas of cultivation, industry, transportation, and urbanization. This comes with many penalties and adverse effects on the surrounding environment. The increase in quality and standards of life and continuously increasing demands of humans resulted in various activities that amplified pollution of air, soil, and aquatic environment. Environmental pollution related to humans activities includes greenhouse gases, particulate pollutants, leachates, oil spills, hazardous wastes, persistent organic pollutants, pesticides, and sludge, as well as nonbiodegradable disposable goods. The problem is aggravated because of the lack of proper facilities for waste disposal and treatment. These pollutants affect human health and its accumulation may make the surrounding environment invalid for supporting life (Gavrilescu 2010; Gavrilescu et al. 2015). Thus, monitoring of environmental contaminants have become a worldwide priority due to their impacts on human health and socioeconomic progress (Justino et al. 2017).

Classical analytical approaches including gas and liquid chromatography and capillary electrophoresis coupled with different types of detectors including mass spectrometric, electrochemical, and optical ones are applied for environmental contaminants' monitoring. However, most of these techniques are harmful to the environment as they include the production of harmful wastes and the disposal of a large volume of organic solvents after each assay. In addition, they require expensive reagents and equipment and laborious sample pretreatment (Lang et al. 2016; Hassani et al. 2017; Justino et al. 2017).

Starting from the opening of the current century, the green chemistry concept has been strongly evolved and there is a solid global trend to make the analytical methods greener with less impact on the environment by minimizing toxic chemicals and solvents, and economizing energy consumption (de la Guardia et al. 1995; Tobiszewski et al. 2010; Guardia and Garrigues 2012). Thus, the green analytical chemistry (GAC) concept has been emerged to achieve these goals. Great efforts were made to optimize analytical approaches capable of direct analysis using miniaturized analytical equipment, reducing solvents consumption, and allowing for on-site analysis (Armenta et al. 2008). The analytical methods dedicated to monitor the environmental pollutants should be designed to be inherently eco-green themselves and to track the rules of GAC (Tobiszewski et al. 2010; Guardia and Garrigues 2012; El-Maghrabey et al. 2019).

There are many approaches for greening analytical methodologies, such as greening classical analytical methods (e.g. HPLC, GC, and CE) or developing new techniques that are themselves green in nature (Farré et al. 2010). Classical methods can be greened through several approaches (El-Shaheny et al. 2015, 2019), however, they are still not effective for on-site measurements that are needed for environmental monitoring especially in cases of accidental release of hazardous wastes or acute poisoning. Thus, rapid, miniaturized and cost-effective methods, and, if applicable, portable devices are strongly needed for environmental monitoring to surpass the increased environmental problems (Justino et al. 2017). For the achievement of this purpose, sensors have attracted the attention of scientists in recent times. A sensor is an electrical subsystem that aims at detecting actions or alterations in its surrounding environment and submits signals either in a simple form as light or complex form as in a computer (Vaz Jr 2018). Thus, many researchers are recently focusing on the development of chemical sensors and biosensors that follow the rules of GAC and allow for rapid and on-site monitoring of environmental contaminants (Justino et al. 2017; El-Maghrabey et al. 2019; Li et al. 2019; Boonkanon et al. 2020). A chemical sensor is an independent analytical device that gives information about the chemical composition of its surrounding environment which is, mainly, a fluidic phase. The information is given as a quantifiable physical signal that is correlated to the concentration of a certain chemical species. On the other hand, biosensors are analytical devices that convert a biological response into a measurable signal (Vaz Jr 2018).

In this chapter, we provided a review for the state of the art in developing chemical sensors that obey the GAC rules for environmental analysis of different contaminants. Regarding biosensors, there are many current reviews about them as recent as 2019, thus, we will provide an umbrella review of these reviews and their concluding remarks.

# 2 Green Chemical Sensors for Environmental Contaminants

Recently, many eco-green chemical sensors were developed for monitoring environmental contaminants. Their green nature varies from one to another where some sensors are synthesized via green routes or using naturally occurring compounds, while others are not green in nature but their sensing mechanism and procedure were optimized and miniaturized to decrease the consumption of solvents, reagents, and energy. Moreover, some of these sensors were evaluated from greenness point of view via comprehensive approaches. In this section, the green approaches for environmental chemical sensors are discussed and briefly presented in Table 1.

## *2.1 Carbon Quantum Dots as Green Sensors for Environmental Contaminants*

Quantum dots (QDs) semiconductors possess exceptional optical properties making them the most leading fluorescent nanoparticles (NPs) in the field of nanoprobe sensing and they were reported many times for sensing environmental toxic pollutants such as mercury (Gonçalves et al. 2010). However, they are toxic to the environment due to their heavy metal components (Hardman 2005). Recently, carbon quantum dots (CQDs) have arisen as a promising substitute for semiconductors QDs with excellent photoluminescent properties. It has many advantages including chemical inertness, thermal stability, water solubility, and exceptional biocompatibility which make CQDs an excellent environmentally friendly alternative to semiconductors QDs (Chandra et al. 2011). Zho et al. utilized ethylenediaminetetraacetic acid (EDTA) as a precursor for the synthesis of CQDs *via* pyrolysis process. The produced CQDs were used unmodified for sensing of the heavy metal environmental toxic pollutant, Hg(II), through fluorescence quenching mechanism (Zhou et al. 2012). Additionally, amino-functionalized fluorescent CQDs were synthesized by Li et al. through the hydrothermal coupling of citric acid as a natural carbon source with ethylenediamine (EDA). The produced CQDs were applied for sensing of Cr(VI) in the wastewater of the vanadium extraction industry via fluorescence quenching mechanism (Li et al. 2018). Despite the use of some synthetic substrates such as EDA and EDTA, these methods are considered eco-friendly as it avoids the use of organic dyes and solvents and could provide a cheap and sensitive sensor for toxic environmental contaminants (Zhou et al. 2012; Li et al. 2018).

In this connection, greener approaches for the synthesis of CQDs using natural substrates or biomasses have been emerged recently. For example, natural compounds including sucrose, citric acid, glycerol, glycol, and glucose (Chandra et al. 2011; Wang et al. 2011; Shi et al. 2012) have been used as main substrates for the synthesis of CQDs via their carbonization hydrothermally or by microwave irradiation. Dong et al. synthesized CQDs capped with polyamine functionalized by pyrolyzing citric acid, as a carbon source, with the water-soluble polyamine cationic

**Table 1** Environmentally green chemical sensors for the monitoring of environmental contaminants

| Analytes | Environmental sample | Sensor (substrate)[a] | Sensing mechanism | LOD | References |
|---|---|---|---|---|---|
| Al(III) | Water | CNOs | Fluorescence quenching | 0.77 μM | Tripathi et al. (2017) |
| Arsenic | Water | Glucose and Au-NPs | Colorimetric | 0.53 ppb | Boruah et al. (2019) |
| Cd(II) | Water | Glutathione and Au-NPs | Colorimetric | 30 nM | Manjumeena et al. (2015) |
| Cr(VI) | Waste water | CQDs (citric acid + EDA) | Fluorescence quenching | 140 nM | Li et al. (2018) |
| Hg(II) | Water | Unmodified CQDs (EDTA) | Fluorescence quenching | 4.2 nM | Zhou et al. (2012) |
| | Lake water | CQDs (sweet potatoes) | Fluorescence quenching | 1 nM | Lu et al. (2012a) |
| | Pond water | CQDs (sweet potatoes) | Fluorescence quenching | 6 nM | Bano et al. (2018) |
| | Lake water | CQDs (pamelo peel) | Fluorescence quenching | 0.22 nM | Lu et al. (2012b) |
| | Lake water | N-CQDs (Phytic acid + Na citrate) | Fluorescence quenching | 3 nM | Huang et al. (2013) |
| | Tap and mineral water | Ag@AgCl-NPs | Colorimetric | 4.2 nM | Karimi and Samimi (2019) |
| Hg(II), Pb(II), and Mn(II) | Water | Ag-NPs and Au-NPs | Colorimetric | 16–53 nM | Annadhasan et al. (2014) |
| Cu(II) | River water | BPEI-CQDs (citric acid) | Fluorescence quenching | 6.0 nM | Dong et al. (2012b) |
| | River water | BPEI-CQDs (bamboo leaves) | Fluorescence quenching | 115 nM | Liu et al. (2014) |
| | River water | N-CQDs (arginine + lemon extract) | Fluorescence quenching | 47 nM | Das et al. (2017) |
| | Water | N-CQDs (phytic acid + Na citrate) | Fluorescence quenching | 1 nM | Yang et al. (2018) |
| $F^-$ | Water | CQDs (citric acid) + Eu(III) | Fluorescence turn off–on | 1 ppm | Singhal et al. (2017) |
| | Water | CQDs (starch) + Fe(III) | Fluorescence turn off–on | NM (not mentioned) | Basu et al. (2015) |
| Chlorine | Tap water | GQDs (citric acid) | Fluorescence quenching | 0.05 μM | Dong et al. (2012a) |
| Bisphenol-A | Tap and mineral water | Ag-NPs | Colorimetric | 45 nM | Khalililaghab et al. (2017) |
| | Wastewater samples | GCE: Fullerene-$C_{60}$ | Electrochemical | 3.7 nM | Rather and De Wael (2013) |
| 4-Aminophenol | Water | GCE:NiO-NPs·CNT NCs | Electrochemical | 15 pM | Hussain et al. (2017) |

(continued)

**Table 1** (continued)

| Analytes | Environmental sample | Sensor (substrate)[a] | Sensing mechanism | LOD | References |
|---|---|---|---|---|---|
| Atrazine pesticide | River water | Boron-doped diamond electrode | Electrochemical | 10 nM | Švorc et al. (2013) |
| Nitrobenzene | Water | GCE:Au-NPs | Electrochemical | 16 nM | Emmanuel et al. (2014) |
| Dichloromethane | Water | SPE:ZnO-NPys | Electrochemical | 17.3 μM | Kim et al. (2019) |
| Pd(II) | River water | Benzofuran-2-boronic acid | Fluorimetric | 9.8 nM | Higashi et al. (2017) |
| Hydrogen sulfide | | HPI-Cu(II) complex | Fluorescence off–on | 5 ppb | El-Maghrabey et al. (2019) |
| Boron | Wastewater | Curcumin nanoparticles supported on starch | Colorimetric | 0.052 ppb | Boonkanon et al. (2020) |

[a]*CNOs* onion-like carbon NPs, *CQDs* carbon quantum dots, *BPEI* branched polyethyleneimine, GCE: glassy carbon electrode, GCE: CNT NCs: carbon nanotube nanocomposites, *SPE* screen-printed electrode, *NPys* nano pyramids, *HPI* 2-(4-Hydroxyphenyl)-4,5-di(2-pyridyl)imidazole

polymer, branched polyethyleneimine (BPEI) (Dong et al. 2012c). They applied the amino capped CQDs for monitoring of Cu(II) in river water samples where Cu(II) quenches the fluorescence of the CQDs through the inner filter effect (Dong et al. 2012b). Alternatively, bamboo leaves were used as a natural biomass precursor for CQDs instead of using chemical substrates. Bamboo leaves were chosen as bio-source due to their richness with complex carbohydrates, which possess many carbonyl, hydroxyl, and carboxylic function groups that are excellent carbon sources for CQDs synthesis. The produced CQDs were capped also with BPEI and were applied for monitoring of Cu(II) (Liu et al. 2014). Sweet potato is another natural mass that has been effectively used as a carbon source for CQDs. The produced sensor from sweet potatoes hydrothermal treatment could detect Hg(II) down to 1 nM (Lu et al. 2012a). Bano et al. utilized another natural biomass, *Tamarindus indica* leaves, which produced CQDS upon hydrothermal treatment. The developed probe could detect Hg(II) in pond water samples down to 6 nM (Bano et al. 2018). Moreover, a huge step forward for waste materials recycling was taken when the same research group used pomelo peel as a carbon source for CQDs. The developed sensor here was extremely sensitive to Hg(II) down to the sub-nanomolar level (Lu et al. 2012b).

Doped carbon dots are known to be more efficient as sensors due to extensive surface functionalization that provides a great binding affinity without the need for capping. Das et al. synthesized N-doped CQDs (N-CQDs) via thermal coupling of lemon extract as a carbon source and L-arginine as N source. The produced N-doped CQDs could efficiently capture Cu(II) ions via the surface amino groups forming cupric ammine complex that lead to fluorescence quenching of the N-CQDs by inner filter effect (Das et al. 2017). Another N-CQDs was synthesized by Hunag et al. via the hydrothermal treatment of strawberry juice providing CQDs with a nitrogen content of 6.9%. The fluorescence of the eco-friendly N-CQDs was dynamically quenched in the presence of Hg(II), thus it was applied for its sensing in lake water samples with sensitivity down to 6 nM. This method is completely green as it used a natural readily available source for carbon and nitrogen (Huang et al. 2013).

Moreover, a green phosphorus-doped CQD (P-CQDs) was synthesized by Yang et al. through a facile hydrothermal coupling of phytic acid and sodium citrate. Phytic acid was used as a phosphorus source due to its hexatomic ring structure and its enrichment with phosphate groups, in addition to being completely natural. The P-CQDs provided a superb sensor for Cu(II) with excellent sensitivity down to 1 nM (Yang et al. 2018).

Additionally, the literature search reveals that carbon nanostructures are the subject of most interest in sensors research (Li et al. 2019). The latest produced carbon nanostructures are onion-like carbon NPs (CNOs). CNOs possess unique physicochemical properties arising from their remarkable edge effect (Sano et al. 2001; Bartelmess and Giordani 2014; Bartelmess et al. 2015). Tripathi et al. synthesized CNOs from natural sources in a green way via pyrolysis of flaxseed oil as the green precursor for carbon followed by oxidation to introduce many carboxyl functionalities to increase its hydrophilicity and stability in water. The unique fluorescence of these CNOs was found to be selectively quenched in the presence of Al(III). Such phenomenon was applied for sensing of Al(III) as an aquatic environment contaminant in water samples with LOD of 0.77 μM (Tripathi et al. 2017). An illustration of the green pathway for synthesis of CQDs and their sensing mechanism for metal environmental pollutant is illustrated in Fig. 1a.

The use of CQDs is not limited to the sensing of toxic metal cations such as Hg(II), Cu(II), Cd(II), and Cr(VI), but it is also extended for analysis of toxic environmental anionic pollutant such as fluoride. CQDs prepared by alkaline hydrothermal treatment of citric acid were developed by Singhal et al. for sensing of $F^-$ through fluorescence turn off–on. The synthesized CQDs fluorescence is turned off via the addition of Eu(III) due to electron transfer and aggregation-induced fluorescence quenching. Upon addition of $F^-$ to this system, fluorescence is regained due to the removal of Eu(III) from the system and formation of $EuF_3$ (Singhal et al. 2017). The same idea was used by Basu et al.; however, they used a more eco-friendly carbon source, tapioca sago starch, for the synthesis of CQDs and they used Fe(III) instead of Eu(III) (Basu et al. 2015). The turn-off–on fluorescence principle for sensing anionic pollutants such as fluoride is illustrated in Fig. 1b.

Graphene quantum dots (GQDs), are a subclass of CQDs that are usually synthesized from graphene or its oxide and their physicochemical properties are very similar to graphene (Pan et al. 2010; Sun et al. 2013). Highly fluorescent GQDs were greenly synthesized through the pyrolysis of citric acid. The fluorescence of the formed GQDs was quenched in the presence of chlorine due to the destruction of its passivated surface by the oxidative power of chlorine. This probe was applied for

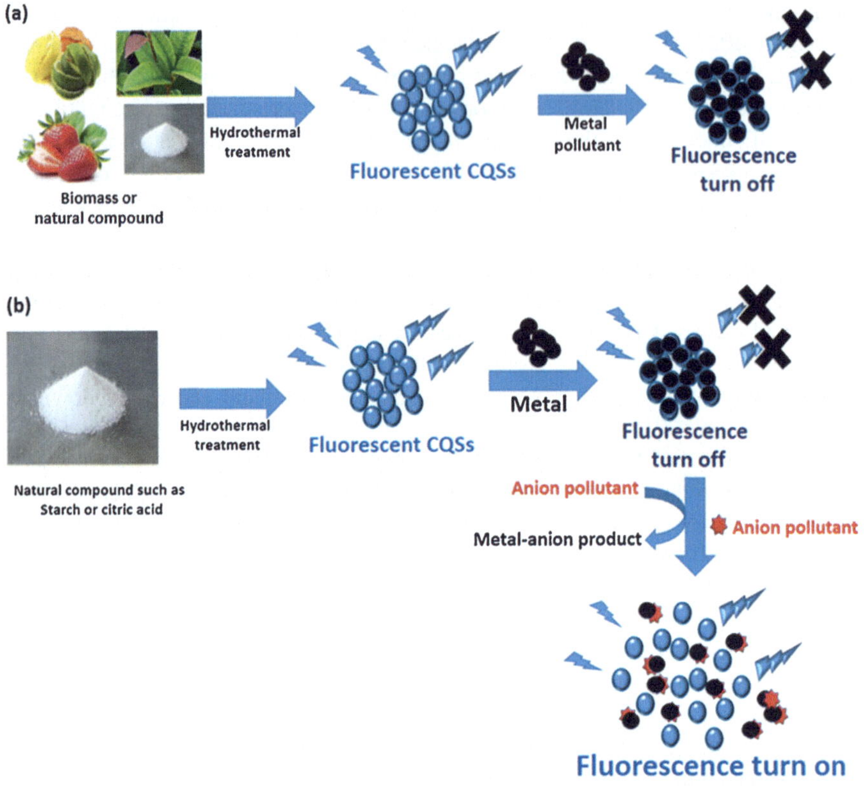


**Fig. 1** Illustration of the general green synthetic pathway of CQDs sensors and their mechanism of sensing environmental pollutants, where (**a**) for sensing metal pollutants and (**b**) for anionic ones

sensing free chlorine in all its forms, including $Cl_2$, HClO, and $ClO^-$, in tap water with sensitivity down to 0.05 μM (Dong et al. 2012a).

Another interesting nanoallotropes of carbon is fullerenes. Fullerenes possess very good electrochemical properties. In addition, they are effectively used as electrocatalysts for several chemical and biochemical reactions. Moreover, their advantageous use for the fabrication of electrochemical sensors is under intensive investigation (Sherigara et al. 2003). Partially reduced fullerene-$C_{60}$ is used for the fabrication of different electrodes providing an excellent working electrode due to its high electroactive surface area and excellent electronic conductivity. Fullerenes are considered green to the environment as they are nontoxic and biocompatible (Csiszar et al. 2001). Rather and De Wael modified glassy carbon electrode (GCE) with fullerene-$C_{60}$ through drop-dry or "casting" method using a suspension of fullerene in a volatile solvent. The fullerene-$C_{60}$ modified GCE was successfully applied for electrochemical detection of the environmental pollutant and endocrine disruptor, bisphenol A (BPA) (Rather and De Wael 2013). BPA is an important environmental pollutant that should be targeted as an analyte due to its wide use in polycarbonate plastics and epoxy resins industry. BPA leaks from these plastic or resin materials to the environment through the acid, base, or heat catalyzed hydrolysis of the ester bond between BPA monomers (Sun et al. 2000). BPA has an estrogenic effect on experimental animals at low doses and the tested subjects showed a high risk of breast cancer incidence that makes it a target for several in-vivo studies (Watanabe et al. 2001; Kuroda et al. 2003; Sun et al. 2004). Thus, BPA environmental monitoring is vital and crucial (Muñoz-de-Toro et al. 2005; Chen et al. 2011). The fullerene-$C_{60}$ modified GCE could detect BPA with very excellent sensitivity down to 3.7 nM (Rather and De Wael 2013).

## 2.2 Greenly Produced Metal Nanoparticles (MNPs) as Sensors for Environmental Contaminants

Metal nanoparticles (MNPs) possess exceptional optical properties originating from oscillations of their charge density. Their unique optical properties are denoted as localized surface plasmon resonance (Wang et al. 2010; Mayer and Hafner 2011). Their unique properties grab increased attention especially in the fields of sensing (Zhou et al. 2015) and imaging (Yang et al. 2016). Thus, many researchers reported different ways for MNPs preparation including wet chemical (Jana et al. 2001), electrochemical (Yamauchi et al. 2012), mechanical (Hanada et al. 2005), and photochemical synthesis (Zhang et al. 2012). However, among these methods, the wet chemical ones are the most commonly used through employing reducing reagents such as citrate (Sener et al. 2013) and sodium borohydride (Chen et al. 2013). Unfortunately, all these aforementioned conventional synthesis methods of NPs have many drawbacks such as high cost, need of laborious and time-consuming steps, and production of unstable NPs that necessitate the use of surfactants (Kumar et al. 2013) that themselves could introduce many cytotoxic effects (Alkilany et al. 2009). Thus, green chemistry-based synthesis of NPs is essentially needed as a better and eco-friendly approach for NPs production (Raveendran et al. 2003; Iravani 2011). Figure 2 illustrates the main principle for green synthesis of MNPs.

Raveendran et al. used glucose and starch for the reduction and stabilization of Ag–Au bimetallic NPs (Raveendran et al. 2006). Also, plant tannin was used for green synthesis of Au-NPs (Huang et al. 2010) and recently trypsin was utilized also for their green synthesis (Zou et al. 2013). Amino acids can play a crucial role in MNPs synthesis through the reduction and stabilization power of their side groups; thus, they can be used for the transformation of reducible metal ions to NPs (Huang et al. 2006). Annadhassan et al. reported a green synthesis approach for gold and silver NPs (Au-NPs and Ag-NPs) as a green sensor for environmental monitoring using L-tyrosine and natural sunlight irradiation. L-tyrosine is used as capping and reducing agent and sunlight irradiation is employed as an external stimulus for controlling NPs switchable behavior. Of course, the use of a natural amino acid such as L-tyrosine and solar irradiation as nontoxic, nonpolluting, and traceless thermal

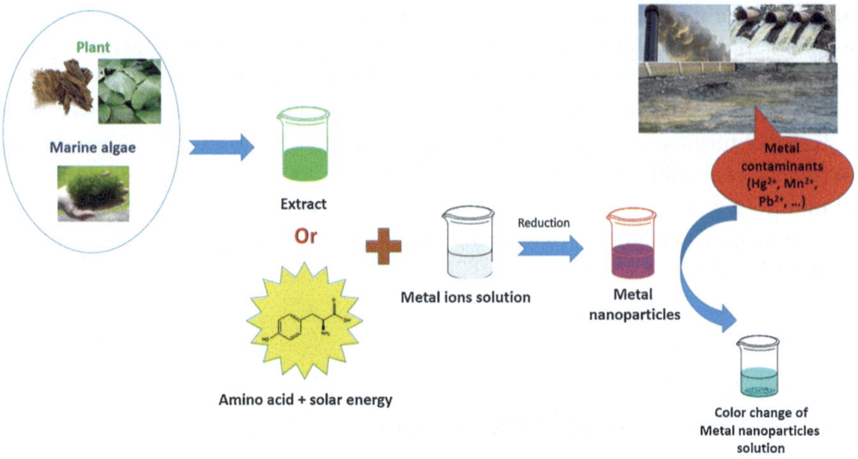


**Fig. 2** Illustration of the general pathway of green synthesis of metal nanoparticles and their mechanism of sensing environmental pollutants

energy substitute made the synthesis process strictly green. The produced Au and Ag-NPs were successfully applied as colorimetric sensors for detecting the ecologically toxic metals Hg(II), Mn(II), and Pb(II) in aquatic samples. The color of Ag-NPs changed from yellow to colorless and brown in the presence of Hg(II) and Mn(II), respectively. Regarding the Au-NPs sensor, its pink color intensity is decreased in presence of Hg(II) and the color was changed to violet in presence of Pb(II) (Annadhasan et al. 2014).

Varma et al. research group has focused on using plant extracts, vitamins, and bio-surfactants in green synthesis of MNPs (Nadagouda and Varma 2006, 2007; Baruwati and Varma 2009; Nadagouda et al. 2009). Emmanuel et al. could greenly synthesize Au-NPs using *Acacia nilotica* twig bark extract at room temperature within a very short time of 10 min upon the introduction of the extract to $HAuCl_4$. The produced Au-NPs were used for modification of GCE producing an electrochemical sensor that could detect the hazardous pollutant nitrobenzene, known for causing Methemoglobinaemia, in water samples with good sensitivity (Emmanuel et al. 2014).

On the other hand, green synthesis of MNPs can be achieved through biological processes using bacteria and fungi. In most of the green synthesis approaches, NPs are formed through reduction as mentioned previously; thus, natural masses like marine algae that abundantly contain natural compounds with reducing ability, like polysaccharides, phenolics, and sugars, are used as reducing and capping reagents for the green production of MNPs. The green synthesis of MNPs adopting algae extract is more advantageous than using bacteria and fungi as it also eliminates the laborious maintaining process for cell culture, and it is more suited for large-scale production processes (Rajeshkumar et al. 2012; Dhas et al. 2014). Khalililaghab et al. developed a green sensor for environmental monitoring based on this approach. They developed a simple and green procedure for the preparation of Ag-NPs by applying the aqueous extract of *Sargassum boveanum* algae to $AgNO_3$. The developed Ag-NPS was applied for colorimetric assay of the environmental pollutant BPA in tap and mineral water. The greenly synthesized Ag-NPs sensor could colorimetrically detect BPA through a color change from yellow to dark purple with good sensitivity down to 45 nM. The sensing mechanism depends on the power of BPA to be strongly adsorbed on the surface of Ag-NPs inducing the formation of aggregates that lead to the color change of the NPs (Khalililaghab et al. 2017). Karimi and Samimi recently reported another example of alga application for green production of MNPs. They added the aqueous crude extract of the green marine alga *Chaetomorpha* sp. to $AgNO_3$ and then boiled the reaction mixture producing Ag@AgCl-NPs. The greenly produced MNPs were applied for the colorimetric detection of Hg(II) in different water samples as the brown color of the MNPs changed from brown to colorless upon its interaction with Hg(II) (Karimi and Samimi 2019).

## 2.3 Greenly Functionalized Metal Nanoparticles (MNPs) as Green Sensors for Environmental Contaminants

Another approach for greening MNPs preparation is to functionalize them using natural compounds such as simple and complex carbohydrates (Shervani and Yamamoto 2011; Boruah et al. 2019), natural amino acids (Si and Mandal 2007), proteins (Wei et al. 2011), and biopolymers such as alginates (Pal et al. 2005). The basic principle of this technique is illustrated in Fig. 3.

A novel environmentally benign colorimetric sensor was developed by Borouah et al. via functionalizing Au-NPs with the biodegradable molecule, glucose. The sensor could successfully detect the major environmental threat, arsenic, down to 0.53 ppb, via changing its color from red to blue. The color change is due to selective binding of arsenic to the hydroxyl groups of glucose which in turn reduces the inter-particle distances altering the color of the NPs from red to blue (Boruah et al. 2019). Another example of the use of functionalized Au-NPs as a green sensor is reported by Manjumeena et al. Au-NPs was photosynthesized in a green way using the aqueous leaves extract of *Rosa indica - wichuraiana* hybrid *Francois Guillot.* Glutathione was utilized for surface modification of Au-NPs and the developed sensor was applied for monitoring of cadmium. The probe color is changed from ruby red into purple when cadmium is present in the environmental sample due to interaction of the metal with the amino, carboxyl, and thiol groups of glutathione changing the inter-particle distances of the NPs (Manjumeena et al. 2015).

## 2.4 Miscellaneous Sensors

Hussain et al. prepared an electrochemical sensor using nanostructures semiconductors composed of nickel oxide-NPs and carbon nanotube nanocomposites (NiO-NPs·CNT NCs) using simplistic wet chemical methods. The wet chemical synthesis is not considered an eco-friendly way for the preparation of MNPs. However, the

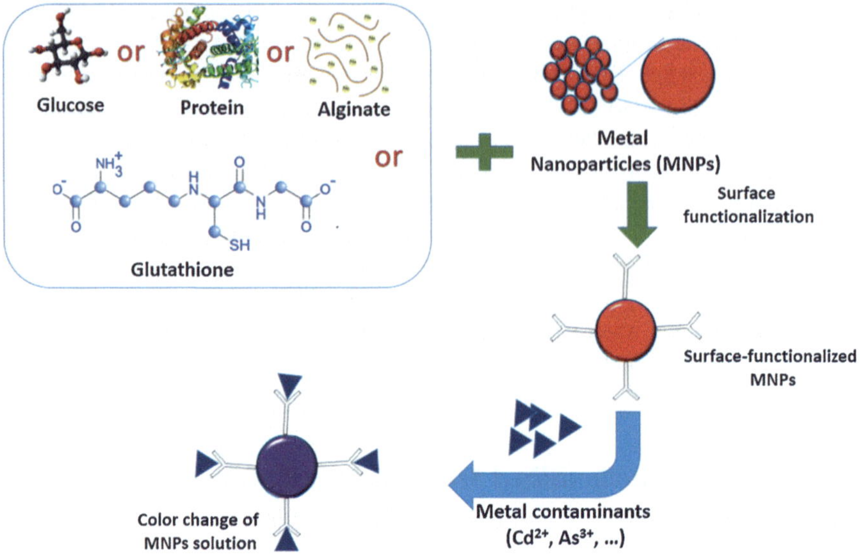


**Fig. 3** The process of formation of green functionalized metal nanoparticles and the mechanism of their sensing for environmental pollutants

mesoporous nature of the NCs allows for their easy recycling without a considerable loss of their efficiency giving them a green nature. Carbon nanotubes were added to NiO electrode in order to enhance its performance because of its large specific surface and high conductivity. Also, due to their unique absorption and adsorption abilities, the hybrid NCs are appropriate for the detection of toxins in the environment. The electrochemical sensor was developed based on fabricating GCE with a thin layer of the hybrid NCs and was applied for the detection of 4-aminophenol (Hussain et al. 2017). 4-Aminophenol is broadly used industrially in addition to being a metabolite of analgesics including acetaminophen. 4-aminophenol is well known to be nephrotoxic and it was proved also that it could cause acute hepatotoxicity in mice (Song and Chen 2001). Thus, monitoring of 4-aminophenol in the environment is very important. The developed green electrochemical sensor can detect 4-aminophenol with exceptional sensitivity down to 15 pM (Hussain et al. 2017). Another greening approach for electrochemical sensors is reported by Švorc et al. through using a boron-doped diamond electrode for square voltammetric detection of pesticides instead of using the extremely environmentally toxic mercury electrodes. Atrazine is one of the most environmentally persistent and toxic herbicides. Moreover, it is a carcinogen and endocrine disruptor even in a small concentration. Thus, Atrazine monitoring in the environment is of great importance. The environmentally green electrode developed by Švorc et al. could detect atrazine in river water samples down to 10 nM (Švorc et al. 2013).

In addition, another green approach for electrochemical sensors was developed through the manufacturing of hexagonal nanopyramids (NPys) of ZnO by hydrothermal treatment of zinc acetate in the presence of oleylamine as a surfactant. The produced ZnO-NPys were used for modifying a screen-printed electrode (SPE) to fabricate a sensitive, precise, and reliable electrochemical sensor for the determination of dichloromethane. Dichloromethane is a commonly used solvent in organic synthesis characterized by its relatively low boiling point (39.6 °C). In addition, it is used frequently for metal surface degreasing and electronics industry. Human intake of dichloromethane, even in traces, could cause fatigue, sleep disturbances, and has a bad influence on the respiratory and central nervous system. Meanwhile, its accidental high intake might cause respiratory failure and coma and very recently, it was listed as a probable human carcinogen. Thus, its analysis in the environment is of great importance for human safety and environment greenness. The developed electrochemical sensor based on ZnO-NPys is capable of detecting dichloromethane down to 17.3 μM (Kim et al. 2019). ZnO is used here due to its wide bandgap, high excitation binding energy, and high-electron communication features that made it a great semiconducting material. Moreover, it is nontoxic and possesses virtuous biocompatibility; thus, its use renders the developed sensor eco-friendly (Ameen et al. 2012, 2013; Kim et al. 2019).

Another approach for greening analytical procedure is miniaturizing the analytical system, for example a miniaturized analytical method based on a microplate-based assay for the determination of Pd(II) in river water samples. Palladium is widely used in several fields including jewelry and dental crowns as well as a catalyst for cross-coupling reactions in the synthesis of many organic compounds including pharmaceutical ones. However, Pd(II) has adverse effects on human health due to its strong binding ability to thiol-containing compounds including glutathione and proteins (e.g., casein), in addition to its interaction with nucleophilic molecules such as DNA and vitamins (e.g., vitamin B6). Thus, Pd(II) traces in pharmaceuticals and environment are considered hazardous. Higashi et al. utilized benzofuran-2-boronic acid as a chemosensor for Pd(II) detection due to the formation of a highly fluorescent benzofuran dimer upon mixing the probe with the target metal at room temperature. The probe showed excellent selectivity for Pd(II) with excellent sensitivity down to 9.88 nM. The measurement was based on fluorescence microplate assay using a small volume of reagents and solvents and allowing for fast assay of a high number of samples simultaneously rendering the method eco-friendly (Higashi et al. 2017). Another method adopting the miniaturization approach was lately developed by El-Maghrabey et al. They developed a green fluorescence turn-on sensor for monitoring of hydrogen sulfide using microplate-based assay. Hydrogen sulfide is a very toxic gas that has no color for its characterization. Humans' exposure to hydrogen sulfide in high amounts can damage the eyes, respiratory and nervous system, and finally can lead to death. Hydrogen sulfide can be found in rivers and wastewaters in the form of $HS^-$ either from natural origin, due to the sulfate-reducing bacteria activity and organic compounds disintegration, or as a contaminant from some industrial activities that use the gas, including petrochemical, paper, and leather industries. Hydrogen sulfide represents a significant pollution index due to its extreme toxicity. Thus, hydrogen sulfide detection and quantification in the occupational susceptible environment are essential. The green sensor developed by El-Maghrabey et al. is based on using the fluorescent lophine analog, 2-(4-hydroxyphenyl)-4,5-di(2-pyridyl)imidazole (HPI), which was synthesized through one-step reaction using a small number of solvents and chemicals. Upon its reaction with Cu(II), HPI loses its fluorescence forming an HPI–Cu(II) complex. This complex is disrupted selectively when reacted with hydrogen sulfide liberating free HPI and regaining its strong fluorescence. The developed sensor, HPI–Cu(II), can sense hydrogen sulfide in river water through fluorescence turn-on mechanism with excellent sensitivity down to 5 ppb adopting microplate-based assay. Using a small volume of chemicals and reagents for the synthesis process and miniaturizing the analytical procedure through using microplate-based assay, together with the restrict avoidance of using any organic solvents in the analysis and the regenerability of the developed HPI–Cu(II) probe renders the assay method environmentally green (El-Maghrabey et al. 2019). Moreover, the authors proved the greenness of their method by using Gałuszka et al. analytical eco-scale score metric (Gałuszka et al. 2012). This analytical metric gives penalty points for any non-green issue in the analytical procedure and subtracts it from 100 to calculate the analytical eco-scale score. The greenness score of this method was above 75 proving the excellent greenness of the assay (El-Maghrabey et al. 2019).

At last, Boonkanon et al. developed a revolutionary approach for environmentally green sensor production for detecting boron in the wastewater. Boron is found naturally in the environment, however, its level is sometimes increased due to its industrial use in glass and ceramic productions and rubberwood treatment and metals fusing. The developed sensor consists of a tapioca starch on which curcumin NPs, extracted from turmeric, are supported and act as a colorimetric sensor. As noted, all the probe materials are natural and environmentally green. However, it is used as thin-film fabricated on a used plastic spoon, but at the same time, the authors said that the spoon can be reused up to 10 times. The color of the film is changed from yellow into reddish-brown after dipping in samples containing boron due to the formation of the rosocyanin complex (Boonkanon et al. 2020).

## 3 Biosensors for Environmental Contaminants

One of the recent approaches of green analytical chemistry (GAC) that has evolved in the twenty-first century is the biosensors. According to the IUPAC description of the biosensor, it is as an integrated device consists of a biochemical receptor in direct contact with a transducer that could provide quantitative or semi-quantitative signals specific to the target analytes (Fig. 4). The biochemical receptor could be enzymes, antibodies, aptamers or nucleic acids, while the transducer part could be electrochemical, optical, or mass sensitive sensor (Farré et al. 2010; Justino et al. 2017).

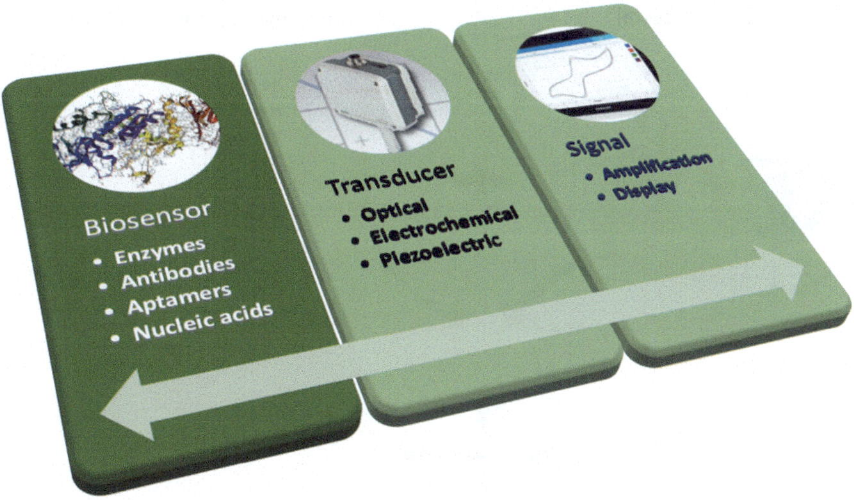


**Fig. 4** General representation for the components of biosensors

## *3.1 Biosensors Mechanism of Sensing*

The basic principle of biosensors depends on using immobilized biological material (enzyme, antibody, aptamer, nucleic acid, etc.) in close contact with the transducer. The target analyte binds to the immobilized biological element forming a bound analyte associated with a specific change that could be converted to a measurable signal by the transducer. A simplified illustration of the principle of environmental biosensors is demonstrated in Fig. 5. Usually, in bio-electrochemistry, the reaction is associated with a quantifiable current (amperometric detection), potential (potentiometric detection), or a change in the conductivity (conductometric detection). The latter two types of electrochemical detectors are the most widely used in environmental biosensing. On the other hand, optical detectors are reliant on several optical phenomena including UV/visible light absorption, fluorescence, phosphorescence, polarization, or rotation. The optical biosensor may comprise the detection of the analyte directly, or indirectly through optical probes. Mass-sensitive sensors depend on measuring small changes in mass. Piezoelectric and surface-acoustic-wave biosensors can be gathered under this class (Farré et al. 2010; Justino et al. 2017).

## *3.2 Greenness of Biosensors*

The development of biosensors for monitoring environmental pollutants contributes greatly to the sustainability of society (Farré et al. 2010). Biosensors help to protect from pollutions while using greener techniques instead of traditional analytical

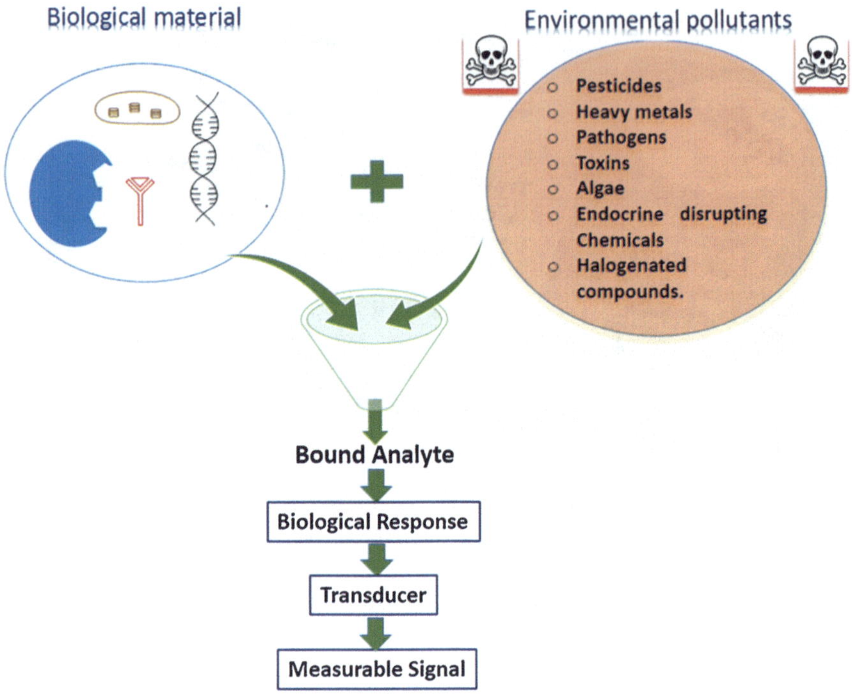


**Fig. 5** Illustration of the sensing principle of environmental biosensors

methodologies (e.g., GC and LC), which require expensive solvents and equipment, and longer time. In addition, biosensors are well-suited for in situ applications and it can be miniaturized to design portable devices. Furthermore, biosensors produce minimal waste and consume limited energy (Justino et al. 2017). In this context, nanotechnology plays a prominent role in the development of biosensing devices since the most recent biosensors use nanomaterials and nanocomposites. This greatly helps to improve analytical features such as sensitivity and detection limits. For example, Au-NPs provides a promising matrix for enzyme immobilization by virtue of their large surface area, ability to stabilize the enzyme by electrostatic attraction, superb biocompatibility, and slight cytotoxicity (Zhao et al. 2013). However, some limitations are facing the commercialization of environmental biosensors, where most biosensors are effectively tested in spiked samples, while during real sample application their performance is greatly influenced by matrix effects (Justino et al. 2017).

### 3.3 *Electrochemical and Enzymatic Biosensors as the Most Frequently Used Types of Biosensors for Environmental Monitoring*

Being on the spot, a vast and increasing number of articles that reported the utilization of biosensors for environmental monitoring is published (Fig. 6). Thus, several reviews already exist to survey biosensors and their applications in environmental monitoring of pollutants, like pesticides, pathogens, heavy metals, and potentially toxic materials, up to recent times. Thus, herein, we found it convenient to briefly overview more than twenty review articles and to provide the concluded remarks from them about the recently published techniques. The features and applications of different biosensors designed for the detection and determination of environmental pollutants have been briefly reviewed by Goradel et al. (Hashemi Goradel et al. 2018). Additionally, the recent advances in the development of biosensors for various environmental contaminants have been deeply reviewed and updated by Justino and colleagues (Justino et al. 2017). The later review article concluded that electrochemical and enzymatic biosensors are the most widely used types in environmental pollutants' detection. In particular, acetylcholinesterase inhibition-based biosensors are the most popular for pesticides determination by virtue of their simplicity, specificity, and cost-effectiveness. Aptamers-based biosensors are advantageous by their thermal stability, the possibility of designing, denaturalizing and rehybridizing their structure, and their distinguishing ability for targets with different functional groups (Justino et al. 2017).

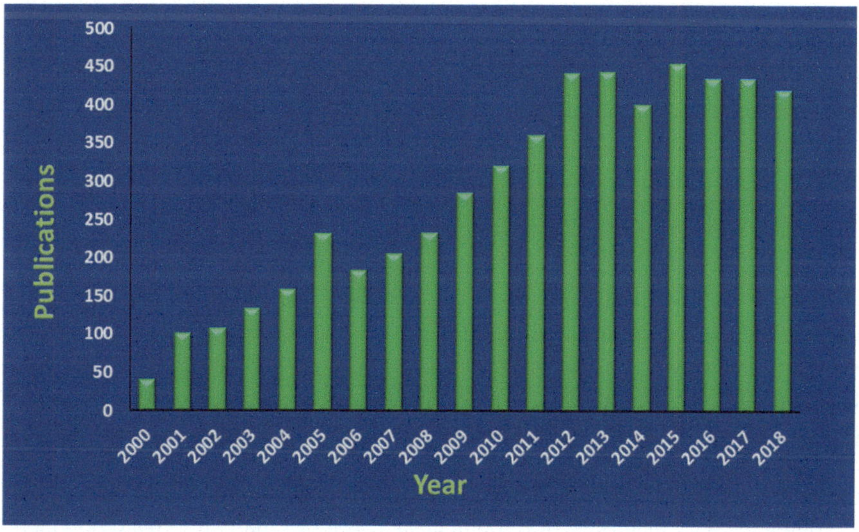


**Fig. 6** The number of publications related to biosensor every year since 2000 according to PubMed search

Hernandez-Vargas et al. reviewed the up-to-date trends for the determination of different environmental pollutants by electrochemical biosensors with a special highlighting of SPE, nanowire-based, and paper-based biosensors. This review pointed out to the fast-growing in the development of electrochemical biosensors and other related fields such as bioelectronics and (bio)-nanotechnology that are expected to have a significant effect on the design of future biosensing strategies (Hernandez-Vargas et al. 2018). Meanwhile, Lee et al. reported a recent review article emphasizing the current trends for the design of paper-based electrochemical biosensors and their applications including environmental monitoring. This type of biosensors provides a cost-effective, easily operated, and sensitive tool with a high capability of customization (Lee et al. 2018). The recent strategies for electrochemical nano biosensors including the principles and applications of protein, antibody and aptamer-based biosensors have been also surveyed and discussed by Mazzei and colleagues (Mazzei et al. 2015).

In addition, Rapini and Marrazza published an article presenting an in-depth review of aptamer-based electrochemical biosensors for different environmental and food contaminants such as toxins, pesticides, drugs, endocrine disruptors, and heavy metals. This review has demonstrated that aptasensors are among the most promising methodologies for biosensor improvements (Rapini and Marrazza 2017). Aptamer-based biosensors have been also reviewed by Meena et al. (2018). Nevertheless, the applications of aptasensors to environmental analysis is still limited due to the inadequate number of existing oligonucleotide sequences (Rapini and Marrazza 2017). Besides, Nigam and Shukla presented a review discussing the enzyme-based biosensors emphasizing the mechanisms of detection of environmental pollutants as well as a survey of the published methods in this area. The authors concluded that, in spite of being ideal tools for environmental monitoring due to their sensitivity, selectivity, and rapidness, enzyme-based biosensors are still in their infancy. Some limitations which need further studies include the encountered interferences from environmental matrix which hinder the detection of the target compounds and the low detection limits (Nigam and Shukla 2015).

### 3.4 Biosensors for Pesticides

Pesticides' environmental monitoring received great attention owing to their acute toxicity and ability to cause long-term harmful effects to humans and the environment. For this reason, a vast number of biosensors have been designed for this goal. In this connection, summaries of different strategies of biosensor technology for pesticide detection has been presented by Verma and Bhardwaj (2015) and Hassani et al. (2017) covering different types of biosensors, such as electrochemical-, optical-, and mass-based, enzyme-based biosensors, immunosensors, aptamers, as well as complementary materials' development such as molecularly imprinted polymers, biochips technology, and nanotechnology.

Furthermore, the recent technological advances in the electrochemical biosensors for organophosphate pesticide chlorpyrifos detection in the environment has been specifically reviewed by Uniyal and Sharma. In this connection, various electrochemical biosensors based on biomolecules, enzymes, and biomimetic molecules were reported. The appearance of biomimetic molecules-based biosensors has sparked a revolution for chlorpyrifos detection with improved analytical performance. Current trends in the use of biosensors based on biomimetic molecules include ones based on nanoparticles (NPs) and molecular imprinting polymerization (Uniyal and Sharma 2018). Further, the different strategies reported for the detection of organophosphates pesticides by electrochemical biosensors have been also documented by Kaur and Prabhakar. The most promising approaches in this connection include aptasensors and non–biomolecule-based biosensors. As well, the use of smartphone-based applications will modernize the rapid, cheap, and convenient techniques for these pollutants' detection (Kaur and Prabhakar 2017). The recent progress in the determination of organophosphorus compounds using organophosphorus hydrolase–based biosensors has been also reviewed (Jain et al. 2019). Songa and Okonkwo reviewed also the recent methods for enhancing enzyme-based biosensors for organophosphorus pesticides in terms of selectivity and sensitivity. Acetylcholinesterase inhibition-based biosensors with enhanced analytical features can be designed with the help of protein engineering for manufacturing tailor-made biorecognition molecules to be integrated into biosensing platforms to improve the biosensor sensitivity (Songa and Okonkwo 2016). A recent review covering the basics and approaches of several organophosphorus pesticides biosensors including optical (fluorescence and surface plasmon resonance), electrochemical (amperometric and potentiometric), thermal and piezoelectric together with microbial and DNA biosensors have been published in 2019 (Pundir and Malik 2019).

## *3.5 Biosensors for Heavy Metals*

Determination of heavy metals as environmental contaminants also received pronounced attention. Thus, a tremendous number of articles reporting biosensors for the detection of heavy metals have been published. In this context, in 2019, Odobašić et al. published a book chapter summarizes the biosensors for determination of heavy metals in water such as Cd(II), Zn(II), Hg(II), Cu(II), Pb(II), Ni(II), Ag(I), U(IV), and Fe(III). In this connection, antibody-, enzyme-, protein-, cell-, immobilized engineered bacteria-based biosensors were utilized for detection of these heavy metal ions (Odobašić et al. 2019). Kanellis reviewed also more than 100 published articles discussing biosensors for monitoring heavy metals in drinking water with a special emphasis on the sensitivity aspect. This later review concluded some clear trends to build on. First, NPs became the main component for the construction of biosensors; second, simple NP receptor molecules are not sufficient but need optimization by surface modification or functionalization; and third, further investigation is needed by researchers to verify the sensitivity and robustness of the biosensors (Kanellis 2018). In addition, a review of the progress in electrochemical biosensors, including enzymatic, immuno, nucleic acid, whole-cell, molecular imprinted and novel nanomaterial modified electrochemical sensors, for heavy metals' detection as pollutants in water has been presented by Dai et al. (2018). Furthermore, affinity-based biosensors for heavy metals in the environment were particularly reviewed. This review covered the protein-based sensors, immunosensors, and aptasensors published in this connection (Antiochia et al. 2013). Very recent, the progress and applications of the oligonucleotide-based biosensors (Verma and Kaur 2019) and protein/enzyme-based biosensors (Osorio-González et al. 2019) for heavy metal contaminants in the environment has been also surveyed.

### 3.6 Biosensors for Toxic Pharmaceuticals

Monitoring and detections of toxic pharmaceuticals in the environment via biosensors attracted also much interest. For this purpose, molecular imprinted polymer-based biosensors were extensively used. Thus, their principle, current status, and applications in these connections have been reviewed in detail by Dhanjai et al. (2019).

## 4 Conclusions

Sensors, including chemical and biosensors, are tremendously used for the detection and determination of a wide variety of environmental pollutants including various heavy metals, pesticides, and toxic pharmaceuticals. As compared to classical analytical techniques, sensors are preferred for environmental monitoring being greener, rapid, cost-effective, and sensitive. The advances in nanotechnology help in improving the simplicity, selectivity, sensitivity, robustness, and green aspects of different types of sensors. CQDS and carbon nanoallotropes including GQDs and fullerenes are very promising tools for manufacturing green sensors for the environmental pollutant. They can be partially or totally synthesized from natural compounds or natural biomasses forming highly fluorescence NPs that lose their fluorescence upon interaction with environmental pollutants such as heavy metals. In addition, MNPs produced or functionalized via green techniques are ideal for this purpose. Also miniaturizing the analytical procedure is a promising way for developing an analytical method that complies with GAC principles. Additionally, a smartphone-based sensor is a perfect match for a portable, cheap, and simple device.

The vast number of applications of biosensors in this field is the proof of their bright future as tools for detection and control of environmental contaminants. In particular, electrochemical and enzymatic biosensors are the most widely applied types for this goal. However, there is still a deficiency in commercially available biosensors which needs more effort from researchers.

## References


Alkilany AM, Nagaria PK, Hexel CR et al (2009) Cellular uptake and cytotoxicity of gold nanorods: molecular origin of cytotoxicity and surface effects. Small 5:701–708

Ameen S, Akhtar MS, Seo H-K et al (2012) Influence of Sn doping on ZnO nanostructures from nanoparticles to spindle shape and their photoelectrochemical properties for dye sensitized solar cells. Chem Eng J 187:351–356

Ameen S, Akhtar MS, Nazim M, Shin H-S (2013) Rapid photocatalytic degradation of crystal violet dye over ZnO flower nanomaterials. Mater Lett 96:228–232

Annadhasan M, Muthukumarasamyvel T, Sankar Babu VR, Rajendiran N (2014) Green synthesized silver and gold nanoparticles for colorimetric detection of $Hg^{2+}$, $Pb^{2+}$, and $Mn^{2+}$ in aqueous medium. ACS Sustain Chem Eng 2:887–896

Antiochia R, Favero G, Conti ME et al (2013) Affinity-based biosensors for heavy metal detection. Int J Environ Health 6:290–303

Armenta S, Garrigues S, de la Guardia M (2008) Green analytical chemistry. TrAC Trends Anal Chem 27:497–511. https://doi.org/10.1016/j.trac.2008.05.003

Bano D, Kumar V, Singh VK, Hasan SH (2018) Green synthesis of fluorescent carbon quantum dots for the detection of mercury(II) and glutathione. New J Chem 42:5814–5821. https://doi.org/10.1039/C8NJ00432C

Bartelmess J, Giordani S (2014) Carbon nano-onions (multi-layer fullerenes): chemistry and applications. Beilstein J Nanotechnol 5:1980–1998. https://doi.org/10.3762/bjnano.5.207

Bartelmess J, Baldrighi M, Nardone V et al (2015) Synthesis and characterization of far-red/NIR-fluorescent BODIPY dyes, solid-state fluorescence, and application as fluorescent tags attached to carbon Nano-onions. Chemistry 21:9727–9732. https://doi.org/10.1002/chem.201500877

Baruwati B, Varma RS (2009) High value products from waste: grape pomace extract—a three-in-one package for the synthesis of metal nanoparticles. ChemSusChem 2:1041–1044

Basu A, Suryawanshi A, Kumawat B et al (2015) Starch (tapioca) to carbon dots: an efficient green approach to an on–off–on photoluminescence probe for fluoride ion sensing. Analyst 140:1837–1841. https://doi.org/10.1039/C4AN02340D

Boonkanon C, Phatthanawiwat K, Wongniramaikul W, Choodum A (2020) Curcumin nanoparticle doped starch thin film as a green colorimetric sensor for detection of boron. Spectrochim Acta Part A Mol Biomol Spectrosc 224:117351. https://doi.org/10.1016/j.saa.2019.117351

Boruah BS, Biswas R, Deb P (2019) A green colorimetric approach towards detection of arsenic(III): a pervasive environmental pollutant. Opt Laser Technol 111:825–829. https://doi.org/10.1016/j.optlastec.2018.09.023

Chandra S, Das P, Bag S et al (2011) Synthesis, functionalization and bioimaging applications of highly fluorescent carbon nanoparticles. Nanoscale 3:1533–1540

Chen X, Wang C, Tan X, Wang J (2011) Determination of bisphenol a in water via inhibition of silver nanoparticles-enhanced chemiluminescence. Anal Chim Acta 689:92–96

Chen Z, Zhang X, Cao H, Huang Y (2013) Chitosan-capped silver nanoparticles as a highly selective colorimetric probe for visual detection of aromatic ortho-trihydroxy phenols. Analyst 138:2343–2349

Csiszar M, Szűcs Á, Tölgyesi M et al (2001) Electrochemical reactions of cytochrome c on electrodes modified by fullerene films. J Electroanal Chem 497:69–74

Dai X, Wu S, Li S (2018) Progress on electrochemical sensors for the determination of heavy metal ions from contaminated water. J Chinese Adv Mater Soc 6:91–111

Das P, Ganguly S, Bose M et al (2017) A simplistic approach to green future with eco-friendly luminescent carbon dots and their application to fluorescent nano-sensor 'turn-off' probe for selective sensing of copper ions. Mater Sci Eng C 75:1456–1464. https://doi.org/10.1016/j.msec.2017.03.045

Dhanjai SA, Mugo SM et al (2019) Molecular imprinted polymer-based biosensors for the detection of pharmaceutical contaminants in the environment. In: Kaur Brar S, Hegde K (eds) Pachapur

techniques and protocols for monitoring environmental contaminants VLBT-T. Elsevier, pp 371–389
Dhas TS, Kumar VG, Karthick V et al (2014) Facile synthesis of silver chloride nanoparticles using marine alga and its antibacterial efficacy. Spectrochim Acta Part A Mol Biomol Spectrosc 120:416–420
Dong Y, Li G, Zhou N et al (2012a) Graphene quantum dot as a green and facile sensor for free chlorine in drinking water. Anal Chem 84:8378–8382. https://doi.org/10.1021/ac301945z
Dong Y, Wang R, Li G et al (2012b) Polyamine-functionalized carbon quantum dots as fluorescent probes for selective and sensitive detection of copper ions. Anal Chem 84:6220–6224. https://doi.org/10.1021/ac3012126
Dong Y, Wang R, Li H et al (2012c) Polyamine-functionalized carbon quantum dots for chemical sensing. Carbon N Y 50:2810–2815. https://doi.org/10.1016/j.carbon.2012.02.046
El-Maghrabey MH, Watanabe R, Kishikawa N, Kuroda N (2019) Detection of hydrogen sulfide in water samples with 2-(4-hydroxyphenyl)-4,5-di(2-pyridyl)imidazole-copper(II) complex using environmentally green microplate fluorescence assay method. Anal Chim Acta 1057:123–131. https://doi.org/10.1016/j.aca.2019.01.006
El-Shaheny RN, El-Maghrabey MH, Belal FF (2015) Micellar liquid chromatography from green analysis perspective. Open Chem 13:877–892. https://doi.org/10.1515/chem-2015-0101
El-Shaheny RN, El-Maghrabey MH, Eid MI, El-Enany NM (2019) Green chromatographic purification of pharmaceuticals. Ind Appl Green Solvents Vol II 54:148–181
Emmanuel R, Karuppiah C, Chen S-M et al (2014) Green synthesis of gold nanoparticles for trace level detection of a hazardous pollutant (nitrobenzene) causing Methemoglobinaemia. J Hazard Mater 279:117–124. https://doi.org/10.1016/j.jhazmat.2014.06.066
Farré M, Pérez S, Gonçalves C et al (2010) Green analytical chemistry in the determination of organic pollutants in the aquatic environment. TrAC Trends Anal Chem 29:1347–1362. https://doi.org/10.1016/j.trac.2010.07.016
Gałuszka A, Migaszewski ZM, Konieczka P, Namieśnik J (2012) Analytical eco-scale for assessing the greenness of analytical procedures. Trends Anal Chem 37:61–72. https://doi.org/10.1016/j.trac.2012.03.013
Gavrilescu M (2010) Environmental biotechnology: achievements, opportunities and challenges. Dyn Biochem Process Biotechnol Mol Biol 4:1–36
Gavrilescu M, Demnerova K, Aamand J et al (2015) Emerging pollutants in the environment: present and future challenges in biomonitoring, ecological risks and bioremediation. New Biotechnol 32:147–156. https://doi.org/10.1016/j.nbt.2014.01.001
Gonçalves HMR, Duarte AJ, Esteves da Silva JCG (2010) Optical fiber sensor for hg(II) based on carbon dots. Biosens Bioelectron 26:1302–1306. https://doi.org/10.1016/j.bios.2010.07.018
Guardia M d l (M d l), Garrigues S (2012) Handbook of green analytical chemistry. Wiley, New York
de la Guardia M, Khalaf KD, Hasan BA et al (1995) In-line, titanium dioxide-catalysed, ultraviolet mineralization of toxic aromatic compounds in the waste stream from a flow injection-based resorcinol analyser. Analyst 120:231–235. https://doi.org/10.1039/AN9952000231
Hanada N, Ichikawa T, Fujii H (2005) Catalytic effect of nanoparticle 3D-transition metals on hydrogen storage properties in magnesium hydride $MgH_2$ prepared by mechanical milling. J Phys Chem B 109:7188–7194
Hardman R (2005) A toxicologic review of quantum dots: toxicity depends on physicochemical and environmental factors. Environ Health Perspect 114:165–172
Hashemi Goradel N, Mirzaei H, Sahebkar A et al (2018) Biosensors for the detection of environmental and urban pollutions. J Cell Biochem 119:207–212
Hassani S, Momtaz S, Vakhshiteh F et al (2017) Biosensors and their applications in detection of organophosphorus pesticides in the environment. Arch Toxicol 91:109–130. https://doi.org/10.1007/s00204-016-1875-8

Hernandez-Vargas G, Sosa-Hernández J, Saldarriaga-Hernandez S et al (2018) Electrochemical biosensors: a solution to pollution detection with reference to environmental contaminants. Biosensors 8:29

Higashi A, Kishikawa N, Ohyama K, Kuroda N (2017) A simple and highly selective fluorescent sensor for palladium based on benzofuran-2-boronic acid. Tetrahedron Lett 58:2774–2778

Huang YF, Lin YW, Chang HT (2006) Growth of various au–Ag nanocomposites from gold seeds in amino acid solutions. Nanotechnology 17:4885–4894

Huang X, Wu H, Liao X, Shi B (2010) One-step, size-controlled synthesis of gold nanoparticles at room temperature using plant tannin. Green Chem 12:395–399

Huang H, Lv J-J, Zhou D-L et al (2013) One-pot green synthesis of nitrogen-doped carbon nanoparticles as fluorescent probes for mercury ions. RSC Adv 3:21691–21696. https://doi.org/10.1039/C3RA43452D

Hussain MM, Rahman MM, Asiri AM (2017) Ultrasensitive and selective 4-aminophenol chemical sensor development based on nickel oxide nanoparticles decorated carbon nanotube nanocomposites for green environment. J Environ Sci 53:27–38. https://doi.org/10.1016/j.jes.2016.03.028

Iravani S (2011) Green synthesis of metal nanoparticles using plants. Green Chem 13:2638–2650

Jain M, Yadav P, Joshi A, Kodgire P (2019) Advances in detection of hazardous organophosphorus compounds using organophosphorus hydrolase based biosensors. Crit Rev Toxicol 3:1–24

Jana NR, Gearheart L, Murphy CJ (2001) Wet chemical synthesis of silver nanorods and nanowires of controllable aspect ratio Electronic supplementary information (ESI) available: UV–VIS spectra of silver nanorods. See http://www.rsc.org/suppdata/cc/b1/b100521i. Chem Commun (2):617–618

Justino CIL, Duarte AC, Rocha-Santos TAP (2017) Recent progress in biosensors for environmental monitoring: a review. Sensors 17:2918

Kanellis VG (2018) Sensitivity limits of biosensors used for the detection of metals in drinking water. Biophys Rev 10:1415–1426

Karimi S, Samimi T (2019) Green and simple synthesis route of Ag@AgCl nanomaterial using green marine crude extract and its application for sensitive and selective determination of mercury. Spectrochim Acta A Mol Biomol Spectrosc 222:117216. https://doi.org/10.1016/j.saa.2019.117216

Kaur N, Prabhakar N (2017) Current scenario in organophosphates detection using electrochemical biosensors. TrAC Trends Anal Chem 92:62–85

Khalililaghab S, Momeni S, Farrokhnia M et al (2017) Development of a new colorimetric assay for detection of bisphenol-a in aqueous media using green synthesized silver chloride nanoparticles: experimental and theoretical study. Anal Bioanal Chem 409:2847–2858. https://doi.org/10.1007/s00216-017-0230-0

Kim E-B, Abdullah AS et al (2019) Environment-friendly and highly sensitive dichloromethane chemical sensor fabricated with ZnO nanopyramids-modified electrode. J Taiwan Inst Chem Eng 102:143–152. https://doi.org/10.1016/j.jtice.2019.05.016

Kumar S, Gradzielski M, Mehta SK (2013) The critical role of surfactants towards CdS nanoparticles: synthesis, stability, optical and PL emission properties. RSC Adv 3:2662–2676

Kuroda N, Kinoshita Y, Sun Y et al (2003) Measurement of bisphenol a levels in human blood serum and ascitic fluid by HPLC using a fluorescent labeling reagent. J Pharm Biomed Anal 30:1743–1749

Lang Q, Han L, Hou C et al (2016) A sensitive acetylcholinesterase biosensor based on gold nanorods modified electrode for detection of organophosphate pesticide. Talanta 156–157:34–41. https://doi.org/10.1016/j.talanta.2016.05.002

Lee VBC, MOHD-NAIM NF, Tamiya E, Ahmed MU (2018) Trends in paper-based electrochemical biosensors: from design to application. Anal Sci 34:7–18

Li H-Y, Li D, Guo Y et al (2018) On-site chemosensing and quantification of Cr(VI) in industrial wastewater using one-step synthesized fluorescent carbon quantum dots. Sensors Actuators B Chem 277:30–38. https://doi.org/10.1016/j.snb.2018.08.157

Li M, Chen T, Gooding JJ, Liu J (2019) Review of carbon and Graphene quantum dots for sensing. ACS Sensors 4:1732–1748. https://doi.org/10.1021/acssensors.9b00514

Liu Y, Zhao Y, Zhang Y (2014) One-step green synthesized fluorescent carbon nanodots from bamboo leaves for copper(II) ion detection. Sensors Actuators B Chem 196:647–652. https://doi.org/10.1016/j.snb.2014.02.053

Lu W, Qin X, Asiri AM et al (2012a) Green synthesis of carbon nanodots as an effective fluorescent probe for sensitive and selective detection of mercury(II) ions. J Nanopart Res 15:1344. https://doi.org/10.1007/s11051-012-1344-0

Lu W, Qin X, Liu S et al (2012b) Economical, green synthesis of fluorescent carbon nanoparticles and their use as probes for sensitive and selective detection of mercury(II) ions. Anal Chem 84:5351–5357. https://doi.org/10.1021/ac3007939

Manjumeena R, Duraibabu D, Rajamuthuramalingam T et al (2015) Highly responsive glutathione functionalized green AuNP probe for precise colorimetric detection of $Cd^{2+}$ contamination in the environment. RSC Adv 5:69124–69133. https://doi.org/10.1039/C5RA12427A

Mayer KM, Hafner JH (2011) Localized surface plasmon resonance sensors. Chem Rev 111:3828–3857

Mazzei F, Favero G, Bollella P et al (2015) Recent trends in electrochemical nanobiosensors for environmental analysis. Int J Environ Health 7:267–291

Meena RK, Kamboj D, Kumar K, Karanwal S (2018) A brief review on Aptamer based biosensors for detection of environmental pollution. Int J Curr Microbiol App Sci 7:1483–1489

Muñoz-de-Toro M, Markey CM, Wadia PR et al (2005) Perinatal exposure to bisphenol-a alters peripubertal mammary gland development in mice. Endocrinology 146:4138–4147

Nadagouda MN, Varma RS (2006) Green and controlled synthesis of gold and platinum nanomaterials using vitamin B2: density-assisted self-assembly of nanospheres, wires and rods. Green Chem 8:516–518

Nadagouda MN, Varma RS (2007) A greener synthesis of core (Fe, Cu)-shell (Au, Pt, Pd and Ag) nanocrystals using aqueous vitamin C. Cryst Growth Des 7:2582–2587

Nadagouda MN, Hoag G, Collins J, Varma RS (2009) Green synthesis of au nanostructures at room temperature using biodegradable plant surfactants. Cryst Growth Des 9:4979–4983

Nigam VK, Shukla P (2015) Enzyme based biosensors for detection of environmental pollutants—a review. J Microbiol Biotechnol 25:1773–1781

Odobašić A, Šestan I, Begić S (2019) Biosensors for determination of heavy metals in waters. In: Environmental biosensors. IntechOpen

Osorio-González CS, Hegde K, Brar SK et al (2019) Advances in protein/enzyme-based biosensors for the detection of metal contaminants in the environment. In: Kaur Brar S, Hegde K (eds) Pachapur techniques and protocols for monitoring environmental contaminants VLBT-T. Elsevier, pp 245–261

Pal A, Esumi K, Pal T (2005) Preparation of nanosized gold particles in a biopolymer using UV photoactivation. J Colloid Interface Sci 288:396–401

Pan D, Zhang J, Li Z, Wu M (2010) Hydrothermal route for cutting Graphene sheets into blue-luminescent Graphene quantum dots. Adv Mater 22:734–738. https://doi.org/10.1002/adma.200902825

Pundir CS, Malik A (2019) Bio-sensing of organophosphorus pesticides: a review. Biosens Bioelectron 140:111348

Rajeshkumar S, Kannan C, Annadurai G (2012) Green synthesis of silver nanoparticles using marine brown algae Turbinaria conoides and its antibacterial activity. Int J Pharma Bio Sci 3:502–510

Rapini R, Marrazza G (2017) Electrochemical aptasensors for contaminants detection in food and environment: recent advances. Bioelectrochemistry 118:47–61

Rather JA, De Wael K (2013) Fullerene-C60 sensor for ultra-high sensitive detection of bisphenol-a and its treatment by green technology. Sensors Actuators B Chem 176:110–117. https://doi.org/10.1016/j.snb.2012.08.081

Raveendran P, Fu J, Wallen SL (2003) Completely "green" synthesis and stabilization of metal nanoparticles. J Am Chem Soc 125:13940–13941

Raveendran P, Fu J, Wallen SL (2006) A simple and '"green"' method for the synthesis of Au, Ag, and Au–Ag alloy nanoparticles. Green Chem 8:34–38

Sano N, Wang H, Chhowalla M et al (2001) Synthesis of carbon "onions" in water. Nature 414:506–507. https://doi.org/10.1038/35107141

Sener G, Uzun L, Denizli A (2013) Lysine-promoted colorimetric response of gold nanoparticles: a simple assay for ultrasensitive mercury (II) detection. Anal Chem 86:514–520

Sherigara BS, Kutner W, D'Souza F (2003) Electrocatalytic properties and sensor applications of fullerenes and carbon nanotubes. Electroanalysis 15:753–772

Shervani Z, Yamamoto Y (2011) Carbohydrate-directed synthesis of silver and gold nanoparticles: effect of the structure of carbohydrates and reducing agents on the size and morphology of the composites. Carbohydr Res 346:651–658

Shi W, Li X, Ma H (2012) A tunable ratiometric pH sensor based on carbon nanodots for the quantitative measurement of the intracellular pH of whole cells. Angew Chem Int Ed 51:6432–6435. https://doi.org/10.1002/anie.201202533

Si S, Mandal TK (2007) Tryptophan-based peptides to synthesize gold and silver nanoparticles: a mechanistic and kinetic study. Chem Eur J 13:3160–3168

Singhal P, Vats BG, Jha SK, Neogy S (2017) Green, water-dispersible photoluminescent on–off–on probe for selective detection of fluoride ions. ACS Appl Mater Interfaces 9:20536–20544. https://doi.org/10.1021/acsami.7b03346

Song H, Chen TS (2001) P-aminophenol-induced liver toxicity: tentative evidence of a role for acetaminophen. J Biochem Mol Toxicol 15:34–40. https://doi.org/10.1002/1099-0461(2001)15:1<34::AID-JBT4>3.0.CO;2-U

Songa EA, Okonkwo JO (2016) Recent approaches to improving selectivity and sensitivity of enzyme-based biosensors for organophosphorus pesticides: a review. Talanta 155:289–304

Sun Y, Wada M, Al-Dirbashi O et al (2000) High-performance liquid chromatography with peroxyoxalate chemiluminescence detection of bisphenol a migrated from polycarbonate baby bottles using 4-(4,5-diphenyl-1H-imidazol-2-yl) benzoyl chloride as a label. J Chromatogr B Biomed Sci Appl 749:49–56

Sun Y, Irie M, Kishikawa N et al (2004) Determination of bisphenol a in human breast milk by HPLC with column-switching and fluorescence detection. Biomed Chromatogr 18:501–507

Sun H, Wu L, Wei W, Qu X (2013) Recent advances in graphene quantum dots for sensing. Mater Today 16:433–442. https://doi.org/10.1016/j.mattod.2013.10.020

Švorc Ľ, Rievaj M, Bustin D (2013) Green electrochemical sensor for environmental monitoring of pesticides: determination of atrazine in river waters using a boron-doped diamond electrode. Sensors Actuators B Chem 181:294–300. https://doi.org/10.1016/j.snb.2013.02.036

Tobiszewski M, Mechlinska A, Namiesnik J (2010) Green analytical chemistry—theory and practice. Chem Soc Rev 39:2869–2878. https://doi.org/10.1039/b926439f

Tripathi KM, Tran TS, Kim YJ, Kim T (2017) Green fluorescent onion-like carbon nanoparticles from flaxseed oil for visible light induced Photocatalytic applications and label-free detection of Al(III) ions. ACS Sustain Chem Eng 5:3982–3992. https://doi.org/10.1021/acssuschemeng.6b03182

Uniyal S, Sharma RK (2018) Technological advancement in electrochemical biosensor based detection of organophosphate pesticide chlorpyrifos in the environment: a review of status and prospects. Biosens Bioelectron 116:37–50

Vaz S Jr (2018) Analytical chemistry applied to emerging pollutants. Springer

Verma N, Bhardwaj A (2015) Biosensor technology for pesticides—a review. Appl Biochem Biotechnol 175:3093–3119

Verma N, Kaur G (2019) Advances in the oligonucleotide-based biosensors for the detection of heavy metal contaminants in the environment. In: Kaur Brar S, Hegde K (eds) Pachapur techniques and protocols for monitoring environmental contaminants VLBT-T. Elsevier, pp 169–185

Wang J, Zhang P, Li JY et al (2010) Adenosine–aptamer recognition-induced assembly of gold nanorods and a highly sensitive plasmon resonance coupling assay of adenosine in the brain of model SD rat. Analyst 135:2826–2831

Wang X, Qu K, Xu B et al (2011) Microwave assisted one-step green synthesis of cell-permeable multicolor photoluminescent carbon dots without surface passivation reagents. J Mater Chem 21:2445–2450. https://doi.org/10.1039/C0JM02963G

Watanabe T, Yamamoto H, Inoue K et al (2001) Development of sensitive high-performance liquid chromatography with fluorescence detection using 4-(4,5-diphenyl-1H-imidazol-2-yl) benzoyl chloride as a labeling reagent for determination of bisphenol a in plasma samples. J Chromatogr B Biomed Sci Appl 762:1–7

Wei H, Wang Z, Zhang J et al (2011) Time-dependent, protein-directed growth of gold nanoparticles within a single crystal of lysozyme. Nat Nanotechnol 6:93–97

Yamauchi Y, Tonegawa A, Komatsu M et al (2012) Electrochemical synthesis of mesoporous Pt–Au binary alloys with tunable compositions for enhancement of electrochemical performance. J Am Chem Soc 134:5100–5109

Yang Y, Wang X, Cui Q et al (2016) Self-assembly of fluorescent organic nanoparticles for iron(III) sensing and cellular imaging. ACS Appl Mater Interfaces 8:7440–7448

Yang F, He X, Wang C et al (2018) Controllable and eco-friendly synthesis of P-riched carbon quantum dots and its application for copper(II) ion sensing. Appl Surf Sci 448:589–598. https://doi.org/10.1016/j.apsusc.2018.03.246

Zhang Y, Yuan X, Wang Y, Chen Y (2012) One-pot photochemical synthesis of graphene composites uniformly deposited with silver nanoparticles and their high catalytic activity towards the reduction of 2-nitroaniline. J Mater Chem 22:7245–7251

Zhao Y, Zhang W, Lin Y, Du D (2013) The vital function of Fe3O4@au nanocomposites for hydrolase biosensor design and its application in detection of methyl parathion. Nanoscale 5:1121–1126. https://doi.org/10.1039/c2nr33107a

Zhou L, Lin Y, Huang Z et al (2012) Carbon nanodots as fluorescence probes for rapid, sensitive, and label-free detection of $Hg^{2+}$ and biothiols in complex matrices. Chem Commun 48:1147–1149. https://doi.org/10.1039/C2CC16791C

Zhou W, Gao X, Liu D, Chen X (2015) Gold nanoparticles for in vitro diagnostics. Chem Rev 115:10575–10636

Zou L, Qi W, Huang R et al (2013) Green synthesis of a gold nanoparticle–nanocluster composite nanostructures using trypsin as linking and reducing agents. ACS Sustain Chem Eng 1:1398–1404

# Index





Inamuddin, A. M. Asiri (eds.), *Nanosensor Technologies for Environmental Monitoring*, Nanotechnology in the Life Sciences,
https://doi.org/10.1007/978-3-030-45116-5

## N

**T**

The manufacturer's authorised representative in the EU is Springer Nature Customer Service Centre GmbH, Europaplatz 3, 69115 Heidelberg, Germany. If you have any concerns regarding our products, please contact ProductSafety@springernature.com

Printed and bound by CPI Group (UK) Ltd, Croydon, CR0 4YY
05/10/2026
02228126-0015